Basic Molecular Quantum Mechanics

Basic Molecular Quantum Mechanics

Steven A. Adelman

CRC Press
Taylor & Francis Group
Boca Raton London New York

CRC Press is an imprint of the
Taylor & Francis Group, an **informa** business

First edition published 2022
by CRC Press
6000 Broken Sound Parkway NW, Suite 300, Boca Raton, FL 33487-2742

and by CRC Press
2 Park Square, Milton Park, Abingdon, Oxon, OX14 4RN

CRC Press is an imprint of Taylor & Francis Group, LLC

Library of Congress Cataloging-in-Publication Data
Names: Adelman, Steven A., author.
Title: Basic molecular quantum mechanics / Steven A. Adelman.
Description: First edition. | Boca Raton : Taylor and Francis, 2021. |
Includes bibliographical references and index.
Identifiers: LCCN 2021001521 (print) | LCCN 2021001522 (ebook) |
ISBN 9781498733991 (hardback) | ISBN 9780429155741 (ebook)
Subjects: LCSH: Quantum theory.
Classification: LCC QC174.12 .A298 2021 (print) | LCC QC174.12 (ebook) |
DDC 530.12—dc23
LC record available at https://lccn.loc.gov/2021001521
LC ebook record available at https://lccn.loc.gov/2021001522

ISBN: 978-1-032-01065-6 (hbk)
ISBN: 978-1-4987-3399-1 (pbk)
ISBN: 978-0-429-15574-1 (ebk)

Typeset in Times
by codeMantra

Contents

Preface

This textbook is designed to provide an introduction to quantum mechanics, emphasizing its applications to chemistry, for advanced undergraduates and beginning graduate students specializing in chemistry, in related fields such as chemical engineering and materials science, and in some areas of biology. Quantum mechanics is a general theory of the motions, structures, properties, and behaviors of particles of atomic and subatomic dimensions, for example, atoms, molecules, and electrons. While quantum mechanics was created in the first third of the twentieth century by a handful of theoretical physicists working on a limited number of problems, it is now further developed and applied by a great number of people working on a vast range of problems in wide areas of science and technology.

This textbook grew out of a graduate course that I taught at Purdue University to students in the fields mentioned above, including students specializing in a field as non-quantitative as pharmacy. Since most students that I taught did not have advanced backgrounds in mathematics or physics, two subjects essential to the understanding of quantum mechanics, this book is carefully written to be easily understood by such students. Thus, it only requires as essential background a first course in calculus which covers as its most advanced topics partial differentiation and multiple integration, although good courses in general chemistry and introductory physics, while not absolutely essential, also provide very helpful background.

Quantum mechanics is an inherently mathematical subject. So the book develops as simply as possible the mathematics beyond basic calculus needed to understand and apply quantum mechanics. Also to comprehend quantum mechanics, one needs a grasp of the elements of classical mechanics, Newton's seventeenth-century theory of motion of macroscopic objects. A simple development of these elements is given in Appendix A. I very strongly recommend that students who are not comfortable with classical mechanics read Appendix A before beginning the study of the main body of this textbook.

The book starts in Chapter 1 by describing the early major developments of quantum mechanics beginning with the seminal contribution of Planck who initiated the quantum theory in 1900 by introducing the revolutionary hypothesis that the harmonic oscillators or resonators in the walls of a black-body radiation cavity have discrete energies rather than the continuous energies predicted by classical mechanics. The book concludes in Chapter 13 by developing the modern computer-based methods for determining molecular electronic structures and for deriving from these structures observable molecular properties such as energies, bond lengths, and bond angles. Comparison of the theoretical results obtained throughout the textbook with observed results is made whenever possible.

The book builds up quantum mechanics systematically with each chapter preparing the student for the next more advanced chapter.

The treatments in this book are rigorous since I recall from my undergraduate courses that precise mathematical derivations in which the student can work through each step of the derivation are far easier to understand and far more convincing than "hand waving" arguments. A small number of the needed derivations, however, are too advanced for this textbook. For these cases, I merely quote the results of the derivation and refer the students to a more advanced textbook where the full derivation is given.

The book often makes use of analogies to help explain abstract quantum concepts. For example, to explain the very important forms of stationary quantum state wave functions, I point out the analogy between these forms and those of the easily visualizable standing wave displacement functions of a stretched string. As a second example, I explain the abstract nature of the measurement of a quantum observable by its analogy to the very simple problem of predicting the outcome (heads or tails) of a coin flip.

At the end of each chapter are one or two Further Readings pages that list references, either books or papers from the literature, which for some of the references students can study alternative presentations of the content of the chapter or for other references can expand their knowledge by studying more extensive or more advanced developments of the content.

Also at the end of each chapter is a Problems set including solutions to some of the problems. The problems range from simple applications of the formulas of the chapter up to often quite advanced developments of new results which extend the core content of the chapter. I strongly recommend that students solve as many of the problems as possible to both test their understanding of the content of the chapter and to develop mathematical skills that are essential if one wants to make applications of quantum mechanics.

This book should be useful for both experimentalists and theoreticians. For experimentalists, the book provides the basic quantum background which when supplemented by more specialized information will permit them to interpret spectroscopic results. It also provides the background needed to intelligently use standard modern quantum chemical computer programs that yield the molecular orbitals and properties of molecules. For theoreticians, the book provides a good departure point for a study of more advanced textbooks on quantum mechanics, for example, the book by Cohen-Tannoudji, Diu, and Laloe or the one by Sakurai. Both of these books are referred to frequently in this textbook, and full references to them may be found in the Further Readings pages of many chapters.

Acknowledgments

I wish to thank my wife Barbara for her patience and encouragement that helped me to carry out the very formidable task of writing this book.

I am also deeply grateful to Mrs. Cindy Everhart of the Purdue University Chemistry Department for her superb typing of this long-and-complex equation-filled textbook. Without the immense amount of challenging work done so capably by Cindy, this book would not have been possible.

Author

Steven A. Adelman was born and grew up in Chicago. He received a Ph.D. from Harvard University in Chemical Physics in 1972. He then became a postdoctoral fellow first at MIT and then at the University of Chicago. In 1975, he joined the Chemistry faculty of Purdue University as an Assistant Professor and was promoted to Full Professor in 1981. He is now Professor Emeritus of Chemistry at Purdue.

Adelman is a theoretical chemist whose research is centered in the field of statistical mechanics, and he has contributed to several areas in that field. His most important contribution is to lay the physical, mathematical, and computational foundations of the new field of chemical reaction dynamics in condensed matter, especially reaction dynamics on solid surfaces and in liquid solutions.

Adelman's theory of solid surface chemical reaction dynamics provides the now-standard conceptualization of the many-body or many-degree-of-freedom influence on these dynamics. It also provides practical tools that permit one to overcome the many-degree-of-freedom roadblock to computer simulations of solid surface chemical reactions and the closely related processes of gas/solid surface energy transfer. These tools have been applied by other workers to obtain molecular insights into the dynamics of a number of solid surface energy transfer and chemical reaction processes.

Adelman's theory of liquid solution chemical reaction dynamics developed out of his theory of solid surface chemical reaction dynamics. But because the physics of the many-degree-of-freedom influence on reaction dynamics is much more complex in liquids than on solids, both the conceptualization and calculation of this influence are much more involved in the former than in the latter theory. In fact, a central part of the development of the liquid solution theory is a new formulation of the statistical mechanical theory of irreversible processes. The theory of irreversible processes was initiated by the Irish-English physicist George Stokes, the co-formulator of the equations of hydrodynamics, in 1851 with his hydrodynamic treatment of the motion of a macroscopic sphere in water, was extended to semi-microscopic particles such as pollen grains by the French physicist Paul Langevin in 1905 yielding the famous Langevin equation that provides the foundation of the theory of Brownian motion, was further developed by other workers including the Dutch physicist Leonard Ornstein who made foundational contributions to the theory of the equilibrium structures of liquids; the Dutch-American physicist George Uhlenbeck, the co-discoverer of electron spin; the Norwegian-American chemical physicist and Nobel Laureate in chemistry Lars Onsager who made major contributions to the theories of the dielectric constant of polar fluids, the conductivity of electrolyte solutions, general, liquid crystal, and liquid helium phase transitions, and moreover, is the discoverer of the celebrated reciprocal relations of transport theory; the Indian-American astrophysicist and Nobel Laureate in physics Subrahmanyan Chandrasekhar while best known for his theories of the evolution of stars also made very significant contributions in a remarkably broad range of fields including atomic physics, hydrodynamic stability, and the general relativistic theory of Black Holes, and was put on a rigorous microscopic foundation by the Japanese physicist Hazime Mori in 1965 to yield the modern theory of generalized Brownian motion.

The problem of the theory of irreversible processes is to determine the motions of a few $n_A \ll N_A$ = Avogadro's number of primary degrees of freedom of a system comprised of many N_A degrees of freedom. The few n_A degrees of freedom are called primary because their motions directly determine the values of observables A of experimental significance. The motions of the $N_A - n_A$ non-primary degrees of freedom only affect the values of the A's indirectly because of their influence on the motions of the primary degrees of freedom. While the values of the A's may be determined by simulating the motions of the full N_A-degree-of-freedom system, this brute force approach is very inconvenient since it requires solving many-degree-of-freedom equations of motion.

The strategy of the theory of irreversible processes is to avoid massive simulations of the full system by developing easily solved *effective equations of motion* for only the few n_A primary degrees of freedom. These effective equations of motion simulate the influence of the non-primary degrees of freedom by including suitably chosen conservative, dissipative, and random force terms (the last needed to account for the Boltzmann spread of the initial conditions of the non-primary degrees of freedom).

For example, in Stokes' theory of the motion of a macroscopic sphere in water, the dissipative term in Stokes' effective equation of motion is approximated by the frictional force exerted by the water on the sphere calculated by solving the Navier–Stokes hydrodynamic equations.

Prior to Adelman's work, all theories of irreversible processes were so-called *slow variable* theories that require the assumption that the primary degrees of freedom have much lower speeds than the non-primary degrees of freedom.

However, for reaction dynamics in liquids the primary degrees of freedom are the Cartesian coordinates or the reaction coordinate of the reacting molecules, while the non-primary degrees of freedom are the Cartesian coordinates or a suitably chosen set of equivalent generalized coordinates of the liquid solvent molecules. Additionally, according to Arrhenius' fundamental principle of chemical kinetics in order for molecules to surmount a chemical activation barrier and hence react, they must be moving at hyperthermal speeds that are much greater than the thermal speeds of the solvent molecules. Thus, to treat chemical reaction dynamics in liquids, one requires a novel fast variable theory of irreversible processes for which the primary degrees of freedom are moving much faster than the non-primary degrees of freedom. Adelman has formulated such a theory. Adelman's fast variable theory yields conservative, dissipative, and random forces whose physical interpretations and mathematical forms are qualitatively different from those of the slow variable theories. These differences yield a physical picture of liquid phase reaction dynamics which is radically different from and much more realistic than the picture of the traditional slow variable Kramers-type theories of reaction dynamics in solution. Adelman's picture has been confirmed (and the Kramers-type picture has been invalidated) by computer simulations of a number of liquid-phase chemical reaction. Thus, Adelman's theory represents a major advance in one of the central questions in chemistry: What is the nature of chemical reactions dynamics in liquid solution?

Adelman has been awarded both Alfred P. Sloan and John Simon Guggenheim Memorial Fellowships. He is a fellow of both the American Physical Society and the American Association for the Advancement of Science. He has been listed in Marquis' Who's Who in America every year since 1985 and has been listed four times in Marquis' Who's Who in the World.

1 Toward Quantum Mechanics

Classical mechanics introduced in Appendix A is the theory of motion developed by Isaac Newton in the last third of the seventeenth century. Classical mechanics successfully describes a tremendous range of phenomena. A few examples include the "break" of a curveball thrown by a major league pitcher, the orbiting of a planet about the sun, the propagation of tidal waves, the vibrations of a violin string, the flow of fluids, and the transmission of sound.

However, by the end of the nineteenth century, experimental results had emerged which revealed phenomena outside of the purview of both classical mechanics and one of the two other pillars of nineteenth-century physics (the second being thermodynamics) Maxwell's theory of the electromagnetic field. From the theoretical analyses of these phenomena and additional phenomena discovered in the early part of the twentieth century as well as from purely theoretical considerations, it eventually became clear that classical mechanics failed when applied to very small particles such as atoms, molecules, and electrons and that a new form of mechanics, *quantum mechanics*, was needed to study the motions of such particles.

The development of quantum mechanics required a radical revision of many of the most basic principles of physics and so it only very gradually emerged over the thirty-year period, ~1900–1930. In this chapter, we describe the first stages of this emergence. We particularly introduce the new ideas that permitted the explanation of four classically inexplicable phenomena: (a) the blackbody radiation spectrum, (b) the photoelectric effect, (c) the low-temperature heat capacities of solids, and most importantly, (d) the line spectra of the hydrogen atom.

We begin by describing how Planck and Einstein explained the first three of these phenomena by introducing the revolutionary concept of *energy quanta*.

1.1 ENERGY QUANTA

We start with blackbody radiation.

To begin, we note that a chunk of metal or any other body continually emits electromagnetic radiation or light. At room temperature, most of this light lies in the infrared region of the electromagnetic spectrum and thus is invisible. But if the chunk of metal is heated, the light becomes visible, and as the temperature increases, the metal will first glow dull red, then bright red, then still brighter yellow, and eventually very bright blue white. What is happening is that as the temperature increases, both the intensities (which determine the brightness) and the frequencies (which determine the colors) of the emitted light increase. While these qualitative features of emission are found for all heated bodies, the details of the emission spectrum vary from body to body.

Thus, it proved convenient to introduce an idealized emitter called a blackbody which completely absorbs light of all frequencies and which emits at all frequencies. The German physicist Gustav Kirchhoff in 1859 showed that the emission spectrum of a blackbody depends only on its Kelvin temperature T and is the same for all blackbodies.

A blackbody can be approximated by a hollow object or cavity with a very small hole through which light can enter or exit. Any radiation that enters the cavity is trapped and reflects off the cavity walls until it is absorbed. The walls constantly emit and absorb radiation, so radiation exists in the cavity in thermodynamic equilibrium with the walls. The spectrum of this radiation can be determined by analyzing the light that exits the cavity. What is measured is $u(\upsilon,T)$, the energy per unit volume and per unit frequency range of cavity radiation of frequency υ. In Figure 1.1, we plot $u(\upsilon,T)$, as determined from experiment, as a function of υ at temperatures $T = 1{,}200$ and $1{,}800\,K$. Note from the plot that both the total amount of energy radiated and the frequency of peak emission increase with T, in qualitative accord with what occurs when heating ordinary bodies.

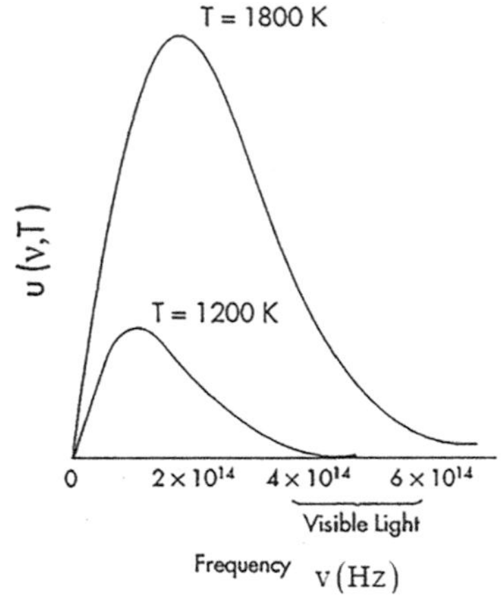

FIGURE 1.1 Experimental blackbody radiation spectra $u(\upsilon,T)$ at temperatures $T = 1{,}200$ and $1{,}800\,K$. The total radiated energies (areas under the $u[\upsilon,T]$ curves) increase sharply with T, and the frequencies of maximum emission (peak positions of the $u[\upsilon,T]$ curves) also increase with T.

While the blackbody spectrum is close to common experience to explain, it theoretically proved to be a tremendous challenge. So even though Kirchhoff studied blackbody radiation theoretically as early as 1859, it was not until 1900 that the blackbody radiation phenomenon was explained quantitatively. The explanation was given by the German physicist Max Planck who derived a formula for the cavity radiation spectrum $u(\upsilon,T)$ which agrees perfectly with experiment for all frequencies υ and temperatures T. To derive this formula, however, Planck was forced to introduce a radically new hypothesis. Namely, he proposed that the harmonic oscillators or resonators in the cavity walls had discrete or quantized energies rather than the continuum of energies predicted by classical mechanics and accepted for centuries. Planck further proposed that light exited the cavity in discrete bursts, rather than continuously, due to emissions of the cavity wall resonators arising from transitions from their higher energy quantized energy levels to their lower energy levels. Planck's proposals initiated the quantum theory and provide an essential part of the basis for all subsequent developments of quantum mechanics.

Despite the seminal nature of Planck's contribution, in this chapter we treat blackbody radiation by a method different from the one used by Planck. The reason is that we believe the formulation used here is more easily understood than Planck's historical formulation. We start by noting that the cavity radiation exists as electromagnetic standing waves that are three-dimensional generalizations of the familiar standing waves of a stretched string. (See Figure 2.3.) For large cavities, one may show that $g(\upsilon)$, the number of electromagnetic standing waves with frequency υ per unit volume and per unit frequency range, is independent of the shape of the cavity and is given by (Cohen-Tannoudji, Diu and Laloe 1977)

$$g(\upsilon) = \frac{8\pi\upsilon^2}{c^3} \tag{1.1}$$

where c is the speed of light.

The energy $u(\upsilon,T)$ we are seeking is just the number of standing waves of frequency υ per unit volume and per unit frequency range times the average energy $U(\upsilon,T)$ of one of these standing waves at temperature T. That is,

$$u(\upsilon,T) = g(\upsilon)U(\upsilon,T) = \frac{8\pi\upsilon^2}{c^3}U(\upsilon,T). \tag{1.2}$$

To determine $U(\upsilon,T)$ and hence $u(\upsilon,T)$, we must use the methods of statistical mechanics. In particular, we need Boltzmann's fundamental result that the probability $P(E)$ that a mechanical system at temperature T has the energy E is proportional to $\exp\left(-\frac{E}{kT}\right)$, where

$$k = 1.38066 \times 10^{-23}\,J\,K^{-1} \tag{1.3}$$

is Boltzmann's constant. We note for future reference that Boltzmann's constant is related to the gas constant R (the same constant that appears in the famous equation $PV = nRT$) and Avogadro's number N_A by

$$R = N_A k. \tag{1.4}$$

Next, note that since the system must have some energy, E $P(E)$ must satisfy the condition

$$\sum_{\text{all } E} P(E) = 1 \tag{1.5}$$

where $\sum_{\text{all } E}$ denotes the sum over all possible energies of the system. The condition of Equation (1.5) is met if $P(E)$ has the form

$$P(E) = \frac{1}{Q(T)} \exp\left(-\frac{E}{kT}\right) \tag{1.6}$$

where $Q(T)$ is the *partition function* of the system defined by

$$Q(T) = \sum_{\text{all } E} \exp\left(-\frac{E}{kT}\right). \tag{1.7}$$

The average energy $U(T)$ of the system is determined from $P(E)$ as

$$U(T) = \sum_{\text{all } E} E\, P(E). \tag{1.8}$$

From Equations (1.6)–(1.8), it follows that $U(T)$ is given in terms of $Q(T)$ by (Problem 1.1)

$$U(T) = kT^2 \frac{d\ell nQ(T)}{dT}. \tag{1.9}$$

We may now return to blackbody radiation. We first note that we may apply statistical mechanics to the blackbody problem since it may be shown that an electromagnetic standing wave of frequency υ for the purpose of determining $U(\upsilon,T)$ is equivalent to a mechanical system, namely a one-dimensional harmonic oscillator of the same frequency υ (Cohen-Tannoudji, Diu, and Laloe 1977, p. 624). Thus, we may calculate the average energy $U(\upsilon,T)$ of a standing wave of frequency υ using the following adaptation of Equation (1.9):

$$U(\upsilon,T) = kT^2 \frac{d\ell nQ(\upsilon,T)}{dT} \tag{1.10}$$

where $Q(\upsilon,T)$ is the partition function of a one-dimensional harmonic oscillator of frequency υ. $Q(\upsilon,T)$ may be found from the following adaptation of Equation (1.7):

$$Q(\upsilon,T)=\sum_{\text{all } E(\upsilon)} \exp\left[-\frac{E(\upsilon)}{kT}\right] \tag{1.11}$$

where the sum is over all possible energies $E(\upsilon)$ of the oscillator.

We first determine $Q(\upsilon,T)$ assuming that classical mechanics is valid and hence that the oscillator can have a continuous range of energies $E(\upsilon)$. A calculation gives the classical form of $Q(\upsilon,T)$ as (Problem 1.2)

$$Q(\upsilon,T)=\frac{kT}{\upsilon}. \tag{1.12}$$

Since $\ell nQ(\upsilon,T)=\ell n\left(\frac{k}{\upsilon}\right)+\ell nT, \frac{d\ell nQ(\upsilon,T)}{dT}=\frac{d\ell nT}{dT}=\frac{1}{T}$, and hence, Equation (1.10) gives the classical average energy of the oscillator or standing wave as

$$U(\upsilon,T)=kT. \tag{1.13}$$

From Equations (1.2) and (1.13), we obtain the classical formula for the blackbody spectrum first derived by the British physicists Rayleigh and Jeans:

$$u_{RJ}(\upsilon,T)=\frac{8\pi kT}{c^3}\upsilon^2. \tag{1.14}$$

The Rayleigh–Jeans equation (Equation 1.14) is compared with experiment in Figure 1.2. Theory and experiment only agree at the very lowest frequencies. At higher frequencies, they are in gross disagreement. Especially at very high frequencies, the experimental spectrum approaches zero, while the Rayleigh–Jeans spectrum diverges as υ^2. This divergence, which absurdly implies that the intensity of the emitted light approaches infinity as the frequency v becomes infinite is referred to as the *ultraviolet catastrophe*.

Since the Rayleigh–Jeans formula is based on a rigorous application of classical physics, its failure presented a paradox. This paradox as already noted was resolved by Planck by introducing

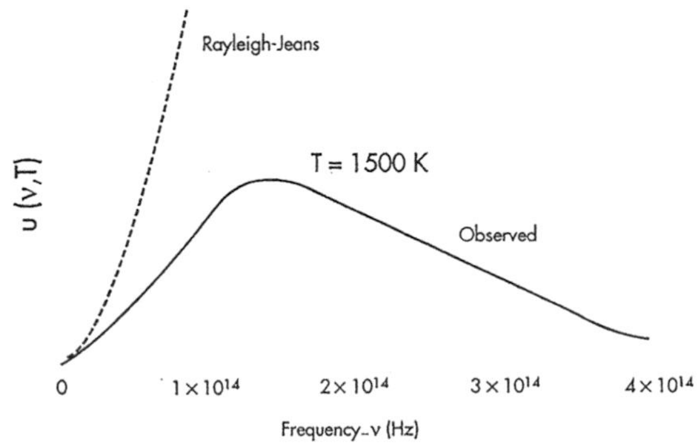

FIGURE 1.2 Comparison of the observed and Rayleigh–Jeans blackbody radiation spectra $u(\upsilon,T)$ at $T=1{,}500\,K$. The Rayleigh–Jeans spectrum $u_{RJ}(\upsilon,T)=\frac{8\pi kT}{c^3}\upsilon^2$ agrees with the experimental spectrum at very low frequencies υ but unphysically diverges at high υ.

a revolutionary hypothesis that is now recognized as the genesis of quantum mechanics. Namely, Planck broke with the classical notion that a physical system has a continuous range of energies and instead postulated that the blackbody cavity standing waves or the harmonic oscillators that represent them could only have discrete or quantized energies. (Planck's actual assumption noted earlier was different but equivalent.) Specifically, he postulated that the oscillator energies $E(\upsilon)$ were restricted to the discrete values

$$E(\upsilon) = nh\upsilon, \quad \text{where } n = 0,1,2,\ldots \tag{1.15}$$

and where h is a proportionality factor now known as Planck's constant. Comparing Equations (1.11) and (1.15) gives the partition function as

$$Q(\upsilon,T) = \sum_{n\times 0}^{\infty} \exp\left(-\frac{nh\upsilon}{kT}\right). \tag{1.16}$$

We may evaluate $Q(\upsilon,T)$ by letting $x = \exp\left(-\frac{h\upsilon}{kT}\right)$ to give $Q(\upsilon,T)$ as $Q(\upsilon,T) = \sum_{n=0}^{\infty} x^n$. The sum is a geometric series that has the value $(1-x)^{-1}$. Thus, the quantum form of $Q(\upsilon,T)$ is

$$Q(\upsilon,T) = \frac{1}{1-\exp\left(-\frac{h\upsilon}{kT}\right)}. \tag{1.17}$$

$U(\upsilon,T)$ may be found from $Q(\upsilon,T)$ using Equation (1.10) as (Problem 1.6)

$$U(\upsilon,T) = \frac{h\upsilon}{\exp\left(\frac{h\upsilon}{kT}\right)-1}. \tag{1.18}$$

Comparing Equations (1.2) and (1.18) gives Planck's blackbody radiation formula

$$u_{Pl}(\upsilon,T) = \frac{8\pi h}{c^3}\frac{\upsilon^3}{\exp\left(\frac{h\upsilon}{kT}\right)-1}. \tag{1.19}$$

Notice that since in the low-frequency limit $\frac{h\upsilon}{kT} << 1$, in this limit $\exp\left(\frac{h\upsilon}{kT}\right)-1 = \left(1+\frac{h\upsilon}{kT}+\cdots\right)-1 = \frac{h\upsilon}{kT}$ and $u_{Pl}(\upsilon,T)$ reduces to the Rayleigh–Jeans law $u_{RJ}(\upsilon,T) = \frac{8\pi kT}{c^3}\upsilon^2$ which agrees with experiment at low frequencies. In the opposite high-frequency limit, $\frac{h\upsilon}{kT} >> 1$ $u_{Pl}(\upsilon,T)$ becomes $\frac{8\pi h}{c^3}\upsilon^3\exp\left(-\frac{h\upsilon}{kT}\right)$, which coincides with Wein's law (derived before Planck's work by the German physicist Max Wien using thermodynamic arguments) which agrees with experiment at high frequencies. What about intermediate frequencies? It turns out that if one takes Planck's constant to have the value (see Problem 1.7)

$$h = 6.62608 \times 10^{-34}\ \text{J.s} \tag{1.20}$$

Planck's blackbody equation (Equation 1.19) perfectly agrees with experiment at all frequencies and temperatures. This agreement provides great support for but does not unequivocally prove Planck's hypothesis of quantized energies.

The next step in establishing the existence of energy quanta was taken by Albert Einstein in 1905. Einstein explained the photoelectric effect, discovered in 1887 by the German physicist Heinrich Hertz, by introducing a concept perhaps even more radical than the one introduced by Planck. In conflict with Maxwell's established conceptualization of light as an electromagnetic wave, Einstein postulated that light is actually a stream of particles that are now called photons. The energy ε_υ of a particle or photon of light of frequency υ was assumed to be $\varepsilon_\upsilon = h\upsilon$, where h is again Planck's constant.

(Notice that our postulate that the energies of the blackbody cavity standing waves are quantized as $E(\upsilon) = nh\upsilon$ emerges naturally from the photon model of light if one reinterprets the standing waves as collections of $n = 0,1,2,\ldots$ photons each of energy $\varepsilon_\upsilon = h\upsilon$.)

We next note that the photoelectric effect is the emission of electrons by a metal due to visible or ultraviolet light striking the surface of the metal. An analogous electron emission phenomenon called photoionization occurs when light of sufficiently high frequency impinges on an atom or a molecule. Since photoionization is easier to visualize than the photoelectric effect, we will actually discuss photoionization.

Consider the simplest case of ultraviolet light impinging on a gas comprised of atoms like lithium with a single easily ionized valence electron. The following is observed:

1. For light of frequencies υ greater than a critical frequency υ_{crit}, atoms are ionized and electrons are observed in the laboratory, but no electrons are observed if υ is less than υ_{crit}*
2. The kinetic energy T_{photo} of the ejected electrons increases linearly with υ for $\upsilon > \upsilon_{crit}$.
3. While the number of electrons ejected by the light increases with the intensity or brightness I of the light, T_{photo} is independent of I.

These experimental results are summarized in Figure 1.3.

As schematically depicted in Figure 1.4a and b, the predictions of the wave model of light conflict with experiment. Namely, the wave model predicts that for fixed υ, T_{photo} increases with the intensity I of the light wave, while for fixed I, it predicts that T_{photo} is independent of υ. On the other hand, as schematically depicted in Figure 1.4c and d, the predictions of the photon model of light qualitatively agree with experiment. Namely, if one assumes that each electron ejection arises from the absorption of a single photon of energy $\varepsilon_\upsilon = h\upsilon$, the photon model predicts that T_{photo} is independent of I, which is proportional to the number of photons present but increases with υ.

The qualitative agreement of the photon model with experiment may be made quantitative by the following argument. When the atom absorbs light and ionizes, energy conservation requires that

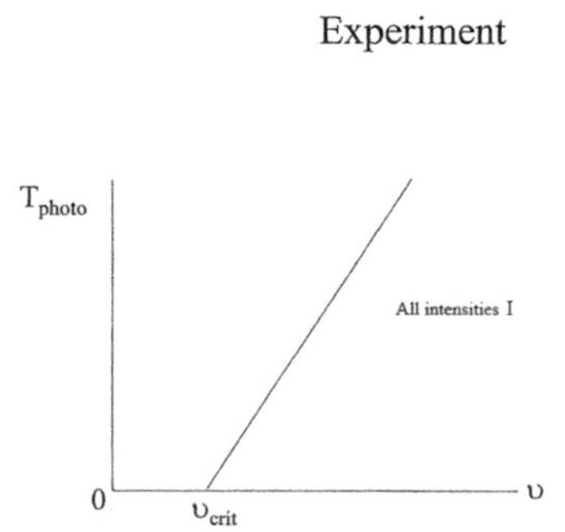

FIGURE 1.3 The kinetic energy T_{photo} of an electron ejected from an atom by ultraviolet light of frequency υ. For υ less than a critical frequency υ_{crit}, ejection does not occur, but for $\upsilon > \upsilon_{crit}$, T_{photo} increases linearly with υ and is independent of the intensity I of the light.

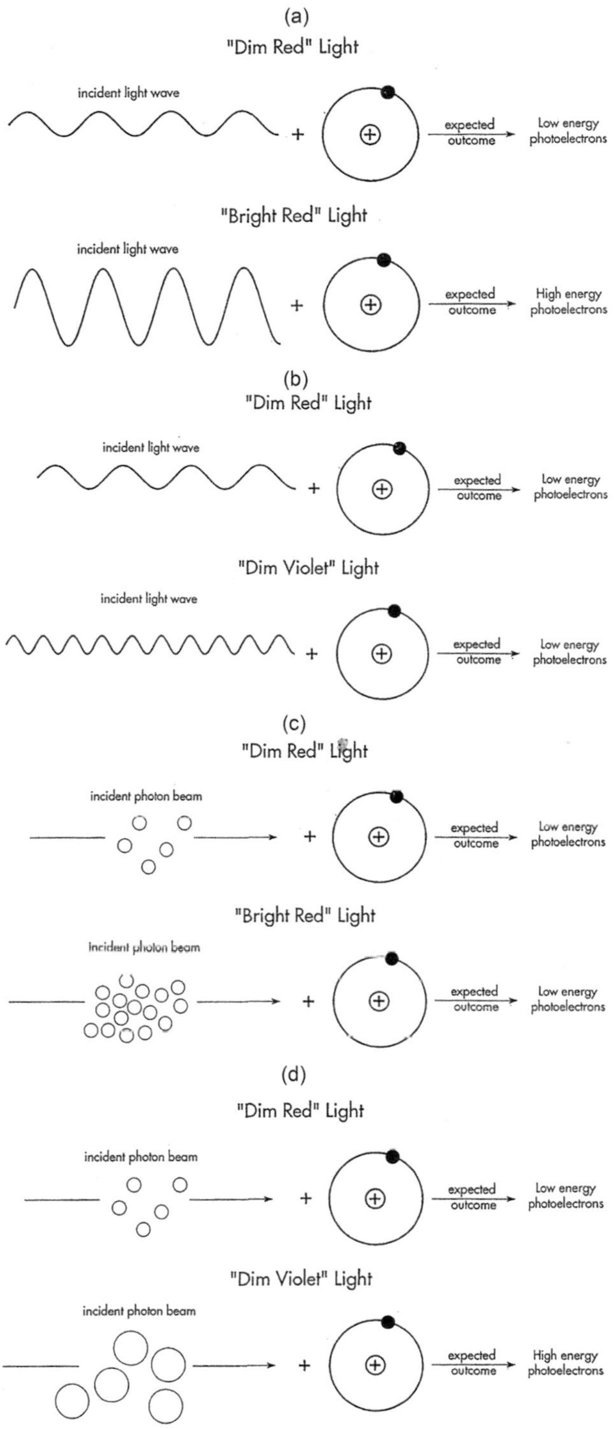

FIGURE 1.4 Photoionization of an atom as predicted by the wave and photon models of light. According to the wave model, (a) for fixed light frequency υ (color), the kinetic energy T_{photo} of the ejected electron increases with the energy E or intensity I (brightness) of the light wave, but (b) for fixed E or I, T_{photo} is independent of υ. According to the photon model, each electron ejection arises from the absorption of a single photon of energy $\varepsilon_\upsilon = h\upsilon$ (c). Therefore, T_{photo} is independent of the number of photons present and hence I, but (d) T_{photo} increases with the amount of energy absorbed per ejection $\varepsilon_\upsilon = h\upsilon$ and hence increases with υ.

$E_i = E_f$, where E_i and E_f are, respectively, the energies of the system before and after the ionization process occurs. Assuming that ionization is due to the absorption of a single photon, it follows that $E_i = E_a + \varepsilon_\upsilon$, where E_a is the energy of the atom to be ionized and where $\varepsilon_\upsilon = h\upsilon$ is the energy of the photon to be absorbed. Also $E_f = E_{ion} + T_{photo}$, where E_{ion} is the energy of the positive ion formed by the ionization process and T_{photo} is the kinetic energy of the ejected electron. Thus, the energy conservation condition $E_i = E_f$ gives $E_a + h\upsilon = E_{ion} + T_{photo}$ or

$$h\upsilon = E_{ion} - E_a + T_{photo}. \tag{1.21}$$

However, $E_{ion} - E_a$ is the energy change for the process atom $\rightarrow$ ion + a stationary electron at infinity. That is, $E_{ion} - E_a$ is the energy needed to remove an electron from the atom and move it to infinity with zero kinetic energy. This energy is called the ionization potential (or ionization energy) of the atom and is denoted by I (not to be confused with the light intensity I). Thus, we have that

$$I = E_{ion} - E_a. \tag{1.22}$$

Comparing Equations (1.21) and (1.22) gives $h\upsilon = I + T_{photo}$ or

$$T_{photo} = h\upsilon - I. \tag{1.23}$$

Equation (1.23) is the basic equation of photoionization and perfectly agrees with what is observed in photoionization experiments, thus confirming the photon model that was used to derive it. For example, let us assume that we have a gas comprised of lithium atoms and we probe the gas by letting ultraviolet light of frequency υ imping on it. Suppose we plot the kinetic energy T_{photo} of the emitted electrons as a function of υ. In complete agreement with Equation (1.23), we will obtain a linear plot like that of Figure 1.3 with a slope equal to Planck's constant h and with a critical frequency (the frequency for which $T_{photo} = 0$) given by

$$\upsilon_{crit} = \frac{I}{h} \tag{1.24}$$

where $I = 5.39$ electron volts (eV) is the ionization potential of lithium.

Next, let us turn to the problem of the low-temperature heat capacities of solids. Following Einstein, we will determine the heat capacity $C(T)$ at Kelvin temperature T of a model solid comprised of Avogadro's number N_A (or one mole) of identical atoms that move like independent three-dimensional harmonic oscillators that execute independent vibrations with the same frequency υ in each of their three directions of motion. Such a system is equivalent to a collection of $3N_A$ one-dimensional harmonic oscillators each of frequency υ.

The heat capacity $C(T)$ of the model solid is given by thermodynamics in terms of its average energy $E(\upsilon,T)$ by

$$C(T) = \frac{\partial E(\upsilon,T)}{\partial T}. \tag{1.25}$$

Let us first determine $E(\upsilon,T)$ and $C(T)$ classically. Recall from Equation (1.13) that the classical average energy of a one-dimensional harmonic oscillator of frequency υ is kT. Since our model solid is equivalent to $3N_A$ of such oscillators, its classical average energy is

$$E(\upsilon,T) = 3N_A kT = 3RT \tag{1.26}$$

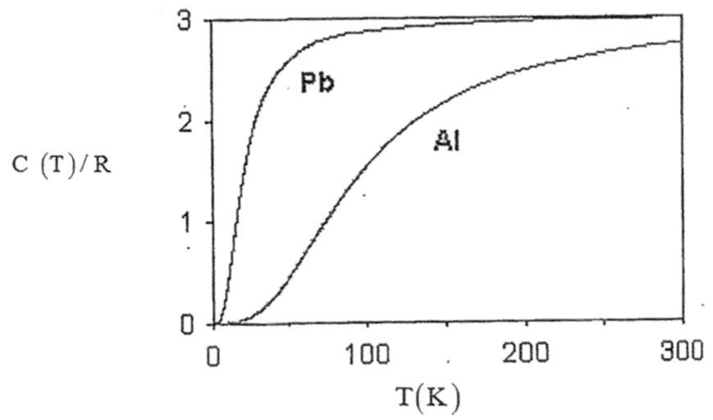

FIGURE 1.5 Measured heat capacities $C(T)$ as a function of Kelvin temperature T for the solids lead and aluminum. At high T, the heat capacities approach the classical Dulong–Petit law $C(T) = 3R$, but as T decreases, the heat capacities also decrease and approach the limit $\lim_{T \to 0K} C(T) = 0$.

where we have used Equation (1.4), $R = N_A k$. Equations (1.25) and (1.26) give that the classical heat capacity of the solid is independent of temperature and equal to

$$C(T) = 3R. \tag{1.27}$$

For historical reasons, the classical equation (Equation 1.27) is called the Dulong–Petit law. (It was first established empirically by Dulong and Petit in the early part of the nineteenth century.) We will see that it is a high-temperature limiting law that turns out to be for many solids a good approximation at room temperature.

We next note that with the development of refrigeration techniques, it became possible to measure the heat capacities of solids at temperatures T well below room temperature. What was observed is that at sufficiently low temperatures, the Dulong–Petit law and hence classical physics fail. Specifically, as illustrated in Figure 1.5, as T decreases, $C(T)$ falls increasingly below its classical Dulong–Petit value 3R and in fact approaches zero as $T \to 0K$.

So the temperature dependence of the heat capacities of solids presents us with still another classically inexplicable phenomenon. Einstein explained this phenomenon in 1907 by applying Planck's concept of energy quanta. Specifically, he postulated that the energies $E(\upsilon)$ of each of the members of the equivalent collection of $3N_A$ one-dimensional harmonic oscillators of frequency υ are quantized as $E(\upsilon) = nh\upsilon$, where $n = 0,1,2,\ldots$. Since this quantization formula is identical to the blackbody quantization equation (Equation 1.15), we may apply results derived earlier for blackbody radiation to the present problem. In particular, we may apply Equation (1.18) for the quantum average energy of a single one-dimensional harmonic oscillator of frequency υ to yield the following expression for the quantum average energy $E(\upsilon, T)$ of the model solid which is equivalent to a collection of $3N_A$ harmonic oscillators of frequency υ:

$$E(\upsilon, T) = \frac{3N_A h\upsilon}{\exp\left(\dfrac{h\upsilon}{kT}\right) - 1}. \tag{1.28}$$

Equations (1.4), (1.25), and (1.28) then yield the quantum heat capacity of the model solid as (Problem 1.16)

$$C(T) = \frac{3R\left(\dfrac{h\upsilon}{kT}\right)^2 \exp\left(-\dfrac{h\upsilon}{kT}\right)}{\left[1 - \exp\left(-\dfrac{h\upsilon}{kT}\right)\right]^2}. \tag{1.29}$$

First, let us find the high-temperature $T \to \infty$ limit of $C(T)$. In this limit, $\frac{h\upsilon}{kT} \to 0$, and hence, $\exp\left(-\frac{h\upsilon}{kT}\right) \to 1$ and $\left[1-\exp\left(-\frac{h\upsilon}{kT}\right)\right]^2 \to \left[1-\left(1-\frac{h\upsilon}{kT}+\cdots\right)\right]^2 = \left(\frac{h\upsilon}{kT}\right)^2$ yielding from Equation (1.29) that

$$\lim_{T\to\infty} C(T) = 3R \tag{1.30}$$

which is the Dulong–Petit law.

Next, let us determine the low-temperature $T \to 0\,K$ limit of $C(T)$. In this limit, $\frac{h\upsilon}{kT} \to \infty$, and Equation (1.29) reduces to

$$\lim_{T\to 0K} C(T) = 3R\left(\frac{h\upsilon}{kT}\right)^2 \exp\left(-\frac{h\upsilon}{kT}\right) \tag{1.31}$$

which in agreement with experiment predicts that $C(T = 0\,K) = 0$.

Additionally, at intermediate temperatures the shapes of Einstein heat capacity curves are qualitatively similar to the shapes of experimental heat capacity curves like those of Figure 1.5 (Problem 1.17). However, the agreement between Einstein's heat capacity formula and experiment is not quantitative because of the error introduced by Einstein's assumption that the atoms of the solid vibrate independently. Especially at very low temperatures, Einstein's formula breaks down qualitatively. Namely, experimentally one finds that $\lim_{T\to 0K} C(T) \alpha T^3$, a result explained theoretically in 1911 by the Dutch physicist and theoretical physical chemist Peter Debye, in qualitative disagreement with Einstein's equation (Equation 1.31).

Summarizing, we have shown in this section that the concept of energy quanta permits the explanation of three phenomena that cannot be explained classically: the blackbody radiation spectrum, the photoelectric effect (or the analogous photoionization effect), and the temperature dependence of the heat capacities of solids.

We next turn the far more central questions that arise when one attempts to understand the line spectra of the hydrogen atom. From 1913 onward, it was the attempts to explain these spectra and the line spectra of more complex atoms (as well as related phenomena) which provided the major driving force for the development of quantum mechanics (van der Waerden 1968).

1.2 CLASSICAL VERSUS LINE SPECTRA

We begin with the planetary model of the atom proposed by the New Zealander-British physicist Ernest Rutherford in 1911. In its modern familiar form, the Rutherford atom is comprised of a tiny positively charged nucleus surrounded by negatively charged orbiting electrons each with a mass m_e that is much less than the nuclear mass m_N. The Rutherford model derived from an analysis of the experiments of Geiger and Marsden who directed a beam of α-particles (helium nuclei) toward a sample of gold foil. Geiger and Marsden found that while most of the α-particles went right through the gold foil, a few were reflected. This result was in conflict with the earlier "plum pudding" model of the atom, but is in agreement with the planetary model, since within this model a few reflections of the α-particles off the gold foil are expected because of rare collisions (rare since the planetary atom consists mainly of empty space) of the α-particles with the tiny gold atomic nuclei. Next, let us specialize the planetary model to the one-electron or hydrogen-like atoms. These atoms include hydrogen, the He^+ ion, the Li^{2+} ion, and so on. For the hydrogen-like atoms, an electron of charge $-e$ where

$$e = 1.6021773 \times 10^{-19}\,C \tag{1.32}$$

orbits a nucleus with charge Ze, where Z is the atomic number, the number of protons in the nucleus of the atom. For H, He^+, Li^{2+}, … $Z = 1, 2, 3, \ldots$.

In this section, we treat the orbiting motion of the electron by classical mechanics and then show that our classical treatment makes predictions in qualitative discord with experiments. In the next section, following the Danish physicist Niels Bohr, we augment our classical treatment with quantum assumptions similar to those of Planck and obtain results for the line spectra of the hydrogen atom in near-perfect agreement with experiment.

We will need the potential energy due to the interaction of the electron and the nucleus. Recall that the Coulomb potential $V(r)$ between two particles with charges q_1 and q_2 depends only on the separation r of the particles and is given in the SI units of joules by

$$V(r) = \frac{q_1 q_2}{4\pi\varepsilon_0 r} \tag{1.33}$$

where

$$4\pi\varepsilon_0 = 1.1126501 \times 10^{-10}\,J^{-1}\,m^{-1}\,C^2 \tag{1.34}$$

is 4π times the vacuum permittivity ε_0. The electron–nucleus potential energy function $U(r)$ is therefore given by

$$U(r) = \frac{(Ze)(-e)}{4\pi\varepsilon_0 r} = -\frac{Ze^2}{4\pi\varepsilon_0 r} \tag{1.35}$$

where r is the distance between the electron and the nucleus.

Further, as indicated, the masses m_N of the nuclei of hydrogen-like atoms are much greater than the electron mass

$$m_e = 9.1093897 \times 10^{-31}\,kg. \tag{1.36}$$

For example, for the hydrogen-like atom with the lightest nucleus hydrogen itself, $m_N \approx 1{,}836\,m_e$. Consequently to an excellent approximation, we may neglect nuclear motion and fix the nucleus at the origin of the coordinate system used to reference the electron motion. Moreover, since $U(r)$ is a central field potential, that is, a potential which depends only on the distance r of the particle from the origin, we may use a result established in every classical mechanics textbook that classical central field motion is planar and choose the coordinate system so that the electron motion occurs in the XY-plane. Then, we may obtain the results from classical mechanics needed to analyze this planar electron motion by adapting results derived in Appendix B from the rigorous analysis of the classical rotational–vibrational motion of a diatomic molecule comprised of one light and one heavy atom, the latter being assumed motionless.

Making this adaptation, we find from Equations (B.4) and (B.5) the following equation of motion for the radial trajectory $r(t)$ of the electron (i.e., for the distance $r[t]$ of the electron from the nucleus as a function of time t):

$$m_e\ddot{r}(t) = -\frac{dU_{eff}\left[r(t)\right]}{dr(t)} \tag{1.37}$$

where the effective potential $U_{eff}\left[r(t)\right]$ is given by

$$U_{eff}\left[r(t)\right] = \frac{L^2}{2m_e r^2(t)} - \frac{Ze^2}{4\pi\varepsilon_0 r(t)}. \tag{1.38}$$

In the above expression for $U_{eff}(r)$, L is the magnitude of the electron's conserved angular momentum vector:

$$\mathbf{L} = \mathbf{r}(t) \times \mathbf{p}(t) \tag{1.39}$$

where $\mathbf{r}(t)$ and $\mathbf{p}(t) = m_e\dot{\mathbf{r}}(t)$ are the position and momentum vectors of the electron. Adapting Equation (B.7), we also find the following result for the conserved energy E of the electron:

$$E = \frac{p_r^{\,2}(t)}{2m_e} + \frac{L^2}{2m_e r^2(t)} - \frac{Ze^2}{4\pi\varepsilon_0 r(t)} \tag{1.40}$$

where $p_r(t) = m_e\dot{r}(t)$ is the electron's radial momentum.

We next specialize our equations to circular orbits of radius r. For such orbits, $r(t)$ has the fixed value r, and thus, $p_r(t) = m_e\dot{r}(t) = 0$ and $\ddot{r}(t) = 0$. Equations (1.37), (1.38), and (1.40) then simplify to, respectively,

$$\frac{dU_{\text{eff}}(r)}{dr} = 0 \tag{1.41}$$

$$U_{\text{eff}}(r) = \frac{L^2}{2m_e r^2} - \frac{Ze^2}{4\pi\varepsilon_0 r} \tag{1.42}$$

and

$$E = \frac{L^2}{2m_e r^2} - \frac{Ze^2}{4\pi\varepsilon_0 r}. \tag{1.43}$$

Comparing Equations (1.41) and (1.42) gives that $\frac{d}{dr}\left(\frac{L^2}{2m_e r^2} - \frac{Ze^2}{4\pi\varepsilon_0 r}\right) = 0$, which in turn yields the following relationship between r and L^2:

$$r_L \equiv r = \frac{4\pi\varepsilon_0}{m_e Z e^2} L^2. \tag{1.44}$$

Using Equation (1.44) to eliminate r from Equation (1.43) gives the energy E in terms of L^2 as

$$E_L \equiv E = -\frac{Z^2 e^4 m_e}{2(4\pi\varepsilon_0)^2 L^2}. \tag{1.45}$$

Additionally, for circular orbits we may easily find the frequency of revolution υ of the electron about the nucleus. The following expression for υ may be obtained by adapting Equation (B.10) and then using Equation (1.44):

$$\upsilon_L \equiv \upsilon = \frac{Z^2 e^4 m_e}{2\pi(4\pi\varepsilon_0)^2 L^3}. \tag{1.46}$$

Next, let us compare the predictions of classical theory with experiment. Classically, the magnitude L of the electron's angular momentum vector **L** can take on any nonnegative value. Thus, from Equation (1.46) the electron frequency υ_L can have any positive value. However, according to classical electromagnetic theory, a charged particle executing uniform circular motion with frequency υ emits light with the same frequency υ. Consequently, classical theory predicts that a

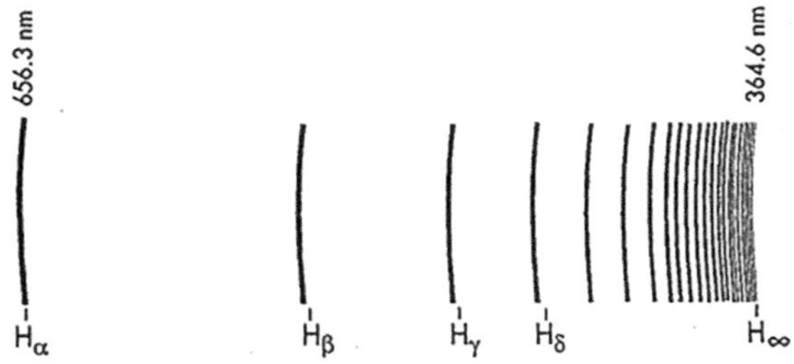

FIGURE 1.6 The Balmer series emission spectrum of the hydrogen atom. The $H_\alpha, H_\beta, H_\gamma$, and H_δ lines occur in the visible region of the electromagnetic spectrum, while the remaining lines occur in the near-ultraviolet region of the electromagnetic spectrum.

gas of hydrogen-like atoms emits at all frequencies (since the atoms have a continuous range of frequencies) and thus will have a continuous emission spectrum. Instead, what is observed is line spectra. For example, the visible emission spectrum of hydrogen consists of four lines called the H_α, H_β, H_γ, and H_δ lines. These lines, shown in Figure 1.6, have discrete wavelengths λ of, respectively, 656.3, 486.1, 434.1, and 410.2 nm.

Classical theory not only fails to explain the line spectra of the hydrogen-like atoms (as well as the line spectra of other atoms) but also predicts that atoms are unstable. This is because according to classical electromagnetic theory orbiting electrons constantly radiate light and thus continually lose energy. Therefore, they will eventually spiral into the nucleus causing the atom to collapse.

Before describing how Bohr's theory of the hydrogen-like atoms eliminates for these atoms the gross deficiencies of classical theory, we discuss in more detail the experimental emission spectra of hydrogen.

In 1885, the Swiss mathematician Johann Balmer realized that the frequencies of the four visible emission lines of hydrogen could be fit to a simple formula that also predicts near-ultraviolet lines that were later observed. In modern notation, Balmer's formula gives the *wave numbers* $\bar{\upsilon} = \frac{\upsilon}{c} = \frac{1}{\lambda}$ of emitted light of frequency υ as

$$\bar{\upsilon} = R\left(\frac{1}{4} - \frac{1}{n_2^2}\right) \tag{1.47}$$

where n_2 is an integer that can take on the values $n_2 = 3, 4, \cdots$ and R is a constant (not to be confused with the gas constant R) with the modern value given in Equation (1.48). For $n_2 = 3, 4, 5$, and 6, Balmer's equation (Equation 1.47) predicts the wave numbers of the visible hydrogen $H_\alpha, H_\beta, H_\gamma$, and H_δ lines. For $n_2 > 6$, it predicts the wave numbers of a sequence of near-ultraviolet lines that were later observed. Spectral lines that conform to Balmer's formula are said to form the Balmer series of hydrogen.

While wave numbers $\bar{\upsilon}$ have the SI units m^{-1}, observed wave numbers are usually reported in cm^{-1} units $(1\,cm^{-1} = 100\,m^{-1})$. R therefore is usually given in cm^{-1} units. The modern value of R is

$$R = 109{,}677.576\ cm^{-1}. \tag{1.48}$$

In 1888, the Swedish physicist Johannes Rydberg generalized Balmer's formula to give an equation that predicts hydrogen atom spectral series additional to the Balmer series. Rydberg's formula is

$$\bar{\upsilon} = R\left(\frac{1}{n_1^2} - \frac{1}{n_2^2}\right) \tag{1.49}$$

where the ranges of n_1 and n_2 are

$$n_1 = 1,2,\ldots \text{ and } n_2 = n_1 + 1, n_1 + 2,\ldots. \quad (1.50)$$

For $n_1 = 2$, Rydberg's formula reduces to Balmer's equation (Equation 1.47). But for $n_1 = 1,3,\ldots$, it predicts new hydrogen spectral series that were later observed. Namely, it predicts the Lyman series $(n_1 = 1)$, Paschen series $(n_1 = 3)$, Brackett series $(n_1 = 4)$, Pfund series $(n_1 = 5)$, and so on. The Lyman series lines occur in the ultraviolet region of the electromagnetic spectrum, while the Paschen, Brackett, and Pfund series lines occur in the infrared region of the electromagnetic spectrum.

We may now turn to Bohr's 1913 theory of the hydrogen-like atoms.

1.3 THE BOHR THEORY OF THE HYDROGEN-LIKE ATOMS

Bohr's theory of the hydrogen-like atoms is based on the following assumptions, some of which generalize to all other atoms:

1. A hydrogen-like atom can only exist in a series of stable states called *stationary states* distinguished by the values of a quantum number $n = 1,2,3\ldots$ and characterized by quantized energies E_n where $E_1 < E_2 < E_3 \cdots$. The stationary state with the lowest energy E_1 and the smallest quantum number $n = 1$ is called the ground state, the stationary state with the next lowest energy E_2 and the next smallest quantum number $n = 2$ is called the first excited state, and so on.
2. Stationary states are stable because in conflict with classical electromagnetic theory, when an atom is in a stationary state the orbital motion of its electron does not induce the emission of electromagnetic radiation.
3. An atom can only emit electromagnetic radiation by making a transition between a stationary state of higher energy E_{n_2} and a stationary state of lower energy E_{n_1} (Figure 1.7). Correspondingly, an atom can only absorb electromagnetic radiation by making a transition from a stationary state of lower energy E_{n_1} to a stationary state of higher energy E_{n_2}.
4. The frequency υ of radiation emitted or absorbed in transitions between stationary states with quantum numbers n_1 and n_2, where $n_2 > n_1$ and hence $E_{n_2} > E_{n_1}$, is given by the *Bohr frequency condition*

$$h\upsilon = E_{n_2} - E_{n_1}. \quad (1.51)$$

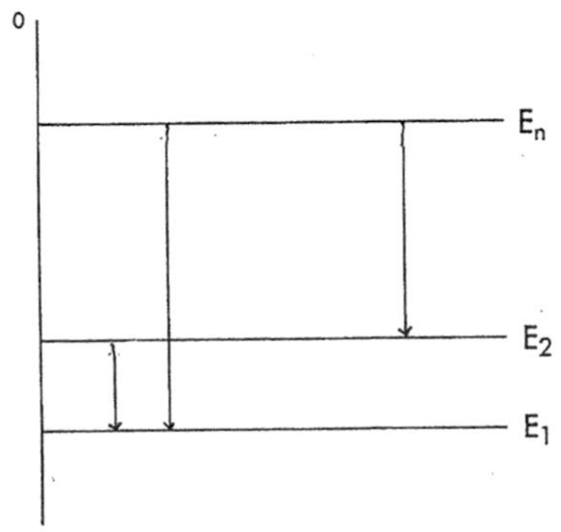

FIGURE 1.7 Schematic depiction of the origin of the emission spectrum of an atom. The emission lines arise from transitions that move the atom from stationary states of higher energy to stationary states of lower energy accompanied by emission of a photon.

Equation (1.51) follows from energy conservation if one assumes that the transitions are caused by emission or absorption of a single photon of energy $\varepsilon_\upsilon = h\upsilon$ (Problem 1.18). (Assumptions 1, 2, and 4 are analogous to the assumptions made by Planck in order to derive his blackbody formula.)

5. In stationary states, the atom's electron moves in the Coulomb potential produced by the nucleus along a classical circular orbit trajectory.
6. The magnitude L of the electron's angular momentum vector **L** when the atom is in the nth stationary state is quantized as

$$L = n\hbar, \quad \text{where } n = 1,2,\ldots \tag{1.52}$$

and where $\hbar$ (read h-bar) is defined by

$$\hbar = \frac{h}{2\pi} = 1.05457 \times 10^{-34}\,\text{J s}. \tag{1.53}$$

The quantization condition of Equation (1.52) is justified later in this section.

We may now obtain the main results of the Bohr theory.

From Equation (1.52) and the classical energy equation (Equation 1.45) whose use is justified by Assumption 5, we find Bohr's result for the quantized energies E_n of a hydrogen-like atom as

$$E_n = -\frac{Z^2 e^4 m_e}{2(4\pi\varepsilon_0)^2 n^2 \hbar^2} \quad \text{where } n = 1,2,\ldots. \tag{1.54}$$

The energy levels E_n are plotted for hydrogen with $Z = 1$ in Figure 1.8. Notice that the levels become increasingly closely spaced as n increases with the level spacing approaching zero as $n \rightarrow \infty$.

Similarly, we find from Equations (1.44) and (1.52) that according to Bohr's theory, the radii of the electron's circular orbits are restricted to the discrete values

$$r_n = \frac{4\pi\varepsilon_0}{m_e Z e^2} n^2 \hbar^2 \quad \text{where } n = 1,2,\ldots. \tag{1.55}$$

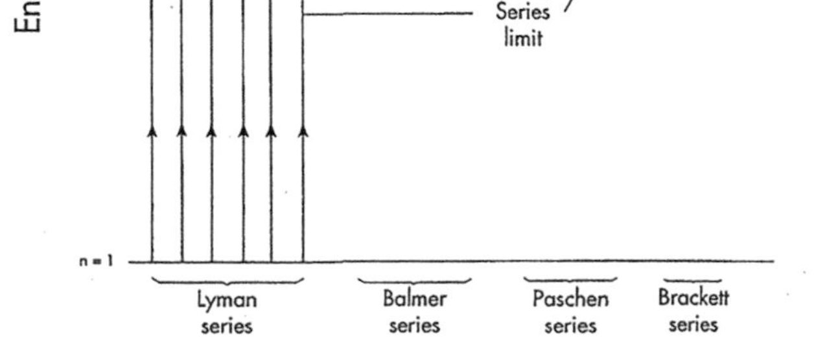

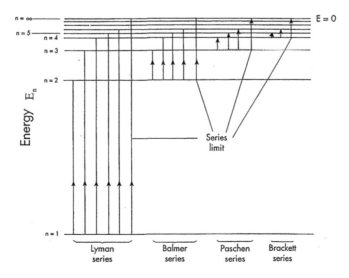

FIGURE 1.8 The transitions that give rise to the absorption spectral series of the hydrogen atom.

The orbit radius of the hydrogen atom in its n = 1 ground state is of special interest since it will turn out to be the natural unit of length in the atomic world. It is called the Bohr radius and is denoted by the symbol a_0. Setting Z = 1 and n = 1 in Equation (1.55) gives the Bohr radius a_0 as

$$a_0 = \frac{4\pi\varepsilon_0}{m_e e^2}\hbar^2 = 5.29177 \times 10^{-11}\,\text{m}. \tag{1.56}$$

Comparing Equations (1.55) and (1.56) yields the following convenient form for the discrete radii r_n (see Figure 1.9):

$$r_n = \frac{n^2}{Z} a_0, \quad \text{where } n = 1, 2, \ldots. \tag{1.57}$$

Equation (1.57) shows that as expected, the atom contracts as the nuclear charge Ze increases.

Next, let us turn to the emission spectra of the hydrogen-like atoms. We wish to determine the wave numbers $\bar{\upsilon}_{n_2 \to n_1}$ of emitted light due to transitions from stationary states of higher quantum numbers n_2 (or higher energies E_{n_2}) to stationary states of lower quantum numbers n_1 (or lower energies E_{n_1}). (For hydrogen, the corresponding absorption transitions are shown in Figure 1.8.)

Since we require wave numbers $\bar{\upsilon} = \frac{\upsilon}{c}$ rather than frequencies υ, we need the Bohr frequency condition of Equation (1.51) in the form

$$\bar{\upsilon}_{n_2 \to n_1} = \frac{E_{n_2} - E_{n_1}}{hc}. \tag{1.58}$$

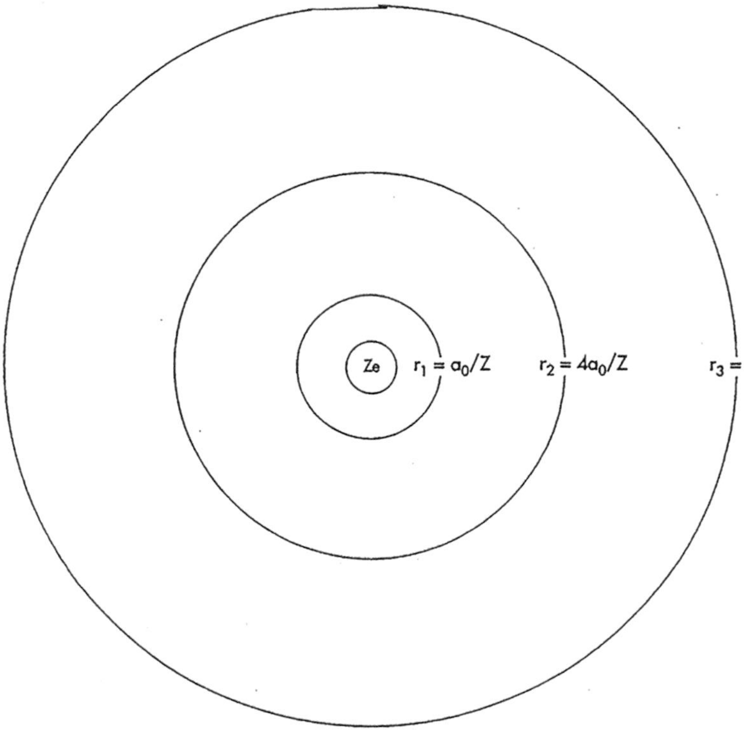

FIGURE 1.9 The radii r_n, as given by Equation (1.57), of the Bohr circular orbits for the first three stationary states of a hydrogen-like atom with atomic number Z.

Comparing this form with Equation (1.54) for the energies E_n gives the wave numbers of the emitted light as

$$\bar{\upsilon}_{n_2 \to n_1} = Z^2 R_\infty \left(\frac{1}{n_1^2} - \frac{1}{n_2^2} \right) \tag{1.59}$$

where the range of the quantum numbers n_1 and n_2 is

$$n_1 = 1, 2, \ldots \text{ and } n_2 = n_1 + 1, n_1 + 2, \cdots \tag{1.60}$$

and where the *Rydberg constant*

$$R_\infty = \frac{2\pi^2 e^4 m_e}{(4\pi\varepsilon_0)^2 h^3 c} = 109{,}737.315\,\text{cm}^{-1}. \tag{1.61}$$

When specialized to the hydrogen atom with $Z = 1$, Bohr's equations (Equations 1.59 and 1.60) for the spectral line positions nearly coincide with Rydberg's empirical equations (Equations 1.49 and 1.50). The only difference occurs because, as may be seen by comparing Equations (1.48) and (1.61), the values of the constants R_∞ and R differ slightly. (Most of the small discrepancy can be eliminated by incorporating nuclear motion into Bohr's treatment. A treatment of the same effect within the modern quantum theory of the hydrogen atom line spectra is given in Section 10.3.)

Thus, Bohr's theory provides an atomic interpretation of the observed hydrogen atom emission line spectra. For example, the Balmer series lines are due to transitions from stationary states with quantum numbers $n_2 = 3, 4, \ldots$ to the stationary state with quantum number $n_1 = 2$. Correspondingly, the Lyman series lines are due to transitions between stationary states with quantum numbers $n_2 = 2, 3, \ldots$ and the stationary state with quantum number $n_1 = 1$. The atomic origins of the Paschen series lines $(n_1 = 3)$, Brackett series lines $(n_1 = 4)$, Pfund series lines $(n_1 = 5)$, and so on are analogous.

We close this section by giving a justification for Bohr's angular momentum quantization hypothesis of Equation (1.52). To start, we note that it follows from the classical energy equation (Equation 1.45) and the Bohr frequency condition that the quantum theory cannot reproduce Rydberg's empirical equation (Equation 1.49) unless L is quantized as

$$L = n \frac{H}{2\pi}, \quad \text{where } n = 1, 2, \ldots \tag{1.62}$$

and where H is a constant to be determined.

Inserting Equation (1.62) into the classical equations (Equations 1.45 and 1.46) and using Equation (1.61) for R_∞ gives the following quantized forms for the energies E_n and the electron frequencies υ_n.

$$E_n = -\frac{hcZ^2 R_\infty}{n^2} \frac{h^2}{H^2} \tag{1.63}$$

and

$$\upsilon_n = \frac{2Z^2 R_\infty}{n^3} \frac{h^3}{H^3} c \tag{1.64}$$

where n = 1,2,....

From the Bohr frequency condition of Equations (1.58) and (1.63) for E_n, we obtain the wave numbers of emitted light as

$$\bar{\upsilon}_{n2\to n1} = Z^2 R_\infty \frac{h^2}{H^2}\left(\frac{1}{n_1^2} - \frac{1}{n_2^2}\right). \tag{1.65}$$

Specializing $\bar{\upsilon}_{n_2\to n_1}$ to the transitions $n_2 = n_1 + 1$ yields

$$\bar{\upsilon}_{n_1+1\to n_1} = Z^2 R_\infty \frac{h^2}{H^2}\left[\frac{2n_1+1}{n_1^2(n_1+1)^2}\right]. \tag{1.66}$$

In the large quantum number limit $n_1 \to \infty$, Equation (1.66) reduces to

$$\lim_{n_1\to\infty} \bar{\upsilon}_{n_1+1\to n_1} = \frac{2Z^2R_\infty}{n_1^3}\frac{h^2}{H^2}. \tag{1.67}$$

To determine H, we will compare $\lim_{n_1\to\infty} \bar{\upsilon}_{n_1+1\to n_1}$ with (see Equation 1.64)

$$\bar{\upsilon}_{n_1} \equiv \frac{\upsilon_{n_1}}{c} = \frac{2Z^2R_\infty}{n_1^3}\frac{h^3}{H^3}. \tag{1.68}$$

The quantity $\bar{\upsilon}_{n_1}$ is the classical wave number for the emission transition $n_1 + 1 \to n_1$ (Problem 1.26). The corresponding quantum wave number is of course $\bar{\upsilon}_{n_1+1\to n_1}$. To proceed further, we need the *Bohr correspondence principle*. According to the correspondence principle, quantum and classical predictions of frequencies or wave numbers for spectroscopic transitions must agree in the limit of large quantum numbers. Applying the correspondence principle to our problem yields

$$\lim_{n_1\to\infty} \bar{\upsilon}_{n_1+1\to n_1} = \bar{\upsilon}_{n_1} \tag{1.69}$$

which gives from Equations (1.67) and (1.68)

$$\frac{2Z^2R_\infty}{n_1^3}\frac{h^2}{H^2} = \frac{2Z^2R_\infty}{n_1^3}\frac{h^3}{H^3} \tag{1.70}$$

which in turn yields $\frac{h^2}{H^2} = \frac{h^3}{H^3}$ or $H = h$. Therefore, Equation (1.62) reduces to Bohr's angular momentum quantization hypothesis of Equation (1.52), thus justifying this hypothesis.

For future reference, let us set $H = h$ to rewrite Equation (1.63) for E_n as

$$\frac{E_n}{hc} = -\frac{Z^2R_\infty}{n^2}. \tag{1.71}$$

Since R_∞ has units of cm^{-1}, Equation (1.71) shows that the energies of the hydrogen-like atoms (as well as all other atoms and also molecules) may be expressed in cm^{-1} units as well as in more conventional units such as joules or electron volts.

We next summarize the steps that led from the Bohr theory to modern quantum mechanics.

1.4 FROM THE BOHR THEORY TO MODERN QUANTUM MECHANICS

Bohr's theory represents a major advance in the understanding of atomic structure and spectra. Most notably, it explains atomic stability by forbidding the emission of electromagnetic radiation by an atom in its ground state, it provides a picture of how the line spectra of atoms emerge as single photon transitions between stable stationary states, it gives a quantitative explanation of these spectra for the one-electron atoms, it provides the Bohr frequency condition which is universally used in modern quantum mechanics to determine atomic and molecular spectral line frequencies, and it provides a natural unit of length for the atomic world, the Bohr radius a_0.

However, Bohr's theory suffers from severe limitations that include the following:

1. An extension of Bohr's ideas known as the old quantum theory (Pauling and Wilson 1985, Chapter 2) which renders these ideas applicable to systems other than the hydrogen-like atoms and to hydrogen-like atoms subject to electric and magnetic fields often yields results only in imperfect agreement with experiment. Most seriously, the old quantum theory fails qualitatively to describe the line spectra of even the simplest multi-electron atom, the helium atom.
2. Both Bohr's theory and the old quantum theory are incomplete. This is because while they predict spectral line positions, they do not predict other spectral features, especially line intensities.

For these and other reasons, the Bohr theory and the old quantum theory were eventually replaced by modern quantum mechanics.

Many of the key developments that culminated in the birth of modern quantum mechanics are documented in the reprint volume edited by van der Waerden (van der Waerden 1968). Here, we will give only the briefest description of some of these developments:

1. The Bohr correspondence principle was greatly generalized by Bohr in 1918. For example, it was extended to include line intensities as well as line frequencies. This reduced the search for quantum mechanics to the problem of extrapolating from the known classical high-quantum-number behavior of a system to its unknown low-quantum-number behavior.
2. The first completely satisfactory way of making this extrapolation was proposed by the German physicist Werner Heisenberg in 1925. Heisenberg's method was based on the following two premises. (a) The equations of the new quantum mechanics must only involve quantities closely related to observables. Thus, in contrast to the Bohr theory and the old quantum theory, the new quantum mechanics must not involve trajectories since as we will see in Section 3.5 for microscopic particles like electrons, trajectories cannot be observed. (b) The equations of classical mechanics are valid when expressed in terms of quantities that are closely related to the observable spectral line frequencies and line intensities. On the basis of these premises, Heisenberg developed a primitive form of modern quantum mechanics and made a few simple applications.
3. Heisenberg's theoretical framework was soon greatly expanded and also was applied to more complex problems like the hydrogen atom by Born, Jordan, Dirac, Pauli, and by Heisenberg himself. The form of modern quantum mechanics which emerged from this work is now called *matrix mechanics*.
4. A second completely independent route to quantum mechanics developed from Einstein's work on the photoelectric effect. Recall that in order to explain the photoelectric effect, Einstein was forced to assume that light has particle-like as well as wave-like properties and thus exhibits a *wave–particle duality*. Motivated by Einstein's work, in 1924 the French physicist Louis de Broglie boldly proposed that the wave–particle duality extended to ordinary particles. That is, de Broglie proposed that particles like electrons have wave-like as well as particle-like properties. De Broglie's hypothesis was soon confirmed experimentally by Davisson and Germer who showed that a beam of electrons displayed the wave interference phenomenon of diffraction.

5. Given that particles have wave-like properties, the following question immediately arises: What is the wave equation which describes the wave-like behavior of particles? This question was answered in 1926 by the Austrian physicist Erwin Schrödinger who developed the required wave equation now known as the Schrödinger equation. By solving his equation, Schrödinger was able to correctly determine the energy levels of the harmonic oscillator, the hydrogen atom, and other simple quantum systems. Schrödinger soon showed that his form of modern quantum mechanics now known as *wave mechanics* is equivalent to matrix mechanics. However, wave mechanics is much more convenient mathematically and conceptually than matrix mechanics. Thus, the Schrödinger equation is now the foundation for nearly all applications of quantum mechanics to chemistry and will be used exclusively in this textbook.

We close this chapter by giving a brief preview of Schrödinger's form of modern quantum mechanics which emphasizes parallel features of the structures of quantum and classical mechanics.

1.5 A PREVIEW OF MODERN QUANTUM MECHANICS

While modern quantum mechanics differs radically from classical mechanics in both outward appearance and essential content, there are some striking parallels between the two theories. We will next summarize these parallels since they aid in the understanding of quantum mechanics.

The parallels are as follows:

1. Both quantum and classical mechanics involve a central quantity X. The quantity X is central since all system observables may be found from X.
2. Both quantum and classical mechanics consist of two types of rules: type 1 rules for determining X and type 2 rules for obtaining the system observables from X.

To illustrate these parallels, we will consider a system comprised of a single particle of mass m with position vector $\mathbf{r} = (x, y, z)$ moving in three dimensions under the influence of a potential energy function $U(\mathbf{r})$.

We first discuss the classical mechanics of the particle. Referring to the discussion of classical mechanics of Appendix A, the central quantity X of the system is the particle's trajectory $\mathbf{r}(t)$, the type 1 rule is Newton's equation of motion for $\mathbf{r}(t)$ $m\ddot{\mathbf{r}}(t) = -\frac{\partial U[\mathbf{r}(t)]}{\partial \mathbf{r}(t)}$, and an example of a type 2 rule is the formula $\mathbf{L}(t) = \mathbf{r}(t) \times \mathbf{p}(t) = m\mathbf{r}(t) \times \dot{\mathbf{r}}(t)$, which determines the particle's observable angular momentum $\mathbf{L}(t)$ from its trajectory $\mathbf{r}(t)$.

Next, let us turn to quantum mechanics. As discussed in detail in Chapter 4, in quantum mechanics the central quantity is the system wave function $\Psi(\mathbf{r}, t)$, the type 1 rule is the time-dependent Schrödinger equation for $\Psi(\mathbf{r}, t)$

$$i\hbar \frac{\partial \Psi(\mathbf{r}, t)}{\partial t} = \left[-\frac{\hbar^2}{2m} \nabla^2 + U(\mathbf{r}) \right] \Psi(\mathbf{r}, t) \tag{1.72}$$

where ∇^2 (read del-squared) is defined by $\nabla^2 = \frac{\partial^2}{\partial x^2} + \frac{\partial^2}{\partial y^2} + \frac{\partial^2}{\partial z^2}$, and a typical type 2 rule is the following formula for the *average* value $\langle \mathbf{L}(t) \rangle$ of the particle's angular momentum at time t

$$\langle \mathbf{L}(t) \rangle = \int \Psi^*(\mathbf{r}, t) \left[\mathbf{r} \times \frac{h}{i} \nabla \Psi(\mathbf{r}, t) \right] d\mathbf{r} \tag{1.73}$$

where $\Psi^*(\mathbf{r},t)$ is the complex conjugate of $\Psi(\mathbf{r},t)$, where ∇ (read del) is defined in terms of the unit vectors $\mathbf{e}_x, \mathbf{e}_y$, and $\mathbf{e}_z$ pointing in, respectively, the X−,Y−, and Z−directions by $\nabla = \mathbf{e}_x \frac{\partial}{\partial x} + \mathbf{e}_y \frac{\partial}{\partial y} + \mathbf{e}_z \frac{\partial}{\partial z}$, and where $\int d\mathbf{r} \equiv \int_{-\infty}^{\infty} dx \int_{-\infty}^{\infty} dy \int_{-\infty}^{\infty} dz$.

One might ask why the quantum type 2 rule gives only the average value of $\mathbf{L}(t)$ rather than its absolute value. This is because in contrast to classical mechanics, quantum mechanics, as is described in detail in sections 4.3 and 4.4 does not in general make definite predictions of the outcomes of experiments that measure the values of observables. Rather, it only predicts the likelihoods or probabilities of each of the many possible outcomes. Consequently, in quantum mechanics the most widely used type 2 rules predict values for observables averaged over the spread of their values found in many different measurements.

In summary, in quantum mechanics the trajectory $\mathbf{r}(t)$ is replaced by the wave function $\Psi(\mathbf{r},t)$, Newton's equation of motion is replaced by the time-dependent Schrödinger equation, and average rather than absolute values are predicted for observables.

FURTHER READINGS

Beiser, Arthur. 1995. *Concepts of modern physics*. 5th ed. New York: McGraw-Hill, Inc.

Cohen-Tannoudji, Claude, Bernard Diu, and Franck Laloe. 1977. *Quantum mechanics*. Vol. 1. Translated by Susan Reid Hemley, Nicole Ostrowsky, and Dan Ostrowsky. New York: John Wiley & Sons.

Dicke, Robert H., and James P. Wittke. 1960. *Introduction to quantum mechanics*. Reading, MA: Addison-Wesley Publishing Company, Inc.

McQuarrie, Donald A. 2008. *Quantum chemistry*. 2nd ed. Sausalito, CA: University Science Books.

Pauling, Linus, and E. Bright Wilson Jr. 1985. *Introduction to quantum mechanics with applications to chemistry*. New York: Dover Publications, Inc.

van der Waerden, B. L., ed. 1968. *Sources of quantum mechanics*. New York: Dover Publications, Inc.

PROBLEMS

1.1 Derive Equation (1.9) for the average energy $U(T)$ from Equations (1.6)–(1.8).

1.2 From Equation (A.22) of Appendix A, the classical energy $E(\upsilon)$ of a one-dimensional harmonic oscillator with coordinate y, momentum p_y, mass m, and frequency υ is $H(y,p_y) = \frac{p_y^2}{2m} + \frac{1}{2} m\omega^2 y^2$, where $\omega = 2\pi\upsilon$. Thus, the classical form of the harmonic oscillator partition function of Equation (1.11) is $Q(\upsilon,T) = \int_{-\infty}^{\infty} dp_y \int_{-\infty}^{\infty} dy \exp\left[-\frac{H(y,p_y)}{kT}\right]$. Using the form of $H(y,p_y)$ just given and the integral $\int_{-\infty}^{\infty} \exp(-\alpha x^2) dx = \left(\frac{\pi}{\alpha}\right)^{1/2}$, show that the classical partition function $Q(\upsilon,T) = \frac{kT}{\upsilon}$.

1.3 The quantity $u(\upsilon,T)$ plotted in Figures 1.1 and 1.2 is the average energy per unit volume and per unit frequency range of blackbody cavity radiation at temperature T. Thus, $u(\upsilon,T)d\upsilon$ is the average energy per unit volume at temperature T of cavity radiation with frequencies between υ and $\upsilon + d\upsilon$. Therefore, $u(\upsilon,T)d\upsilon$ has the units $J\,m^{-3}$. Show from Equations (1.14) and (1.19) that both $u_{RJ}(\upsilon,T)d\upsilon$ and $u_{PI}(\upsilon,T)d\upsilon$ have the units $J\,m^{-3}$.

1.4 The Planck blackbody radiation formula can be expressed in terms of the wavelengths $\lambda = \frac{c}{\upsilon}$ of the cavity radiation as well as its frequencies υ. Let $\rho_{PI}(\lambda,T)d\lambda$ be the Planck theory expression for the average energy per unit volume at temperature T of cavity radiation with wavelengths between λ and $\lambda + d\lambda$. Then, it follows that $\rho_{PI}(\lambda,T)d\lambda = -u_{PI}(\upsilon,T)d\upsilon$. (The minus sign occurs because a decrease in wavelength corresponds to an increase in frequency.) (a) From this relation and Equation (1.19) for $u_{Pl}(\upsilon,T)$, show that $\rho_{Pl}(\lambda,T)=\frac{8\pi hc}{\lambda^s}\frac{1}{\exp\left(\frac{hc}{\lambda kT}\right)-1}$. (b) Plot $\rho_{Pl}(\lambda,T)$ vs λ in nm at $T=1{,}000$, 5,000, and 10,000 K.

1.5 Before the work of Planck, Wein showed by thermodynamic arguments that the wavelength λ_{max} of maximum blackbody emission is related to the temperature T of the blackbody by the Wien displacement law $\lambda_{max}=\frac{b}{T}$, where b is a constant determined empirically as $b=2.89777\times10^6\,\text{nm K}$. Notice that in accord with your plots of Problem 1.4b, λ_{max} decreases as T increases. In this problem, you will derive the Wien displacement law from Planck's blackbody theory. You will need Planck's blackbody energy density $\rho_{Pl}(\lambda,T)$ whose form was derived in Problem 1.4a as $\rho_{Pl}(\lambda,T)=\frac{8\pi hc}{\lambda^5}\frac{1}{\exp\left(\frac{hc}{\lambda kT}\right)-1}$. In particular, you will take λ_{max} as the wavelength λ which maximizes $\rho_{Pl}(\lambda,T)$ and thus is determined by the condition $\left[\frac{d\rho_{Pl}(\lambda,T)}{d\lambda}\right]_{\lambda=\lambda_{max}}=0$. (a) From this condition and the form of $\rho_{Pl}(\lambda,T)$, show that $x\equiv\frac{hc}{\lambda_{max}kT}$ satisfies the equation $\frac{1}{5}x+e^{-x}=1$. (b) Solve this equation iteratively on a calculator or computer (taking as your first guess $x=5$) to show that $x=4.96511$. (c) Thus, show that $\lambda_{max}=\frac{hc}{kx}\frac{1}{T}=\frac{2.89777\times10^6\,\text{nm K}}{T}$ in perfect accord with Wien's empirical displacement law. (d) Compute λ_{max} in nm at the temperatures 1,000, 5,000, and 10,000 K, and referring to Table 7.2, determine what region of the electromagnetic spectrum λ_{max} lies in at each of these temperatures.

1.6 Using Equation (1.10), derive Equation (1.18) from Equation (1.17).

1.7 In Problem 1.5, it was shown that Planck's theory gives the following result for the wavelength λ_{max} of maximum blackbody emission at temperature T: $\lambda_{max}=\frac{hc}{kx}\frac{1}{T}$, where $x=4.96511$. Compare this result with Wien's empirical displacement law $\lambda_{max}=\frac{2.89777\times10^6\,\text{nm K}}{T}$ to determine from experiment the numerical value of Planck's constant h.

1.8 To estimate the surface temperatures of stars, it is often assumed that their observed emission spectra may be approximated by blackbody spectra. With this assumption, the surface temperature T of a star may be estimated from its wavelength of maximum emission λ_{max} given by Wien's displacement law (see Problem 1.5) as $T=\frac{2.9\times10^6\,\text{nm K}}{\lambda_{max}}$. For the sun, $\lambda_{max}=500\,\text{nm}$. Estimate the surface temperature of the sun.

1.9 One of the hottest stars with surface temperature $T\approx40{,}000\,\text{K}$ is Eta Carinae located about 7,500 light-years from the sun. Assume that the emission spectrum of Eta Carinae may be approximated by a blackbody spectrum. Then, referring to Problem 1.8, estimate the wavelength of maximum emission λ_{max} of Eta Carinae. Also referring to Table 7.2, determine the region of the electromagnetic spectrum λ_{max} lies in.

1.10 This problem deals with the Stefan–Boltzmann law and its derivation from the Planck blackbody theory. The Stefan–Boltzmann law that was derived by thermodynamic arguments before Planck's work is as follows: $R = \frac{c}{4}u(T) = \sigma T^4$, where R is the total energy radiated by the blackbody per unit area per unit time at temperature T, where $u(T) = \int_0^{\infty} u(\upsilon, T)\, d\upsilon$ is the total energy per unit volume of the blackbody, and where σ is the Stefan–Boltzmann constant that has the empirical value $\sigma = 5.670367 \times 10^{-8}\, J\, m^{-2}\, s^{-1}\, K^{-4}$. You will derive the Stefan–Boltzmann law by first setting $u(\upsilon, T)$ equal to $u_{Pl}(\upsilon, T) = \frac{8\pi h}{c^3}\frac{\upsilon^3}{\exp\left(\frac{h\upsilon}{kT}\right) - 1}$, then evaluating $u(T)$ as $u_{Pl}(T) \equiv \int_0^{\infty} u_{Pl}(\upsilon, T) d\upsilon$, and finally determining R as $R = \frac{c}{4}u_{Pl}(T)$. This derivation will give σ in terms of fundamental constants. To carry through the derivation, you will need the integral $\int_0^{\infty} \frac{x^3 dx}{e^x - 1} = \frac{\pi^4}{15}$. (a) Show that $u_{Pl}(T) = \frac{8\pi^5 k^4}{15 c^3 h^3} T^4$ (b) Thus, show that $R = \frac{c}{4}u_{Pl}(T) = \sigma T^4$ where $\sigma = \frac{2\pi^5 k^4}{15 c^2 h^3}$ (c) Numerically evaluate the theoretical formula for σ found in part (b) and compare your result with the experimental value $\sigma = 5.670367 \times 10^{-8}\, J\, m^{-2}\, s^{-1}\, K^{-4}$.

1.11 What is the energy E in J of a photon of ultraviolet light with wavelength $\lambda = \frac{c}{\upsilon} = 200$ nm.

1.12 Cesium has the ionization potential $I = 3.8939$ eV. What is the wavelength λ_{max} in nm of the longest wavelength light that can photoionize cesium? (Note $1\ eV = 1.60218 \times 10^{-19}$ J.)

1.13 Referring to Problem 1.12, if a gas of cesium atoms is irradiated with ultraviolet light of wavelength $\lambda = 200\,nm$, what is (a) the kinetic energy T_{photo} of the ejected electron in J and (b) the speed v of the ejected electron in $m\,s^{-1}$?

1.14 It is found that if a sample of rubidium gas is irradiated with ultraviolet light with wavelengths less than a critical wavelength $\lambda_{max} = 297.24$, ejected electrons appear in the laboratory. What is the ionization potential I of rubidium in eV?

1.15 A potassium gas irradiated with ultraviolet light of wavelengths $\lambda = 200$ and $250\,nm$ releases electrons with respective speeds v of $8.085 \times 10^5\, m\, s^{-1}$ and $4.665 \times 10^5\, m\, s^{-1}$. From this information, determine Planck's constant h in Js and the ionization potential I of potassium in eV.

1.16 Derive Einstein's heat capacity equation (Equation 1.29) from the formula for $E(\upsilon, T)$ of Equation (1.28).

1.17 The Einstein heat capacity formula of Equation (1.29) may be written as $\frac{C(T)}{3R} = \frac{\left(\frac{\Theta_E}{T}\right)^2 \exp\left(-\frac{\Theta_E}{T}\right)}{\left[1 - \exp\left(-\frac{\Theta_E}{T}\right)\right]^2}$, where the *Einstein temperature* Θ_E is defined by $\Theta_E = \frac{h\upsilon}{k}$.

(Θ_E is not a real temperature. It is just the frequency υ expressed in Kelvin units.) Plot $\frac{C(T)}{3R}$ versus $\frac{T}{\Theta_E}$ and note that your curve is qualitatively similar in shape to the experimental curves of Figure 1.5.

1.18 Derive using energy conservation arguments the Bohr frequency condition of Equation (1.51):

(a) for emission assuming transitions from a stationary state of higher energy E_{n_2} to a state of lower energy E_{n_1} are accompanied by emission of a single photon of energy $\varepsilon_\upsilon = h\upsilon$ and (b) for absorption assuming transitions from a state of lower energy E_{n_1} to a state of higher energy E_{n_2} are accompanied by the absorption of a single photon of energy $\varepsilon_\upsilon = h\upsilon$.

1.19 Determine the ground state energy of the hydrogen atom in (a) J, (b) eV, and (c) cm^{-1}.

1.20 Hydrogen atoms of high quantum number have been observed in space. (a) Find the quantum number of a hydrogen atom whose orbit radius is 10^{-5} m. (b) What is the energy of that atom in eV.

1.21 Use Equation (1.59) to find the maximum and minimum wavelengths $\lambda = \bar{\upsilon}^{-1}$ in nm of the lines in the Lyman series $(n_1 = 1)$ of hydrogen.

1.22 Use Equation (1.59) to find the wavelengths $\lambda = \bar{\upsilon}^{-1}$ in nm of the first four ultraviolet lines of the Balmer series $(n_1 = 2)$ of hydrogen.

1.23 Use Equation (1.59) to determine the smallest and largest wave numbers $\bar{\upsilon}$ in cm^{-1} of (a) the Paschen $(n_1 = 3)$, (b) Brackett $(n_1 = 4)$, and (c) Pfund $(n_1 = 5)$ series of hydrogen.

1.24 From Equations (1.43), (1.44), and (1.52), show that according to the Bohr theory, the kinetic energy KE and the potential energy PE for an electron in a hydrogen-like atom are given by $KE = \dfrac{Z^2e^4m_e}{2(4\pi\varepsilon_0)^2 n^2\hbar^2}$ and $PE = -\dfrac{Z^2e^4m_e}{(4\pi\varepsilon_0)^2 n^2\hbar^2}$. Thus, prove the virial theorem $PE = -2KE$ (which as shown in Chapter 10 also holds in modern quantum mechanics).

1.25 From your result for the kinetic energy KE in Problem 1.24, find the speed v in ms^{-1} of the electron for the ground states of (a) the hydrogen atom and (b) the He^+ ion.

1.26 This problem deals with the Bohr correspondence principle. From Equation (1.66), it follows (since H = h) that for the hydrogen atom, the frequency $\upsilon_{n_1+1\to n_1}$ of the light emitted in the transition $n_1+1 \to n_1$ is given by $\upsilon_{n_1+1\to n_1} = cR_\infty\left[\dfrac{2n_1+1}{n_1^2(n_1+1)^2}\right]$. Correspondingly, the classical approximation $\upsilon_{n_1+1\to n_1}$ to which is the classical frequency of revolution υ_{n_1} is from Equation (1.68) given by $\upsilon_{n_1} = \dfrac{2cR_\infty}{n_1^3}$. (a) Show in accord with the Bohr correspondence principle that $\lim_{n_1\to\infty} \upsilon_{n_1+1\to n_1} = \upsilon_{n_1}$. Compute the percent error $\dfrac{|\upsilon_{n_1+1\to n_1} - \upsilon_{n_1}|}{\upsilon_{n_1+1\to n_1}} \times 100\% = \left|1 - \dfrac{2}{n_1}\dfrac{(n_1+1)^2}{2n_1+1}\right| \times 100\%$ in the classical approximation to $\upsilon_{n_1+1\to n_1}$ for (b) $n_1 = 1$, (c) $n_1 = 10$, (d) $n_1 = 100$, and (e) $n_1 = 1{,}000$.

1.27 Referring to Problem 1.26, compute for the frequencies of hydrogen $\upsilon_{n_1+1\to n_1}$, $\upsilon_{n_1=1}$, and $\upsilon_{n_1=2}$ in s^{-1} and show that the value of $\upsilon_{n_1+1\to n_1}$ lies between the values of the classical frequencies $\upsilon_{n_1=1}$ and $\upsilon_{n_1=2}$.

1.28 Determine the ionization potentials in eV of (a) the hydrogen atom and (b) the He^+ ion. (The ionization potential is the energy needed to take the atom from the ground state with $n_1 = 1$ to the state with $n_2 = \infty$.)

1.29 In Problem A.12, the following relation is derived for the classical period T_{rot} of a particle of mass m and magnitude J of its rotational angular momentum vector **J** moving in a circular orbit of radius r under the influence of a central field potential $U(r)$: $T_{rot} = \dfrac{2\pi mr^2}{J}$. Specialize to the problem of classical electron motion in a circular orbit of radius r in a hydrogen-like atom to derive the following formula for the classical frequency of revolution of the electron $\upsilon_n = \dfrac{1}{T_n}$ where T_n is the period (the time needed for one electron revolution) when the atom is in the nth stationary state υ_n: $n = \dfrac{2^{1/2}}{\pi}\dfrac{(4\pi\varepsilon_0)}{Z_e^2 m_e^{1/2}}(-E_n)^{3/2}$. This is a challenging problem. The formula for υ_n just given is a key relation appearing in Bohr's classic 1913 paper in which he derived Equation (1.59) and thus theoretically explained Rydberg's empirical equation (Equation 1.49) for the line spectra of the hydrogen atom.

SOLUTIONS TO SELECTED PROBLEMS

1.5 (d) $\lambda_{max}(T = 1{,}000\,K) = 2{,}898\,nm$ (infrared), $\lambda_{max}(T = 5{,}000\,K) = 579\,nm$ (visible), and $\lambda_{max}(T = 10{,}000\,K) = 289.8\,nm$ (ultraviolet)

1.7 $h = 6.62617\ J\ s^{-1}$.

1.8 $T \approx 5{,}800\,K$.

1.9 $\lambda_{max} \approx 73\,nm$ (ultraviolet)

1.11 $E = 9.9322 \times 10^{-19}\ J$.

1.12 $\lambda_{max} = 318.40\,nm$.

1.13 (a) $T_{photo} = 3.6934 \times 10^{-19}\,J$ and (b) $v = 9.005 \times 10^5\,m\,s^{-1}$.

1.14 $I = 4.1771\ eV$.

1.15 $h = 6.625 \times 10^{-34}\,J\,s$ and $I = 4.346\,eV$.

1.19 (a) -21.7969×10^{-19} J, (b) $-13.6044\,eV$, and (c) $-109{,}737\,cm^{-1}$

1.20 (a) $n = 435$ and (b) $-7.19 \times 10^{-5}\,eV$.

1.21 $\lambda_{max} = 121.503\,nm$ and $\lambda_{min} = 91.127\,nm$.

1.22 $\lambda = 396.909, 388.808, 383.443, 379.696\,nm$.

1.23 (a) $\bar{\upsilon}_{smallest} = 5{,}334.44\,cm^{-1}$ and $\bar{\upsilon}_{largest} = 12{,}193.0\,cm^{-1}$.
(b) $\bar{\upsilon}_{smallest} = 2{,}469.08\,cm^{-1}$ and $\bar{\upsilon}_{largest} = 6{,}858.56\,cm^{-1}$.
(c) $\bar{\upsilon}_{smallest} = 1{,}341.23\,cm^{-1}$ and $\bar{\upsilon}_{largest} = 4{,}389.48\,cm^{-1}$.

1.25 (a) $v = 2.186 \times 10^6\,m\,s^{-1}$ and (b) $v = 4.371 \times 10^6\,m\,s^{-1}$.

1.26 (b) 167%, (c) 15.2%, (d) 1.50%, and (e) 0.150%.

1.27 For $n_l = 1$, $\upsilon_{n_l+1 \to n_l} = 2.47 \times 10^{15}\,s^{-1}$, $\upsilon_{n_l=1} = 6.58 \times 10^{15}\,s^{-1}$, and $\upsilon_{n_l=2} = 0.82 \times 10^{15}\,s^{-1}$.

1.28 (a) 13.6044 eV and (b) 54.4176 eV.

2 Mathematics for Quantum Mechanics

As noted in Section 1.5, the average values of the observables of a quantum system are determined by its wave function Ψ. In this chapter, we introduce some of the mathematical concepts and techniques needed to analyze quantum wave functions. Especially, we will develop the theory of *Hermitian eigenvalue problems*. Such problems as we will see occur nearly universally in applications of quantum mechanics to chemical problems.

The wave functions Ψ of chemical systems typically depend on more than one variable. However, in this chapter, for simplicity we will restrict ourselves to functions $\Psi(x)$ of a single variable x. This is not a serious limitation since the multivariable characteristics of actual wave functions are natural extensions of the single-variable attributes of the functions $\Psi(x)$.

The functions $\Psi(x)$ like real quantum wave functions are assumed to be single-valued, finite everywhere, continuous with a continuous first derivative, and, as will be described later, square integrable. That is, the functions $\Psi(x)$ are assumed to be *well-behaved*. We will also as in quantum mechanics exclude the trivial function $\Psi(x)=0$. Additionally, like actual wave functions, we will assume that the functions $\Psi(x)$ are in general *complex functions*; that is, we will assume that they can take on complex values for real values of their arguments x.

We thus begin with a discussion of complex functions.

2.1 COMPLEX FUNCTIONS

We start with a brief review of some of the elementary properties of complex numbers.

Recall that a complex number z is defined in terms of two real numbers x and y by

$$z = x + iy \tag{2.1}$$

where $i = \sqrt{-1}$ is the imaginary unit. The number x is called the real part of z and is written as Re z. Correspondingly, the number y is called the imaginary part of z and is written Im z. The complex conjugate z^* of z is defined by

$$z^* = x - iy. \tag{2.2}$$

Notice that $\left(z^*\right)^* = z$ and that if $z^* = z$, then z is real. Also notice that the product z^*z is a nonnegative real number given by

$$z^*z = x^2 + y^2 \geq 0 \tag{2.3}$$

which only vanishes if $x = y = 0$. Additionally, z^*z is often written as $z^*z = |z|^2$, where $|z| = \left(z^*z\right)^{1/2} = \left(x^2 + y^2\right)^{1/2}$ is called the *absolute value* of z.

We may now define complex functions. Complex functions are of the form

$$\Psi(x) = f(x) + ig(x) \tag{2.4}$$

where $f(x)$ and $g(x)$ are ordinary real-valued functions of x. The complex conjugate function

$\Psi^*(x)$ is defined in analogy to the definition of z^* by

$$\Psi^*(x) = f(x) - ig(x). \tag{2.5}$$

In quantum mechanics, we will often require functions like

$$P(x) = \Psi^*(x)\Psi(x) = |\Psi(x)|^2 = f^2(x) + g^2(x) \geq 0. \tag{2.6}$$

This is because such functions will turn out to give the probabilities of finding the particles of a quantum system at particular positions. Notice that $P(x)$ is a nonnegative real function as is required for a probability.

A simple example of a complex function is the complex exponential function $\exp(i\alpha x)$, where α is a real constant. In analogy to the power series expansion of the ordinary real exponential function $\exp(\alpha x)$

$$\exp(\alpha x) = 1 + \alpha x + \frac{\alpha^2 x^2}{2!} + \frac{\alpha^3 x^3}{3!} + \cdots \tag{2.7a}$$

the complex exponential function $\exp(i\alpha x)$ is defined as the power series

$$\exp(i\alpha x) = 1 + i\alpha x + \frac{(i\alpha)^2}{2!} x^2 + \frac{(i\alpha)^3 x^3}{3!} + \cdots. \tag{2.7b}$$

By comparing Equation (2.7b) with the power series expansions of the functions $\cos\alpha x$ and $\sin\alpha x$, one may show that (Problem 2.2)

$$\exp(i\alpha x) = \cos\alpha x + i\sin\alpha x. \tag{2.8}$$

Notice that $\exp(i\alpha x)$ shares the form of the general complex function of Equation (2.4) with $\cos\alpha x = f(x)$ and $\sin\alpha x = g(x)$.

The complex conjugate of $\exp(i\alpha x)$ is

$$[\exp(i\alpha x)]^* = \cos\alpha x - i\sin\alpha x. \tag{2.9}$$

Thus, from Equations (2.8) and (2.9), for the complex exponential function, $P(x)$ of Equation (2.6) specializes to

$$P(x) = [\exp(i\alpha x)]^*[\exp(i\alpha x)] = \cos^2\alpha x + \sin^2\alpha x = 1 > 0. \tag{2.10}$$

Additionally, it follows from Equations (2.8) and (2.9) that

$$[\exp(i\alpha x)]^* = \exp(-i\alpha x). \tag{2.11}$$

Comparing the previous two equations yields that

$$\exp(-i\alpha x)\exp(i\alpha x) = 1. \tag{2.12}$$

Equation (2.12) is of the same form as the real exponential relation $\exp(-\alpha x) \times \exp(\alpha x) = 1$, suggesting that complex exponentials obey the same mathematical rules as real exponentials. This is in fact the case. For example, one may verify that (Problem 2.3)

$$\exp(i\alpha x)\exp(i\beta x) = \exp[i(\alpha + \beta)x] \tag{2.13}$$

$$\frac{d\exp(i\alpha x)}{dx} = i\alpha \exp(i\alpha x) \tag{2.14}$$

and

$$\int \exp(i\alpha x)dx = \frac{1}{i\alpha}\exp(i\alpha x) + C \tag{2.15}$$

where C is an arbitrary constant. Finally, we note that to obtain the complex conjugate of an arbitrary expression formed from complex functions $\Psi_i(x)$, complex constants C_i, and the imaginary unit i, one merely replaces all complex quantities in the expression by their complex conjugates (Problem 2.4). For example, the complex conjugate of the function $\Psi(x) = [\Psi_1(x) + iC_1\Psi_2(x)]\ell n[1 + C_2\Psi_3(x)]$ is $\Psi^*(x) = [\Psi_1^*(x) - iC_1^*\Psi_2^*(x)]\ell n[1 + C_2^*\Psi_3^*(x)]$.

We next introduce *operators.*

2.2 OPERATORS

An operator $\hat{O}$ is a rule that converts an original complex function $\Psi(x)$ into a second complex function $\Phi(x)$. This statement may be written symbolically as

$$\Phi(x) = \hat{O}\ \Psi(x). \tag{2.16}$$

Operators typically act on sets of original complex functions $\{\Psi(x)\}$ rather than on individual functions to give whole sets of second complex functions $\{\Phi(x)\}$.

Some common examples of operators are listed in Table 2.1. Two of the listed operators are $\hat{x}$, which multiplies a function $\Psi(x)$ by x, and $\frac{\hbar}{i}\frac{d}{dx}$, which differentiates a function $\Psi(x)$ with respect to x and then multiplies the result by the complex constant $\frac{\hbar}{i}$. For example, if the original complex function is $\Psi(x) = \exp(i\pi x)$, then for the operator $\hat{x}$ the second complex function is $x\exp(i\pi x)$, while for the operator $\frac{\hbar}{i}\frac{d}{dx}$ the second complex function is $\pi\hbar\exp(i\pi x)$. (For more examples, see Problem 2.5.)

We next define *linear operators* (Problem 2.6).

Let $C_1, C_2, \ldots$ and $\Psi_1(x), \Psi_2(x), \ldots$ be, respectively, a set of complex constants and complex functions. Then, the *linear superposition function* $\Psi(x)$ formed from these quantities is defined as follows:

$$\Psi(x) = C_1\Psi_1(x) + C_2\Psi_2(x) + \cdots. \tag{2.17}$$

An operator $\hat{O}$ is linear if it acts on a linear superposition function $\Psi(x)$ in the following manner:

$$\begin{aligned} \hat{O}\Psi(x) &= \hat{O}[C_1\Psi_1(x) + C_2\Psi_2(x) + \cdots] \\ &= C_1\hat{O}\Psi_1(x) + C_2\hat{O}\Psi_2(x) + \cdots. \end{aligned} \tag{2.18}$$

It is easily verified that all of the operators listed in Table 2.1 including $\hat{x}$ and $\frac{\hbar}{i}\frac{d}{dx}$ are linear. However, not all operators are linear (Problem 2.6). For example, the square root operator $\sqrt{\ }$ that converts a complex function into its square root is not linear since when it acts on a linear superposition function $\Psi(x)$, it gives $\sqrt{\Psi(x)} = \sqrt{C_1\Psi_1(x) + C_2\Psi_2(x) + \cdots} \neq C_1\sqrt{\Psi_1(x)} + C_2\sqrt{\Psi_2(x)} + \cdots$.

TABLE 2.1
Some Common Operators

Operation	Symbol for Operator	Original Complex Function	Second Complex Function
Multiplication by a complex constant γ	$\hat{\gamma}$	$\Psi(x)$	$\gamma\Psi(x)$
Multiplication by the real variable x	$\hat{x}$	$\Psi(x)$	$x\Psi(x)$
Multiplication by the real function $U(x)$	$\hat{U}(x)$	$\Psi(x)$	$U(x)\Psi(x)$
Differentiation with respect to x followed by multiplication by the complex constant $\frac{\hbar}{i}$	$\frac{\hbar}{i}\frac{d}{dx}$	$\Psi(x)$	$\frac{\hbar}{i}\frac{d\Psi(x)}{dx}$
Second differentiation with respect to x	$\frac{d^2}{dx^2}$	$\Psi(x)$	$\Psi''(x)=\frac{d^2\Psi(x)}{dx^2}$
Second differentiation with respect to x followed by multiplication by the real constant $-\frac{\hbar^2}{2m}$ added to multiplication by the real function $U(x)$	$-\frac{\hbar^2}{2m}\frac{d^2}{dx^2}+\hat{U}(x)$	$\Psi(x)$	$-\frac{\hbar^2}{2m}\frac{d^2}{dx^2}\Psi(x)+U(x)\Psi(x)$

However, except at very advanced levels, all operators occurring in quantum mechanics are linear. Consequently, all operators we will encounter are linear and so we will henceforth use the terms operator and linear operator interchangeably.

We next turn to operator addition and multiplication.

The sum $\hat{O}=\hat{O}_1+\hat{O}_2$ of two operators $\hat{O}_1$ and $\hat{O}_2$ which is also an operator is defined by its action on an arbitrary complex function $\Psi(x)$ as follows:

$$\Phi(x)=\hat{O}\Psi(x)=\left(\hat{O}_1+\hat{O}_2\right)\Psi(x)=\hat{O}_1\Psi(x)+\hat{O}_2\Psi(x). \tag{2.19}$$

The product $\hat{O}_{12}=\hat{O}_1\hat{O}_2$ of two operators $\hat{O}_1$ and $\hat{O}_2$ which is also an operator is formed by successively applying the operators to a complex function in the order $\hat{O}_2$ first and then $\hat{O}_1$. Correspondingly, the product $\hat{O}_{21}=\hat{O}_2\hat{O}_1$ which is also an operator is formed by successively applying the operators to a complex function in the order $\hat{O}_1$ first and then $\hat{O}_2$. Explicitly, the products $\hat{O}_{12}$ and $\hat{O}_{21}$ are defined by their action an arbitrary complex function $\Psi(x)$ as follows:

$$\Phi_{12}(x)\equiv\hat{O}_{12}\Psi(x)=\hat{O}_1\hat{O}_2\Psi(x)=\hat{O}_1\Psi_2(x) \tag{2.20a}$$

where $\Psi_2(x)=\hat{O}_2\Psi(x)$ and

$$\Phi_{21}(x)\equiv\hat{O}_{21}\Psi(x)=\hat{O}_2\hat{O}_1\Psi(x)=\hat{O}_2\Psi_1(x) \tag{2.20b}$$

where $\Psi_1(x)=\hat{O}_1\Psi(x)$.

The operator multiplication rules just given may be easily extended to more than two operators. For example, the rule for the product $\hat{O}_1\hat{O}_2\hat{O}_3$ is to apply the operators successively $\hat{O}_3$ first, then $\hat{O}_2$, and finally $\hat{O}_1$.

Operator multiplication involves one complication that does not occur for operator addition; namely, operator multiplication can be *non-commutative*. What this means is that for some but not all operator pairs $\hat{O}_1$ and $\hat{O}_2$, the products $\hat{O}_{12} = \hat{O}_1\hat{O}_2$ and $\hat{O}_{21} = \hat{O}_2\hat{O}_1$ are *different operators* in the sense that the functions $\Phi_{12}(x) = \hat{O}_{12}\Psi(x)$ and $\Phi_{21}(x) = \hat{O}_{21}\Psi(x)$ are different functions.

The following terminology is used to describe the commutivity and non-commutivity of operators. If for *all* well-behaved functions $\Psi(x)$ $\Phi_{12}(x) = \Phi_{21}(x)$, then it is said that the operators $\hat{O}_1$ and $\hat{O}_2$ *commute* and this is written as $\hat{O}_1\hat{O}_2 = \hat{O}_2\hat{O}_1$. But if for even some well-behaved functions $\Psi(x)$ $\Phi_{12}(x) \neq \Phi_{21}(x)$, then it is said that the operators $\hat{O}_1$ and $\hat{O}_2$ *do not commute* and this is written as $\hat{O}_1\hat{O}_2 \neq \hat{O}_2\hat{O}_1$.

To determine whether or not two operators $\hat{O}_1$ and $\hat{O}_2$ commute, it is convenient to evaluate their *commutator* $\left[\hat{O}_1, \hat{O}_2\right]$ defined as the operator

$$\left[\hat{O}_1, \hat{O}_2\right] = \hat{O}_1\hat{O}_2 - \hat{O}_2\hat{O}_1. \tag{2.21}$$

If $\left[\hat{O}_1, \hat{O}_2\right] = \hat{O}$, where $\hat{O}$ is the *null operator* multiply by zero, $\hat{O}_1$ and $\hat{O}_2$ commute, but if $\left[\hat{O}_1, \hat{O}_2\right] \neq \hat{O}$, then $\hat{O}_1$ and $\hat{O}_2$ do not commute.

To illustrate the use of commutators, we determine the commutator $\left[\hat{x}, \frac{\hbar}{i}\frac{d}{dx}\right]$ of the operators $\hat{x}$ and $\frac{\hbar}{i}\frac{d}{dx}$. To make this determination, we evaluate $\left[\hat{x}, \frac{\hbar}{i}\frac{d}{dx}\right]\Psi(x)$, where $\Psi(x)$ is an arbitrary well-behaved complex function. The evaluation proceeds as follows: $\left[\hat{x}, \frac{\hbar}{i}\frac{d}{dx}\right]\Psi(x) =$ $\left[\hat{x}\frac{\hbar}{i}\frac{d}{dx} - \frac{\hbar}{i}\frac{d}{dx}\hat{x}\right]\Psi(x) = \frac{\hbar}{i}\left[\hat{x}\Psi'(x) - \frac{d}{dx}x\Psi(x)\right] = \frac{\hbar}{i}\left[x\Psi'(x) - x\Psi'(x) - \Psi(x)\right] = -\frac{\hbar}{i}\Psi(x) = i\hbar\Psi(x)$. Thus, we have shown that $\left[\hat{x}, \frac{\hbar}{i}\frac{d}{dx}\right]\Psi(x) = i\hbar\Psi(x)$. Since this is true for all well-behaved functions $\Psi(x)$, we have proven the *commutator identity*

$$\left[\hat{x}, \frac{\hbar}{i}\frac{d}{dx}\right] = i\hbar \tag{2.22}$$

and thus have shown that the commutator $\left[\hat{x}, \frac{\hbar}{i}\frac{d}{dx}\right]$ is non-null and that consequently the operators $\hat{x}$ and $\frac{\hbar}{i}\frac{d}{dx}$ do not commute.

We close this section by introducing another important concept, the notion of *linearly independent complex functions* (Problem 2.8). First, consider the simplest case of two functions $\Psi_1(x)$ and $\Psi_2(x)$. These are linearly independent if they differ by more than a complex constant. For example, the functions $\exp(i\alpha x)$ and $(2+ix)\exp(i\alpha x)$ are linearly independent, but the functions $\exp(i\alpha x)$ and $(2+i)\exp(i\alpha x)$ are not linearly independent. For the general case of a set of n functions $\Psi_1(x), \Psi_2(x), \ldots, \Psi_n(x)$, the functions are linearly independent if the linear superposition function $\Psi(x) = C_1\Psi_1(x) + C_2\Psi_2(x) + \cdots + C_n\Psi_n(x)$ vanishes only if all of the coefficients $C_1, C_2, \ldots, C_n$ vanish. Equivalently, the set of n functions $\Psi_1(x), \Psi_2(x), \ldots, \Psi_n(x)$ is linearly independent if no one of the functions can be expressed as a linear superposition of the other $n-1$ functions in the set.

We may now turn to *operator eigenvalue problems*.

2.3 OPERATOR EIGENVALUE PROBLEMS

To start, we return to Equation (2.16) $\Phi(x) = \hat{O}\Psi(x)$. Let us suppose that for some choices $\phi(x)$ of the original complex function $\Psi(x)$, the second complex function $\Phi(x)$ is a complex constant λ times $\phi(x)$. That is, let us assume that if

$$\Psi(x) = \phi(x), \quad \text{then}\, \Phi(x) = \lambda\phi(x). \tag{2.23}$$

The constant λ is said to be an *eigenvalue* of the operator $\hat{O}$, and $\phi(x)$ is referred to as an *eigenfunction* of $\hat{O}$ associated with the eigenvalue λ (Problem 2.9).

Equations (2.16) and (2.23) may be combined into the single *eigenvalue equation*

$$\hat{O}\phi(x) = \lambda\phi(x). \tag{2.24}$$

Eigenvalue equations like Equation (2.24) are typically differential equations with many solutions.

For future reference, notice that since $\hat{O}$ is a linear operator for all complex constants C, $\hat{O}[C\phi(x)] = C\hat{O}\phi(x)$, and hence, from Equation (2.24)

$$\hat{O}[C\phi(x)] = \lambda[C\phi(x)]. \tag{2.25}$$

Thus, if $\phi(x)$ is an eigenfunction of $\hat{O}$ with eigenvalue λ, $C\phi(x)$ is also an eigenfunction of $\hat{O}$ with the same eigenvalue λ.

We additionally note that for many of the operators $\hat{O}$ we will encounter in quantum mechanics for some of their eigenvalues λ, more than one linearly independent eigenfunction $\phi(x)$ is associated with λ. If $q > 1$ linearly independent eigenfunctions are associated with an eigenvalue λ, then that eigenvalue is said to be *q-fold degenerate*. On the other hand, if only one eigenfunction is associated with an eigenvalue λ, then that eigenvalue is said to be *non-degenerate*.

Next, we turn to some examples of these concepts.

First, consider the operator $\hat{O} = \frac{\hbar}{i}\frac{d}{dx}$. Its eigenvalue equation is

$$\frac{\hbar}{i}\frac{d\phi(x)}{dx} = \lambda\phi(x). \tag{2.26}$$

As may be verified by substitution, the eigenfunctions $\phi(x)$ of the operator $\frac{\hbar}{i}\frac{d}{dx}$ are

$$\phi(x) = \exp\left(\frac{i}{\hbar}\lambda x\right) \tag{2.27}$$

where λ can be any complex constant. Also for each eigenvalue λ, there is only a single eigenfunction $\phi(x)$. Thus, the operator $\frac{\hbar}{i}\frac{d}{dx}$ has a continuum of non-degenerate eigenvalues λ since λ can be any complex constant. (We assume in this textbook that the mathematical rules obeyed by functions of a real variable are also obeyed if the variable is complex, an assumption which we have only proven for complex exponential functions.)

As a second example, consider the operator $\hat{O} = \frac{d^2}{dx^2}$ which converts a function $\Psi(x)$ into its second derivative $\Psi''(x)$. Its eigenvalue equation is

$$\frac{d^2}{dx^2}\phi(x) = \lambda\phi(x). \tag{2.28}$$

Substitution gives that for each eigenvalue $-k^2 \equiv \lambda$, there are two linearly independent eigenfunctions:

$$\phi(x) = \cos kx \text{ and } \phi(x) = \sin kx. \tag{2.29}$$

Also k can take on any complex value. Thus, the operator $\frac{d^2}{dx^2}$ has a continuum of two-fold degenerate eigenvalues $\lambda = -k^2$.

We next introduce the centrally important mathematical concept of an *orthonormal set* of functions and in the process describe a mathematical technique often used in quantum mechanics, *Gram–Schmidt orthogonalization*.

2.4 ORTHONORMAL SETS AND THE GRAM–SCHMIDT METHOD

Consider a set $\{\bar{\phi}(x)\}$ of linearly independent complex functions that are defined over the range of x values $a \le x \le b$. We will assume that most members of the set are not normalized. This means that for a typical member $\bar{\phi}_i(x)$ of the set, the integral

$$\int_a^b \bar{\phi}_i^*(x)\bar{\phi}_i(x)dx \neq 1. \tag{2.30}$$

We will also assume that most members of the set are not mutually orthogonal. This means that for a typical pair of *different* members of the set $\bar{\phi}_i(x)$ and $\bar{\phi}_j(x)$, the integral

$$\int_a^b \bar{\phi}_i^*(x)\bar{\phi}_j(x)dx \neq 0. \tag{2.31}$$

We wish to convert the set $\{\bar{\phi}(x)\}$ into an *orthonormal set* $\{\Psi(x)\}$. For the orthonormal set, every member of the set $\Psi_i(x)$ is normalized. That is, for all i

$$\int_a^b \Psi_i^*(x)\Psi_i(x)dx = 1. \tag{2.32}$$

Also for the orthonormal set, every pair of different members of the set $\Psi_i(x)$ and $\Psi_j(x)$ is orthogonal. That is, for all $i \neq j$

$$\int_a^b \Psi_i^*(x)\Psi_j(x)dx = 0. \tag{2.33}$$

The proceeding two equations may be condensed into a single relation if we introduce the *Kronecker delta* δ_{ij} defined by

$$\delta_{ij} = \begin{cases} 1 & \text{if } i = j \\ 0 & \text{if } i \neq j \end{cases} \tag{2.34}$$

In terms of the Kronecker delta, Equations (2.32) and (2.33) become the single statement that for all i and j

$$\int_a^b \Psi_i^*(x)\Psi_j(x)dx = \delta_{ij}. \tag{2.35}$$

Equation (2.35) is referred to as the *orthonormality relation* for the orthonormal set $\{\Psi(x)\}$.

To convert the original set $\{\bar{\phi}(x)\}$ into the orthonormal set $\{\Psi(x)\}$, we proceed in two steps. (1) We use the Gram–Schmidt orthogonalization method to transform the set $\{\bar{\phi}(x)\}$ into the set $\{\phi(x)\}$ comprised of mutually orthogonal but unnormalized functions. (2) We then normalize the functions in the set $\{\phi(x)\}$ to convert that set into the orthonormal set $\{\Psi(x)\}$.

To carry out Step 1, we require the Gram–Schmidt algorithm that permits one to recursively determine the functions $\phi_1(x), \phi_2(x), \cdots$ which comprise the mutually orthogonal set from the known functions $\bar{\phi}_1(x), \bar{\phi}_2(x), \cdots$ which comprise the original set. The Gram–Schmidt algorithm is as follows:

$$\phi_1(x) = \bar{\phi}_1(x) \tag{2.36}$$

$$\phi_2(x) = \bar{\phi}_2(x) - \frac{\int_a^b \bar{\phi}_2(x)\phi_1^*(x)dx}{\int_a^b \phi_1^*(x)\phi_1(x)dx}\phi_1(x) \tag{2.37}$$

$$\phi_3(x) = \bar{\phi}_3(x) - \frac{\int_a^b \bar{\phi}_3(x)\phi_1^*(x)dx}{\int_a^b \phi_1^*(x)\phi_1(x)dx}\phi_1(x) - \frac{\int_a^b \bar{\phi}_3(x)\phi_2^*(x)dx}{\int_a^b \phi_2^*(x)\phi_2(x)dx}\phi_2(x) \tag{2.38}$$

$$\begin{aligned}\phi_4(x) = \bar{\phi}_4(x) &- \frac{\int_a^b \bar{\phi}_4(x)\phi_1^*(x)dx}{\int_a^b \phi_1^*(x)\phi_1(x)dx}\phi_1(x) - \frac{\int_a^b \bar{\phi}_4(x)\phi_2^*(x)dx}{\int_a^b \phi_2^*(x)\phi_2(x)dx}\phi_2(x) \\ &- \frac{\int_a^b \bar{\phi}_4(x)\phi_3^*(x)dx}{\int_a^b \phi_3^*(x)\phi_3(x)dx}\phi_3(x)\end{aligned} \tag{2.39}$$

and so on.

From the proceeding Gram–Schmidt equations, one may show that the functions $\phi_1(x), \phi_2(x), \phi_3(x), \cdots$ are mutually orthogonal (Problem 2.10). For example, it follows immediately from Equation (2.37) that

$$\int_a^b \phi_2^*(x)\,\phi_1(x)dx = 0. \tag{2.40}$$

To normalize the functions $\phi_i(x)$ and thus convert the mutually orthogonal set $\{\phi(x)\}$ into the orthonormal set $\{\Psi(x)\}$ is easy. We simply multiply each of the functions $\phi_i(x)$ by the *normalization factor* $\left[\int_a^b \phi_i^*(x)\phi_i(x)dx\right]^{-1/2}$ to obtain the functions

$$\Psi_i(x) = \left[\frac{1}{\int_a^b \phi_i^*(x)\phi_i(x)dx}\right]^{1/2} \phi_i(x). \tag{2.41}$$

The functions $\Psi_i(x)$ are mutually orthogonal since they are proportional to the mutually orthogonal functions $\phi_i(x)$. Also they are normalized since it follows from Equation (2.41) that

$$\int_a^b \Psi_i^*(x)\Psi_i(x)dx = 1. \tag{2.42}$$

Thus, the set of functions $\Psi_i(x)$ comprise an orthonormal set $\{\Psi(x)\}$ characterized by the orthonormality relation of Equation (2.35).

We close this section with a simple application of the ideas just described. We consider two linearly independent functions $\bar{\phi}_1(x) = 1$ and $\bar{\phi}_2(x) = x$ defined on the range $0 \le x \le 1$. We wish to convert $\bar{\phi}_1(x)$ and $\bar{\phi}_2(x)$ into an orthonormal pair $\Psi_1(x)$ and $\Psi_2(x)$ defined on the same range. We start by specializing the Gram–Schmidt equations (Equations 2.36 and 2.37) to real functions defined on the range $0 \le x \le 1$. This specialization gives that

$$\phi_1(x) = \bar{\phi}_1(x) \tag{2.43}$$

and

$$\phi_2(x) = \bar{\phi}_2(x) - \frac{\int_0^1 \bar{\phi}_2(x)\phi_1(x)dx}{\int_0^1 \phi_1(x)\phi_1(x)dx}\phi_1(x). \tag{2.44}$$

For $\bar{\phi}_1(x) = 1$ and $\bar{\phi}_2(x) = x$, the above relations give that

$$\phi_1(x) = 1 \text{ and } \phi_2(x) = x - \frac{\int_0^1 x\,dx}{\int_0^1 dx} = x - \frac{1}{2}. \tag{2.45}$$

We next normalize $\phi_1(x)$ and $\phi_2(x)$ to obtain the orthonormal pair $\Psi_1(x)$ and $\Psi_2(x)$. From Equation (2.41), to perform this normalization, we must evaluate the integrals $\int_0^1 \phi_1(x)\phi_1(x)dx$ and $\int_0^1 \phi_2(x)\phi_2(x)dx$. However, from Equation (2.45), $\phi_1(x) = 1$ and $\phi_2(x) = x - \frac{1}{2}$. Thus, $\int_0^1 \phi_1(x)\phi_1(x)dx = 1$ and $\int_0^1 \phi_2(x)\phi_2(x)dx = \frac{1}{12}$. Therefore, from Equations (2.41) and (2.45), the orthonormal pair is $\Psi_1(x) = 1$ and $\Psi_2(x) = 12^{1/2}\left(x - \frac{1}{2}\right)$.

We next turn to *Hermitian operators*. Such operators will be of great importance later since we will see in Section 4.2 in quantum mechanics observables like position, momentum, angular momentum, and energy are represented by Hermitian operators.

2.5 HERMITIAN OPERATORS

We will generically denote Hermitian operators by $\hat{A}$ instead of by $\hat{O}$.

Consider a set of linearly independent complex functions $\{\Psi(x)\}$ defined on the range $a \le x \le b$ and a linear operator $\hat{A}$ that acts on the members $\Psi_i(x)$ of this set. We require the *matrix elements* A_{ij} of $\hat{A}$ for the set $\{\Psi(x)\}$. These are defined by

$$A_{ij} = \int_a^b \Psi_i^*(x)\left[\hat{A}\Psi_j(x)\right]dx = \int_a^b \Psi_i^*(x)\hat{A}\Psi_j(x)dx \tag{2.46}$$

where $\Psi_i(x)$ and $\Psi_j(x)$ are any two members of the set. The operator $\hat{A}$ is said to be Hermitian for the set $\{\Psi(x)\}$ on the range $a \le x \le b$ if for *all* functions $\Psi_i(x)$ and Ψ_j in the set

$$A_{ij} = A_{ji}^*. \tag{2.47}$$

We will refer to Equation (2.47) as the *Hermitian condition* for the operator $\hat{A}$. Using the definition for A_{ij} of Equation (2.46), the Hermitian condition may be written explicitly as

$$\int_a^b \Psi_i^*(x)\hat{A}\Psi_j(x)dx = \left[\int_a^b \Psi_j^*(x)\hat{A}\Psi_i(x)dx\right]^*. \tag{2.48}$$

or

$$\int_a^b \Psi_i^*(x)\hat{A}\Psi_j(x)dx = \int_a^b \left[\hat{A}\Psi_i^*(x)\right]^* \Psi_j(x)dx. \tag{2.49}$$

Notice that since the integrals in the above equation depend on a and b as well as on the functions $\Psi_i(x)$ and $\Psi_j(x)$, whether an operator is Hermitian depends on more than just its form. It also depends on both the function set $\{\Psi(x)\}$ and the values of the range boundary points a and b. Thus, an operator $\hat{A}$ may be Hermitian for a function set $\{\Psi(x)\}$ on the range $a \le x \le b$ but not on a different range $a' \le x \le b'$. Similarly, $\hat{A}$ may be Hermitian for a function set $\{\Psi(x)\}$ on the range $a \le x \le b$ but not for a different function set $\{\Phi(x)\}$ on either the same range $a \le x \le b$ or a different range $a' \le x \le b'$.

Let us illustrate these points for the operator $\hat{A} = \frac{\hbar}{i}\frac{d}{dx}$. For this operator, $\left[\hat{A}\Psi_i(x)\right]^* = \left[\frac{\hbar}{i}\frac{d\Psi_i(x)}{dx}\right]^* = -\frac{\hbar}{i}\frac{d\Psi_i^*(x)}{dx}$, and thus, the Hermitian condition of Equation (2.49) reduces to

$$\int_a^b \Psi_i^*(x)\frac{d\Psi_j(x)}{dx} = -\int_a^b \frac{d\Psi_i^*(x)}{dx}\Psi_j(x)dx. \tag{2.50}$$

We will first show from Equation (2.50) that the operator $\frac{\hbar}{i}\frac{d}{dx}$ is Hermitian on the range $0 \le x \le 2\pi$ for the function set $\{\Psi_n(x)\}$ comprised of the complex exponentials $\Psi_n(x) = \exp(inx)$, where $n = 0, \pm 1, \ldots$. That is, we will show for all n and m that

$$\int_0^{2\pi} \Psi_n^*(x)\frac{d\Psi_m(x)}{dx}dx = -\int_0^{2\pi} \frac{d\Psi_n^*(x)}{dx}\Psi_m(x)dx. \tag{2.51}$$

Denoting the left- and right-hand sides of Equation (2.51) by, respectively, I_{nm} and J_{nm}, we must show that for all n and m, $I_{nm} = J_{nm}$. But $I_{nm} = \int_0^{2\pi} \Psi_n^*(x)\frac{d\Psi_m(x)}{dx}dx = \int_0^{2\pi} \exp(-inx)\frac{d\exp(imx)}{dx}dx = im\int_0^{2\pi} \exp\left[i(m-n)x\right]dx$. For $n = m$ $I_{nm} = 2\pi im = 2\pi in$. For $n \ne m$, $I_{nm} = \frac{m}{m-n}\{\exp\left[2\pi i(m-n)\right]-1\} = \frac{m}{n-m}\{\cos\left[2\pi(m-n)\right] + i\sin\left[2\pi(m-n)\right]-1\}$ which vanishes since, because n and m are integers, $\cos\left[2\pi(m-n)\right] = 1$ and $\sin\left[2\pi(m-n)\right] = 0$. Thus, $I_{nm} = 2\pi in\delta_{nm}$, where δ_{nm} is the Kronecker delta. Similarly, we may show that $J_{nm} = 2\pi in\delta_{nm}$. Thus, we have proven for all n and m that $I_{nm} = J_{nm}$ and hence have shown that the operator $\frac{\hbar}{i}\frac{d}{dx}$ is Hermitian on the range $0 \le x \le 2\pi$ for the function set $\{\Psi_n(x)\}$.

We next show that the operator $\frac{\hbar}{i}\frac{d}{dx}$ is *not* Hermitian for the function set $\{\Psi_n(x)\}$ on the different range $0 \le x \le \pi$. To do this, we must show for at least some values of n and m that

$$\int_0^{\pi} \Psi_n^*(x)\frac{d\Psi_m(x)}{dx}dx \neq -\int_0^{\pi} \frac{d\Psi_n^*(x)}{dx}\Psi_m(x)dx. \tag{2.52}$$

Denoting the left-and right-hand sides of Equation (2.52) by I_{nm} and J_{nm}, we thus must show that for at least some values of n and m, $I_{nm} \neq J_{nm}$. An evaluation of I_{nm} and J_{nm} gives the following (Problem 2.11): For $n = m$ $I_{nm} = J_{nm} = in\pi$, and for $n \neq m$ and $n - m$ even $I_{nm} = J_{nm} = 0$, but for $n - m$ odd $I_{nm} = -\frac{2m}{m-n}$ and $J_{nm} = -\frac{2n}{m-n}$. Hence, for $n - m$ odd $I_{nm} \neq J_{nm}$, showing that the operator $\frac{\hbar}{i}\frac{d}{dx}$ is not Hermitian for the function set $\{\Psi_n(x)\}$ on the range $0 \le x \le \pi$.

One may similarly show for the function set $\{\Psi_{n/2}(x)\}$ comprised of the complex exponentials $\exp\left(i\frac{n}{2}x\right)$ where $n = 0, \pm 1, \ldots$ that the operator $\frac{\hbar}{i}\frac{d}{dx}$ is not Hermitian on either of the ranges $0 \le x \le 2\pi$ or $0 \le x \le \pi$ (Problem 2.12).

This method for checking whether an operator $\hat{A}$ is Hermitian for a particular function set and range which we have illustrated for $\hat{A} = \frac{\hbar}{i}\frac{d}{dx}$ while rigorous is very inconvenient. Instead, what is done in practice is that the Hermitian condition of Equation (2.49) is replaced by equivalent but much more convenient *boundary conditions* (i.e., conditions that require knowledge of the functions $\Psi_i[x]$ only at the boundary points a and b). We next illustrate how this replacement is made taking the operator $\hat{A} = \frac{\hbar}{i}\frac{d}{dx}$ as an example. For this operator, recall that the general Hermitian condition of Equation (2.49) reduces to the condition of Equation (2.50).

Let us consider the left-hand side of Equation (2.50) $\int_a^b \Psi_i^*(x)\frac{d}{dx}\Psi_j(x)dx$. Integrating this left-hand side by parts gives

$$\int_a^b \Psi_i^*(x)\frac{d}{dx}\Psi_j(x)dx = \Psi_i^*(x)\Psi_j(x)]_a^b - \int_a^b \frac{d}{dx}\Psi_i^*(x)\Psi_j(x)dx. \tag{2.53}$$

Comparing Equations (2.50) and (2.53) shows that the Hermitian condition of Equation (2.50) is satisfied if

$$\Psi_i^*(x)\Psi_j(x)]_a^b = \Psi_i^*(b)\Psi_j(b) - \Psi_i^*(a)\Psi_j(a) = 0. \tag{2.54}$$

Thus, if Equation (2.54) holds for all pairs of functions $\Psi_i(x)$ and $\Psi_j(x)$ in the function set $\{\Psi(x)\}$, the operator $\frac{\hbar}{i}\frac{d}{dx}$ is Hermitian for that set on the range $a \le x \le b$. However, Equation (2.54) does indeed hold for all pairs $\Psi_i(x)$ and $\Psi_j(x)$ in the set $\{\Psi(x)\}$ if *all* functions $\Psi_k(x)$ in the set satisfy the boundary condition

$$\Psi_k(b) = \Psi_k(a) \tag{2.55}$$

since then $\Psi_i^*(b) = \Psi_i^*(a)$ and $\Psi_j(b) = \Psi_j(a)$ rendering $\Psi_i^*(b)\Psi_j(b) - \Psi_i^*(a)\Psi_j(a) = \Psi_i^*(a)\Psi_j(a) - \Psi_i^*(a)\Psi_j(a) = 0$.

In summary, we have shown *that if all members* $\Psi_k(x)$ *of a function set* $\{\Psi(x)\}$ *obey the boundary condition* $\Psi_k(b) = \Psi_k(a)$, *then the operator* $\frac{\hbar}{i}\frac{d}{dx}$ *is Hermitian for that set on the range* $a \le x \le b$.

To illustrate these principles, we will using the boundary condition of Equation (2.55) again prove that the operator $\frac{\hbar}{i}\frac{d}{dx}$ is Hermitian for the set $\{\Psi_n(x)\}$ of complex exponentials $\Psi_n(x) = \exp(inx)$ on the range $0 \le x \le 2\pi$ but not on the range $0 \le x \le \pi$. For the range $0 \le x \le 2\pi$, the boundary condition of Equation (2.55) becomes $\Psi_k(2\pi) = \Psi_k(0)$. However, for $\Psi_k(x) = \Psi_n(x) = \exp(inx)$, $\Psi_k(0) = 1$ and $\Psi_k(2\pi) = \exp(2in\pi) = \cos 2n\pi + i \sin 2n\pi = 1$, thus showing that the boundary condition $\Psi_k(2\pi) = \Psi_k(0)$ is obeyed for all k = n and hence that $\frac{\hbar}{i}\frac{d}{dx}$ is Hermitian for the set $\{\Psi_n(x)\}$ on the range $0 \le x \le 2\pi$. Next, for the range $0 \le x \le \pi$, the boundary condition becomes $\Psi_k(\pi) = \Psi_k(0)$. However, for $\Psi_k(x) = \Psi_n(x) = \exp(inx)\Psi_k(0) = 1$ and $\Psi_k(\pi) = \exp(in\pi) = \cos n\pi + i \sin n\pi = (-1)^n$, showing that for odd n, the boundary condition $\Psi_k(\pi) = \Psi_k(0)$ is not obeyed and hence $\frac{\hbar}{i}\frac{d}{dx}$ is not Hermitian for the set $\{\Psi_n(x)\}$ on the range $0 \le x \le \pi$.

We next turn to Hermitian eigenvalue problems, namely to the problems of determining the eigenfunctions and eigenvalues of Hermitian operators $\hat{A}$. Hermitian eigenvalue problems are of paramount importance in quantum mechanics since their solutions are the quantized energies and stationary state (i.e., stable state) wave functions of quantum systems.

2.6 HERMITIAN EIGENVALUE PROBLEMS

We will see that (except in applications too advanced for this book) the eigenvalues of Hermitian operators are quantized. We thus denote the eigenvalues of a Hermitian operator $\hat{A}$ by a_n, where n is a quantum number. For simplicity, we will assume that all of the eigenvalues a_n are non-degenerate but all of our results follow even if this assumption is lifted (Problem 2.13).

In analogy to Equation (2.24), the eigenvalue equation for $\hat{A}$ is

$$\hat{A}\Psi_n(x) = a_n\Psi_n(x). \tag{2.56}$$

Equation (2.56) alone does not specify a Hermitian eigenvalue problem. Rather, to have a Hermitian eigenvalue problem, one must supplement Equation (2.56) with the condition that $\hat{A}$ be Hermitian for its set of eigenfunctions $\{\Psi_n(x)\}$ on some range $a \le x \le b$. From Equation (2.49), the required Hermitian condition is that for all n and m,

$$\int_a^b \Psi_n^*(x)\hat{A}\Psi_m(x)dx = \int_a^b \left[\hat{A}\Psi_n(x)\right]^* \Psi_m(x)dx. \tag{2.57}$$

Equations (2.56) and (2.57) together define a Hermitian eigenvalue problem. What occurs in practice is that Equation (2.56) has a continuum of solutions. Most of these solutions, however, are rejected because they fail to satisfy the Hermitian condition of Equation (2.57). The surviving solutions that are called the Hermitian eigenfunctions and eigenvalues of $\hat{A}$ have quantized eigenvalues a_n and associated discrete eigenfunctions $\Psi_n(x)$.

In practice, the Hermitian condition of Equation (2.57) is replaced by equivalent boundary conditions. As we will see later in this section, it is the process of enforcing the boundary conditions which yields quantized eigenvalues a_n.

Before showing this, we first prove that Hermitian eigenvalue problems have the following two properties:

1. The quantized Hermitian eigenvalues a_n are real numbers.
2. The discrete Hermitian eigenfunctions $\Psi_n(x)$ and $\Psi_m(x)$ associated with *different* eigenvalues a_n and a_m are orthogonal. That is, if we define S_{nm} by

$$S_{nm} = \int_a^b \Psi_n^*(x)\Psi_m(x)dx \tag{2.58}$$

then

$$S_{nm} = 0 \quad \text{if } n \neq m. \tag{2.59}$$

To prove these results, we first note from Equation (2.56) that $\hat{A}\Psi_m(x) = a_m\Psi_m(x)$ and $\left[\hat{A}\Psi_n(x)\right]^* = a_n^*\Psi_n^*(x)$. Inserting these results into Equation (2.57) and using Equation (2.58) yield

$$a_m S_{nm} = a_n^* S_{nm}. \tag{2.60}$$

Next, setting $n = m$, the above relation becomes

$$a_n S_{nn} = a_n^* S_{nn} \tag{2.61}$$

where $S_{nn} = \int_a^b \Psi_n^*(x)\Psi_n(x)dx$. However, since $\Psi_n^*(x)\Psi_n(x) \geq 0$ and since we are excluding the trivial function $\Psi_n(x) = 0$, it follows that $S_{nn} > 0$. Moreover, as noted at the start of the chapter, we are restricting ourselves to square-integrable functions $\Psi(x)$, that is, to those for which $\int_a^b \Psi^*(x)\Psi(x)dx < \infty$. This restriction implies that $S_{nn} < \infty$. Thus, it follows that

$$0 < S_{nn} < \infty. \tag{2.62}$$

Given Equation (2.62), S_{nn} may be canceled from Equation (2.61) yielding

$$a_n^* = a_n \tag{2.63}$$

the condition that a_n is real. Thus, we have proven property 1.

Next, using Equation (2.63), we may rewrite Equation (2.60) as

$$a_m S_{nm} = a_n S_{nm}. \tag{2.64}$$

For $n \neq m$ and hence $a_n \neq a_m$, the above relation can be satisfied only if $S_{nm} = \int_a^b \Psi_n^*(x)\Psi_m(x)dx = 0$ which is the condition for the orthogonality of the eigenfunctions $\Psi_n(x)$ and $\Psi_m(x)$. Thus, we have proven property 2.

However, we have not shown that linearly independent eigenfunctions associated with the *same* degenerate eigenvalue a_n are orthogonal. In general, they are not. However, such eigenfunctions can be orthogonalized by the Gram–Schmidt method of Section 2.4. Thus, the normalized Hermitian eigenfunctions of an operator $\hat{A}$ may always be taken as an orthonormal set.

As an example of these concepts, we next determine the Hermitian eigenfunctions and eigenvalues of the operator $\hat{A} = \frac{\hbar}{i}\frac{d}{dx}$ on the range $0 \leq x \leq 2\pi$. We first ignore the Hermitian condition of Equation (2.57) and simply deal with the eigenvalue problem

$$\frac{\hbar}{i}\frac{d\Psi_n(x)}{dx} = a_n\Psi_n(x). \tag{2.65}$$

Solving Equation (2.65), one finds that the operator $\frac{\hbar}{i}\frac{d}{dx}$ has a continuum of eigenfunctions

$$\Psi_n(x) = \exp\left(\frac{i}{\hbar}a_n x\right) \tag{2.66}$$

where a_n can take on any complex value.

We next impose the Hermitian condition or rather the equivalent boundary condition that from Equation (2.55) is

$$\Psi_n(2\pi) = \Psi_n(0). \tag{2.67}$$

Equations (2.65) and (2.67) together define a Hermitian eigenvalue, and thus, its solution should yield real eigenvalues and orthogonal eigenfunctions. We next show that this is indeed the case. Requiring the eigenfunctions of Equation (2.66) conform to the boundary condition of Equation (2.67) yields that $\exp\left[\frac{i}{\hbar}2\pi a_n\right] = 1$ or equivalently that

$$\cos\left(\frac{2\pi a_n}{\hbar}\right) + i\sin\left(\frac{2\pi a_n}{\hbar}\right) = 1. \tag{2.68}$$

The above condition is satisfied only if the eigenvalues a_n are restricted to the real quantized values

$$a_n = n\hbar, \quad \text{where } n = 0, \pm 1, \ldots. \tag{2.69}$$

The eigenfunctions $\Psi_n(x)$ associated with the quantized eigenvalues a_n are from Equations (2.66) and (2.69)

$$\Psi_n(x) = \exp(inx), \quad \text{where } n = 0, \pm 1, \ldots. \tag{2.70}$$

We close this section by showing that the Hermitian eigenfunctions $\Psi_n(x)$ are mutually orthogonal. We thus consider $S_{nm} = \int_0^{2\pi} \Psi_n^*(x)\Psi_m(x)dx = \int_0^{2\pi} \exp[i(m-n)x]dx$. For $n \neq m$, $S_{nm} = \frac{1}{i(m-n)}\{\exp[2\pi i(m-n)] - 1\} = \frac{1}{i(m-n)}\{\cos[2\pi(m-n)] + i\sin[2\pi(m-n)] - 1\} = 0$, thus proving orthogonality.

Next, we introduce another mathematical concept that is essential in quantum mechanics, the concept of *complete orthonormal expansions.*

2.7 COMPLETE ORTHONORMAL EXPANSIONS

To start, we convert the set of Hermitian eigenfunctions $\{\Psi_n(x)\}$ of an arbitrary Hermitian operator $\hat{A}$ into an orthonormal set. Since the eigenfunctions $\Psi_n(x)$ are already orthogonal to make this conversion, it is only necessary to normalize each eigenfunction. This is easily done if we recall Equation (2.25), which implies that the eigenfunctions $\Psi_n(x)$ are undetermined to within a complex constant. Denoting the constant, called the normalization factor, by N_n, we have that the function

$$\bar{\Psi}_n(x) = N_n\Psi_n(x) \tag{2.71}$$

like $\Psi_n(x)$ is an eigenfunction of $\hat{A}$ with the eigenvalue a_n. We determine N_n by requiring that $\bar{\Psi}_n(x)$ be normalized and thus satisfy the condition

$$\int_a^b \bar{\Psi}_n^*(x)\bar{\Psi}_n(x)dx = 1. \tag{2.72}$$

Equations (2.71) and (2.72) give the absolute value $|N_n| = (N_n^* N_n)^{1/2}$ of N_n as

$$|N_n| = \left[\int_a^b \Psi_n^*(x)\Psi_n(x)dx\right]^{-1/2}. \tag{2.73}$$

Usually, N_n may be chosen to be real. Then, $|N_n| = N_n$, and hence, one obtains the normalized eigenfunction $\bar{\Psi}_n(x)$ from Equations (2.71) and (2.73) as

$$\bar{\Psi}_n(x) = \left[\int_a^b \Psi_n^*(x)\Psi_n(x)dx\right]^{-1/2} \Psi_n(x). \tag{2.74}$$

To illustrate this normalization procedure, we find the normalized Hermitian eigenfunctions $\bar{\Psi}_n(x)$ of the operator $\frac{\hbar}{i}\frac{d}{dx}$ on the range $0 \le x \le 2\pi$. We have already shown that the corresponding unnormalized Hermitian eigenfunctions are $\Psi_n(x) = \exp(inx)$. Thus, on the range $0 \le x \le 2\pi$ the integral $\int_a^b \Psi_n^*(x)\Psi_n(x)dx$ in Equation (2.74) specializes to $\int_0^{2\pi} \exp(-inx)\exp(inx)dx = 2\pi$, and hence, the normalized Hermitian eigenfunctions are

$$\bar{\Psi}_n(x) = \left(\frac{1}{2\pi}\right)^{1/2} \exp(inx). \tag{2.75}$$

We will henceforth drop the bar on the normalized Hermitian eigenfunctions $\bar{\Psi}_n(x)$ and denote these functions simply as $\Psi_n(x)$. Since the functions $\Psi_n(x)$ are both normalized and orthogonal, the set $\{\Psi_n(x)\}$ comprised of these functions is an orthonormal set and thus satisfies for all n and m the orthonormality relation

$$\int_a^b \Psi_n^*(x)\Psi_m(x)dx = \delta_{nm}. \tag{2.76}$$

We may now define *complete orthonormal sets*. An orthonormal set $\{\Psi_n(x)\}$ is said to be complete if any well-behaved function $\Psi(x)$ that satisfies the same boundary conditions as the members $\Psi_n(x)$ of the set may be exactly expanded as the following linear superposition of these members:

$$\Psi(x) = \sum_{\text{all } n} C_n \Psi_n(x) \tag{2.77}$$

where, as we will soon show, the expansion coefficients C_n are given by

$$C_n = \int_a^b \Psi_n^*(x)\Psi(x)dx. \tag{2.78}$$

The expansion of Equation (2.77) is referred to as a complete orthonormal expansion.

We next derive Equation (2.78) from the orthonormality relation for the set $\{\Psi_n(x)\}$. To start, we first change the dummy summation index in Equation (2.77) from n to m to yield the expansion as $\Psi(x)=\sum_{\text{all } m} C_m\Psi_m(x)$. Multiplying this expansion on the left by $\Psi_n^*(x)$, integrating, and then using the orthonormality relation of Equation (2.76) give $\int_a^b \Psi_n^*(x)\Psi(x)dx = \sum_{\text{all } m} C_m \int_a^b \Psi_n^*(x)\Psi_m(x)dx = \sum_{\text{all } m} C_m\delta_{mn}$. But since $\delta_{mn}=0$ if $m \neq n$ and $\delta_{mn}=1$ if $m=n$, $\sum_{\text{all } m} C_m\delta_{mn} = C_n\delta_{nn} = C_n$. Therefore, $\int_a^b \Psi_n^*(x)\Psi(x)dx = C_n$, which is Equation (2.78), thus proving that important relation.

An example of a complete orthonormal expansion will be given in the next section.

Before proceeding to that section, we note that the assumption that the set of orthonormal eigenfunctions of a Hermitian operator is complete is unproven for arbitrary Hermitian operators. So here we will simply follow convention and postulate that the normalized eigenfunctions of all quantum Hermitian operators form complete sets.

To illustrate the mathematical concepts developed in this chapter as well as to introduce some additional ideas that will prove important in quantum mechanics, we next study a simple prototype problem for quantum mechanics, wave motion on a stretched string.

2.8 WAVE MOTION ON A STRETCHED STRING: A PROTOTYPE PROBLEM FOR QUANTUM MECHANICS

The following discussion refers to Figure 2.1. In these figures, we draw a tightly stretched string of length a which runs along the X-axis between the points $x=0$ and $x=a$. The string is displaced vertically at time $t=0$ to produce one-dimensional wave motion for $t>0$. The wave motion is described by a *wave displacement function* $\Psi(x,t)$ whose value at time t is the vertical displacement of the string at point x. Since the string is pinned down at its endpoints 0 and a for all times t, the wave displacement function must obey the boundary conditions

$$\Psi(0,t)=0 \text{ and } \Psi(a,t)=0. \tag{2.79}$$

Our problem is to determine the possible wave motions $\Psi(x,t)$ of the string.

To make this determination, we must develop a classical equation of motion for the string, namely its wave equation. Then, we must solve the wave equation subject to the boundary conditions of Equation (2.79) and also subject to the initial conditions developed shortly which specify the wave displacement function at $t=0$.

Because the string is a continuous dynamical system, its Newton's equation of motion cannot be immediately written down from the classical mechanical results for point particle systems of Appendix A. Fortunately, an effective method exists for treating continuous systems, namely approximating them by point particle systems with nearest-neighbor Hooke's law forces and then taking the continuum limit for which the number of particles $N \to \infty$ and the interparticle separation $\Delta \to 0$. This method is developed in detail by Goldstein, Poole, and Safko (2002, Chapter 13). Here, we will merely summarize their results.

The main result of Goldstein, Poole, and Safko is that in the continuum limit, the Newton's equations of motion for the N-particle system become the following wave equation for $\Psi(x,t)$:

$$\frac{1}{v_0^2}\frac{\partial^2\Psi(x,t)}{\partial t^2}=\frac{\partial^2\Psi(x,t)}{\partial x^2}. \tag{2.80}$$

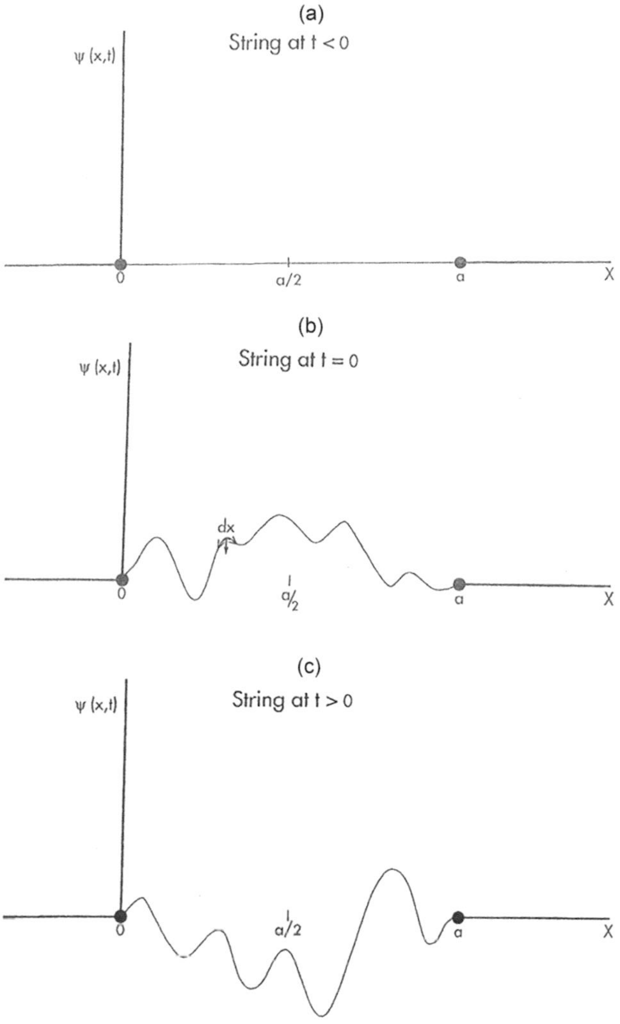

FIGURE 2.1 Wave motion on a stretched string of length a with fixed endpoints. Shown is the wave displacement function $\Psi(x,t)$ for times (a) $t<0$ when string is at equilibrium and hence $\Psi(x,t)=0$, (b) $t=0$ when wave motion is started by an initial displacement $\Psi(x,0)=\Psi(x)$, and (c) $t>0$ while wave motion is taking place.

The parameter v_0 in the wave equation is a property of the string which depends on its mass per unit length and on its elasticity. From Equation (2.80), it is evident that v_0 has units of velocity. In fact, v_0 would be equal to the propagation speed of a traveling wave on the string if the string were of infinite length. For this reason, we will call v_0 the *wave velocity* of the string.

Notice that the wave equation is a second-order differential equation in both x and t. Since it is second order in t in analogy to our discussion of Newton's equation of motion in Appendix A, the wave equation requires two initial conditions, each pair of distinct initial conditions generating a unique wave motion $\Psi(x,t)$.

The initial conditions for the wave equation are found by taking the continuum limit of the initial conditions of the N-particle system. This limiting process yields the initial conditions for the wave equation as

$$\Psi(x,0)=\Psi(x) \text{ and } \dot{\Psi}(x,0)=\dot{\Psi}(x) \tag{2.81}$$

where $\Psi(x)$ and $\dot{\Psi}(x)$ are, respectively, the initial vertical displacement and initial vertical velocity of the string at point x.

For simplicity, we will henceforth assume that the string is initially motionless and therefore that $\dot{\Psi}(x,0)=\dot{\Psi}(x)=0$. Thus, we will restrict ourselves to the subclass of wave motions generated by the initial conditions

$$\Psi(x,0)=\Psi(x) \text{ and } \dot{\Psi}(x,0)=\dot{\Psi}(x)=0. \tag{2.82}$$

Before solving the wave equation (Equation 2.80) subject to the boundary conditions of Equation (2.79) and the initial conditions of Equation (2.82), we note that two qualitatively different types of solutions of the wave equation are possible, and which type occurs depends on the form of the initial wave displacement function $\Psi(x)=\Psi(x,0)$. The first type occurs if $\Psi(x)$ has an arbitrary form like that depicted in Figure 2.1b. Then, $\Psi(x,t)$ has a complex time dependence that, as illustrated in Figure 2.1c, does not preserve the shape of $\Psi(x)$. The second type occurs when $\Psi(x)=\phi(x)$, where $\phi(x)$ is a sine function of the type plotted in Figure 2.2. Then, $\Psi(x,t)$ describes simple *standing wave motion* like that shown in Figure 2.3 which preserves the shape of $\Psi(x)$.

We first treat standing wave motion. For standing wave motion, $\Psi(x,t)$ has the shape preserving factorized form

$$\Psi(x,t)=\phi(x)\psi(t) \tag{2.83}$$

and the initial conditions of Equation (2.82) specialize to

$$\Psi(x,0)=\phi(x) \text{ and } \dot{\Psi}(x,0)=0. \tag{2.84}$$

We next solve the wave equation for the standing waves by a method frequently used in quantum mechanics, the *method of separation of variables*. To carry out this solution, we first insert Equation (2.83) for $\Psi(x,t)$ into the wave equation (Equation 2.80) to obtain $\frac{1}{v_0^2}\phi(x)\frac{d^2\psi(t)}{dt^2}=\psi(t)\frac{d^2\phi(x)}{dx^2}$ or dividing through by $\phi(x)\psi(t)$

$$\frac{1}{v_0^2}\frac{1}{\psi(t)}\frac{d^2\psi(t)}{dt^2}=\frac{1}{\phi(x)}\frac{d^2\phi(x)}{dx^2}. \tag{2.85}$$

The left-hand side of the above relation depends only on t, while the right-hand depends only on x Thus, if we vary, say, t keeping x constant, the right-hand side remains constant and so the left-hand side despite its apparent dependence on t must also remain constant, implying that it is actually independent of t and hence equal to a constant which we will denote by λ. But since the right-hand side is equal to the left-hand side, the right-hand side must also be equal to the constant λ. Thus, we have the relations $\frac{1}{v_0^2}\frac{1}{\psi(t)}\frac{d^2\psi(t)}{dt^2}=\lambda$ and $\frac{1}{\phi(x)}\frac{d^2\phi(x)}{dx^2}=\lambda$, which imply that $\psi(t)$ and $\phi(x)$ satisfy the *independent* equations

$$\frac{d^2\psi(t)}{dt^2}=v_o^2\lambda\psi(t) \tag{2.86}$$

and

$$\frac{d^2\phi(x)}{dx^2}=\lambda\phi(x). \tag{2.87}$$

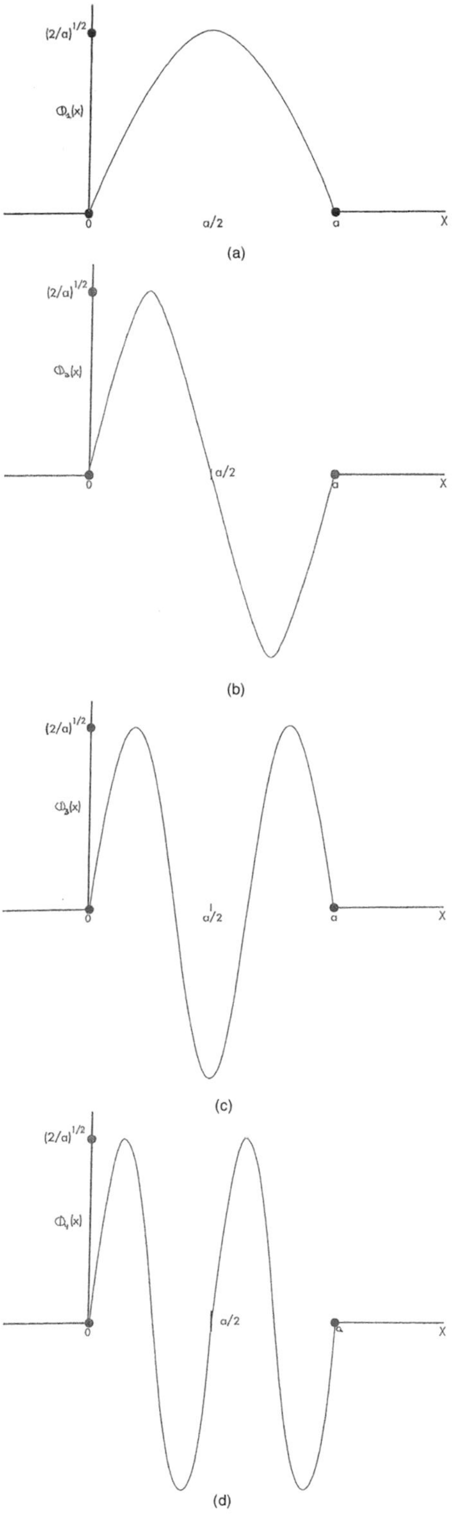

FIGURE 2.2 The normalized standing wave eigenfunctions $\phi_n(x) = (2/a)^{1/2} \sin\left(\frac{n\pi x}{a}\right)$ for (a) $n = 1$, (b) $n = 2$, (c) $n = 3$, and (d) $n = 4$.

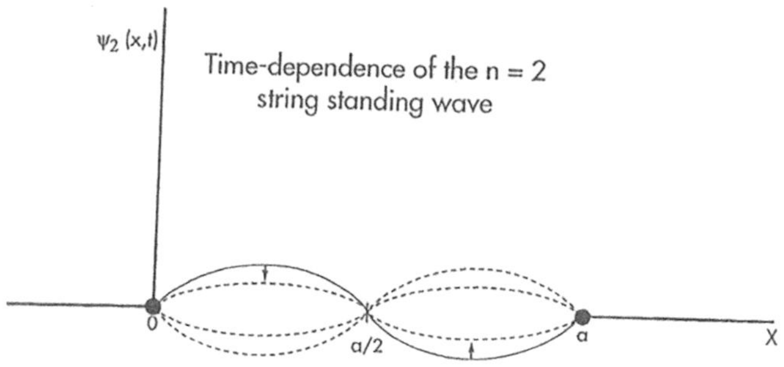

FIGURE 2.3 The $n=2$ stretched string standing wave. While arbitrary initial wave displacements $\Psi(x)$ (Figure 2.1b) give rise to complex wave motions $\Psi(x,t)$ (Figure 2.1c), the special initial displacement $\phi_2(x)=(2/a)^{1/2}\sin\left(\frac{2\pi x}{a}\right)$ (solid curve) produces simple standing wave motions $\Psi_2(x,t)=\cos\left(\frac{2\pi v_0 t}{a}\right)\phi_2(x)$ (dashed curves) which preserve the shape of $\phi_2(x)$.

Notice that both of the above equations are one-variable *ordinary* differential equations and are therefore much simpler than the wave equation (Equation 2.80), which is a two-variable *partial* differential equation.

To proceed further, we first discuss the $\phi(x)$ equation (Equation 2.87). Notice that it is an eigenvalue equation for the operator $\frac{d^2}{dx^2}$ which as may be verified by substitution has two linearly independent solutions $\phi(x)=\cos kx$ and $\phi(x)=\sin kx$, where $-k^2=\lambda$. The general solution of Equation (2.87) is thus

$$\phi(x)=A\cos kx+B\sin kx \tag{2.88}$$

where A and B are arbitrary constants.

So far, k and hence λ can take on any complex value rather than just the real values expected for wave motion. This discordant feature occurs since the eigenvalue equation (Equation 2.87) alone does not define a Hermitian eigenvalue problem. We will convert it into a Hermitian eigenvalue problem by supplementing it with appropriate boundary conditions. From Equations (2.79) and (2.83), these boundary conditions are $\Psi(0,t)=\phi(0)\psi(t)=0$ and $\Psi(a,t)=\phi(a)\psi(t)=0$ or canceling the $\psi(t)$ factors

$$\phi(0)=0 \text{ and } \phi(a)=0. \tag{2.89}$$

One may prove that the operator $\frac{d^2}{dx^2}$ is Hermitian on the range $0\le x\le a$ for function sets all of whose members satisfy the boundary conditions of Equation (2.89). (See Problem 2.14.) This result is sufficient to show that Equations (2.87) and (2.89) together define a Hermitian eigenvalue problem and therefore that the eigenvalues $\lambda=-k^2$ are real and quantized and the normalized eigenfunctions $\phi(x)$ form an orthonormal set.

We next show explicitly that this is indeed the case. First, applying the boundary condition $\phi(0)=0$ to the general solution of Equation (2.88) yields $\phi(0)=A\cos 0=0$, which since $\cos 0=1$ gives $A=0$. Thus, the general solution reduces to

$$\phi(x)=B\sin kx. \tag{2.90}$$

Then, applying the boundary condition $\phi(a)=0$ to Equation (2.90) gives $B\sin ka=0$, which has three solutions $B=0$, $k=0$, or $k=\frac{n\pi}{a}$ where $n=1,2,\ldots$. The choices $B=0$ and $k=0$ give the trivial solution $\phi(x) = 0$ (no wave motion) and thus must be rejected. From Equation (2.90), the allowed choice $k = k_n = \frac{n\pi}{a}$ yields for $n = 1,2,\ldots$ the discrete eigenfunctions

$$\phi(x) = \phi_n(x) = B_n \sin\left(\frac{n\pi x}{a}\right). \tag{2.91}$$

Associated with these eigenfunctions are the real quantized eigenvalues

$$\lambda = \lambda_n = -k_n^2 = -\frac{n^2\pi^2}{a^2}. \tag{2.92}$$

To complete the determination of the eigenfunctions $\phi_n(x)$, we must evaluate the constant B_n in Equation (2.91). This may be done by requiring that the eigenfunctions $\phi_n(x)$ be normalized on the interval $0 - a$. This requirement gives B_n as $B_n = \left(\frac{2}{a}\right)^{1/2}$ (Problem 2.15a). Hence, the normalized eigenfunctions are

$$\phi_n(x) = \left(\frac{2}{a}\right)^{1/2} \sin\left(\frac{n\pi x}{a}\right). \tag{2.93}$$

The normalized eigenfunctions $\phi_n(x)$ are plotted in Figure 2.2 for $n = 1-4$.

Further, one may readily verify that the normalized eigenfunctions form a real orthonormal set and thus for all n and m conform to the real orthonormality relation (Problem 2.15b)

$$\int_0^a \phi_n(x)\phi_m(x)dx = \delta_{nm}. \tag{2.94}$$

To complete the treatment of standing wave motion, we next turn to the $\psi(t)$ equation (Equation 2.86) specialized to the nth standing wave (the wave motion generated by the initial wave displacement function $\Psi[x] = \phi_n[x]$). For the nth standing wave, we denote $\psi(t)$ by $\psi_n(t)$ and take $\lambda = \lambda_n = -k_n^2$ to write Equation (2.86) as

$$\frac{d^2\psi_n(t)}{dt^2} = -\omega_n^2\psi_n(t) \tag{2.95}$$

where

$$\omega_n = v_0 k_n = v_0\left(\frac{n\pi}{a}\right). \tag{2.96}$$

For the nth standing wave, Equations (2.83) and (2.84) become, respectively, $\Psi_n(x,t) = \phi_n(x)\psi_n(t)$ and $\Psi_n(x,0) = \phi_n(x)$ and $\dot{\Psi}_n(x,0) = 0$. Comparing these relations gives $\Psi_n(x,0) = \phi_n(x)\psi_n(0) = \phi_n(x)$ and $\dot{\Psi}_n(x,0) = \phi_n(x)\dot{\psi}_n(0) = 0$, which yield the following initial conditions for $\psi(t)$

$$\psi_n(0) = 1 \text{ and } \dot{\psi}_n(0) = 0. \tag{2.97}$$

It is easily verified that the solution of Equation (2.95) that satisfies the initial conditions of Equation (2.97) is

$$\psi_n(t) = \cos\omega_n t. \tag{2.98}$$

Thus, specializing Equation (2.83) to the nth standing wave gives the full form of the wave displacement function $\Psi_n(x,t)$ of that standing wave as

$$\Psi_n(x,t) = \cos\omega_n t\phi_n(x). \tag{2.99}$$

$\Psi_n(x,t)$ for $n = 2$ is schematically depicted in Figure 2.3.

We may now treat general wave motion which recall is generated by an arbitrary initial wave displacement function $\Psi(x) = \Psi(x, t = 0)$. It is shown in Problem 2.16 that the solution $\Psi(x,t)$ of the wave equation generated by $\Psi(x)$ is the following linear superposition of the standing wave solutions $\cos\omega_n t\phi_n(x)$

$$\Psi(x,t) = \sum_{n\neq 1}^{\infty} C_n \cos\omega_n t\phi_n(x) \tag{2.100}$$

where the expansion coefficients C_n are given in Equation (2.102).

To derive Equation (2.102), we set $t = 0$ in Equation (2.100) to give

$$\Psi(x) = \sum_{n=1}^{\infty} C_n\phi_n(x). \tag{2.101}$$

Since the functions $\phi_n(x)$ form a real orthonormal set, Equation (2.101) is a real complete orthonormal expansion of the function $\Psi(x)$. Thus, in analogy to Equation (2.78), the coefficients

$$C_n = \int_0^a \phi_n(x)\Psi(x)dx. \tag{2.102}$$

We conclude this section by giving a numerical illustration of the complete orthonormal expansion of Equation (2.101). We choose the initial wave displacement function to have the form

$$\Psi(x) = \sin^2\left(\frac{2\pi x}{a}\right) \tag{2.103}$$

plotted in Figure 2.4. Since $\Psi(x)$ satisfies the boundary conditions $\Psi(0) = 0$ and $\Psi(a) = 0$ obeyed by the orthonormal eigenfunctions $\phi_n(x) = \left(\frac{2}{a}\right)^{1/2} \sin\left(\frac{n\pi x}{a}\right)$, it may indeed be expanded in terms

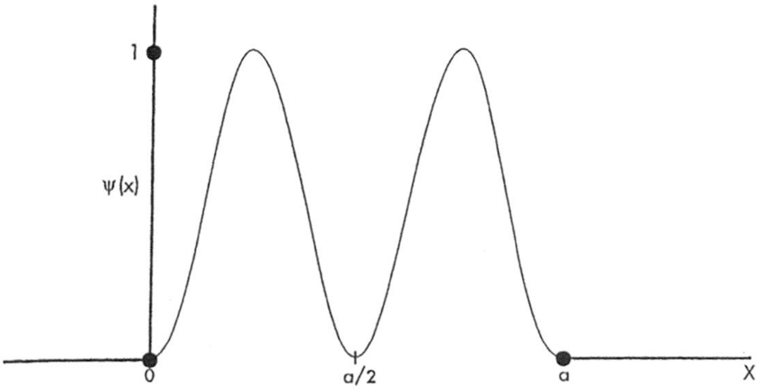

FIGURE 2.4 Initial wave displacement function $\Psi(x) = \sin^2\left(\frac{2\pi x}{a}\right)$ of Equation (2.103).

of these eigenfunctions. Written in terms of the functions $\sin\left(\frac{n\pi x}{a}\right)=\left(\frac{a}{2}\right)^{1/2}\phi_n(x)$, the complete orthonormal expansion of Equation (2.101) of $\Psi(x)=\sin^2\left(\frac{2nx}{a}\right)$ is

$$\Psi(x)=\sum_{n=1}^{\infty} D_n \sin\left(\frac{n\pi x}{a}\right) \tag{2.104}$$

where from Equations (2.102) and (2.103)

$$D_n=\frac{2}{a}\int_0^a \sin\left(\frac{n\pi x}{a}\right)\sin^2\left(\frac{2\pi x}{a}\right)dx. \tag{2.105}$$

Evaluation of D_n gives (Problem 2.17)

$$D_n=\begin{cases} 0 & \text{for n even} \\ \frac{2}{\pi}\left[\frac{1}{n}-\frac{1}{2(n+4)}-\frac{1}{2(n-4)}\right] & \text{for n odd.} \end{cases} \tag{2.106}$$

The non-vanishing values of D_n for n = 1–15 are listed in Table 2.2, and a study of the convergence of the expansion of Equation (2.104) is given in Table 2.3. It is found that the approximation

$$\Psi_{15}(x)=\sum_{n=1}^{15} D_n \sin\left(\frac{n\pi x}{a}\right) \tag{2.107}$$

reproduces $\Psi(x)$ to within 0.3%.

Finally, the time dependence of $\Psi(x,t)$ for $\Psi(x)=\sin^2\left(\frac{2\pi x}{a}\right)$ is studied in Problem 2.19.

TABLE 2.2
The First Eight Nonzero Expansion Coefficients D_n of Equation (2.106)

N	D_n
1	0.6791
3	0.4850
5	−0.2264
7	−0.0441
9	−0.0174
11	−0.0088
13	−0.0051
15	−0.0032

TABLE 2.3

Convergence in N of the Expansions $\Psi_N(x)=\sum_{n=1}^{N} D_n \sin\left(\frac{n\pi x}{a}\right)$ (see Table 2.2 for D_n Values) Obtained by Terminating the Complete Orthonormal Expansion of $\Psi(x)=\sin^2\left(\frac{2\pi x}{a}\right)$ Given in Equation (2.104) at N Terms

$\frac{x}{a}$	$\Psi_1(x)$	$\Psi_3(x)$	$\Psi_5(x)$	$\Psi_7(x)$	$\Psi_{15}(x)$	$\Psi(x)$
0.0	0	0	0	0	0	0
0.1	0.210	0.602	0.376	0.340	0.345	0.346
0.2	0.399	0.860	0.860	0.902	0.903	0.905
0.3	0.549	0.699	0.926	0.912	0.903	0.905
0.4	0.646	0.361	0.361	0.335	0.346	0.346
0.5	0.679	0.194	−0.032	0.012	0.001	0

FURTHER READINGS

Arfken, George B., Hans J. Weber, and Frank E. Harris. 2012. *Mathematical methods for physicists*. 7th ed. Waltham, MA: Academic Press.

Goldstein, Herbert, Charles Poole, and John Safko. 2002. *Classical mechanics*. 3rd ed. San Francisco: Pearson Education, Inc.

Ward, James B., and Ruel V. Churchill. 1993. *Fourier series and boundary value problems*. 5th ed. New York: McGraw-Hill.

PROBLEMS

2.1 (a) Show that $i=\exp\left(\frac{i\pi}{2}\right)$. Evaluate the real and imaginary parts of the complex numbers (b) $z=i^{1/2}$ and (c) $z=i^{1/3}$.

2.2 Prove Equation (2.8).

2.3 (a) Prove Equation (2.13) by showing that it is equivalent to the standard trigonometric identities $\cos(\alpha+\beta)x=\cos\alpha x\cos\beta x-\sin\alpha x\sin\beta x$ and $\sin(\alpha+\beta)x=\sin\alpha x\cos\beta x+\sin\beta x\cos\alpha x$. From Equation (2.8), prove (b) Equation (2.14) and (c) Equation (2.15).

2.4 Let $\Psi_1(x),\Psi_2(x),\Psi_3(x)$ and C_1,C_2,C_3 be complex functions and complex constants. By multiplying out real and imaginary parts, prove that (a) if $\Psi(x)=\Psi_1(x)\Psi_2(x)\Psi_3(x)$, then $\Psi^*(x)=\Psi_1^*(x)\Psi_2^*(x)\Psi_3^*(x)$, and (b) if $\Psi(x)=C_1\Psi_1(x)+C_2\Psi_2(x)+C_3\Psi_3(x)$, then $\Psi^*(x)=C_1^*\Psi_1^*(x)+C_2^*\Psi_2^*(x)+C_3^*\Psi_3^*(x)$.

2.5 Referring to Equation (2.16), find $\Phi(x)$ if $\Psi(x)=\exp(i\pi x)$, and if $\hat{O}$ is the rule, (a) $[\ ]^2$ = square $\Psi(x)$, (b) $\sqrt{\ }$ = take the square root of $\Psi(x)$, (c) apply $\frac{d^2}{dx^2}+3\frac{d}{dx}+2$ to $\Psi(x)$, (d) $[\ ]^*$ = take the complex conjugate of $\Psi(x)$, (e) $[\ ]^{-1}$ = take the reciprocal of $\Psi(x)$, (f) ℓn = take the logarithm of $\Psi(x)$, and (g) $\int_{-1/2}^{1/2}dx$ = evaluate the integral of $\Psi(x)$ between the limits $x=-\frac{1}{2}$ and $x=\frac{1}{2}$.

2.6 Referring to the definitions of Problem 2.5, determine which of the following operators are linear or non-linear: (a) $[\ ]^2$, (b) $\sqrt{\ }$, (c) $\frac{d}{dx^2}+3\frac{d}{dx}+2$, (d) $[\ ]^*$, (e) $[\ ]^{-1}$, (f) ln, and (g) $\int_{-\frac{1}{2}}^{1/2}dx$.

2.7 (a) Prove the commutator identity $\left[\hat{x}^2, \frac{\hbar}{i}\frac{d}{dx}\right] = 2i\hbar\hat{x}$. (b) Show that the operators $\frac{\hbar}{i}\frac{d}{dx}, \frac{d^2}{dx^2}$, and $\frac{d^2}{dx^2} + 3\frac{d}{dx} + 2$ mutually commute.

2.8 Show that (a) the functions $\exp(i\alpha x), \exp(-i\alpha x), \cos\alpha x, \sin\alpha x$ are not linearly independent, but (b) the functions $1, 2x, 4x^2 - 2, 8x^3 - 12x$ are linearly independent.

2.9 (a) Show that the functions $\exp(\pm i\alpha x)$ are eigenfunctions of the operator $\frac{d^2}{dx^2} + 3\frac{d}{dx} + 2$, and give the values of the two non-degenerate eigenvalues. (b) Show that the functions $\cos\alpha x$ and $\sin\alpha x$ are eigenfunctions of the operator $\frac{d^4}{dx^4} + 2\frac{d^2}{dx^2} + 1$, and give the value of the two-fold degenerate eigenvalue.

2.10 This problem deals with the Gram–Schmidt orthogonalization method. (a) Prove from Equations (2.36)–(2.38) that the real functions $\phi_1(x), \phi_2(x)$, and $\phi_3(x)$ are mutually orthogonal. (b) Write down the form of the Gram–Schmidt transformation relation for $\phi_n(x)$ where n is arbitrary as suggested by the forms of Equations (2.37)–(2.39). (c) Show that on the range $-1 \le x \le 1$, the linearly independent set of functions $1, x, x^2$, and x^3 may be converted into the orthonormal set $\frac{1}{\sqrt{2}}, \sqrt{\frac{3}{2}}x, \sqrt{\frac{5}{8}}(3x^2 - 1)$, and $\sqrt{\frac{7}{8}}(5x^3 - 3x)$ by using Equations (2.36)–(2.39) and then normalizing. (d) Prove that on the range $-1 \le x \le 1$, the functions $\sqrt{\frac{5}{8}}(3x^2 - 1)$ and $\sqrt{\frac{7}{8}}(5x^2 - 1)$ derived in part (c) are orthogonal.

2.11 Denote the left-hand and right-hand sides of Equation (2.52) by, respectively, I_{nm} and J_{nm}. Assuming $\Psi_n(x) = \exp(inx)$ and $\Psi_m(x) = \exp(imx)$, show that for $n = m$ $I_{nm} = J_{nm} = in\pi$, for $n \ne m$ and $n - m$ even $I_{nm} = J_{nm} = 0$, and for $n - m$ odd $I_{nm} = -\frac{2m}{m - n}$ and $J_{nm} = -\frac{2n}{n - m}$.

2.12 (a) From the Hermitian condition of Equation (2.50), show for the function set $\{\Psi_{n/2}(x)\}$ comprised of the complex exponentials $\exp\left(\frac{inx}{2}\right)$ where $n = 0, \pm 1, \ldots$ that the operator $\frac{\hbar}{i}\frac{d}{dx}$ is not Hermitian on either of the ranges $0 \le x \le 2\pi$ or $0 \le x \le \pi$. (b) Repeat your proofs of part (a) except base your arguments on the boundary condition $\Psi_k(b) = \Psi_k(a)$ of Equation (2.55).

2.13 Show that the eigenvalues a_n of a Hermitian operator $\hat{A}$ are real and that the eigenfunctions of $\hat{A}$ associated with different eigenvalues are orthogonal even when some or all of the eigenvalues a_n are degenerate.

2.14 (a) Show that for the operator $\frac{d^2}{dx^2}$, on the range $0 \le x \le a$ the general Hermitian condition of Equation (2.49) when applied to the function set $\{\phi(x)\} = \phi_1(x), \phi_2(x), \ldots$ specializes to $\int_0^a \phi_i^*(x)\frac{d^2\phi_j(x)}{dx^2}dx = \int_0^a \frac{d^2\phi_i^*(x)}{dx^2}\phi_j(x)dx$. (b) By integrating the left-hand side of the Hermitian condition of part (a) by parts twice, show that $\int_0^a \phi_i^*(x)\frac{d^2\phi_j(x)}{dx^2}dx = \phi_i^*(x)\phi_j'(x)\Big|_0^a - \left[\phi_i^*(x)\right]'\phi_j(x)\Big|_0^a + \int_0^a \frac{d^2\phi_i^*(x)}{dx^2}\phi_j(x)dx$. (c) By comparing you results in parts (a) and (b), show that the condition that the operator $\frac{d^2}{dx^2}$ be Hermitian for the function set $\{\phi(x)\}$ on the range $0 \le x \le a$ is that all members $\phi_i(x)$ and $\phi_j(x)$ of the set conform to the relation $\phi_i^*(x)\phi_j'(x)\Big|_0^a - \left[\phi_i^*(x)\right]'\phi_j(x)\Big|_0^a = 0$.

(c) From this relation, show that the operator $\frac{d^2}{dx^2}$ is Hermitian for the function set $\{\phi(x)\}$ on the range $0 \le x \le a$ if each member $\phi_k(x)$ of the set obeys the boundary conditions $\phi_k(0) = 0$ and $\phi_k(a) = 0$.

2.15 (a) Prove that the normalization factor B_n in Equation (2.91) is given by $B_n = \left(\frac{2}{a}\right)^{1/2}$

(b) Prove the orthonormality relation of Equation (2.94). (Hint: The trigonometric identities $\cos(A \pm B) = \cos A \cos B \mp \sin A \sin B$ will prove helpful.)

2.16 Show that $\Psi(x,t)$ of Equation (2.100) is the solution of the wave equation (Equation 2.80) generated by the initial wave displacement function $\Psi(x)$. To do this, show that $\Psi(x,t)$ is a solution of the wave equation, satisfies the boundary conditions of Equation (2.79), and conforms to the initial conditions of Equation (2.82).

2.17 Derive from Equations (2.105) and (2.106) for the expansion coefficients D_n. (Hint: Use the trigonometric identities $\cos[A \pm B] = \cos A \cos B \mp \sin A \sin B$ and $\sin[A \pm B] = \sin A \cos B \pm \sin B \cos A$.)

2.18 In Section 2.8, we applied the complete orthonormal expansion equations $\Psi(x) = \sum_{n=1}^{\infty} D_n \sin\left(\frac{n\pi x}{a}\right)$ and $D_n = \frac{2}{a}\int_0^a \sin\left(\frac{n\pi x}{a}\right)\Psi(x)dx$ to the initial wave displacement function $\Psi(x) = \sin^2\left(\frac{2\pi x}{a}\right)$. In this problem, you will apply the expansion equations to another initial wave displacement function $\Psi(x)$. This $\Psi(x)$ is V-shaped. Specifically, it ascends linearly from $x = 0$ to the string midpoint $x = \frac{a}{2}$ where it has its maximum value $\Psi\left(\frac{a}{2}\right) = 1$ and then descends linearly to $x = a$. The functional form for this wave displacement function is $\Psi(x) = \begin{cases} \frac{2}{a}x & \text{for } 0 \le x \le \frac{a}{2} \\ \frac{2}{a}(a-x) & \text{for } \frac{a}{2} < x \le a \end{cases}$. (a) Show for $\Psi(x)$ that $D_n = I_{1n} + I_{2n} + I_{3n}$ where $I_{C1n} = \frac{4}{a^2}\int_0^{a/2} x \sin\left(\frac{n\pi x}{a}\right)dx, I_{2n} = \frac{4}{a}\int_{a/2}^{a} \sin\left(\frac{n\pi x}{a}\right)dx$, and $I_{3n} - = \frac{4}{a^2}\int_{a/2}^{a} x \sin\left(\frac{n\pi x}{a}\right)dx$.

(b) Show using integration by parts that $I_{1n} = \frac{4}{n^2\pi^2}\sin\left(\frac{n\pi}{2}\right) - \frac{2}{n\pi}\cos\left(\frac{n\pi}{2}\right)$, $I_{2n} = \frac{4}{n\pi}\cos\left(\frac{n\pi}{2}\right) - \frac{4}{n\pi}\cos n\pi$, and $I_{3n} = \frac{4}{n^2\pi^2}\sin\left(\frac{n\pi}{2}\right) + \frac{4}{n\pi}\left(\cos n\pi - \frac{2}{n\pi}\right) + \cos\left(\frac{n\pi}{2}\right)$.

(c) Thus, show that $D_n = -\frac{8}{n^2\pi^2}\sin\left(\frac{n\pi}{2}\right)$. (d) Show that D_n may be rewritten as $D_n = \frac{8}{n^2\pi^2}\begin{cases} 0 & \text{for n even} \\ (-1)^{\frac{n-1}{2}} & \text{for n odd} \end{cases}$. (e) Using this result for D_n, compute the approximations $\Psi_N\left(\frac{a}{2}\right) = \sum_{n=1}^{N} D_n \sin\left(\frac{n\pi}{2}\right)$ for $N = 1,3,5,7$, and 9 and compare with the exact result $\Psi\left(\frac{a}{2}\right) = 1$.

2.19 In this problem, you will study the time dependence of $\Psi(x,t)$ generated by $\Psi(x) = \sin^2\left(\frac{n\pi x}{a}\right)$. Referring to Equation (2.100), plot for $\Psi(x) = \sin^2\left(\frac{2\pi x}{a}\right)$ the approximation to $\Psi(x,t)$ $\Psi_{15}(x,t)=\sum_{n=1}^{15} C_n \cos\omega_n t \sin\left(\frac{n\pi x}{a}\right)$ versus x/a at dimensionless times $\tau = v_0 t/a = 0.00, 0.25, 0.50, 0.75$, and 1.00. Use the D_n values listed in Table 2.2.

2.20 Consider two Hermitian operators $\hat{A}$ and $\hat{B}$. Show that the matrix elements of the product operator $\hat{A}\hat{B}$ satisfy the relationships $\int_a^b \Psi_i^*(x)\hat{A}\hat{B}\Psi_j(x)dx = \left[\int_a^b \Psi_j^*(x)\hat{B}\hat{A}\Psi_i(x)dx\right]^*$. Notice that this result shows that $\hat{A}\hat{B}$ is Hermitian only if $\hat{A}$ and $\hat{B}$ commute.

2.21 Consider the standing wave eigenfunctions $\Psi_n(x)=\left(\frac{2}{a}\right)^{1/2}\sin\left(\frac{n\pi x}{a}\right), n=1,2,\ldots$ Show that these eigenfunctions are symmetric (do not change sign) under reflection through the string midpoint $x=\frac{a}{2}$ if n is odd and are antisymmetric (change sign) under the same reflection if n is even.

SOLUTIONS TO SELECTED PROBLEMS

2.1 (b) $\mathrm{Re}\,z=\cos\left(\frac{\pi}{4}\right)=0.707$ and $\mathrm{Im}\,z=\sin\left(\frac{\pi}{4}\right)=0.707$. (c) $\mathrm{Re}\,z=\cos\left(\frac{\pi}{6}\right)=0.866$ and $\mathrm{Im}\,z=\sin\left(\frac{\pi}{6}\right)=0.500$.

2.5 (a) $\exp(2i\pi x)$, (b) $\exp\left(\frac{i\pi x}{2}\right)$, (c) $\left(-\pi^2+3i\pi+2\right)\exp(i\pi x)$, (d) $\exp(-i\pi x)$, (e) $\exp(-i\pi x)$, (f) $i\pi x$, and (g) $2/\pi$.

2.6 (a) Non-linear, (b) non-linear, (c) linear, (d) non-linear, (e) non-linear, (f) non-linear, and (g) linear.

2.9 (a) $-\alpha^2+3i\alpha+2$ and (b) $\alpha^4-2\alpha^2+1$.

2.10 (b) $\phi_n(x)=\bar{\phi}_n(x)-\sum_{i=1}^{n-1}\frac{\int_a^b \bar{\phi}_n(x)\phi_i^*(x)dx}{\int_a^b \phi_i^*(x)\phi_i(x)dx}$.

3 The Schrödinger Equation and the Particle-in-a-Box

In Chapters 4–6, we lay out the full formal basis of quantum mechanics, and in the succeeding chapters, we describe increasingly advanced applications of this basis. However, in order to convey some feeling for quantum mechanics before developing its full formal machinery in this chapter, we introduce the most useful quantum equation, the *time-independent Schrödinger equation*, and then apply it to one of the simplest quantum systems, the instructive one-dimensional particle-in-a-box system.

The time-independent Schrödinger equation, as we will show in Section 3.2, may be derived from the more fundamental *time-dependent Schrödinger equation*, already touched on in Section 1.5. So we begin with the time-dependent Schrödinger equation. This equation cannot be derived from anything more fundamental. Rather, like Newton's equation of motion, it is best viewed as a postulate that is accepted because it successfully predicts a vast range of phenomena.

While the time-dependent Schrödinger equation cannot be derived, several non-rigorous plausibility arguments for its form exist. We next give one of these.

3.1 A HEURISTIC "DERIVATION" OF THE TIME-DEPENDENT SCHRÖDINGER EQUATION

In our discussion of the photoelectric effect in Section 1.1, we noted that Einstein discovered that light exhibits a wave–particle duality, namely that a light wave of frequency υ or wavelength $\lambda = \frac{c}{\upsilon}$, where c is the speed of light, could also be viewed as a stream of particles called photons each with an energy E and momentum p. Einstein postulated that the particle properties of light E and p were related to its wave properties υ and λ as follows:

$$E = h\upsilon \text{ and } p = \frac{h}{\lambda}. \tag{3.1}$$

We further noted in Section 1.4 that de Broglie later hypothesized that ordinary particles also exhibit a wave–particle duality. Namely, de Broglie hypothesized that associated with a particle is a wave with *de Broglie wavelength* λ. In analogy to Einstein's photon relation $p = \frac{h}{\lambda}$, de Broglie postulated that the wavelength of the matter wave associated with a particle of momentum p is given by

$$\lambda = \frac{h}{p}. \tag{3.2}$$

Because of the minute magnitude of Planck's constant h, the de Broglie wavelength of a macroscopic particle is too small for the particle's wave properties to be observable (Problem 3.2). But for microscopic particles like electrons, the de Broglie wavelength can be large enough for the wave-like behavior of the particles to be detectable (Problem 3.3).

To proceed further, we return to the problem of wave motion of a stretched string studied in Section 2.8. In that section, we assumed a string that is pinned down at its endpoints and thus dealt with standing waves and their superpositions. Here in contrast, we will consider a traveling wave with wavelength λ moving with the wave velocity v_0 on an infinite string in the positive X-direction (Figure 3.1).

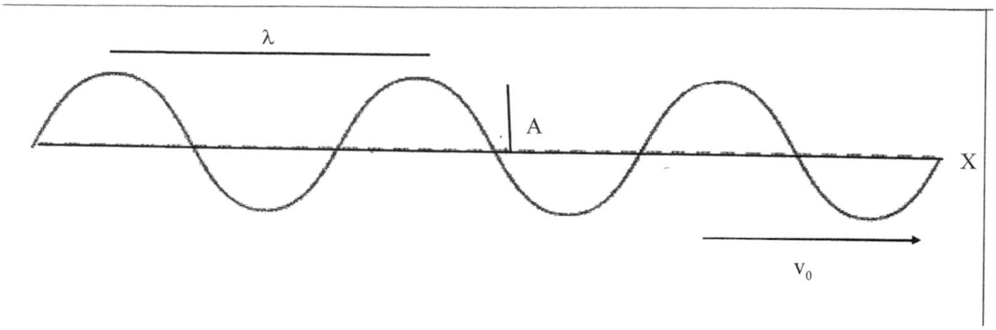

FIGURE 3.1 A traveling wave on a string moving in the X-direction with wave displacement function (see text) $\Psi(x,t) = A\cos\left[2\pi\left(\frac{x}{\lambda} - \frac{v_0 t}{\lambda}\right)\right]$, where A is the wave amplitude, λ is the wavelength, and v_0 is the wave velocity.

As in Section 2.8, we will characterize the wave by a wave displacement function $\Psi(x,t)$ whose value at time t is the vertical displacement of the string at point x. We will again denote the initial wave displacement function $\Psi(x,t=0)$ by $\Psi(x)$. Since the wave displacement must repeat itself at intervals of the wavelength λ, $\Psi(x)$ must satisfy the condition

$$\Psi(x+n\lambda) = \Psi(x) \quad \text{where } n = 1,2,\ldots \tag{3.3}$$

Also since the shape of the wave does not change as it travels along the string, an additional condition must be satisfied by $\Psi(x,t)$; namely,

$$\Psi(x,t) = \Psi(x - v_0 t) \tag{3.4}$$

where (Problem 3.7) $\Psi(x - v_0 t)$ is $\Psi(x)$ evaluated at the point $x - v_0 t$. Notice from Equation (3.4) that the full form of $\Psi(x,t)$ may be immediately found from its initial form $\Psi(x)$.

Perhaps the simplest choice of $\Psi(x)$ which conforms to Equation (3.3) and therefore is acceptable is the cosine function

$$\Psi(x) = A\cos\left(\frac{2\pi x}{\lambda}\right) \tag{3.5}$$

where A is the wave amplitude. However, for many purposes the following more general choice of $\Psi(x)$

$$\Psi(x) = A\exp\left(\frac{2\pi i x}{\lambda}\right) \tag{3.6}$$

which also conforms to Equation (3.3) is preferred since it is mathematically very convenient. (The proceeding form $\Psi(x)$ is complex while actual initial wave displacement functions are of course real. This is not a serious problem since what is done in practice is to use the complex form as the starting point of a wave calculation and then retain only the real part of the final result.)

From Equation (3.4), for $\Psi(x)$ of Equation (3.6)

$$\Psi(x,t) = A\exp\left[2\pi i\left(\frac{x}{\lambda} - \frac{v_0 t}{\lambda}\right)\right]. \tag{3.7}$$

However, the wavelength λ, frequency υ, and wave velocity v_0 are related by (Problem 3.9)

$$\upsilon = \frac{v_0}{\lambda}. \tag{3.8}$$

Consequently, we may rewrite Equation (3.7) as

$$\Psi(x,t) = A\exp\left[2\pi i\left(\frac{x}{\lambda} - \upsilon t\right)\right]. \tag{3.9}$$

However, the proceeding form for $\Psi(x,t)$ not only describes wave motion on a string but also represents a simplified description of the propagation of a light wave. For a light wave, however, we may use Einstein's relation $E = h\upsilon$ to express $\Psi(x,t)$ in terms of the photon energy E as

$$\Psi(x,t) = A\exp\left[2\pi i\left(\frac{x}{\lambda} - \frac{E}{h}t\right)\right]. \tag{3.10}$$

We next assume (non-rigorously) that $\Psi(x,t)$ of Equation (3.10) describes the wave motion of the de Broglie matter wave of a particle. We thus reinterpret λ and E as, respectively, the particle's de Broglie wavelength and energy. Then, given Equation (3.2) for the de Broglie wavelength λ, $\Psi(x,t)$ may be written as

$$\Psi(x,t) = A\exp\left[\frac{2\pi i}{h}(px - Et)\right] \tag{3.11}$$

where p is the momentum of the particle. Additionally, the energy E of a particle of mass m moving subject to a potential energy function $U(x)$ is given by

$$E = \frac{p^2}{2m} + U(x). \tag{3.12}$$

Next, we differentiate $\Psi(x,t)$ of Equation (3.11) with respect to t and rearrange to give

$$i\hbar\frac{\partial\Psi(x,t)}{\partial t} = E\Psi(x,t) \tag{3.13}$$

where recall $\hbar = \frac{h}{2\pi}$. Then, we differentiate $\Psi(x,t)$ twice with respect to x and rearrange to yield

$$-\frac{\hbar^2}{2m}\frac{\partial^2\Psi(x,t)}{\partial x^2} = \frac{p^2}{2m}\Psi(x,t). \tag{3.14}$$

But from Equation (3.12), $\frac{p^2}{2m} = E - U(x)$. Hence, Equation (3.14) may be rewritten as

$$-\frac{\hbar^2}{2m}\frac{\partial^2\Psi(x,t)}{\partial x^2} + U(x)\Psi(x,t) = E\Psi(x,t). \tag{3.15}$$

However, the left-hand sides of both Equations (3.13) and (3.15) are equal to $E\Psi(x,t)$ and thus are equal to each other. Equating these left-hand sides yields

$$i\hbar\frac{\partial\Psi(x,t)}{\partial t} = -\frac{\hbar^2}{2m}\frac{\partial^2\Psi(x,t)}{\partial x^2} + U(x)\Psi(x,t). \tag{3.16}$$

Equation (3.16) is the time-dependent Schrödinger equation for the wave function $\Psi(x,t)$ of a particle of mass m moving in one dimension under the influence of a potential energy function $U(x)$.

Equation (3.16) is readily extended from one to three dimensions. Consider a particle of mass m with position vector $\mathbf{r}=(x,y,z)$ subject to a potential energy function $U(\mathbf{r})$. Then, the time-dependent Schrödinger equation for the particle's wave function $\Psi(\mathbf{r},t)$ is, as in Equation (1.72)

$$i\hbar\frac{\partial\Psi(\mathbf{r},t)}{\partial t}=-\frac{\hbar^2}{2m}\nabla^2\Psi(\mathbf{r},t)+U(\mathbf{r})\Psi(\mathbf{r},t) \tag{3.17}$$

where the operator ∇^2 is defined by

$$\nabla^2=\frac{\partial^2}{\partial x^2}+\frac{\partial^2}{\partial y^2}+\frac{\partial^2}{\partial z^2}. \tag{3.18}$$

Since the time-dependent Schrödinger equation (Equation 3.16) is first order in time, it requires only a single initial condition, which we write as

$$\Psi(x,t=0)=\Psi(x) \tag{3.19}$$

where the initial wave function $\Psi(x)$ can be any of an infinity of well-behaved and (usually) normalized wave functions. Each choice of $\Psi(x)$ generates a distinct time-dependent wave function $\Psi(x,t)$ that in turn determines a unique set of average values (or, as touched on in Section 3.3 and described in detail in Chapter 4, more properly probability distributions) for the system's observables.

We next derive the time-independent Schrödinger equation for a particle moving in one dimension from its time-dependent Schrödinger equation (Equation 3.16).

3.2 STATIONARY STATE WAVE FUNCTIONS AND THE TIME-INDEPENDENT SCHRÖDINGER EQUATION

The derivation is based on the analogy between the problem of finding the standing wave solutions of the wave equation (Equation 2.80) and the problem of determining the stationary state solutions (to be described shortly) of the time-dependent Schrödinger equation (Equation 3.16).

To start, we note that in analogy to what we found in Section 2.8 for wave motion on a stretched string, we expect two qualitatively different types of solutions of the time-dependent Schrödinger equation. We further expect that which type occurs depends on the form of the initial wave function $\Psi(x)$.

Namely, recall for wave motion on a stretched string that arbitrary initial wave displacement functions $\Psi(x)$ generate time-dependent wave displacement functions $\Psi(x,t)$ with complex time dependencies which do not preserve the shape of $\Psi(x)$. However, initial wave displacement functions $\Psi(x)$ that are eigenfunctions $\phi_n(x)$ of the operator $\frac{d^2}{dx^2}$ generate standing wave displacement functions $\Psi_n(x,t)$ with simple cosine time dependencies which preserve the shape of $\Psi(x)$.

Analogously in quantum mechanics, arbitrary initial wave functions $\Psi(x)$ generate time-dependent wave functions $\Psi(x,t)$ with complex time dependencies which do not preserve the form of $\Psi(x)$. However, initial wave functions $\Psi(x)$ that are eigenfunctions $\Psi_n(x)$ of the *Hamiltonian operator*

$$\hat{H}=-\frac{\hbar^2}{2m}\frac{d^2}{dx^2}+U(x) \tag{3.20}$$

generate *stationary state wave functions* $\Psi_n(x,t)$ with simple complex exponential time dependencies which preserve the form of $\Psi(x)$.

Before determining the form of the stationary state wave functions, we note that these wave functions are of paramount importance in the quantum theory of atoms and molecules. This is because stationary state wave functions are the wave functions of a quantum system when it is in one of its *stationary states* that are the normal stable quantum states of the system.

To determine the stationary state wave functions, we first use Equation (3.20) for the Hamiltonian operator $\hat{H}$ to write the time-dependent Schrödinger equation (Equation 3.16) concisely as

$$i\hbar\frac{\partial\Psi(x,t)}{\partial t} = \hat{H}\Psi(x,t). \tag{3.21}$$

To proceed further, guided by our development of the standing wave solutions of the wave equation, we assume that the stationary state wave functions $\Psi(x,t)$ have the following factorized form:

$$\Psi(x,t) = \Psi(x)\Phi(t) \tag{3.22}$$

where $\Psi(x)$ is the initial wave function that we will soon see is an eigenfunction of $\hat{H}$.

We next derive independent equations for $\Psi(x)$ and $\Phi(t)$ by the separation-of-variables technique used in Section 2.8 to determine the standing wave solutions of the wave equation. Thus, we insert Equation (3.22) into Equation (3.21) and note that $\hat{H}$ is free of any reference to the time t to obtain $i\hbar\Psi(x)\frac{d\Phi(t)}{dt} = \Phi(t)\hat{H}\Psi(x)$ or equivalently

$$i\hbar\frac{1}{\Phi(t)}\frac{d\Phi(t)}{dt} = \frac{1}{\Psi(x)}\hat{H}\Psi(x). \tag{3.23}$$

Since the left-hand side of Equation (3.23) depends only on t while the right-hand side depends only on x, we may make the separation-of-variables argument and equate both sides to a constant. We will see in Section 4.4 that this constant is the energy of the system and so we will denote it by E. Then, Equation (3.23) separates into two independent equations

$$i\hbar\frac{d\Phi(t)}{dt} = E\Phi(t) \tag{3.24}$$

and

$$\hat{H}\Psi(x) = E\Psi(x). \tag{3.25}$$

Equation (3.25), which is the eigenvalue equation for the Hamiltonian operator $\hat{H}$, is called the time-independent Schrödinger equation or often just the Schrödinger equation. Its solutions are the energies E and the wave functions $\Psi(x)$. for the stationary states, which as noted are the normal stable states of a quantum system. For this reason, the Schrödinger equation (Equation 3.25) and its multi-dimensional generalizations will provide the basis for most of our later applications of quantum mechanics.

However, while the energies of quantum systems are real and quantized, the eigenvalues E of $\hat{H}$ are in general complex and continuous. This incompatibility arises since the Schrödinger equation alone does not define a Hermitian eigenvalue problem and, thus in analogy to what we found for the standing wave equation (Equation 2.87), must be supplemented by boundary conditions. The needed boundary conditions are

$$\lim_{x\to\infty}\Psi(x) = 0 \text{ and } \lim_{x\to\infty}\Psi(x) = 0. \tag{3.26}$$

These boundary conditions arise from the requirement that the eigenfunctions $\Psi(x)$ must be square integrable and thus must satisfy the condition

$$\int_{-\infty}^{\infty} \Psi^*(x)\Psi(x)dx < \infty. \tag{3.27}$$

The Schrödinger equation (Equation 3.25) and the boundary conditions of Equation (3.26) together define a Hermitian eigenvalue problem (Problem 3.11). Consequently, solving this problem yields real eigenvalues E_n that are the quantized energies of the system and discrete orthogonal eigenfunctions $\Psi_n(x)$ which if normalized conform to the orthonormality relation

$$\int_{-\infty}^{\infty} \Psi_n^*(x)\Psi_m(x)dx = \delta_{nm}. \tag{3.28}$$

To emphasize these features, we rewrite the Schrödinger equation as

$$\hat{H}\Psi_n(x) = E_n\Psi_n(x). \tag{3.29}$$

Before proceeding further, we note that while within the Planck, Bohr, and Einstein theories of Chapter 1 quantized energies were postulated in order to explain particular experiments, in modern quantum mechanics they arise naturally by imposing boundary conditions (or other restrictions like those described in Appendices C and D) on the solutions of the Schrödinger equation that is a fundamental relation not tied to any specific experiment.

To complete the determination of the stationary state wave functions $\Psi(x,t)$, let us solve Equation (3.24) for $\Phi(t)$, which for the nth stationary state becomes

$$i\hbar\frac{d\Phi_n(t)}{dt} = E_n\Phi_n(t). \tag{3.30}$$

Equation (3.30) is a first-order differential equation in time and thus requires only a single initial condition. This initial condition may be found by specializing Equation (3.22) to give the full wave function $\Psi_n(x,t)$ for the nth stationary state as

$$\Psi_n(x,t) = \Psi_n(x)\Phi_n(t). \tag{3.31}$$

But $\Psi_n(x,0) = \Psi_n(x)$ yielding the initial condition for $\Phi_n(t)$ as $\Phi_n(0) = 1$. Solving Equation (3.30) subject to this initial condition gives $\Phi_n(t)$ as

$$\Phi_n(t) = \exp\left(-\frac{i}{\hbar}E_n t\right). \tag{3.32}$$

Comparing Equations (3.31) and (3.32) yields the final form for $\Psi_n(x,t)$

$$\Psi_n(x,t) = \exp\left(-\frac{i}{\hbar}E_n t\right)\Psi_n(x). \tag{3.33}$$

Both $\Psi_n(x,t)$ and $\Psi_n(x)$ are referred to as stationary state wave functions.

Finally, notice that in contrast to determining arbitrary system wave functions, to determine stationary state wave functions $\Psi_n(x,t)$ it is not necessary to deal with the time-dependent Schrödinger equation (Equation 3.16). Instead, since $\Psi_n(x,t)$ depends only on E_n and $\Psi_n(x)$, it is only necessary to solve the much simpler time-independent Schrödinger equation (Equation 3.29).

Next, we introduce the probability interpretation of quantum wave functions developed much more completely in Sections 4.3 and 4.4.

3.3 THE PROBABILITY INTERPRETATION OF THE WAVE FUNCTION AND THE DETERMINATION OF AVERAGE VALUES OF OBSERVABLES

As indicated in Section 1.5, in contrast to classical mechanics, quantum mechanics (typically but not always) does not make a definite prediction of the outcome of a single measurement of the value of an observable. Rather, as described in detail in Sections 4.3 and 4.4, quantum mechanics only predicts probability distributions that give the likelihoods of each of the many possible outcomes. The information in the probability distributions is then usually condensed to give values for the observable averaged over the spread of its values found in many individual measurements.

As a simple example of these principles, let us consider the problem of determining at time t the position x of a particle moving in one dimension. Classical mechanics predicts from the particle's trajectory $x(t)$ the precise value of the position. In contrast, quantum mechanics predicts from the particle's wave function $\Psi(x,t)$ a probability distribution $P(x,t)$ that gives the likelihood of measuring a particular position from among the range of positions which potentially could be measured. Quantum mechanics also predicts from the wave function the average value $\langle x(t)\rangle$ of the position determined by averaging over the spread of results found from many position measurements.

The probability distribution $P(x,t)$ has the following interpretation:

$$P(x,t)dx = \text{probability that a measurement of the particle's position at time t will give a value between x and } x+dx. \quad (3.34)$$

Since the particle must have some position in the range $-\infty < x < \infty$, it follows from Equation (3.34) that

$$\int_{-\infty}^{\infty} P(x,t)\,dx = 1. \quad (3.35)$$

Additionally, it follows from Equation (3.34) that $\langle x(t)\rangle$ is given by

$$\langle x(t)\rangle = \int_{-\infty}^{\infty} xP(x,t)\,dx. \quad (3.36)$$

To proceed further, we need a relation that determines $P(x,t)$ from the system wave function $\Psi(x,t)$. The required relation was proposed by the German physicist Max Born in 1926 and which we will call the *Born rule* is

$$P(x,t) = \Psi^*(x,t)\Psi(x,t) \geq 0. \quad (3.37)$$

Using the Born rule, Equations (3.35) and (3.36) become

$$\int_{-\infty}^{\infty} \Psi^*(x,t)\Psi(x,t)dx = 1 \quad (3.38)$$

and

$$\langle x(t)\rangle = \int_{-\infty}^{\infty} x\Psi^*(x,t)\Psi(x,t)dx. \quad (3.39)$$

Note that Equation (3.38) shows that the Born rule requires that the wave function $\Psi(x,t)$ must be normalized.

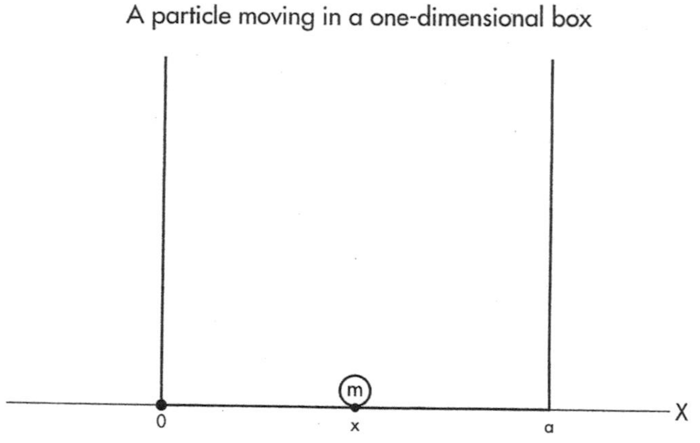

FIGURE 3.2 A particle of mass m confined to a one-dimensional box of length a.

We next specialize our results to stationary states. We will show that for such states, the Born rule probability distribution $P(x,t) = \Psi^*(x,t)\Psi(x,t)$, the normalization requirement on $P(x,t)$, and the average value $\langle x(t)\rangle$ of x at time t are all *time independent*. This time independence is the basis of the term stationary state.

We assume that the system is in the nth stationary state. From Equation (3.33), for the wave function for this state and Equation (3.37), for the nth stationary state $P(x,t) = \Psi_n^*(x,t)\Psi_n(x,t) = \exp\left(i\frac{E_n}{\hbar}t\right)\exp\left(-\frac{i}{\hbar}E_n t\right)\Psi_n^*(x)\Psi_n(x)$. Then, canceling the complex exponential factors yields the following time-independent form for $P(x,t)$:

$$P(x,t) = P_n(x) = \Psi_n^*(x)\Psi_n(x) \geq 0. \tag{3.40}$$

Similarly, the normalization requirement of Equation (3.38) and the formula for $\langle x(t)\rangle$ of Equation (3.39) reduce to the time-independent forms

$$\int_{-\infty}^{\infty} \Psi_n^*(x)\Psi_n(x) = 1 \tag{3.41}$$

and

$$\langle x(t)\rangle = \langle x\rangle_n = \int_{-\infty}^{\infty} x\Psi_n^*(x)\Psi_n(x)dx. \tag{3.42}$$

Notice that Equation (3.41) shows that $\Psi_n(x)$ like $\Psi_n(x,t)$ must be normalized.

We are now ready to determine the stationary state wave functions $\Psi_n(x)$ and the stationary state properties of the one-dimensional particle-in-a-box system depicted in Figure 3.2.

3.4 THE STATIONARY STATES OF THE ONE-DIMENSIONAL PARTICLE-IN-A-BOX SYSTEM

To start, using Equation (3.20) for $\hat{H}$, we write out the Schrödinger equation (Equation 3.29) explicitly as

$$-\frac{\hbar^2}{2m}\frac{d^2\Psi_n(x)}{dx^2} + U(x)\Psi_n(x) = E_n\Psi_n(x). \tag{3.43}$$

For the one-dimensional particle-in-a-box system of Figure 3.2, the potential energy function $U(x)$ is zero inside of the box and infinite outside of the box and thus has the form

$$U(x) = \begin{cases} 0 & \text{for } 0 \leq x \leq a \\ \infty & \text{for } x < 0 \text{ or } x > a \end{cases} \tag{3.44}$$

where a is the length of the box.

Since inside the box $U(x) = 0$, inside the box the Schrödinger equation (Equation 3.43) reduces to

$$-\frac{\hbar^2}{2m}\frac{d^2\Psi_n(x)}{dx^2} = E_n\Psi_n(x) \quad \text{for } 0 \leq x \leq a. \tag{3.45}$$

Outside of the box $U(x)$ is infinite and so the Schrödinger equation (Equation 3.43) includes an infinite term unless the wave functions $\Psi_n(x)$ vanish outside of the box. Thus, to avoid an infinite term, we require that

$$\Psi_n(x) = 0 \quad \text{for } x < 0 \text{ and } x > a \tag{3.46}$$

Therefore, since we know that the wave functions vanish outside of the box, it is only necessary to solve Equation (3.45) for the wave functions inside of the box.

However, the Schrödinger equation (Equation 3.45) does not yet define a Hermitian eigenvalue problem. To convert it into a Hermitian eigenvalue problem, we must supplement it with the following boundary conditions:

$$\Psi_n(0) = 0 \text{ and } \Psi_n(a) = 0. \tag{3.47}$$

These boundary conditions follow since (a) the wave functions vanish outside of the box and (b) the wave functions must be continuous, implying that they must vanish just inside the box, that is, at $x = 0$ and $x = a$.

To proceed further, let us rewrite the Schrödinger equation (Equation 3.45) as

$$\frac{d^2\Psi_n(x)}{dx^2} = -k_n{}^2\Psi_n(x) \tag{3.48}$$

where $k_n{}^2 = \frac{2mE_n}{\hbar^2}$ or equivalently

$$E_n = \frac{k_n{}^2\hbar^2}{2m}. \tag{3.49}$$

We next note that (except for notation) Equation (3.48) and its boundary conditions of Equation (3.47) are identical to the stretched string standing wave equation (Equation 2.87) and its boundary conditions of Equation (2.89). Thus, we may immediately write down the solutions of the particle-in-a-box Schrödinger equation from the results of Section 2.8 for the stretched string standing wave eigenvalue problem. (Also see Problem 3.15.) We first find that k_n is restricted to the discrete values

$$k_n = \frac{n\pi}{a} \quad \text{where } n = 1, 2, \ldots. \tag{3.50}$$

From Equations (3.49) and (3.50), we then find that the particle-in-a-box energies have the quantized values

$$E_n = \frac{n^2\pi^2\hbar^2}{2ma^2} = \frac{n^2h^2}{8ma^2}, \quad \text{where } n = 1,2,\ldots. \tag{3.51}$$

The ground $(n = 1)$ and first two excited state $(n = 2 \text{ and } 3)$ particle-in-a-box energy levels are depicted in Figure 3.3. Further from Equation (2.93) for the normalized standing wave eigenfunctions, the normalized particle-in-a-box wave functions are real and are given inside the box by

$$\Psi_n(x) = \left(\frac{2}{a}\right)^{1/2} \sin\left(\frac{n\pi x}{a}\right), \quad \text{where } n = 1,2,\ldots. \tag{3.52}$$

Additionally, from Equation (2.94) the normalized wave functions $\Psi_n(x)$ conform for all n and m to the real orthonormality relation

$$\int_0^a \Psi_n(x)\Psi_m(x)dx = \delta_{nm}. \tag{3.53}$$

The particle-in-a-box wave functions $\Psi_n(x)$ and the corresponding Born rule probability distributions for the particle's position x

$$P_n(x) = \Psi_n(x)\Psi_n(x) = \frac{2}{a}\sin^2\left(\frac{n\pi x}{a}\right) \tag{3.54}$$

are plotted for $n = 1 - 3$ in Figure 3.4.

The plots of the wave functions illustrate an important general principle. Namely, if one defines a *node* of a stationary state wave function as a point other than one of the boundary points (which are $x = 0$ and $x = a$ for the particle-in-a-box system) where the wave function vanishes, then the ground state wave function has zero nodes, the first excited state wave function has one node, and so on. In accord with this general principle, we see from Figure 3.4 that $\Psi_1(x)$ is nodeless, that $\Psi_2(x)$ has a single node at $x = \frac{a}{2}$, and that $\Psi_3(x)$ has two nodes: one at $x = \frac{a}{3}$ and the other at $x = \frac{2a}{3}$.

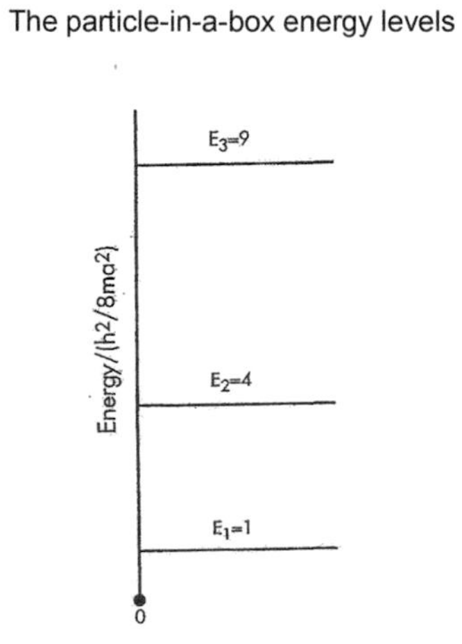

FIGURE 3.3 The ground and first two excited state energy levels of a one-dimensional particle-in-a-box system.

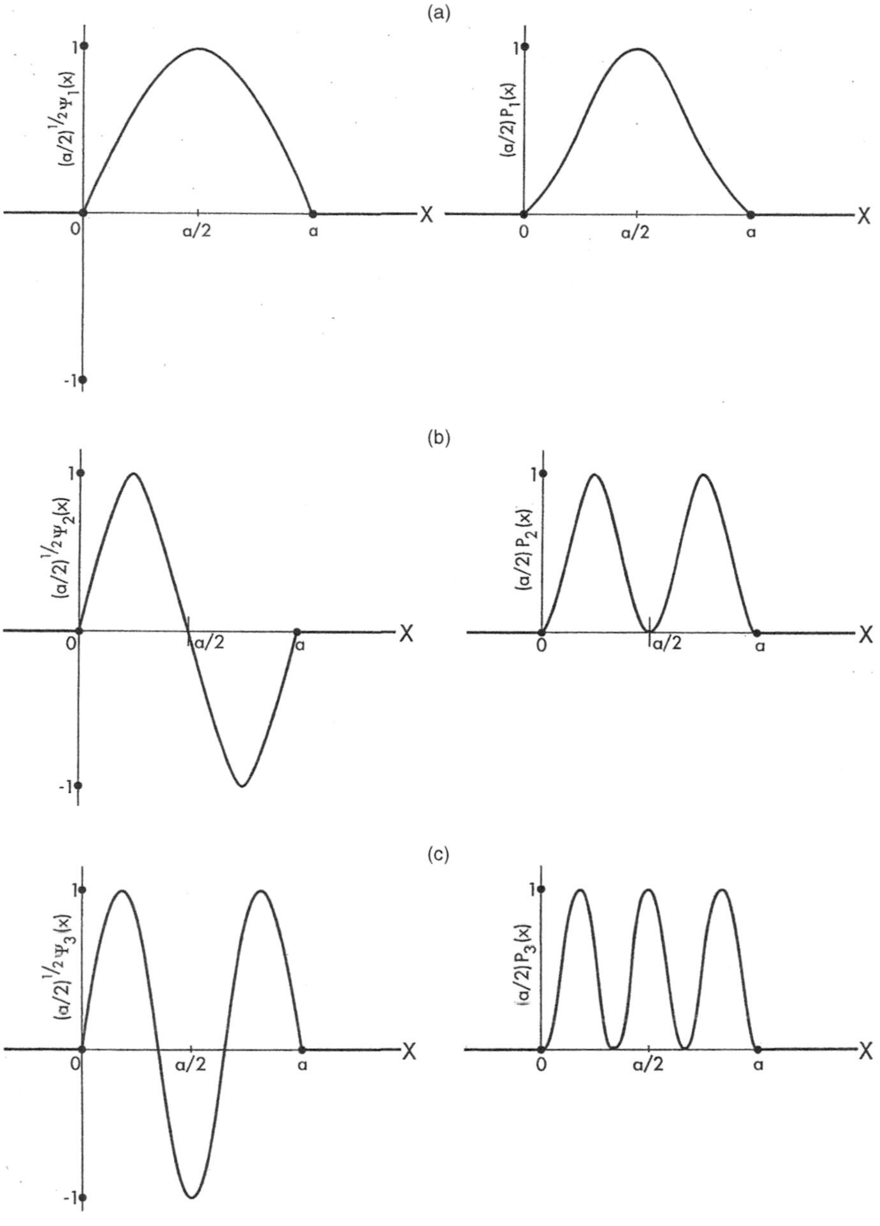

FIGURE 3.4 The one-dimensional particle-in-a-box wave functions $\Psi_n(x)$ and Born rule probability distributions $P_n(x) = \Psi_n(x)\Psi_n(x)$ for (a) n = 1, (b) n = 2, and (c) n = 3.

The plots of the probability distributions $P_n(x)$ illustrate how radically behavior in the quantum world differs from classical intuition. For example, from the plot of $P_2(x)$ in Figure 3.4b, one sees that $P_2(x)$ vanishes at the box midpoint $x = \frac{a}{2}$, and therefore, the particle has zero probability of being at the midpoint. Consequently, the particle can apparently "magically" move from a point to the left of the midpoint to a point to the right of it without ever passing through the midpoint. The resolution of this paradox of "quantum magic" as we will soon see is that particle trajectories cannot be defined in quantum mechanics.

We next turn to the determination of average values of observables for the particle-in-a-box system. This will lead us to the concept of *uncertainties* and to the *Heisenberg uncertainty relation.*

3.5 UNCERTAINTIES AND THE HEISENBERG UNCERTAINTY RELATION FOR THE PARTICLE-IN-A-BOX SYSTEM

We begin by specializing Equation (3.42) to give the average value $\langle x \rangle_n$ of the particle's position x in the nth particle-in-a-box stationary state as

$$\langle x \rangle_n = \int_0^a x\Psi_n(x)\Psi_n(x)dx. \tag{3.55}$$

Evaluation of $\langle x \rangle_n$ using Equation (3.52) for the particle-in-a-box wave functions $\Psi_n(x)$ gives (Problem 3.18a)

$$\langle x \rangle_n = \frac{a}{2}. \tag{3.56}$$

The above expression for $\langle x \rangle_n$ is expected since the particle is equally likely to be a distance δ from either the left or the right of the box midpoint $\frac{a}{2}$.

Equation (3.55) is easily generalized to give the average value $\langle g(x) \rangle_n$ of an arbitrary function $g(x)$ of x as

$$\langle g(x) \rangle_n = \int_0^a g(x)\Psi_n(x)\Psi_n(x)dx. \tag{3.57}$$

A particularly important choice of $g(x)$ is $(x - \langle x_n \rangle)^2$. We will denote the average of $(x - \langle x \rangle_n)^2$ by $(\Delta x)_n^2$. From Equation (3.57) with $g(x) = (x - \langle x_n \rangle)^2$

$$(\Delta x)_n^{\ 2} = \int_0^a (x - \langle x \rangle_n)^2 \Psi_n(x)\Psi_n(x)dx. \tag{3.58}$$

We next turn to the *very important* interpretation of $(\Delta x)_n$ noting the following points. As indicated earlier, quantum mechanics does not make a definite prediction of the outcome of any specific measurement of the particle's position x. Rather, quantum mechanics only predicts that if many measurements of x are made all with the system having the wave function $\Psi_n(x)$, a spread of values of x will be obtained with the probability of finding each value of x in the spread determined by the probabilities distribution $P_n(x) = \Psi_n(x)\Psi_n(x)$. Because of this spread of possible measured values of x when a particular measurement of x is made, the value found will typically deviate from the average value $\langle x \rangle_n$.

$(\Delta x)_n$ is this deviation averaged over many measurements. If $(\Delta x)_n$ is large, the spread of possible outcomes is broad, and hence, the uncertainty of the outcome of a particular measurement of x is large. On the other hand, if $(\Delta x)_n$ is small, most of the measurements will give values of x close to $\langle x \rangle_n$, and therefore, the spread will be narrow. In this case, the uncertainty of the outcome of any particular measurement of x is small.

For these reasons, $(\Delta x)_n$ is referred to as the *uncertainty of x*.

Next, we obtain a form for $(\Delta x)_n^{\ 2}$ which is much more convenient than the form of Equation (3.58). Noting the normalization condition $\int_0^a \Psi_n(x)\Psi_n(x)dx = 1$ for the particle-in-a-box wave functions

$\Psi_n(x)$, we obtain this more convenient form as follows. From Equation (3.58), $(\Delta x)_n^2 = \int_0^a (x - \langle x \rangle_n)^2 \times \Psi_n(x)\Psi_n(x)dx$, which may be written as $(\Delta x)_n^2 = \int_0^a (x^2 - 2x\langle x \rangle_n + \langle x \rangle_n^2)\Psi_n(x)\Psi_n(x)dx = \langle x^2 \rangle_n - 2\langle x \rangle_n^2 + \langle x \rangle_n^2$ yielding the convenient form as

$$(\Delta x)_n^2 = \langle x^2 \rangle_n - \langle x \rangle_n^2. \tag{3.59}$$

Notice that we have used Equation (3.57) to write

$$\int_0^a x^2 \Psi_n(x)\Psi_n(x) = \langle x \rangle_n^2. \tag{3.60}$$

Also notice from Equation (3.59) that since $(\Delta x)_n^2 > 0$ $\langle x^2 \rangle_n > \langle x \rangle_n^2$.

While we have derived Equation (3.59) for the particle-in-a-box system, it holds generally.

Next, using Equation (3.52) for $\Psi_n(x)$, $\langle x^2 \rangle_n$ may be evaluated from Equation (3.60) as (Problem 3.18b)

$$\langle x^2 \rangle_n = \frac{a^2}{3} - \frac{a^2}{2n^2\pi^2}. \tag{3.61}$$

Then, using $\langle x \rangle_n = \frac{a}{2}$,$(\Delta x)_n$ may be evaluated from Equations (3.59) and (3.61) as

$$(\Delta x)_n = \left(\langle x^2 \rangle_n - \langle x \rangle_n^2\right)^{1/2} = a\left(\frac{1}{12} - \frac{1}{2n^2\pi^2}\right)^{1/2}. \tag{3.62}$$

Next, let us consider the particle's X-component of momentum p_x. The uncertainty $(\Delta p_x)_n$ of p_x is defined in analogy to Equation (3.59) for $(\Delta x)_n^2$ as

$$(\Delta p_x)_n^2 = \langle p_x^2 \rangle_n - \langle p_x \rangle_n^2. \tag{3.63}$$

The interpretation of $(\Delta p_x)_n$ is analogous to that of $(\Delta x)_n$.

To determine $(\Delta p_x)_n$ from Equation (3.63), we must evaluate $\langle p_x \rangle_n$ and $\langle p_x^2 \rangle_n$. However, these averages cannot be found in the same way that we determined $\langle x \rangle_n$ and $\langle x^2 \rangle_n$. This is because we do not know p_x as a function of x and therefore guesses like $\langle p_x \rangle_n \stackrel{?}{=} \int_{-\infty}^{\infty} p_x \Psi_n^*(x)\Psi_n(x)dx$ are wrong. To obtain correct expressions for $\langle p_x \rangle_n$ and $\langle p_x^2 \rangle_n$, we must draw on results developed in Chapter 4. Specifically, specializing Equation (4.50) to the one-dimensional particle-in-a-box system yields the correct expressions as

$$\langle p_x \rangle_n = \frac{\hbar}{i}\int_0^a \Psi_n(x)\frac{d}{dx}\Psi_n(x)dx \tag{3.64}$$

and

$$\langle p_x^2 \rangle_n = -\hbar^2 \int_0^a \Psi_n(x)\frac{d^2}{dx^2}\Psi_n(x)dx. \tag{3.65}$$

Evaluating the integrals in the proceeding two relations using Equation (3.52) for $\Psi_n(x)$ yields (Problem 3.19)

$$\langle p_x \rangle_n = 0 \text{ and } \langle p_x^2 \rangle_n = \left(\frac{n\pi\hbar}{a}\right)^2 \tag{3.66}$$

From the above equation, the uncertainty $(\Delta p_x)_n$ is given by

$$(\Delta p_x)_n = \left(\langle p_x^2 \rangle_n - \langle p_x \rangle^2\right)^{1/2} = \frac{n\pi\hbar}{a}. \tag{3.67}$$

Comparing Equation (3.62) for $(\Delta x)_n$ and Equation (3.67) for $(\Delta p_x)_n$ then gives

$$(\Delta x)_n (\Delta p_x)_n = \frac{\hbar}{2}\left(\frac{n^2\pi^2}{3} - 2\right)^{1/2}. \tag{3.68}$$

Since for all n $\left(\frac{n^2\pi^2}{3} - 2\right)^{1/2} > 1$, it follows that

$$(\Delta x)_n (\Delta p_x)_n > \frac{\hbar}{2}. \tag{3.69}$$

Equation (3.69), which we derived for the particle-in-a-box system, but as described in Section 5.3 actually holds for all systems and is a major result of quantum mechanics known as the Heisenberg uncertainty relation. Its meaning is as follows. Suppose that the system is in a quantum state α for which measurements of the position x give a narrow spread of results and hence for state $\alpha \Delta x$ is small. Then, in order that the Heisenberg uncertainty relation of Equation (3.69) be satisfied, when the system is in state $\alpha \Delta p_x$ must be large and hence in state α measurements of the momentum p_x give a broad spread of results. Conversely, if the system is in a different state β for which Δp_x is small and hence measurements of p_x give a narrow spread of results, the uncertainty relation requires that in state $\beta \Delta x$ be large and hence measurements of x give a broad spread of results. Consequently, due to the uncertainty relation, it is impossible to find quantum states for which measurements of x and p_x will both give a narrow spread of results. In other words, it is impossible to find quantum states for which both x and p_x are known precisely (which conflicts with the Heisenberg uncertainty relation).

However, as discussed in Section A.1, specification of a particle trajectory requires simultaneous precise knowledge of the initial values of both x and p_x. Consequently, trajectories are not defined in quantum mechanics.

FURTHER READINGS

Atkins, Peter and Ronald Friedman. 2011. *Molecular quantum mechanics.* 5th ed. New York: Oxford University Press.

Beiser, Arthur. 1995. *Concepts of modern physics.* 5th ed. New York: McGraw-Hill, Inc.

Levine, Ira N. 2013. *Quantum chemistry.* 7th ed. New York: Prentice-Hall.

McQuarrie, Donald A. 2008. *Quantum chemistry.* 2nd ed. Sausalito, CA: University Science Books.

Pauling, Linus, and E. Bright Wilson Jr. 1985. *Introduction to quantum mechanics with applications to chemistry.* New York: Dover Publications, Inc.

PROBLEMS

3.1 Determine in $kgms^{-1}$ the momentum ρ of a photon of light with wavelength $\lambda = 200\,nm$.

3.2 Compute in meters the de Broglie wavelength λ of a baseball of mass $m = 0.15\,kg$ traveling at 95 miles $hour^{-1}$. By comparing your results for λ with the radius $r \approx 4\,cm$ of the baseball, explain why the wave properties of the baseball are not observable.

3.3 Determine in meters the de Broglie wavelength λ of an electron moving in the $n = 1$ Bohr circular orbit of the hydrogen atom. Show that your result for λ is of the order of atomic dimensions by comparing it with the value of the Bohr radius $a_0 = 5.29 \times 10^{-11}$ m. (Hint: See Problem 1.25.)

3.4 Consider an electron with de Broglie wavelength $\lambda = 3.0 \times 10^{-10}$ m and a light wave with the same wavelength. Determine in joules the kinetic energy $T = \frac{p^2}{2m_e}$ of the electron, and compare it with the energy $E = h\upsilon$ of one photon of the light.

3.5 In this problem, you will derive Equation (3.51) for the energies of a particle-in-a-box of length a from an argument based on de Broglie matter waves. (a) First, show that the energy E of a particle of mass m confined to a one-dimensional box may be expressed in terms of its de Broglie wavelength λ as $E = \frac{h^2}{2m\lambda^2}$. The condition for the stable states of the particle is that its matter waves must be such that an integral or half-integral number of their de Broglie wavelengths λ fit into the box. (b) Show that the de Broglie wavelengths of the allowed matter waves are thus restricted to the discrete values $\lambda_n = \frac{2a}{n}$, where $n = 1, 2, \ldots$. (c) From your results in parts (a) and (b), derive Equation (3.51).

3.6 The Bohr theory of the hydrogen-like atoms may be derived from an argument based on de Broglie matter waves. Namely, a circular electron orbit of radius r can only be stable if the electron's matter wave is such that an integral number $n = 1, 2, \ldots$ of its de Broglie wavelengths λ fit into the circumference $2\pi r$ of the orbit. Derive from this requirement Bohr's angular momentum quantization formula $L = n\hbar$.

3.7 Give an argument which shows that the wave displacement function $\Psi(x,t)$ of a traveling wave on a string moving in the positive X-direction with velocity v_0 must have the form given in Equation (3.4).

3.8 Show that the wave displacement function $\Psi(x,t)$ of Equation (3.4) solves the wave equation (Equation 2.80).

3.9 Prove the relation between υ, v_0, and λ of Equation (3.8).

3.10 By analogy to Equations (3.16) and (3.17), write down the time-dependent Schrödinger equation for a system comprised of two particles with masses m_1 and m_2 and position vectors $r_1 = (x_1, y, z_1)$ and $r_2 = (x_2, y_2, z_2)$ interacting via a potential energy function $U(r_1, r_2)$.

3.11 Show that Equations (3.25) and (3.26) together define a Hermitian eigenvalue problem. (Hint: See Problem 2.14.)

3.12 Classically, the minimum energy of a particle of mass m in a box of length a is $E_{cl,min} = 0$. Quantum mechanically, the minimum energy called the *zero point energy* is the ground state energy given from Equation (3.51) as $E_{qu,min} = E_1 = \frac{h^2}{8ma^2}$. Explain why the Heisenberg uncertainty relation forbids a particle-in-a-box to have an energy of zero and thus requires the zero point energy.

3.13 For macroscopic particles, the zero point energy defined in Problem 3.12 is in agreement with classical intuition too small to be detectable. Show that this is true by determining in joules the zero point energy of a marble of mass 1.0 g confined to a one-dimensional box of length a = 5 cm.

3.14 For microscopic particles, the zero point energy defined in Problem 3.12 can be appreciable. To show that this is true, determine the zero point energy in eV of an electron confined to a one-dimensional box of length $a = 5a_0$ where a_0 is the Bohr radius, and then compare your result to the ionization potential of the hydrogen atom I = 13.6 eV.

3.15 Without referring to the standing wave eigenvalue problem, solve the particle-in-a-box Schrödinger equation (Equation 3.48) subject to the boundary conditions of Equation (3.47) to obtain Equations (3.51) and (3.52) for the quantized energies E_n and normalized wave functions $\Psi_n(x)$.

3.16 (a) Show that the particle-in-a-box wave function $\Psi_n(x) = \sqrt{\frac{2}{a}}\sin\left(\frac{n\pi x}{a}\right)$ has $n-1$ nodes occurring at the points $x_m = \frac{ma}{n}$ where $m = 1,2,\ldots n-1$. (b) Show that the particle-in-a-box probability distribution for position $P_n(x) = \frac{2}{a}\sin^2\frac{n\pi x}{a}$ has n maxima occurring at the points $x_m = \frac{ma}{2n}$ where $m = 1,3,5,\ldots,2n-1$.

3.17 Referring to Equation (3.54), the probability $P_n(x_1)$ that a particle in a box of length a in the nth stationary state is located between the points x_1 and x_2 is given by $P_n(x_1,x_2) = \frac{2}{a}\int_{x_1}^{x_2}\sin^2\left(\frac{n\pi x}{a}\right)dx$. (a) Show that $P_n(x_1,x_2) = \frac{x_2 - x_1}{a} + \frac{1}{2n\pi}\left[\sin\left(\frac{2n\pi x_1}{a}\right) - \sin\left(\frac{2n\pi x_2}{a}\right)\right]$. (b) Show from your result in part (a) that as expected, $P_n(0,a) = 1$ and $P_n\left(0,\frac{a}{2}\right) = \frac{1}{2}$. (c) Show that $P_n\left(0,\frac{a}{4}\right) = \frac{1}{4} - \frac{1}{2n\pi}\begin{Bmatrix} 0 & \text{for n even} \\ (-1)^{\frac{n^{(-1)}}{2}} & \text{for n odd} \end{Bmatrix}$ and that (d) $P_n\left(\frac{a}{4},\frac{3a}{4}\right) = \frac{1}{2} + \frac{1}{n\pi}\begin{Bmatrix} 0 & \text{for n even} \\ (-1)^{\frac{n^{(-1)}}{2}} & \text{for n odd} \end{Bmatrix}$.

3.18 (a) Derive Equation (3.56) for $\langle x\rangle_n$. (b) Derive Equation (3.61) for $\langle x^2\rangle_n$. (Hint: To perform the integrals in parts [a] and [b], use the trigonometric identity $\sin^2\theta = \frac{1-\cos 2\theta}{2}$ and then integrate by parts.)

3.19 Derive Equation (3.66) for $\langle p_x\rangle_n$ and $\langle p_x^2\rangle_n$. (Hint: The trigonometric identities $\sin 2\theta = 2\sin\theta\cos\theta$ and $\sin^2\theta = \frac{1-\cos 2\theta}{2}$ will be helpful.)

3.20 Consider a classical particle confined to a box of length a. The particle will have equal probability for being at any point inside the box and zero probability for being outside of the box. (a) Show that this implies that the classical probability distribution for position $P_{cl}(x)$ is given by $P_{cl}(x) = \begin{cases} \frac{1}{2} & \text{inside the box} \\ 0 & \text{outside the box} \end{cases}$. (b) Show from this form for $P_{cl}(x)$ that the classical uncertainty in position $(\Delta x)_{cl} = a\left(\frac{1}{12}\right)^{1/2}$. (c) Show from this result for $(\Delta x)_{cl}$ and from Equation (3.62) for the quantum uncertainty $(\Delta x)_n$ that $\lim_{n\to\infty}(\Delta x)_n = (\Delta x)_{cl}$.

3.21 In the nearly free electron model, molecular spectra that arise from pi electron transitions are crudely treated using particle-in-a-box approximations. In this problem, we use the nearly free electron model to study the ultraviolet–visible absorption spectrum of the six-π electron-conjugated dye molecule hexatriene $CH_2 = CH - CH = CH - CH = CH_2$. In our treatment, we approximate the π orbital energy levels of hexatriene by the particle-in-a-box energy levels of Equation (3.51). We will choose the length a of the box as the length of the molecule plus the length of one $C-C$ single bond. We assume that the $C-C$ and $C=C$ bond lengths are, respectively, 154 and 135 pm. To obtain the ground state π electron configuration, in accord with the Pauli exclusion principle, we fill the three lowest particle-in-a-box energy levels so that each is occupied by two π electrons.

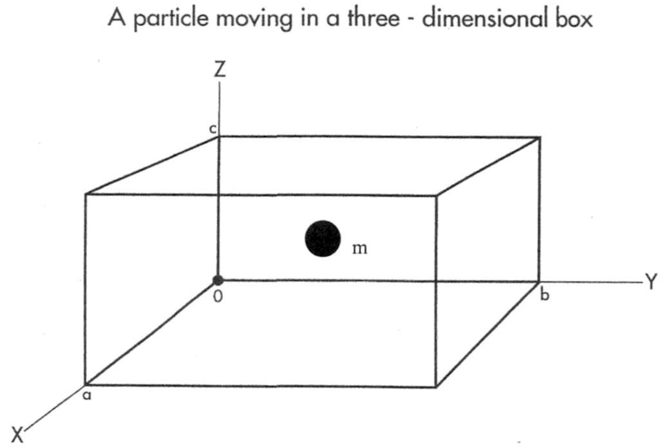

FIGURE 3.5 A particle of mass m confined to a three-dimensional box with sides of lengths a, b, and c.

Compute in cm^{-1} units (a) the energy of the highest occupied level, (b) the energy of the lowest unoccupied level, and (c) the energy difference between the lowest unoccupied level and the highest occupied level. Compare the energy difference determined in part (c) with the experimental wave number $\bar{\upsilon} = 37{,}310\ cm^{-1}$ of the longest wavelength absorption band in the ultraviolet–visible spectrum of hexatriene.

3.22 Consider a particle of mass m confined to the *three-dimensional* box with sides of lengths a, b, and c as shown in Figure 3.5. The potential energy function of the particle $U(x,y,z)$ vanishes inside of the box and is infinite outside of the box. That is, $U(x,y,z)$ conforms to the following three-dimensional generalization of Equation (3.44):

$$U(x,y,z) = \begin{cases} 0 & \text{for } 0 \le x \le a \text{ and } 0 \le y \le b \text{ and } 0 \le z \le c \\ \infty & \text{for } x < 0 \text{ or } y < 0 \text{ or } z < 0 \text{ or } x > a \text{ or } y > b \text{ or } z > c \end{cases}.$$

Inside the box, the wave functions $\Psi_{n_x n_y n_z}(x,y,z)$, which depend on three quantum numbers n_x, n_y, and n_z one for each of the three directions of motion, satisfy the Schrödinger equation $-\frac{\hbar^2}{2m}\left(\frac{\partial^2}{\partial x^2} + \frac{\partial^2}{\partial y^2} + \frac{\partial^2}{\partial z^2}\right)\Psi_{n_x n_y n_z}(x,y,z) = E_{n_x n_y n_z}\Psi_{n_x n_y n_z}(x,y,z)$.

In order that the wave functions be continuous at the box walls, they must obey boundary conditions that are the following extensions of those of Equation (3.47): $\Psi_{n_x n_y n_z}(0,y,z) = \Psi_{n_x n_y n_z}(a,y,z) = 0$, $\Psi_{n_x n_y n_z}(x,0,z) = \Psi_{n_x n_y n_z}(x,b,z) = 0$, and $\Psi_{n_x n_y n_z}(x,y,0) = \Psi_{n_x n_y n_z}(x,y,c)$. (a) Assume that the wave functions factorize as $\Psi_{n_x n_y n_z}(x,y,z) = \Psi_{n_x}(x)\Psi_{n_y}(y)\Psi_{n_z}(z)$, then insert this factorized form into the Schrödinger equation inside the box, and use the separation-of-variables technique to derive the following one-dimensional Schrödinger equations for $\Psi_{n_x}(x), \Psi_{n_y}(y)$, and $\Psi_{n_z}(z)$:

$$-\frac{\hbar^2}{2m}\frac{d^2\Psi_{n_x}(x)}{dx^2} = E_{n_x}\Psi_{n_x}(x), -\frac{\hbar^2}{2m}\frac{d^2\Psi_{n_y}(y)}{dy^2} = E_{n_y}\Psi_{n_y}(y), \text{ and} -\frac{\hbar^2}{2m}\frac{d^2\Psi_{n_z}(z)}{dz^2} = E_{n_z}\Psi_{n_z}$$

(z) where $E_{n_x n_y n_z} = E_{n_x} + E_{n_y} + E_{n_z}$. (b) Show that given the factorization of the wave function of part (a), the boundary conditions on the wave function reduce to $\Psi_{n_x}(0) = \Psi_{n_x}(a) = 0, \Psi_{n_y}(0) = \Psi_{n_y}(b)$, and $\Psi_{n_z}(0) = \Psi_{n_z}(c)$. (c) Explain why your results of parts (a) and (b) show that the $\Psi_{n_x}(x), \Psi_{n_y}(y)$, and $\Psi_{n_z}(z)$ are the wave functions for one-dimensional particle-in-a-box systems with respective lengths a, b, and c and that E_{n_x}, E_{n_y}, and E_{n_z} are the corresponding energies for these systems. (d) Show from your results in part (c) that the normalized wave functions and energies for the particle-in-a-box system of Figure 3.5 are $\Psi_{n_x n_y n_z}(x,y,z) = \left(\frac{8}{V}\right)^{1/2} \sin\left(\frac{n_x \pi}{a} x\right) \times \sin\left(\frac{n_y \pi}{b} y\right) \sin\left(\frac{n_z \pi}{c} z\right)$ and $E_{n_x n_y n_z} = \frac{h^3}{8m}\left(\frac{n_x^2}{a^2} + \frac{n_y^2}{b^2} + \frac{n_z^2}{c^2}\right)$ where $V = abc$ is the volume of the box and where the ranges of the quantum numbers n_x, n_y, and n_z are $n_x = 1,2,\ldots, n_y = 1,2,\ldots$, and $n_z = 1,2,\ldots$.

3.23 Consider a particle of mass m moving in a cubic box with sides of length a. From part (d) of Problem 3.22, the energy levels of this system are given by $E_{n_x n_y n_z} = \frac{\left(n_{x^2} + n_{y^2} + n_{z^2}\right)h^2}{8ma^2}$ where the ranges of the quantum numbers are $n_x = 1,2,\ldots, n_y = 1,2,\ldots$, and $n_z = 1,2,\ldots$. For the particle-in-a-cubic-box system, (a) show that the ground state is non-degenerate. (b) Find the energies of the first three excited states and show that all are threefold degenerate.

SOLUTIONS TO SELECTED PROBLEMS

3.1 $3.313 \times 10^{-27}\ \mathrm{kg\,m\,s^{-1}}$.

3.2 $1.040 \times 10^{-34}\ \mathrm{m}$.

3.3 $2.66 \times 10^{-10}\ \mathrm{m}$.

3.4 Electron kinetic energy $= 2.68 \times 10^{-18}$ J, and photon energy $= 9.93 \times 10^{-16}$ J.

3.10 $i\hbar \frac{\partial \Psi(r_1, r_2, t)}{\partial t} = -\frac{\hbar^2}{2m_1}\nabla_1^2 \Psi(\mathbf{r}_1, \mathbf{r}_2, t) - \frac{\hbar^2}{2m_2}\nabla_2^2 \Psi(\mathbf{r}_1, \mathbf{r}_2, t) + U(\mathbf{r}_1, \mathbf{r}_2)\Psi(\mathbf{r}_1, \mathbf{r}_2, t)$ where for $i = 1 \text{ and } 2$, $\nabla_i^2 = \frac{\partial}{\partial x_i^2} + \frac{\partial}{\partial y_i^2} + \frac{\partial}{\partial z_i^2}$.

3.13 1.46×10^{-63} J.

3.14 5.37 eV.

3.21 (a) $36{,}315\ \mathrm{cm^{-1}}$, (b) $64{,}560\ \mathrm{cm^{-1}}$, and (c) $28{,}245\ \mathrm{cm^{-1}}$.

3.23 (b) In units of $\frac{h^2}{8ma^2}$, energies are 6, 9, and 11.

4 Wave Functions and Experimental Outcomes

In this chapter, we develop some of the most central principles of quantum mechanics. We especially develop the quantum probability rules that predict from the wave function of a quantum system the likelihoods of each of the possible outcomes of experiments which measure the values of the observables of the system.

Before starting, we note that most of the quantum principles described in this chapter are best viewed as postulates whose validity is accepted because of the vast number of experimental confirmations of their consequences.

We begin with a discussion of the system wave function and its equation of motion, the time-dependent Schrödinger equation.

4.1 THE WAVE FUNCTION AND ITS EQUATION OF MOTION

We postulate the following concerning the wave function.

> Postulate. *For all times* $t \geq 0$, *the normalized system wave function* $\Psi(t)$, *which is a complex function of the system's configuration and the time, determines all that can be observed about the system. Concisely stated, at time* $t\Psi(t)$ *completely specifies the quantum state of the system.* (4.1)

For example, at time t the square of the absolute value $|\Psi(t)|$; namely $\Psi^*(t)\Psi(t) \geq 0$, determines the probability that a system will have a particular configuration from among the range of its possible configurations. As a second example, at time t the wave function $\Psi(t)$ completely determines the average values $\langle A \rangle_t$ and uncertainties $(\Delta A)_t = \left(\langle A^2 \rangle_t - \langle A \rangle_t^2\right)^{1/2}$ of all system observables A.

For the present, we will restrict ourselves to systems comprised of spinless particles. Then, the system's configuration is specified by the particles' positions. Hence, $\Psi(t)$ is a function of either the particles' Cartesian coordinates or some equivalent set of non-Cartesian or *generalized coordinates* and the time t.

The following question immediately arises: What set of coordinates should one choose to specify wave functions $\Psi(t)$? The answer is that the wave functions of a quantum system are best taken as functions of the same coordinates which would most naturally describe the system's trajectories if the system behaved classically. However, as shown in Appendices A and B, for many systems the natural coordinates for trajectories are generalized rather than Cartesian coordinates. While in later chapters we will deal with the complications that arise from generalized coordinate specifications of system wave functions, in this chapter for simplicity we will take all wave functions to be functions of the particles' Cartesian coordinates and the time t.

For example, for a general system comprised of N particles each moving in three dimensions with particle position vectors $\mathbf{r}_1 = (x_1, y_1, z_1), \ldots, \mathbf{r}_N = (x_N, y_N, z_N)$, the wave functions are taken to have the form

$$\Psi(t) = \Psi(\mathbf{r}_1, \cdots, \mathbf{r}_N, t). \tag{4.2}$$

Special cases of a general system are systems comprised of a single particle moving either in three dimensions for which $\Psi(t) = \Psi(\mathbf{r}, t)$ or in one dimension for which $\Psi(t) = \Psi(x, t)$.

However, not all functions of the particle coordinates and the time are acceptable wave functions. For a function to be acceptable, it must also be *well-behaved.* That is, it must be single-valued, finite everywhere, continuous with all first derivatives also continuous, and square integrable. The first two of these restrictions are necessary for validity of the Born rule probability interpretation of the wave function described in Section 3.3. The continuity requirements are necessary in order that the wave function can be a solution of the time-dependent Schrödinger equation. The square integrability condition is required so that the wave function can be normalized.

For a general system, the square integrability condition is

$$\int d\mathbf{r}_1 \ldots \int d\mathbf{r}_N \Psi^*(\mathbf{r}_1, \ldots, \mathbf{r}_N, t)\Psi(\mathbf{r}_1, \ldots, \mathbf{r}_N, t) < \infty \tag{4.3}$$

where for all $i = 1, \ldots, N$ we use the shorthand notation

$$\int d\mathbf{r}_i = \int_{-\infty}^{\infty} dx_i \int_{-\infty}^{\infty} dy_i \int_{-\infty}^{\infty} dz_i. \tag{4.4}$$

We next turn to the time dependence of $\Psi(t)$ which determines the time dependence of the quantum state of the system. The time dependence of $\Psi(t)$ is generated by the time-dependent Schrödinger equation that was already discussed for one-dimensional systems in Chapter 3. Specifically, in Section 3.1 we gave a non-rigorous plausibility argument for the form of the one-dimensional time-dependent Schrödinger equation. Here, we give the time-dependent Schrödinger equation for a general system as a postulate without attempting to justify its form.

Postulate. *For a system comprised of* N *spinless particles with masses* $m_1, \ldots, m_N$ *each moving in three dimensions under the influence of a potential energy function* $U(\mathbf{r}_1, \ldots, \mathbf{r}_N)$, *the time dependence of the system wave function* $\Psi(\mathbf{r}_1, \ldots, \mathbf{r}_N, t)$ *is generated by the following time-dependent Schrödinger equation:*

$$i\hbar \frac{\partial \Psi(\mathbf{r}_1, \ldots, \mathbf{r}_N, t)}{\partial t} = \sum_{i=1}^{N} -\frac{\hbar^2}{2m_i} \nabla_i^2 \Psi(\mathbf{r}_1, \ldots, \mathbf{r}_N, t) + U(\mathbf{r}_1, \ldots, \mathbf{r}_N)\Psi(\mathbf{r}_1, \ldots, \mathbf{r}_N, t) \tag{4.5}$$

where the operator ∇_i^2 (read del i-squared) is given by

$$\nabla_i^2 = \frac{\partial^2}{\partial x_i^2} + \frac{\partial^2}{\partial y_i^2} + \frac{\partial^2}{\partial z_i^2}. \tag{4.6}$$

The time-dependent Schrödinger equation (Equation 4.5) is a first-order differential equation in time and thus requires only a single initial condition which we write as

$$\Psi(\mathbf{r}_1, \ldots, \mathbf{r}_N, t = 0) = \Psi(\mathbf{r}_1, \ldots, \mathbf{r}_N) \tag{4.7}$$

where the initial wave function $\Psi(\mathbf{r}_1, \ldots, \mathbf{r}_N)$ can be any well-behaved and usually normalized function. Each of the infinity of distinct choices of $\Psi(\mathbf{r}_1, \ldots, \mathbf{r}_N)$ generates a unique well-behaved time-dependent wave function $\Psi(\mathbf{r}_1,\ldots,\mathbf{r}_N,t)$ which we prove in Section 6.5 is normalized if $\Psi(\mathbf{r}_1,\ldots,\mathbf{r}_N)$ is normalized. $\Psi(\mathbf{r}_1,\ldots,\mathbf{r}_N,t)$ if normalized in turn determines a unique set of probability distributions for system observables like position, momentum, energy, and angular momentum.

Before going on, we note for simplicity of notation that in the remainder of this chapter, instead of considering a general system, we will consider a system comprised of a single particle with mass m and position vector $\mathbf{r} = (x,y,z)$ moving in three dimensions under the influence of a potential energy function $U(\mathbf{r})$. The wave functions for this system will be denoted by $\Psi(\mathbf{r},t)$. All of our results for the single-particle three-dimensional system are readily extended to corresponding results for a general system.

We next turn to a new topic. To start, we note that in Section A.2, we defined a classical observable $A(t)$ of a particle system as any quantity whose value is determined by the trajectory of the system. However, as noted at the conclusion of Section 3.5, because of the Heisenberg uncertainty relation, trajectories cannot be defined in quantum mechanics. So for quantum systems, we need a new definition of an observable. In quantum mechanics, classical observables $A(t)$ are represented by linear operators $\hat{A}$ which transform function sets $\{\Psi\}$ comprised of linearly independent well-behaved functions of the particle coordinates. Moreover, the operators $\hat{A}$ must be Hermitian for the function sets $\{\Psi\}$ on the ranges of the particle coordinates. The Hermitian requirement on the operators $\hat{A}$ is necessary since, as we will see in Section 4.4, the eigenvalues a_n of these operators are the possible measured values of the corresponding classical observables $A(t)$ and thus must be real numbers, a restriction which is only guaranteed if the operators $\hat{A}$ are Hermitian.

Rules are available for determining the quantum Hermitian operators $\hat{A}$ for each of the corresponding classical observables $A(t)$. While in advanced treatments of quantum mechanics these rules are typically derived from fundamentals (Cohen-Tannoudji, Diu, and Laloe 1977, pp. 144–150), here we will merely give the rules as postulates.

4.2 THE HERMITIAN OPERATOR REPRESENTATIVES OF OBSERVABLES

We will first consider the simplest type of system, a single particle of mass m and coordinate x moving in one dimension subject to a potential energy function $U(x)$. To start, we note, as pointed out in Section A.2, that the basic classical observables of such a system are

$$x(t) \text{ and } p_x(t) \tag{4.8}$$

where the position observable $x(t)$ is the position of the particle at time t and where the momentum observable $p_x(t) = m\dot{x}(t)$ is the momentum of the particle at time t.

Complex classical observables are functions

$$A(t) = A\left[x(t), p_x(t)\right] \tag{4.9}$$

of the basic observables. Two simple examples of complex classical observables are the kinetic energy observable $T(t) = \frac{p_x^2(t)}{2m}$ and the potential energy observable $U(t) = U\left[x(t)\right]$. The most important complex classical observable is the Hamiltonian observable $H(t)$ or more simply just the Hamiltonian. $H(t)$ is defined by

$$H(t) = H\left[x(t), p_x(t)\right] = \frac{p_x^2(t)}{2m} + U\left[x(t)\right]. \tag{4.10}$$

Notice that the Hamiltonian $H(t)$ is the sum of the kinetic energy observable $T(t)$ and the potential energy observable $U(t)$ and thus is equal to the conserved total energy E of the system.

We may now give the rules for converting classical observables $A(t)$ into quantum Hermitian operators $\hat{A}$. We start with the rules for converting the basic classical observables, the position observable $x(t)$ and the momentum observable $p_x(t)$, into the basic quantum operators, the position operator $\hat{x}$ and the momentum operator $\hat{p}_x$. These rules are

$$x(t) \to \hat{x} = \text{multiply by x and } p_x(t) \to \hat{p}_x = \frac{\hbar}{i}\frac{\partial}{\partial x}. \tag{4.11}$$

To form the quantum operator $\hat{A}$ that represents an arbitrary complex classical observable $A(t) = A[x(t), p_x(t)]$, we usually merely have to replace the arguments $x(t)$ and $p_x(t)$ of $A[x(t), p_x(t)]$ by the corresponding quantum operators $\hat{x}$ and $\hat{p}_x$ to yield the rule

$$A(t) = A[x(t), p_x(t)] \to \hat{A} = A(\hat{x}, \hat{p}_x). \tag{4.12}$$

However, if $A[x(t), p_x(t)]$ includes terms like $x(t)p_x(t)$, this simple procedure leads to ambiguity and therefore is not satisfactory. This ambiguity arises because the two apparently equally valid correspondences $x(t)p_x(t) \to \hat{x}\hat{p}_x$ and $x(t)p_x(t) \to \hat{p}_x\hat{x}$ are not identical since $\hat{x}$ and $\hat{p}_x$ do not commute and thus $\hat{x}\hat{p}_x \neq \hat{p}_x\hat{x}$. Moreover, neither $\hat{x}\hat{p}_x$ nor $\hat{p}_x\hat{x}$ are Hermitian operators. Both of these difficulties may be resolved if we modify the rule of Equation (4.12) by making the symmetrized correspondence $x(t)p_x(t) \to \frac{1}{2}(\hat{x}\hat{p}_x + \hat{p}_x\hat{x})$ (Cohen-Tannoudji, Diu. and Laloe 1977, p. 263).

We next specialize to the case that the classical observable $A(t)$ is the Hamiltonian $H(t)$ of Equation (4.10) in order to obtain the all-important (recall Chapter 3) Hamiltonian operator $\hat{H}$ Since $H(t)$ does not contain terms like $x(t)p_x(t)$, no symmetrization is necessary and so we can obtain the rule for forming $\hat{H}$ by straightforwardly specializing Equation (4.12) to yield

$$H(t) = \frac{p_x^2(t)}{2m} + U[x(t)] \to \hat{H} = \frac{\hat{p}_x^2}{2m} + U(\hat{x}). \tag{4.13}$$

We may obtain a more convenient form for $\hat{H}$ than the one just given by noting that $\frac{\hat{p}_x\hat{p}_x}{2m} = \frac{\left(\frac{\hbar}{i}\frac{\partial}{\partial x}\right)\left(\frac{\hbar}{i}\frac{\partial}{\partial x}\right)}{2m} = -\frac{\hbar^2}{2m}\frac{\partial^2}{\partial x^2}$, yielding from Equation (4.13) the more convenient form as

$$\hat{H} = -\frac{\hbar^2}{2m}\frac{\partial^2}{\partial x^2} + U(\hat{x}). \tag{4.14}$$

Next, recall that the role of operators is to transform functions. Thus, let us see how the quantum operators we have just developed transform an arbitrary well-behaved complex function $\Psi(x,t)$. First, applying the basic operators $\hat{x}$ and $\hat{p}_x$ to $\Psi(x,t)$ yields

$$\hat{x}\Psi(x,t) = x\Psi(x,t) \text{ and } \hat{p}_x\Psi(x,t) = \frac{\hbar}{i}\frac{\partial\Psi(x,t)}{\partial x}. \tag{4.15}$$

Assuming that no symmetrization is necessary, the generic complex operator $\hat{A}$ is given by $\hat{A} = A(\hat{x}, \hat{p}_x)$. Hence, applying $\hat{A}$ to $\Psi(x,t)$ yields a new function

$$\hat{A}\Psi(x,t) = A(\hat{x}, \hat{p}_x)\Psi(x,t). \tag{4.16}$$

To develop the explicit form of the new function, one must expand $A(\hat{x}, \hat{p}_x)$ in a double-operator power series in $\hat{x}$ and $\hat{p}_x$ (symmetrizing all products of non-commuting operators) and then apply this power series to $\Psi(x,t)$ using Equation (4.11) repeatedly to determine the action of each of its terms on $\Psi(x,t)$.

This power series method yields a very simple result if we take $\hat{A}$ to be the Hamiltonian operator $\hat{H}$. To see this, first let us apply $\hat{H}$ of Equation (4.14) to $\Psi(x,t)$ to yield

$$\hat{H}\Psi(x,t) = -\frac{\hbar^2}{2m}\frac{\partial^2}{\partial x^2}\Psi(x,t) + U(\hat{x})\Psi(x,t). \tag{4.17}$$

While evaluating the term $-\frac{\hbar^2}{2m}\frac{\partial^2}{\partial x^2}\Psi(x,t)$ in the proceeding expression is straightforward, how to evaluate the term $U(\hat{x})\Psi(x,t)$ is not immediately evident since $U(\hat{x})$ is a *function of an operator*, namely $\hat{x}$. However, we will shortly show by the operator power series method that

$$U(\hat{x})\Psi(x,t) = U(x)\Psi(x,t) \tag{4.18}$$

a result which eliminates the difficulty since $U(x)$ is just the simple multiplication operator multiply by $U(x)$. Given this result, Equation (4.17) greatly simplifies to

$$\hat{H}\Psi(x,t) = -\frac{\hbar^2}{2m}\frac{\partial^2\Psi(x,t)}{\partial x^2} + U(x)\Psi(x,t) \tag{4.19}$$

a form which may be straightforwardly evaluated.

To verify Equation (4.18) and therefore prove Equation (4.19) for $\hat{H}\Psi(x,t)$, we proceed as follows. (a) We first expand the classical observable $U[x(t)]$ as the Maclaurin series $U[x(t)] = \sum_{n=0}^{\infty}\frac{u^{(n)}}{n!}x^n(t)$ where $u^{(n)} = \left\{\frac{d^nU[x(t)]}{dx(t)^n}\right\}_{x(t)=0} = \left[\frac{dU^n(x)}{dx^n}\right]_{x=0}$. (b) Next, we make the classical to quantum correspondence $x^n(t) \to \hat{x}^n$ = multiply by x^n which follows from the basic correspondence $x(t) \to \hat{x}$ = multiply by x to convert the classical Maclaurin series expansion of $U[x(t)]$ in (a) into the following operator power series expansion of $U(\hat{x})$; $U(\hat{x}) = \sum_{n=0}^{\infty}\frac{u^{(n)}}{n!}\hat{x}^n$. (c) Then, we apply this operator power series expansion to $\Psi(x,t)$ to obtain $U(\hat{x})\Psi(x,t) = \sum_{n=0}^{\infty}\frac{u^{(n)}}{n!}\hat{x}^n\Psi(x,t) = \sum_{n=0}^{\infty}\frac{u^{(n)}}{n!}x^n\Psi(x,t) = \left[\sum_{n=0}^{\infty}\frac{u^{(n)}}{n!}x^n\right]\Psi(x,t)$. (d) Finally, we note that $\sum_{n=0}^{\infty}\frac{u^{(n)}}{n!}x^n = U(x)$, and hence, our result in (c) reduces to $U(\hat{x})\Psi(x,t) = U(x)\Psi(x,t)$, which is the relation of Equation (4.18), thus validating that relation and consequently establishing our result for $\hat{H}\Psi(x,t)$ of Equation (4.19).

The results developed so far in this section are summarized in Table 4.1.

Finally, we turn to the question of extending the one-dimensional expressions listed in Table 4.1 to corresponding results for systems comprised of a single particle moving in three dimensions. However, since the three-dimensional results are natural extensions of our one-dimensional expressions, we omit their detailed development and merely summarize them in Table 4.1.

We next turn to a new topic. Specifically, we present a detailed development of the Born rule, already described in Section 3.3 for one-dimensional single-particle systems, for the probability distribution, or more properly the probability density, of the position observable **r** of a single particle moving in three dimensions. The Born rule is the simplest quantum probability rule for predicting experimental outcomes from the normalized system wave function $\Psi(\mathbf{r},t)$. Additionally and very importantly, the Born rule provides the physical interpretation of the wave function $\Psi(\mathbf{r},t)$. Namely, since the wave function is a complex quantity, it cannot be straightforwardly related to observables that are real-valued, and therefore, its physical interpretation is not immediately evident.

TABLE 4.1

Classical Observables $\mathbf{A(t)}$ and Corresponding Quantum Hermitian Operators $\hat{\mathbf{A}}$ for One- and Three-Dimensional Single-Particle Systems

System	$A(t)$	$\hat{A}$	$\hat{A}\Psi(t)$
A particle of mass m moving in one dimension with coordinate x, potential energy function $U(x)$, and wave functions $\Psi(t)=\Psi(x,t)$	$x(t)$	$\hat{x}$ = multiply by x	$x\Psi(x,t)$
	$p_x(t)$	$\hat{p}_x = \frac{\hbar}{i}\frac{\partial}{\partial x}$	$\frac{\hbar}{i}\frac{\partial}{\partial x}\Psi(x,t)$
	$A(t)=A[x(t),p(t)]$	$\hat{A}=A[\hat{x},\hat{p}_x]$	$A(\hat{x},\hat{p}_x)\Psi(x,t)$
	$H(t)=\frac{p_x^2(t)}{2m}+U[x(t)]$	$\hat{H}=\frac{\hat{p}_x^2}{2m}+U(\hat{x})$	$-\frac{\hbar^2}{2m}\frac{\partial^2}{\partial x^2}\Psi(x,t)+U(x)\Psi(x,t)$
A particle of mass m moving in three dimensions with position vector $r=(x,y,z)$, potential energy function $U(r)$, and wave functions $\Psi(t)=\Psi(\mathbf{r},t)$	$\mathbf{r}(t)=[x(t),y(t),z(t)]$	$\hat{\mathbf{r}}$ = multiply by $\mathbf{r}=(x,y,z)$	$[x\Psi(\mathbf{r},t),y\Psi(\mathbf{r},t),z\Psi(\mathbf{r},t)]$
	$\mathbf{p}(t)=[p_x(t),p_y(t),p_z(t)]$	$\hat{p}=\left(\frac{\hbar}{i}\frac{\partial}{\partial x},\frac{\hbar}{i}\frac{\partial}{\partial y},\frac{\hbar}{i}\frac{\partial}{\partial z}\right)$	$\left[\frac{\hbar}{i}\frac{\partial}{\partial x}\Psi(\mathbf{r},t),\frac{\hbar}{i}\frac{\partial}{\partial y}\Psi(\mathbf{r},t),\frac{\hbar}{i}\frac{\partial}{\partial z}\Psi(\mathbf{r},t)\right]$
	$A(t)=A[\mathbf{r}(t),\mathbf{p}(t)]$	$\hat{A}=A(\hat{\mathbf{r}},\hat{\mathbf{p}})$	$A(\hat{\mathbf{r}},\hat{\mathbf{p}})\Psi(\mathbf{r},t)$
	$H(t)=\frac{\mathbf{p}(t)\cdot\mathbf{p}(t)^{(a)}}{2m}+U[\mathbf{r}(t)]$	$\hat{H}=\frac{\hat{\mathbf{p}}\cdot\hat{\mathbf{p}}^{(a)}}{2m}+U(\hat{\mathbf{r}})$	$-\frac{\hbar^2}{2m}\nabla^2\Psi(\mathbf{r},t)+U(\mathbf{r})\Psi(\mathbf{r},t)$
	$L_x(t)^{(b)}=y(t)p_z(t)-z(t)p_y(t)$	$\hat{L}_x^{(b)}=\frac{\hbar}{i}\left(\hat{y}\frac{\partial}{\partial z}-z\frac{\partial}{\partial y}\right)$	$\frac{\hbar}{i}\left[y\frac{\partial}{\partial z}\Psi(\mathbf{r},t)-z\frac{\partial}{\partial y}\Psi(\mathbf{r},t)\right]$
	$L_y(t)=z(t)p_x(t)-x(t)p_z(t)$	$\hat{L}_y=\frac{\hbar}{i}\left(\hat{z}\frac{\partial}{\partial x}-\hat{x}\frac{\partial}{\partial z}\right)$	$\frac{\hbar}{i}\left[z\frac{\partial}{\partial x}\Psi(\mathbf{r},t)-x\frac{\partial}{\partial z}\Psi(\mathbf{r},t)\right]$
	$L_z(t)=x(t)p_y(t)-y(t)p_x(t)$	$\hat{L}_z=\frac{\hbar}{i}\left(\hat{x}\frac{\partial}{\partial y}-\hat{y}\frac{\partial}{\partial x}\right)$	$\frac{\hbar}{i}\left[x\frac{\partial}{\partial y}\Psi(\mathbf{r},t)-y\frac{\partial}{\partial z}\Psi(\mathbf{r},t)\right]$

(a) The dot product $\mathbf{p}(t)\cdot\mathbf{p}(t)=p_x^2(t)+p_y^2(t)+p_z^2(t)$. The dot product $\hat{\mathbf{p}}\cdot\hat{\mathbf{p}}=\hat{p}_x^2+\hat{p}_y^2+\hat{p}_z^2=-\hbar^2\left(\frac{\partial^2}{\partial x^2}+\frac{\partial^2}{\partial y^2}+\frac{\partial^2}{\partial z^2}\right)=-\hbar^2\nabla^2$.

(b) $L_x(t), L_y(t)$, and $L_z(t)$ are the Cartesian components of the classical angular momentum vector $\mathbf{L}(t)$. $\hat{L}_x, \hat{L}_y$, and $\hat{L}_z$ are the Cartesian components of the Hermitian operator $\hat{\mathbf{L}}$ which represents $\mathbf{L}(t)$.

The interpretation of the wave function is provided by the Born rule formula $P(\mathbf{r},t) = \Psi^*(\mathbf{r},t)\Psi(\mathbf{r},t) \geq 0$ since $P(\mathbf{r},t)$ is the probability per unit volume of finding the particle at point $\mathbf{r}$ at time t.

To proceed further, we give a detailed discussion of probability distributions, which were already touched on in Section 3.3, and the related concept of *random variables.*

4.3 RANDOM VARIABLES AND THE BORN PROBABILITY RULE FOR THE POSITION OBSERVABLE r

As in classical mechanics, in quantum mechanics the position vector $\mathbf{r}$ of a particle is a measurable quantity whose Cartesian components x, y, and z can take on all values between $-\infty$ and ∞. However, while in classical mechanics positions may be determined precisely from the particle's trajectory $\mathbf{r}(t)$, in quantum mechanics only a probability distribution that gives the likelihood that the particle has a particular position may be determined from the particle's normalized wave function $\Psi(\mathbf{r},t)$.

Thus, we next discuss probability distributions. This discussion is facilitated by the introduction of the concept of a random variable and so we will begin by defining random variables.

A random variable X is a (real) measurable quantity with the following two properties:

1. X can take on only a limited range of values. The range of possible values of X can be either discrete, in which case X is a discrete random variable, or continuous, in which case X is a continuous random variable.
2. If a measurement of X is made with certainty, one of its possible values will be found, but which of these possible values is actually found depends solely on chance and therefore cannot be predicted in advance of the measurement.

Let us illustrate this definition for one of the simplest random variables X, the discrete coin face random variable. For this choice, X can take on only two values: x_1 = heads and x_2 = tails. If a measurement of X is made, which is a flip of the coin, which of its possible values x_1 or x_2 is found depends solely on chance and therefore cannot be predicted in advance of the coin flip.

While the outcome of any particular coin flip cannot be predicted in advance, the coin face random variable probability distribution $P(X)$, which gives the probabilities of each of the two possible outcomes, can be determined. This determination may be made by making many coin flips. If the coin is fair, we will find after many flips that x_1 and x_2 are found with equal frequency. We will thus conclude that the probability $P(X = x_1) = P(x_1)$ for finding heads in any single coin flip is $\frac{1}{2}$, while the probability $P(X = x_2) = P(x_2)$ for finding tails in any single flip is also $\frac{1}{2}$. The pair of probabilities $P(x_1)$ and $P(x_2)$ comprise the coin face random variable probability distribution $P(X)$ if the coin is fair. If the coin is not fair, $P(x_1)$ and $P(x_2)$ still comprise the coin face random variable probability distribution, but in this case, $P(x_1) \neq P(x_2)$.

(A somewhat more complex random variable X is the discrete die face random variable. For this random variable, the possible values of X are the number of dots on each of the six die faces. Moreover, a measurement of X consists of a roll of the die. In analogy to determining $P(X)$ for the coin face random variable, the die face random variable probability distribution may be found by making many rolls of the die, thus determining the six probabilities for finding each of the die faces in any single roll.)

Let us now return to quantum mechanics.

Just as one cannot predict in advance the outcome of any single coin flip (or die roll), one cannot predict in advance the outcome among the many possible outcomes of an experiment designed to measure the value of a position observable or in fact of any quantum observable (within the very important exceptions discussed in Section 4.5).

However, just as for the coin (or die) face random variables, one can predict statistical information about the outcomes of measurements of the values of quantum observables. Especially, one can predict probability distributions that are the collections of probabilities of each of the possible outcomes. One can also predict the values of quantum observables averaged over many measurements.

We may now return to the Born rule.

To start, we note that while in classical mechanics the Cartesian components x, y, and z of the particle's position vector $\mathbf{r}$ are deterministic quantities, within quantum mechanics they comprise a set of three independent continuous random variables each with the range of possible values $-\infty$ to ∞. Thus, if x, y, and z are measured, the values found for each of these random variables will lie in the range of their possible values $-\infty$ to ∞. However, which set of values for x, y, and z in this range is actually found depends solely on chance and therefore cannot be predicted in advance. Stated slightly differently, if the position vector $\mathbf{r}$ is measured, any vector of finite magnitude can potentially be found, but which vector is actually found depends solely on chance and therefore cannot be predicted in advance.

While the outcome of any particular measurement of $\mathbf{r}$ cannot be predicted in advance, what can be predicted is a probability density (which is the probability distribution of a continuous random variable) $P(\mathbf{r}, t)$ which determines the probabilities of each of the continuum of possible outcomes.

Next, using the abbreviated notation $d\mathbf{r} = dxdydz$, we may give the interpretation of the probability density $P(\mathbf{r}, t)$ as

$$\begin{aligned} P(\mathbf{r},t)d\mathbf{r} = {} & \text{the probability that if the particle's postition} \\ & \text{vector } \mathbf{r} \text{ is measured at time t, its Cartesian} \\ & \text{components x, y, and z will be found to lie} \\ & \text{between x and x + dx, y + dy, and z + dz.} \end{aligned} \tag{4.20}$$

We conclude this section by giving as a postulate the Born rule for $P(\mathbf{r}, t)$.

Postulate. *The probability density* $P(\mathbf{r}, t)$ *for the position vector* $\mathbf{r}$ *of a particle moving in three dimensions is given in terms of the particle's normalized wave function* $\Psi(\mathbf{r}, t)$ *by the Born rule*

$$P(\mathbf{r},t) = \Psi^*(\mathbf{r},t)\Psi(\mathbf{r},t). \tag{4.21}$$

We next turn to the quantum probability rules for arbitrary observables $A = A(t)$. As shown in advanced treatments of quantum mechanics, the Born rule for position observables is a special case of the general rules for arbitrary observables (Cohen-Tannoudji, Diu, and Laloe, 1977, pp. 218 and 225).

4.4 THE QUANTUM RULES FOR THE PROBABILITY DISTRIBUTIONS OF ARBITRARY OBSERVABLES

Consider the eigenvalues a_n and associated eigenfunctions Ψ_n^A of the Hermitian operator $\hat{A}$ which represents an arbitrary classical observable A. For simplicity, we will assume that the eigenvalues a_n are non-degenerate and form a discrete set. Further and very importantly, we note that while in classical mechanics an observable A can take on any value, in quantum mechanics A can take on only a restricted range of values. The precise nature of this restriction is the content of the following quantum postulate which holds even if our assumption that the eigenvalues form a non-degenerate discrete set is violated.

$$\begin{aligned} & \text{Postulate. } \textit{The only possible values of an observable} \\ & \textit{A are the (real) eigenvalues } a_n \textit{ of its Hermitian operator} \\ & \textit{representative } \hat{A}. \end{aligned} \tag{4.22}$$

For example, the only possible values of a quantum system's Hamiltonian H or equivalently of its energy E are the eigenvalues E_n of its Hamiltonian operator $\hat{H}$.

Next and also very importantly, we note that while in classical mechanics an observable A is a deterministic quantity, in quantum mechanics (given our assumption that the eigenvalues a_n form a discrete set) A is a discrete random variable whose possible values are from the postulate of Equation (4.22) the eigenvalues a_n of $\hat{A}$. Consequently, if a measurement of the value of A is made when the system is in a specified quantum state, which can be any quantum state, with certainty one of the eigenvalues a_n will be found, but which eigenvalue is actually found depends solely on chance and therefore cannot be predicted in advance of the measurement. (Within the die face random variable analogy, a specified quantum state is analogous to a specific die loading. Just as changing the die loading changes the probability of finding a specific die face in a single roll of the die, changing the specified quantum state changes the probability of finding a specific eigenvalue a_n in a single measurement of the observable A.)

While quantum mechanics cannot predict the outcome of any particular measurement of the value of A (conditional on the exceptions described in the next section), it does predict a probability distribution $P(A, t)$ which is the specified quantum state-dependent collection of probabilities $P(a_n,t)$ for finding each of the eigenvalues a_n in any single measurement of the value of A performed at time t.

In Equation (4.27), we give as a postulate the rule for the form of $P(a_n,t)$. In Section 4.6, we derive from this form an expression that determines from the normalized system wave function $\Psi(\mathbf{r},t)$ the value $\langle A\rangle_t$ of A averaged over the spread of eigenvalues a_n found from many measurements of the value of A all performed at the same time t and all performed with the system in the same specified quantum state. (The spread occurs because A is a random variable and therefore different measurements of its value usually give different results.)

To start the development of the rule for $P(A,t)$, We note from the postulate of Equation (4.22) that this probability distribution must have the following form:

$$P(A,t)=\begin{cases}0 & \text{if A is not an eigenvalue of } \hat{A}\\ P(a_n,t) & \text{if A is the eigenvalue } a_n \text{ of } \hat{A}.\end{cases} \tag{4.23}$$

Thus, to determine $P(A,t)$ it is only necessary to find the form of the probabilities $P(a_n,t)$.

To develop this form, we first consider the eigenvalue problem for $\hat{A}$ which for our single-particle three-dimensional system is

$$\hat{A}\Psi_n^A(\mathbf{r})=a_n\Psi_n^A(\mathbf{r}). \tag{4.24}$$

Since $\hat{A}$ is Hermitian, its normalized eigenfunctions $\Psi_n^A(\mathbf{r})$ form an orthonormal set which we will assume is complete. Then, the specified state-dependent wave function of the system $\Psi(\mathbf{r},t)$ has the complete orthonormal expansion

$$\Psi(\mathbf{r},t)=\sum_{\text{all } n} C_n^A(t)\Psi_n^A(\mathbf{r}). \tag{4.25}$$

The expansion coefficients $C_n^A(t)$ are given in analogy to Equation (2.78) by

$$C_n^A(t)=\int\left[\Psi_n^A(\mathbf{r})\right]^*\Psi(\mathbf{r},t)d\mathbf{r}. \tag{4.26}$$

We will shortly see that the larger the contribution of the eigenfunction $\Psi_n^A(\mathbf{r})$ to $\Psi(\mathbf{r},t)$, as measured by the magnitude of $\left|C_n^A(t)\right|$ the larger is the probability $P(a_n,t)$ of measuring the eigenvalue a_n associated with $\Psi_n^A(\mathbf{r})$.

We may now give the rule for the probabilities $P(a_n, t)$ as a postulate.

$$\text{Postulate. } P(a_n,t)=\left|C_n^A(t)\right|^2=\left[C_n^A(t)\right]^*\left[C_n^A(t)\right]\geq 0 \text{ if } \Psi(\mathbf{r},t) \text{ is normalized} \tag{4.27}$$

Notice from the proceeding two equations that, as indicated earlier, $P(a_n, t)$ and hence $P(A, t)$ depend on the choice of the specified quantum state because of their dependence on the normalized system wave function $\Psi(\mathbf{r}, t)$.

We next discuss the special quantum states for which in many measurements of the value of an observable A in each measurement the same value is found for A rather than the spread of values found for most quantum states.

4.5 THE SPECIAL QUANTUM STATES FOR WHICH AN OBSERVABLE A HAS A DEFINITE VALUE

To begin, we note that if the system is in a typical quantum state specified by some normalized wave function $\Psi(\mathbf{r}, t)$ if many measurements of the value of an observable A are made all performed at the same time t since A is a random variable, a spread of the eigenvalues a_n of the representative $\hat{A}$ of A will be found.

However, special quantum states exist, which are specific to the observable A, for which in many measurements of the value of A no spread of the eigenvalues a_n is found. Rather, when the system is in one of these special states in *every* measurement of the value of A, the *same* eigenvalue of $\hat{A}$ is found. Additionally, when the system is in the special state for which the eigenvalue a_m of $\hat{A}$ is always found, we say that the observable A has the *definite value* a_m.

For simplicity, we well assume that for the special states, the system wave functions and hence the probability distributions for all observables are *time independent*. Thus, when the system is in the special state for which the observable A has the definite value a_m, we will write the normalized system wave function $\Psi(\mathbf{r}, t)$ as $\Psi_m(\mathbf{r})$ and the probability distribution $P(A, t)$ for A as $P_m(A)$. Also when the system is in this special state in any single measurement of the value of A, the value found is *with certainty* a_m. Therefore, for this special state the probability distribution $P_m(A)$ is given by

$$P_m(A)=\begin{cases} P_m(a_n)=0 & \text{if} \quad n\neq m \\ P_m(a_n)=1 & \text{if} \quad n=m \end{cases} \tag{4.28}$$

where $P_m(a_n)$ is the probability of finding the eigenvalue a_n in any single measurement of the value of A. The content of Equation (4.28) may be written concisely as

$$P_m(a_n)=\delta_{nm}. \tag{4.29}$$

We next show that for the special state for which A has the definite value a_m, the normalized system wave function $\Psi_m(\mathbf{r})$ is given by

$$\Psi_m(\mathbf{r})=\Psi_m^A(\mathbf{r}) \tag{4.30}$$

where $\Psi_m^A(\mathbf{r})$ is the normalized eigenfunction of $\hat{A}$ associated with the eigenvalue a_m.

To prove this assertion, we assume Equation (4.30) $\Psi_m(\mathbf{r})=\Psi_m^A(\mathbf{r})$ and then show that this assumption implies the condition that the system is in the special state for which A has the definite value a_m, namely the condition $P_m(a_n)=\delta_{nm}$ of Equation (4.29). To show this, we specialize Equation (4.26) for the coefficients $C_n^A(t)$ and the postulate of Equation (4.27) for the probabilities

$P(a_n,t)$ to the case that the system wave function $\Psi(\mathbf{r},t)$ is the time-independent function $\Psi_m^A(\mathbf{r})$. So specializing Equation (4.26) gives the expansion coefficients as

$$C_n(t) = C_n^A = \int \left[\Psi_n^A(\mathbf{r})\right]^* \left[\Psi_m^A(\mathbf{r})\right] d\mathbf{r} = \delta_{nm}. \tag{4.31}$$

Given the above relation, Equation (4.27) yields the probabilities as

$$P(a_n,t) = P(a_n) = \delta_{nm}^* \delta_{nm} = \delta_{nm}^2 = \delta_{nm} \tag{4.32}$$

which is the condition $P(a_m) = \delta_{nm}$ which we set out to prove.

In summary, we have proven the following:

The normalized system wave function which specifies the special quantum state for which an observable A has the definite value a_m is the normalized eigenfunction $\Psi_m^A(\mathbf{r})$ of the representative $\hat{A}$ of A associated with the eigenvalue a_m.

We next note that when the system is in a quantum state for which an observable A has a definite value, other system observables B usually do not have definite values. For example, suppose the system is in the state for which its energy E has the definite value E_m (which we will see in Chapter 6 is the stationary state with energy E_m), and therefore, the system wave function $\Psi_m(r)$ is the eigenfunction of the system's Hamiltonian operator $\hat{H}$ associated with the eigenvalue E_m. Then, the position observable r does not have a definite value. Thus, if many measurements are made of the values of x, y, and z, the Cartesian components of r, a spread of values for each component will be found. The spread is determined quantitatively by the Born rule probability density for the state $\Psi_m^*(r)\Psi_m(r)$. This is the situation for all of the applications we will make of quantum mechanics in Chapters 7–13. (That is, in these applications E always has a definite value but **r** never does.)

A very important exception occurs for two observables A and B whose Hermitian operator representatives $\hat{A}$ and $\hat{B}$ commute. For this case, it is possible to find quantum states for which both A and B have definite values. This follows since as shown in Section 5.2 the representatives $\hat{A}$ and $\hat{B}$ have a complete orthonormal set of common eigenfunctions $\Psi_{a_n,b_m}(\mathbf{r})$ which are the solutions of the simultaneous eigenvalue problem $\hat{A}\Psi_{a_n,b_m}(\mathbf{r}) = a_n\Psi_{a_n,b_m}(\mathbf{r})$ and $\hat{B}\Psi_{a_n,b_m}(\mathbf{r}) = b_m\Psi_{a_n,b_m}(\mathbf{r})$. Then, by a straightforward extension of the argument just given for a single observable A, one may show that if the system is in the quantum state specified by $\Psi_{a_n,b_m}(\mathbf{r})$, both A and B have definite values, a_n for A and b_m for B.

In other words, when the system is in the quantum state specified by the common eigenfunction $\Psi_{a_n,b_m}(\mathbf{r})$ if many simultaneous measurements of the values of A and B are made in each of these measurements, the value found for A will be the eigenvalue a_n and the value found for B will be the eigenvalue b_m.

We next turn to the development of an expression which determines from the normalized system wave function $\Psi(\mathbf{r},t)$ the average or *expected* value $\langle A \rangle_t$ of an arbitrary observable A with Hermitian operator representative $\hat{A}$.

4.6 THE QUANTUM FORMULA FOR THE EXPECTED VALUE OF AN ARBITRARY OBSERVABLE A

To start, we note that perhaps surprisingly, we usually *cannot* determine the expected value $\langle A \rangle_t$ as the average value of A found from many measurements of this value all performed at (nearly the same) time t and all performed with the system in the quantum state specified by $\Psi(\mathbf{r},t)$ if these measurements are made on a *single system*. This is because quantum (but not classical) systems are so "delicate" that even *one* measurement of the value of A usually spoils the system for subsequent measurements of this value. This spoiling occurs because the measurement changes the system's quantum state from the state specified by $\Psi(\mathbf{r},t)$ to a new state. This new state is identified by the following quantum postulate.

Postulate. *If up to some time t the normalized system wave function is* $\Psi(\mathbf{r},t)$ *and if at time t a measurement of the value of an observable A is made yielding the eigenvalue* a_n *of its representative* $\hat{A}$*, the system wave function instantaneously after the measurement changes from* $\Psi(\mathbf{r},t)$ *to the normalized eigenfunction* $\Psi_n^A(\mathbf{r})$ *of* $\hat{A}$ *associated with the measured eigenvalue* a_n*. Consequently, at time t the system's quantum state abruptly changes from the state specified by* $\Psi(\mathbf{r},t)$ *to the state specified by* $\Psi_n^A(\mathbf{r})$.

Therefore, to determine $\langle A \rangle_t$, the expected value of A when the system wave function is $\Psi(\mathbf{r},t)$, in order to avoid the spoiling of the system by the measurement one must average over the spread of eigenvalues of $\hat{A}$ found from $N \to \infty$ measurements of the value of A all performed at time t, but each performs on *different but identical* systems all with the wave function $\Psi(\mathbf{r},t)$.

This measurement process yields $\langle A \rangle_t$ as the average of the eigenvalues a_n of $\hat{A}$ found in the $N \to \infty$ measurements of the value of A performed on different but identical systems. Namely, $\langle A \rangle_t$ is the following average of the values of A found in the $N \to \infty$ measurements

$$\langle A \rangle_t = \lim_{N\to\infty} \frac{1}{N} S \tag{4.33}$$

where

$$S = \begin{pmatrix} \text{the value of} & & \text{the value of} & & \text{the value of} \\ \text{A found} & + & \text{A found} & +\cdots+ & \text{A found} \\ \text{in the first} & & \text{in the second} & & \text{in the Nth} \\ \text{measurement} & & \text{measurement} & & \text{measurement} \end{pmatrix} \tag{4.34}$$

We evaluate the sum of Equation (4.34) in the following manner. Let us define $N(a_n,t)$ as the number of times the eigenvalue a_n of $\hat{A}$ is the value of A found in the $N \to \infty$ measurements. With this definition, there are $N(a_1,t)$ terms in the sum for which the eigenvalue a_1 is measured. Adding up these terms gives the partial sum $a_1N(a_1,t)$. Similarly, adding up all terms in the sum for which the eigenvalue a_2 is measured gives the second partial sum $a_2N(a_2,t)$. Continuing in manner, we obtain a partial sum for each eigenvalue of $\hat{A}$. Adding up all of these partial sums permits us to express the total sum S as

$$S = \sum_{\text{all n}} a_n N(a_n,t). \tag{4.35}$$

Comparing Equations (4.33) and (4.35) then yields $\langle A \rangle_t$ as

$$\langle A \rangle_t = \lim_{N\to\infty} \frac{1}{N} \sum_{\text{all n}} a_n N(a_n,t). \tag{4.36}$$

However, $\lim_{N\to\infty} \frac{1}{N} N(a_n,t)$ is the fraction of measurements for which the eigenvalue a_n is found. But this fraction is equal to $P(a_n,t)$, the probability of finding the eigenvalue a_n in any single measurement of the value of the observable A. Thus, we may rewrite Equation (4.36) for $\langle A \rangle_t$ as

$$\langle A \rangle_t = \sum_{\text{all n}} a_n P(a_n,t). \tag{4.37}$$

We next derive from Equation (4.37) an expression which determines $\langle A\rangle_t$ from the normalized system wave function $\Psi(\mathbf{r},t)$. To start, we use the postulate of Equation (4.27) $P(a_n,t)=\left[C_n^A(t)\right]^*\left[C_n^A(t)\right]$ to rewrite Equation (4.37) for $\langle A\rangle_t$ as

$$\langle A\rangle_t=\sum_{\text{all n}} a_n\left[C_n^A(t)\right]^*\left[C_n^A(t)\right]. \tag{4.38}$$

Next, taking the complex conjugate of Equation (4.26) yields

$$\left[C_n^A(t)\right]^*=\int \Psi_n^A(\mathbf{r})\Psi^*(\mathbf{r},t)d\mathbf{r}. \tag{4.39}$$

Since $\hat{A}\Psi_n^A(\mathbf{r})=a_n\Psi_n^A(\mathbf{r})$, from the proceeding equation it follows that

$$a_n\left[C_n^A(t)\right]^*=\int a_n\Psi_n^A(\mathbf{r})\Psi^*(\mathbf{r},t)d\mathbf{r}=\int \hat{A}\Psi_n^A(\mathbf{r})\Psi^*(\mathbf{r},t)d\mathbf{r}. \tag{4.40}$$

Using the previous equation to eliminate $a_n\left[C_n^A(t)\right]^*$ from the Equation (4.38) yields

$$\langle A\rangle_t=\int \Psi^*(\mathbf{r},t)\sum_{\text{all n}} C_n^A(t)\hat{A}\Psi_n^A(\mathbf{r})d\mathbf{r}. \tag{4.41}$$

However, since $\hat{A}$ is a linear operator, the above relation may be reexpressed as

$$\langle A\rangle_t=\int \Psi^*(\mathbf{r},t)\hat{A}\sum_{\text{all n}} C_n^A(t)\Psi_n^A(\mathbf{r}). \tag{4.42}$$

But since from Equation (4.25) $\sum_{\text{all n}} C_n^A(t)\Psi_n^A(\mathbf{r})=\Psi(\mathbf{r},t)$, Equation (4.42) may be rewritten as

$$\langle A\rangle_t=\int \Psi^*(\mathbf{r},t)\hat{A}\Psi(\mathbf{r},t)d\mathbf{r} \tag{4.43}$$

which is the expression which determines $\langle A\rangle_t$ in terms of the normalized system wave function $\Psi(\mathbf{r},t)$ which we have been seeking.

By making the transposition $\hat{A}\to\hat{A}^k, k=2,3,\ldots$, our expression for $\langle A\rangle_t$ generalizes to the following relation for the expected values $\langle A^k\rangle_t$ of powers of A:

$$\langle A^k\rangle_t=\int \Psi^*(\mathbf{r},t)\hat{A}^k\Psi(\mathbf{r},t)d\mathbf{r}. \tag{4.44}$$

The proceeding relation may be readily extended to yield the following result for a general system:

$$\langle A^k\rangle_t=\int d\mathbf{r}_1\ldots\int d\mathbf{r}_n\Psi^*(\mathbf{r}_1,\ldots,\mathbf{r}_N,t)\hat{A}^k\Psi(\mathbf{r}_1,\ldots,\mathbf{r}_N,t). \tag{4.45}$$

The above expression greatly simplifies when specialized to one-dimensional systems with normalized wave functions $\Psi(x,t)$. Making this specialization yields $\langle A^k\rangle_t$ as

$$\langle A^k\rangle_t=\int_{-\infty}^{\infty} \Psi^*(x,t)\hat{A}^k\Psi(x,t)dx. \tag{4.46}$$

Two common observables for one-dimensional systems are the position observable x and the momentum observable p_x with respective Hermitian operator representatives $\hat{x}$ = multiply by x and $\hat{p}_x = \frac{\hbar}{i}\frac{\partial}{\partial x}$. Given the form for the representative $\hat{x}$, specializing Equation (4.46) to the observables x and x^2 yields the results expected from the Born rule

$$\langle x \rangle_t = \int_{-\infty}^{\infty} x\Psi^*(x,t)\Psi(x,t)dx \quad \text{and} \quad \langle x^2 \rangle_t = \int_{-\infty}^{\infty} x^2\Psi^*(x,t)\Psi(x,t)dx. \tag{4.47}$$

Similarly, specializing Equation (4.46) to the observables p_x and p_x^2 gives the new results

$$\langle p_x \rangle_t = \frac{\hbar}{i}\int_{-\infty}^{\infty} \Psi^*(x,t)\frac{\partial\Psi(x,t)dx}{\partial x} \quad \text{and} \quad \langle p_x{}^2 \rangle_t = -\hbar^2\int_{-\infty}^{\infty} \Psi^*(x,t)\frac{\partial^2\Psi(x,t)dx}{\partial x^2}. \tag{4.48}$$

Finally, if the system is in a stationary state with normalized wave function $\Psi(x,t) = \exp\left(-\frac{i}{\hbar}E_n t\right)\Psi_n(x)$, Equations (4.47) and (4.48) simplify to the time-independent forms

$$\langle x \rangle_n = \int_{-\infty}^{\infty} x\Psi_n^*(x)\Psi_n(x)dx \quad \text{and} \quad \langle x^2 \rangle_n = \int_{-\infty}^{\infty} x^2\Psi_n^*(x)\Psi_n(x)dx. \tag{4.49}$$

and

$$\langle p_x \rangle_n = \frac{\hbar}{i}\int_{-\infty}^{\infty} \Psi_n^*(x)\frac{d\Psi_n(x)}{dx} \quad \text{and} \quad \langle p_x \rangle_n = -\hbar^2\int_{-\infty}^{\infty} \Psi_n^*(x)\frac{d^2\Psi_n(x)dx}{dx^2}. \tag{4.50}$$

FURTHER READINGS

Atkins, Peter and Ronald Friedman. 2011. *Molecular quantum mechanics.* 5th ed. New York: Oxford University Press.

Cohen-Tannoudji, Claude, Bernard Diu, and Franck Laloe. 1977. *Quantum mechanics.* Vol. 1. Translated by Susan Reid, Nicole Ostrowsky, and Dan Ostrowsky. New York: John Wiley & Sons.

Dicke, Robert H., and James P. Wittke. 1960. *Introduction to quantum mechanics.* Reading, MA: Addison-Wesley Publishing Company, Inc.

Levine, Ira N. 2013. *Quantum chemistry.* 7th ed. New York: Prentice-Hall.

McQuarrie, Donald A. 2008. *Quantum chemistry.* 2nd ed. Sausalito, CA: University Science Books.

Merzbacher, Eugen. 1961. *Quantum mechanics.* New York: John Wiley & Sons, Inc.

Pauling, Linus, and E. Bright Wilson Jr. 1985. *Introduction to quantum mechanics with applications to chemistry.* New York: Dover Publications, Inc.

Sakurai, J. J. 1985. *Modern quantum mechanics.* Rev. ed. Edited by San Fu Tuan. Reading, MA: Addison-Wesley Publishing Co.

PROBLEMS

4.1 Which of the following functions are square integrable?

(a) $x^{1/2}$

(b) x^{-2}

(c) $\sin kx \quad k\,\text{real}$

(d) $\sin kx \exp(-|x|) \quad k\,\text{real}$

(e) $\sin kx \exp(-|x|) \quad \text{Re}\,k > 1\ \text{Im}\,k > 1$

(f) $\sin kx \exp(-x^2)$

4.2 Which of the following functions are well-behaved?

(a) $x^{-1/2}$ $-\infty < x < \infty$

(b) x for $x \geq 0$ zero otherwise

(c) $\exp(-x)$ $-\infty < x < \infty$

(d) $\exp(-x^2)$ $-\infty < x < \infty$

(e) $\exp(-|x|)$ $-\infty < x < \infty$

(f) $\left(x - \frac{1}{2}\right)^{-1}$ $-\infty < x < \infty$

4.3 Consider a particle of mass m moving in three dimensions under the influence of a potential energy function of the form $U(\mathbf{r}) = U(x,y,z) = U(x) + U(y) + U(z)$. Using the separation-of-variables technique, show that the particle's three-dimensional time-dependent Schrödinger equation separates into three independent one-dimensional time-dependent Schrödinger equations with respective potential energy functions $U(x), U(y)$, and $U(z)$.

4.4 Show that the position operator $\hat{x}$ is Hermitian for all function sets comprised of well-behaved functions $\Psi(x,t)$ over all ranges.

4.5 Show that the momentum operator $\hat{p}_x = \frac{\hbar}{i}\frac{\partial}{\partial x}$ is Hermitian for the function set comprised of all well-behaved functions $\Psi(x,t)$ over the range $-\infty$ to ∞. (Hint: Integrate by parts.)

4.6 Show that the Hamiltonian operator $\hat{H} = \frac{\hat{p}_x^{\,2}}{2m} + U(\hat{x})$ is Hermitian for the function set comprised of all well-behaved functions $\Psi(x,t)$ over the range $-\infty$ to ∞.

4.7 Show that the angular momentum operator $\hat{L}_z$ given in Table 4.1 is Hermitian for the function set comprised of all well-behaved functions $\Psi(\mathbf{r},t)$ over the ranges $-\infty < x < \infty, -\infty < y < \infty$, and $-\infty < z < \infty$.

4.8 Extend the argument which led from Equations (4.17) to (4.19) from one to three dimensions to show that the expression $\hat{H}\Psi(\mathbf{r},t) = \left[-\frac{\hbar^2}{2m}\nabla^2\Psi(\mathbf{r},t) + U(\hat{\mathbf{r}})\right]\Psi(\mathbf{r},t)$ may be rigorously transformed into the simpler expression $\hat{H}\Psi(\mathbf{r},t) = \left[-\frac{\hbar^2}{2m}\nabla^2\Psi(\mathbf{r},t) + U(\mathbf{r})\right]\Psi(\mathbf{r},t)$.

4.9 Work out how the operator $\hat{H}^2 = \left[\frac{\hat{p}_x}{2m} + U(\hat{x})\right]^2$ transforms an arbitrary well-behaved function $\Psi(x,t)$.

4.10 Fill in Table 4.1 for a general system comprised of N particles moving in three dimensions with masses $m_1,\ldots,m_N$.

4.11 This problem and the two that follow are concerned with a discrete random variable X with M possible values $x_1,\ldots,x_M$ and with a probability distribution $P(X)$ which is the collection of M probabilities $P(x_i)$ for measuring each of the possible values of X. If $N \to \infty$ measurements of X are made and the value x_i is found N_i times, what is the probability $P(x_i)$ of finding the value x_i in a single measurement of X?

4.12 This problem generalizes the discussion of uncertainties given in Section 3.5 for the particle-in-a-box system. We will assume as in Problem 4.11 a discrete random variable X with the M possible values $x_1,\ldots,x_M$ characterized by a probability distribution $P(X)$ which is the collection of M probabilities $P(x_i)$ for measuring each of the possible values x_i of X.

(a) Show by generalizing the argument which led to Equation (4.37) that, with a change in notation, for $k = 1,2,\ldots \langle X^k \rangle = \sum_{i=1}^{M} x_i^k P(x_i)$. Next, consider the uncertainty ΔX of X which is defined by $(\Delta X)^2 = \sum_{i=1}^{M} (x_i - \langle X \rangle)^2 P(x_i)$. (b) Show that $(\Delta X)^2$ may be written in the more convenient form $(\Delta X)^2 = \langle X^2 \rangle - \langle X \rangle^2$.

4.13 This problem is concerned with the same random variable X dealt with in Problem 4.11. Here, however we will assume that X can take on only $M = 9$ possible values $x_1,\ldots,x_9$. We will also assume that these nine values are distributed symmetrically as $x_1 = x_9 = 0.04$, $x_2 = x_8 = 0.08$, $x_3 = x_7 = 0.12$, $x_4 = x_6 = 0.16$, and $x_5 = 0.20$, and additionally for all $i = 1-9$, $P(x_i) = x_i$. Then, using the result $\langle X^k \rangle = \sum_{i>1}^{M} x_i^k P(x_i)$ derived in Problem 4.12, numerically evaluate the average value $\langle X \rangle$ of X and the uncertainty $\Delta X = \left(\langle X^2 \rangle - \langle X \rangle^2\right)^{1/2}$ of X.

4.14 Suppose that at time t, measurements are made of the energy E of $N \to \infty$ identical systems each with the same normalized wave function $\Psi(t)$. What can you say about $\Psi(t)$ if the same value E_n is found for E in all of the measurements?

4.15 This problem and the next are concerned with some aspects of the quantum theory of the one-dimensional harmonic oscillator dealt with in detail in Chapter 7. (For a classical picture of the oscillator, see Figure A.2.) Denote the coordinate of the oscillator by y, its mass by m, and its frequency (the number of vibrations the oscillator would make per second if it moved classically) by υ. Define the circular frequency ω of the oscillator by $\omega = 2\pi\upsilon$. In terms of ω, the potential energy of the oscillator is $U(y) = \frac{1}{2} m\omega^2 y^2$. Also the oscillator's kinetic energy is $\frac{p_y^2}{2m}$ where $p_y = m\dot{y}$. Therefore, the oscillator's classical Hamiltonian is $H(t) = \frac{p_y^2(t)}{2m} + \frac{1}{2} m\omega^2 y^2(t)$.

(a) Using the rules of Table 4.1, write down the oscillator's Hamiltonian operator $\hat{H}$. The oscillator's ground state wave function $\Psi_0(y)$ and energy E_0 may be found by solving the Schrödinger equation $\hat{H}\Psi_0(y) = E_0\Psi_0(y)$. As shown in Appendix C, this solution gives $\Psi_0(y)$ as $\Psi_0(y) = N_0 \exp\left(-\frac{1}{2}\frac{m\omega}{\hbar} y^2\right)$ where N_0 is a normalization factor.

(b) By inserting this form for $\Psi_0(y)$ into the Schrödinger equation $\hat{H}\Psi_0(y) = E_0\Psi_0(y)$, derive an expression for E_0.

(c) Derive the normalized form of $\Psi_0(y)$ by determining N_0. You will need for $n = 0$ the integral $G_n(\alpha) = \int_{-\infty}^{\infty} y^n \exp(-\alpha y^2) dy$ whose form is derived for n even in Chapter 7 as $G_n(\alpha) = \frac{n!}{2^n (n/2)!} \frac{\pi^{1/2}}{\alpha^{(n+L)/2}}$. Note that $0! = 1$.

4.16 This problem concerns an important quantum result, the virial theorem. A virial theorem is a relationship between the average kinetic energy $\langle KE \rangle$ and the average potential energy $\langle PE \rangle$ of a quantum system. Virial theorems hold for many types of systems, but the form of the virial theorem is specific to the type of system. For example, the hydrogen atom virial theorem is $\langle PE \rangle = -2\langle KE \rangle$, while the harmonic oscillator virial theorem is $\langle PE \rangle = \langle KE \rangle$. Prove the harmonic oscillator virial theorem for the ground state of the one-dimensional harmonic oscillator of Problem 4.15. That is, prove $\frac{1}{2}m\omega^2\int_{-\infty}^{\infty}\Psi_0(y)y^2\Psi_o(y)dy = \frac{1}{2m}\int_{-\infty}^{\infty}\Psi_0(y)\hat{p}_y^2\Psi_o(y)dy$. You will need the result for $\Psi_o(y)$ derived in Problem 4.15 and the result for $G_n(\alpha)$ given in Problem 4.15.

4.17 Consider a particle with wave function $\Psi(t)$. Suppose that at time t, $\Psi(t)$ is such that one of the particle's observables A has the definite value a. (a) Show that at time t, the average value and uncertainty of A are $\langle A\rangle_t = a$ and $(\Delta A)_t = \left(\langle A^2\rangle_t - \langle A^2\rangle_t\right)^{1/2} = 0$. (b) Why are results in part (a) to be expected?

4.18 Consider a particle moving in one dimension. Suppose that at time τ, the particle's energy is measured and the value E_n is always found. Further suppose that a time τ_- just before the measurement, the particle is in a stationary state with normalized wave function $\Psi_n(x)$. What is the particle's normalized wave function $\Psi(x,t)$ for all times $t > \tau$ after the measurement?

4.19 Consider a particle of mass m confined to a one-dimensional box of length a. The particle's Hamiltonian operator is $\hat{H} = -\frac{\hbar}{2m}\frac{d^2}{dx^2}$. Assume that the particle's wave function is $\phi(x) = Nx(x-a)$ where N is a normalization factor. (a) Explain why $\phi(x)$ is a possible wave function of the particle. (b) Show that the quantum state specified by $\phi(x)$ is a non-stationary state, and further explain why the particle does not have a definite energy when it is in this non-stationary state. (c) Show that after normalization, $\phi(x) = \frac{30^{1/2}}{a^{5/2}}x(x-a)$. While when the particle is in the state specified by $\phi(x)$ it does not have a definite energy, it does have an average energy $\langle E\rangle$. (d) Show that $\langle E\rangle = \frac{10}{\pi^2}E_1 = 1.0132E_1$ where $E_1 = \frac{h^2}{8ma^2}$ is the particle-in-a-box ground stationary state energy. (e) Why do you think $\langle E\rangle$ is so close to E_1?

4.20 In this problem, we return to the particle-in-a-box system of Problem 4.19. From Section 3.4, the particle's normalized stationary state wave functions and stationary state energies are $\Psi_n(x) = \left(\frac{2}{a}\right)^{1/2}\sin\frac{n\pi x}{a}$ and $E_n = \frac{n^2h^2}{8ma^2}$ where $n = 1,2,\ldots$. As in Problem 4.19, assume that the normalized wave function of the particle is $\phi(x) = \frac{30^{1/2}}{a^{5/2}}x(x-a)$. As noted in Problem 4.19, when the system is in the quantum state specified by $\phi(x)$, it does not have a definite energy. It does however have a probability distribution $P(E)$ which is the collection of probabilities $P(E_n)$ of measuring each of the stationary state energies E_n. $P(E_n)$ may be determined by adapting Equations (4.26) and (4.27) to give $P(E_n) = C_n^2$ where $C_n = \int_0^a \Psi_n(x)\phi(x)dx$. (a) From this information, show $P(E_n) = \frac{960}{n^6\pi^6}$ for n odd and $P(E_n) = 0$ for n even. (b) Explain using a symmetry argument why $P(E_n) = 0$ for n even. Note that $P(E_1) = 0.999361$. (c) Given this value for $P(E_1)$, what can you infer about $\phi(x)$?

4.21 Again consider the particle-in-a-box system of Problems 4.19 and 4.20 with normalized wave function $\phi(x) = \frac{30^{1/2}}{a^{5/2}}x(x-a)$. Calculate the uncertainty product of the particle $(\Delta x)(\Delta p_x)$, and show that it conforms to the Heisenberg uncertainty relation $(\Delta x)(\Delta p_x) \geq \frac{\hbar}{2}$.

SOLUTIONS TO SELECTED PROBLEMS

4.1 (b), (c), and (f)

4.2 (d)

4.9 $\hat{H}^2\Psi(x,t) = \frac{\hbar^2}{8m^2}\frac{\partial^2}{\partial x^2}\Psi(x,t) - \frac{\hbar^2}{2m}U(x)\frac{\partial^2}{\partial x^2}\Psi(x,t) - \frac{\hbar^2}{2m}\frac{U(x)}{\partial x^2} - U^2(x)\Psi(x,t).$

4.11 $P(x_i) = \lim_{N\to\infty}\frac{N_i}{N}.$

4.13 $\langle X\rangle = 0.136$ and $\Delta X = 0.055$.

4.14 At time t, $\hat{H}\Psi(t) = E_n\Psi(t)$ where $\hat{H}$ is the system's Hamiltonian operator.

4.15 (a) $\hat{H} = -\frac{\hbar^2}{2m}\frac{d^2}{dy^2} + \frac{1}{2}m\omega^2\hat{y}^2$, (b) $E_0 = \frac{1}{2}\hbar\omega$, and (c) $\Psi_0(y) = \left(\frac{m\omega}{\pi\hbar}\right)^{1/4}\exp\left(-\frac{1}{2}\frac{m\omega}{\hbar}y^2\right).$

4.17 (b) Since a is measured every time, $\langle A\rangle_t = a$, and there is no uncertainty in the outcome of a measurement, so $(\Delta A)_t = 0$.

4.18 $\Psi(x,t) = \Psi_n(x)\exp\left(-\frac{i}{\hbar}E_n[t-\tau]\right).$

4.19 $\phi(x)$ is well-behaved and satisfies the particle-in-a-box bounding conditions since $\phi(0) = \phi(a) = 0$. (e) Probably because $\phi(x)$ is a good approximation to the particle's ground stationary state wave function $\Psi_1(x)$.

4.20 (b) $\phi(x)$ is even under reflection through the box midpoint $\frac{a}{2}$, while for n even $\Psi_n(x)$ is odd under this reflection. So for n even, C_n vanishes. (c) $\phi(x)$ is nearly identical to $\Psi_1(x)$.

4.21 $(\Delta x)(\Delta p_x) = \left(\langle x^2\rangle - \langle x\rangle^2\right)^{1/2}\left(\langle p_x\rangle^2 - \langle p_x^2\rangle\right)^{1/2} = (3.162)(0.189)\hbar = 0.598\hbar > \frac{\hbar}{2}.$

5 Commutation Rules and Uncertainty Relations

In this chapter, we develop results that will prove to be very important later. First, we show that if two observables A and B are *compatible*, that is, if their Hermitian operator representatives $\hat{A}$ and $\hat{B}$ commute, then these representatives have a complete orthonormal set of common eigenfunctions $\Psi_{a_n,b_m}(\mathbf{r})$ which are the solutions of the simultaneous eigenvalue problem $\hat{A}\Psi_{a_n,b_m}(\mathbf{r}) = a_n\Psi_{a_n,b_m}(\mathbf{r})$ and $\hat{B}\Psi_{a_n,b_m}(\mathbf{r}) = b_n\Psi_{a_n,b_m}(\mathbf{r})$. Consequently, from the discussion at the end of Section 4.5, when the system is in a quantum state specified by one of the common eigenfunctions $\Psi_{a_n,b_m}(\mathbf{r})$, both A and B have definite values, a_n for A and b_m for B. In contrast, we show that if two observables A and B are *incompatible*, that is, if their representatives $\hat{A}$ and $\hat{B}$ do not commute, then these representatives have few if any common eigenfunctions. As a consequence rarely if ever can one find quantum states for which two incompatible observables A and B both have definite values rather than spreads of values. Therefore, two incompatible observables cannot (with rare exceptions) be simultaneously precisely measured. Rather, in analogy to the Heisenberg uncertainty relation $\Delta x \Delta p_x \geq \frac{\hbar}{2}$ which exists between the incompatible position and momentum observables, a Heisenberg uncertainty relation exists between *any* pair of incompatible observables A and B.

We start with a summary and extension of the discussion of commuting and non-commuting operators given in Section 2.2.

5.1 COMMUTING AND NON-COMMUTING OPERATORS

Recall that the commutator $[\hat{A},\hat{B}]$ of two operators $\hat{A}$ and $\hat{B}$ is defined by

$$[\hat{A},\hat{B}] = \hat{A}\hat{B} - \hat{B}\hat{A}. \tag{5.1}$$

Consider a set comprised of all well-behaved and linearly independent complex functions. Let us apply the operator $[\hat{A},\hat{B}]$ to a member $\phi_i(\mathbf{r})$ of the set. If for *all* functions $\phi_i(\mathbf{r})$ in the set

$$[\hat{A},\hat{B}]\phi_i(\mathbf{r}) = 0 \tag{5.2}$$

then we say that $\hat{A}$ and $\hat{B}$ commute and write this as either

$$\hat{A}\hat{B} = \hat{B}\hat{A} \text{ or } [\hat{A},\hat{B}] = \hat{O} \tag{5.3}$$

where recall $\hat{O}$ is the null operator that multiplies all functions $\phi_i(\mathbf{r})$ in the set by zero. On the other hand, if for even *some* functions $\phi_i(\mathbf{r})$ in the set

$$[\hat{A},\hat{B}]\phi_i(\mathbf{r}) \neq 0 \tag{5.4}$$

then we say that $\hat{A}$ and $\hat{B}$ do not commute and write this as either

$$\hat{A}\hat{B} \neq \hat{B}\hat{A} \text{ or } [\hat{A},\hat{B}] \neq \hat{O}. \tag{5.5}$$

An elementary example of commuting operators occurs when $\hat{B} = \hat{A}$. Then, for all $\hat{A}$

$$\left[\hat{A},\hat{B}\right]=\left[\hat{A},\hat{A}\right]=\hat{O}. \tag{5.6}$$

A simple example of non-commuting operators occurs when $\hat{A}$ is the position operator $\hat{x}$ = multiply by x and $\hat{B}$ is the momentum operator $\hat{p}_x = \frac{\hbar}{i}\frac{\partial}{\partial x}$. Following steps like those that led to Equation (2.22) yields the commutator of $\hat{x}$ and $\hat{p}_x$ as $\left[\hat{x},\hat{p}_x\right] = i\hbar$ which is non-null showing that $\hat{x}$ and $\hat{p}_x$ do not commute. Analogously, we may show for the operator pairs $\hat{y}$ = multiply by $y, \hat{p}_y = \frac{\hbar}{i}\frac{\partial}{\partial y}$ and $\hat{z}$ = multiply by $z, \hat{p}_z = \frac{\hbar}{i}\frac{\partial}{\partial z}$ that the commutator for each pair is non-null and equal to $i\hbar$ and therefore each pair of operators do not commute. In summary,

$$\left[\hat{x},\hat{p}_x\right]=\left[\hat{y},\hat{p}_y\right]=\left[\hat{z},\hat{p}_z\right]=i\hbar. \tag{5.7}$$

We may also easily show that except for the commutators of Equation (5.7), all commutators formed from the Cartesian components $\hat{x}, \hat{y}$, and $\hat{z}$ of the vector position operator $\hat{\mathbf{r}}$ and/or from the Cartesian components $\hat{p}_x, \hat{p}_y$, and $\hat{p}_z$ of the vector momentum operator $\hat{\mathbf{p}}$ are null; for example,

$$\left[\hat{x},\hat{p}_y\right]=\left[\hat{x},\hat{y}\right]=\left[\hat{p}_x,\hat{p}_y\right]=\hat{O}, \tag{5.8}$$

and therefore, all Cartesian component operator pairs other than $\hat{x}$ and $\hat{p}_x, \hat{y}$ and $\hat{p}_y$, and $\hat{z}$ and $\hat{p}_z$ commute.

We next turn to our main topic, the common eigenfunctions of the representatives of compatible observables and the lack of common eigenfunctions of the representatives of incompatible observables.

5.2 COMMON EIGENFUNCTIONS FOR COMPATIBLE OBSERVABLES AND THE LACK OF COMMON EIGENFUNCTIONS FOR INCOMPATIBLE OBSERVABLES

To start, let us consider two observables A and B and their respective Hermitian operator representatives $\hat{A}$ and $\hat{B}$. We assume that $\hat{A}$ and $\hat{B}$ have a complete orthonormal set of common eigenfunctions $\Psi_{a_n,b_m}(\mathbf{r})$ which are the solutions of the simultaneous eigenvalue problem

$$\hat{A}\Psi_{a_n,b_m}(\mathbf{r}) = a_n\Psi_{a_n,b_m}(\mathbf{r}) \tag{5.9a}$$

and

$$\hat{B}\Psi_{a_n,b_m}(\mathbf{r}) = b_m\Psi_{a_n,b_m}(\mathbf{r}). \tag{5.9b}$$

We next show that Equation (5.9) imply that $\hat{A}$ and $\hat{B}$ commute, and therefore, their commutator

$$\left[\hat{A},\hat{B}\right]=\hat{O}. \tag{5.10}$$

The proof of this assertion is very simple. First, we multiply Equation (5.9a) by $\hat{B}$ and Equation (5.9b) by $\hat{A}$ and then use both Equation (5.9a) and Equation (5.9b) to yield

$$\hat{B}\hat{A}\Psi_{a_n,b_m}(\mathbf{r}) = a_n\hat{B}\Psi_{a_n,b_m}(\mathbf{r}) = a_n b_m \Psi_{a_n,b_m}(\mathbf{r}) \tag{5.11a}$$

and

$$\hat{A}\hat{B}\Psi_{a_n,b_m}(\mathbf{r}) = b_m\hat{A}\Psi_{a_n,b_m}(\mathbf{r}) = a_n b_m \Psi_{a_n,b_m}(\mathbf{r}). \tag{5.11b}$$

Next, subtracting Equation (5.11a) from Equation (5.11b) yields

$$\left(\hat{A}\hat{B} - \hat{B}\hat{A}\right)\Psi_{a_n,b_m}(\mathbf{r}) = \left[\hat{A},\hat{B}\right]\Psi_{a_n,b_m}(\mathbf{r}) = 0. \tag{5.12}$$

However, since Equation (5.12) holds for each member $\Psi_{a_n,b_m}(\mathbf{r})$ of the complete set of common eigenfunctions, it implies that Equation (5.10) holds and hence that $\hat{A}$ and $\hat{B}$ commute.

We have just shown that if the representatives $\hat{A}$ and $\hat{B}$ of two observables A and B have a complete set of common eigenfunctions, then these representatives commute and hence the observables A and B are compatible. The more powerful converse of this theorem is also true. Namely, for two compatible observables A and B with commuting Hermitian operator representatives $\hat{A}$ and $\hat{B}$, these representatives have a complete orthonormal set of common eigenfunctions.

We next prove this converse theorem. We start with the eigenvalue equation for $\hat{A}$

$$\hat{A}\Psi_{a_n,b_m}(\mathbf{r}) = a_n\Psi_{a_n,b_m}(\mathbf{r}). \tag{5.13}$$

Notice that using foresight, we have written the eigenfunctions of $\hat{A}$ as $\Psi_{a_n,b_m}(\mathbf{r})$ rather than as, say, $\Psi_n(\mathbf{r})$. For simplicity, we assume that the eigenvalues a_n of $\hat{A}$ are non-degenerate. However, the converse theorem is also true for the more realistic case for which some of the eigenvalues a_n are degenerate (Sakurai 1985, pp. 30–31). To proceed further, we multiply the eigenvalue equation (Equation 5.13) by $\hat{B}$ to obtain

$$\hat{B}\hat{A}\Psi_{a_n,b_m}(\mathbf{r}) = a_n\hat{B}\Psi_{a_n,b_m}(\mathbf{r}) \tag{5.14}$$

However, given our assumption that $\hat{A}$ and $\hat{B}$ commute, the previous equation may be rewritten as $\hat{A}\hat{B}\Psi_{a_n,b_m}(\mathbf{r}) = a_n\hat{B}\Psi_{a_n,b_m}(\mathbf{r})$ or alternatively as

$$\hat{A}\left[\hat{B}\Psi_{a_n,b_m}(\mathbf{r})\right] = a_n\left[\hat{B}\Psi_{a_n,b_m}(\mathbf{r})\right]. \tag{5.15}$$

Notice that the above relation shows that $\hat{B}\Psi_{a_n,b_m}(\mathbf{r})$ is an eigenfunction of $\hat{A}$ associated with the eigenvalue a_n. However, $\hat{B}\Psi_{a_n,b_m}(\mathbf{r})$ cannot be a genuinely new eigenfunction of $\hat{A}$ in the sense that it cannot be linearly independent of the original eigenfunction $\Psi_{a_n,b_m}(\mathbf{r})$ of $\hat{A}$. This is because if $\hat{B}\Psi_{a_n,b_m}(\mathbf{r})$ and $\Psi_{a_n,b_m}(\mathbf{r})$ were linearly independent, they would comprise a set of two linearly independent eigenfunctions of $\hat{A}$ each associated with the eigenvalue a_n, a result which conflicts with our assumption that the eigenvalue a_n is non-degenerate. Therefore, the functions $\hat{B}\Psi_{a_n,b_m}(\mathbf{r})$ and $\Psi_{a_n,b_m}(\mathbf{r})$ must be linearly dependent, implying as noted at the close of Section 2.2 that $\hat{B}\Psi_{a_n,b_m}(\mathbf{r})$ must be proportional to $\Psi_{a_n,b_m}(\mathbf{r})$. Denoting the proportionality constant by b_m yields the proportionality relation between $\hat{B}\Psi_{a_n,b_m}(\mathbf{r})$ and $\Psi_{a_n,b_m}(\mathbf{r})$ as

$$\hat{B}\Psi_{a_n,b_m}(\mathbf{r}) = b_m\Psi_{a_n,b_m}(\mathbf{r}). \tag{5.16}$$

However, Equation (5.16) is much more than a mere proportionality relation. Rather, it is also a statement that not only is $\Psi_{a_n,b_m}(\mathbf{r})$ an eigenfunction of $\hat{A}$ with eigenvalue a_n, it is also an eigenfunction of $\hat{B}$ with eigenvalue b_m. Thus, we have arrived at the very important conclusion that $\Psi_{a_n,b_m}(\mathbf{r})$ is a common eigenfunction of $\hat{A}$ and $\hat{B}$. This common eigenfunction is a solution of the simultaneous eigenvalue problem defined by Equations (5.13) and (5.16).

Moreover, since $\hat{A}$ is Hermitian, its normalized eigenfunctions $\Psi_{a_n,b_m}(\mathbf{r})$ comprise a complete orthonormal set. Given the proceeding discussion, the eigenfunctions $\Psi_{a_n,b_m}(\mathbf{r})$ form a complete orthonormal set of common eigenfunctions of $\hat{A}$ and $\hat{B}$, which is the converse theorem that we set out to prove.

So far we have only considered compatible observables with commuting Hermitian operative representatives. We now turn to incompatible observables A and B whose representatives $\hat{A}$ and $\hat{B}$ do not commute and therefore conform to the inequality

$$\hat{A}\hat{B} \neq \hat{B}\hat{A}. \tag{5.17}$$

We next prove for incompatible observables another important theorem. Specifically, we prove that the representatives $\hat{A}$ and $\hat{B}$ of two incompatible observables A and B *do not* have a complete set of common eigenfunctions. In fact, in practice $\hat{A}$ and $\hat{B}$ typically have few if any common eigenfunctions.

The proof of the theorem proceeds by contradiction. Namely, we assume that $\hat{A}$ and $\hat{B}$ do have a complete set of common eigenfunctions. But we have already shown that if the representatives $\hat{A}$ and $\hat{B}$ of two observables A and B have a complete set of common eigenfunctions, then these representatives commute; that is,

$$\hat{A}\hat{B} = \hat{B}\hat{A}. \tag{5.18}$$

The previous relation however contradicts the true equation (Equation 5.17), thus showing that the assumption that $\hat{A}$ and $\hat{B}$ have a complete set of common eigenfunctions is false, therefore proving the theorem.

We next turn to a new topic, namely the Heisenberg uncertainty relations for incompatible observables A and B.

5.3 HEISENBERG UNCERTAINTY RELATIONS FOR INCOMPATIBLE OBSERVABLES

In Section 3.5, we derived the position x and momentum p_x uncertainty relation for the limited case of the stationary states of the one-dimensional particle-in-a-box system. Here, we describe the general Heisenberg uncertainty relation derived, for example, by Sakurai (Sakurai 1985, pp. 35–36) which holds for any two incompatible observables A and B and for any quantum state of any system.

This general Heisenberg uncertainty relation is

$$(\Delta A)(\Delta B) \geq \frac{1}{2}\left[\left\langle\left[\hat{A},\hat{B}\right]\right\rangle^{*}\left\langle\left[\hat{A},\hat{B}\right]\right\rangle\right]^{1/2}. \tag{5.19}$$

In Equation (5.19), $\hat{A}$ and $\hat{B}$ are the Hermitian operator representatives of A and B,

$$\left\langle\left[\hat{A},\hat{B}\right]\right\rangle = \int \Psi^{*}(\mathbf{r})\left[\hat{A},\hat{B}\right]\Psi(\mathbf{r})d\mathbf{r} \tag{5.20}$$

where $\Psi(\mathbf{r})$ is any well-behaved normalized wave function and ΔA and ΔB are the uncertainties of A and B given by

$$\Delta A = \left[\left\langle A^2\right\rangle - \langle A\rangle^2\right]^{1/2} \text{ and } \Delta B = \left[\left\langle B^2\right\rangle - \langle B\rangle^2\right]^{1/2}. \tag{5.21}$$

The uncertainties depend on the wave function $\Psi(\mathbf{r})$ since

$$\left\langle A^2\right\rangle = \int \Psi^*(\mathbf{r})\hat{A}^2\Psi(\mathbf{r})d\mathbf{r} \text{ and } \langle A\rangle = \int \Psi^*(\mathbf{r})\hat{A}\Psi(\mathbf{r})d\mathbf{r} \tag{5.22}$$

with corresponding expressions for $\left\langle B^2\right\rangle$ and $\langle B\rangle$.

Notice that the quantity $\left\langle\left[\hat{A},\hat{B}\right]\right\rangle$ of Equation (5.20) as well as the uncertainties ΔA and ΔB depends on the wave function $\Psi(\mathbf{r})$. Thus, irrespective of the wave function, the Heisenberg uncertainty relation puts a lower bound on the uncertainty product $(\Delta A)(\Delta B)$. However, the exact value of the product does depend on the wave function $\Psi(\mathbf{r})$.

Next, let us assume that A and B are the position and momentum observables x and p_x. Then, specializing Equations (5.19) and (5.20) to these observables, making use of the commutator identity $\left[\hat{x},\hat{p}_x\right] = i\hbar$, the general Heisenberg uncertainty relation (Equation 5.19) reduces to the familiar position–momentum uncertainty relation

$$(\Delta x)(\Delta p_x) \geq \frac{1}{2}\hbar. \tag{5.23}$$

Like the general Heisenberg uncertainty relation, the position–momentum uncertainty relation only provides a lower bound to the uncertainty product. The actual value of the uncertainty product $(\Delta x)(\Delta p_x)$ depends on the system wave function $\Psi(\mathbf{r})$ or equivalently on the quantum state of the system. This dependence is illustrated at the close of Section 7.10 where it is shown that $(\Delta x)(\Delta p_x)$ for the harmonic oscillator stationary states increases very rapidly with the degree of excitation of the oscillator.

Finally, we note, as already emphasized in Section 3.5, that the position–momentum uncertainty relation shows that if Δx is small, Δp_x must be large, and vice versa. This point is illustrated in Figure 5.1, and its implication, that the classical trajectory concept is invalid in quantum mechanics, is emphasized at the close of Chapter 3.

FURTHER READINGS

Cohen-Tannoudji, Claude, Bernard Diu, and Franck Laloe. 1977. *Quantum mechanics*. Vol. 1. Translated by Susan Reid, Nicole Ostrowsky, and Dan Ostrowsky. New York: John Wiley and Sons.

Merzbacher, Eugen. 1961. *Quantum mechanics*. New York: John Wiley & Sons, Inc.

Sakurai, J. J. 1985. *Modern quantum mechanics*. Rev. ed. Edited by San Fu Tuan. Reading, MA: Addison-Wesley Publishing Co.

PROBLEMS

5.1 Prove the following commutator identities:

(a) $\left[\hat{A},\hat{B}\right] = -\left[\hat{B},\hat{A}\right]$

(b) $\left[\hat{A},\hat{B}+\hat{C}\right] = \left[\hat{A},\hat{B}\right] + \left[\hat{A},\hat{C}\right]$

(c) $\left[\hat{A}\hat{B},\hat{C}\right] = \hat{A}\left[\hat{B},\hat{C}\right] + \left[\hat{A},\hat{C}\right]\hat{B}$

(d) $\left[\hat{A}\hat{B},\hat{C}\hat{D}\right] = \hat{A}\hat{C}\left[\hat{B},\hat{D}\right] + \hat{A}\left[\hat{B},\hat{C}\right]\hat{D} + \hat{C}\left[\hat{A},\hat{D}\right]\hat{B} + \left[\hat{A},\hat{C}\right]\hat{D}\hat{B}.$

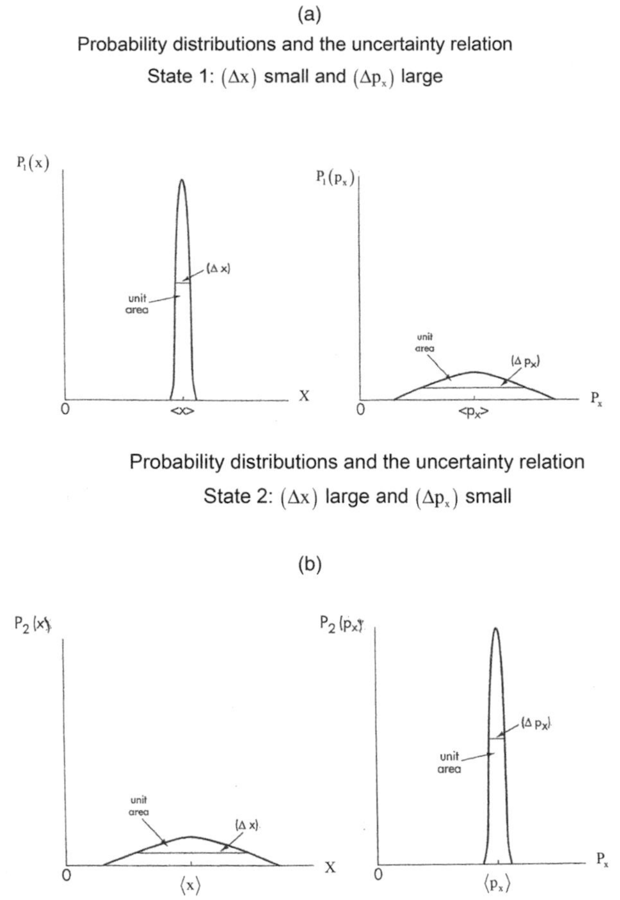

FIGURE 5.1 Mutual variation of the position probability distribution $P(x)$ and the momentum probability distribution $P(p_x)$ for a particle moving in one dimension. Since for the same quantum state $P(x)$ and $P(p_x)$ derive from the same wave function and since x and p_x are incompatible observables, these probability distributions are not independent but rather are related. The basis of the relation is the Heisenberg uncertainty relation $(\Delta x)(\Delta p_x) \geq \frac{\hbar}{2}$. The uncertainty relation implies that if (Δx) is small, (Δp_x) is large, and vice versa. (a) Quantum state 1: (Δx) is small and hence (in order to guarantee that $\int_0^\infty P(x)dx = 1$) $P(x)$ is tall and narrow, while (Δp_x) is large and hence $P(p_x)$ is short and broad. (b) Quantum state 2: (Δx) is large and hence $P(x)$ is short and broad, while (Δp_x) is small and hence $P(p_x)$ is tall and narrow.

(Hint: Use the result of part [c] to prove part [d].)

5.2 Show that $[\hat{x}, \hat{p}_y] = [\hat{x}, \hat{p}_z] = [\hat{y}, \hat{p}_x] = \hat{O}$.

5.3 Show that $[\hat{x}, \hat{p}_x] = [\hat{y}, \hat{p}_y] = [\hat{z}, \hat{p}_z] = i\hbar$.

5.4 Using the commuter identities of Problems 5.1d and 5.3 and the fact that all commutators involving $\hat{x}, \hat{y}, \hat{z}$ and $\hat{p}_x, \hat{p}_y, \hat{p}_z$ except those of Problems 5.3 are null, evaluate the commutator $[\hat{x}\hat{p}_y, \hat{y}\hat{p}_x]$.

5.5 Consider the angular momentum operators $\hat{L}_x = \hat{y}\hat{p}_z - \hat{z}\hat{p}_y, \hat{L}_y = \hat{z}\hat{p}_x - x\hat{p}_z$, and $\hat{L}_z = \hat{x}\hat{p}_y - \hat{y}\hat{p}_x$. Using the results of Problems 5.1 and 5.3 and results like those of Problems 5.2, one may prove the following commutation rules for $\hat{L}_x, \hat{L}_y$, and $\hat{L}_z$: $[\hat{L}_x, \hat{L}_y] = i\hbar\hat{L}_z, [\hat{L}_y, \hat{L}_z] = i\hbar\hat{L}_x$, and $[\hat{L}_z, \hat{L}_x] = i\hbar\hat{L}_y$.

(a) Using these commutation rules and the commutator identity of Problem 5.1c, prove that $\left[\hat{L}^2, \hat{L}_z\right] = \hat{O}$ where $\hat{L}^2 = \hat{L}_x^2 + \hat{L}_y^2 + \hat{L}_z^2$ is the Hermitian operator that represents the square of the classical angular momentum vector $L^2 = L \cdot L$.

(b) Do the operators $\hat{L}^2$ and $\hat{L}_z$ have a common complete set of eigenfunctions?

(c) Do the operators $\hat{L}_x, \hat{L}_y$, and $\hat{L}_z$ have a common complete set of eigenfunctions?

5.6 Derive the familiar form of the Heisenberg uncertainty relation of Equation (5.23) from the general form of Equation (5.19).

5.7 In this problem, you will verify the Heisenberg uncertainty relation $(\Delta x)(\Delta p_x) \geq \frac{1}{2}\hbar$ for the ground state of the one-dimensional harmonic oscillator. The uncertainty relation only gives a lower bound to the uncertainty product $(\Delta x)(\Delta p_x)$. The exact result for the product depends on the system wave function $\Psi(x)$. Since $\Psi(x)$ can be any well-behaved normalized function, many results for $(\Delta x)(\Delta p_x)$ consistent with the uncertainty relation are possible, one for each choice of $\Psi(x)$. (For example, see Equation [7.77].)

Here, we will choose $\Psi(x)$ to be the normalized ground state harmonic oscillator wave function given in Equation (7.15) as $\Psi_0(x) = \left(\frac{m\omega}{\pi\hbar}\right)^{1/4} \exp\left[-\frac{1}{2}\left(\frac{m\omega}{\hbar}\right)x^2\right]$. In this problem, you will determine $\Delta x = \left[\langle x^2\rangle - \langle x\rangle^2\right]^{1/2}$ and $\Delta p_x = \left[\langle p_x^2\rangle - \langle p_x\rangle^2\right]^{1/2}$ for the wave function $\Psi_0(x)$. You will use the general formula given in Equation (4.46) $\langle A^k\rangle = \int_{-\infty}^{\infty} \Psi^*(x)\hat{A}^k \Psi(x)dx$ valid for any one-dimensional observable A and any wave function $\Psi(x)$.

(a) Apply the general formula for $\langle A^k\rangle$ to give expressions for Δx and Δp_x for the harmonic oscillator ground state wave function $\Psi_0(x)$.

(b) From your result in part (a), show that $\langle x\rangle = \langle p_x\rangle = 0$ so that the harmonic oscillator ground state formulas for Δx and Δp_x reduce to $\Delta x = \langle x^2\rangle^{1/2}$ and $\Delta p_x = \langle p_x^2\rangle^{1/2}$.

(c) Evaluate $\Delta x = \langle x^2\rangle^{1/2}$ and $\Delta p_x = \langle p_x^2\rangle^{1/2}$ using the wave function $\Psi_0(x)$. You will need the integrals $\int_{-\infty}^{\infty} \exp(-\alpha x^2)dx = \left(\frac{\pi}{\alpha}\right)^{1/2}$ and $\int_{-\infty}^{\infty} x^2 \exp(-\alpha x^2)dx = \frac{1}{2}\frac{\pi^{1/2}}{\alpha^{3/2}}$.

(d) From your result in part (c), evaluate $(\Delta x)(\Delta p_x)$ and compare your result with the Heisenberg uncertainty relation $(\Delta x)(\Delta p_x) \geq \frac{1}{2}\hbar$.

5.8 Give a reason why the uncertainty relation of Equation (5.22) does not hold for compatible observables A and B.

SOLUTIONS TO SELECTED PROBLEMS

5.4 $i\hbar(\hat{y}\hat{p}_y - \hat{x}\hat{p}_x)$.

5.5 (b) Yes and (c) no.

5.7 (a) $\Delta x = \left\{\int_{-\infty}^{\infty} x^2\Psi_0^2(x)dx - \left[\int_{-\infty}^{\infty} x\Psi_0^2(x)\right]^2\right\}^{1/2}$,

$$\Delta p_x = \left\{-\hbar^2\int_{-\infty}^{\infty} \Psi_0(x)\frac{d^2}{dx^2}\Psi_0(x)dx - \int_{-\infty}^{\infty}\left[\Psi_0(x)\frac{\hbar}{i}\frac{d\Psi_0(x)}{dx}\right]^2\right\}^{1/2}. \text{ and}$$

(d) $(\Delta x)(\Delta p_x) = \frac{\hbar}{2} \geq \frac{\hbar}{2}$.

5.8 For quantum states specified by the eigenfunctions $\Psi_{a_n,b_m}(\mathbf{r})\Delta A = \Delta B = 0$.

6 Stationary and Non-stationary Quantum States

In Section 3.2, we introduced stationary quantum states. In this chapter, we expand this introduction and also derive some basic results for non-stationary states. Before beginning our mathematical treatments of stationary and non-stationary quantum states, we briefly review the stationary states of the Bohr theory and also qualitatively describe their similarities to and differences from the stationary states of modern quantum mechanics.

6.1 BOHR VERSUS MODERN QUANTUM STATIONARY STATES

Recall from Section 1.3 that Bohr introduced the concept of stationary states as the stable states of atoms with discrete or quantized energies. Fusing this picture of the atom with the principle of conservation of energy (Bohr frequency condition), Bohr was able to quantitatively explain the line spectra of the hydrogen-like atoms (and also qualitatively explain the line spectra of more complex atoms) as arising from single photon transitions between discrete stationary state energy levels.

As already shown in Section 3.2, modern quantum mechanics predicts the existence of special quantum states, also called stationary states. The modern quantum stationary states, as we will show in Section 6.4, like those of the Bohr theory have definite energies. For this reason, like the Bohr theory, modern quantum mechanics predicts that the absorption and emission of electromagnetic radiation by atoms (and molecules) is due to transitions between stable stationary state energy levels.

There are however many significant differences between the Bohr and modern quantum stationary states. For example, while within the Bohr theory stationary states are the only possible type of quantum state, as we will see in this chapter in modern quantum mechanics stationary states are only a special type of quantum state and other types are possible.

Despite this, as already noted in Section 3.2, stationary states are of premier importance in the modern quantum theory of atoms and molecules. This is because they are the normal stable states of quantum systems.

We may now begin the quantitative treatment of both stationary and non-stationary quantum states. For simplicity, we will study a system comprised of a single particle of mass m with position vector $r = (x, y, z)$ moving in three dimensions under the influence of a potential energy function $U(r)$. We will denote the wave functions of this system by $\Psi(r,t)$. The results that we will develop for our single-particle system are readily extended to corresponding results for a general system comprised of N particles each moving in three dimensions.

Our treatments of both stationary and non-stationary quantum states are based on the time-dependent Schrödinger equation for our single-particle three-dimensional system. We next turn to this equation.

6.2 THE TIME-DEPENDENT SCHRÖDINGER EQUATION FOR A SINGLE PARTICLE MOVING IN THREE DIMENSIONS

The time-dependent Schrödinger equation for our system is the following specialization of Equation (4.5) for a general system:

$$i\hbar\frac{\partial\Psi(r,t)}{\partial t} = -\frac{\hbar^2}{2m}\nabla^2\Psi(r,t) + U(r)\Psi(r,t). \tag{6.1}$$

Equation (6.1) must be solved subject to the initial condition

$$\Psi(r,t=0)=\Psi(r) \tag{6.2}$$

where the initial wave function $\Psi(r)$ can be any well-behaved and usually normalized function. The time-dependent Schrödinger equation (Equation 6.1) may be written more concisely as

$$i\hbar\frac{\partial\Psi(r,t)}{\partial t}=\hat{H}\Psi(r,t) \tag{6.3}$$

where

$$\hat{H}=-\frac{\hbar^2}{2m}\nabla^2+U(r) \tag{6.4}$$

is the Hamiltonian operator of our single-particle system.

We next turn to the derivation as a relation for the forms of both the stationary and non-stationary state wave functions of our system and then specialize that relation to stationary state wave functions.

6.3 DERIVATION OF THE FORM OF STATIONARY STATE WAVES FUNCTIONS

The form of stationary state wave functions was already derived for one-dimensional systems in Section 3.2 by applying the separation-of-variables technique to the time-dependent Schrödinger equation (Equation 3.16) and is given in Equation (3.33). Here, we rederive this form for our single-particle three-dimensional system by a more general technique which is also applicable to non-stationary state wave functions.

We start with the Schrödinger equation for our system which is the following eigenvalue equation for the Hamiltonian operator $\hat{H}$ of Equation (6.4):

$$\hat{H}\Psi_m(r)=E_m\Psi_m(r). \tag{6.5}$$

For simplicity, we assume that the energy eigenvalues E_m are non-degenerate.

Since $\hat{H}$ is Hermitian, its normalized eigenfunctions $\Psi_m(r)$ form an orthonormal set which we assume is complete. With this assumption, the system wave function $\Psi(r,t)$ has the following complete orthonormal expansion in terms of the eigenfunctions $\Psi_m(r)$ of $\hat{H}$:

$$\Psi(r,t)=\sum_{\text{all m}}C_m(t)\Psi_m(r) \tag{6.6}$$

where the expansion coefficients $C_m(t)$ are given by

$$C_m(t)=\int\Psi_m^*(\mathbf{r})\Psi(\mathbf{r},t)d\mathbf{r}. \tag{6.7}$$

Next, let us differentiate the expansion of Equation (6.6) with respect to t and then multiply the result by $i\hbar$ to yield

$$i\hbar\frac{\partial\Psi(r,t)}{\partial t}=i\hbar\sum_{\text{all m}}\frac{dC_m(t)}{dt}\Psi_m(r). \tag{6.8}$$

Then, let us multiply Equation (6.6) by $\hat{H}$ and use the Schrödinger equation (Equation 6.5) to obtain

$$\hat{H}\Psi(\mathbf{r},t) = \sum_{\text{all m}} C_m(t)\hat{H}\Psi_m(\mathbf{r}) = \sum_{\text{all m}} C_m(t)E_m\Psi_m(\mathbf{r}). \tag{6.9}$$

Next, comparing Equations (6.8) and (6.9) with the time-dependent Schrödinger equation (Equation 6.3) $i\hbar\frac{\partial\Psi(r,t)}{\partial t} = \hat{H}\Psi(r,t)$ shows that the left-hand sides of these equations are equal and hence their right-hand sides are also equal. Equating these right-hand sides yields the relation

$$i\hbar\sum_{\text{all m}} \frac{dC_m(t)}{dt}\Psi_m(\mathbf{r}) = \sum_{\text{all m}} C_m(t)E_m\Psi_m(\mathbf{r}). \tag{6.10}$$

Equating the coefficients of $\Psi_m(\mathbf{r})$ on both sides of this relation yields the following differential equation for $C_m(t)$:

$$\frac{dC_m(t)}{dt} = -\frac{i}{\hbar}C_m(t)E_m. \tag{6.11}$$

The differential equation (Equation 6.11) must be solved subject to the initial condition $C_m(t=0) = C_m(0)$. As may be seen by substitution, the solution of the differential equation (Equation 6.11) in accord with its initial condition is

$$C_m(t) = C_m(0)\exp\left(-\frac{i}{\hbar}E_m t\right). \tag{6.12}$$

Substituting this form for $C_m(t)$ into the expansion of Equation (6.6) yields the system wave function $\Psi(r,t)$ as

$$\Psi(r,t) = \sum_{\text{all m}} C_m(0)\exp\left(-\frac{i}{\hbar}E_m t\right)\Psi_m(r). \tag{6.13}$$

Equation (6.13) is the quantum analogue of the expansion of Equation (2.100) of an arbitrary stretched string wave displacement function in terms of the string's standing wave displacement functions. The above representation of $\Psi(r,t)$, which provides the basis of our work in the remainder of this chapter, holds for both stationary and non-stationary state wave functions $\Psi(r,t)$.

However, while for stationary state wave functions only one term in the sum in Equation (6.13) is non-vanishing, for non-stationary state wave functions at least two and often an infinity of terms are non-vanishing. The reason why for stationary state wave functions only a single term in the sum survives derives from the fact, soon to be proven, that for an arbitrary nth stationary state wave function $\Psi(r,t) = \Psi_n(r,t)$, the coefficients $C_m(0)$ have the values $C_m(0) = \delta_{mn}$. Inserting this still-unproved form into Equation (6.13) yields the nth stationary state wave function as $\Psi_n(r,t) = \exp\left(-\frac{i}{\hbar}E_n t\right)\Psi_n(r)$. While for stationary state wave functions the coefficients $C_m(0)$ have the very simple forms $C_m(0) = \delta_{mn}$, for non-stationary state wave functions these coefficients have more complex forms with unique forms for each of the infinity of linearly independent non-stationary state wave functions $\Psi(r,t)$.

We may now complete the derivation of the forms of stationary state wave functions. To do this, we prove the correctness of our assumed form $C_m(0) = \delta_{mn}$ for $C_m(0)$ which as already shown yields the form of an arbitrary nth stationary state wave function as

$$\Psi_n(\mathbf{r},t) = \exp\left(-\frac{i}{\hbar}E_n t\right)\Psi_n(\mathbf{r}). \tag{6.14}$$

To prove that $C_m(0) = \delta_{mn}$, we first evaluate Equation (6.13) at $t = 0$ to obtain

$$\Psi(\mathbf{r}) = \sum_{\text{all m}} C_m(0)\Psi_m(\mathbf{r}) \tag{6.15}$$

where $\Psi(\mathbf{r}) = \Psi(\mathbf{r}, t = 0)$. We next note that Equation (6.15) is the complete orthonormal expansion of $\Psi(\mathbf{r})$ in terms of the eigenfunctions $\Psi_m(\mathbf{r})$ of $\hat{H}$. Consequently, the expansion coefficients $C_m(0)$ are given by

$$C_m(0) = \int \Psi_m^*(\mathbf{r})\Psi(\mathbf{r})d\mathbf{r}. \tag{6.16}$$

To obtain the nth stationary state wave function $\Psi_n(\mathbf{r}, t)$, we choose the initial wave function $\Psi(\mathbf{r})$ to be

$$\Psi(\mathbf{r}) = \Psi_n(\mathbf{r}) \tag{6.17}$$

where $\Psi_n(\mathbf{r})$ is the nth normalized eigenfunction of $\hat{H}$. However, since both $\Psi_m(\mathbf{r})$ and $\Psi_n(\mathbf{r})$ are normalized eigenfunctions of $\hat{H}$ and therefore form an orthonormal pair, it follows from Equations (6.16) and (6.17) that for the nth stationary state,

$$C_m(0) = \int \Psi_m^*(\mathbf{r})\Psi_n(\mathbf{r})d\mathbf{r} = \delta_{mn}. \tag{6.18}$$

Thus, we have proven that $C_m(0) = \delta_{mn}$ and therefore have established Equation (6.14) as the form of the nth stationary state wave function.

We next derive two very important properties of stationary states from the forms of their wave functions.

6.4 TWO PROPERTIES OF STATIONARY STATES

Specifically, in this section we establish the following two points:

Point 1. For all times $t \geq 0$, the energy E of a quantum system when it is in its nth stationary state with wave function $\Psi_n(\mathbf{r}, t) = \exp\left(-\frac{i}{\hbar}E_n t\right)\Psi_n(\mathbf{r})$ has the definite value E_n.

Point 2. When a quantum system is in an arbitrary stationary state, say the kth stationary state, all of its observable properties are time independent.

To establish Point 1, we first note, as is easily verified, that since $\Psi_n(\mathbf{r})$ is normalized, $\Psi_n(\mathbf{r}, t)$ is also normalized. The normalization of $\Psi_n(\mathbf{r}, t)$ is essential to our proof of Point 1 since this proof hinges on the postulate of Equation (4.27) which requires that the system wave function $\Psi(\mathbf{r}, t)$, which is here $\Psi_n(\mathbf{r}, t)$, be normalized. (For the same reason, the normalization of $\Psi_n[\mathbf{r}, t]$ is essential to the proof of Point 2.)

To prove Point 1, we specialize Equations (4.26) and (4.27) to the case that the arbitrary observable A, with Hermitian operator representative $\hat{A}$, is the system's energy E, with representative $\hat{H}$, and the system wave function $\Psi(\mathbf{r}, t)$ is the nth stationary state wave function $\Psi_n(\mathbf{r}, t) = \exp\left(-\frac{i}{\hbar}E_n t\right)\Psi_n(\mathbf{r})$.

To start, we specialize Equation (4.26) for the coefficients $C_m^A(t)$. To make this specialization, we replace in Equation (4.26) $\Psi(\mathbf{r}, t)$ by $\Psi_n(\mathbf{r}, t)$ and also replace the eigenfunctions $\Psi_m^A(\mathbf{r})$ of $\hat{A}$

by the eigenfunctions $\Psi_m(\mathbf{r})$ of $\hat{H}$. With these replacements, Equation (4.26) becomes the following expression:

$$C_m(t) = \int \Psi_m^*(\mathbf{r})\Psi_n(\mathbf{r},t)d\mathbf{r} = \exp\left(-\frac{i}{\hbar}E_n t\right)\int \Psi_m^*(\mathbf{r})\Psi_n(\mathbf{r})d\mathbf{r} \tag{6.19}$$

for the coefficients $C_m(t)$ of the complete orthonormal expansion of $\Psi_n(\mathbf{r},t)$ in terms of the eigenfunctions $\Psi_m(\mathbf{r})$ of $\hat{H}$. However, since both $\Psi_m(\mathbf{r})$ and $\Psi_n(\mathbf{r})$ are normalized eigenfunctions of $\hat{H}$,

$$\int \Psi_m^*(\mathbf{r})\Psi_n(\mathbf{r})d\mathbf{r} = \delta_{mn} \tag{6.20}$$

and hence, $C_m(t)$ simplifies to

$$C_m(t) = \exp\left(-\frac{i}{\hbar}E_n t\right)\delta_{mn}. \tag{6.21}$$

Next, let us determine the probability $P(E_m,t)$ of finding the eigenvalue E_m of $\hat{H}$ in a measurement of the energy E performed at time t when the system wave function is $\Psi_n(\mathbf{r},t)$. Specializing Equation (4.27) gives this probability as

$$P(E_m,t) = C_m^*(t)C_m(t). \tag{6.22}$$

Finally, comparing Equations (6.21) and (6.22) yields $P(E_m,t)$ as

$$P(E_m,t) = \exp\left(\frac{i}{\hbar}E_n t\right)\exp\left(-\frac{i}{\hbar}E_n t\right)\delta_{mn}^*\delta_{mn} = \delta_{mn}. \tag{6.23}$$

The proceeding equation shows that for any time $t \geq 0$, the probability of measuring any eigenvalue E_m of $\hat{H}$ other than E_n is zero, while the probability of measuring E_n is unity. This is the criterion for the system's energy E to have the definite value E_n. Thus, we have proven Point 1.

To prove Point 2, which is the statement that when the system is in arbitrary kth stationary state with wave function

$$\Psi_k(\mathbf{r},t) = \exp\left(-\frac{i}{\hbar}E_k t\right)\Psi_k(\mathbf{r}) \tag{6.24}$$

all of its observable properties are time independent, it is sufficient to show that the probability distributions $P(A,t)$ for all observables A are time independent.

To prove this result, recall that each distribution $P(A,t)$ is the collection of probabilities $P(a_n,t)$ of measuring at time t the eigenvalues a_n of the representative $\hat{A}$ of the observable A. So to establish Point 2, it is sufficient to show that when the system is in the kth stationary state, all probabilities $P(a_n,t)$ are time independent and therefore conform to the relation

$$P(a_n,t) = P(a_n,t=0). \tag{6.25}$$

Before proving this fundamental relation, we illustrate how it suggests that for a stationary state, the observable properties of a quantum system are time independent. Specifically, we show that if the relation of Equation (6.25) is true, then the expected value $\langle A\rangle_t$ of an arbitrary observable A is time independent; that is,

$$\langle A\rangle_t = \langle A\rangle_{t=0}. \tag{6.26}$$

To show this, we note that Equation (6.26) follows immediately from the relation $\langle A \rangle_t = \sum_{\text{all n}} a_n P(a_n, t)$ of Equation (4.37) and from the time independence of $P(a_n, t)$ given in Equation (6.25).

We next derive the fundamental equation (Equation 6.25).

Given our assumption that the system is in the kth stationary state, its wave function $\Psi(r,t)$ is $\Psi_k(r,t)$. Then, Equation (4.26) for the coefficients $C_n^A(t)$ specializes to

$$C_n^A(t) = \int \left[\Psi_n^A(\mathbf{r})\right]^* \Psi_k(\mathbf{r},t)d\mathbf{r}. \tag{6.27}$$

With Equation (6.24) for $\Psi_k(r,t)$, the proceeding relation becomes

$$C_n^A(t) = \exp\left(-\frac{i}{\hbar}E_k t\right)\int \left[\Psi_n^A(\mathbf{r})\right]^* \Psi_k(\mathbf{r})d\mathbf{r}. \tag{6.28}$$

Finally, substituting Equation (6.28) for $C_n^A(t)$ into Equation (4.27) for $P(a_n, t)$ and canceling the complex exponential factors yield

$$P(a_n,t) = \left\{\int \left[\Psi_n^A(\mathbf{r})\right]^* \Psi_k(\mathbf{r})d\mathbf{r}\right\}^* \left\{\int \left[\Psi_n^A(\mathbf{r})\right]^* \Psi_k(\mathbf{r})d\mathbf{r}\right\}. \tag{6.29}$$

But since the right-hand side of the proceeding equation is time independent, $P(a_n, t)$ is time independent and hence Equation (6.25) holds. Thus, we have proven Equation (6.25) and therefore have established Point 2.

This concludes our discussion of stationary states. We next turn to non-stationary states.

We first note that while a stationary state occurs when a system's initial wave function $\Psi(\mathbf{r})$ is an eigenfunction of its Hamiltonian operator $\hat{H}$ if $\Psi(\mathbf{r})$ is not an eigenfunction of $\hat{H}$, then the derived full wave function $\Psi(\mathbf{r},t)$ specifies a non-stationary state.

To begin our study of non-stationary states, we prove that non-stationary state wave functions $\Psi(\mathbf{r},t)$ are normalized if their initial values $\Psi(\mathbf{r})$ are normalized. In the course of this proof, we introduce a powerful mathematical construct that is widely used in quantum mechanics, the *Dirac δ – function*. The Dirac δ – function is the continuum analogue of the Kronecker delta δ_{ij}.

6.5 THE NORMALIZATION OF NON-STATIONARY STATE WAVE FUNCTIONS AND THE DIRAC δ – FUNCTION.

To start, let us consider the normalization integral of an arbitrary non-stationary state wave function $\Psi(\mathbf{r},t)$. From Equation (6.13), this integral is $\int \Psi^*(\mathbf{r},t)\Psi(\mathbf{r},t)d\mathbf{r} = \sum_{\text{all n}}\sum_{\text{all m}} C_n^*(0)C_m(0)\times$

$$\exp\left[-\frac{i}{\hbar}(E_m - E_n)t\right]\int \Psi_n^*(\mathbf{r})\Psi_m(\mathbf{r})d\mathbf{r}. \tag{6.30}$$

But since both $\Psi_n(\mathbf{r})$ and $\Psi_m(\mathbf{r})$ are normalized eigenfunctions of $\hat{H}$,

$$\int \Psi_n^*(\mathbf{r})\Psi_m(\mathbf{r})d\mathbf{r} = \delta_{nm} \tag{6.31}$$

and hence, Equation (6.30) simplifies to

$$\int \Psi^*(\mathbf{r},t)\Psi(\mathbf{r},t)d\mathbf{r} = \sum_{\text{all m}} C_m^*(0)C_m(0). \tag{6.32}$$

Next, using Equation (6.16) for $C_m(0)$, the previous equation may be rewritten as

$$\int \Psi^*(\mathbf{r},t)\Psi(\mathbf{r},t)d\mathbf{r} = \int d\mathbf{r}' \int d\mathbf{r}\Psi^*(\mathbf{r}')\Psi(\mathbf{r})\sum_{\text{all m}} \Psi^*_m(\mathbf{r})\Psi_m(\mathbf{r}'). \tag{6.33}$$

To proceed further, we make a mathematical digression. Consider an arbitrary well-behaved complex function $\Phi(\mathbf{r})$. It may be expanded in terms of the complete orthonormal set of eigenfunctions $\Psi_m(\mathbf{r})$ of $\hat{H}$ as

$$\Phi(\mathbf{r}) = \sum_{\text{all m}} C_m \Psi_m(\mathbf{r}) \tag{6.34}$$

where the expansion coefficients C_m are given by

$$C_m = \int \Psi^*_m(\mathbf{r}')\Phi(\mathbf{r}')d\mathbf{r}'. \tag{6.35}$$

Using Equation (6.35) to eliminate C_m from Equation (6.34) yields $\Phi(\mathbf{r})$ as

$$\Phi(\mathbf{r}) = \int d\mathbf{r}'\Phi(\mathbf{r}')\sum_{\text{all m}} \Psi^*_m(\mathbf{r}')\Psi_m(\mathbf{r}). \tag{6.36}$$

Next, in analogy to the Kronecker delta relation $f_i = \sum_{\text{all j}} f_j\delta_{ij}$, let us define the Dirac $\delta-$ function $\delta(\mathbf{r}-\mathbf{r}')$ by

$$f(\mathbf{r}) = \int d\mathbf{r}' f(\mathbf{r}')\delta(\mathbf{r}-\mathbf{r}') \tag{6.37}$$

where $f(\mathbf{r})$ is an arbitrary well-behaved complex function. For $f(\mathbf{r}) = \Phi(\mathbf{r})$, the definition of Equation (6.37) becomes

$$\Phi(\mathbf{r}) = \int d\mathbf{r}'\Phi(\mathbf{r}')\delta(\mathbf{r}-\mathbf{r}'). \tag{6.38}$$

Comparing Equations (6.36) and (6.38) yields the following representation of the Dirac $\delta-$ function:

$$\delta(\mathbf{r}-\mathbf{r}') = \sum_{\text{all m}} \Psi^*_m(\mathbf{r}')\Psi_m(\mathbf{r}). \tag{6.39}$$

The previous equation is called the *closure relation*. Since $\Phi(\mathbf{r})$ may be expanded in terms of the complete orthonormal set of eigenfunctions of any Hermitian operator, the closure relation holds for any complete orthonormal set of eigenfunctions, not just the eigenfunctions of $\hat{H}$.

We may now return to the normalization integral of Equation (6.33). Noting that the complex conjugate of the closure relation is $\delta^*(r-r) = \sum_{\text{all m}} \Psi^*_m(r)\Psi_m(r')$, the normalization integral may be recast as

$$\int \Psi^*(r,t)\Psi(r,t)dr = \int dr' \int dr\Psi^*(r')\Psi(r)\delta^*(r-r'). \tag{6.40}$$

However, $\delta(\mathrm{r}-\mathrm{r}')$ may be exactly represented by the following limiting Gaussian function (Cohen-Tannoudji, Diu, and Laloe 1977): $\lim_{\sigma\to 0}\left(\frac{1}{2\pi\sigma^2}\right)^{3/2}\exp\left[-\frac{1}{2}\frac{(\mathrm{r}-\mathrm{r}')\cdot(\mathrm{r}-\mathrm{r}')}{\sigma^2}\right]$, which like $\delta(\mathrm{r}-\mathrm{r}')$ is infinitely sharply peaked at $\mathrm{r}=\mathrm{r}'$ and normalized to unity. But since the limiting Gaussian is real, $\delta(\mathrm{r}-\mathrm{r}')$ is also real. Thus, Equation (6.40) may be rewritten as

$$\int \Psi^*(\mathbf{r},t)\Psi(\mathbf{r},t)d\mathbf{r} = \int d\mathbf{r}'\int d\mathbf{r}\Psi^*(\mathbf{r}')\Psi(\mathbf{r})\delta(\mathbf{r}-\mathbf{r}'). \tag{6.41}$$

But setting $f(\mathbf{r})=\Psi^*(\mathbf{r})$ in Equation (6.37) yields $\int d\mathbf{r}'\Psi^*(\mathbf{r}')\delta(\mathbf{r}-\mathbf{r}')=\Psi^*(\mathbf{r})$. Thus, Equation (6.41) reduces to

$$\int \Psi^*(\mathbf{r},t)\Psi(\mathbf{r},t)d\mathbf{r} = \int \Psi^*(\mathbf{r})\Psi(\mathbf{r})d\mathbf{r}. \tag{6.42}$$

Equation (6.42) is the result we have been seeking; namely, if the initial wave function $\Psi(\mathbf{r})$ is normalized, the derived time-dependent wave function $\Psi(\mathbf{r},t)$ is also normalized. We henceforth assume that $\Psi(\mathrm{r})$ and therefore $\Psi(\mathbf{r},t)$ are normalized.

We may now turn to a new topic.

In Section 6.4, we showed that for a system in a stationary state, the probability distributions $P(A,t)$ for all observables A are time independent. We will soon see that the situation is more complex if the system is in a non-stationary state.

To show this, we must first summarize some aspects of classical mechanics described in more detail in Appendix A. Consider a classical system which we assume for simplicity is comprised of only a single particle moving in three dimensions. Along a trajectory of the particle, its position vector $\mathbf{r}(t)$ and its momentum vector $\mathbf{p}(t)$ of course change with time. Therefore, one might expect that since the values of the system's observables $A(t)=A[\mathbf{r}(t),\mathbf{p}(t)]$ depend on $\mathbf{r}(t)$ and $\mathbf{p}(t)$ along a trajectory, these values are time dependent. This is usually but not always true. Rather, for many systems, special observables $A(t)$ exist which are time independent in the sense that they maintain their initial values along a trajectory. Such special observables are called *constants of the motion.* The most important constant of the motion is the Hamiltonian $H(t)$ of a system whose forces are derived from a potential energy function. Such a system is referred to as a *conservative system.* The time independence of the Hamiltonian of a conservative system is called *the law of conservation of energy.*

We may now consider the question of the time dependence of non-stationary state probability distributions $P(A,t)$. We will show in Sections 6.6 and 6.7 that the time dependence of probability distributions for observables that classically are time dependent is qualitatively different from the time dependence of probability distributions for observables that classically are constants of the motion.

6.6 THE PROBABILITY DISTRIBUTIONS $P(A,t)$ ASSOCIATED WITH ORDINARY TIME-DEPENDENT CLASSICAL OBSERVABLES $A(t)$

We prove in this section that for an ordinary time-dependent classical observable $A(t)$, the corresponding non-stationary state quantum probability distribution $P(A,t)$ is also time dependent. Thus, for this case, expected values $\langle A\rangle_t$ of $A(t)$ are time dependent.

To make this proof, we again need Equation (4.26) and the postulate of Equation (4.27). We note that we may use Equation (4.27), which requires normalized wave functions, since we showed in Section 6.5 that for all times $t\geq 0$, non-stationary state wave functions $\Psi(\mathbf{r},t)$ may be taken as normalized.

To start, recall from Equation (4.27) that for any normalized system wave function $\Psi(\mathbf{r},t)$, the probability $P(a_m,t)$ that at time t the measured value of an arbitrary observable A is equal to the eigenvalue a_m of its representative $\hat{A}$ is given by

$$P(a_m,t)=\left[C_m^A(t)\right]^*\left[C_m^A(t)\right]. \tag{6.43}$$

From Equation (4.26), $C_m^A(t)$ is given in terms of $\Psi(\mathbf{r},t)$ and the normalized eigenfunction $\Psi_m^A(\mathbf{r})$ of $\hat{A}$ associated with the eigenvalue a_m by

$$C_m^A(t)=\int\left[\Psi_m^A(\mathbf{r})\right]^*\Psi(\mathbf{r},t)d\mathbf{r}. \tag{6.44}$$

Next, substituting Equation (6.13) for $\Psi(\mathbf{r},t)$ into the previous relation yields $C_m^A(t)$ as

$$C_m^A(t)=\sum_{\text{all n}}C_n(0)\exp\left(-\frac{i}{\hbar}E_nt\right)\int\left[\Psi_m^A(\mathbf{r})\right]^*\Psi_n(\mathbf{r})d\mathbf{r} \tag{6.45}$$

where recall $\Psi_n(\mathbf{r})$ is the nth normalized eigenfunction of the system's Hamiltonian operator $\hat{H}$ and E_n is the associated energy eigenvalue. Then, taking the complex conjugate of the above form for $C_m^A(t)$ and changing the dummy summation index from n to n′ give this complex conjugate as

$$\left[C_m^A(t)\right]^*=\sum_{\text{all n}'}C_{n'}^*(0)\exp\left(\frac{i}{\hbar}E_{n'}t\right)\int\left[\Psi_m^A(\mathbf{r})\right]\Psi_{n'}^*(\mathbf{r})d\mathbf{r}. \tag{6.46}$$

Inserting Equations (6.45) and (6.46) into Equation (6.43) yields our final result for $P(a_m,t)$

$$P(a_m,t)=\sum_{\text{all n}'}\sum_{\text{all n}}C_{n'}^*(0)C_n(0)\exp\left[\frac{i}{\hbar}(E_{n'}-E_n)t\right]\int\left[\Psi_m^A(\mathbf{r})\right]\Psi_{n'}^*(\mathbf{r})d\mathbf{r}\int\left[\Psi_m^A(\mathbf{r})\right]^*\Psi_n(\mathbf{r})d\mathbf{r}. \tag{6.47}$$

Notice that $P(a_m,t)$ is time dependent because of its oscillatory complex exponential factors. Similarly, the probabilities for measuring all of the other eigenvalues of $\hat{A}$ are time dependent. Consequently, $P(A,t)$ is time dependent. Thus, we have proven the desired result. If a classical observable $A(t)$ is time dependent, then the corresponding non-stationary state quantum probability distribution $P(A,t)$ is also time dependent.

We next determine the time dependence of non-stationary state probability distributions $P(A,t)$ which correspond to classical observables $A(t)$ which are constants of the motion.

6.7 THE PROBABILITY DISTRIBUTIONS $P(A,t)$ ASSOCIATED WITH CLASSICAL OBSERVABLES $A(t)$ WHICH ARE CONSTANTS OF THE MOTION

In this section, we show that if a classical observable $A(t)$ is a time-independent constant of the motion, then the corresponding non-stationary state quantum probability distribution $P(A,t)$ is also time independent. Consequently, the expected values $\langle A\rangle_t$ of a constant of the motion $A(t)$ are time independent.

To start, we note that the results of Problem 6.7 suggest that if $A(t)$ is a constant of the motion, its representative $\hat{A}$ must commute with the system's Hamiltonian operator $\hat{H}$. Therefore, for a constant of the motion, we must have that

$$\left[\hat{A},\hat{H}\right]=\hat{O}. \tag{6.48}$$

However, in Section 5.2 we proved a theorem that states that if the representatives $\hat{A}$ and $\hat{B}$ of two classical observables A and B commute, these representatives have a complete orthonormal set of common eigenfunctions. Applying this theorem to the present problem, we conclude that if a classical observable $A(t)$ is a constant of motion, then since its representative $\hat{A}$ commutes with $\hat{H}, \hat{A}$ and $\hat{H}$ have a complete orthonormal set of common eigenfunctions. However, within our assumption that all eigenvalues E_m of $\hat{H}$ are non-degenerate, the set of common eigenfunctions of $\hat{A}$ and $\hat{H}$ is may be taken as the complete orthonormal set of eigenfunctions $\Psi_m(\mathbf{r})$ of $\hat{H}$. Thus, to obtain the probability $P(a_m,t)$ for the case that $A(t)$ is a constant of the motion, one merely replaces in Equation (6.47) the eigenfunctions $\Psi_m^A(\mathbf{r})$ of $\hat{A}$ by the eigenfunctions $\Psi_m(\mathbf{r})$ of $\hat{H}$. These replacements yield the probability $P(a_m,t)$ as

$$P(a_m,t)=\sum_{\text{all } n'}\sum_{\text{all } n} C_{n'}^*(0)C_n(0)\exp\left[\frac{i}{\hbar}(E_{n'}-E_n)t\right]\int \Psi_m(\mathbf{r})\Psi_{n'}^*(\mathbf{r})d\mathbf{r}\int \Psi_m^*(\mathbf{r})\Psi_n(\mathbf{r})d\mathbf{r}. \quad (6.49)$$

But since $\Psi_m(\mathbf{r}), \Psi_{n'}(\mathbf{r})$, and $\Psi_n(\mathbf{r})$ are all normalized eigenfunctions of $\hat{H}$, one may easily evaluate the integrals in the proceeding equation to obtain the probability $P(a_m,t)$ as

$$P(a_m,t)=\sum_{\text{all } n'}\sum_{\text{all } n} C_{n'}^*(0)C_n(0)\exp\left[\frac{i}{\hbar}(E_{n'}-E_n)t\right]\delta_{mn'}\delta_{mn}. \quad (6.50)$$

However, since $\delta_{mn'}\delta_{mn}$ is zero unless $n' = n = m$, only one term in the above double sum is non-vanishing, the term for which $n' = n = m$. Thus, our final result for $P(a_m,t)$ has the very simple form

$$P(a_m,t)=C_m^*(0)C_m(0) \quad (6.51)$$

which is time independent. Similarly, the probabilities for measuring all other eigenvalues of $\hat{A}$ are time independent. Hence, the probability distribution $P(A,t)$ is time independent.

Therefore, we have established the sought-after result. For a classical observable $A(t)$ which is a time-independent constant of the motion, the corresponding non-stationary state quantum probability distribution $P(A,t)$ is also time independent.

FURTHER READINGS

Cohen-Tannoudji, Claude, Bernard Diu, and Franck Laloe. 1977. *Quantum mechanics*. Translated by Susan Reid Hemley, Nicole Ostrowsky, and Dan Ostrowsky. New York: John Wiley & Sons Inc.

Merzbacher, Eugen. 1961. *Quantum mechanics*. New York: John Wiley & Sons, Inc.

Sakurai, J. J. 1985. *Modern quantum mechanics*. Rev. ed. Edited by San Fu Tuan. Reading, MA: Addison-Wesley Publishing Co.

PROBLEMS

6.1 For a system in the nth stationary state with wave function $\Psi(\mathbf{r},t)=\exp\left(-\frac{i}{\hbar}E_n t\right)\Psi_n(\mathbf{r})$, show that Equation (4.44) for the expected value $\langle A^k\rangle_t$ reduces to the time-independent form $\langle A^k\rangle_t = \langle A^k\rangle_n = \int \Psi_n^*(\mathbf{r})\hat{A}^k\Psi_n(\mathbf{r})d\mathbf{r}$.

6.2 Derive the differential equation (Equation 6.11) for $C_m(t)$ from Equation (6.10) using only the orthonormality of the eigenfunctions $\Psi_m(\mathbf{r})$ of $\hat{H}$.

6.3 The Born rule for the probability density $P(\mathbf{r},t)$ for the position observable $\mathbf{r}$ is $P(\mathbf{r},t)=\Psi^*(\mathbf{r},t)\Psi(\mathbf{r},t)$ where $\Psi(\mathbf{r},t)$ is the normalized system wave function.

(a) Show that if the system is in a general non-stationary state with wave function $\Psi(\mathbf{r},t)$ given by Equation (6.13), then $P(\mathbf{r},t)$ has the time-dependent form $P(\mathbf{r},t)=\sum_{\text{all m}'}\sum_{\text{all m}} C^*_{m'}(0)C_m(0)\exp\left[\frac{i}{\hbar}(E_{m'}-E_m)t\right]\Psi^*_{m'}(\mathbf{r})\Psi_m(\mathbf{r})$.

(b) Show that if the system is in the kth stationary state with wave function $\Psi_k(\mathbf{r},t)=\exp\left(-\frac{i}{\hbar}E_k t\right)\Psi_k(\mathbf{r})$, the Born rule probability density has the time-independent form $P(\mathbf{r},t)=\Psi^*_k(\mathbf{r})\Psi_k(\mathbf{r})$.

(c) Find the values of the coefficients $C_{m'}(0)$ and $C_m(0)$ in the non-stationary state form for $P(\mathbf{r},t)$ derived in part (a) which reduce that form to the stationary state form for $P(\mathbf{r},t)$ derived in part (b).

6.4 Consider an arbitrary observable B with Hermitian operator representative $\hat{B}$. When the system wave function is $\Psi(\mathbf{r},t)$, the expected value $\langle B\rangle_t$ of B is $\int\Psi^*(\mathbf{r},t)\hat{B}\Psi(\mathbf{r},t)d\mathbf{r}$.

(a) If the system is in the ground stationary state with wave function $\Psi_1(\mathbf{r},t)=\exp\left(\frac{i}{\hbar}E_1 t\right)\Psi_1(\mathbf{r})$, show that $\langle B\rangle_t$ has the time-independent value $\langle B\rangle_t=\int\Psi^*_1(\mathbf{r})B\Psi_1(\mathbf{r})d\mathbf{r}$.

(b) If the system is in a non-stationary state with wave function $\Psi(\mathbf{r},t)= C_1(0)\exp\left(-\frac{i}{\hbar}E_1 t\right)\Psi_1(\mathbf{r})+C_2(0)\exp\left(-\frac{i}{\hbar}E_2 t\right)\Psi_2(\mathbf{r})$, show that $\langle B\rangle_t$ has the time-dependent value $\langle B\rangle_t=C^*_1(0)C_1(0)\int\Psi^*_1(\mathbf{r})\hat{B}\Psi_1(\mathbf{r})d\mathbf{r}+C^*_2(0)C_2(0)\times \int\Psi^*_2(\mathbf{r})\hat{B}\Psi_2(\mathbf{r})d\mathbf{r}+2\mathrm{Re}\left\{C^*_1(0)C_2(0)\exp\left[\frac{i}{\hbar}(E_1-E_2)t\right]\int\Psi^*_1(\mathbf{r})\hat{B}\Psi_2(\mathbf{r})d\mathbf{r}\right\}$.

(c) Show that if the system is in a general non-stationary state with wave function $\Psi(\mathbf{r},t)$ given by Equation (6.13), $\langle B\rangle_t$ has the time-dependent form $\langle B\rangle_t=\sum_{\text{all n}}\sum_{\text{all n}'} C^*_{n'}(0)C_n(0)\exp\left[\frac{i}{\hbar}(E_{n'}-E_n)t\right]\int\Psi^*_{n'}(\mathbf{r})\hat{B}\Psi_n(\mathbf{r})d\mathbf{r}$.

6.5 This problem deals with the extension of the classical principle of conservation of energy described in Appendix A to quantum mechanics. You will show that on the average, energy is conserved for the three systems of Problem 6.4. Specifically, you will show that the average energies $\langle E\rangle_t$ of these systems are time independent and hence are constants of the motion. To do this, you will specialize the results of Problem 6.4 by taking the arbitrary observable B to be the energy E and the Hermitian operator $\hat{B}$ to be the system's Hamiltonian operator $\hat{H}$.

(a) Show for the system of Problem 6.4a that $\langle E\rangle_t$ has the time-independent value E_1.

(b) Show for the system of Problem 6.4b that $\langle E\rangle_t$ has the time-independent value $E_1C^*_1(0)C_1(0)+E_2C^*_2(0)C_2(0)$.

(c) Show for system of Problem 6.4c that $\langle E\rangle_t$ has the time-independent value $\sum_{\text{all n}}E_nC^*_n(0)C_n(0)$.

(d) Why is it expected for the systems of parts (b) and (c) that $\langle E\rangle_t$ is time independent?

6.6 In this problem, you will show for an arbitrary observable A with Hermitian operator representative $\hat{A}$ that when the normalized system wave function is $\Psi(\mathbf{r},t)$, the sum of all probabilities $P(a_n,t)$ for measuring the eigenvalues a_n of $\hat{A}$ at time t is as expected $\sum_{\text{all n}}P(a_n,t)=1$. Prove this result using Equations (4.26), (4.27), and (6.39).

6.7 In this problem, you will develop the concept of a constant of the motion in quantum mechanics. Consider an arbitrary observable A with Hermitian operator representative $\hat{A}$. If the system wave function is $\Psi(\mathbf{r},t)$, the expected value of A is $\langle A\rangle_t = \int \Psi^*(\mathbf{r},t)\hat{A}\Psi(\mathbf{r},t)d\mathbf{r}$.

(a) Differentiate $\langle A\rangle_t$ with respect to t and use the time-dependent Schrödinger equation (Equation 6.3) to show that $\frac{d\langle A\rangle_t}{dt} = \frac{1}{i\hbar}\left[\int \Psi^*(\mathbf{r},t)\hat{A}\hat{H}\Psi(\mathbf{r},t)d\mathbf{r} - \int \hat{H}\Psi^*(\mathbf{r},t)\hat{A}\Psi(\mathbf{r},t)d\mathbf{r}\right]$.

(b) Use the fact that $\hat{H}$ is Hermitian to show that $\int \hat{H}\Psi^*(\mathbf{r},t)\hat{A}\Psi(\mathbf{r},t)d\mathbf{r} = \int \Psi^*(\mathbf{r},t)\hat{H}\hat{A}\Psi(\mathbf{r},t)d\mathbf{r}$.

(c) Combine your results in parts (a) and (b) to show that $\frac{d\langle A\rangle_t}{dt} = \frac{1}{i\hbar}\int \Psi^*(\mathbf{r},t)\times \left[\hat{A},\hat{H}\right]\Psi(\mathbf{r},t)d\mathbf{r} = \frac{1}{i\hbar}\left\langle\left[\hat{A},\hat{H}\right]\right\rangle_t$. Notice that $\frac{d\langle A\rangle_t}{dt} = 0$ if $\left[\hat{A},\hat{H}\right] = \hat{O}$, and hence, on the average the observable A is a constant of the motion if $\hat{A}$ and $\hat{H}$ commute.

6.8 This problem deals with another aspect of conservation of energy in quantum mechanics. In particular, you will show that if the energy E of a system is measured at time t, the probability $P(E_m,t)$ of finding the eigenvalue E_m of $\hat{H}$ is time independent even for a system in a non-stationary state.

(a) By taking the observable A to be E, show from Equation (6.51) that $P(E_m,t)$ has the time-independent form $P(E_m,t) = C_m^*(0)C_m(0)$.

(b) Why is it permissible to use Equation (6.51)?

(c) From your result in part (a), derive that energy conservation law of Problem 6.5c $\langle E\rangle_t = \sum_{\text{all } m} E_m C_m^*(0)C_m(0)$.

6.9 This problem deals with Ehrenfest's theorem that provides a link between classical and quantum mechanics. To start, we specialize the equation of motion of part (c) of Problem 6.7 $\frac{d\langle A\rangle_t}{dt} = \frac{1}{i\hbar}\left\langle\left[\hat{A},\hat{H}\right]\right\rangle_t$ to the cases that the observable A is the position observable $\mathbf{r} = (x,y,z)$ or the momentum observable $\mathbf{p} = (p_x,p_y,p_z)$. This specialization yields the equations of motion $\frac{d\langle \mathbf{r}\rangle_t}{dt} = \frac{1}{i\hbar}\left\langle\left[\hat{\mathbf{r}},\hat{H}\right]\right\rangle_t$ and $\frac{d\langle \mathbf{p}\rangle_t}{dt} = \frac{1}{i\hbar}\left\langle\left[\hat{\mathbf{p}},\hat{H}\right]\right\rangle_t$. Then, note that the Hamiltonian operator $\hat{H}$ of Equation (6.4) may be rewritten as $\hat{H} = \frac{\hat{\mathbf{p}}\cdot\hat{\mathbf{p}}}{2m} + U(r)$ where the momentum operator $\hat{\mathbf{p}} = \frac{\hbar}{i}\nabla$ and where the vector operator $\nabla \equiv \left(\frac{\partial}{\partial x},\frac{\partial}{\partial y},\frac{\partial}{\partial z}\right)$.

(a) From this form for $\hat{H}$, show that $\left\langle\left[\hat{\mathbf{r}},\hat{H}\right]\right\rangle_t = \left\langle\left[\hat{\mathbf{r}},\frac{\hat{\mathbf{p}}\cdot\hat{\mathbf{p}}}{2m}\right]\right\rangle_t$ and $\left\langle\left[\hat{\mathbf{p}},\hat{H}\right]\right\rangle = \left\langle\left[\hat{\mathbf{p}},U(\mathbf{r})\right]\right\rangle$

(b) Using $\hat{\mathbf{r}} = (\hat{x},\hat{y},\hat{z})$ and $\hat{\mathbf{p}} = (\hat{p}_x,\hat{p}_y,\hat{p}_z)$, show that $\left[\hat{\mathbf{r}},\frac{\hat{\mathbf{p}}\cdot\hat{\mathbf{p}}}{2m}\right] = \frac{1}{2m}\times \left(\left[\hat{x},\hat{p}_x^2\right]+\left[\hat{x},\hat{p}_y^2\right]+\left[\hat{x},\hat{p}_z^2\right]+\left[\hat{y},\hat{p}_x^2\right]+\left[\hat{y},\hat{p}_y^2\right]+\left[\hat{y},\hat{p}_z^2\right]+\left[\hat{z},\hat{p}_x^2\right]+\left[\hat{z},\hat{p}_y^2\right]+\left[\hat{z},\hat{p}_z^2\right]\right)$.

(c) Using the commutator identity of Problem 5.1c and the fact that the only non-null commutators of the components of $\hat{\mathbf{r}}$ and $\hat{\mathbf{p}}$ are those of Problem 5.3, show that the expression for $\left[\hat{\mathbf{r}},\frac{\hat{\mathbf{p}}\cdot\hat{\mathbf{p}}}{2m}\right]$ of part (b) simplifies to $\left[\hat{\mathbf{r}},\frac{\hat{\mathbf{p}}\cdot\hat{\mathbf{p}}}{2m}\right] = \frac{i\hbar}{m}\hat{\mathbf{p}}$.

(d) Additionally, show that $\left[\hat{p}_x, U(x)\right] = \frac{\hbar}{i}\frac{dU(x)}{dx}$.

(e) Extend your result in part (d) to three dimensions to show that $\left[\hat{\mathbf{p}}, U(\mathbf{r})\right] = \frac{\hbar}{i}\nabla U(\mathbf{r})$. From the expressions you developed in this problem, derive Ehrenfest's equations of motion $\frac{d}{dt}\langle\mathbf{r}\rangle_t = \frac{1}{m}\langle\mathbf{p}\rangle_t$ and $\frac{d}{dt}\langle\mathbf{p}\rangle_t = -\langle\nabla U(\mathbf{r})\rangle_t$. Next, consider Hamilton's classical equations of motion given for one-dimensional systems in Equation (A.41). Notice that on the average, Ehrenfest's quantum equations of motion are three-dimensional generalizations of Hamilton's one-dimensional equations of motion.

(f) To further see the link between classical and quantum mechanics, recast Ehrenfest's equations of motion to obtain the equation of motion $m\frac{d^2}{dt^2}\langle\mathbf{r}\rangle_t = -\langle\nabla U(\mathbf{r})\rangle_t$. Notice that on the average, this last quantum equation of motion is identical to Newton's three-dimensional classical equation of motion.

6.10 This problem concerns the time dependence of the expected value of the coordinate of a one-dimensional harmonic oscillator in a non-stationary state. As in Chapter 7, where the quantum theory of the one-dimensional harmonic oscillator is developed in detail, we assume an oscillator of mass m, vibrational displacement y, and circular frequency ω. (The circular frequency $\omega = 2\pi\upsilon$, where υ is the oscillator's actual frequency, which is the number of vibrations the oscillator would make per second if it moved classically.) We assume the oscillator is in the non-stationary state specified by the wave function $\Psi(y,t) = N\left[\exp\left(-\frac{i}{\hbar}E_n t\right)\Psi_0(y) + \exp\left(-\frac{i}{\hbar}E_1 t\right)\Psi_1(y)\right]$ where N is a normalization factor and where $\Psi_0(y)$ and $\Psi_1(y)$ are, respectively, the oscillator's normalized ground and first excited stationary state wave functions and where E_0 and E_1 are the corresponding ground and first excited state oscillator energies.

(a) Show that the normalized form of $\Psi(y,t)$ is $\Psi(y,t) = \left(\frac{1}{2}\right)^{1/2}\left[\exp\left(-\frac{i}{\hbar}\frac{E_0 t}{\hbar}\right)\Psi_0(y) + \exp\left(-\frac{i}{\hbar}E_1 t\right)\Psi_1(y)\right]$. In Equations (7.15), (7.16), and (7.10), we give that $\Psi_0(y) = \left(\frac{m\omega}{\pi\hbar}\right)^{1/4}\exp\left(-\frac{1}{2}\frac{m\omega}{\hbar}y^2\right)$, $\Psi_1(y) = \left(\frac{m\omega}{\pi\hbar}\right)^{1/4}\left(\frac{2m\omega}{\hbar}\right)^{1/2} y\exp\left(-\frac{1}{2}\frac{m\omega}{\hbar}y^2\right)$, $E_0 = \frac{1}{2}\hbar\omega$, and $E_1 = \frac{3}{2}\hbar\omega$. To solve this problem, you will also need the value of the integral $G_0(\alpha) = \int_{-\infty}^{\infty} y^n \exp(-\alpha y^2)dy$ where n is a nonnegative integer and where α is a positive constant. In Section 7.9, we show that for n even $G_n(\alpha) = \frac{n!}{2^n(n/2)!}\frac{\pi^{1/2}}{\alpha(n+1)^{1/2}}$, while for n odd $G_n(\alpha) = 0$

(b) On the basis of this information, show that the expected value $\langle y\rangle_t = \int_{-\infty}^{\infty} y\Psi^*(y,t)\Psi(y,t)dy = \left(\frac{\hbar}{2m\omega}\right)^{1/2}\cos\omega t$.

SOLUTIONS TO SELECTED PROBLEMS

6.3 (c) $C_{m'}(0) = \delta_{m'k}$ and $C_m(0) = \delta_{mk}$.

6.5 (c) Because $H(t)$ is a constant of the motion.

6.8 (b) Since $\left[\hat{H}, \hat{H}\right] = \hat{O}$.

7 The Harmonic Oscillator

In Chapters 2 and 4–6, we developed the mathematical and physical foundations of quantum mechanics. In this and the remaining chapters, we describe successively advanced applications of these foundations.

In this chapter, we study the stationary states of the one-dimensional harmonic oscillator which after the one-dimensional particle-in-a-box is perhaps the simplest quantum system. We base our study on the analysis of Appendix C in which the harmonic oscillator Schrödinger equation is solved exactly to yield the oscillator's stationary state wave functions $\Psi_v(y)$ and definite quantized energies E_v. In particular, in this chapter we describe the wave functions and energy levels, develop some of the properties of the wave functions, and very importantly apply our results to the problems of the infrared (IR) spectra of diatomic and polyatomic molecules.

We start by writing down the Schrödinger equation of the oscillator.

7.1 THE TIME-INDEPENDENT SCHRÖDINGER EQUATION OF THE ONE-DIMENSIONAL HARMONIC OSCILLATOR

To do this, we must outline some aspects of the classical treatment of the one-dimensional harmonic oscillator described in more detail in Section A.1. In that section, we note that the oscillator is an idealized rather than a real system. But we also point out that a real system that approximates a one-dimensional harmonic oscillator is a particle of mass m and coordinate x which is bound to a wall by a spring (Figure A.2a). The spring produces a restoring force $F_x(x)$ on the particle which resists vibrational displacements $y = x - x_e$ of x from its equilibrium value x_e (Figure A.2a and b) and thus can induce vibratory motions of the particle about x_e. The real system of Figure A.2 becomes a one-dimensional harmonic oscillator if one takes the restoring force to have the idealized *Hooke's law* form

$$F_x(x) = -k(x - x_e) = -ky \tag{7.1}$$

where the *force constant* k is a property of the spring which increases with the spring's "stiffness."

Next, let us define the *circular frequency* ω by

$$\omega = \left(\frac{k}{m}\right)^{1/2}. \tag{7.2}$$

The circular frequency ω is related to the oscillator's actual frequency υ by

$$\omega = 2\pi\upsilon. \tag{7.3}$$

υ is the number of vibrations the oscillator would make per second if it moved classically. We have introduced ω because the oscillator's wave functions $\Psi_v(y)$ and energies E_v are more simply expressed in terms of ω than in terms of k.

From Equations (7.1) and (7.2), the Hooke's law restoring force expressed in terms of ω is

$$F_x(x) = -m\omega^2(x - x_e) = -m\omega^2 y. \tag{7.4}$$

We may now write down the potential energy function $U(x)$ of the oscillator which is defined by $F_x(x) = -\frac{dU(x)}{dx}$. Choosing the arbitrary zero of energy so $U(x_e) = 0$ and using Equation (7.4) for $F_x(x)$, we find that

$$U(x) = \frac{1}{2}m\omega^2(x - x_e)^2 = \frac{1}{2}m\omega^2y^2. \tag{7.5}$$

The classical Hamiltonian of the oscillator is thus $H = \frac{p_x^2}{2m} + U(x) = \frac{p_x^2}{2m} + \frac{1}{2}m\omega^2y^2$, where $p_x = m\dot{x}$ is the momentum of the particle. However, $p_x = m\dot{x} = m(\dot{x} - \dot{x}_e) = m\dot{y} \equiv p_y$. Thus, H can be written totally in terms of the vibrational coordinate y and the vibrational momentum p_y as

$$H = \frac{p_y^{\ 3}}{2m} + \frac{1}{2}m\omega^2y^2. \tag{7.6}$$

However, the oscillator's classical Hamiltonian H may using the rules of Table 4.1 be easily converted into its quantum Hamiltonian operator

$$\hat{H} = -\frac{\hbar^2}{2m}\frac{d^2}{dy^2} + \frac{1}{2}m\omega^2\hat{y}^2 \tag{7.7}$$

where the operator $\hat{y}$ = multiply by y. Using the above form for $\hat{H}$, the oscillator's Schrödinger equation $\hat{H}\Psi_v(y) = E_v\Psi_v(y)$ may be written down explicitly as

$$-\frac{\hbar^2}{2m}\frac{d^2}{dy^2}\Psi_v(y) + \frac{1}{2}m\omega^2y^2\Psi_v(y) = E_v\Psi_v(y) \tag{7.8}$$

The Schrödinger equation (Equation 7.8) must be supplemented by the boundary condition

$$\lim_{y\to\pm\infty} \Psi_v(y) = 0 \tag{7.9}$$

which guarantees that the wave functions $\Psi_v(y)$ are square integrable.

Equations (7.8) and (7.9) are the main results of this section.

We next compare the harmonic oscillator stationary state wave functions and energies derived in Appendix C with the corresponding particle-in-a-box quantities derived in Section 3.4.

7.2 PARTICLE-IN-A-BOX VERSUS HARMONIC OSCILLATOR STATIONARY STATES

In Figure 7.1, we briefly preview the forms of the harmonic oscillator wave functions $\Psi_v(y)$ and energies E_v and compare these quantities with the particle-in-a-box wave functions $\Psi_n(x)$ and energies E_n. From Figure 7.1b and c, it is evident that the particle-in-a-box and harmonic oscillator wave functions share some similarities. Most notably, for both systems, in accord with general principles the ground state wave functions have zero nodes and the first excited state wave functions have one node. There are also important differences in the wave functions. These differences derive from the very different forms of the particle-in-a-box and harmonic oscillator potentials shown in

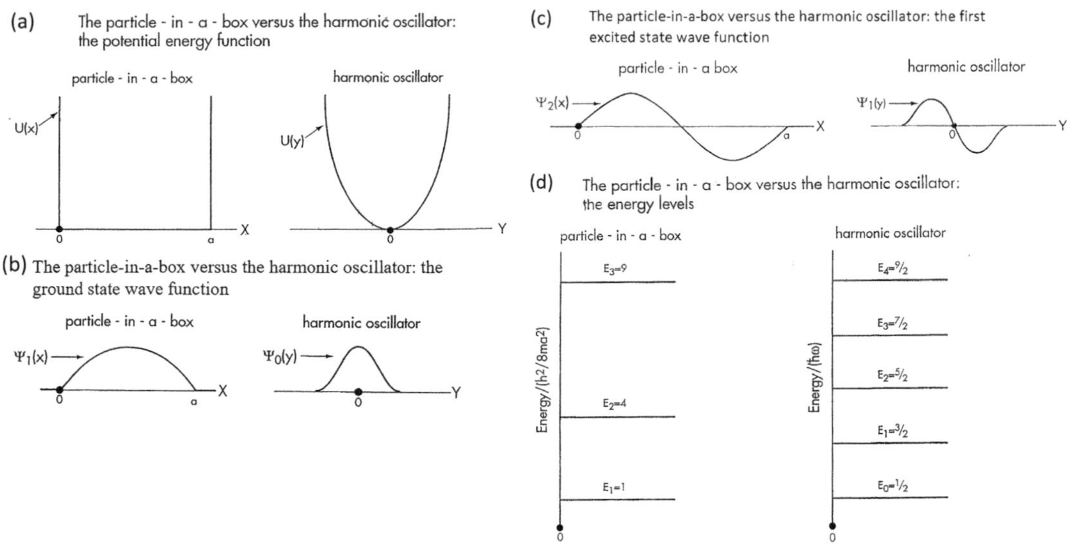

FIGURE 7.1 Four comparisons of the one-dimensional particle-in-a-box and harmonic oscillator systems. (a) The particle-in-a-box potential $U(x)$ and the harmonic oscillator potential $U(y)$. (b) The ground state wave functions $\Psi_1(x)$ and $\Psi_0(y)$ for the particle-in-a-box and harmonic oscillator systems. (c) The first excited state wave functions $\Psi_2(x)$ and $\Psi_1(y)$ for the particle-in-a-box and harmonic oscillator systems. (d) Particle-in-a-box and harmonic oscillator energy levels.

Figure 7.1a. For example, since the particle-in-a-box potential $U(x)$ rises abruptly to infinity at the box edges, it rigidly confines the particle to the interior of the box. This reflects itself in wave functions $\Psi_n(x)$ which rigorously vanish outside of the box, and which thus have discontinuous first derivatives at the box edges. The harmonic oscillator potential $U(y)$ in contrast rises gradually to infinity as $y \to \pm\infty$. This gradual rise reflects itself in wave functions $\Psi_v(y)$ which decay to zero smoothly as $y \to \pm\infty$.

Additionally, from Figure 7.1d the particle-in-a-box and harmonic oscillator energy levels differ radically with the spacing of the particle-in-a-box levels increasing rapidly with the quantum number n while the harmonic oscillator levels are equally spaced.

We next summarize the complete results for the harmonic oscillator energies and wave functions derived in Appendix C.

7.3 SUMMARY OF THE RESULTS OF THE SOLUTION OF THE HARMONIC OSCILLATOR SCHRÖDINGER EQUATION

As shown in Appendix C, imposition of the requirement that the wave functions be well-behaved at $y = \pm\infty$ yields discrete values 0,1, … for the quantum number v and energy levels which are non-degenerate and take the quantized form

$$E_v = \left(v + \frac{1}{2}\right)\hbar\omega = \left(v + \frac{1}{2}\right)h\upsilon. \tag{7.10}$$

As illustrated in Figure 7.2, the energy levels are equally spaced with spacing

$$E_{v+1} - E_v = \frac{1}{2}\hbar\omega. \tag{7.11}$$

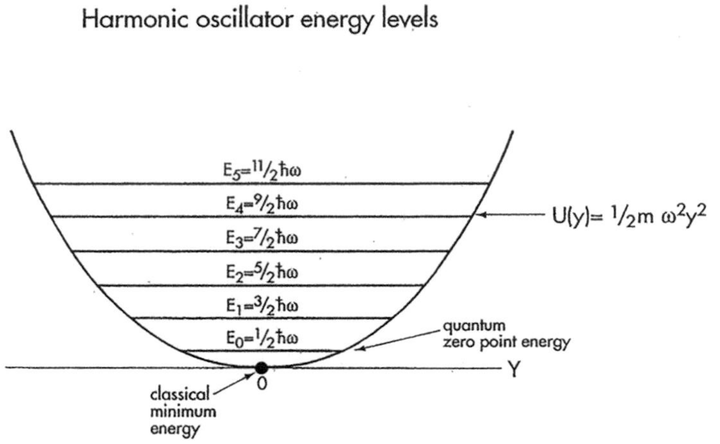

FIGURE 7.2 The equally spaced energy levels of a one-dimensional harmonic oscillator of mass and circular frequency ω.

Finally, the wave functions are real and of the form

$$\Psi_v(y) = N_v \exp\left[-1/2\left(\frac{m\omega}{\hbar}\right)y^2\right]h_v(y) \tag{7.12}$$

and thus are products of three factors: (a) a normalization factor N_v, (b) a Gaussian function of $y, \exp\left[-\frac{1}{2}\left(\frac{m\omega}{\hbar}\right)y^2\right]$, and (c) a polynomial function $h_v(y)$ of y. These three factors govern different aspects of the behavior of the wave functions. The normalization factor

$$N_v = \left(\frac{m\omega}{\pi\hbar}\right)^{1/4}\left(\frac{1}{2^v v!}\right)^{1/2} \tag{7.13}$$

insures that the wave functions are properly normalized to unity. The Gaussian factor determines the large y behavior of $\Psi_v(y)$ and especially enforces the boundary conditions of Equation (7.9). The polynomial factor $h_v(y)$ guarantees in accord with general principles that $\Psi_v(y)$ has v nodes and also insures in agreement with classical mechanics that the vibrational amplitude of the oscillator grows as E_v increases.

The polynomials $h_v(y)$ turn out to be simply related to standard functions, the Hermite functions or Hermite polynomials $H_n(x)$. Specifically,

$$h_v(y) = H_{n=v}\left(\alpha^{1/2}y\right) \tag{7.14}$$

where $\alpha = \frac{m\omega}{\hbar}$. The Hermite functions are polynomials of order $n = 0,1,\ldots$ as may be seen from Table 7.1 where the first nine are listed. Thus, given Table 7.1 and Equations (7.12)–(7.14), we may write down the explicit forms of the wave functions $\Psi_v(y)$. For example, the ground and first two excited state wave functions are, respectively,

$$\Psi_0(y) = \left(\frac{m\omega}{\pi\hbar}\right)^{1/4} \exp\left[-1/2\left(\frac{m\omega}{\hbar}\right)y^2\right] \tag{7.15}$$

TABLE 7.1
The First Nine Hermite Polynomials $H_n(x)$

n	$H_n(x)$
0	1
1	$2x$
2	$4x^2 - 2$
3	$8x^3 - 12x$
4	$16x^4 - 48x^2 + 12$
5	$32x^5 - 160x^3 + 120x$
6	$64x^6 - 480x^4 + 720x^2 - 120$
7	$128x^7 - 1{,}344x^5 + 3{,}360x^3 - 1{,}680x$
8	$256x^8 - 3{,}584x^6 + 13{,}440x^4 - 13{,}440x^2 + 1{,}680$

$$\Psi_1(y) = \left(\frac{m\omega}{\pi\hbar}\right)^{1/4}\left(\frac{2m\omega}{\hbar}\right)^{1/2} y \exp\left[-1/2\left(\frac{m\omega}{\hbar}\right)y^2\right] \tag{7.16}$$

and

$$\Psi_2(y) = \left(\frac{m\omega}{\pi\hbar}\right)^{1/4}\left(\frac{1}{2}\right)^{1/2}\left(2\left(\frac{m\omega}{\hbar}\right)y^2 - 1\right)\exp\left[-1/2\left(\frac{m\omega}{\hbar}\right)y^2\right]. \tag{7.17}$$

Let us verify that $\Psi_0(y)$ is a solution of the Schrödinger equation (Equation 7.8). Since the normalization factor $\left(\frac{m\omega}{\pi\hbar}\right)^{1/4}$ of $\Psi_0(y)$ cancels from both sides of Equation (7.8), we may ignore it and thus only need to show that $\hat{H}\left\{\exp\left[-\frac{1}{2}\left(\frac{m\omega}{\hbar}\right)y^2\right]\right\} = \frac{1}{2}\hbar\omega\exp\left[-\frac{1}{2}\frac{m\omega}{\hbar}y^2\right]$ or

$$-\frac{\hbar^2}{2m}\frac{d^2}{dy^2}\left\{\exp\left[-\frac{1}{2}\left(\frac{m\omega}{\hbar}\right)y^2\right]\right\} + \frac{1}{2}m\omega^2 y^2\exp\left(-\frac{1}{2}\frac{m\omega}{\hbar}y^2\right) = \frac{1}{2}\hbar\omega\exp\left(-\frac{1}{2}\frac{m\omega}{\hbar}y^2\right) \tag{7.18}$$

where we have used $E_0 = \frac{1}{2}\hbar\omega$. But

$$-\frac{\hbar^2}{2m}\frac{d^2}{dy^2}\left\{\exp\left[-\frac{1}{2}\left(\frac{m\omega}{\hbar}\right)y^2\right]\right\} = \left(\frac{1}{2}\hbar\omega - \frac{1}{2}m\omega^2 y^2\right)\exp\left[-\frac{1}{2}\left(\frac{m\omega}{\hbar}\right)y^2\right]. \tag{7.19}$$

Inserting the above equation into Equation (7.18) proves the latter equation is an identity, thus establishing that $\Psi_0(y)$ solves Equation (7.8).

Next, we discuss some qualitative aspects of the harmonic oscillator stationary states starting with the *zero point energy*.

7.4 THE ZERO POINT ENERGY

Within classical mechanics, the minimum energy of a one-dimensional harmonic oscillator is zero. This follows from the classical equation (Equation A.22) that shows that the oscillator takes on a minimum energy of zero when its coordinate and momentum have the values $x = x_e$ and $p_x = 0$.

In contrast, within quantum mechanics the minimum energy of an oscillator is the positive ground state energy $E_0 = E_{v=0} = \frac{1}{2}\hbar\omega$. The classical minimum energy $E = 0$ is ruled out by the Heisenberg uncertainty relation since the conditions $x = x_e$ and $p_x = 0$ imply in conflict with this relation that both the oscillator's position and momentum are simultaneously known with certainty. In summary, while in classical mechanics a harmonic oscillator can be at rest at its equilibrium position x_e and thus have a minimum energy $E = 0$, in quantum mechanics the oscillator "jitters" about x_e in order to conform to the uncertainty relation and thus has a minimum energy $E_0 = \frac{1}{2}\hbar\omega > 0$ called the *zero point energy*. (Also see Problem 3.12.)

Next, we examine the ground state harmonic oscillator Born rule probability distribution

$$P_0(y) = \Psi_0(y)\Psi_0(y). \tag{7.20}$$

In the process, we introduce the concept of *classically forbidden behavior.*

7.5 THE GROUND STATE BORN RULE PROBABILITY DISTRIBUTION $P_0(y)$ AND CLASSICALLY FORBIDDEN BEHAVIOR

Equations (7.15) and (7.20) yield that the ground state Born rule probability distribution $P_0(y)$ has the Gaussian form

$$P_0(y) = \left(\frac{m\omega}{\pi\hbar}\right)^{1/2} \exp\left[-\left(\frac{m\omega}{\hbar}\right)y^2\right]. \tag{7.21}$$

$P_0(y)$ is compared with the corresponding classical probability distribution $P_{cl}(y)$ in Figure 7.3 computed assuming the oscillator's classical energy E is the ground state energy $E_0 = \frac{1}{2}\hbar\omega$. Notice that $P_0(y)$ and $P_{cl}(y)$ differ qualitatively making evident the highly non-classical behavior of the harmonic oscillator ground state. The qualitative form of $P_{cl}(y)$ may be inferred from the classical energy conservation arguments of Section A.3, especially Figure A.3. In brief, the classical energy of the oscillator $E = \frac{p_y^2}{2m} + \frac{1}{2}m\omega^2 y^2$ is purely kinetic at the equilibrium point $y=0$. Thus, the oscillator is moving fastest near the equilibrium point and hence spends the least amount of time in its vicinity. Consequently, $P_{cl}(y)$ has its minimum value at $y=0$. In contrast, $P_{cl}(y)$ has its maximum value at the right- and left-hand classical turning points given by Equation (A.27) as

$$y_{R_0} = x_{R_0} - x_e = \left(\frac{2E_0}{m\omega^2}\right)^{1/2} = \left(\frac{\hbar}{m\omega}\right)^{1/2} \text{ and } y_{L_0} = x_{L_0} - x_e = -\left(\frac{2E_0}{m\omega^2}\right)^{1/2} = -\left(\frac{\hbar}{m\omega}\right)^{1/2} \tag{7.22}$$

since as discussed in Section A.3 the oscillator is motionless at the turning points and thus spends the greatest amount of time in their neighborhoods. Notice from Figure 7.3, the behavior of $P_0(y)$ is nearly the opposite of that of $P_{cl}(y)$. Namely, $P_0(y)$ has its maximum value, not its minimum value, at $y=0$ and instead of approaching infinity at the turning points it decays smoothly at zero as $|y| \to \infty$. Classical mechanics predicts that the oscillator cannot have stretches in which $y > y_{R_0}$ or compressions in which $y < y_{L_0}$. This is because from Figure A.4 during such stretches and compressions the oscillator's potential energy $U(y) = 1/2\, m\omega^2 y^2$ would exceed its total energy E_0 and hence by energy conservation its kinetic energy would become negative and its velocity would take on imaginary values. In contrast, from Figure 7.3 it is evident that quantum mechanics permits oscillator stretches for which $y > y_{R_0}$ and oscillator compressions for which $y < y_{L_0}$. Such stretches and compressions are referred to as classically forbidden behavior. The total probability P_0^{CF} that the oscillator is in the classically forbidden region is the area of the shaded regions in Figure 7.3; namely,

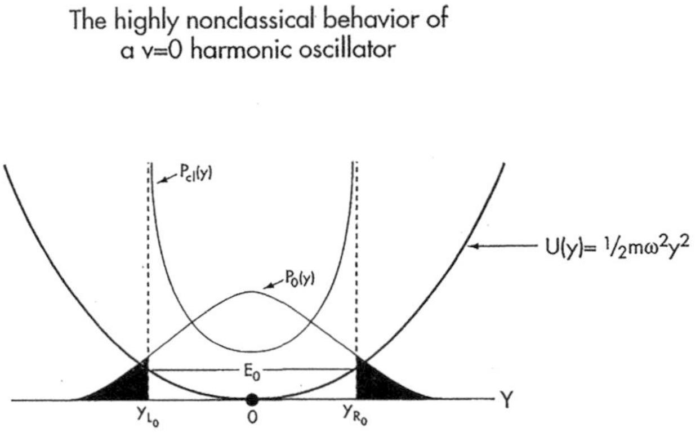

FIGURE 7.3 Ground state quantum $P_0(y)$ and classical $P_{cl}(y)$ position probability distributions for the harmonic oscillator. $P_0(y)$ is as in Equation (7.21), while $P_{cl}(y)$ is computed for energy $E = E_0 = 1/2\hbar\omega$ (see text). The classical turning points y_{R_0} and y_{L_0} are given in Equation (7.22), and the shaded area is the classically forbidden probability $P_0^{CF} = 0.1573$.

$$P_0^{CF} = \int_{-\infty}^{y_{L0}} P_0(y)dy + \int_{y_{R0}}^{\infty} P_0(y)dy. \tag{7.23}$$

Using Equations (7.21) and (7.22), Equation (7.23) may be recast as (Problem 7.5)

$$P_0^{CF} = 1 - \text{erf}(1.00) \tag{7.24}$$

where erf (z) is a well-studied function, the error function, defined by

$$\text{erf}(z) = \frac{2}{\pi^{1/2}} \int_0^z \exp(-t^2)dt. \tag{7.25}$$

However, from standard tabulations (Abramovitz and Stegun 1972, Table 7.1) $\text{erf}(1.0) = 0.8427$. Thus, P_0^{CF} of Equation (7.24) has the value

$$P_0^{CF} = 0.1573. \tag{7.26}$$

Therefore, the oscillator when it is in its ground state has a nearly 16% probability of being found in the region $y > y_{R_0}$ and $y < y_{L_0}$ where it is forbidden classically.

Next, we discuss the wave functions and Born rule probability distributions for the $v = 0$ ground state and the $v = 1 \text{ and } 2$ excited states.

7.6 THE WAVE FUNCTIONS AND BORN RULE PROBABILITY DISTRIBUTIONS FOR THE $v = 0 - 2$ STATES

The wave functions and Born rule probability distributions for these states are plotted in Figure 7.4a–c. Notice from these figures that the ground state wave function has zero nodes, the first excited state wave function has one node, and the second excited state wave function has two nodes. In general, the vth state has v nodes. This follows since the Hermite polynomial $H_n(x)$ has n real zeroes and so from Equations (7.12) and (7.14) $\Psi_v(y)$ has v nodes corresponding to these zeroes, in accord with general principles.

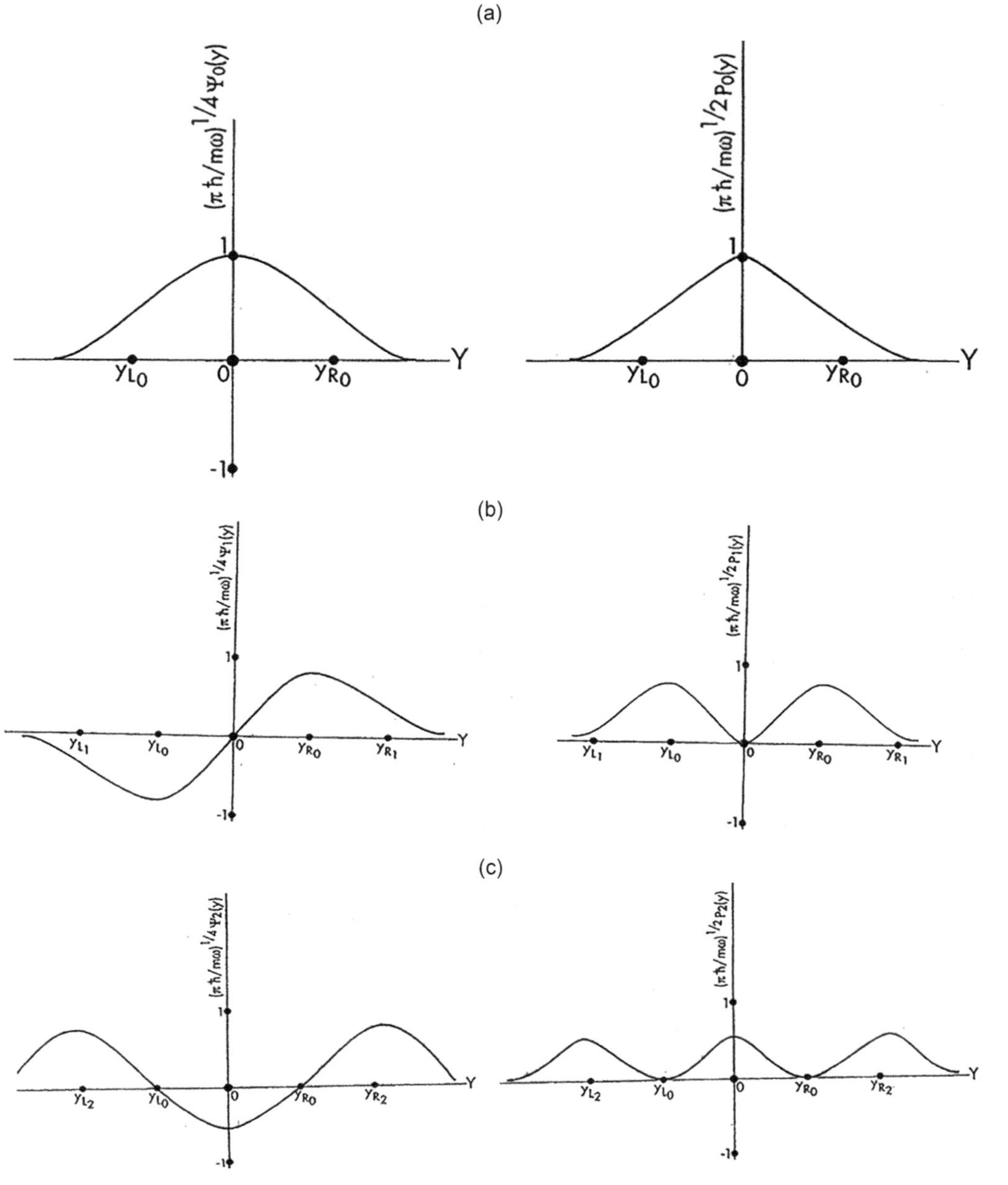

FIGURE 7.4 Ground and low-lying excited state wave functions (left-side plots) and position probability distributions (right-side plots) for the harmonic oscillator. Wave functions $\Psi_v(y)$ for $v = 0-2$ are determined from Equations (7.15)–(7.17). The corresponding Born rule position probability distributions $P_v(y) = \Psi_v(y)\Psi_v(y)$. (a) The ground $v = 0$ state. (b) The first excited $v = 1$ state. (c) The second excited $v = 2$ state. Note that y_{R_v} and y_{L_v} for $v = 0-2$ are the classical turning points for the energies $E_v = \left(v + \frac{1}{2}\right)\hbar\omega$.

The excited state probability distributions $P_1(y)$ and $P_2(y)$ depicted in Figure 7.4b and c vanish at the nodal points of the corresponding wave functions. This vanishing from the classical standpoint seems "magical." Consider, for example, the first excited state probability distribution $P_1(y)$ plotted in Figure 7.4b. Notice that $P_1(y)$ is positive for either $y > 0$ or $y < 0$ but vanishes for $y = 0$. That is, the oscillator has a positive probability of being either stretched $(y > 0)$ or compressed

$(y<0)$ but has zero probability of being found at equilibrium $(y=0)$. Thus, the oscillator can apparently pass from a stretched configuration to a compressed configuration without ever passing through its equilibrium configuration! This "quantum magic," which we earlier encountered for the particle-in-a-box system, reflects the fact that trajectories are not a valid quantum concept.

We next examine the Born rule probability distribution $P_{12}(y)$ for the highly excited $v=12$ state. In contrast to the highly non-classical behavior of the $v=0$ state, we will see the emergence of classical behavior for the $v=12$ state.

7.7 THE BORN RULE PROBABILITY DISTRIBUTION $P_{12}(y)$ AND THE EMERGENCE OF CLASSICAL BEHAVIOR

In Figure 7.5, we compare the twelfth excited state Born rule probability distribution

$$P_{12}(y)=\Psi_{12}(y)\Psi_{12}(y) \tag{7.27}$$

with the corresponding classical distribution $P_{cl}(y)$. $P_{cl}(y)$ is found by assuming the oscillator's classical energy is equal to its $v=12$ quantum energy $E_{12}=\left(12+\frac{1}{2}\right)\hbar\omega=\frac{25}{2}\hbar\omega$. Notice from Figure 7.5 that $P_{12}(y)$ is highly oscillatory. This is due to the twelve zeroes of $P_{12}(y)$ arising from the twelve nodes of $\Psi_{12}(y)$. Despite this oscillatory behavior, $P_{12}(y)$ in accord with the Bohr correspondence principle (the Bohr correspondence principle, as discussed in Sections 1.3 and 1.4, originally referred to the approach of spectral line frequencies and intensities to their classical values in the limit of high quantum numbers n. Now the term is used to describe the approach of *all* quantum properties to their classical values in the limit of high n.) has a close connection with the smooth classical distribution $P_{cl}(y)$. This is because the envelope (average value) of $P_{12}(y)$ is very similar to that of $P_{cl}(y)$. Moreover, in analogy to Equation (7.23) for P_0^{CF}, the probability P_{12}^{CF} that a $v=12$ oscillator is in the classically forbidden region is

$$P_{12}^{CF}=\int_{-\infty}^{y_{L12}}P_{12}(y)\,dy+\int_{y_{R12}}^{\infty}P_{12}(y)\,dy \tag{7.28}$$

where $y_{R_{12}}$ and $y_{L_{12}}$ are classical turning points for the $v=12$ state.

$$y_{R_{12}}=\left(\frac{2E_{12}}{m\omega^2}\right)^{1/2}=5\left(\frac{\hbar}{m\omega}\right)^{1/2} \text{ and } y_{L_{12}}=-\left(\frac{2E_{12}}{m\omega^2}\right)^{1/2}=-5\left(\frac{\hbar}{m\omega}\right)^{1/2}. \tag{7.29}$$

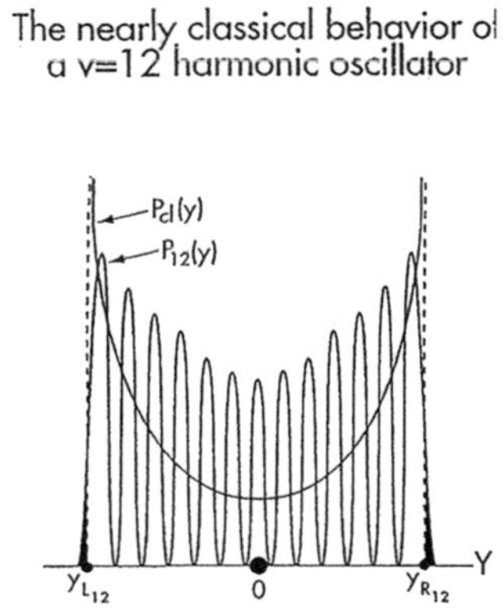

FIGURE 7.5 Quantum $P_{12}(y)$ and classical $P_{cl}(y)$ position probability distributions for the highly excited $v=12$ state of the harmonic oscillator. Note that in accord with the Bohr correspondence principle, the envelope (average value) of $P_{12}(y)$ nearly coincides with $P_{cl}(y)$.

However, comparing Figures 7.3 and 7.5 shows that $P_{12}^{CF} \ll P_0^{CF} \approx 16\%$. This result is again in accord with the Bohr correspondence principle which requires that

$$\lim_{v \to \infty} P_v^{CF} = 0. \tag{7.30}$$

We next turn to quantitative aspects of the stationary state wave functions $\Psi_v(y)$. We first note that some harmonic oscillator integrals vanish due to the symmetry properties of the wave functions.

7.8 EVEN OR ODD PROPERTY OF $\Psi_v(y)$ AND THE VANISHING OF CERTAIN HARMONIC OSCILLATOR INTEGRALS

Even functions $f_E(y)$ and odd functions $f_O(y)$ of a variable y are those that obey, respectively, the conditions

$$f_E(-y) = f_E(y) \text{ and } f_O(-y) = -f_O(y). \tag{7.31}$$

Examples of even functions are y^2, cos y, and y sin y. Examples of odd functions are y^3, $\sin y$ and $\tan y$. We will be interested in the following three integrals over even and odd functions:

$$S_{EO} = \int_{-b}^{b} f_E(y) f_O(y) dy \tag{7.32}$$

$$S_{EE} = \int_{-b}^{b} f_E^2(y) dy \text{ and } S_{OO} = \int_{-b}^{b} f_O^2(y) dy. \tag{7.33}$$

The product of an even and odd function is an odd function, and the integral of an odd function over symmetric limits –b to b vanishes. So as shown in Figure 7.6,

$$S_{EO} = 0. \tag{7.34}$$

Similarly, the products of two even functions or two odd functions are even functions. Therefore,

$$S_{EE} = 2\int_{0}^{b} f_E^2(y) dy \text{ and } S_{OO} = 2\int_{0}^{b} f_O^2(y) dy. \tag{7.35}$$

The vanishing of integrals of the form $\int_{-b}^{b} f_E(y)\, f_O(y) dy$

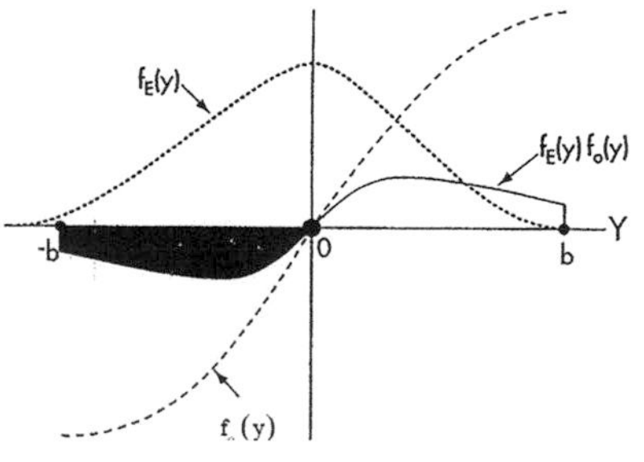

FIGURE 7.6 The integrals $S_{EO} = \int_{-b}^{b} f_E(y) f_O(y) dy$ of Equation (7.32) vanish since they are equal to the total area under the curve $f_E(y) f_O(y)$, which is the sum of two canceling contributions: the shaded negative area to the left of the origin and the unshaded positive area to the right of the origin.

These results may be applied to evaluate harmonic oscillator integrals since the harmonic oscillator wave functions

$$\Psi_v(y) = N_v \exp\left[-\frac{1}{2}\left(\frac{m\omega}{\hbar}\right)y^2\right] H_v\left[\left(\frac{m\omega}{\hbar}\right)^{1/2} y\right] \tag{7.36}$$

are even in y if v is even and odd in y if v is odd. This may be seen as follows. Since the Gaussian factor $\exp\left[-\frac{1}{2}\frac{m\omega}{\hbar}y^2\right]$ is even in y, the evenness or oddness of $\Psi_v(y)$ is that of $H_v\left[\left(\frac{m\omega}{\hbar}\right)^{1/2} y\right] = H_v(x)$. But from Table 7.1 and Problem 7.13, $H_v(x)$ is even in x for v even and is odd in x for v odd, implying the same symmetry for $\Psi_v(y)$.

An example of an integral that vanishes on the basis of symmetry is $\langle y \rangle_v$, the expected value of y in the vth stationary state. From Problem 6.1

$$\langle y \rangle_v = \int_{-\infty}^{\infty} y\Psi_v^2(y)dy \tag{7.37}$$

which vanishes because $\Psi_v^2(y)$ as the square of either an even or odd function is even in y while y is odd in y. Also from Problem 6.1,

$$\langle p_y \rangle = \int_{-\infty}^{\infty} \Psi_v(y)\hat{p}_y\Psi_v(y)\,dy = \frac{\hbar}{i}\int_{-\infty}^{\infty} \Psi_v(y)\frac{d}{dy}\Psi_v(y)\,dy \tag{7.38}$$

vanishes since if $\Psi_v(y)$ is even in y, $\frac{d\Psi_v(y)}{dy}$ is odd in y, and vice versa (Problem 7.6c).

We next turn to the *Gaussian integrals.*

7.9 THE GAUSSIAN INTEGRALS

Because of the polynomial times Gaussian form of the wave functions $\Psi_v(y)$, many harmonic oscillator expected values and matrix elements may be expressed in terms of the Gaussian integrals $G_n(\alpha)$ defined by

$$G_n(\alpha) = \int_{-\infty}^{\infty} y^n \exp\left(-\alpha y^2\right)dy \tag{7.39}$$

where n is a nonnegative integer and α is a positive constant. Since the Gaussian function $\exp\left(-\alpha y^2\right)$ is even in y and the power function y^n is odd in y if n is odd and even in y if n is even, the integral $G_n(\alpha)$ has the form S_{EO} for n odd and S_{EE} for n even. Thus, $G_n(\alpha)$ simplifies to

$$G_n(\alpha) = \begin{cases} \int_{-\infty}^{\infty} y^n \exp(-\alpha y^2)\,dy = 0 & \text{for n odd} \\ 2\int_{0}^{\infty} y^n \exp(-\alpha y^2)\,dy & \text{for n even.} \end{cases} \tag{7.40}$$

From Equation (7.40), we only need to evaluate $G_n(\alpha)$ for n even. This evaluation requires the following result for $G_0(\alpha)$ derived in Problem 7.8:

$$G_0(\alpha) = 2\int_{0}^{\infty} \exp\left(-\alpha y^2\right)\,dy = \left(\frac{\pi}{\alpha}\right)^{1/2}. \tag{7.41}$$

To start, we evaluate $G_2(\alpha)$ as follows. Differentiating the first two members of Equation (7.41) with respect to α gives

$$\frac{dG_0(\alpha)}{d\alpha} = -2\int_0^\infty y^2 \exp(-\alpha y^2)dy = -G_2(\alpha). \tag{7.42}$$

Next, differentiating the first and third members of Equation (7.41) with respect to α gives

$$\frac{dG_0(\alpha)}{d\alpha} = -\frac{1}{2}\frac{\pi^{1/2}}{\alpha^{3/2}}. \tag{7.43}$$

Comparing Equations (7.42) and (7.43) yields $G_2(\alpha)$ as

$$G_2(\alpha) = 2\int_0^\infty y^2 \exp(-\alpha y^2)dy = \frac{1}{2}\frac{\pi^{1/2}}{\alpha^{3/2}}. \tag{7.44}$$

$G_4(\alpha)$ may be analogously evaluated from $G_2(\alpha)$ as

$$G_4(\alpha) = 2\int_0^\infty y^4 \exp(-\alpha y^2)dy = \frac{1}{2}\frac{3}{2}\frac{\pi^{1/2}}{\alpha^{5/2}}. \tag{7.45}$$

Continuing in this manner, we find for $n = 2,4\ldots$ that

$$G_n(\alpha) = 2\int_0^\infty y^n \exp(-\alpha y^2)\, dy = \left(\frac{1}{2}\right)\left(\frac{3}{2}\right)\cdots\frac{(n-1)}{2}\frac{\pi^{1/2}}{\alpha^{(n+1)/2}} \tag{7.46}$$

which may be recast as (Problem 7.7)

$$G_n(\alpha) = \frac{n!}{2^n (n/2)!}\frac{\pi^{1/2}}{\alpha^{(n+1)/2}} \tag{7.47}$$

Comparing Equations (7.40) and (7.47) then yields the following final form for the Gaussian integrals:

$$G_n(\alpha) = \begin{cases} 0 & \text{for n odd} \\ \dfrac{n!}{2^n (n/2)!}\dfrac{\pi^{1/2}}{\alpha^{(n+1)/2}} & \text{for n even} \end{cases} \tag{7.48}$$

We next use the Gaussian integrals to determine $\langle y^2 \rangle$ and $\langle p_y^2 \rangle$ for the harmonic oscillator ground and first excited states. This determination will permit us to prove the harmonic oscillator virial theorem and verify the Heisenberg uncertainty relation for these states.

7.10 VIRIAL THEOREM AND HEISENBERG UNCERTAINTY RELATION FOR THE HARMONIC OSCILLATOR

Let us start with the ground state for which the wave function $\Psi_0(y)$ is given by Equation (7.15). From Problem 6.1 and Equation (7.15) for the ground state,

$$\langle y^2 \rangle_0 = \int_{-\infty}^\infty y^2 \Psi_0(y)\Psi_0(y)dy = \left(\frac{m\omega}{\pi\hbar}\right)^{1/2}\int_{-\infty}^\infty y^2 \exp\left[-\left(\frac{m\omega}{\hbar}\right)y^2\right]dy. \tag{7.49}$$

Thus, from Equation (7.44)

$$\langle y^2\rangle_0 = \left(\frac{m\omega}{\pi\hbar}\right)^{1/2} G_2\left(\frac{m\omega}{\hbar}\right). \tag{7.50}$$

Similarly,

$$\langle p_y{}^2\rangle_0 = \int_{-\infty}^{\infty} \Psi_0(y)\left(\frac{\hbar}{i}\right)^2 \frac{d^2}{dy^2}\Psi_0(y)\,dy \tag{7.51}$$

may be expressed as

$$\langle p_y{}^2\rangle_0 = -\hbar^2\left(\frac{m\omega}{\pi\hbar}\right)^{1/2}\left[-\frac{m\omega}{\hbar}G_0\left(\frac{m\omega}{\hbar}\right)+\left(\frac{m\omega}{\hbar}\right)^2 G_2\left(\frac{m\omega}{\hbar}\right)\right]. \tag{7.52}$$

Equations (7.41) and (7.44) for $G_0(\alpha)$ and $G_2(\alpha)$ then give for $\langle y^2\rangle_0$ and $\langle P_y^2\rangle_0$

$$\langle y^2\rangle_0 = \frac{1}{2}\frac{\hbar}{m\omega} \tag{7.53}$$

and

$$\langle p_y{}^2\rangle_0 = \frac{1}{2}\hbar\omega m. \tag{7.54}$$

An analogous calculation for the first excited state based on Equation (7.16) for $\Psi_1(y)$ gives

$$\langle y^2\rangle_1 = \int_{-\infty}^{\infty} y^2\Psi_1(y)\Psi_1(y)\,dy = \left(\frac{m\omega}{\pi\hbar}\right)^{1/2}\left(\frac{2m\omega}{\hbar}\right)G_4\left(\frac{m\omega}{\hbar}\right)^2 \tag{7.55}$$

and

$$\langle p_y^2\rangle_1 = \int_{-\infty}^{\infty} \Psi_1(y)\left(\frac{\hbar}{i}\right)^2 \frac{d^2}{dy^2}\Psi_1(y)\,dy = -\hbar^2\left(\frac{m\omega}{\pi\hbar}\right)^{1/2}$$
$$\times\left(\frac{2m\omega}{\hbar}\right)\left[-\frac{3m\omega}{\hbar}G_2\left(\frac{m\omega}{\hbar}\right)+\left(\frac{m\omega}{\hbar}\right)^2 G_4\left(\frac{m\omega}{\hbar}\right)\right]. \tag{7.56}$$

Using Equations (7.44) and (7.45) for $G_2(\alpha)$ and $G_4(\alpha)$ gives the following results for $\langle y^2\rangle_1$ and $\langle p_y^2\rangle_1$:

$$\langle y^2\rangle_1 = \frac{3}{2}\frac{\hbar}{m\omega} \tag{7.57}$$

and

$$\langle p_y^2\rangle_1 = \frac{3}{2}m\hbar\omega. \tag{7.58}$$

We next turn to the virial theorem for the harmonic oscillator. Virial theorems are relationships between average kinetic energies $\langle KE\rangle$ and average potential energies $\langle PE\rangle$ which exist for many types of systems. For example, for the hydrogen atom as we will see in Section 10.4, the virial theorem is $\langle PE\rangle = -2\langle KE\rangle$. For the harmonic oscillator, it is

$$\langle PE\rangle = \langle KE\rangle. \tag{7.59}$$

A consequence of the harmonic oscillator virial theorem is that the total energy $E = \langle KE \rangle + \langle PE \rangle = 2\langle KE \rangle = 2\langle PE \rangle$ or

$$\langle PE \rangle = \frac{1}{2}E \tag{7.60}$$

and

$$\langle KE \rangle = \frac{1}{2}E. \tag{7.61}$$

The harmonic oscillator virial theorem is easily verified for the ground state. Using $\langle PE \rangle_0 = \frac{1}{2}m\omega^2\langle y^2 \rangle_0$ and Equation (7.53) $\langle y^2 \rangle_0 = \frac{1}{2}\frac{\hbar}{m\omega}$ gives

$$\langle PE \rangle_0 = \frac{1}{2}\frac{1}{2}\hbar\omega = \frac{1}{2}E_0. \tag{7.62}$$

Similarly, using $\langle KE \rangle_0 = \frac{\langle p_y{}^2 \rangle_0}{2m}$ and Equation (7.54) $\langle p_y{}^2 \rangle_0 = \frac{1}{2}\hbar\omega m$ gives

$$\langle KE \rangle_0 = \frac{1}{2}\frac{1}{2}\hbar\omega = \frac{1}{2}E_0. \tag{7.63}$$

For the first excited state, a similar argument based on the forms for $\langle y^2 \rangle_1$ and $\langle p_y^2 \rangle_1$ of Equations (7.57) and (7.58) gives the viral theorem for the first excited state

$$\langle PE \rangle_1 = \frac{1}{2}\frac{3}{2}\hbar\omega = \frac{1}{2}E_1 \tag{7.64}$$

$$\langle KE \rangle_1 = \frac{1}{2}\frac{3}{2}\hbar\omega = \frac{1}{2}E_1. \tag{7.65}$$

If we conjecture that the virial theorem holds for the general v^{th} excited state, then we may find the values of $\langle y_v^2 \rangle$ and $\langle p_y^2 \rangle_v$. This is done as follows. If the virial theorem holds for the vth state, then from Equations (7.60) and (7.61)

$$\langle PE \rangle_v = \frac{1}{2}E_v = \frac{1}{2}\left(v + \frac{1}{2}\right)\hbar\omega \tag{7.66}$$

and

$$\langle KE \rangle_v = \frac{1}{2}E_v = \frac{1}{2}\left(v + \frac{1}{2}\right)\hbar\omega. \tag{7.67}$$

Using $\langle PE \rangle_v = \frac{1}{2}m\omega^2\langle y^2 \rangle_v$ and $\langle KE \rangle_v = \frac{\langle p_y{}^2 \rangle_v}{2m}$ in Equations (7.66) and (7.67) yields

$$\langle y^2 \rangle_v = \left(v + \frac{1}{2}\right)\frac{\hbar}{m\omega} \tag{7.68}$$

and

$$\langle p_y^2 \rangle_v = \left(v + \frac{1}{2}\right) m\hbar\omega. \tag{7.69}$$

We rigorously prove Equations (7.68) and (7.69) in Section 7.12 and Problem 7.16.

We next verify the Heisenberg uncertainty relation for the harmonic oscillator. From Equation (5.23), the uncertainty relation for the vth harmonic oscillator state is

$$(\Delta y)_v (\Delta p_y)_v \geq \frac{\hbar}{2} \tag{7.70}$$

where

$$(\Delta y)_v = \left[\langle y^2 \rangle_v - \langle y \rangle_v^2\right]^{1/2} \tag{7.71}$$

and

$$(\Delta p_y)_v = \left[\langle p_y^2 \rangle_v - \langle p_y \rangle_v^2\right]^{1/2}. \tag{7.72}$$

However, earlier we have shown by symmetry arguments that $\langle y \rangle_v = \langle p_y \rangle_v = 0$. Hence,

$$(\Delta y)_v = \langle y^2 \rangle_v^{1/2} \tag{7.73}$$

and

$$(\Delta p_y)_v = \langle p_y^2 \rangle_v^{1/2}. \tag{7.74}$$

Using Equations (7.68) and (7.69), $(\Delta y)_v$ and $(\Delta p_y)_v$ thus become

$$(\Delta y)_v = \left(v + \frac{1}{2}\right)^{1/2} \left(\frac{\hbar}{m\omega}\right)^{1/2} \tag{7.75}$$

and

$$(\Delta p_y)_v = \left(v + \frac{1}{2}\right)^{1/2} (m\hbar\omega)^{1/2}. \tag{7.76}$$

Consequently, the uncertainty product (which depends on the oscillator's quantum state in accord with the discussion of Section 5.3) is

$$(\Delta y)_v (\Delta p_y)_v = \left(v + \frac{1}{2}\right)\hbar \geq \frac{\hbar}{2} \tag{7.77}$$

in agreement with the uncertainty relation of Equation (7.70).

Next, we turn to some additional Hermite polynomial relations that will prove useful later.

7.11 ADDITIONAL RELATIONS FOR THE HERMITE POLYNOMIALS FROM THE GENERATING FUNCTION

The forms for the Hermite polynomials $H_n(x)$ listed in Table 7.1 are derived in Appendix C as solutions to Hermite's differential equation (Equation C.11). Here, we use an equivalent but more convenient method of determination of the Hermite polynomials based on their *generating function*

$$g(t,x) = \exp(-t^2 + 2tx). \tag{7.78}$$

The utility of the generating function $g(t,x)$ follows since one may show that (Arfken, Weber, and Harris 2012)

$$g(t,x) = \sum_{n=0}^{\infty} H_n(x)\frac{t^n}{n!}. \tag{7.79}$$

From Equations (7.78) and (7.79), one may derive many useful relations for the Hermite polynomials. For example, $g(t,x)$ may be expanded in a power series in the variable t to give

$$g(t,x) = \sum_{n=0}^{\infty}\left[\frac{\partial^n g(t,x)}{\partial t^n}\right]_{t=0}\frac{t^n}{n!}. \tag{7.80}$$

Comparing Equations (7.79) and (7.80) gives $H_n(x)$ as

$$H_n(x) = \left[\frac{\partial^n g(t,x)}{\partial t^n}\right]_{t=0}. \tag{7.81}$$

Next, rewriting $g(t,x)$ of Equation (7.78) as

$$g(t,x) = \exp(x^2)\exp\left[-(t-x)^2\right] \tag{7.82}$$

it follows that

$$\frac{\partial^n g(t,x)}{\partial t^n} = \exp(x^2)\frac{\partial^n}{\partial t^n}\exp\left[-(t-x)^2\right]. \tag{7.83}$$

However, it is easily verified that

$$\frac{\partial^n}{\partial t^n}\exp\left[-(t-x)^2\right] = (-1)^n\frac{\partial^n}{\partial x^n}\exp\left[-(t-x)^2\right]. \tag{7.84}$$

Comparing the preceding two relations yields

$$\frac{\partial^n g(t,x)}{\partial t^n} = (-1)^n\exp(x^2)\frac{\partial^n}{\partial x^n}\exp\left[-(t-x)^2\right]. \tag{7.85}$$

Evaluating Equation (7.85) at $t = 0$ gives

$$\left[\frac{\partial^n g(t,x)}{\partial t^n}\right]_{t=0} = (-1)^n\exp(x^2)\frac{d^n}{dx^n}\exp(-x^2). \tag{7.86}$$

Finally, comparing Equations (7.81) and (7.86) then gives the following *Rodrigues representation* for $H_n(x)$:

$$H_n(x) = (-1)^n \exp(x^2) \frac{d^n}{dx^n} \exp(-x^2). \tag{7.87}$$

Equation (7.87) provides a convenient way of generating the Hermite polynomials of Table 7.1 (which is much simpler than the power series method of determining these polynomials described in Appendix C). Thus, for n=0 it yields $H_0(x) = (-1)^0 \exp(x^2) \frac{d^0}{dx^0} \exp(-x^2) = \exp(x^2)\exp(-x^2) = 1$. For n=1, it gives $H_1(x) = (-1)^1 \exp(x^2) \frac{d}{dx} \exp(-x^2) = (-1)(-2x)\exp(x^2)\exp(-x^2) = 2x$, and so on.

From the generating function, one may also prove the very useful Hermite polynomial recursion relation

$$H_{n+1}(x) = 2xH_n(x) - 2nH_{n-1}(x). \tag{7.88}$$

Equation (7.88) is proven as follows. Differentiating Equation (7.78) with respect to t gives

$$\frac{\partial g(t,x)}{\partial t} = -2(t-x)\exp(-t^2 + 2tx) = -2(t-x)g(x,t). \tag{7.89}$$

Also differentiating Equation (7.79) with respect to t gives

$$\frac{\partial g(t,x)}{\partial t} = \sum_{n=1}^{\infty} \frac{nH_n(x)t^{n-1}}{n!} = \sum_{n=1}^{\infty} \frac{H_n(x)t^{n-1}}{(n-1)!}. \tag{7.90}$$

Equating the last terms on the right-hand sides of both Equations (7.89) and (7.90) yields

$$-2(t-x)g(t,x) = \sum_{n=1}^{\infty} \frac{H_n(x)t^{n-1}}{(n-1)!}. \tag{7.91}$$

Next, using Equation (7.79) to evaluate $-2(t-x)g(t,x)$, Equation (7.91) becomes

$$-2\sum_{n=0}^{\infty} H_n(x)\frac{t^{n+1}}{n!} + 2x\sum_{n=0}^{\infty} H_n(x)\frac{t^n}{n!} = \sum_{n=1}^{\infty} \frac{H_n(x)t^{n-1}}{(n-1)!}. \tag{7.92}$$

However, Equation (7.92) may be recast as (Problem 7.11)

$$\sum_{n=0}^{\infty} [2xH_n(x) - 2nH_{n-1}(x) - H_{n+1}(x)]\frac{t^n}{n!} = 0. \tag{7.93}$$

Equation (7.93) can only hold for all t if the coefficients of t^n vanish for all n, that is, if

$$H_{n+1}(x) = 2xH_n(x) - 2nH_{n-1}(x). \tag{7.94}$$

But Equation (7.94) is the recursion relation of Equation (7.88), thus proving this important relation. It is easy to show that the first few Hermite polynomials obey the recursion relation (Problem 7.12).

A second recursion relation for the Hermite polynomials which can be derived by differentiating the generating function with respect to x is (Problem 7.14)

$$H_n'(x) = 2nH_{n-1}(x). \tag{7.95}$$

7.12 DERIVATION OF THE FORM OF $\langle y^2 \rangle_v$ FROM THE RECURSION RELATION

In Equations (7.68) and (7.69), we conjectured the forms of $\langle y^2 \rangle_v$ and $\langle p_y^2 \rangle_v$. We may now rigorously derive these forms from the two Hermite polynomial relations. We next show that the form for $\langle y^2 \rangle_v$ may be derived from the recursion relation of Equation (7.88). The form for $\langle p_y^2 \rangle_v$ may be similarly derived from both recursion relations, those of Equations (7.88) and (7.95) (Problem 7.16).

We begin by introducing a new variable

$$\xi = \alpha^{1/2} y \quad \text{where } \alpha = \left(\frac{m\omega}{\hbar}\right). \tag{7.96}$$

From Equations (7.12) and (7.14), normalized harmonic oscillator wave functions may be written in terms of ξ as

$$\Psi_v(y) = N_v \exp\left(-\frac{1}{2}\xi^2\right) H_v(\xi). \tag{7.97}$$

Further, since the harmonic oscillator Hamiltonian operator $\hat{H}$ of Equation (7.7) is Hermitian, the functions $\Psi_v(y)$ form an orthonormal set and thus satisfy the orthonormality relation

$$\int_{-\infty}^{\infty} \Psi_{v'}(y)\Psi_v(y)dy = \delta_{vv'}. \tag{7.98}$$

Next, using Equations (7.96) and (7.97), Equation (7.98) may be written as

$$N_{v'}N_v\alpha^{-1/2}\int_{-\infty}^{\infty} \exp(-\xi^2)H_{v'}(\xi)H_v(\xi)d\xi = \delta_{vv'}. \tag{7.99}$$

Additionally, from Equation (7.96)

$$\langle \xi^2 \rangle_v = \alpha \langle y^2 \rangle_v. \tag{7.100}$$

However,

$$\begin{aligned} \langle \xi^2 \rangle_v &= \int_{-\infty}^{\infty} \xi^2 \Psi_v(y)\Psi_v(y)dy \\ &= N_v^2\alpha^{-1/2}\int_{-\infty}^{\infty} \xi^2 \exp(-\xi^2)H_v(\xi)H_v(\xi)d\xi \end{aligned} \tag{7.101}$$

where we have used Equation (7.97) for $\Psi_v(y)$.

To simplify Equation (7.101), we use the recursion relation of Equation (7.88) which when evaluated for $x = \xi$ and $n = v$ is $H_{v+1}(\xi) = 2\xi H_v(\xi) - 2vH_{v-1}(\xi)$
or

$$\xi H_v(\xi) = vH_{v-1}(\xi) + \frac{1}{2}H_{v+1}(\xi). \tag{7.102}$$

Multiplying the above equation by ξ gives

$$\xi^2 H_v(\xi) = v\xi H_{v-1}(\xi) + \frac{1}{2}\xi H_{v+1}(\xi). \tag{7.103}$$

Evaluating Equation (7.102) for, respectively, $v \rightarrow v-1$ and $v+1$ gives

$$\xi H_{v-1}(\xi) = (v-1)H_{v-2}(\xi) + \frac{1}{2}H_v(\xi) \tag{7.104}$$

and

$$\xi H_{v+1}(\xi) = (v+1)H_v(\xi) + \frac{1}{2}H_{v+2}(\xi). \tag{7.105}$$

Using Equations (7.104) and (7.105) to eliminate $\xi H_{v-1}(\xi)$ and $\xi H_{v+1}(\xi)$ from Equation (7.103) yields

$$\xi^2 H_v(\xi) = v(v-1)H_{v-2}(\xi) + \left(v + \frac{1}{2}\right)H_v(\xi) + \frac{1}{4}H_{v+2}(\xi). \tag{7.106}$$

Employing Equation (7.106) to eliminate $\xi^2 H_v(\xi)$ from Equation (7.101) gives

$$\left\langle \xi^2 \right\rangle_v = N_v^{\,2}\alpha^{-1/2}\int_{-\infty}^{\infty} \exp(-\xi^2)H_v(\xi)\left[v(v-1)H_{v-2}(\xi) + \left(v + \frac{1}{2}\right)H_v(\xi) + \frac{1}{4}H_{v+2}(\xi)\right]d\xi. \tag{7.107}$$

Finally, given the orthonormality relation of Equation (7.99), the above relation simplifies to

$$\left\langle \xi^2 \right\rangle_v = v + \frac{1}{2}. \tag{7.108}$$

However, from Equation (7.100) $\left\langle y^2 \right\rangle_v = \alpha^{-1}\left\langle \xi^2 \right\rangle_v$ and since $\alpha = \frac{m\omega}{\hbar}$

$$\left\langle y^2 \right\rangle_v = \left(v + \frac{1}{2}\right)\frac{\hbar}{m\omega} \tag{7.109}$$

which is our conjecture of Equation (7.68), thus establishing that conjecture.

Next, let us turn to a new topic, specifically to show that the harmonic oscillator approximation provides a zeroth-order model for the IR spectra and other vibrational properties of diatomic molecules.

7.13 THE HARMONIC OSCILLATOR APPROXIMATION FOR THE VIBRATIONAL MOTIONS AND IR SPECTRUM OF A DIATOMIC MOLECULE

Consider a diatomic molecule comprised of atoms with masses m_1 and m_2. We showed within classical mechanics in Section B.6 that the vibrational motions of the molecule in the absence of rotations are rigorously those of a fictitious particle moving in the molecule's potential energy function $U(r)$ (described in detail in Section B.1) where r is the internuclear separation (the distance between the diatomic's nuclei)) with mass equal to the molecule's reduced mass

$$\mu = \frac{m_1 m_2}{m_1 + m_2}. \tag{7.110}$$

As established in Section 9.1, the same simplification holds within quantum mechanics. Thus, we may immediately write down the Schrödinger equation for a non-rotating diatomic's vibrational wave functions $\Psi_v(r)$ and energies E_v

$$-\frac{\hbar^2}{2\mu}\frac{d^2\Psi_v(r)}{dr^2} + U(r)\Psi_v(r) = E_v\Psi_v(r). \tag{7.111}$$

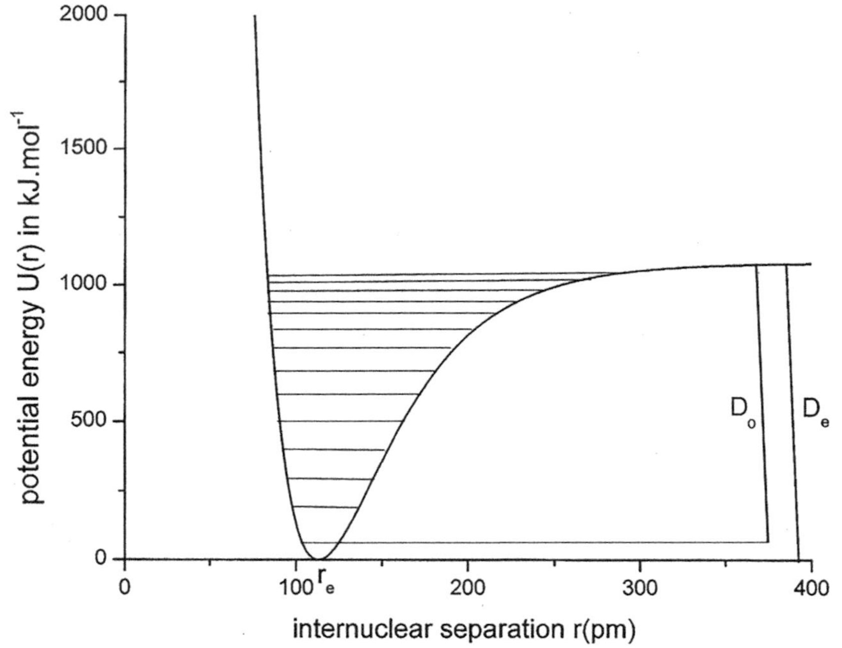

FIGURE 7.7 The potential energy function $U(r)$ and its energy levels for a generic diatomic molecule.

In Figure 7.7, U(r) is plotted for a generic diatomic molecule with several of the diatomic's vibrational energy levels E_v included. Notice that as energy increases, the energy levels become increasingly closely paced, a consequence of the breakdown of the harmonic oscillator approximation to the diatomic's motion (which we will develop shortly) which predicts equally spaced levels. Also in Figure 7.7, two dissociation energies D_e and D_0 are indicated. D_e is the *experimentally inaccessible* energy needed to dissociate the diatomic given that it is initially sitting motionless at the minimum of U(r), an impossibility because of the uncertainty relation which is why D_e is experimentally inaccessible. D_0 is the *experimentally accessible* energy required to dissociate the molecule when it is in its ground state. We will refer to D_e as the *classical dissociation energy* and D_0 as the *quantum dissociation energy*. From Figure 7.7, D_0 and D_e are related by

$$D_0 = D_e - E_0 \tag{7.112}$$

where E_0 is the ground state energy of the molecule.

We next make the harmonic oscillator approximation for $U(r)$. To start, we define the vibrational displacement $y = r - r_e$ where r_e is the molecule's equilibrium internuclear separation (the point at which $U[r]$ has its minimum) or bond length and write $U(r) = U(y + r_e)$. Assuming for characteristic vibrational displacements y that $y \ll r_e$, we expand $U(y + r_e)$ in a power series in y and keep terms only up to quadratic order to obtain

$$U(y + r_e) \doteq U_{HO}(y) \equiv U(r_e) + U'(r_e)y + \frac{1}{2}U''(r_e)y^2. \tag{7.113}$$

From Figure 7.7, we have chosen the zero of energy so that $U(r_e) = 0$. Also since r_e corresponds to the minimum of $U(r)$, $U'(r_e) = 0$. Thus, Equation (7.113) simplifies to

$$U_{HO}(y) = \frac{1}{2}U''(r_e)y^2. \tag{7.114}$$

The harmonic oscillator restoring force $F_y = -\frac{dU_{H0}(y)}{dy} = -U''(r_e)y$ is of the Hooke's law form $F_y(y) = -ky$ with force constant

$$k = U''(r_e). \tag{7.115}$$

Thus, $U_{H0}(y)$ may be written as

$$U_{H0}(y) = \frac{1}{2}ky^2. \tag{7.116}$$

We next define the circular frequency ω by

$$\omega = \left(\frac{k}{\mu}\right)^{1/2} = \left[\frac{U''(r_e)}{\mu}\right]^{1/2}. \tag{7.117}$$

We then have from Equation (7.116) that

$$U_{H0}(y) = \frac{1}{2}\mu\omega^2 y^2. \tag{7.118}$$

Since $U_{H0}(y)$ is identical to the harmonic oscillator potential of Equation (7.5) except for the replacement $m \to \mu$, we may immediately write down the harmonic oscillator approximation to the diatomic's vibrational energy levels as

$$E_v = \left(v + \frac{1}{2}\right)\hbar\omega = \left(v + \frac{1}{2}\right)h\upsilon, \tag{7.119}$$

where υ is the diatomic's classical vibrational frequency.

In Figure 7.8, we plot $U(r)$ of Figure 7.7 and its harmonic oscillator fit $U_{H0}(y)$. The two potentials differ qualitatively above the dissociation limit since $U_{H0}(y)$ does not permit dissociation. At lower but still high energies, $U(r)$ and $U_{H0}(y)$ differ appreciably. However, $U_{H0}(y)$ provides an excellent approximation $U(r)$ at the lowest energies. Hence, the first few harmonic oscillator vibrational energy levels provide a good approximation to the true levels obtained by solving Equation (7.111). As an additional consequence, one may replace with little error in Equation (7.112) the diatomic's true ground state energy E_0 by the harmonic oscillator zero point energy $\frac{1}{2}\hbar\omega$. This gives D_0 as $D_0 = D_e - \frac{1}{2}\hbar\omega$. We will express D_0 and D_e in $kJ\,mol^{-1}$. Then, the zero point energy must be multiplied by Avogadro's number N_A to give

$$D_0 = D_e - \frac{1}{2}\hbar\omega N_A. \tag{7.120}$$

We next determine within the harmonic oscillator approximation the vibrational absorption spectrum of a diatomic molecule. This spectrum occurs in the IR region of the electromagnetic spectrum (see Table 7.2) and arises from transitions between initial vibrational states with quantum numbers v to final vibrational states with quantum numbers $v' > v$. While transitions between all v and all $v' > v$ are in principle possible, in practice experimental spectra (except at very high temperatures) are dominated by the $v = 0 \to 1$ transition. This is because at room temperature most diatomic molecules are in their $v = 0$ ground vibrational states, as we will see later in this section, and because

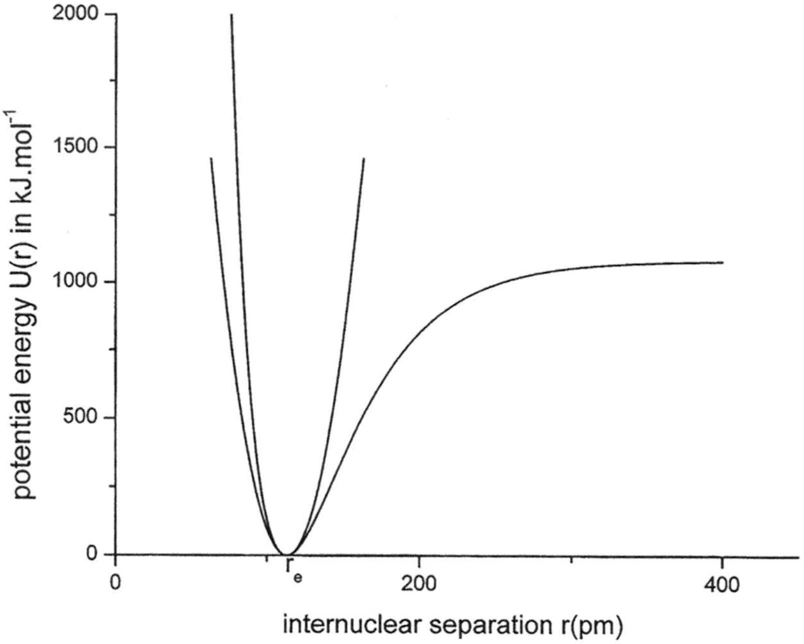

FIGURE 7.8 The potential energy function $U(r)$ of Figure 7.7 and its parabolic harmonic oscillator fit.

TABLE 7.2
Regions of the Electromagnetic Spectrum and Associated Molecular Transitions

Spectral Regions	Wavelength $\lambda(m)$	Frequency $\upsilon(Hz = s^{-1})$	Wave Number $\bar{\upsilon}(cm^{-1})$	Molecular Transition
Microwave	$1.0 - 10^{-3}$	$3.0\times10^{8} - 3.0\times10^{11}$	$0.01 - 10.0$	Rotational
Far infrared	$10^{-3} - 1.5\times10^{-5}$	$3.0\times10^{11} - 2.0\times10^{13}$	$10.0 - 667$	Rotational and vibrational
Near infrared	$1.5\times10^{-5} - 0.8\times10^{-6}$	$2.0\times10^{13} - 3.75\times10^{14}$	$667 - 12,500$	Vibrational
Visible and ultraviolet	$0.8\times10^{-6} - 10^{-7}$	$3.75\times10^{14} - 3.0\times10^{15}$	$12,500 - 100,000$	Electronic

there is a *selection rule* for strictly harmonic molecules, to be derived in Section 7.14, which only allows transitions for which $\Delta v = v' - v = 1$.

We will denote the frequency of electromagnetic radiation absorbed by υ_{IR}. Assuming that the spectrum is due to the $v = 0 \rightarrow 1$ transition, υ_{IR} is given by the Bohr frequency condition as

$$\upsilon_{IR} = h^{-1}(E_1 - E_0). \tag{7.121}$$

Using the harmonic oscillator equation (Equation 7.119) to approximate E_0 and E_1, one has for υ_{IR} that

$$\upsilon_{IR} = \upsilon. \tag{7.122}$$

To proceed further, we define the wave number of the light by

$$\bar{\upsilon}_{IR} = \frac{\upsilon_{IR}}{c} \tag{7.123}$$

TABLE 7.3
Vibrational Wave Numbers $\bar{\omega}$, Frequencies υ, and Force Constants k, determined experimentally from IR line positions, for Selected Diatomic Molecules

Molecule	$\bar{\omega}(cm^{-1})$	$\upsilon(ps^{-1})$	$k(Nm^{-1})$
$^{1}H_2$	4,158.5	124.7	513
$^{1}H^{35}Cl$	2,885.3	86.5	480
$^{12}C^{16}O$	2,143.2	64.3	1,856
$^{16}O_2$	1,556.2	46.7	1,142
$^{35}Cl_2$	555.4	16.7	319
$^{23}Na_2$	157.7	4.7	34

and the vibrational wave number of the molecule by

$$\bar{\omega} = \frac{\upsilon}{c} = \frac{\omega}{2\pi c}. \tag{7.124}$$

where c is the speed of light. Comparing Equations (7.122)–(7.124) yields

$$\bar{\upsilon}_{IR} = \bar{\omega}. \tag{7.125}$$

Equation (7.125) determines a basic molecular parameter $\bar{\omega}(\text{or}\,\upsilon)$ in terms of the position of the IR absorption line $\bar{\upsilon}_{IR}$. Thus, it is a fundamental equation of IR spectroscopy. In Table 7.3, we list for several diatomic molecules vibrational wave numbers $\bar{\omega}$ determined from experimental line positions $\bar{\upsilon}_{IR}$ and also give derived quantities, the vibrational frequencies υ and the force constants k.

The force constants are expressed in terms of ω by Equation (7.117) written as

$$k = \mu\omega^2. \tag{7.126}$$

Since from Equation (7.124) $\omega = 2\pi c\bar{\omega}$, k may be rewritten in terms of $\bar{\omega}$ as

$$k = 4\pi^2\mu c^2\bar{\omega}^2. \tag{7.127}$$

We may now determine k, which is of considerable interest since it provides valuable information about U(r) near its minimum and also determines the "stiffness" of the diatomic's chemical bond, from the position of the IR absorption line $\bar{\upsilon}_{IR}$. Let us consider as an example the $^{12}C^{16}O$ molecule. Its reduced mass is (Note that u is an abbreviation for atomic mass unit) $\mu = \frac{m_{^{12}C} m_{^{16}O}}{m_{^{12}C} + m_{^{16}O}} u \times 1.66054\times10^{-27}\,\text{kg}\,u^{-1} = \frac{(12)(15.9945)}{12+15.9945} \times 1.66054\times10^{-27}\,\text{kg} = 1.1385\times10^{-26}\,\text{kg}$. For $^{12}C^{16}O$, the IR spectrum consists of an intense line at $\bar{\upsilon}_{IR} = \bar{\omega} = 2{,}143.2\,\text{cm}^{-1}$. Thus, from Equation (7.127) the $^{12}C^{16}O$ force constant is $k = (4.0)\pi^2(1.1385\times10^{-26}\,\text{kg})(2.998\times10^{8}\,\text{m}\,\text{s}^{-1})^2 \times (2.1432\times10^{3}\,\text{cm}^{-1})^2\left(10^2\frac{\text{m}^{-1}}{\text{cm}^{-1}}\right)^2 = 1{,}856\,\text{kg}\,\text{s}^{-2} = 1{,}856\,\text{N}\,\text{m}^{-1}$.

We next turn to isotope effects on IR spectra. The basic principle on which the treatment of isotope effects rests derived in Section 13.1 is that

the potential energy function

$U(r)$ is independent of isotope. (7.128)

Consequences are that r_e, D_e, and $k = U''(r_e)$ are independent of isotope. However, from Equation (7.117), ω and hence $\bar{\omega}$ depend on isotope through their dependence on μ. From Equation (7.125), the isotope dependence of $\bar{\omega}$ reflects itself in a corresponding isotope dependence of the position $\bar{\upsilon}_{IR}$ of the IR absorption line. Next from Equation (7.127), the relation between the $\bar{\omega}$ values of arbitrary isotopes 1 and 2 is $\frac{\bar{\omega}_2^2}{\bar{\omega}_1^2} = \frac{\frac{k}{4\pi^2\mu_2 c^2}}{\frac{k}{4\pi^2\mu_1 c^2}} = \frac{\mu_1}{\mu_2}$

or

$$\bar{\omega}_2 = \bar{\omega}_1\left(\frac{\mu_1}{\mu_2}\right)^{1/2}. \tag{7.129}$$

Let us apply Equation (7.129) to obtain the value of $\bar{\omega}$ for ${}^{16}O^2H$ from that for ${}^{16}O^1H$. To do this, we must compute the reduced masses $\mu_{^{16}O^1H} = \frac{m_{^{16}O} m_{^1H}}{m_{^{16}O} + m_{^1H}} = \frac{(15.9945)(1.00794)}{(15.9945+1.00794)} = 0.948u$ and $\mu_{^{16}O^2H} = \frac{m_{^{16}O} m_{^2H}}{m_{^{16}O} + m_{^2H}} = \frac{(15.9945)(2.014)}{(15.9945+2.014)} = 1.789u$. Then, using Equation (7.129) and the value of $\bar{\omega}$ for ${}^{16}O^1H$ of 3568.0 cm^{-1} gives $\bar{\omega}_{^{16}O^2H} = (3{,}568.0\,cm^{-1})\left(\frac{0.948}{1.789}\right)^{1/2} = 2{,}597.3\,cm^{-1}$, a substantial isotope shift from $3{,}568.0\,cm^{-1}$ to $2.507.3\,cm^{-1}$.

Next, from the following rewritten form of Equation (7.120)

$$D_0 = D_e - \frac{1}{2}hc\bar{\omega}N_A, \tag{7.130}$$

one may determine the experimentally inaccessible D_e, a very important quantity since it provides valuable information about the potential U(r), from the experimentally accessible D_0.

Let us compute D_e for HCl from Equation (7.130). Note from Equation (7.130) that D_0 since it depends on $\bar{\omega}$ is isotope dependent. For our determination of D_e, we will thus use the experimental values $D_0 = 427.8\,kJ\,mol^{-1}$ and $\bar{\omega} = 2{,}885.3\,cm^{-1}$ for the ${}^1H^{35}Cl$ isotope. Using Equation (7.130) written as $D_e = D_0 + \frac{1}{2}hc\bar{\omega}N_A$, we have for the isotope-independent D_e that $D_e = 427.8\,kJ\,mol^{-1} + (0.5)(6.626\times10^{-34}\,J\,s)(2.998\times10^8\,m\,s^{-1})(2{,}885.3\,cm^{-1})(6.022\times10^{23}\,mol^{-1})\times \left(\frac{100\,m^{-1}}{cm^{-1}}\right)\left(\frac{10^{-3}\,kJ}{J}\right) = 427.8 + 17.3\,kJ\,mol^{-1}$ or $D_e = 445.1$ kJ mol^{-1} . In Table 7.4 we list D_e and D_0 for several diatomic molecules with D_e computed, as just described, from experimental D_0 values and the $\bar{\omega}$ values in Table 7.3. For all molecules, the zero point energy $\frac{1}{2}hc\bar{\omega}N_A$ is small compared to D_0 and so the differences between the D_0 and D_e values are small.

Next, while D_e is independent of isotope, D_0 depends on isotope. To illustrate how to determine the isotope dependence of D_0, we will evaluate $D_{0,^2H^{35}Cl}$ from $D_{0,^1H^{35}Cl}$. To do this, we use Equation (7.130) to write $D_{0,^2H^{35}Cl} = D_e - \frac{1}{2}hc\bar{\omega}_{^2H^{35}Cl}N_A$ and $D_{0,^1H^{35}Cl} = D_e - \frac{1}{2}hc\bar{\omega}_{^1H^{35}Cl}N_A$. Then, subtracting these two equations gives

$$D_{0,^2H^{35}Cl} = D_{0,^1H^{35}Cl} - \frac{1}{2}hc\left(\bar{\omega}_{^2H^{35}Cl} - \bar{\omega}_{^1H^{35}Cl}\right)N_A. \tag{7.131}$$

TABLE 7.4
Classical D_e and Quantum D_0 Dissociation Energies for Selected Diatomic Molecules

Molecule	$D_e\left(\text{kJ mol}^{-1}\right)$	$D_0\left(\text{kJ mol}^{-1}\right)$
$^1H_2$	456.9	432.1
$^1H^{35}Cl$	445.1	427.8
$^{12}C^{16}O$	1,082.3	1,070.2
$^{16}O_2$	502.9	493.6
$^{35}Cl_2$	242.5	239.2
$^{23}Na_2$	72.1	71.1

From Tables 7.3 and 7.4, we have $\bar{\omega}_{1_{H^{35}Cl}} = 2{,}885.3\,\text{cm}^{-1}$ and $D_{0,^1H^{35}Cl} = 427.8\,\text{kJ.mol}^{-1}$. Thus, to implement Equation (7.131), we only need $\bar{\omega}_{2_{H^{35}Cl}}$, which may be computed from $\bar{\omega}_{1_{H^{35}Cl}}$ using the procedure described earlier. We find $\omega_{2_{H^{35}Cl}} = 2{,}070.1\,\text{cm}^{-1}$. Therefore, $\bar{\omega}_{2_{H^{35}Cl}} - \bar{\omega}_{1_{H^{35}Cl}} = 2{,}070.1 - 2{,}885.3\ \text{cm}^{-1} = -815.2\ \text{cm}^{-1}$. Equation (7.131) therefore yields

$$D_{0,^2H^{35}Cl} = 427.8\,\text{kJ}\,\text{mol}^{-1} + \frac{1}{2}hc\left(815.2\,\text{cm}^{-1}\right)N_A. \tag{7.132}$$

Evaluating the zero point energy term gives $D_{0,^2H^{35}Cl} = 427.8 + 4.88\,\text{kJ}\,\text{mol}^{-1} = 432.7\,\text{kJ.mol}^{-1}$, showing a small shift from $D_{0,^1H^{35}Cl} = 427.8\,\text{kJ}\,\text{mol}^{-1}$.

We next test the validity of the harmonic oscillator approximation for the $^1H^{35}CI$ ground state. This validity requires that characteristic displacements y_{ch} of the oscillator be small compared to r_e. That is, the validity condition is $y_{ch} \ll r_{e^-}$. To test this validity condition, we assume that y_{ch} is the ground state right-hand classical turning point y_{R_0}. This may be determined by letting $m \to \mu$ in Equation (7.22) to yield

$$y_{R_0} = \left(\frac{\hbar}{\mu\omega}\right)^{1/2}. \tag{7.133}$$

Since $\omega = 2\pi c\bar{\omega}$, y_{R_0} may be written as

$$y_{R_0} = \left(\frac{h}{4\pi^2 c\mu\bar{\omega}}\right)^{1/2}. \tag{7.134}$$

Let us compute y_{R_0} for $^1H^{35}Cl$. The reduced mass of $^1H^{35}Cl$ is $\mu = 1.6268 \times 10^{-27}\,\text{kg}$ and $\bar{\omega} = 2{,}885.3\,\text{cm}^{-1}$. Then, from Equation (7.134) for $^1H^{35}Cl$

$$y_{R_0} = \left[\frac{6.626\times10^{-34}\,\text{J}\,\text{s}^{-1}}{\left(4\pi^2\right)\left(2.998\times10^{8}\,\text{m}\,\text{s}^{-1}\right)\left(1.6268\times10^{-27}\,\text{kg}\right)\left(2885.3\,\text{cm}^{-1}\right)\left(\frac{10^2\,\text{m}^{-1}}{\text{cm}^{-1}}\right)}\right]^{1/2} = 11.12\times10^{-12}\,\text{m} = 11.12\,\text{pm}.$$

For HCl, $r_e = 127.46$ pm. Thus, $y_{R_0} \ll r_e$, and hence for our choice $y_{ch} = y_{R_0}$, the validity condition for the harmonic oscillator approximation is well satisfied for the $^1H^{35}Cl$ ground state.

Finally, let us justify our statement that at room temperature, most diatomic molecules are in their ground vibrational states. Let us assume for simplicity that the molecules comprise a gas at Kelvin temperature T. Each molecule has a set of energy levels $E_0, E_1, \ldots$. Because the molecules experience constant collisions, which energy level a molecule will be in at a particular time cannot be known in advance. Instead, only a probability distribution for the energy levels is known. As in Section 1.1, this probability distribution is the Boltzmann distribution that takes the form (compare Equation [1.6])

$$P_T(E_i) = \frac{g_i \exp\left(-\frac{E_i}{kT}\right)}{Q(T)} \tag{7.135}$$

where k is Boltzmann's constant given by

$$k = 1.38066 \times 10^{-23}\,\mathrm{J\,K^{-1}}. \tag{7.136}$$

The factor g_i is the degeneracy of level i and accounts for the fact that $P_T(E_i)$ is proportional to the number of states included in level i. $Q(T)$ is the partition function chosen so that $P_T(E_i)$ is normalized to unity. That is, the partition function is chosen to satisfy the condition $\sum_{\text{all } i} P_T(E_i) = 1$ which from Equation (7.135) gives

$$\sum_{\text{all } i} P_T(E_i) = \frac{\sum_{\text{all } i} g_i \exp(-E_i/kT)}{Q(T)} = 1 \tag{7.137}$$

which in turn yields $Q(T)$ as

$$Q(T) = \sum_{\text{all } i} g_i \exp(-E_i/kT). \tag{7.138}$$

Let us now specialize to the harmonic oscillator with the non-degenerate energy levels $E_v = \left(v + \frac{1}{2}\right)h\upsilon$. For the harmonic oscillator, Equations (7.135) and (7.138) become

$$P_T(E_v) = \frac{\exp\left[-\left(v+\frac{1}{2}\right)\frac{h\upsilon}{kT}\right]}{Q(T)} = \frac{\exp\left[-\left(v+\frac{1}{2}\right)\frac{h\upsilon}{kT}\right]}{\sum_{v=0}^{\infty}\exp\left[-\left(v+\frac{1}{2}\right)\frac{h\upsilon}{kT}\right]}. \tag{7.139}$$

Canceling the factor $\exp\left(-\frac{1}{2}\frac{h\upsilon}{kT}\right)$ from the numerator and denominator in Equation (7.139) becomes

$$P_T(E_v) = \frac{\exp\left(-\frac{vh\upsilon}{kT}\right)}{\sum_{v=0}^{\infty}\exp\left(-\frac{vh\upsilon}{kT}\right)}. \tag{7.140}$$

To evaluate the sum in Equation (7.140) as in the evaluation of the blackbody harmonic oscillator partition function of Equation (1.16), let

$$x = \exp\left(-\frac{h\upsilon}{kT}\right). \tag{7.141}$$

Then, the sum becomes the geometric series

$$\sum_{v=0}^{\infty} x^{v} = \frac{1}{1-x} \tag{7.142}$$

yielding from Equation (7.141)

$$\sum_{v=0}^{\infty} \exp\left(-\frac{vh\upsilon}{kT}\right) = \frac{1}{1-\exp\left(-\frac{h\upsilon}{kT}\right)}. \tag{7.143}$$

Combining Equations (7.140) and (7.143) gives

$$P_T(E_v) = \exp\left(-\frac{vh\upsilon}{kT}\right)\left[1-\exp\left(-\frac{h\upsilon}{kT}\right)\right]. \tag{7.144}$$

From Equation (7.144), the probability that the molecule is in the v = 0 ground state is

$$P_T(E_0) = \left[1-\exp\left(-\frac{h\upsilon}{kT}\right)\right]. \tag{7.145}$$

For most diatomics, at room temperature $h\upsilon \gg kT$ and so $P(E_0) \doteq 1$. At much higher temperatures, $h\upsilon/kT$ decreases and molecules can have a substantial probability of being in excited states.

Equation (7.145) may be put in a more convenient form by defining the *vibrational temperature* Θ_{VIB} by

$$\Theta_{VIB} = \frac{hc}{k}\bar{\omega}. \tag{7.146}$$

Since $c\bar{\omega} = \upsilon$, expressed in terms of Θ_{VIB}, Equation (7.145) becomes

$$P_T(E_0) = 1-\exp\left(-\frac{\Theta_{VIB}}{T}\right). \tag{7.147}$$

From Equation (7.146), Θ_{VIB} may be immediately evaluated from $\bar{\omega}$ via the relation

$$\Theta_{VIB}(K) = 1.439\,K\,cm\,\bar{\omega}\left(cm^{-1}\right) \tag{7.148}$$

Let us do an example. For the $^{12}C^{16}O$ molecule, $\bar{\omega} = 2{,}143.2\,cm^{-1}$. We then ask: What are $P_{300K}(E_0)$ and $P_{1,000K}(E_0)$? From Equation (7.148), for $^{12}C^{16}O\ \Theta_{VIB} = (1.439\,K\,cm)\times(2{,}143.2\,cm^{-1})$ $= 3{,}084.1\,K$. So $P_{300K}(E_0) = 1-\exp\left(-\frac{3{,}084.1}{300}\right)$, showing that nearly all $^{12}C^{16}O$ molecules are in the ground state at $T = 300\,K$. However, for $^{12}C^{16}O$, $P_{1,000K}(E_0) = 1-\exp\left(\frac{-3084.1}{1{,}000K}\right) = 0.95422$. So at $T = 1{,}000K$, nearly 5% of the molecules are in excited states.

The values for $P_T(E_0)$ at $T = 300K$ and $T = 2{,}000K$ calculated from Equation (7.147) are listed for several diatomic molecules in Table 7.5. At $T = 300K$, for all but the lowest frequency diatomics, the $v = 0$ state is almost exclusively populated. Consequently, at $T = 300K$ the IR absorption is almost entirely due to the $v = 0 \rightarrow 1$ *fundamental band* transition. At $T = 2{,}000K$, excited states have

TABLE 7.5
Boltzmann Probabilities $P_T(E_0)$, Calculated from Equations (7.147) and (7.148) with the $\bar{\omega}$ Values of Table 7.3 for Occupation of the Ground Vibrational State at $T = 300\,K$ and $T = 2{,}000\,K$

Molecule	$\Theta_{VIB}(K)$	$P_{300K}(E_0)$	$P_{2,000K}(E_0)$
$^1H_2$	5,984.1	1.0	0.9498
$^1H^{35}Cl$	4,151.9	1.0	0.8746
$^{12}C^{16}O$	3,084.1	1.0	0.7861
$^{16}O_2$	2,239.4	0.9994	0.6736
$^{35}Cl_2$	799.2	0.9303	0.3294
$^{23}Na_2$	266.9	0.5306	0.1073

a non-negligible probability of being populated. Consequently, at $T = 2{,}000\,K$ the $v = 1 \rightarrow 2$ (or even the $v = 2 \rightarrow 3$) "*hot bands*" can contribute significantly to the IR absorption.

We next turn to the derivation of the harmonic oscillator selection rules for diatomic IR transitions.

7.14 DERIVATION OF THE HARMONIC OSCILLATOR SELECTION RULES FOR IR TRANSITIONS OF DIATOMIC MOLECULES

Consider a diatomic molecule subject to IR radiation whose electric field vector points along the $Z-$axis. The condition that the probability of a transition $v \rightarrow v'$ be non-vanishing is that

$$M_z \equiv \int_{-\infty}^{\infty} \Psi_{v'}(y)\mu_z(r)\Psi_v(y)dr \neq 0 \tag{7.149}$$

where $\mu_z(r)$ is the $Z-$component of the permanent electric dipole moment $\mu(r)$ of the molecule. Note from Equation (7.149) that a molecule must have a permanent dipole moment in order to have an IR spectrum. Thus, heteronuclear but not homonuclear diatomics absorb and emit in the IR. Also the r-dependence of $\mu(r)$ arises since the charge density of the molecule depends on r.

We next expand $\mu_z(r) = \mu_z(r_e + y)$ as a power series in y. Since we are assuming $y \ll r_e$, we terminate this expansion at linear order to obtain $\mu_z(r) \doteq \mu_z(r_e) + \left[\frac{d\mu_z(r)}{dr}\right]_{r=r_e} y$. Inserting this approximation into Equation (7.149) gives M_z as

$$M_z = \mu_z(r_e)\int_{-\infty}^{\infty} \Psi_{v'}(y)\Psi_v(y)dy + \left[\frac{d\mu_z(r)}{dr}\right]_{r=r_e} \int_{-\infty}^{\infty} \Psi_{v'}(y)y\Psi_v(y)dy. \tag{7.150}$$

The first term on the right-hand side of Equation (7.150) vanishes due to the orthogonality of the harmonic oscillator wave functions $\Psi_{v'}(y)$ and $\Psi_v(y)$. Thus, to linear order in y

$$M_z = \left[\frac{d\mu_z(r)}{dr}\right]_{r=r_e} I_{v'v} \tag{7.151}$$

where

$$I_{v'v} = \int_{-\infty}^{\infty} \Psi_{v'}(y)y\Psi_v(y)dy. \tag{7.152}$$

Equation (7.151) shows that (a) in order for a molecule to have an IR spectrum in addition to needing a permanent dipole moment, it must also have a nonzero dipole moment derivative $\left[\frac{d\mu_z(r)}{dr}\right]_{r=r_e}$ and (b) if $I_{v'v} \neq 0$ the transition $v \rightarrow v'$ occurs or is *allowed*, while if $I_{v'v} = 0$ the transition $v \rightarrow v'$ does not occur or is *forbidden*.

Next, let us use Equations (7.96) and (7.97) to write $I_{v'v}$ as

$$I_{v'v} = N_{v'}N_{v'}\alpha^{-1}\int_{-\infty}^{\infty} \exp(-\xi^2)H_{v'}(\xi)\xi H_v(\xi)d\xi. \tag{7.153}$$

Employing the recursion relation of Equation (7.102) to eliminate $\xi H_v(\xi)$ from Equation (7.153) gives

$$I_{v'v} = N_{v'}N_v\alpha^{-1}\left[v\int_{-\infty}^{\infty} \exp(-\xi^2)H_{v'}(\xi)H_{v-1}(\xi)d\xi + \frac{1}{2}\int_{-\infty}^{\infty} \exp(-\xi^2)H_{v'}(\xi)H_{v+1}(\xi)d\xi\right] \tag{7.154}$$

From Equation (7.13) (with $m \rightarrow \mu$) for N_v and the orthonormality integral of Equation (7.99), one may evaluate $I_{v'v}$ as (Problem 7.29)

$$I_{v'v} = \left(\frac{\hbar}{2\mu\omega}\right)^{1/2}\left[(v+1)^{1/2}\delta_{v'v+1} + v^{1/2}\delta_{v'v-1}\right]. \tag{7.155}$$

The content of Equation (7.155) is that an absorption transition is allowed only if $v' = v + 1$ and an emission transition is allowed only if $v' = v - 1$.

Summarizing, for a perfectly harmonic diatomic molecule to have an IR spectrum in addition to conforming to the requirement that the molecule have a permanent dipole moment, it must also obey the following selection rules

$$\left[\frac{d\mu_z(r)}{dr}\right]_{r=r_e} \neq 0 \tag{7.156}$$

and

$$\Delta v = \begin{cases} v' - v = +1 & \text{for absorption} \\ v' - v = -1 & \text{for emission.} \end{cases} \tag{7.157}$$

We next turn to the final topic of this chapter, the vibrations of polyatomic molecules.

7.15 THE VIBRATIONAL PROPERTIES AND IR SPECTRA OF POLYATOMIC MOLECULES

The number of degrees of freedom of a dynamical system is equal to the number of independent coordinates needed to specify its classical trajectory. For a polyatomic molecule with N atoms, 3N coordinates are needed to specify its classical trajectory and thus the molecule has 3N degrees of freedom.

The coordinates are most conveniently chosen as the three center of mass coordinates associated with the translational degrees of freedom, as the two for a linear molecule or three for a non-linear

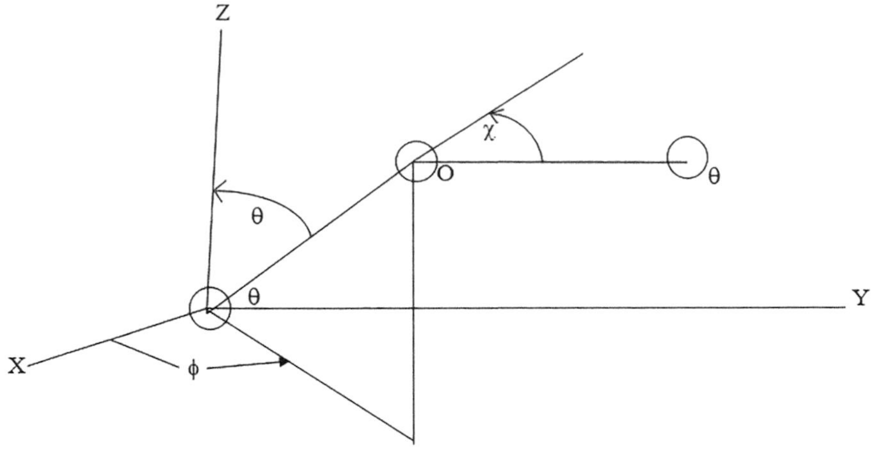

FIGURE 7.9 The angles θ, ϕ, and χ needed to describe the orientation of a non-linear molecule. The angles θ and ϕ specify the orientation of an axis through two of the atoms, and the angle χ describes the orientation of the molecule about that axis.

TABLE 7.6
Number of Translational, Rotational, and Vibrational Degrees of Freedom for Linear and Non-linear Molecules Containing N Atoms

	Translational	Rotational	Vibrational
Linear	3	2	3N − 5
Non-linear	3	3	3N − 6

molecule (Figure 7.9) orientational coordinates associated with the rotational degrees of freedom, and as the internal coordinates associated with the vibrational degrees of freedom.

We may now easily count the number of vibrational degrees of freedom as the total number of degrees of freedom − the number of translational degrees of freedom − the number of rotational degrees of freedom = 3N − 3 − 2 (linear molecule) or 3N − 3 − 3 (non-linear molecule). Summarizing,

$$\text{the number of vibrational degrees of freedom} = 3N - 6(5). \tag{7.158}$$

with six for a non-linear molecule and 5 for a linear molecule. The famous $3N - 6(5)$ rule of Equation (7.158) is highlighted in Table 7.6 and illustrated for several molecules in Table 7.7. As an example, consider methanol CH_3OH. For methanol, N = 6 and the molecule is non-linear, so the number of vibrational degrees of freedom $= 3(6) - 6 = 12$.

We next show that if the vibrational motions of a polyatomic molecule are of sufficiently small amplitude, the $3N - 6(5)$ vibrational (internal) coordinates that make these complex motions may be rigorously transformed into $3N - 6(5)$ new coordinates called *normal mode coordinates* which each execute independent one-dimensional harmonic oscillator motions, an enormous simplification!

We will develop the theory of small amplitude polyatomic vibrations in analogy to the development of the theory of small amplitude diatomic vibrations in Section 7.13. To start, we note that as for a diatomic molecule, the potential energy function U of a polyatomic molecule is independent of the molecule's translational and rotational coordinates and depends only on its $3N - 6(5)$ vibrational coordinates. Denoting the vibrational coordinates by $r_1, \ldots, r_{3N-6(5)}$, we thus have that $U = \left(r_1, \ldots, r_{3N-6[5]}\right)$. Further, let us denote the equilibrium values of the vibrational coordinates by $r_{e1}, \ldots, r_{e3N-6(5)}$.

TABLE 7.7
Number of Translational, Rotational, and Vibrational Degrees of Freedom for Selected Molecules Containing N Atoms

Molecule	N	Linearity	Translational	Rotational	Vibrational
HCl	2	Linear	3	2	1
CO_2	3	Linear	3	2	4
H_2O	3	Non-linear	3	3	3
C_2H_2	4	Linear	3	2	7
NH_3	4	Non-linear	3	3	6
CCl_4	5	Non-linear	3	3	9
C_6H_6	12	Non-linear	3	3	30

Then, in analogy to our definition $y = r - r_e$ of diatomic vibrational displacements from equilibrium, we define $3N-6(5)$ polyatomic vibrational displacements from equilibrium by $y_1 = r_1 - r_{e1}, \ldots, y_{3N-6(5)} = r_{3N-6(5)} - r_{e3N-6(5)}$. Thus, in analogy to the diatomic relation $U(r) = U(y_e + r_e)$, we have the polyatomic relation $U\left(r_1,\ldots,r_{3N-6[5]}\right) = U\left(y_1 + r_{e1},\ldots,y_{3N-6[5]} + r_{e3N-6[5]}\right)$. Finally, in analogy to diatomic quadratic order expansion of the potential energy function $U(y + r_e)$ about r_e

$$U(y + r_e) = \frac{1}{2} k y^2 \tag{7.159}$$

where the force constant

$$k = \frac{d^2U(r_e)}{dr_e^{\,2}} \tag{7.160}$$

we have the polyatomic quadratic order expansion of the potential energy function $U\left(y_1 + r_{e1},\ldots,y_{3N-6[5]} + r_{e3N-6[5]}\right)$ about $r_{e1},\ldots,r_{e3N-6[5]}$

$$U\left(y_1 + r_{e1},\ldots,y_{3N-6[5]} + r_{e3N-6[5]}\right) = \frac{1}{2}\sum_{i=1}^{3N-6(5)}\sum_{j=1}^{3N-6(5)} k_j y_i y_j \tag{7.161}$$

where the force constants

$$k_{ij} = \frac{\partial^2 U\left(r_{e1},\ldots,r_{e3N-6[5]}\right)}{\partial r_{ei}\,\partial r_{ej}}. \tag{7.162}$$

Since in general for $i \neq j$ the force constants k_{ij} are non-vanishing, the potential energy function U of Equation (7.161) involves cross terms (terms for which $i \neq j$).

Because of these cross terms, the classical equations of motion for the polyatomic vibrational displacements y_i are *coupled* or *non-separable*. What this means is that the motion of each displacement y_i depends on the motion of all other displacements y_j. Thus, the coupling renders the determination of polyatomic vibrational motions very complex.

Fortunately, as shown classically in Appendices A and B, for some systems non-separability problems may be effectively dealt with by transforming from the original set of coordinates used to describe the system's motion (in the present case, the vibrational displacements y_i) to a set of generalized coordinates judiciously chosen so that their motions are much simpler than those of the original coordinates. For polyatomic vibrations, the simplifying set of generalized coordinates is the molecule's $3N-6(5)$ normal mode coordinates which we will denote by ξ_k.

The theory of the normal mode transformation (which hinges on the diagonalization of the matrix whose elements are the force constants k_{ij}) is a bit advanced. So here we will omit the development of the theory but will merely summarize its main results. (a) The normal mode coordinates ξ_k are linear superpositions of the vibrational displacements y_i. (b) Expressed in terms of the normal mode coordinates, the polyatomic potential U is free of cross terms and has the form

$$U=\frac{1}{2}\sum_{k=1}^{3N-6(5)}\omega_k^2\xi_k^2 \quad (7.163)$$

where the circular frequencies ω_k^2 of the normal modes are functions of the force constants k_{ij}. Notice that in contrast to ordinary harmonic oscillators with potentials $\frac{1}{2}m\omega^2y^2$, for the normal modes the mass factors $m_k=1$. Also because U of Equation (7.163) is free of cross terms, the normal mode coordinates ξ_k execute independent rather than coupled motions. (c) The force on the kth normal mode coordinate ξ_k is $F_k=-\frac{\partial U}{\partial \xi_k}=-\omega_k^2\xi_k$, so the classical equation of motion for ξ_k has the remarkably simple form

$$\ddot{\xi}_k(t)=-\omega_k^2\xi_k(t) \quad (7.164)$$

which is just the analytically solvable equation of motion for a classical one-dimensional harmonic oscillator (see Section A.1) with coordinate ξ_k, mass $m_k=1$, and circular frequency ω_k.

In summary, within classical mechanics the normal mode transformation reduces the apparently very complex problem of small amplitude polyatomic vibrations to the simple task of analytically solving for the motions of $3N-6(5)$ *independent one-dimensional harmonic oscillators.*

We next show that analogous simplifications occur in quantum mechanics. To start, we note that Equation (7.163) for U implies that when expressed in terms of the normal mode coordinates ξ_k and their conjugate (associated) momenta $p_k=\dot{\xi}_k$, the polyatomic's classical Hamiltonian is

$$H=\sum_{k=1}^{3N-6(5)}\left(\frac{p_k^2}{2}+\frac{1}{2}\omega_k^2\xi_k^2\right). \quad (7.165)$$

Next, using the correspondences of Table 4.1, the above classical Hamiltonian H may be converted into the following quantum Hamiltonian operator:

$$\hat{H}=\sum_{k=1}^{3N-6(5)}-\frac{\hbar}{2}\frac{\partial^2}{\partial\xi_k^2}+\frac{1}{2}\omega_k^2\hat{\xi}_k^2. \quad (7.166)$$

Given Equation (7.166) for $\hat{H}$, the Schrödinger equation for the vibrational wave functions $\Psi_{v1,\ldots,v_{3N-6(5)}}(\xi_1,\cdots,\xi_{3N-6[5]})$ is

$$\sum_{k=1}^{3N-6(5)}\left(-\frac{\hbar^2}{2}\frac{\partial^2}{\partial\xi_k^2}+\frac{1}{2}\omega_k^2\right)\Psi_{v_1,\ldots,v_{3N-6(5)}}(\xi_1,\ldots,\xi_{3N-6[5]})$$
$$=E_{v_1,\ldots,v_{N-6(5)}}\Psi_{v_1,\ldots,v_{N-6(5)}}(E_1,\ldots,\xi_{N-6[5]}). \quad (7.167)$$

However, since $\hat{H}$ of Equation (7.166) is the sum of independent single normal mode Hamiltonian operators $\hat{H}_k = -\frac{\hbar^2}{2}\frac{\partial^2}{\partial \xi_k^2} + \frac{1}{2}\omega_k^2\hat{\xi}_k^2$, the problem of solving the vibrational Schrödinger equation (Equation 7.167) may be greatly simplified. To see how this simplification occurs, first note that in Problem 3.22 by making two successive applications of the separation-of-variables technique, we reduced the three-dimensional particle-in-a-box Schrödinger equation to three independent one-dimensional particle-in-a-box Schrödinger equations. Similarly, by making $3N-6(5)-1$ successive applications of the separation-of-variables technique, we may reduce the $3N-6(5)$–dimensional vibrational Schrödinger equation (Equation 7.167) to $3N-6(5)$ independent single normal mode Schrödinger equations (Problem 7.25). The results of this reduction are that the polyatomic vibrational wave functions simplify to the products

$$\Psi_{v1,\ldots,v_{3N-6(5)}}\left(\xi_1,\ldots,\xi_{3N-6[5]}\right) = \Psi_{v_1}\left(\xi_1\right)\ldots\Psi_{v_{3N-6(5)}}\left(\xi_{3N-6[5]}\right) \tag{7.168}$$

and the polyatomic vibrational energies simplify to the sums

$$E_{v_1,\ldots,v_{3N-6(5)}} = E_{v_1} + \cdots + E_{v3N-6(5)} \tag{7.169}$$

where for $k = 1,\ldots,3N-6(5)\Psi_{v_k}\left(\xi_{v_k}\right)$ and E_{v_k} are the solutions of the one-dimensional harmonic oscillator Schrödinger equation

$$-\frac{\hbar^2}{2}\frac{\partial^2}{\partial \xi_k^2}\Psi_{v_k}\left(\xi_k\right) + \frac{1}{2}\omega_k^2\Psi_{v_k}\left(\xi_k\right) = E_{v_k}\Psi_{v_k}\left(\xi_k\right). \tag{7.170}$$

However, the solutions of the Schrödinger equation (Equation 7.170) are known from the results of Section 7.3. Thus, one may immediately write down the wave functions $\Psi_{v_k}\left(\xi_k\right)$ and the energies E_{v_k}. Especially, the energies when exposed in terms of the normal mode wave numbers $\bar{\omega}_k = \frac{\omega_k}{2\pi c}$ are $E_{vk} = \left(v_k + \frac{1}{2}\right)hc\bar{\omega}_k$. With this form for the energies from Equation (7.169), the polyatomic's total vibrational energy in the small amplitude vibration limit is

$$E_{v_1,\ldots,v_{3N-6(5)}} = \left(v_1 + \frac{1}{2}\right)hc\bar{\omega}_1 + \cdots + \left(v_{3N-6[s]} + \frac{1}{2}\right)hc\bar{\omega}_{3N-6(5)}. \tag{7.171}$$

Let us now specialize to triatomic molecules. By the $3N-6(5)$ rule, linear triatomics have four normal modes while non-linear triatomics have three normal modes. As illustrated in Figure 7.10 for both types of molecules, the normal modes are the *symmetric stretch*, *symmetric bend*, and *asymmetric stretch* modes. We will label these three types of normal modes by the respective subscripts 1, 2, and 3. As an example, consider the four normal modes of the; linear CO_2 isotope $^{12}C^{16}O^{12}C$ (Figure 7.10a). The classical motions of its two stretching mode coordinates ξ_1 and ξ_3 change the CO bond lengths, while the classical motions of its two bending mode coordinates ξ_2 (we imprecisely denote both bending mode coordinates by the same symbol ξ_2) change the OCO bond angle. For the symmetric stretch mode with wave number $\bar{\omega}_1 = 1{,}333\ \text{cm}^{-1}$, the two CO bonds lengthen and shorten synchronously. For the asymmetric stretch mode with wave number $\bar{\omega}_3 = 2{,}349\,\text{cm}^{-1}$, when one CO bond shortens, the other synchronously lengthens. The two bending modes are degenerate each having the wave number $\bar{\omega}_2 = 667\,\text{cm}^{-1}$. The motions of the two bending modes are identical except that one bending mode coordinate moves in the plane of Figure 7.10a while the other moves perpendicular to that plane.

The experimental values of $\bar{\omega}_1$, $\bar{\omega}_2$, and $\bar{\omega}_3$ are listed for several linear and non-linear triatomics in Table 7.8.

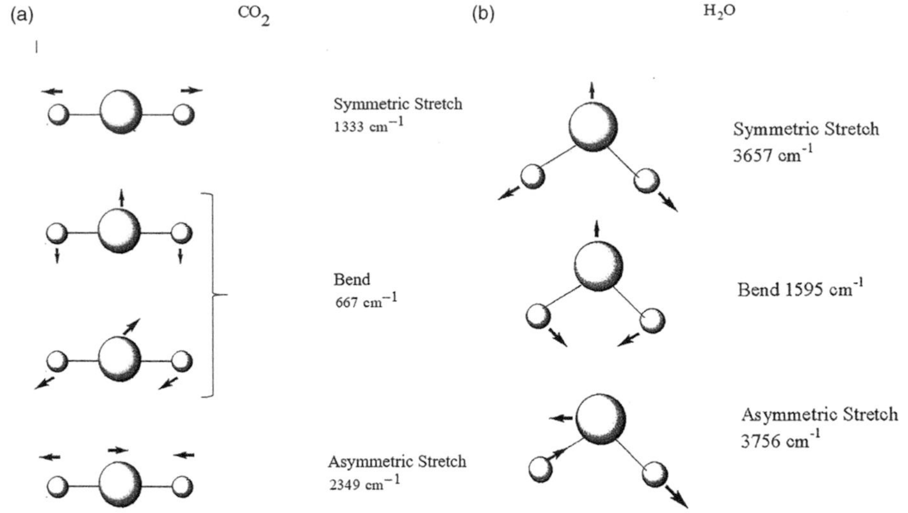

FIGURE 7.10 Normal modes of (a) the linear molecule CO_2 and (b) the non-linear molecule H_2O.

TABLE 7.8
Vibrational Wave Numbers $\bar{\omega}$ for Triatomic Molecules for Symmetric Stretch $(\bar{\omega}_1)$, Symmetric Bend $(\bar{\omega}_2)$, and Asymmetric Stretch $(\bar{\omega}_3)$ Normal Modes

Molecule	Linearity	$\bar{\omega}_1\left(cm^{-1}\right)$	$\bar{\omega}_2\left(cm^{-1}\right)$	$\bar{\omega}_3\left(cm^{-1}\right)$
O−C−O	Linear	1,333	667	2,349
O−C−S	Linear	859	520	2,062
N−O−N	Linear	2,224	589	1,285
H−C−N	Linear	3,311	713	2,089
O−S−O	Non-linear	1,151	518	1,362
H−S−H	Non-linear	2,615	1,183	2,626
D−S−H	Non-linear	1,896	855	1,999
H−O−H	Non-linear	3,657	1,595	3,756
H−O−D	Non-linear	2,727	1,402	3,707
D−O−D	Non-linear	2,671	1,178	2,788

Given our labeling convention for triatomic normal modes, we will indicate the quantum state of a triatomic molecule by grouping together the quantum numbers of each of its normal modes as (v_1, v_2, v_3). (For linear triatomics strictly, we should group four rather than three quantum numbers, but this refinement is often ignored.) For example, with this notation, $(0,0,0)$ indicates the triatomic's ground vibrational state, while for linear triatomics, $(0,1,0)$ indicates the excited state for which the two stretch modes are in their ground states while the two bend modes are in their first excited states.

Next, let us consider triatomic spectroscopic transitions

$$(v_1, v_2, v_3) \rightarrow (v_1', v_2', v_3'). \tag{7.172}$$

From Equation (7.171) specialized to a triatomic molecule, the Bohr frequency condition gives the wave number $\bar{v}_{IR}$ for the transition as

$$\bar{v}_{IR} = (v_1' - v_1)\bar{\omega}_1 - (v_2' - v_2)\bar{\omega}_2 + (v_3' - v_3)\bar{\omega}_3 \tag{7.173}$$

The transition of course only appears in the spectrum if it obeys the selection rules for polyatomic molecules. Within the harmonic oscillator approximation, there are two selection rules. (a) In a transition, only a single normal mode may change its quantum number and that change must be by one unit. (b) A normal mode can only make a transition of the dipole moment of the molecule changes during a vibration of the mode.

Next, let us look at examples of these rules. Suppose a triatomic makes a transition starting from its $(0,0,0)$ ground states. If only the first selection rule held, for all triatomics the harmonic oscillator spectrum would consist of three lines occurring at wave numbers $\bar{\omega}_1, \bar{\omega}_2$, and $\bar{\omega}_3$ arising from the respective transitions $(0,0,0) \rightarrow (1,0,0), (0,0,0) \rightarrow (0,1,0)$, and $(0,0,0) \rightarrow (0,0,1)$. These three lines actually occur for non-linear molecules like H_2O and non-symmetric linear molecules like OCS. However, for symmetric linear molecules like CO_2, the line at $\bar{\omega}_1$ does not occur since the symmetric stretch transition $(0,0,0) \rightarrow (1,0,0)$ is forbidden for such molecules by the second selection rule. (See the schematic IR spectra of Figure 7.11.)

Next, let us turn to the terminology used to describe both the IR spectral lines allowed by the harmonic oscillator selection rules (Figure 7.11) and the weaker lines forbidden by these rules but which appear in the spectrum due to breakdown of the harmonic oscillator approximations.

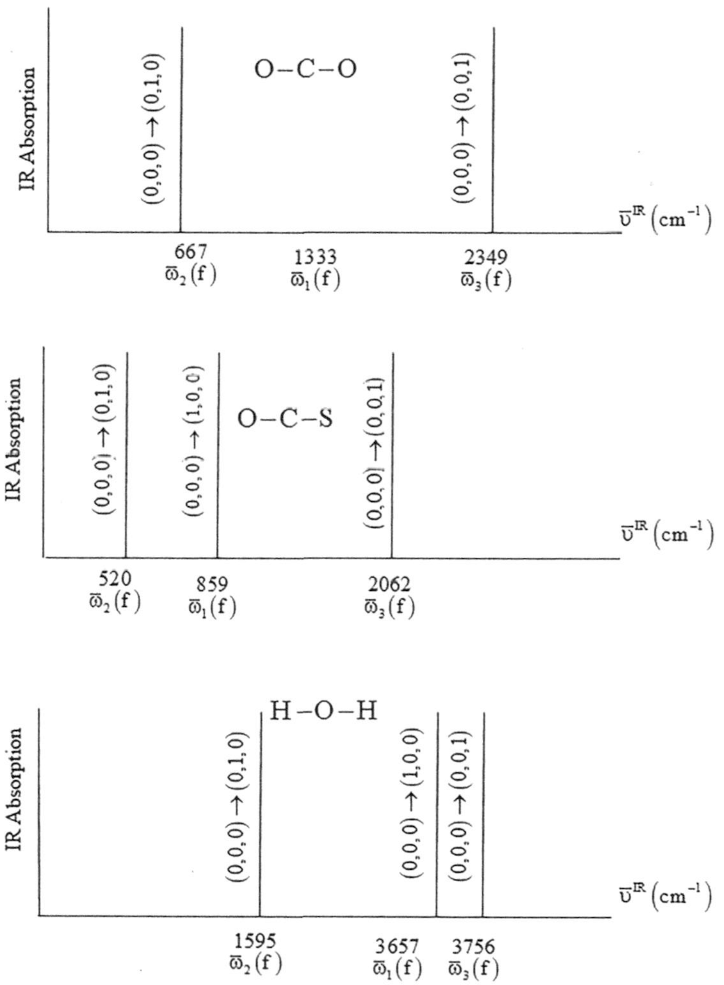

FIGURE 7.11 Schematic fundamental band infrared spectra. Top: CO_2. Middle: OCS. Bottom: H_2O. Since CO_2 is symmetric and linear, it does not have a symmetric stretch fundamental band IR absorption.

The lines at $\bar{\omega}_1, \bar{\omega}_2$, and $\bar{\omega}_3$ which are allowed by the first harmonic oscillator selection rule are called *fundamental bands*. In spectroscopic notation, these three lines are denoted by $\omega_1(f), \omega_2(f)$, and $\omega_3(f)$, with the f's designating fundamental. Analogous notation is used to describe the weaker IR lines present in the spectrum due to anharmonic breakdown of the harmonic oscillator selection rule. For example, lines arising from transitions for which the quantum number of a single normal mode increases by two or more units are called *overtone bands*. As an example of the spectroscopic notion used to describe these lines, the overtone line arising from the transition $(0,0,0) \rightarrow (0,3,0)$ is denoted by $3\omega_2(o)$. Similarly, the lines arising from transitions for which two or more normal modes change their quantum numbers are called *combination bands*. As an example of the spectroscopic notation used to describe combination bands, the line arising from the transition $(1,0,0) \rightarrow (0,0,1)$ is denoted by $\omega_3(c) - \omega_1(c)$. (The minus sign on $\omega_1[c]$ indicates that in the transition, the quantum number v_1 of the symmetric stretch mode decreases by one unit.)

To complete this chapter, we estimate the IR wave number $\bar{\nu}_{IR}$ of the $^{16}O^{12}C^{16}O(1,0,0) \rightarrow (0,0,1)$ combination band. We use the $^{16}O^{12}C^{16}O$ normal mode wave numbers $\bar{\omega}_1 = 1{,}333\,\text{cm}^{-1}$ and $\bar{\omega}_3 = 2{,}349\,\text{cm}^{-1}$. To estimate the $(1,0,0) \rightarrow (0,0,1)$ wave number $\bar{\nu}_{IR}$, we simply set in Equation (7.173) the initial and final symmetric stretch quantum numbers to $v_1 = 1$ and $v_1' = 0$, the initial and final symmetric bend quantum numbers to $v_2 = 0$ and $v_2' = 0$, and the initial and final asymmetric stretch quantum numbers to $v_3 = 0$ and $v_3' = 1$. This gives the estimate $\bar{\nu}_{IR} = \bar{\omega}_3 - \bar{\omega}_1 = 2{,}349 - 1{,}333\,\text{cm}^{-1} = 1{,}016\,\text{cm}^{-1}$. This estimate is slightly higher than the experimental wave number for the $(1,0,0) \rightarrow (0,0,1)$ transition $\bar{\nu}_{IR} = 945\,\text{cm}^{-1}$. The error derives from the fact that the wave numbers $\bar{\omega}_1, \bar{\omega}_2$, and $\bar{\omega}_3$ are obtained from fundamental band measurements.

FURTHER READINGS

Abramovitz, Milton, and Irene A. Stegun, eds. 1972. *Handbook of mathematical functions with formulas, graphs, and mathematical tables*. New York: Dover Publications.

Arfken, George B., Hans J. Weber, and Frank E. Harris. 2012. *Mathematical methods for physicists*. 7th ed. Waltham, MA: Academic Press.

Atkins, Peter, and Ronald Friedman. 2011. *Molecular quantum mechanics*. 5th ed. New York: Oxford University Press.

Banwell, C. N. 1966. *Fundamentals of molecular spectroscopy*. New York: McGraw-Hill.

Bernath, Peter F. 2005. *Spectra of atoms and molecules*. 2nd ed. Oxford: Oxford University Press.

Cohen-Tannoudji, Claude, Bernard Diu, and Franck Laloe. 1977. *Quantum mechanics*. Vol. 1. Translated by Susan Reid Hemley, Nicole Ostrowsky, and Dan Ostrowsky. New York: John Wiley & Sons.

Goswami, Amit. 1997. *Quantum mechanics*. 2nd ed. Dubuque, IA: Wm. C. Brown Publishers.

Herzberg, Gerhard. 1939. *Molecular spectra and molecular structure I. diatomic molecules*. New York: Prentice-Hall, Inc.

Hollas, J. Michael. 2002. *Basic atomic and molecular spectroscopy*. Exeter, Great Britain: Wiley Interscience.

Levine, Ira N. 2013. *Quantum chemistry*. 7th ed. New York: Prentice-Hall.

McQuarrie, Donald A. 2000. *Statistical mechanics*. Mill Valley, CA: University Science Books.

McQuarrie, Donald A. 2008. *Quantum chemistry*. 2nd ed. Sausalito, CA: University Science Books.

Pauling, Linus, and E. Bright Wilson Jr. 1985. *Introduction to quantum mechanics with applications to chemistry*. New York: Dover Publications, Inc.

PROBLEMS

7.1 Show by direct substitution that $\Psi_1(y)$ and $\Psi_2(y)$ solve the Schrödinger equation (Equation 7.8).

7.2 Using Table 7.1 and Equations (7.12)–(7.14), write down the form of $\Psi_3(y)$.

7.3 (a) Show using Gaussian integrals that (a) the harmonic oscillator wave functions $\Psi_v(y)$ for $v = 0 - 2$ are normalized and (b) $\Psi_0(y)$ and $\Psi_2(y)$ are orthogonal.

7.4 (a) Using Table 7.1, find the zeroes of the Hermite polynomials $H_1(x), H_2(x)$, and $H_3(x)$. (b) From these zeroes, find the nodes of the wave functions $\Psi_1(y), \Psi_2(y)$, and $\Psi_3(y)$.

7.5 Prove Equation (7.24) from Equations (7.21)–(7.23) and (7.25).

7.6 (a) Prove from Equation (7.31) that the product of an even and odd function is an odd function. (b) Prove for an odd function $f_0(x) \int_{-b}^{b} f_0(x)dx = 0$ and thus show analytically that S_{EO} of Equation (7.32) vanishes. (c) Show that if $f(x)$ is even, $\frac{df(x)}{dx}$ is odd, and vice versa.

7.7 Derive Equation (7.47) from Equation (7.46).

7.8 In this problem, you will prove Equation (7.41) for $G_0(\alpha)$. Proceed as follows. (a) From Equation (7.39) evaluated for n=0, show that $G_0^2(\alpha) = \int_{-\infty}^{\infty}\int_{-\infty}^{\infty} \exp\left[-\alpha\left(x^2+y^2\right)\right]dxdy$. (b) The integral in part (a) may be regarded as a two-dimensional integral in the XY-plane with x and y being interpreted as the Cartesian coordinates of a point in the plane. Hence, it may be rewritten as an integral over the polar coordinates r and ϕ given in terms of the Cartesian coordinates x and y by Equation (A.36). Using this idea and the results of Problem 8.1, show that $G_0^2(\alpha)$ may be recast as $G_0^2(\alpha) = \int_0^{2\pi} d\phi \int_0^{\infty} r\exp\left(-\alpha r^2\right)dr$. (c) Evaluate the integral in part (b) to yield $G_0^2(\alpha) = \pi/\alpha$, thus proving Equation (7.41).

7.9 This is a challenging problem. A virial theorem for the harmonic oscillator exists in classical mechanics as well as in quantum mechanics. Thus, if for arbitrary functions $f(t)$ we define the average $\langle f\rangle_T$ as $\langle f\rangle_T = T^{-1}\int_0^T f(t)dt$, where $T = \frac{2\pi}{\omega}$ is the period, the virial theorem states $\langle PE\rangle_T = \langle KE\rangle_T$ or $\langle PE\rangle_T = \frac{1}{2}E$ and $\langle KE\rangle_T = \frac{1}{2}E$. To prove this virial theorem, we start with Equation (A.11) for the classical trajectory of the oscillator $y(t) = \cos\omega t\, y + \frac{\sin\omega t}{m\omega}p_y$ and the corresponding equation for the momentum $p_y(t) = m\dot{y}(t) = -m\omega\sin\omega t y + \cos\omega t p_y$. From these equations for $y(t)$ and $p_y(t)$, we prove in Section A.2 the conservation of energy theorem $E = \frac{p_y^2(t)}{2m} + \frac{1}{2}m\omega^2 y^2(t) = \frac{p_y^2}{2m} + \frac{1}{2}m\omega^2 y^2$. From these results, prove the classical virial theorem. (a) First, prove $\langle PE\rangle_T = \frac{1}{2}E$. To start, show that $PE(t) = \frac{1}{2}\times m\omega^2 y^2(t) = \frac{1}{2}m\omega^2\cos^2\omega t y^2 + \frac{1}{2}\sin^2\omega t\frac{p_y^2}{m} - 2m\omega\sin\omega t\cos\omega t y p_x$. (b) Then, using this result for $PE(t)$ and trigonometric identities, prove $\langle PE\rangle_T = \frac{1}{2}E$. (c) Similarly, prove $\langle KE\rangle_T = \frac{1}{2}E$.

7.10 Using the Rodrigues representation of Equation (7.87), derive the forms for the Hermite polynomials $H_n(x)$ for $n = 0-5$ and compare the results with those of Table 7.1.

7.11 Derive Equation (7.93) from Equation (7.92).

7.12 Assuming $H_o(x) = 1$, derive the forms of $H_n(x)$ for $n = 1-5$ from the recursion relation of Equation (7.88) and compare the results with those of Table 7.1.

7.13 The harmonic oscillator wave functions $\Psi_\nu(y)$ have the symmetry property $\Psi_\nu(-y) = (-1)^\nu\Psi_\nu(y)$ if the Hermite polynomials have the symmetry property $H_n(-x) = (-1)^n H_n(x)$. Show that the Hermite polynomials have this symmetry property. (Hint: Show that the generating function $g[t,x]$ of Equation [7.78] obeys the relation $g[-t,-x] = g[t,x]$ and then evaluate Equation [7.79] for $g[-t,-x]$.)

7.14 Prove the recursion relation of Equation (7.95) by differentiating the generating function of Equation (7.78) with respect to *x* to obtain $\frac{\partial g(t,x)}{\partial x} = 2tg(t,x)$ and then use Equation (7.78).

7.15 The recursion relation of Equation (7.95) may be integrated to yield $H_n(x) = H_n(0) + 2n\int_0^x H_{n-1}(z)dz$. Using values of $H_n(0)$ from Table 7.1 and assuming $H_0(x) = 1$, derive from this equation the forms for $H_n(x)$ for $n = 1-5$ and compare your results with those of Table 7.1.

7.16 Derive the form for $\langle p_y^2 \rangle_v$ of Equation (7.69) using the recursion relations of Equations (7.88) and (7.95). Use a method similar to that of Section 7.12.

7.17 In this challenging problem, you will prove the harmonic oscillator orthonormality relation of Equation (7.98) from the Hermite polynomial generating function. The proof is essentially that of Pauling and Wilson and proceeds as follows. (a) Using Equations (7.12) and (7.18), rewrite Equation (7.98) as $(\hbar/m\omega)^{1/2} N_v N_{v'} I_{vv'} = \delta_{vv'}$ where $I_{vv'} \equiv \int_{-\infty}^{\infty} \exp(-x^2) H_v(x) H_{v'}(x)dx$. (b) Using Equations (7.78) and (7.79), show that $\int_{-\infty}^{\infty} \exp(-x^2) g(t,x)\, g(s,x)dx = \sum_{n=0}^{\infty}\sum_{m=0}^{\infty} I_{nm} \frac{t^n}{n!}\frac{s^m}{m!}$ where $I_{nm} \equiv \int_{-\infty}^{\infty} \exp(-x^2) H_n(x) H_m(x)dx$. (c) Show that $\int_{-\infty}^{\infty} \exp(-x^2) g(t,x) g(s,x)dx = \exp(2ts) \int_{-\infty}^{\infty} \exp\left[-(x-t-s)^2\right]dx$ $= \exp(2ts)\int_{-\infty}^{\infty} \exp(-u^2)du = \pi^{1/2}\exp(2ts)$. (d) Show that $\pi^{1/2}\exp(2ts) = \pi^{1/2}\sum_{n=0}^{\infty}\frac{2^n t^n s^n}{n!}$.

(e) Combine the results of parts (b)–(d) to show that $\sum_{n=0}^{\infty}\sum_{m=0}^{\infty} I_{nm}\frac{t^n}{n!}\frac{s^m}{m!} = \pi^{1/2}\sum_{n=0}^{\infty}\frac{2^n t^n s^n}{n!}$.

(f) Equate like powers of *t* and *s* in the result of part (e) to show that $I_{nm} = \pi^{1/2} 2^n n!\delta_{nm}$. (g) Use the result of part (f) to show that $I_{vv'}$ defined in part (a) may be written as $I_{vv'} = \pi^{1/2} 2^v v!\delta_{vv'} = \pi^{1/2}(2^v v!)^{1/2}(2^{v'} v'!)^{1/2}\delta_{vv'}$. (h) Finally, combine Equation (7.13) and the result of part (g) to derive the form of Equation (7.98) obtained in part (a), thus proving Equation (7.98).

7.18 Consider the reduced mass $\mu = \frac{m_1 m_2}{m_1 + m_2}$. (a) Show that if $m_1 = m_2 = m, \mu = \frac{m}{2}$. (b) Show $\mu < m_1$ and $\mu < m_2$. This is why μ is called the reduced mass.

7.19 The molecule $^7Li^1H$ has a strong IR absorption line at $\bar{\upsilon}_{IR} = 1{,}359.3\,cm^{-1}$. Compute the vibrational frequency υ in ps^{-1} and the force constant k in $N\,m^{-1}$ of $^7Li^1H$.

7.20 (a) The vibrational wave number $\bar{\omega}$ of $^{16}O^1H$ is $3{,}568.0\,cm^{-1}$. What is the force constant k of $^{16}O^1H$ in $N\ m^{-1}$? (b) The vibrational wave number $\bar{\omega}$ of $^{16}O^2H$ is $2{,}632.1\,cm^{-1}$. What is the force constant k of $^{16}O^2H$ in $N\,m^{-1}$?

7.21 The vibrational wave number $\bar{\omega}$ of $^1H^{35}Cl$ is 2885.3 cm^{-1}. Compute in cm^{-1} the vibrational wave numbers for (a) $^2H^{35}Cl$ and (b) $^1H^{37}Cl$. (c) Explain why the shift in $\bar{\omega}$ is much larger for $^2H^{35}Cl$ than for $^1H^{37}Cl$.

7.22 For $^1H^{79}$, $Br\,\bar{\omega} = 2{,}558.0\,cm^{-1}$. Compute in $kJ\,mol^{-1}$ the zero point energies of (a) $^1H^{79}Br$ and (b) $^2H^{79}Br$. The quantum dissociation energy D_0 of $^1H^{79}Br$ is $362.6\,kJ\,mol^{-1}$. (c) Compute the classical dissociation energy D_e of HBr in $kJ\,mol^{-1}$. (d) Compute the quantum dissociation energy of $^2H^{79}Br$ in $kJ\,mol^{-1}$.

7.23 The $^{35}Cl_2$ molecule has a vibrational wave number $\bar{\omega} = 555.4\,\text{cm}^{-1}$ and a bond length $r_e = 198.79\,\text{pm}$. Compute y_{R_0} in pm and compare it to r_e. What do you conclude about the validity of the harmonic oscillator approximation for the Cl_2 ground vibrational state?

7.24 Compute the number of normal modes for the non-linear molecules (a) $t-$butyl amine $C_4H_{11}N$, (b) pyridine C_5H_5N, (c) anthracene $C_{14}H_{10}$, and (d) aniline C_6H_7N.

7.25 Using $3N-6(5)$ successive applications of the separation-of-variables technique, show that the solution of the Schrödinger equation (Equation 7.167) is given by Equations (7.168) and (7.169).

7.26 Referring to Table 7.8, draw schematic IR spectra like those of Figure 7.11 for the molecules $N-O-N, H-C-N, O-S-O$, and $H-S-H$.

7.27 Referring to the entry for CO_2 in Table 7.8, assign the transition and estimate the wave number for the $\omega_3 - 2\omega_2(c)$ combination band in the IR spectrum of CO_2. Compare with the experimental wave number of $1,044\,\text{cm}^{-1}$.

7.28 For a triatomic molecule, a normal mode is IR active if the dipole moment of the molecule changes during a vibration. From this statement, show that (a) for a symmetric linear molecule like CO_2, the symmetric stretch mode is IR inactive and the symmetric bend and asymmetric stretch modes are IR active; (b) for a non-symmetric linear molecule like COS, all four normal modes are IR active; and (c) for a non-linear molecule like H_2S, all three normal modes are IR active. (Refer to Figure 7.10.)

7.29 Prove Equation (7.155).

7.30 $^1H^{127}I$ absorbs in the IR at a wave number $\bar{\upsilon}_{IR} = 2,229.7\,\text{cm}^{-1}$. Let $I(0 \rightarrow 1)$ and $I(1 \rightarrow 2)$ be the intensities of the $0 \rightarrow 1$ fundamental band and $1 \rightarrow 2$ hot band transitions. (a) Compute $I(1 \rightarrow 2)/I(0 \rightarrow 1)$ at temperatures (a) 300 K and (b) 2,000 K. (b) Repeat for $^2H^{127}I$ at temperatures (c) 300 K and (d) 2,000 K.

SOLUTIONS TO SELECTED PROBLEMS

7.2 $\Psi_3(y) = \left(\frac{m\omega}{\pi\hbar}\right)^{1/4}\left(\frac{m\omega}{\hbar}\right)^{1/2}\left(\frac{1}{3}\right)^{1/2}\left(\frac{2m\omega}{\hbar}y^3 - 3y\right)\exp\left(-\frac{1}{2}\frac{m\omega}{\hbar}y^2\right).$

7.4 (a) $0, \pm\left(\frac{1}{2}\right)^{1/2}, \pm\left(\frac{3}{2}\right)^{1/2}$ and (b) $0, \pm\alpha^{-1/2}\left(\frac{1}{2}\right)^{1/2}, \pm\alpha^{-1/2}\left(\frac{3}{2}\right)^{1/2}$.

7.19 $40.8\,\text{ps}^{-1}$ and $96\,\text{N m}^{-1}$.

7.20 (a) $711\,\text{N m}^{-1}$ and (b) $711\,\text{N m}^{-1}$.

7.21 (a) $2,069.2\,\text{cm}^{-1}$ and (b) $2,883.5\,\text{cm}^{-1}$.

7.22 (a) $15.2\,\text{kJ mol}^{-1}$, (b) $10.8\,\text{kJ mol}^{-1}$, (c) $377.8\,\text{kJ mol}^{-1}$, and (d) $367.0\,\text{kJ mol}^{-1}$.

7.23 19.8 pm.

7.24 (a) 42, (b) 27, (c) 66, and (d) 36.

7.27 $(0,2,0) \rightarrow (0,0,3), 1,015\,\text{cm}^{-1}$.

7.30 (a) 2.27×10^{-5}, (b) 0.20, (c) 5.03×10^{-4}, and (d) 0.32.

8 Rigid Rotations and Rotational Angular Momentum

In this chapter, we continue the analysis begun in Chapters 3 and 7 of the simplest quantum systems. However, in contrast to the systems studied in these chapters, we deal here with a system, like those dealt with classically in Appendices A and B, whose Cartesian coordinate equations are non-separable and which thus must be treated using generalized coordinates, in this case spherical polar coordinates. Specifically, in this chapter we determine the stationary state wave functions and definite energies of a diatomic molecule in the rigid rotor approximation within which the diatomic's internuclear separation r is fixed at its equilibrium value r_e.

In contrast to real diatomic molecules that both vibrate and rotate, rigid rotor diatomics execute only rotational motions. Despite this lack of realism, a determination of the rigid rotor stationary states is of interest for at least four reasons:

1. The rigid rotor diatomic is perhaps the simplest quantum system that requires a generalized coordinate treatment. Thus, our analysis of the rigid rotor diatomic provides a background for the study of much more complex systems that also require generalized coordinates, for example, the hydrogen-like atoms dealt with in Chapter 10.
2. Our analysis of the rigid rotor system provides an excellent starting point for the determination, given in Chapter 9, of the energy levels and spectra of real diatomic molecules.
3. Our treatment of the rigid rotor diatomic requires a detailed study of rotational angular momentum. It thus provides an introduction to the broad problem of angular momentum in quantum mechanics.
4. The rigid rotor model provides the basis for a first approximation treatment of the pure rotational spectra of diatomic molecules.

It will be easier to develop the theory of rigid rotor diatomics if we first overview the full problem of real rotating and vibrating diatomic molecules.

8.1 OVERVIEW OF THE STATIONARY STATES OF ROTATING AND VIBRATING DIATOMIC MOLECULES

We will consider in this overview the generic diatomic molecule depicted in Figure B.2 with atomic masses m_1 and m_2, atomic Cartesian coordinates $\mathbf{r}_1 = (x_1, y_1, z_1)$ and $\mathbf{r}_2 = (x_2, y_2, z_2)$, and potential energy function $U(r)$.

We begin by briefly summarizing the classical treatment of diatomic motion given in Section B.5. Our motivation is the observation made in Section 4.1 that the wave functions of a quantum system are best taken as functions of the same coordinates which would most naturally describe the trajectories of the system if it behaved classically.

We note three points established in Section B.5:

1. The classical motion of the diatomic molecule cannot be easily analyzed in terms of the Cartesian coordinates $\mathbf{r}_1$ and $\mathbf{r}_2$. However, it is very conveniently analyzed in terms of the diatomic's *relative coordinates* $\mathbf{r} = \mathbf{r}_1 - \mathbf{r}_2$ and its *center-of-mass coordinates* $\mathbf{R} = \dfrac{m_1\mathbf{r}_1 + m_2\mathbf{r}_2}{m_1 + m_2}$. The relative coordinates execute the three-dimensional central field

motion of a fictitious particle of mass $\mu = \frac{m_1 m_2}{m_1 + m_2}$ = reduced mass of the diatomic in the potential $U(r)$. The center of mass coordinates execute the three-dimensional motion of a fictitious free particle (i.e., a particle with no forces acting on it) of mass $M = m_1 + m_2 =$ total mass of the diatomic.
2. Classically, the uninteresting motion of the center of mass coordinates **R** (which are the overall translational motions of the diatomic) may be ignored when treating the motion of the relative coordinates **r**.
3. The rotational–vibrational trajectories of the diatomic molecule are most naturally described in terms of the spherical polar coordinates r, θ, and ϕ of **r** shown in Figure B.8. It is important to note that the spherical polar coordinate r is identical to the internuclear separation r.

Points 1 and 2 extend to quantum mechanics. Point 3 implies that the diatomic's rotational–vibrational stationary state wave functions are best taken as functions of r, θ, and ϕ and are thus written as

$$\Psi_{vJM} = \Psi_{vJM}(r,\theta,\phi). \tag{8.1}$$

The wave functions Ψ_{vJM} since they depend on three coordinates are labeled by three quantum numbers. The quantum number v is associated with the coordinate r, while the quantum numbers J and M are associated, respectively, with the coordinates θ and ϕ. Additionally, as shown in Appendix B, within classical mechanics r may be interpreted as the diatomic's vibrational coordinate and θ and ϕ may be interpreted as its rotational coordinates. The same interpretations hold in quantum mechanics. Consequently, we will refer to v as the diatomic's vibrational quantum number and to J and M as its rotational quantum numbers.

Given this background, we are almost ready to write down the Schrödinger equation for the rotational–vibrational wave functions Ψ_{vJM}. First, however, we must develop the form for the diatomic's rotational–vibrational Hamiltonian operator which is its relative coordinate Hamiltonian operator $\hat{H}$ expressed in terms of the spherical polar coordinates r, θ, and ϕ.

We start with $\hat{H}$ expressed in terms of the Cartesian components x, y, and z of the relative coordinate vector **r**; namely,

$$\hat{H} = -\frac{\hbar^2}{2\mu}\left(\frac{\partial^2}{\partial x^2} + \frac{\partial^2}{\partial y^2} + \frac{\partial^2}{\partial z^2}\right) + U\left[\left(x^2 + y^2 + z^2\right)^{1/2}\right]. \tag{8.2}$$

In the above equation, the operators $-\frac{\hbar^2}{2\mu}\frac{\partial^2}{\partial x^2}, -\frac{\hbar^2}{2\mu}\frac{\partial^2}{\partial y^2}$, and $-\frac{\hbar^2}{2\mu}\frac{\partial^2}{\partial z^2}$ represent, respectively, the kinetic energies arising from the motions of the x-, y- and z-coordinates and $U\left[\left(x^2 + y^2 + z^2\right)^{1/2}\right]$ is the potential energy function $U(r)$ expressed in terms of x, y, and z.

We next introduce the transformation relations between the Cartesian and spherical polar representations of **r**. These are

$$x = r\sin\theta\cos\phi, y = r\sin\theta\sin\phi, \text{ and } z = r\cos\theta. \tag{8.3}$$

These transformation relations permit us to use the chain rule of partial differential calculus (see Problem 8.4) to convert the Cartesian partial derivatives in Equation (8.2) into spherical polar partial derivatives yielding the diatomic's rotational–vibrational Hamiltonian operator. We finally arrive at the diatomic's rotational–vibrational Hamiltonian operator

$$\hat{H} = -\frac{\hbar^2}{2\mu}\frac{1}{r^2}\frac{\partial}{\partial r}\left(r^2\frac{\partial}{\partial r}\right) - \frac{\hbar^2}{2\mu r^2}\frac{1}{\sin\theta}\frac{\partial}{\partial\theta}\left(\sin\theta\frac{\partial}{\partial\theta}\right) - \frac{\hbar^2}{2\mu r^2\sin^2\theta}\frac{\partial^2}{\partial\phi^2} + U(r). \tag{8.4}$$

We next turn to the physical interpretation of $\hat{H}$ of Equation (8.4). To make this interpretation, we note that the terms of $\hat{H}$ are in one-to-one correspondence with the terms of the classical rotational–vibrational Hamiltonian H of Equation (B.36)

$$H = \frac{p_r^2(t)}{2\mu} + \frac{p_\theta^2(t)}{2\mu r^2(t)} + \frac{p_\phi^2(t)}{2\mu r^2(t)\sin^2\theta(t)} + U[r(t)]. \tag{8.5}$$

Comparing $\hat{H}$ and H yields that the terms in $\hat{H}$ are to be interpreted as the operators which, respectively, represent the diatomic's vibrational kinetic energy (the term proportional to p_r^2), the θ and ϕ components of its rotational kinetic energy (the terms proportional to p_θ^2, and p_ϕ^2), and its potential energy.

We may now write down the diatomic's Schrödinger equation for its rotational and vibrational motions

$$\hat{H}\Psi_{vJM}(r,\theta,\phi) = E_{vJ}\Psi_{vJM}(r,\theta,\phi). \tag{8.6}$$

where $\hat{H}$ is given by Equation (8.4). (*Note*: Using foresight, we have anticipated that the energy E is independent of the quantum number M.) Equation (8.6) must be solved subject to boundary conditions at the boundary points of the ranges of the spherical polar coordinates. These ranges are

$$0 \le r < \infty \quad 0 \le \theta \le \pi \quad 0 \le \phi \le 2\pi. \tag{8.7}$$

Thus, the boundary points are as follows: for $r, 0$ and ∞, for θ, 0 and π, and for ϕ, 0 and 2π. The boundary conditions are as follows (Pauling and Wilson 1985):

$$\Psi_{vJM}(0,\theta,\phi) \text{ is finite and } \lim_{r\to\infty} \Psi_{vJM}(r,\theta,\phi) = 0 \tag{8.8a}$$

$$\Psi_{vJM}(r,0,\phi) \text{ is finite and } \Psi_{vJM}(r,\pi,\phi) \text{ is finite} \tag{8.8b}$$

and

$$\Psi_{vJM}(r,\theta,0) = \Psi_{vJM}(r,\theta,2\pi). \tag{8.8c}$$

In addition to obeying boundary conditions, the rotational–vibrational wave functions must conform to the following normalization condition:

$$\int_{-\infty}^{\infty}\int_{-\infty}^{\infty}\int_{-\infty}^{\infty} \Psi_{vJM}^*(r,\theta,\phi)\Psi_{vJM}(r,\theta,\phi)\,dxdydz = 1 \tag{8.9}$$

where the triple integral is over the Cartesian components x, y, and z of **r**. Equation (8.9) however is not yet a satisfactory expression. This is because it mixes two different coordinate representations of r, the Cartesian representation x, y, and z and the spherical polar representation r, θ, and ϕ. Thus, to proceed further, we must convert Equation (8.9) from a mixed expression to a pure spherical polar expression. This may be done using the following spherical polar form for the volume element dxdydz (Problem 8.1):

$$dxdydz = r^2 \sin\theta\, dr d\theta d\phi. \tag{8.10}$$

Comparing Equations (8.9) and (8.10) and noting the ranges of the spherical polar coordinates given in Equation (8.7) yield the following pure spherical polar expression for the normalization condition on the wave functions:

$$\int_0^\infty \int_0^\pi \int_0^{2\pi} r^2 \sin\theta \Psi^*_{vJM}(r,\theta,\phi)\Psi_{vJM}(r,\theta,\phi)\,dr d\theta d\phi = 1. \tag{8.11}$$

Solving Equation (8.6) subject to the boundary and normalization conditions gives a complete set of rotational–vibrational eigenfunctions which obey the orthonormality condition

$$\int_0^\infty \int_0^\pi \int_0^{2\pi} r^2 \sin\theta \Psi^*_{v'J'M'}(r,\theta,\phi)\Psi_{vJM}(r,\theta,\phi)\,dr d\theta d\phi = \delta_{v'v}\delta_{J'J}\delta_{M'M}. \tag{8.12}$$

We next turn to the rigid rotor model for diatomic rotations and its relation to rotational angular momentum.

8.2 THE RIGID ROTOR STATIONARY STATES AND THE ROTATIONAL ANGULAR MOMENTUM EIGENVALUE PROBLEM

In the remainder of this chapter, we will analyze the rigid rotor diatomic depicted in Figure 8.1 and related rotational angular momentum problems. We begin by developing the forms for the rigid rotor Hamiltonian operator $\hat{H}_{RR}$, the rigid rotor Schrödinger equation, and so forth from the corresponding rotating–vibrating diatomic molecule forms. This is straightforward since the rigid rotor approximate equations may be obtained by setting $r = r_e$ and dropping all references to r and partial derivatives with respect to r, as well as dropping the quantum number v in the corresponding rotating-vibrating diatomic molecule equations. Thus, the rigid rotor Hamiltonian operator $\hat{H}_{RR}$ follows from Equation (8.4) as

$$\hat{H}_{RR} = -\frac{\hbar^2}{2\mu r_e^2}\frac{1}{\sin\theta}\frac{\partial}{\partial\theta}\left(\sin\theta\frac{\partial}{\partial\theta}\right) - \frac{\hbar^2}{2\mu r_e^2 \sin^2\theta}\frac{\partial^2}{\partial\phi^2} \tag{8.13}$$

where we have set the zero of energy so that $U(r_e) = 0$. The rigid rotor Schrödinger equation

$$\hat{H}_{RR}\Psi_{JM}(\theta,\phi) = E_J \Psi_{JM}(\theta,\phi) \tag{8.14}$$

similarly follows from Equation (8.6). The rigid rotor boundary, normalization, and orthonormality conditions follow from Equations (8.8b and c), (8.11), and (8.12) as

$$\Psi_{JM}(0,\phi) \text{ is finite and } \Psi_{JM}(\pi,\phi) \text{ is finite} \tag{8.15a}$$

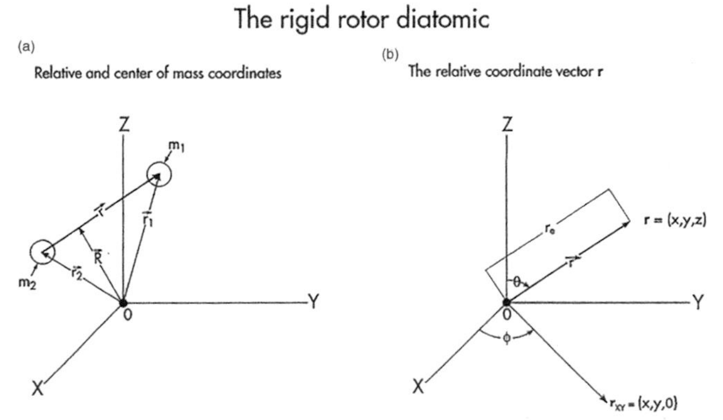

FIGURE 8.1 Generalized coordinates for rigid rotor diatomic motion. (a) As for a rotating–vibrating diatomic (Figure B.7), the motion of a rigid rotor diatomic is best expressed in terms of its relative and center of mass coordinate vectors **r** and **R**. (b) For the rigid rotor diatomic, the spherical polar coordinate r of **r** is fixed at r_e, and thus, the motion of **r** (and, hence, the rotational motion of the diatomic) is exactly expressed in terms of the spherical polar angles θ and ϕ of **r**.

$$\Psi_{JM}(\theta,0)=\Psi_{JM}(\theta,2\pi) \tag{8.15b}$$

$$\int_0^{\pi}\int_0^{2\pi}\sin\theta\Psi^*_{JM}(\theta,\phi)\Psi_{JM}(\theta,\phi)\,d\theta d\phi=1 \tag{8.16}$$

and

$$\int_0^{\pi}\int_0^{2\pi}\sin\theta\Psi^*_{J'M'}(\theta,\phi)\Psi_{JM}(\theta,\phi)\,d\theta d\phi=\delta_{JJ'}\delta_{MM'}. \tag{8.17}$$

We next connect the rigid rotor model to rotational angular momentum. To start, we return to the classical Hamiltonian H for a rotating–vibrating diatomic molecule previously written in the form of Equation (B.36) but now written in the form of Equation (B.37)

$$H=\frac{p_r^2(t)}{2\mu}+\frac{J^2}{2\mu r^2(t)}+U[r(t)] \tag{8.18}$$

where $p_r(t)=\mu\dot{r}(t)$ is the diatomic's classical vibrational momentum and where $J^2=\mathbf{J}\cdot\mathbf{J}$ is the square of the magnitude of the diatomic's classical rotational angular momentum vector **J** defined below in Equation (8.26). (Note that $J=[\mathbf{J}\cdot\mathbf{J}]^{1/2}$must not be confused with the rotational quantum number J.)

We next reduce H to the diatomic's classical rigid rotor Hamiltonian H_{RR}. This is done by setting $r(t)=r_e$ and $p_r(t)=\mu\dot{r}_e=0$ in Equation (8.18) and by recalling that we have set the zero of energy such that $U(r_e)=0$. These substitutions yield H_{RR} as

$$H_{RR}=\frac{J^2}{2\mu r_e^{\,2}}. \tag{8.19a}$$

The corresponding quantum operator is the following new representation of $\hat{H}_{RR}$:

$$\hat{H}_{RR}=\frac{\hat{J}^2}{2\mu r_e^{\,2}} \tag{8.19b}$$

Comparing our earlier representation of $\hat{H}_{RR}$ (Equation 8.13) and our new representation equation (Equation 8.19b) gives $\hat{J}^2$ as

$$\hat{J}^2=-\hbar^2\frac{1}{\sin\theta}\frac{\partial}{\partial\theta}\left(\sin\theta\frac{\partial}{\partial\theta}\right)-\frac{\hbar^2}{\sin^2\theta}\frac{\partial^2}{\partial\phi^2}. \tag{8.20}$$

Next, consider the eigenvalue problem

$$\hat{J}^2Y_{JM}(\theta,\phi)=\lambda_{JM}\hbar^2Y_{JM}(\theta,\phi). \tag{8.21}$$

The eigenfunctions $Y_{JM}(\theta,\phi)$ are standard functions called the *spherical harmonics*. We will develop their forms later in this chapter. We will also show that the eigenvalues $\lambda_{JM}\hbar^2$ are independent of M and take on the quantized values

$$\lambda_{JM}\hbar^2=J(J+1)\hbar^2,\quad \text{where } J=0,1,\ldots. \tag{8.22}$$

With Equation (8.22), the $\hat{J}^2$ eigenvalue problem of Equation (8.21) becomes

$$\hat{J}^2 Y_{JM}(\theta,\phi) = J(J+1)\hbar^2 Y_{JM}(\theta,\phi). \tag{8.23}$$

Since from Equation (8.19b) $\hat{H}_{RR}$ and $\hat{J}^2$ differ only by the constant factor $\frac{1}{2\mu r_e^{\,2}}$, it follows that solving the rigid rotor Schrödinger equation (Equation 8.14) is equivalent to solving the $\hat{J}^2$ eigenvalue problem. One consequence is that the rigid rotor wave functions are just the spherical harmonics; that is,

$$\Psi_{JM}(\theta,\phi) = Y_{JM}(\theta,\phi). \tag{8.24}$$

A second consequence is the form of the rigid rotor energies E_J. To obtain this form, we write the rigid rotor Schrödinger equation $\hat{H}_{RR}\Psi_{JM}(\theta,\phi) = E_J\Psi_{JM}(\theta,\phi)$ as $\frac{\hat{J}^2}{2\mu r_e^{\,2}} Y_{JM}(\theta,\phi) = E_J Y_{JM}(\theta,\phi)$. However, since from Equation (8.23) $\hat{J}^2 Y_{JM}(\theta,\phi) = J(J+1)\hbar^2 Y_{JM}(\theta,\phi)$, the rigid rotor Schrödinger equation reduces to the relation $\frac{J(J+1)\hbar^2}{2\mu r_e^{\,2}} Y_{JM}(\theta,\phi) = E_J Y_{JM}(\theta,\phi)$ which immediately yields the rigid rotor energies as

$$E_J = \frac{J(J+1)\hbar^2}{2\mu r_e^{\,2}}, \quad \text{where } J = 0,1,2,\ldots. \tag{8.25}$$

The ground and the three lowest lying excited state energy levels E_J are plotted in Figure 8.2.

We have introduced quantum angular momentum in the context of the rigid rotor stationary states. But questions of angular momentum in quantum mechanics far transcend the rigid rotor problem. For example, we will deal with two other types of angular momenta, electron orbital and electron

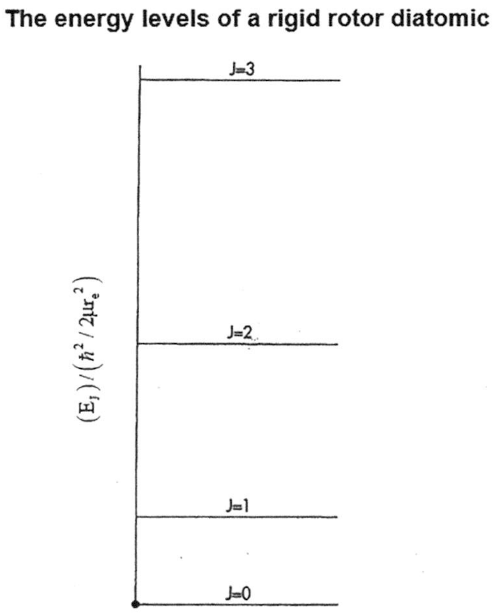

FIGURE 8.2 The ground $(J = 0)$ and low-lying excited state $(J = 1-3)$ energies of a rigid rotor diatomic found from Equation (8.25).

spin, in later chapters. Since the theories of all types of angular momenta are very similar, much of the following discussion of rotational angular momentum will illustrate the general theory of quantum angular momentum.

8.3 THE ROTATIONAL ANGULAR MOMENTUM OPERATORS AND THEIR COMMUTATION RULES

In Equation (B.38), we define a diatomic's classical rotational angular momentum vector $\mathbf{J}(t)$ in terms of its relative coordinate vector $\mathbf{r}(t)=[x(t),y(t),z(t)]$ and its relative momentum vector $\mathbf{p}(t)=\mu\dot{r}(t)=[p_x(t),p_y(t)p_z(t)]$ as the vector cross product

$$\mathbf{J}(t)=r(t)\times p(t). \tag{8.26}$$

From the definition of the cross product, Equation (8.26) may be rewritten as the following three scalar equations for the Cartesian components $J_x(t), J_y(t)$, and $J_z(t)$ of $\mathbf{J}(t)$:

$$J_x(t)=y(t)p_z(t)-z(t)p_y(t) \tag{8.27a}$$

$$J_y(t)=z(t)p_x(t)-x(t)p_z(t) \tag{8.27b}$$

and

$$J_z(t)=x(t)p_y(t)-y(t)p_x(t). \tag{8.27c}$$

Finally, the classical observable $J^2(t)=\mathbf{J}(t)\bullet\mathbf{J}(t)$ may be expressed in terms of $J_x(t), J_y(t)$, and $J_z(t)$ as

$$J^2(t)=J_x^2(t)+J_y^2(t)+J_z^2(t). \tag{8.28}$$

We next move to quantum mechanics. We first convert the classical angular momentum observables $J_{x(t)}, J_{y(t)}$, and $J_z(t)$ to the corresponding quantum Hermitian operators $\hat{J}_x, \hat{J}_y$, and $\hat{J}_z$ by applying the correspondences of Table 4.1 to Equation (8.27) to yield

$$\hat{J}_x=\frac{\hbar}{i}\left(\hat{y}\frac{\partial}{\partial z}-\hat{z}\frac{\partial}{\partial y}\right) \tag{8.29a}$$

$$\hat{J}_y=\frac{\hbar}{i}\left(\hat{z}\frac{\partial}{\partial x}-\hat{x}\frac{\partial}{\partial z}\right) \tag{8.29b}$$

and

$$\hat{J}_z=\frac{\hbar}{i}\left(\hat{x}\frac{\partial}{\partial y}-\hat{y}\frac{\partial}{\partial x}\right). \tag{8.29c}$$

Equation (8.37) expresses $\hat{J}_x$, $\hat{J}_y$, and $\hat{J}_z$ as explicit Cartesian partial differential operators. We may similarly obtain an explicit Cartesian partial differential form for the operator $\hat{J}^2$ which represents J^2. This is done as follows. First, converting Equation (8.28) for $J^2(t)$ yields the following symbolic operator form for $\hat{J}^2$:

$$\hat{J}^2=\hat{J}_x^2+\hat{J}_y^2+\hat{J}_z^2. \tag{8.30}$$

Then, combining Equations (8.29) and (8.30) yields the required Cartesian partial differential operator form for $\hat{J}^2$ (Problem 8.2a).

In Problem 8.3, we prove the basic commutation rules for $\hat{J}_x, \hat{J}_y$, and $\hat{J}_z$. These are

$$\left[\hat{J}_x, \hat{J}_y\right] = i\hbar\hat{J}_z \tag{8.31a}$$

$$\left[\hat{J}_y, \hat{J}_z\right] = i\hbar\hat{J}_x \tag{8.31b}$$

and

$$\left[\hat{J}_z, \hat{J}_x\right] = i\hbar\hat{J}_y \tag{8.31c}$$

From Equation (8.31), we may prove the basic commutation rules for $\hat{J}^2$

$$\left[\hat{J}^2, \hat{J}_x\right] = \hat{O} \ \left[\hat{J}^2, \hat{J}_y\right] = \hat{O} \text{ and} \left[\hat{J}^2, \hat{J}_z\right] = \hat{O} \tag{8.32}$$

where recall $\hat{O}$ is the null operator multiply by zero. Actually, here we will prove only the commutation rule $\left[\hat{J}^2, \hat{J}_z\right] = \hat{O}$. The proofs of the rules $\left[\hat{J}^2, \hat{J}_x\right] = \hat{O}$ and $\left[\hat{J}^2, \hat{J}_y\right] = \hat{O}$ follow parallel arguments.

To prove that $\left[\hat{J}^2, \hat{J}_z\right] = \hat{O}$, we proceed as follows. First, noting Equation (8.30) we rewrite $\left[\hat{J}^2, \hat{J}_z\right]$ as

$$\left[\hat{J}^2, \hat{J}_z\right] = \left[\hat{J}_x^2, \hat{J}_z\right] + \left[\hat{J}_y^2, \hat{J}_z\right] + \left[\hat{J}_z^2, \hat{J}_z\right]. \tag{8.33}$$

Then, recalling the general commutator identity proven in Problem 5.1c

$$\left[\hat{A}\hat{B}, \hat{C}\right] = \hat{A}\left[\hat{B}, \hat{C}\right] + \left[\hat{A}, \hat{C}\right]\hat{B} \tag{8.34}$$

we rewrite $\left[\hat{J}_x^2, J_z\right]$ as

$$\left[\hat{J}_x^2, \hat{J}_z\right] = \hat{J}_x\left[\hat{J}_x, \hat{J}_z\right] + \left[\hat{J}_x, \hat{J}_z\right]\hat{J}_x. \tag{8.35}$$

However, from Equation (8.31c)

$$\left[\hat{J}_x, \hat{J}_z\right] = -\left[\hat{J}_z, \hat{J}_x\right] = -i\hbar\hat{J}_y. \tag{8.36}$$

Equations (8.35) and (8.36) then yield

$$\left[\hat{J}_x^2, \hat{J}_z\right] = -i\hbar\left(\hat{J}_x\hat{J}_y + \hat{J}_y\hat{J}_x\right). \tag{8.37}$$

Similar steps yield

$$\left[\hat{J}_y^2, J_z\right] = i\hbar\left(\hat{J}_x\hat{J}_y + \hat{J}_y\hat{J}_x\right). \tag{8.38}$$

Analogously,

$$\left[\hat{J}_z^2, \hat{J}_z\right] = \hat{J}_z\left[\hat{J}_z, \hat{J}_z\right] + \left[\hat{J}_z, \hat{J}_z\right]\hat{J}_z = \hat{O}. \tag{8.39}$$

Finally, comparing Equations (8.33) and (8.37)–(8.39) gives

$$\left[\hat{J}^2,\hat{J}_z\right] = -i\hbar\left(\hat{J}_x\hat{J}_y + \hat{J}_y\hat{J}_x\right) + i\hbar\left(\hat{J}_x\hat{J}_y + \hat{J}_y\hat{J}_x\right) + \hat{O} = \hat{O} \tag{8.40}$$

thus proving the commutation rule $\left[\hat{J}^2,\hat{J}_z\right] = \hat{O}$.

Many key results for rotational angular momentum may be derived from the commutation relations of Equations (8.31) and (8.32) *alone*. However, the commutation rules for all types of angular moments are identical in form to those of Equations (8.31) and (8.32). Thus, an analysis of the many consequences of the rotational angular momentum commutation rules yields broadly applicable results.

One of the most remarkable consequences of the commutation rules is that while in classical mechanics the angular momentum vector **J** can point in any direction, in quantum mechanics the direction of **J** is restricted to discrete values. This quantum restriction of the possible directions of **J** is called *space quantization* (Atkins, de Paula, and Friedman 2014, Figure 14.4).

However, for our present problem of determining the rigid rotor wave functions or equivalently the spherical harmonics, a very different consequence of the commutation rules is essential: the simultaneous eigenvalue problem for $\hat{J}^2$ and $\hat{J}_z$.

8.4 THE COMMUTATION RULES AND THE SIMULTANEOUS EIGENVALUE PROBLEM FOR $\hat{J}^2$ AND $\hat{J}_z$

To begin, consider two observables A and B which are incompatible and thus have non-commuting Hermitian operator representatives $\hat{A}$ and $\hat{B}$. From Section 5.2, because of their non-commutivity, $\hat{A}$ and $\hat{B}$ have few if any common eigenfunctions. Therefore, for incompatible observables A and B, it is usually impossible to find quantum states for which both A and B have definite values.

In contrast, two compatible observables A and B have commuting Hermitian operator representatives $\hat{A}$ and $\hat{B}$. As proven in Section 5.2, because of their commutivity, $\hat{A}$ and $\hat{B}$ have a complete orthonormal set of common eigenfunctions which we denote by Ψ_{a_n,b_m}. These common eigenfunctions are the solutions of the simultaneous eigenvalue problem $\hat{A}\Psi_{a_n,b_m} = a_n\Psi_{a_n,b_m}$ and $\hat{B}\Psi_{a_n,b_m} = b_m\Psi_{a_n,b_m}$. Moreover, as indicated at the close of Section 4.5 if the system is in the quantum state specified by Ψ_{a_n,b_m}, both observables A and B have definite values, a_n for A and b_m for B. In other words, when the system is in the state specified by the common eigenfunction Ψ_{a_n,b_m} if many simultaneous measurements of the values of A and B are made on identical but different systems, in each of these measurements the eigenvalue a_n will be found for A and the eigenvalue b_m will be found for B.

Next, let us apply these concepts to rotational angular momentum. First, let us consider the angular momentum observables J_x, J_y, and J_z and their respective Hermitian operator representatives $\hat{J}_x, \hat{J}_y$, and $\hat{J}_z$. No two of these observables are compatible because from Equation (8.31) no two of their representatives commute. For example, J_x and J_y are incompatible because from Equation (8.31a) their representatives $\hat{J}_x$ and $\hat{J}_y$ do not commute. Because of this non-commutivity, $\hat{J}_x$ and $\hat{J}_y$ do not have common eigenfunctions, and therefore, it is impossible to find quantum states for which both J_x and J_y have definite values. Similarly, for the other two pairs of observables J_x and J_z and J_y and J_z, it is impossible to find quantum states for which both members of a pair have definite values. These considerations imply that it is impossible to find quantum states for which more than one of the observables J_x, J_y, and J_z has a definite value.

(There is however one important exception. This exception is a consequence of the fact that the $J = 0$, $M = 0$ spherical harmonic $Y_{00}[\theta,\phi]$ is unique because it is a constant, namely $\left[\frac{1}{4\pi}\right]^{1/2}$. Because $Y_{00}[\theta,\phi]$ is a constant, letting the angular momentum operators $\hat{J}_x, \hat{J}_y$, and $\hat{J}_z$ act on it shows that it is a common eigenfunction of these operators with the common eigenvalue zero. Consequently, when the system is in the quantum state specified by $Y_{00}[\theta,\phi]$, which is the rigid rotor ground state, the observables J_x, J_y, and J_z all have the definite value zero.)

Let us next turn from incompatible to compatible angular momentum observables. Specifically, let us consider the three pairs of observables $(J^2, J_x), (J^2, J_y)$, and (J^2, J_z). Each of these pairs is comprised of compatible observables. This is because by Equation (8.32) each of the corresponding Hermitian operator pairs $(\hat{J}^2, \hat{J}_x), (\hat{J}^2, \hat{J}_y)$, and $(\hat{J}^2, \hat{J}_z)$ commutes. Moreover, because the operators that make up each pair commute, the operators for each pair have a complete set of common eigenfunctions (which is unique to that pair and therefore, for example, is one set for the pair $[\hat{J}^2, \hat{J}_x]$ and a different set for the pair $[\hat{J}^2, \hat{J}_y]$) which are determined by solving the simultaneous eigenvalue problem for that pair. Additionally, if the system is in a quantum state specified by, say, one of the common eigenfunctions of $\hat{J}^2$ and $\hat{J}_z$, then the observables J^2 and J_z both have definite values which are, respectively, the eigenvalues of $\hat{J}^2$ and $\hat{J}_z$ for that common eigenfunction.

Finally, we note that by convention, the component $\hat{J}_z$ of $\hat{\mathbf{J}}$ has been selected as preferred (even though from a theoretical standpoint, the components $\hat{J}_x$ and $\hat{J}_y$ are just as satisfactory).

Thus, henceforth we will be concerned with the simultaneous eigenvalue problem for $\hat{J}^2$ and $\hat{J}_z$ and with the definite values for the corresponding observables J^2 and J_z.

We further note that since from Equation (8.21) the eigenfunctions of $\hat{J}^2$ are the spherical harmonics, we will assume (and later prove) that the common eigenfunctions of $\hat{J}^2$ and $\hat{J}_z$ are also the spherical harmonics. Thus, for most of the remainder of this chapter, we will be concerned with deriving the forms for the spherical harmonics by solving the simultaneous eigenvalue problem for $\hat{J}^2$ and $\hat{J}_z$

This simultaneous eigenvalue problem is

$$\hat{J}^2 Y_{JM}(\theta,\phi) = \lambda_{JM}\hbar^2 Y_{JM}(\theta,\phi) \tag{8.41a}$$

and

$$\hat{J}_z Y_{JM}(\theta,\phi) = \mu_{JM}\hbar Y_{JM}(\theta,\phi). \tag{8.41b}$$

The main results which we will obtain later by solving Equation (8.41) are as follows:

1. The quantum number J is restricted to the integral values

$$J = 0, 1, 2, \ldots. \tag{8.42a}$$

2. For each J, the quantum number M takes on all integral values from −J to J and thus has the range

$$M = -J, -J+1, \ldots, 0, \ldots, J-1, J. \tag{8.42b}$$

Therefore, for each J value there are $2J+1$ M values and hence $2J+1$ spherical harmonics.

3. The eigenvalues $\lambda_{JM}\hbar^2$ of $\hat{J}^2$ are independent of M and thus depend only on J. Consequently, there are $2J+1$ spherical harmonic eigenfunctions $Y_{JM}(\theta,\phi)$ each with the same J value but with different M values associated with each eigenvalue $\lambda_{JM}\hbar^2$. Therefore, the eigenvalues $\lambda_{JM}\hbar^2$ are $(2J+1)$−fold degenerate. Moreover, these eigenvalues have the quantized forms

$$\lambda_{JM}\hbar^2 = J(J+1)\hbar^2. \tag{8.43}$$

4. The eigenvalues $\mu_{JM}\hbar$ of $\hat{J}_z$ are independent of J and have the quantized forms

$$\mu_{JM}\hbar = M\hbar. \tag{8.44}$$

5. The spherical harmonics $Y_{JM}(\theta,\phi)$ have the form given in Equation (8.105).

We next solve the simultaneous eigenvalue problem for $\hat{J}^2$ and $\hat{J}_z$ of Equations (8.41) and thus determine the spherical harmonics $Y_{JM}(\theta,\phi)$. To start, we must write down the eigenvalue equations (Equation 8.41) as explicit spherical polar coordinate partial differential equations, since the spherical polar equations but not their Cartesian equivalents are separable. This requires the spherical polar forms for the angular momentum operators.

8.5 THE SIMULTANEOUS EIGENVALUE PROBLEM FOR $\hat{J}^2$ AND $\hat{J}_z$ IN SPHERICAL POLAR COORDINATES

The spherical polar forms for $\hat{J}_x, \hat{J}_y$, and $\hat{J}_z$ are

$$\hat{J}_x = -\frac{\hbar}{i}\left(\sin\phi\frac{\partial}{\partial\theta} + \cos\phi\cot\theta\frac{\partial}{\partial\phi}\right) \tag{8.45a}$$

$$\hat{J}_y = \frac{\hbar}{i}\left(\cos\phi\frac{\partial}{\partial\theta} - \sin\phi\cot\theta\frac{\partial}{\partial\phi}\right) \tag{8.45b}$$

and

$$\hat{J}_z = \frac{\hbar}{i}\frac{\partial}{\partial\phi}. \tag{8.45c}$$

Equations (8.45) may be derived from the Cartesian forms of Equations (8.29) using the transformation relations between Cartesian and spherical polar coordinates of Equation (8.3) (Problems 8.4 and 8.5). Moreover, Equation (8.20) for $\hat{J}^2$ which we will write as

$$\hat{J}^2 = -\hbar^2\left(\frac{\partial^2}{\partial\theta^2} + \cot\theta\frac{\partial}{\partial\theta} + \frac{1}{\sin^2\theta}\frac{\partial^2}{\partial\phi^2}\right) \tag{8.46}$$

while originally derived heuristically by specializing the classical Hamiltonian for the rotating–vibrating diatomic molecule to the rigid rotor model may now be derived rigorously from Equations (8.45) and the relation of Equation (8.30) $\hat{J}^2 = \hat{J}_x^2 + \hat{J}_y^2 + \hat{J}_z^2$ (Problem 8.6).

To begin the solution of the simultaneous eigenvalue problem for $\hat{J}^2$ and $\hat{J}_z$ of Equations (8.41), we use Equations (8.45c) and (8.46) to rewrite this problem as

$$\left(\frac{\partial^2}{\partial\theta^2} + \cot\theta\frac{\partial}{\partial\theta} + \frac{1}{\sin^2\theta}\frac{\partial^2}{\partial\phi^2}\right)Y_{JM}(\theta,\phi) = -\lambda_{JM}Y_{JM}(\theta,\phi) \tag{8.47a}$$

and

$$\frac{\partial Y_{JM}(\theta,\phi)}{\partial\phi} = i\mu_{JM}Y_{JM}(\theta,\phi). \tag{8.47b}$$

Equations (8.47) must be solved subject to boundary and normalization conditions satisfied by the spherical harmonics. These conditions turn out to be identical to those satisfied by the rigid rotor wave functions and thus are given by

$$Y_{JM}(0,\phi)\,\text{is finite and}\,Y_{JM}(\pi,\phi)\text{ is finite} \tag{8.48a}$$

and

$$Y_{JM}(\theta,0)=Y_{JM}(\theta,2\pi) \tag{8.48b}$$

and

$$\int_0^{\pi}\int_0^{2\pi}\sin\theta Y_{JM}^{*}(\theta,\phi)Y_{JM}(\theta,\phi)\,d\theta d\phi=1. \tag{8.49}$$

We next simplify the eigenvalue equation (Equation 8.47a) so that it is free of the spherical polar coordinate ϕ and thus depends only on the spherical polar coordinate θ and partial derivatives with respect to θ. To make this simplification, we note from Equation (8.47b) that $\frac{\partial^2 Y_{JM}(\theta,\phi)}{\partial\phi^2}=i\mu_M\frac{\partial Y_{JM}(\theta,\phi)}{\partial\phi}=(i\mu_{JM})(i\mu_{JM})Y_{JM}(\theta,\phi)$ or $\frac{\partial^2 Y_{JM}(\theta,\phi)}{\partial\phi^2}=-\mu_{JM}^2 Y_{JM}(\theta,\phi)$. Using this result to eliminate the term $\frac{\partial^2 Y_{JM}(\theta,\phi)}{\partial\phi^2}$ from Equation (8.47a) gives the required simplified equation

$$\left(\frac{\partial^2}{\partial\theta^2}+\cot\theta\frac{\partial}{\partial\theta}-\frac{\mu_{JM}^2}{\sin^2\theta}\right)Y_{JM}(\theta,\phi)=-\lambda_{JM}Y_{JM}(\theta,\phi). \tag{8.50}$$

Thus, our problem is to solve the coupled partial differential equations (Equations 8.47b and 8.50) for the spherical harmonics $Y_{JM}(\theta,\phi)$. Our next step is to make a separation-of-variables factorization assumption. With this assumption, we will be able to reduce the coupled partial differential equations to a pair of independent ordinary differential equations (Equations 8.52 and 8.53), an enormous simplification.

8.6 SEPARATION OF VARIABLES FOR THE SPHERICAL HARMONICS

We assume that the spherical harmonics factorize as

$$Y_{JM}(\theta,\phi)=\Theta_{JM}(\theta)\Phi_M(\phi) \tag{8.51}$$

where $\Theta_{JM}(\theta)$ and $\Phi_M(\phi)$ are new functions which we will eventually determine. Notice that while the spherical harmonics $Y_{JM}(\theta,\phi)$ depend on two variables the new functions in Equation (6.51) each depend on only one variable. This fact is the basis of the huge simplification that arises from our factorization assumption.

Also notice that we have anticipated that the functions $\Phi_M(\phi)$ do not depend on the J quantum number. This expectation will soon be validated.

We may now easily derive independent eigenvalue equations for $\Phi_M(\phi)$ and $\Theta_{JM}(\theta)$. First, inserting the factorization equation (Equation 8.51) into Equation (8.47b) gives $\Theta_{JM}(\theta)\frac{d\Phi_{JM}(\phi)}{d\phi}=i\mu_{JM}\Theta_{JM}(\theta)\Phi_M(\phi)$. Then, canceling the $\Theta_{JM}(\theta)$ factor yields the following independent eigenvalue equation for $\Phi_M(\phi)$:

$$\frac{d\Phi_M(\phi)}{d\phi}=i\mu_{JM}\Phi_M(\phi). \tag{8.52}$$

Similarly, inserting Equation (8.51) into Equation (8.50) and then canceling the $\Phi_M(\phi)$ factor give the following independent eigenvalue equation for $\Theta_{JM}(\theta)$:

$$\left[\frac{d^2}{d\theta^2} + \cot\theta\frac{d}{d\theta} - \frac{\mu_{JM}^2}{\sin^2\theta}\right]\Theta_{JM}(\theta) = -\lambda_{JM}\Theta_{JM}(\theta). \tag{8.53}$$

Equations (8.52) and (8.53) must be supplemented by boundary and normalization conditions for both $\Phi_M(\phi)$ and $\Theta_{JM}(\theta)$. From the factorization equation (Equation 8.51) and from the boundary and normalization conditions for $Y_{JM}(\theta,\phi)$ of Equations (8.48b) and (8.49), we find for $\Phi_M(\phi)$ the conditions

$$\Phi_M(0) = \Phi_M(2\pi) \tag{8.54}$$

and

$$\int_0^{2\pi} \Phi_M^*(\phi)\Phi_M(\phi)d\phi = 1 \tag{8.55}$$

where to obtain Equation (8.55) we have assumed that $\Phi_M(\phi)$ and $\Theta_{JM}(\theta)$ are separately normalized to unity. Similarly, we find the following boundary and normalization conditions for $\Theta_{JM}(\theta)$:

$$\Theta_{JM}(0) \text{ is finite and } \Theta_{JM}(\pi) \text{ is finite} \tag{8.56}$$

and

$$\int_0^{\pi} \sin\theta\Theta_{JM}^*(\theta)\Theta_{JM}(\theta)d\theta = 1. \tag{8.57}$$

We next determine the functions $\Phi_M(\phi)$ by solving their eigenvalue equation (Equation 8.52) subject to the boundary and normalization conditions of Equations (8.54) and (8.55).

8.7 SOLUTION OF THE OF $\Phi_M(\phi)$ PROBLEM

The general solution of Equation (8.52) for $\Phi_M(\phi)$ is

$$\Phi_M(\phi) = N\exp(i\mu_{JM}\phi) \tag{8.58}$$

where N is a normalization factor. Applying the boundary condition for $\Phi_M(\phi)$ of Equation (8.54) to Equation (8.58) yields the following restriction on μ_{JM}:

$$\exp(2\pi i\mu_{JM}) = 1 \tag{8.59}$$

or equivalently $\cos(2\pi\mu_{JM}) + i\sin(2\pi\mu_{JM}) = 1$. The proceeding restriction can only be satisfied if μ_{JM} is an integer which we will call M. That is, in order to satisfy the boundary condition of Equation (8.54), μ_{JM} must be restricted to the integer values

$$\mu_{JM} = M = \ldots -2,-1,\ldots,1,2,\ldots. \tag{8.60}$$

With Equation (8.60), $\Phi_M(\phi)$ of Equation (8.58) simplifies to

$$\Phi_M(\phi) = N\exp(iM\phi). \tag{8.61}$$

Notice as mentioned earlier that $\Phi_M(\phi)$ is independent of the quantum number J. To complete the determination of $\Phi_M(\phi)$, we must determine the normalization factor N. Assuming that N is real, a brief calculation based on Equations (8.55) and (8.61) gives that $N = \left(\frac{1}{2\pi}\right)^{1/2}$. Thus, the final form of $\Phi_M(\phi)$ is

$$\Phi_M(\phi) = \left(\frac{1}{2\pi}\right)^{1/2} \exp(iM\phi). \tag{8.62}$$

Finally, it is easy to show from Equation (8.62) that the functions $\Phi_M(\phi)$ are orthonormal on the range 0–2π and thus conform to the orthonormality relation

$$\int_0^{2\pi} \Phi^*_{M'}(\phi)\Phi_M(\phi)d\phi = \delta_{MM'}. \tag{8.63}$$

We next turn to the challenging problem of determining the functions $\Theta_{JM}(\theta)$ by solving the eigenvalue equation (Equation 8.53) subject to the boundary and normalization conditions of Equations (8.56) and (8.57).

8.8 SOLUTION OF THE $\Theta_{JM}(\theta)$ PROBLEM

To start, using Equation (8.60) for $\mu_{JM} = M$, we rewrite Equation (8.53) as

$$\left[\frac{d}{d\theta^2} + \cot\theta\frac{d}{d\theta} + \lambda_{JM} - \frac{M^2}{\sin^2\theta}\right]\Theta_{JM}(\theta) = 0. \tag{8.64}$$

To solve the above equation, we will relate the eigenfunctions $\Theta_{JM}(\theta)$ to standard functions, the *associated Legendre functions* (Arfken, Weber, and Harris, 2012) which are denoted by $P_J^{|M|}(x)$ where J and $|M|$ can take on only nonnegative integer values. (The functions $P_J^{|M|}(x)$ are independent of the sign of M since their defining differential equation (Equation 8.65) depends on M only as M^2.)

The associated Legendre functions are defined on the range $-1 \le x \le 1$ as those solutions of the associated Legendre equation

$$\left(1-x^2\right)\frac{d^2P_J^{|M|}(x)}{dx^2} - 2x\frac{dP_J^{|M|}(x)}{dx} + \left[J(J+1) - \frac{M^2}{1-x^2}\right]P_J^{|M|}(x) = 0 \tag{8.65}$$

which satisfy the boundary conditions

$$P_J^{|M|}(1)\,\text{is finite and}\,P_J^{|M|}(-1)\,\text{is finite} \tag{8.66}$$

and a normalization condition given later in Equation (8.100).

To relate the eigenfunctions $\Theta_{JM}(\theta)$ to the associated Legendre functions $P_J^{|M|}(x)$, we begin by recalling that in Equation (8.43), we stated without proof that $\lambda_{JM} = J(J+1)$. If this result is true, and we will soon show that it is, then it is straightforward to show that the eigenfunctions $\Theta_{JM}(\theta)$ are proportional to the functions $P_J^{|M|}(x)$ evaluated for $x = \cos\theta$. That is,

$$\Theta_{JM}(\theta) = N_{JM}P_J^{|M|}(\cos\theta) \tag{8.67}$$

where N_{JM} is a proportionality factor which we will determine in Section 8.9.

We next prove that the proportionality relation of Equation (8.67) holds if $\lambda_{JM} = J(J+1)$. To do this, we first note that the associated Legendre boundary conditions of Equation (8.66) when evaluated for $x = \cos\theta$ are identical to the boundary conditions for $\Theta_{JM}(\theta)$ of Equation (8.56). Thus, to prove the proportionality equation (Equation 8.67), it is only necessary to show that Equation (8.65) for $P_J^{|M|}(x)$ is equivalent to Equation (8.64) for $\Theta_{JM}(\theta)$ conditional that $\lambda_{JM} = J(J+1)$.

To show this, we first rewrite Equation (8.65) as

$$\frac{d}{dx}\left[\left(1-x^2\right)\frac{dP_J^{|M|}(x)}{dx}\right]+\left[J(J+1)-\frac{M^2}{1-x^2}\right]P_J^{|M|}(x)=0. \tag{8.68}$$

We next express Equation (8.68) in terms of the variable $\theta=\cos^{-1}x$. To make this transformation, we note that since $x=\cos\theta, P_J^{|M|}(x)=P_J^{|M|}(\cos\theta), 1-x^2=\sin^2\theta$, and $\frac{d}{dx}=\frac{d}{d\cos\phi}=-\frac{1}{\sin\theta}\frac{d}{d\theta}$. Hence, when expressed in terms of θ, Equation (8.68) becomes $\frac{1}{\sin\theta}\frac{d}{d\theta}\left[\sin\theta\frac{d}{d\theta}P_J^{|M|}(\cos\theta)\right]+\left[J(J+1)-\frac{M^2}{\sin^2\theta}\right]P_J^{|M|}(\cos\theta)=0$ or equivalently

$$\left[\frac{d^2}{d\theta^2}+\cot\theta\frac{d}{d\theta}+J(J+1)-\frac{M^2}{\sin^2\theta}\right]P_J^{M}(\cos\theta)=0. \tag{8.69}$$

Note that Equation (8.64) for $\Theta_{JM}(\theta)$ and Equation (8.69) for $P_J^{|M|}(\cos\theta)$ are identical if $\lambda_{JM} = J(J+1)$. Thus, we have established Equation (8.67) conditional that $\lambda_{JM} = J(J+1)$.

We next prove that λ_{JM} is indeed equal to $J(J+1)$ and also verify Equation (8.42) for the ranges of the rotational quantum numbers J and M. To do this, we solve Equation (8.64) for $\Theta_{JM}(\theta)$. We start by defining a new function $P_{JM}(x)=\Theta_{JM}(\theta)$ which is just $\Theta_{JM}(\theta)$ expressed in terms of the variable $x=\cos\theta$. Thus, solving the differential equation (Equation 8.70) for $P_{JM}(x)$, we next derive, is equivalent to solving Equation (8.64) for $\Theta_{JM}(\theta)$.

To derive Equation (8.70), we start with Equation (8.64) for $\Theta_{JM}(\theta)$ and then invert the steps which had from Equations (8.65) to (8.69). This procedure yields the following differential equation for $P_{JM}(x)$ which is analogous to Equation (8.65):

$$\left(1-x^2\right)\frac{d^2P_{JM}(x)}{dx^2}-2x\frac{dP_{JM}(x)}{dx}+\left[\lambda_{JM}-\frac{M^2}{1-x^2}\right]P_{JM}(x)=0. \tag{8.70}$$

We solve Equation (8.70) by a method which is similar to that used in Appendix C to determine the Hermite polynomials. Thus, we will merely outline our method of solution. We begin by defining the new function $Q(x)$ as follows:

$$P_{JM}(x)=\left(1-x^2\right)^{|M|/2}Q(x). \tag{8.71}$$

Next, inserting Equation (8.71) into Equation (8.70) yields the following differential equation for $Q(x)$ (Problem 8.11):

$$\left(1-x^2\right)\frac{d^2Q(x)}{dx^2}-2(|M|+1)x\frac{dQ(x)}{dx}+\left[\lambda_{JM}-|M|(|M|+1)\right]Q(x)=0. \tag{8.72}$$

Next, we make the following power series expansion of $Q(x)$:

$$Q(x)=\sum_{n=0}^{\infty}a_nx^n. \tag{8.73}$$

Inserting Equation (8.73) into Equation (8.72) and proceeding as in Appendix C yields the following recursion relation for the coefficients a_n (Problem 8.12):

$$a_{k+2} = \frac{(k+|M|)(k+|M|+1)-\lambda_{JM}}{(k+1)(k+2)} a_k. \tag{8.74}$$

From Equations (8.73) and (8.74), for arbitrary values of λ_{JM} $Q(x)$ is an infinite series that diverges so severely at $x = \pm 1$ that $P_{JM}(x)$ diverges at $x = \pm 1$ and thus $\Theta_{JM}(\theta)$ diverges at $\theta = 0$ and π, in conflict with the boundary conditions on $\Theta_{JM}(\theta)$ of Equation (8.56).

The remedy is similar to that used in Appendix C to determine the Hermite polynomials. Namely, we cut off the series of Equation (8.73) at the kth term where k can have any of the values $0,1,2\ldots$. With this cutoff, $Q(x)$ becomes a kth-order polynomial that is finite at $x = \pm 1$, thus eliminating the violation of the boundary conditions on $\Theta_{JM}(\theta)$. To make the cutoff, we choose λ_{JM} so that $a_{k+2} = 0$. From Equation (8.74), the proper choice of λ_{JM} is

$$\lambda_{JM} = (k+|M|)(k+|M|+1). \tag{8.75}$$

We next define a new quantum number J by

$$J = k+|M|. \tag{8.76}$$

Since k can take on any nonnegative integral value and since by Equation (8.60) $|M|$ can only take on the values $0,1,2\ldots$, the possible values of are

$$J = 0,1,2,\ldots. \tag{8.77}$$

Thus, if we properly identify J of Equation (8.76) with the rotational quantum number J, then Equation (8.77) verifies the range of J stated without proof in Equation (8.42a). Moreover, from Equations (8.75) and (8.76), we have that

$$\lambda_{JM} = J(J+1). \tag{8.78}$$

Equation (8.78), however, is just the validity condition that we have been seeking for the proportionality relation of Equation (8.67) between the eigenfunctions $\Theta_{JM}(\theta)$ and the associated Legendre functions $P_J^{|M|}(\cos\theta)$. Additionally, since $J = k+|M|$ and since $k \geq 0$, it follows that $J \geq |M|$ or $|M| \leq J$. Moreover, since $|M|$ is restricted to the values $|M| = 0,1,\ldots$ and since we have just shown that $|M| \leq J$, from Equation (8.77) the possible values of $|M|$ are $|M| = 0,1,\ldots,J-1,J$. Consequently, the range of M is

$$M = -J,-J+1,\ldots,0,\ldots,J-1,J. \tag{8.79}$$

Equation (8.79) establishes the previously unproven equation (Equation 8.42b).

We have just proven Equation (8.67) and therefore have shown that to within a constant N_{JM}, the eigenfunctions $\Theta_{JM}(\theta)$ and the associated Legendre functions $P_J^{|M|}(\cos\theta)$ are identical. Additionally, from the factorization equation (Equation 8.51), the functions $\Theta_{JM}(\theta)$ and hence the functions $P_J^{|M|}(\cos\theta)$ determine the θ–dependence of the spherical harmonics $Y_{JM}(\theta,\phi)$. Therefore, it is not surprising that in order to determine the spherical harmonics, we need to further study the associated Legendre functions.

This study is greatly facilitated if we introduce new functions, the *Legendre polynomials.* The Legendre polynomials are simpler than the associated Legendre functions but are closely related to these functions. For example, many of the properties of the associated Legendre functions are natural generalizations of simpler properties of the Legendre polynomials. Moreover, since the Legendre polynomials are easily generated from either their Rodrigues representation or their recursion relation, the best way to obtain the associated Legendre functions is to derive their forms from those of the Legendre polynomials. Given these comments, we next turn to a discussion of the Legendre polynomials and the associated Legendre functions.

8.9 THE LEGENDRE POLYNOMIALS AND THE ASSOCIATED LEGENDRE FUNCTIONS

The Legendre polynomials $P_J(x)$ are defined as the special case or the associated Legendre functions $P_J^{|M|}(x)$

$$P_J(x) = P_J^{|M|=0}(x). \tag{8.80}$$

Setting $M = 0$ in the associated Legendre equation (Equation 8.65) gives the following Legendre differential equation for the Legendre polynomials:

$$\left(1-x^2\right)\frac{d^2P_J(x)}{dx^2} - 2x\frac{dP_J(x)}{dx} + J(J+1)P_J(x) = 0. \tag{8.81}$$

The first nine Legendre polynomials are listed in Table 8.1. It is easy to verify that the first few are solutions of the Legendre equation (Equation 8.81). (See Problem 8.29.)

We showed in Section 7.11 that many properties of the Hermite polynomials may be derived from a generating function. The same is true for the Legendre polynomials. The Legendre polynomial generating function is denoted by $h(t,x)$ and is given by

$$h(t,x) = \left(1-2xt+t^2\right)^{-1/2}. \tag{8.82}$$

The value of the generating function $h(t,x)$ arises since in many applied mathematics textbooks it is shown that it is related to the Legendre polynomials $P_J(x)$ as follows:

$$h(t,x) = \sum_{J=0}^{\infty} P_J(x)t^J. \tag{8.83}$$

TABLE 8.1
The First Nine Legendre Polynomials

J	$P_J(x)$
0	1
1	x
2	$\frac{1}{2}\left(3x^2-1\right)$
3	$\frac{1}{2}\left(5x^3-3x\right)$
4	$\frac{1}{8}\left(35x^4-30x^2+3\right)$
5	$\frac{1}{8}\left(63x^5-70x^3+15x\right)$
6	$\frac{1}{16}\left(231x^6-315x^4+105x^2-5\right)$
7	$\frac{1}{16}\left(429x^7-693x^5+315x^3-35x\right)$
8	$\frac{1}{128}\left(6435x^8-12012x^6+6930x^4-1260x^2+35\right)$

One of the valuable relations that may be derived from Equations (8.82) and (8.83) is the following Rodrigues representation of the Legendre polynomials:

$$P_J(x) = \frac{1}{2^J J!} \frac{d^J}{dx^J}\left(x^2 - 1\right)^J. \tag{8.84}$$

The Legendre polynomials may be easily obtained from the above Rodrigues representation (Problem 8.14). An especially important result that may be derived from the generating function $h(t,x)$ is the Legendre polynomial recursion relation

$$(J+1)P_{J+1}(x) = (2J+1)xP_J(x) - JP_{J-1}(x). \tag{8.85}$$

If one takes $P_0(x)$ to have the value one, then the Legendre polynomials $P_J(x)$ for $J > 0$ may be readily obtained from the above recursion relation (Problem 8.18).

We next derive the recursion relation of Equation (8.85) from the generating function $h(t,x)$. We start by differentiating both Equations (8.82) and (8.83) with respect to t to obtain the relation

$$\frac{\partial h(t,x)}{\partial t} = \frac{x-t}{\left(1-2xt+t^2\right)^{3/2}} = \sum_{J=0}^{\infty} JP_J(x)t^{J-1}. \tag{8.86}$$

Next, since $\frac{1}{\left(1-2xt+t^2\right)^{3/2}} = \frac{1}{\left(1-2xt+t^2\right)} \frac{1}{\left(1-2xt+t^2\right)^{1/2}} = \frac{1}{\left(1-2xt+t^2\right)} h(t,x) = \frac{1}{\left(1-2xt+t^2\right)} \times \sum_{J=0}^{\infty} P_J(x)t^J$, Equation (8.86) may be rewritten as

$$(x-t)\sum_{J=0}^{\infty} P_J(x)t^J = \left(1-2xt+t^2\right)\sum_{J=0}^{\infty} JP_J(x)t^{J-1}. \tag{8.87}$$

By multiplying out the terms in Equation (8.87) and then grouping like powers of t, Equation (8.87) may be recast as

$$\sum_{J=0}^{\infty}(2J+1)xP_J(x)t^J - \sum_{J=0}^{\infty} JP_J(x)t^{J-1} - \sum_{J=0}^{\infty}(J+1)P_J(x)t^{J+1} = 0. \tag{8.88}$$

However (Problem 8.16),

$$\sum_{J=0}^{\infty} JP_J(x)t^{J-1} = \sum_{J=0}^{\infty}(J+1)P_{J+1}(x)t^J \tag{8.89a}$$

and

$$\sum_{J=0}^{\infty}(J+1)P_J(x)t^{J+1} = \sum_{J=0}^{\infty} JP_{J-1}(x)t^J. \tag{8.89b}$$

Substituting Equation (8.89) into Equation (8.88) yields

$$\sum_{J=0}^{\infty}\left[(2J+1)xP_J(x) - (J+1)P_{J+1}(x) - JP_{J-1}(x)\right]t^J = 0. \tag{8.90}$$

However, Equation (8.90) can hold for all t only if the coefficients of each power of t vanish. This requirement yields the relation

$$(2J+1)xP_J(x) = (J+1)P_{J+1}(x) + JP_{J-1}(x) \tag{8.91}$$

which is just a slightly rearranged form of the Legendre polynomial recursion relation of Equation (8.85). Thus, we have proven the Legendre polynomial recursion relation.

Next, an examination of the $J = 0-8$ Legendre polynomials listed in Table 8.1 shows that for all nine,

$$P_J(1) = 1. \tag{8.92}$$

To further illustrate the utility of the generating functions $h(t,x)$, we use it to prove that Equation (8.92) holds for all J and thus is a general property of the Legendre polynomials.

We start by evaluating $h(t,x)$ for $x = 1$. From Equations (8.82) and (8.83), this evaluation yields

$$\left(1-2t+t^2\right)^{-1/2} = \sum_{J=0}^{\infty} P_J(1)t^J. \tag{8.93}$$

However, since $\left(1-2t+t^2\right) = (1-t)^2, \left(1-2t+t^2\right)^{-1/2} = (1-t)^{-1}$, which is equal to the geometric series $1+1+t+t^2+\cdots = \sum_{J=0}^{\infty} t^J = \sum_{J=0}^{\infty} (1)t^J$, and therefore, Equation (8.93) may be rewritten as

$$\sum_{J=0}^{\infty} (1)t^J = \sum_{J=0}^{\infty} P_J(1)t^J. \tag{8.94}$$

Equating like powers of t on the left- and right-hand sides of Equation (8.94) gives Equation (8.92). Thus, we have proven that Equation (8.92) is a general identity obeyed by all Legendre polynomials.

Another important property of the Legendre polynomials which may be derived from $h(t,x)$ is the symmetry relation (Problem 8.19)

$$P_J(-x) = (-1)^J P_J(x). \tag{8.95}$$

Thus, in accord with the list in Table 8.1, $P_J(x)$ is even in x if J is even and odd in x if J is odd.

A very important property of the Legendre polynomials which may be derived from the Legendre differential equation (Equation 8.81) (and hence indirectly from $h[t,x]$) is their "orthonormality" relation, which may be written as either

$$\int_{-1}^{1} P_{J'}(x)P_J(x)dx = \frac{2\delta_{J'J}}{2J+1} \tag{8.96a}$$

or in terms of $\theta = \cos^{-1} x$

$$\int_0^{\pi} \sin\theta P_{J'}(\cos\theta)P_J(\cos\theta)d\theta = \frac{2\delta_{J'J}}{2J+1}. \tag{8.96b}$$

(We have put quotation marks on the word orthonormality since it is evident from Equation [8.96] that while the Legendre polynomials are mutually orthogonal, they are not normalized to unity.)

We may now return to the associated Legendre functions $P_J^{|M|}(x)$. One may show that by differentiating Equation (8.81) $|M|$ times to obtain Equation (8.65), they may be obtained from the Legendre polynomials $P_J(x)$ via the relation

$$P_J^{|M|}(x)=\left(1-x^2\right)^{|M|/2}\frac{d^{|M|}}{dx^{|M|}}P_J(x). \tag{8.97}$$

The associated Legendre functions may be conveniently obtained from the Legendre polynomials via Equation (8.97) (see Problem 8.13). The associated Legendre functions so obtained are listed in Table 8.2 for $J=1-4$ and for $|M|\le J$.

As noted earlier, many properties of the associated Legendre functions are natural generalizations of the corresponding properties of the Legendre polynomials. For example, the associated Legendre recursion relation is the following generalization of the Legendre polynomial recursion relation (Equation 8.85)

$$\left(J-|M|+1\right)P_{J+1}^{|M|}(x)=(2J+1)xP_J^{|M|}(x)-\left(J+|M|\right)P_{J-1}^{|M|}(x), \tag{8.98}$$

Similarly, the associated Legendre "orthonormality" relation

$$\int_{-1}^{1}P_{J'}^{|M|}(x)P_J^{|M|}(x)dx=\frac{2}{2J+1}\frac{\left(J+|M|\right)!}{\left(J-|M|\right)!}\delta_{J'J} \tag{8.99a}$$

TABLE 8.2
The Associated Legendre Functions for $J=0-4$ and $|M|\le J$

JJ	$\lvert M\rvert$	$P_J^{\lvert M\rvert}(x)$
0	0	1
1	0	x
	1	$\left(1-x^2\right)^{1/2}$
2	0	$\frac{1}{2}\left(3x^2-1\right)$
	1	$3x\left(1-x^2\right)^{1/2}$
	2	$3\left(1-x^2\right)$
3	0	$\frac{1}{2}\left(5x^3-3x\right)$
	1	$\frac{3}{2}\left(5x^2-1\right)\left(1-x^2\right)^{1/2}$
	2	$15x\left(1-x^2\right)$
	3	$15\left(1-x^2\right)^{3/2}$
4	0	$\frac{1}{8}\left(35x^4-30x^2+3\right)$
	1	$\frac{5}{2}\left(7x^3-3x\right)\left(1-x^2\right)^{1/2}$
	2	$\frac{15}{2}\left(7x^2-1\right)\left(1-x^2\right)$
	3	$105x\left(1-x^2\right)^{3/2}$
	4	$105\left(1-x^2\right)^2$

or equivalently

$$\int_0^{\pi} \sin\theta\, P_{J'}^{|M|}(\cos\theta) P_J^{|M|}(\cos\theta)\, d\theta = \frac{2}{2J+1}\frac{(J+|M|)!}{(J-|M|)!}\delta_{J'J} \tag{8.99b}$$

is a natural extension of the Legendre polynomial "orthonormality" relation of Equation (8.96).

We will shortly require the normalization condition for the associated Legendre functions. Setting $J' = J$ in Equation (8.99b) gives this condition as

$$\int_0^{\pi} \sin\theta\, P_J^{|M|}(\cos\theta) P_J^{|M|}(\cos\theta)\, d\theta = \frac{2}{2J+1}\frac{(J+|M|)!}{(J-|M|)!}. \tag{8.100}$$

We may now determine the proportionality factor N_{JM} of Equation (8.67). To do this, note from Equation (8.67) that

$$\int_0^{\pi} \sin\theta\, \Theta_{JM}(\theta)\Theta_{JM}(\theta)\, d\theta = N_{JM}^2 \int_0^{\pi} \sin\theta\, P_J^{|M|}(\cos\theta) P_J^{|M|}(\cos\theta)\, d\theta. \tag{8.101}$$

However, noting from Equation (8.57) that on the interval 0–π the functions $\Theta_{JM}(\theta)$ are normalized to unity and also noting Equation (8.100), it follows that Equation (8.101) reduces to

$$1 = N_{JM}^2 \frac{2}{2J+1}\frac{(J+|M|)!}{(J-|M|)!} \tag{8.102}$$

and hence

$$N_{JM} = \left[\left(\frac{2J+1}{2}\right)\frac{(J-|M|)!}{(J+|M|)!}\right]^{1/2}. \tag{8.103}$$

The explicit relation that we have been seeking between the eigenfunctions $\Theta_{JM}(\theta)$ and the associated Legendre functions $P_J^{|M|}(\cos\theta)$ then follows immediately from Equations (8.67) and (8.103) as

$$\Theta_{JM}(\theta) = \left[\left(\frac{2J+1}{2}\right)\frac{(J-|M|)!}{(J+|M|)!}\right]^{1/2} P_J^{|M|}(\cos\theta). \tag{8.104}$$

The functions $\Theta_{JM}(\theta)$ found from Equation (8.104) with the associated Legendre functions obtained from Table 8.2 are listed in Table 8.3 for $J = 0-4$. with $|M| \le J$.

8.10 THE SPHERICAL HARMONICS AND THE RIGID ROTOR STATIONARY STATE WAVE FUNCTIONS

We are now finally ready to write down the solution of our original problem, determining the normalized rigid rotor wave functions or equivalently the spherical harmonics. From the separation-of-variables equation (Equation 8.51), from the form of $\Phi_M(\phi)$ of Equation (8.62), and from Equation (8.104) for $\Theta_{JM}(\theta)$, the spherical harmonics are given by

$$Y_{JM}(\theta,\phi) = \left(\frac{1}{2\pi}\right)^{1/2}\left[\left(\frac{2J+1}{2}\right)\frac{(J-|M|)!}{(J+|M|)!}\right]^{1/2} P_J^{|M|}(\cos\theta)\exp(iM\phi) \tag{8.105}$$

where the associated Legendre functions are found from Table 8.2 or more generally from Equation (8.97) and where recall that we have earlier shown that the quantum numbers J and M are restricted to the values

$$J = 0,1,2\ldots \text{ and } M = -J,-J+1,\ldots,0,\ldots,J-1,J. \tag{8.106}$$

TABLE 8.3
The Eigenfunctions $\Theta_{JM}(\theta)$ of Equation (8.104) for $J = 0 - 4$ and $|M| \le J$

JJ	|M|	$\Theta_{JM}(\theta)$
0	0	$\left(\frac{1}{2}\right)^{1/2}$
1	0	$\left(\frac{3}{2}\right)^{1/2}\cos\theta$
	1	$\left(\frac{3}{4}\right)^{1/2}\sin\theta$
2	0	$\left(\frac{5}{8}\right)^{1/2}\left(3\cos^2\theta - 1\right)$
	1	$\left(\frac{15}{4}\right)^{1/2}\sin\theta\cos\theta$
	2	$\left(\frac{15}{16}\right)^{1/2}\sin^2\theta$
3	0	$\left(\frac{63}{8}\right)^{1/2}\left(\frac{5}{3}\cos^3\theta - \cos\theta\right)$
	1	$\left(\frac{21}{32}\right)^{1/2}\sin\theta\left(5\cos^2\theta - 1\right)$
	2	$\left(\frac{105}{16}\right)^{1/2}\sin^2\theta\cos\theta$
	3	$\left(\frac{35}{32}\right)^{1/2}\sin^3\theta$
4	0	$\left(\frac{81}{128}\right)^{1/2}\left(\frac{35}{3}\cos^4\theta - 10\cos^2\theta + 1\right)$
	1	$\left(\frac{405}{32}\right)^{1/2}\sin\theta\left(\frac{7}{3}\cos^3\theta - \cos\theta\right)$
	2	$\left(\frac{45}{64}\right)^{1/2}\sin^2\theta\left(7\cos^2\theta - 1\right)$
	3	$\left(\frac{315}{32}\right)^{1/2}\sin^3\theta\cos\theta$
	4	$\left(\frac{315}{256}\right)^{1/2}\sin^4\theta$

Often the spherical harmonics are redefined so that the right-hand side of Equation (8.105) is multiplied by a physically insignificant term $\exp[i\delta]$ where δ is real. This term is called a *phase factor*. We will use here the Condon–Shortley choice of phase factor. With this choice, the spherical harmonics are given by Equation (8.105) except that those with odd and positive M are given by the negative of Equation (8.105). The spherical harmonics for $J = 0 - 3$ with the Condon–Shortley phase factor are listed in Table 8.4.

Note from Table 8.4 that for $J = 0, M = 0$ and there is only a single spherical harmonic; for $J = 1, M = -1, 0,$ or 1 and there are three spherical harmonics; for $J = 2, M = -2, -1, 0, 1,$ or 2 and there are five spherical harmonics; and so on. This pattern is reminiscent of that of the hydrogen atom orbitals since there is one s orbital, three p orbitals, five d orbitals, and so on. This similarity between

TABLE 8.4
The s, p, d, and f Spherical Harmonics $Y_{JM}(\theta,\phi)$

Type	Degeneracy $2J+1$	J	M	$Y_{JM}(\theta,\phi)$
S	1	0	0	$\left(\frac{1}{4\pi}\right)^{1/2}$
P	3	1	−1	$\left(\frac{3}{8\pi}\right)^{1/2}\sin\theta\exp(-i\phi)$
			0	$\left(\frac{3}{4\pi}\right)^{1/2}\cos\theta$
			1	$-\left(\frac{3}{8\pi}\right)^{1/2}\sin\theta\exp(i\phi)$
d	5	2	−2	$\left(\frac{15}{32\pi}\right)^{1/2}\sin^2\theta\exp(-2i\phi)$
			−1	$\left(\frac{15}{8\pi}\right)^{1/2}\sin\theta\cos\theta\exp(-i\phi)$
			0	$\left(\frac{5}{16\pi}\right)^{1/2}(3\cos^2\theta-1)$
			1	$-\left(\frac{15}{8\pi}\right)^{1/2}\sin\theta\cos\theta\exp(i\phi)$
			2	$\left(\frac{15}{32\pi}\right)^{1/2}\sin^2\theta\exp(2i\phi)$
f	7	3	−3	$\left(\frac{35}{64\pi}\right)^{1/2}\sin^3\theta\exp(-3i\phi)$
			−2	$\left(\frac{105}{32\pi}\right)^{1/2}\sin^2\theta\cos\theta\exp(-2i\phi)$
			−1	$\left(\frac{21}{64\pi}\right)^{1/2}(5\cos^2\theta-1)\sin\theta\exp(-i\phi)$
			0	$\left(\frac{63}{16\pi}\right)^{1/2}\left(\frac{5}{3}\cos^3\theta-\cos\theta\right)$
			1	$-\left(\frac{21}{64\pi}\right)^{1/2}(5\cos^2\theta-1)\sin\theta\exp(i\phi)$
			2	$\left(\frac{105}{32\pi}\right)^{1/2}\sin^2\theta\cos\theta\exp(2i\phi)$
			3	$-\left(\frac{35}{64\pi}\right)^{1/2}\sin^3\theta\exp(3i\phi)$

the apparently very different rigid rotor and the hydrogen atom systems is not an accident since as we will see in Chapter 10 the angular parts of the hydrogen atom orbitals are the spherical harmonics. Thus, in accord with Table 8.4, we will henceforth refer to the J = 0,1,2,... spherical harmonics as s,p,d... functions.

Additionally, three-dimensional plots of the spherical harmonic Born rule probability densities $Y^*_{JM}(\theta,\phi)Y_{JM}(\theta,\phi)$, which determine the orientational probabilities of a rigid rotor diatomic, reveal that the characteristic shapes of the hydrogen atom orbitals (spherical for s orbitals, dumbbell-like for p orbitals, and so on) follow from the forms of the spherical harmonics.

We further note that the spherical harmonics form an orthonormal set and thus conform to the orthonormality relation

$$\int_0^{\pi}\int_0^{2\pi} \sin\theta Y^*_{J'M'}(\theta,\phi)\, Y_{JM}(\theta,\phi)\, d\theta d\phi = \delta_{J'J}\delta_{M'M}. \tag{8.107}$$

The spherical harmonic orthonormality relation follows from the forms of the spherical harmonics given in Equation (8.105), from the orthonormality relation of Equation (8.63) for the function $\Phi_M(\phi) = \left(\frac{1}{2\pi}\right)^{1/2} \exp(M\phi)$, and from the associated Legendre "orthonormality" relation of Equation (8.99b).

Furthermore, since the normalized rigid rotor stationary state wave functions are identical to the spherical harmonics, the spherical harmonic orthonormality relation also follows from the orthonormality of the rigid rotor wave functions. The rigid rotor orthonormality in turn follows because the rigid rotor Hamiltonian operator $\hat{H}_{RR}$ is Hermitian.

Finally, for future reference, we note from Equations (8.41) and results derived in Sections 8.7 and 8.8 that the spherical harmonics solve the simultaneous eigenvalue problem

$$\hat{J}^2 Y_{JM}(\theta,\phi) = J(J+1)\hbar^2 Y_{JM}(\theta,\phi) \tag{8.108a}$$

and

$$\hat{J}_z Y_{JM}(\theta,\phi) = M\hbar Y_{JM}(\theta,\phi). \tag{8.108b}$$

In the remainder of this chapter, we will present applications of the results developed so far. Specifically, in the next section we will derive the rigid rotor approximation to the pure rotational absorption spectrum of a diatomic molecule. In the final section (Section 8.12), we will derive the rigid rotor selection rules for the rotational transitions which give rise to this spectrum.

8.11 THE RIGID ROTOR PURE ROTATIONAL SPECTRA OF DIATOMIC MOLECULES

We next turn to the pure rotational transitions of a diatomic molecule. The transitions depending on the molecule lead to absorptions and emissions in either the microwave or far-infrared regions of the electromagnetic spectrum. (See Table 7.2.) However, for convenience we will refer to both microwave and far-infrared spectra as pure rotational or rotational spectra. Additionally, microwave frequencies are conventionally reported in units of $\text{GHz}\left(1\text{GHz} = 10^9\,\text{Hz}\right)$, while far-infrared frequencies are conventionally reported in units of cm^{-1}. The conversion factor between these two units is $1\,\text{cm}^{-1} = 29.979\,\text{GHz}$.

We will determine the rotational absorption spectra of diatomic molecules within the rigid rotor approximation. We start with the rigid rotor energy level equation (Equation 8.25) written in terms of the moment of inertia I of the diatomic as

$$E_J = \frac{J(J+1)\hbar^2}{2I} \tag{8.109}$$

where we have used the familiar diatomic molecule form for I (Atkins, de Paula, and Friedman 2014, p. 104)

$$I = \mu r_e^2 \tag{8.110}$$

where recall that r_e is the equilibrium internuclear separation of the molecule which is identical to its bond length.

Let us next consider a transition due to absorption of a photon of frequency $\upsilon_{J\rightarrow J'}$ which takes the system from an initial state with energy E_J to a final state with energy $E_{J'} > E_J$. Using Equation (8.109), the Bohr frequency condition gives for $\upsilon_{J\rightarrow J'}$

$$\upsilon_{J\rightarrow J'} = \frac{E_{J'} - E_J}{h} = \frac{\left[J'(J'+1) - J(J+1)\right]h}{8\pi^2 I}. \tag{8.111}$$

Defining the rotational wave number by $\bar{\upsilon}_{J\rightarrow J'} = c^{-1}\upsilon_{J\rightarrow J'}$, we have from Equation (8.111) that

$$\bar{\upsilon}_{J\rightarrow J'} = \frac{\left[J'(J'+1) - J(J+1)\right]h}{8\pi^2 Ic}. \tag{8.112}$$

Note that $\bar{\upsilon}_{J\rightarrow J'}$ has units of cm^{-1} if c is measured in units of $cm\,s^{-1}$.

In Section 8.12, we derive the following rigid rotor selection rules for rotational absorption transitions:

$$\Delta J = J' - J = 1 \tag{8.113a}$$

and

$$\Delta M = M' - M = 0. \tag{8.113b}$$

The selection rule for J requires that $J' = J+1$. Consequently, Equation (8.112) for the rotational absorption wave numbers becomes

$$\bar{\upsilon}_{J\rightarrow J+1} = \frac{\left[(J+1)(J+2) - J(J+1)\right]}{8\pi^2 Jc}h = 2\bar{B}(J+1), \text{ where } J = 0, 1, 2, \ldots \tag{8.114}$$

and where the *rotational constant* $\bar{B}$ is defined by

$$\bar{B} = \frac{h}{8\pi^2 Ic}. \tag{8.115}$$

From Equation (8.114), the rigid rotor pure rotational spectrum of a diatomic molecule consists of a series of equally spaced lines occurring at $2\bar{B}(J=0), 4\bar{B}(J=1), 6\bar{B}(J=2)$, and so on (Figure 8.3). From the observed spectrum, one can therefore obtain the line spacing

$$2\bar{B} = \frac{h}{4\pi^2 Ic} = \frac{h}{4\pi^2 \mu r_e^2 c} \tag{8.116}$$

and thus determine experimentally both the diatomic's moment of inertia I and its bond length r_e.

Next, let us look at an example. To a good approximation, the $^1H^{35}CI$ pure rotational absorption spectrum consists of a series of equally spaced lines with line spacing $2\bar{B} = 20.90\,cm^{-1}$. Let us determine the rigid rotor model predictions for the moment of inertia and bond length of $^1H^{35}CI$.

First, let us consider I. From Equation (8.116), I is given in terms of $2\bar{B}$ by $I = \frac{h}{4\pi^2(2\bar{B})c}$. Thus, for $^1H^{35}CI$,

$$I = \frac{6.626\times10^{-34}\,J.s}{(4\pi^2)(20.90\ cm^{-1})(2.998\times10^8 m.s^{-1})\left(\frac{100m^{-1}}{cm^{-1}}\right)} = 2.68\times10^{-47}\,kg.m^2.$$

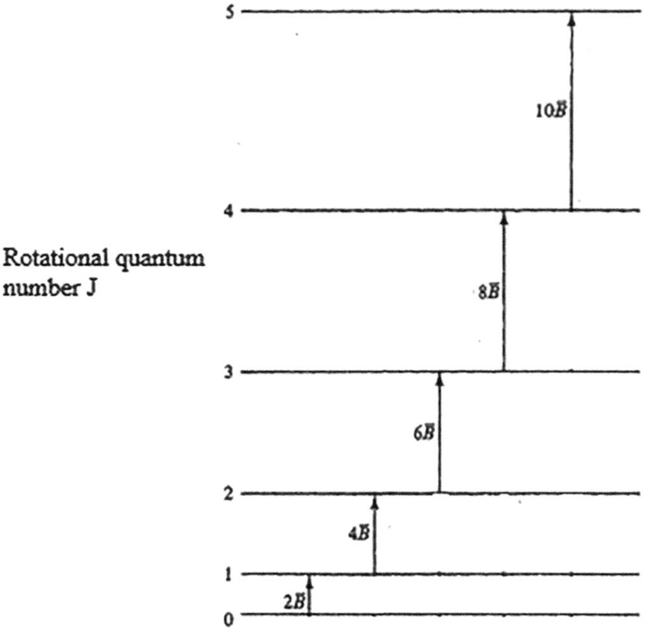

FIGURE 8.3 Rotational absorption spectrum of a diatomic molecule within the rigid rotor approximation. The spectrum consists of a series of equally spaced lines occurring at $2\bar{B}$ arising from the transition $J = 0 \rightarrow J = 1$, at $4\bar{B}$ arising from the transition $J = 1 \rightarrow 2$, and so on.

Next, let us determine $r_e = \left(\frac{I}{\mu}\right)^{1/2}$. For ${}^1H^{35}CI$,

$$\mu = \frac{m_{^1H} m_{^{35}CI}}{m_{^1H} + m_{^{35}CI}} u \times 1.661\,\text{kg}\,u^{-1} = \frac{(1.0078)(34.97)}{1.0078 + 34.97} \times 1.661\,\text{kg} = 1.63 \times 10^{-27}\,\text{kg}.$$

Therefore, for ${}^1H^{35}CI$ $r_e = \left(\frac{2.68 \times 10^{-47}\,\text{kg}\,\text{m}^2}{1.63 \times 10^{-27}\,\text{kg}}\right)^{1/2} = 1.28 \times 10^{-10}\,\text{m}$ or $r_e = 128$ pm.

This rigid rotor approximation for r_e is in excellent agreement with the most precise experimental value 127.4552 pm.

Finally, let us look at isotope effects. Consider two isotopes of a diatomic molecule labeled, respectively, by 1 and 2. Suppose that $2\bar{B}_1$ is known, but $2\bar{B}_2$ is not known. We wish to determine $2\bar{B}_2$ from $2\bar{B}_1$. Recalling that r_e is independent of isotope, we have from Equation (8.116) that $2\bar{B}_2 = \left(\frac{\mu_1}{\mu_2}\right) 2\bar{B}_1$. Next, let us apply this relation to the HCI isotopes ${}^2H^{35}CI$ for which we assume that $2\bar{B}$ is unknown and ${}^1H^{35}CI$ for which $2\bar{B} = 20.90\,\text{cm}^{-1}$. The reduced mass of ${}^2H^{35}CI$ is

$$\mu_{^2H^{35}CI} = \frac{m_{^2H} m_{^{35}CI}}{m_{^2H} + m_{^{35}CI}} u = \frac{(2.014)(34.97)}{2.104 + 34.97} u = 1.904u$$

while the reduced mass of ${}^1H^{35}CI$ is

$$\mu_{^1H^{35}CI} = \frac{m_{^1H} m_{^{35}CI}}{m_{^1H} + m_{^{35}CI}} u = \frac{(1.0078)(34.97)}{1.0078 + 34.97} u = 0.9804u.$$

Therefore, $2\bar{B}_{{}^2H^{35}Cl} = \left(\dfrac{0.98044}{1.9044}\right)\left(2\bar{B}_{{}^1H^{35}Cl}\right) = (0.515)\left(20.90\,\text{cm}^{-1}\right) = 10.76\,\text{cm}^{-1}$.

Thus, the rigid rotor pure rotational spectrum of ${}^2H^{35}Cl$ consists of a series of equally spaced lines with line spacing $2\bar{B}_{{}^2H^{35}Cl} = 10.76\,\text{cm}^{-1}$.

Next, we derive the rigid rotor absorption selection rules of Equation (8.113).

8.12 DERIVATION OF THE RIGID ROTOR SELECTION RULES FOR DIATOMIC PURE ROTATIONAL SPECTRA

In Section 7.14, we derived the selection rules for the infrared spectrum of a diatomic molecule within the harmonic oscillator approximation. As for the harmonic oscillator, the gross selection rule for the rigid rotor is that in order for the diatomic molecule to absorb or emit electromagnetic radiation, it must have a permanent dipole moment $\boldsymbol{\mu}$. Moreover, for radiation whose polarization vector points along the Z-axis, the condition that the probability of an absorption transition $JM \rightarrow J'M'$ where $J' > J$ be non-vanishing is that the transition dipole moment M_z defined in the next equation be non-vanishing, that is, that

$$M_z \equiv \int_0^{\pi}\int_0^{2\pi} \sin\theta Y^*_{J'M'}(\theta,\phi)\,\mu_z Y_{JM}(\theta,\phi)\,d\theta d\phi \neq 0. \tag{8.117}$$

In Equation (8.117), μ_z is the Z-component of the dipole moment vector $\boldsymbol{\mu}$ given by

$$\mu_z = \mu\cos\theta \tag{8.118}$$

where μ is the magnitude of the vector $\boldsymbol{\mu}$. Given Equation (8.118), Equation (8.117) becomes

$$M_z = \mu\int_0^{\pi}\int_0^{2\pi} \sin\theta\cos\theta Y^*_{J'M'}(\theta,\phi) Y_{JM}(\theta,\phi)\,d\theta d\phi \neq 0. \tag{8.119}$$

However, from Equations (8.103) and (8.105), the spherical harmonics may be written as

$$Y_{JM}(\theta,\phi) = N_{JM}P_J^{|M|}(\cos\theta)\left(\frac{1}{2\pi}\right)^{1/2}\exp(iM\phi) \tag{8.120}$$

and thus, the condition of Equation (8.119) may be expressed as

$$M_z = \mu N_{J'M'}N_{JM} \times$$

$$\int_0^{\pi}\int_0^{2\pi} \sin\theta\cos\theta P_{J'}^{|M'|}(\cos\theta)P_J^{|M|}(\cos\theta)\frac{\exp\left[i(M-M')\phi\right]}{2\pi}\,d\theta\, d\phi \neq 0. \tag{8.121}$$

When expressed in terms of the variable $x = \cos\theta$, Equation (8.121) transforms to

$$M_z = \mu N_{J'M'}N_{JM}\int_{-1}^{1} P_{J'}^{|M'|}(x)\,x P_J^{|M|}(x)\,dx \int_0^{2\pi}\frac{\exp\left[i(M-M')\phi\right]}{2\pi}d\phi \neq 0. \tag{8.122}$$

But it is easily verified that

$$\int_0^{2\pi} \frac{\exp\left[i(M-M')\phi\right]}{2\pi} d\phi = \delta_{M'M}. \tag{8.123}$$

Inserting Equation (8.123) into Equation (8.122) shows that M_z is non-vanishing and hence an absorption transition is possible only if $M' = M$ or $\Delta M = M' - M = 0$, thus verifying the second part of the absorption selection rule of Equation (8.113). To establish the first part of the selection rule, first we set $M' = M$ in Equation (8.122) to obtain the condition

$$M_z = \mu N_{J'M} N_{JM} \int_{-1}^{1} P_{J'}^{|M|}(x)\, x P_J^{|M|}(x)\, dx \neq 0. \tag{8.124}$$

Then, to eliminate the factor $xP_J^{|M|}(x)$ in Equation (8.124), we insert the recursion relation of Equation (8.98) written as

$$xP_J^{|M|}(x) = \frac{(J-|M|+1)}{2J+1} P_{J+1}^{|M|}(x) + \frac{J-|M|}{2J+1} P_{J-1}^{|M|}(x) = 0 \tag{8.125}$$

into Equation (8.124) to obtain

$$M_z = \mu N_{J'M} N_{JM} \times \left[\frac{(J-|M|+1)}{2J+1} \int_{-1}^{1} P_{J'}^{|M|}(x) P_{J+1}^{|M|}(x)\, dx + \frac{(J-|M|)}{2J+1} \int_{-1}^{1} P_{J'}^{|M|}(x) P_{J-1}^{|M|}(x)\, dx \right] \neq 0. \tag{8.126}$$

But according to Equation (8.99a), associated Legendre functions $P_J^{|M|}(x)$ with the same $|M|$ values but different J values are orthogonal on the interval $-1 \le x \le 1$. Thus, since for absorption $J' > J$, the second integral on the right-hand side of Equation (8.126) vanishes. For the first integral to be non-vanishing and thus for the absorption transition $JM \rightarrow J'M$ to be allowed, we must similarly have that $J' = J+1$ or $\Delta J = J' - J = 1$, which is the first part of the selection rule of Equation (8.113).

In summary, we have proven the absorption selection rules $\Delta J = J' - J = 1$ and $\Delta M = M' - M = 0$ of Equation (8.113). The emission selection rules $\Delta J = J' - J = -1$ and $\Delta M = M' - M = 0$ may be established similarly.

FURTHER READINGS

Abramovitz, Milton, and Irene A. Stegun, eds. 1965. *Handbook of mathematical functions with formulas, graphs, and mathematical tables.* New York: Dover Publications.

Arfken, George B., Hans J. Weber, and Frank E. Harris. 2012. *Mathematical methods for physicists.* 7th ed. Waltham, MA: Academic Press.

Atkins, Peter, Julio de Paula, and Ronald Friedman. 2014. *Physical Chemistry: quanta, matter, and change.* 2nd ed. Great Britain: Oxford University Press.

Atkins, Peter, and Ronald Friedman. 2011. *Molecular quantum mechanics.* 5th ed. New York: Oxford University Press.

Cohen-Tannoudji, Claude, Bernard Diu, and Franck Laloe. 1977. *Quantum mechanics.* Vol. 1. Translated by Susan Reid Hemley, Nicole Ostrowsky, and Dan Ostrowsky. New York: John Wiley & Sons.

Jeffreys, Harold, and Bertha Swirles. 1956. *Methods of mathematical physics.* 3rd ed. New York: Cambridge University Press.

Levine, Ira N. 2015. *Quantum chemistry.* 7th ed. Englewood Cliffs, NJ: Prentice-Hall Inc.

McQuarrie, Donald A. 2008. *Quantum chemistry.* 2nd ed. Sausalito, CA: University Science Books.

Merzbacher, Eugen. 1961. *Quantum mechanics.* New York: John Wiley & Sons, Inc.

Pauling, Linus, and E. Bright Wilson Jr. 1985. *Introduction to quantum mechanics with applications to chemistry.* New York: Dover Publications, Inc.
Sakurai, J. J. 1985. *Modern quantum mechanics.* Rev. ed. Edited by San Fu Tuan. Reading, MA: Addison-Wesley Publishing Co.

PROBLEMS

8.1 In this problem, we will be concerned with the origin of generalized coordinate expressions for integrals like those of Equations (8.11) and (8.12). In our development, we will consider a general f-degree-of-freedom system with Cartesian coordinates $x_1, y_1, \ldots$ and generalized coordinates $q_1, \ldots, q_f$ linked by transformation relations of the form $x_1 = x_1(q_1, \ldots, q_f), y_1 = y_1(q_1, \ldots, q_f)$, and so on. The focus of our discussion will be the generalized coordinate form of the system's volume element $dx_1dy_1 \ldots$. This form is developed in applied mathematics texts (Jeffreys and Swirles 1956, sec. 5.052) as $dx_1dy_1 \ldots = \partial(x_1, y_1, \ldots)/\partial(q_1, \ldots, q_f)\ dq_1, dq_f \ldots,$ where $\partial(x_1, y_1, \ldots)/\partial(q_1, \ldots, q_f)$ is the *Jacobian* of $x_1, y_1, \ldots$ with respect to $q_1, \ldots, q_f$. The Jacobian is an extension of the concept of a partial derivative and is defined as the following determinant of partial derivatives: $\begin{vmatrix} \frac{\partial x_1}{\partial q_1} & \frac{\partial y_1}{\partial q_1} \cdots \\ \vdots & \vdots \\ \frac{\partial x_1}{\partial q_f} & \frac{\partial y_1}{\partial q_f} \cdots \end{vmatrix}$.

The partial derivatives $\partial x_1/\partial q_1$ and so forth in the Jacobian may be found explicitly if the transformation relations linking $x_1, y_1, \ldots$ and $q_1, \ldots, q_f$ are known. We next consider applications of these ideas. (a) First, consider a single-degree-of-freedom system. Such a system is specified by a single coordinate, and thus, the transformation relations reduce to $x = x(q)$. Show for this system that the Jacobian $\partial(x_1, y_1, \ldots)/\partial(q_1, \ldots, q_f) = dx(q)/dq$ and hence the relation $dx_1dy_1 \ldots = \partial(x_1, y_1, \ldots)/\partial(q_1, \ldots, q_f)dq_1 \ldots dq_f$ reduces to the elementary calculus result $dx = [dx(q)/dq]dq$. (b) Denote the stationary state wave functions of the single-degree-of-freedom system by $\Psi_n(q)$. In analogy to Equation (8.9), the mixed expression for the normalization condition of $\Psi_n(q)$ is $\int_{-\infty}^{\infty} \Psi_n^*(q)\Psi_n(q)dx = 1$. Using your result in part (a), show that the pure generalized coordinate expression for this condition is $\int_{\text{all } q} \Psi_n^*(q)\Psi_n(q)[dx(q)/dq]dq = 1$ where $\int_{\text{all } q} dq$ indicates an integral over the full range of q. (c) Next, consider a two-degree-of-freedom system. Show for this system that the extension of your result for the generalized coordinate firm of the normalization integral in part (b) is $\int_{\text{all } q_1}\int_{\text{all } q_2} \Psi_{n_1n_2}^*(q_1, q_2)\Psi_{n_1n_2}(q_1, q_2)[(\partial x/\partial q_1)(\partial y/\partial q_2) - (\partial y/\partial q_1)(\partial x/\partial q_2)]dq_1dq_2 = 1$. (d) Next, specialize the results of part (c) to the two-degree-of-freedom system depicted in Figure A.5. For this system, the generalized coordinates q_1 and q_2 are the polar coordinates r and ϕ. Using the transformation relations of Equation (A.36), show for this system that the Jacobian $(\partial x/\partial q_1)(\partial y/\partial q_2) - (\partial y/\partial q_1)(\partial x/\partial q_2) = r$ and hence the normalization integral of part (c) reduces to $\int_0^{\infty}\int_0^{2\pi} r\Psi_{n_1n_2}^*(r, \phi)\Psi_{n_1n_2}(r, \phi)drd\phi = 1$. (e) The Jacobian for the transformation from

Cartesian to spherical polar coordinates is $\partial(x,y,z)/\partial(r,\theta,\phi)=\begin{vmatrix}\frac{\partial x}{\partial r} & \frac{\partial y}{\partial r} & \frac{\partial z}{\partial r}\\ \frac{\partial x}{\partial \theta} & \frac{\partial y}{\partial \theta} & \frac{\partial z}{\partial \theta}\\ \frac{\partial x}{\partial \phi} & \frac{\partial y}{\partial \phi} & \frac{\partial y_1}{\partial \phi}\end{vmatrix}$. Using the spherical polar to Cartesian coordinate transformation equations (Equation 8.3) and the standard rule for evaluating third-order determinants (Arfken, Weber, and Harris 2012), show that $\partial(x,y,z)/\partial(r,\theta,\phi)=r^2\sin\theta$ and hence prove Equation (8.10).

8.2 (a) Derive the Cartesian partial differential form of $\hat{J}^2$ from Equations (8.29) and (8.30). (b) Compare this Cartesian form with the corresponding spherical polar form of Equation (8.46). Does this comparison suggest which choice of coordinates, Cartesian or spherical polar, is better suited for dealing with the $\hat{J}^2$ eigenvalue problem?

8.3 Consider the commutator $\left[\hat{A}\hat{B},\hat{C}\hat{D}\right]$ where $\hat{A}$ and so on are generic quantum operators. In Problem 5.1d, we showed that $\left[\hat{A}\hat{B},\hat{C}\hat{D}\right]=\hat{A}\hat{C}\left[\hat{B},\hat{D}\right]+\hat{A}\left[\hat{B},\hat{C}\right]\hat{D}+\hat{C}\left[\hat{A},\hat{D}\right]\hat{B}+\left[\hat{A},\hat{C}\right]\hat{D}\hat{B}$. Prove Equation (8.31a) by writing $\left[\hat{J}_x,\hat{J}_y\right]=\left[\hat{y}\hat{p}_z,\hat{z}\hat{p}_x\right]+$ three similar terms and finally by evaluating $\left[\hat{y}\hat{p}_z,\hat{z}\hat{p}_x\right]$ and so on using the above commutator identity and the fact that the only non-null commutators of the components of $\hat{\mathbf{r}}$ and $\hat{\mathbf{p}}$ are those of Problem 5.3.

8.4 In this problem, we will outline the derivation of the spherical polar coordinate expression $\hat{J}_z=\frac{\hbar}{i}\frac{\partial}{\partial\phi}$ of Equation (8.45c) from the Cartesian expression $\hat{J}_z=\frac{\hbar}{i}\left(x\frac{\partial}{\partial y}-y\frac{\partial}{\partial x}\right)$. We will need the following relations that follow from the Cartesian to spherical polar coordinate transformation equations of Equation (8.3): $r^2=x^2+y^2+z^2(1)$, $\cos\theta=\frac{z}{\left(x^2+y^2+z^2\right)^{1/2}}(2)$, $\tan\phi=\frac{y}{x}(3)$, $x=r\sin\theta\sin\phi(4)$, $y=r\sin\theta\sin\phi(5)$, and $z=r\cos\theta(6)$. We will use these relations to express $\frac{\partial}{\partial y}$ and $\frac{\partial}{\partial x}$ in terms of spherical polar partial derivatives. To start, we use the chain rule of partial differential calculus to obtain $\frac{\partial}{\partial y}=\frac{\partial r}{\partial y}\frac{\partial}{\partial r}+\frac{\partial\theta}{\partial y}\frac{\partial}{\partial\theta}+\frac{\partial\phi}{\partial y}\frac{\partial}{\partial\phi}(7)$ and $\frac{\partial}{\partial x}=\frac{\partial r}{\partial x}\frac{\partial}{\partial r}+\frac{\partial\theta}{\partial x}\frac{\partial}{\partial\theta}+\frac{\partial\phi}{\partial x}\frac{\partial}{\partial\phi}(8)$. We next evaluate $\frac{\partial r}{\partial y},\frac{\partial\theta}{\partial y}$, and so on in (7) and (8). We first evaluate $\frac{\partial r}{\partial y}$. Using (1), $\frac{\partial r^2}{\partial y}=2r\frac{\partial r}{\partial y}=2y$. So $\frac{\partial r}{\partial y}=\frac{y}{r}$, which from (5) becomes $\frac{\partial r}{\partial y}=\frac{r\sin\theta\sin\phi}{r}=\sin\theta\sin\phi$.

(a) Similarly, show that $\frac{\partial r}{\partial x}=\frac{x}{r}=\sin\theta\cos\phi$. Next, let us evaluate $\frac{\partial\theta}{\partial y}$. Using (2), $\frac{\partial\cos\theta}{\partial y}=-\sin\theta\frac{\partial\theta}{\partial y}=-\frac{zy}{\left(x^2+y^2+z^2\right)^{3/2}}=-\frac{r\cos\theta r\sin\theta\sin\phi}{r^3}$. Therefore, $\frac{\partial\theta}{\partial y}=\frac{\cos\theta\sin\phi}{r}$.

(b) Similarly, show that $\frac{\partial\theta}{\partial x}=\frac{\cos\theta\cos\phi}{r}$. Next, let us evaluate $\frac{\partial\phi}{\partial y}$. Differentiating $\tan\phi$ with respect to y and using (3) gives

$\frac{d\tan\phi}{d\phi}\frac{\partial\phi}{\partial y} = \frac{1}{\cos^2\phi}\frac{\partial\phi}{\partial y} = \frac{1}{x}$. So $\frac{\partial\phi}{\partial y} = \frac{\cos^2\phi}{x} = \frac{\cos^2\phi}{r\cos\phi\sin\theta} = \frac{\cos\phi}{r\sin\theta}$. (c) Similarly, show that $\frac{\partial\phi}{\partial x} = -\frac{\sin\phi}{r\sin\theta}$. Pulling these results together, we find that $\frac{\partial}{\partial y}$ of (7) and $\frac{\partial}{\partial x}$ of (8) are given by $\frac{\partial}{\partial y} = \sin\theta\sin\phi\frac{\partial}{\partial r} + \frac{\cos\theta\sin\phi}{r}\frac{\partial}{\partial\theta} + \frac{\cos\phi}{r\sin\theta}\frac{\partial}{\partial\phi}$. and $\frac{\partial}{\partial x} = \sin\theta\cos\phi\frac{\partial}{\partial r} + \frac{\cos\theta\cos\phi}{r}\frac{\partial}{\partial\theta} - \frac{\sin\phi}{r\sin\theta}\frac{\partial}{\partial\phi}$. (d) Use these results in the Cartesian expression for $\hat{J}_z$ to obtain the spherical polar expression for $\hat{J}_z$.

8.5 Using the method of Problem 8.4, derive the spherical polar forms for $\hat{J}_x$ and $\hat{J}_y$ given in Equations (8.45a) and (8.45b).

8.6 Derive the form for $\hat{J}^2$ of Equation (8.46) from the spherical polar forms for $\hat{J}_x, \hat{J}_y, \hat{J}_z$ of Equations (8.45) and also from (8.30).

8.7 Derive the angular momentum commutation rules of Equation (8.31) from the spherical polar forms of Equation (8.45).

8.8 Derive the commutation rule $\left[\hat{J}^2, \hat{J}_z\right] = \hat{O}$ from the spherical polar forms for $\hat{J}_z$ and $\hat{J}^2$ given in Equations (8.45c) and (8.46).

8.9 Prove that Equations (8.52) and (8.54) define a Hermitian eigenvalue problem.

8.10 Using Table 8.4, verify the spherical harmonic identity $\sum_{M=-J}^{J} \left|Y_{JM}(\theta,\phi)\right|^2 = \frac{2J+1}{4\pi}$ for $J = 0, 1,$ and 2.

8.11 Derive Equation (8.72) for $Q(x)$ by inserting the form for $Q(x)$ of Equation (8.71) into Equation (8.70).

8.12 By inserting the power series expansion of Equation (8.73) into Equation (8.72), derive the recursion relation of Equation (8.74).

8.13 From the $P_J(x)$ values for $J = 0-3$ listed in Table 8.1 using Equation (8.97), determine $P_J^{|M|}(x)$ for $J = 0-3$ and $|M| \le J$ and compare your results with those of Table 8.2.

8.14 From the Rodrigues representation of Equation (8.84), determine $P_J(x)$ for $J = 0-3$ and compare your results with those of Table 8.1. Use the result $0! = 1$ derived in the next problem.

8.15 For positive integral n, n! may be defined as $n! = \int_0^\infty t^n \exp(-t)dt$. (a) Prove that for positive integral n, $I_n = \int_0^\infty t^n \exp(-t)dt = n(n-1) \ldots 1 = n!$ (Hint: Integrate I_n by parts n times.) The definition of z! for arbitrary z is $z! = \int_0^\infty t^z \exp(-t)dt$. (b) Use this definition to show $0! = 1$. (c) Show $\frac{1}{2}! = G_2(\alpha = 1) = \frac{1}{2}\pi^{1/2}$ where $G_2(\alpha)$ is the Gaussian integral defined in Equation (7.44).

8.16 Verify Equation (8.89).

8.17 Derive the forms of $P_J(x)$ for $J = 0-2$ from the generating function of Equations (8.82) and (8.83) and its first two time derivatives evaluated at $t = 0$. Compare your results with those of Table 8.1.

8.18 Given that $P_0(x) = 1$, determine $P_J(x)$ for $J = 1-4$ from the recursion relation of Equation (8.85) and compare your results with those of Table 8.1.

8.19 From the generating function $h(t, x)$ defined in Equations (8.82) and (8.83), prove the symmetry relation for the Legendre polynomials given in Equation (8.95). (Hint: Note that $(h[-t, -x] = h[t, x])$.)

8.20 Using the forms for $P_J(x)$ listed in Table 8.1, prove in agreement with Equation (8.96a) that (a) $\int_{-1}^{1} P_2(x)P_2(x)dx = \frac{2}{5}$, (b) $\int_{-1}^{1} P_1(x)P_3(x)dx = 0$, and (c) $\int_{-1}^{1} P_2(x)P_3(x)dx = 0$. (d) Why is your result in part (c) also a consequence of Equation (8.95)?

8.21 Using the forms for $P_J^{|M|}(x)$ listed in Table 8.2, prove in agreement with Equation (8.99a) that (a) $\int_{-1}^{1} P_1^1(x)P_1^1(x)dx = \frac{4}{3}$ and (b) $\int_{-1}^{1} P_1^1(x)P_2^1(x)dx = 0$.

8.22 In this problem, you will verify the orthonormality equation (Equation 8.107) for a few spherical harmonics. Use Table 8.4 to show that (a) $\int_0^{\pi} \sin\theta d\theta \int_0^{2\pi} d\phi Y_{21}^*(\theta,\phi)Y_{21}(\theta,\phi) = 1$, (b) $\int_0^{\pi} \sin\theta d\theta \int_0^{2\pi} d\phi Y_{21}^*(\theta,\phi)Y_{06}(\theta,\phi) = 0$, and (c) $\int_0^{\pi} \sin\theta d\theta \int_0^{2\pi} d\phi Y_{31}^*(\theta,\phi)Y_{21}(\theta,\phi) = 0$.

8.23 (a) Show that $Y_{11}(\theta,\phi)$ given in Table 8.4 is an eigenfunction of $\hat{J}^2$ of Equation (8.46) with eigenvalue $2\hbar^2$. (b) Given Equation (8.45a) for $\hat{J}_x$, $Y_{11}(\theta,\phi)$ is not an eigenfunction of $\hat{J}_x$.

8.24 For all spherical harmonics $Y_{JM}(\theta,\phi)\langle J_x\rangle \equiv \int_0^{\pi}\int_0^{2\pi} \sin\theta Y_{JM}^*(\theta,\phi)\, \hat{J}_x Y_{JM}(\theta,\phi)d\theta d\phi = 0$ and $\langle J_y\rangle \equiv \int_0^{\pi}\int_0^{2\pi} \sin\theta Y_{JM}^*(\theta,\phi)\hat{J}_x Y_{JM}(\theta,\phi)d\theta d\phi = 0$ where $\hat{J}_x$ and $\hat{J}_y$ are given in Equations (8.45a) and (8.45b). Prove these two results. (Hint: Write $Y_{JM}[\theta,\phi]$ as $Y_{JM}[\theta,\phi] = \left[\frac{1}{2\pi}\right]^{1/2} \Theta_{JM}[\theta]\exp[iM\phi]$ and perform the integral over ϕ first.)

8.25 (a) Compute $2\bar{B}$ for $^1H^{37}CI$ using the value $2\bar{B} = 20.90\,cm^{-1}$ for $^1H^{35}CI$. Notice that the isotopic shift relative to $^1H^{35}CI$ is much smaller than the corresponding shift for $^2H^{35}CI$. (b) Explain why this is the case.

8.26 The $J = 0 \rightarrow J = 1$ transition in the rotational spectrum of $^{12}C^{16}O$ occurs at frequency 115.27121 GHz. (a) Compute the moment of inertia of $^{12}C^{16}O$. (b) Compute the bond length r_e of CO. (c) Compute the frequency in GHz of the $J = 8 \rightarrow J = 9$ transition of $^{12}C^{16}O$.

8.27 The bond length of HBr is 141.4pm. (a) Compute the moment of inertia for $^1H^{79}Br$. (b) Compute the rotational spectrum absorption wave number $\bar{\upsilon}_{8\rightarrow 9}$ for $^1H^{79}Br$. (c) Compute the moment of inertia for $^2H^{79}Br$. (d) Compute the rotational spectrum absorption wave number $\bar{\upsilon}_{8\rightarrow 9}$ for $^2H^{79}$ Br.

8.28 Two adjacent lines in the rotational spectrum of a diatomic occur at 14.67 and $16.30\,cm^{-1}$. What is the transition $J \rightarrow J+1$ for the lower frequency line?

8.29 Referring to Table 8.1, show that the first four Legendre polynomials satisfy Legendre's equation (Equation 8.81).

SOLUTIONS TO SELECTED PROBLEMS

8.25 (a) $20.87\,cm^{-1}$.
8.26 (a) $1.456\times10^{-46}\,kg\,m^2$, (b) 113.1 pm, and (c) $1.0374409\times10^3\,GH_z$.
8.27 (a) $3.308\times10^{-47}\,kg\,m^2$, (b) $152.4\,cm^{-1}$, (c) $6.522\times10^{-47}\,kg\,m^2$, and (d) $77.25\,cm^{-1}$.
8.28 $J = 8 \rightarrow 9$.

9 Diatomic Rotational–Vibrational Spectroscopy

In Sections 7.13 and 8.11, we gave preliminary accounts of the IR and pure rotational spectra of diatomic molecules. Here, we give more detailed descriptions. We start by deriving the fundamental equation that underlies our treatments, the Schrödinger equation for the vibrational motion of a diatomic in the presence of rotation.

9.1 THE VIBRATIONAL TIME-INDEPENDENT SCHRÖDINGER EQUATION

Our starting point is the diatomic's rotational–vibrational Hamiltonian operator $\hat{H}$ of Equation (8.4) written in terms of $\hat{J}^2$ of Equation (8.20)

$$\hat{H} = -\frac{\hbar^2}{2\mu}\frac{1}{r^2}\frac{\partial}{\partial r}\left(r^2\frac{\partial}{\partial r}\right) + \frac{\hat{J}^2}{2\mu r^2} + U(r). \tag{9.1}$$

With the above form for $\hat{H}$, the rotational–vibrational Schrödinger equation (Equation 8.6) becomes

$$\left[-\frac{\hbar^2}{2\mu}\frac{1}{r^2}\frac{\partial}{\partial r}\left(r^2\frac{\partial}{\partial r}\right) + \frac{\hat{J}^2}{2\mu r^2} + U(r)\right]\Psi_{vJM}(r,\theta,\phi) = E_{vJ}\Psi_{vJM}(r,\theta,\phi). \tag{9.2}$$

We seek solutions of the form

$$\Psi_{vJM}(r,\theta,\phi) = R_{vJ}(r)Y_{JM}(\theta,\phi) \tag{9.3}$$

where $R_{vJ}(r)$ will turn out to be a vibrational wave function and where $Y_{JM}(\theta,\phi)$ is a spherical harmonic. Inserting Equation (9.3) into Equation (9.2) and recalling Equation (8.108a) $\hat{J}^2 Y_{JM}(\theta,\phi) = J(J+1)\hbar^2 Y_{JM}(\theta,\phi)$ gives $Y_{JM}(\theta,\phi)\left[-\frac{\hbar^2}{2\mu}\frac{1}{r^2}\frac{d}{dr}\left(r^2\frac{d}{dr}\right) + \frac{J(J+1)\hbar^2}{2\mu r^2} + U(r)\right]R_{VJ}(r) = E_{vJ}Y_{JM}(\theta,\phi)R_{vJ}(r)$.

Canceling the $Y_{JM}(\theta,\phi)$ factor then yields the vibrational Schrödinger equation

$$\left[-\frac{\hbar^2}{2\mu}\frac{1}{r^2}\frac{d}{dr}\left(r^2\frac{d}{dr}\right) + \frac{J(J+1)\hbar^2}{2\mu r^2} + U(r)\right]R_{vJ}(r) = E_{vJ}R_{vJ}(r). \tag{9.4}$$

Equation (9.4) takes on an especially useful form when written in terms of the functions $\Psi_{vJ}(r)$ defined by $\Psi_{vJ}(r) = rR_{vJ}(r)$. In terms of $\Psi_{vJ}(r)$, Equation (9.4) becomes (Problem 9.1)

$$-\frac{\hbar^2}{2\mu}\frac{d^2}{dr^2}\Psi_{vJ}(r) + \left[\frac{J(J+1)\hbar^2}{2\mu r^2} + U(r)\right]\Psi_{vJ}(r) = E_{vJ}\Psi_{vJ}(r). \tag{9.5}$$

Defining the *effective potential* $U_{eff}(r)$ by (compare with the classical effective potential of Equation [B.40])

$$U_{eff}(r) = \frac{J(J+1)\hbar^2}{2\mu r^2} + U(r) \tag{9.6}$$

Equation (9.5) may then be written as

$$-\frac{\hbar^2}{2\mu}\frac{d^2}{dr^2}\Psi_{vJ}(r) + U_{eff}(r)\Psi_{vJ}(r) = E_{vJ}\Psi_{vJ}(r). \tag{9.7}$$

Notice that if we set J = 0 in Equation (9.6), $U_{eff}(r)$ reduces to the true potential $U(r)$ and Equation (9.7) becomes the Schrödinger equation (Equation 7.111) for the vibrational wave functions of a non-rotating diatomic molecule. However, this quantum equation (Equation 7.111) was originally arrived at non-rigorously by analogy the corresponding classical equation. The present derivation provides a rigorous justification for the central Equation (7.111).

Also notice that Equation (9.7) is identical in form to a *one-dimensional* Schrödinger equation for a particle of mass μ = diatomic's reduced mass and coordinate r = diatomic's internuclear separation subject to the potential energy function $U_{eff}(r)$. Equation (9.7) however is not a true one-dimensional Schrödinger equation since for such an equation the range of the particle's coordinate x is $-\infty < x < \infty$ while the range of the internuclear separation r is $0 \le r < \infty$. Correspondingly, while the boundary conditions on the solutions $\Psi_n(x)$ of a true one-dimensional Schrödinger equation are $\lim_{x\to\pm\infty} \Psi_n(x) = 0$, the boundary conditions on the solutions $\Psi_{vJ}(r)$ of Equation (9.7) are (Problem 9.2)

$$\lim_{r\to 0}\Psi_{vJ}(r) = 0 \text{ and } \lim_{r\to\infty}\Psi_{vJ}(r) = 0. \tag{9.8}$$

These differences in boundary conditions can yield solutions $\Psi_{vJ}(r)$ of Equation (9.7) which are radically different from the solutions $\Psi_n(x)$ of a true one-dimensional Schrödinger equation. For this reason, Equation (9.7) is referred to as a *pseudo-one-dimensional* Schrödinger equation.

We next turn to the harmonic oscillator rigid rotor approximation to the pseudo-one-dimensional Schrödinger equation (Equation 9.7) and the rotational–vibrational energy levels which emerge from solving it.

9.2 THE HARMONIC OSCILLATOR RIGID ROTOR APPROXIMATION TO THE ROTATIONAL–VIBRATIONAL ENERGY LEVELS OF A DIATOMIC MOLECULE

The harmonic oscillator rigid rotor approximation combines the harmonic oscillator approximation of Section 7.13 and the rigid rotor approximation of Chapter 8. The approximation, like the pure harmonic oscillator approximation, is based on the assumption that the characteristic displacements $y = r - r_e$ of the internuclear separation r from its equilibrium value r_e are of small amplitude. In accord with this assumption, within harmonic oscillator rigid rotor approximation the true potential $U(r)$ is replaced by the harmonic oscillator potential $U_{MO}(y) = \frac{1}{2}\mu\omega^2 y^2$ of Equation (7.118). Additionally, the rotational term $\frac{J(J+1)\hbar^2}{2\mu r^2}$ in $U_{eff}(r)$ is replaced by its rigid rotor approximation $\frac{J(J+1)\hbar^2}{2\mu r_e^2}$. The rigid rotor approximation to the rotational term may also be written as $\frac{J(J+1)\hbar^2}{2I}$, where $I = \mu r_e^2$ is the moment of inertia of the diatomic molecule. Thus, within the harmonic oscillator rigid rotor approximation the effective potential of Equation (9.6) has the approximate form

$$U_{eff}(r) \doteq \frac{1}{2}\mu\omega^2 y^2 + \frac{J(J+1)\hbar^2}{2I}. \tag{9.9}$$

Equation (9.9) yields the harmonic oscillator rigid rotor approximation to the pseudo-one-dimensional Schrödinger equation (Equation 9.7)

$$-\frac{\hbar^2}{2\mu}\frac{d^2}{dr^2}\Psi_{vJ}(r) + \frac{1}{2}\mu\omega^2 y^2 \Psi_{vJ}(r) + \frac{J(J+1)\hbar^2}{2I}\Psi_{vJ}(r) = E_{vJ}\Psi_{vJ}(r). \tag{9.10}$$

However, since the term $\frac{J(J+1)\hbar^2}{2I}$ in Equation (9.10) is an r-independent constant, it does not affect the forms of the vibrational wave functions $\Psi_{vJ}(r)$. That is, these wave functions are the same for all values of the quantum number J and thus may be written as $\Psi_v(r)$. Therefore, Equation (9.10) simplifies to $-\frac{\hbar^2}{2\mu}\frac{d^2}{dr^2}\Psi_v(r) + \frac{1}{2}\mu\omega^2 y^2 \Psi_v(r) + \frac{J(J+1)\hbar^2}{2I}\Psi_v(r) = E_{vJ}\Psi_v(r)$ or equivalently

$$-\frac{\hbar^2}{2\mu}\frac{d^2\Psi_v(r)}{dr^2} + \frac{1}{2}\mu\omega^2 y^2 \Psi_v(r) = \hat{E}_{vJ}\Psi_v(r) \tag{9.11}$$

where

$$\hat{E}_{vJ} \equiv E_{vJ} - \frac{J(J+1)\hbar^2}{2I}. \tag{9.12}$$

Defining $\tilde{\Psi}_v(y)$ by $\tilde{\Psi}_v(y) = \tilde{\Psi}_v(r - r_e) = \Psi_v(r)$, Equation (9.11) becomes

$$-\frac{\hbar^2}{2\mu}\frac{d^2}{dy^2}\tilde{\Psi}_v(y) + \frac{1}{2}\mu\omega^2 y^2 \tilde{\Psi}_v(y) = \hat{E}_{vJ}\tilde{\Psi}_v(y). \tag{9.13}$$

Equation (9.13) is identical to the Schrödinger equation for a harmonic oscillator of mass μ, circular frequency ω, vibrational displacement y, and energy $\hat{E}_{vJ}$. However, the boundary conditions on the solutions $\tilde{\Psi}_v(y)$ of Equation (9.13) are not the same as those on the solutions $\Psi_v(y)$ of a true harmonic oscillator Schrödinger equation

$$\lim_{y\to-\infty}\Psi_v(y) = 0 \text{ and } \lim_{y\to\infty}\Psi_v(y) = 0. \tag{9.14a}$$

Rather, since for $r = 0$ $y = -r_e$, from Equation (9.8) the boundary conditions on the functions $\tilde{\Psi}_v(y)$ are

$$\lim_{y\to-r_e}\tilde{\Psi}_v(y) = 0 \text{ and } \lim_{y\to\infty}\tilde{\Psi}_v(y) = 0. \tag{9.14b}$$

Notice that the left-most boundary conditions in Equations (9.14a) and (9.14b) differ. Thus, the wave functions $\tilde{\Psi}_v(y)$ which we require and the true harmonic oscillator wave functions $\Psi_v(y)$ are not identical. However, when the vibrational displacement y of the true oscillator is $-r_e$, the true harmonic oscillator is so far into its classically forbidden region that its wave function $\Psi_v(-r_e)$

is of negligible magnitude. Thus, $\lim_{y\to -r_e}\Psi_v(y) \doteq \lim_{y\to -\infty}\Psi_v(y) = 0$. Consequently, the exact boundary conditions on $\Psi_v(y)$ of Equation (9.14a) may be replaced with negligible error by the approximate boundary conditions $\lim_{y\to -r_e}\Psi_v(y) = 0$ and $\lim_{y\to -\infty}\Psi_v(y) = 0$. But these approximate boundary conditions are identical to the boundary conditions on $\tilde{\Psi}_v(y)$ of Equation (9.14b). Therefore, with negligible error the solutions $\tilde{\Psi}_v(y)$ and $\hat{E}_{vJ}$ of Equation (9.13) may be taken as the true harmonic oscillator wave functions $\Psi_v(y)$ and the true harmonic oscillator energies $E_v = \left(v + \frac{1}{2}\right)\hbar\omega = \left(v + \frac{1}{2}\right)h\upsilon$. Especially, to an excellent approximation

$$\hat{E}_{vJ} = \left(v + \frac{1}{2}\right)\hbar\omega = \left(v + \frac{1}{2}\right)h\upsilon. \tag{9.15}$$

Comparing Equations (9.12) and (9.15) then gives the harmonic oscillator rigid rotor approximation to the diatomic's rotational–vibrational energy levels for E_{vJ} as

$$E_{vJ} = \left(v + \frac{1}{2}\right)h\upsilon + \frac{J(J+1)\hbar^2}{2I}, \text{ where } v = 0, 1, \ldots \text{ and } J = 0, 1, \ldots. \tag{9.16}$$

Notice that within the harmonic oscillator rigid rotor approximation, E_{vJ} is the sum of the energy $E_v = \left(v + \frac{1}{2}\right)h\upsilon$ of an isolated harmonic oscillator and the energy $E_J = \frac{J(J+1)\hbar^2}{2I}$ of an isolated rigid rotor.

Next, let us determine the IR absorption spectrum of a diatomic molecule within the harmonic oscillator rigid rotor approximation.

9.3 THE IR ABSORPTION SPECTRUM OF A DIATOMIC MOLECULE WITHIN THE HARMONIC OSCILLATOR RIGID ROTOR APPROXIMATION

In Section 7.13, we discussed the low-resolution IR absorption spectrum of a diatomic molecule. We found that at low resolution, only a single line is detectable. Neglecting rotations and assuming that the vibrations of the diatomic are harmonic and also that only the $v = 0 \to v = 1$ transition contributes to the line, we found that the position $\bar{\upsilon}_{IR}$ of the line coincides with the diatomic's vibrational wave number $\bar{\omega}$. That is, we found that (Equation 7.125)

$$\bar{\upsilon}_{IR} = \bar{\omega}. \tag{9.17}$$

However, our discussion of Section 7.13 is incomplete since rotational transitions (nearly) always accompany changes in vibrational quantum number. These rotational transitions produce fine structure in the IR spectrum which splits the low-resolution line into two series of narrowly spaced lines which are detectable at high resolution. Next, we treat this fine structure within the harmonic oscillator rigid rotor approximation.

To start, we rewrite Equation (9.16) as

$$\frac{E_{vJ}}{hc} = \left(v + \frac{1}{2}\right)\bar{\omega} + J(J+1)\bar{B} \tag{9.18}$$

where as in Equations (7.124) and (8.115) the diatomic's vibrational wave number $\bar{\omega}$ and rotational constant $\bar{B}$ are given by

$$\bar{\omega} = \frac{\upsilon}{c} \text{ and } \bar{B} = \frac{h}{8\pi^2 Ic}. \tag{9.19}$$

We next note for *all* diatomics

$$\bar{B} \ll \bar{\omega}. \tag{9.20}$$

For example, for $^1H^{35}Cl$ $\bar{B} = 10.59\,cm^{-1}$ and $\bar{\omega} = 2{,}885\,cm^{-1}$, while for $^{12}C^{16}O$ $\bar{B} = 1.93\,cm^{-1}$ and $\bar{\omega} = 2{,}143\,cm^{-1}$. Equation (9.20) follows because the classical frequency of rotation of a diatomic is much less than its classical frequency of vibration (Problem 9.3).

In Figure 9.1, we show a schematic energy level diagram in accord with Equations (9.18) and (9.20). Notice that the energy level diagram is comprised of widely spaced vibrational levels enclosing many much more narrowly spaced rotational sublevels.

We next turn to the spectroscopic transitions between the energy levels of Equation (9.18). Let us consider an absorption transition from an initial level with quantum numbers vJ to a final level with quantum numbers v′J′. The wave number of IR radiation absorbed in this transition, from the Bohr frequency condition is

$$\bar{\upsilon}_{vJ \to v'J'} = (v' - v)\bar{\omega} + \left[J'(J'+1) - J(J+1)\right]\bar{B}. \tag{9.21}$$

Equation (9.21) must be supplemented by selection rules. For absorption within the harmonic oscillator rigid rotor approximation, these are

$$\Delta v = v' - v = 1 \tag{9.22a}$$

and

$$\Delta J = J' - J = 1 \text{ or } -1. \tag{9.22b}$$

Notice the transition for which $\Delta J = 0$ is forbidden by Equation (9.22b). Thus, as noted, changes in rotational energy must accompany changes in vibrational energy. Inserting Equation (9.22) into Equation (9.21) yields two series of lines, the *R-branch* lines for which $\Delta J = 1$ occurring at

$$\bar{\upsilon}^R(J) = \bar{\omega} + 2(J+1)\bar{B}, \text{ where } J = 0, 1, \ldots \tag{9.23a}$$

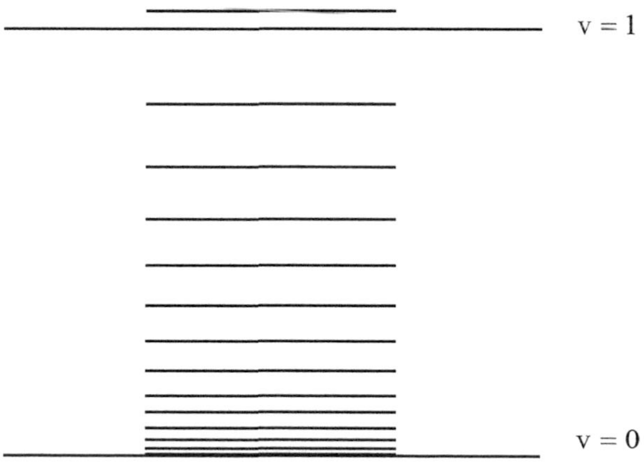

FIGURE 9.1 The energies of the rotational sublevels of the v = 0 vibrational state of a diatomic molecule. Notice that the spacing between adjacent rotational sublevels is much less than the v = 0-to-v = 1 spacing.

and the *P-branch* lines for which $\Delta J = -1$ occurring at wave numbers

$$\upsilon^{P}(J) = \bar{\omega} - 2J\bar{B}, \text{ where } J = 1, 2, \ldots \tag{9.23b}$$

The first three R- and P-branch transitions are shown in Figure 9.2.

Now, we may use experimental information and Equation (9.23) to construct harmonic oscillator rigid rotor high-resolution fundamental band $(v = 0 \rightarrow v = 1)$ IR spectra of diatomic molecules. For example, $^{1}H^{35}Cl$ has a low-resolution infrared absorption spectrum that consists of an intense line at $\bar{\upsilon}_{IR} = 2{,}885.3\,cm^{-1}$. Thus, from Equation (9.17), $\bar{\omega} = 2{,}885.3\,cm^{-1}$. Also from the pure rotational absorption spectrum of $^{1}H^{35}Cl$, one finds that for $^{1}H^{35}Cl$, $\bar{B} = 10.6\,cm^{-1}$.

Thus, from Equation (9.23a), for $^{1}H^{35}Cl$ the R-branch consists of a series of lines occurring at wave numbers $2{,}885.3\,cm^{-1} + 2(10.6)\,cm^{-1} = 2{,}906.5\,cm^{-1}$, $2{,}885.3\,cm^{-1} + 4(10.6)\,cm^{-1} = 2{,}927.7\,cm^{-1}$, and so on. Similarly, from Equation (9.23b) the P-branch consists of a series of lines occurring at wave numbers $2{,}885.3\,cm^{-1} - 2(10.6)\,cm^{-1} = 2{,}864.1\,cm^{-1}$, $2{,}885.3\,cm^{-1} - 4(10.6)\,cm^{-1} = 2{,}842.9\,cm^{-1}$, and so on.

A fundamental band harmonic oscillator rigid rotor IR absorption spectrum for $^{1}H^{35}Cl$ determined in this manner is depicted in Figure 9.3. The spectrum consists of two series of equally spaced lines, the R- and the P-branch lines, both series having the same line spacing $2\bar{B}$. The two series are separated by a gap of $4\bar{B}$. At the midpoint of this gap at wave number $\bar{\omega}$ is a "missing line" that corresponds to the $\Delta J = 0$ forbidden transition.

The intensities of the lines in the harmonic oscillator rigid rotor spectrum of Figure 9.3 were chosen to duplicate those of the experimental fundamental band $T = 300\,K$ $^{1}H^{35}Cl$ IR absorption spectrum shown in Figure 9.4.

We next show that the variations of these experimental intensities are roughly determined by the Boltzmann distribution of the populations of the diatomic's rotational sublevels.

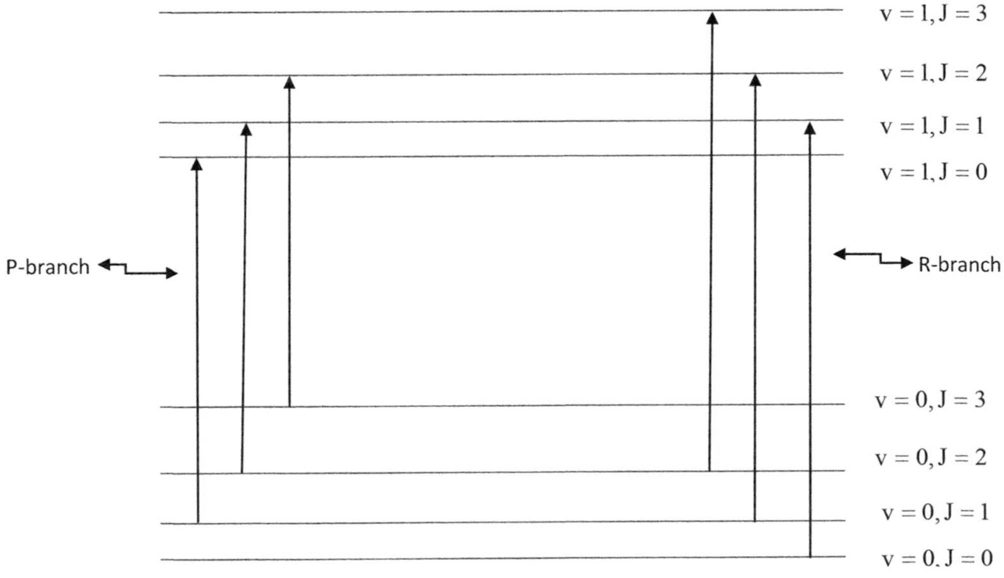

FIGURE 9.2 The first three P- and R-branch transitions in the fundamental band IR absorption spectrum of a diatomic molecule. For clarity, the spacing between the rotational sublevels is greatly exaggerated.

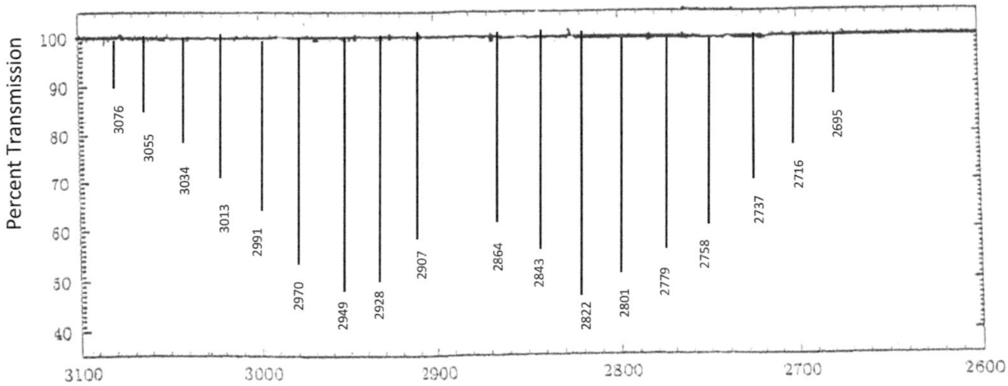

FIGURE 9.3 T = 300 K harmonic oscillator rigid rotor fundamental band IR spectrum of $^1H^{35}Cl$. Intensities are those of the experimental spectrum of Figure 9.4.

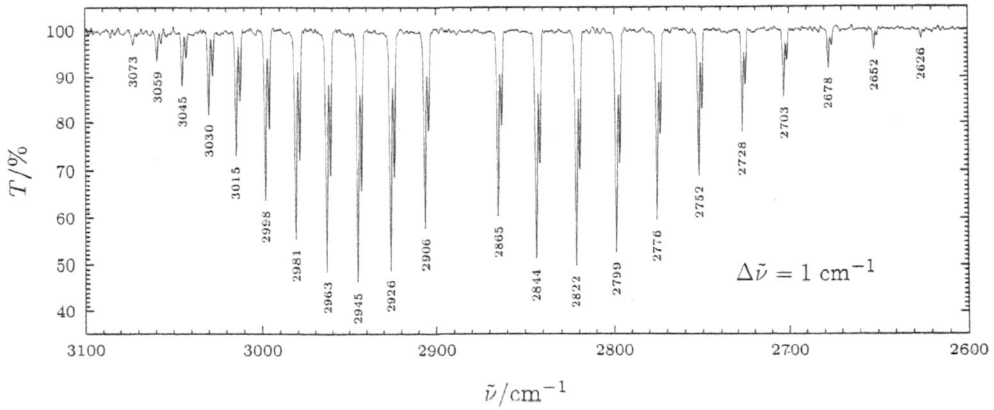

FIGURE 9.4 Experimental fundamental band T = 300 K $^1H^{35}Cl$ IR spectrum. The satellite lines are due to the absorption of $^1H^{37}Cl$. (This spectrum appears on the Wikipedia web page https://upload.wikimedia.org/wikipedia/commons/2/29/Ir_hcl_rot_vib_mrtz.svg released under the http://creativecommons.org/licenses/by-sa/3.0/ Creative Commons Attribution-Share-Alike License 3.0.)

9.4 INTENSITIES OF THE LINES IN THE IR ABSORPTION SPECTRUM OF A DIATOMIC MOLECULE

To see why this is so, consider the number of molecules N_J in a pure gas with rotational quantum number J or equivalently with rigid rotor rotational energy $E_J = \frac{J(J+1)\hbar^2}{2I} = hc\bar{B}J(J+1)$. Since the intensity of a spectral line increases with the number of molecules which can absorb to produce that line, it is evident that the intensities of both the $J \rightarrow J+1$ R-branch line and the $J \rightarrow J-1$ P-branch line increase with N_J. In fact, it is reasonable to assume that these intensities are proportional to or are at least approximately proportional to N_J. If this is so, the rotational Boltzmann distribution must influence the J-dependence of the intensities since $N_J = NP_T(E_J)$, where N is the total number of molecules in the sample and where $P_T(E_J)$ is the Boltzmann probability that at Kelvin temperature T, a molecule has the rotational energy E_J.

$P_T(E_J)$ may be found by specializing the general Boltzmann distribution of Equation (7.135). Noting that the rigid rotor eigenvalue E_J is $(2J+1)$–fold degenerate (since the $2J+1$ rigid rotor states with the same J value but with different M values all have the energy E_J), this specialization yields $P_T(E_J)$ as

$$P_T(E_J) = \frac{(2J+1)\exp\left[-\frac{hc}{kT}\bar{B}J(J+1)\right]}{Q(T)} \tag{9.24}$$

where $Q(T) = \sum_{J=0}^{\infty}(2J+1)\exp\left[-\frac{hc}{kT}\bar{B}J(J+1)\right]$ is the rotational partition function.

To proceed further, we require the ratio $\frac{N_J}{N_{J=0}}$ which is the population N_J of the Jth rotational sublevel relative to the population $N_{J=0}$ of the rotational ground state. Since $\frac{N_J}{N_{J=0}} = \frac{NP_T(E_J)}{NP_T(E_{J=0})} = \frac{P_T(E_J)}{P_T(E_{J=0})}$, from Equation (9.24)

$$\frac{N_J}{N_{J=0}} = (2J+1)\exp\left[-\frac{hc}{kT}\bar{B}J(J+1)\right]. \tag{9.25}$$

We will approximately relate the relative intensities of the infrared absorption lines to the population ratios $\frac{N_J}{N_{J=0}}$ which will then let us approximately evaluate the J-dependence of these intensities from the Boltzmann distribution of Equation (9.25). For definiteness, we will consider the intensities $I_{J\to J+1}$ of the $J \to J+1$ R-branch lines, but similar comments hold for the intensities of the P-branch lines.

To start, we note that the intensity $I_{J\to J+1}$ is determined by two independent factors. The first factor, as already emphasized, is the magnitude of the population N_J. The second factor is the intrinsic line intensity (which is proportional to the probability of the transition J to J + 1 for a single molecule irrespective of the population N_J) which we will denote by $i_{J\to J+1}$. The intrinsic intensity is proportional to the square of the absolute value of M_z, the transition dipole moment of Equation (8.117), evaluated for $J' = J+1$ and $M' = M$. That is, $i_{J\to J+1} \alpha \left|\int_0^{\pi}\int_0^{2\pi} \sin\theta Y^*_{J+1M}(\theta,\phi) M_z Y_{JM}(\theta,\phi)\, d\theta\, d\phi\right|^2$. The variation of experimental infrared line intensities, evident, for example, from Figure 9.4, arises from both the (temperature-dependent) J-dependence of the populations N_J and the (temperature-independent) J-dependence of the intrinsic intensities $i_{J\to J+1}$.

Let us assume that the effects of the J-dependence of the intrinsic intensities $i_{J\to J+1}$ are small and therefore that the variations of the actual intensities $I_{J\to J+1}$ are largely determined by the J-dependence of the populations N_J. If this is so, the relative intensities $\frac{I_{J\to J+1}}{I_{J=0\to 1}}$ may be reasonably assumed to be approximately equal to $\frac{N_J}{N_{J=0}}$ and therefore may be estimated from the Boltzmann distribution of Equation (9.25).

To test our assumption, we compare our ideal relation $\frac{I_{J\to J+1}}{I_{J=0\to 1}} = \frac{N_J}{N_{J=0}}$ with experimental $T = 300\,K$ $^1H^{35}Cl$ relative intensities determined from Figure 9.4. This test is facilitated if in analogy to our definition of the vibrational temperature Θ_{VIB} of Equation (7.146) we define the *rotational temperature* Θ_{ROT} by

$$\Theta_{ROT} = \frac{hc}{k}\bar{B} \tag{9.26}$$

or

$$\Theta_{ROT}(K) = 1.439\,K\,cm\,\bar{B}\left(cm^{-1}\right). \tag{9.27}$$

In terms of Θ_{ROT}, Equation (9.25) becomes

$$\frac{N_J}{N_{J=0}} = (2J+1)\exp\left[-\frac{\Theta_{ROT}}{T}J(J+1)\right]. \quad (9.28)$$

For most diatomic molecules, near room temperature, $\Theta_{VIB} \gg T$, and therefore, nearly all molecules are in the ground vibrational state near $T = 300\,K$ (Table 7.5). However, for all diatomics near $T = 300K$, $\Theta_{ROT} \ll T$, and thus, many excited rotational states are significantly populated at room temperature. For example, for $^1H^{35}CI$

$$\Theta_{ROT} = 1.439\,K\,cm \times 10.59\,cm^{-1} = 15.24\,K \ll 300\,K \quad (9.29)$$

and therefore unsurprisingly the $T = 300\,K$ plot of Equation (9.28) for $^1H^{35}CI$ of Figure 9.5 shows that rotational levels with J as large as ten have non-negligible populations.

Next, let us compare the experimental intensities of Figure 9.4 with the plot of $\frac{N_J}{N_{J=0}}$ of Figure 9.5. Notice that the variations of the experimental relative intensities for both the R- and P-branch spectra and the shape of the $\frac{N_J}{N_{J=0}}$ curve are qualitatively similar. Especially, the experimental relative intensities first increase with J until they reach maxima and then decrease with J and approach zero as $J \to \infty$. Correspondingly, the ratio $\frac{N_J}{N_{J=0}}$ also first increases with J (due to the $2J+1$ degeneracy factor in Equation [9.28]) until it reaches a maximum and then decreases with J and approaches zero as $J \to \infty$ (due to the $\exp\left[-\frac{\Theta_{rot}}{J}J(J+1)\right]$ Boltzmann factor in Equation [9.28]). Moreover, the experimental R- and P-branch lines of maximum intensity have J values of, respectively, 2 and 3. The maximum of $\frac{N_J}{N_{J=0}}$ occurs at $J_{max} = 2.75$ in qualitative agreement with the experimental maxima. However, for $J < J_{max}$ $\frac{N_J}{N_{J=0}}$ falls off much more slowly as J decreases than do the experimental

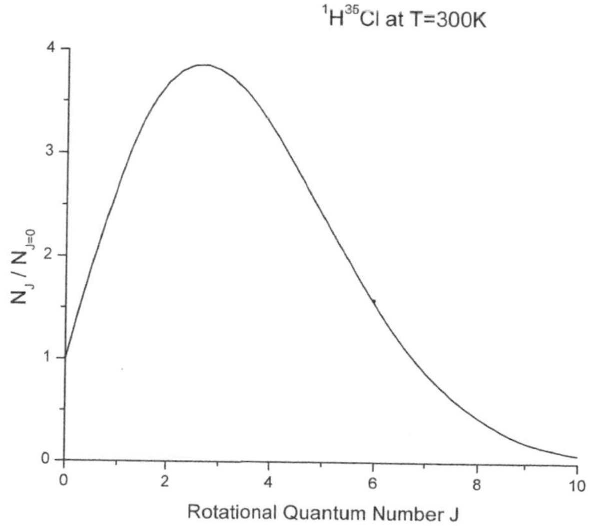

FIGURE 9.5 Relative populations of the rotational levels of $^1H^{35}Cl$ at $T = 300\,K$ with $\Theta_{rot=15.24\,K}$ according to the Boltzmann distribution of Equation (9.28).

relative line intensities. For example, $\frac{N_{J=2}}{N_{J=0}} = 3.7$, while from Figure 9.4 $\frac{I_{J=2\to J=3}}{I_{J=0\to J=1}} = 1.25$, showing that our ideal relation $\frac{I_{J=2\to J=3}}{I_{J=0\to J=1}} = \frac{N_{J=2}}{N_{J=0}}$ is in poor agreement with experiment.

In summary, the J-dependence of the relative populations $\frac{N_J}{N_{J=0}}$ as determined by the Boltzmann distribution gives a qualitative but not quantitative account of the variation of experimental relative IR line intensities.

Finally, let us compare the harmonic oscillator rigid rotor IR spectrum of Figure 9.3 with the experimental spectrum of Figure 9.4. At first glance, we are struck with the veracity of the harmonic oscillator rigid rotor approximation. Despite its simplicity, it qualitatively reproduces the most prominent features of the experimental spectrum. However, a careful look at the experimental spectrum reveals a breakdown of the harmonic oscillator rigid rotor approximation most evident at high J. Namely, in contrast to the equally spaced lines of the approximate spectrum, the lines in the R-branch of the true spectrum become increasingly narrowly spaced as J increases, while the lines in the P-branch become increasingly widely spaced as J increases. The explanation of this phenomenon is developed in Problem 9.15. Next, we turn to a less subtle manifestation of the breakdown of the harmonic oscillator rigid rotor approximation, the existence of *overtone bands* in the IR spectra of diatomic molecules.

9.5 OVERTONE BANDS IN THE IR SPECTRA OF DIATOMIC MOLECULES

As in Section 7.13, we will consider only low-resolution IR spectra, so rotations of the diatomic may be ignored. Here however, we will not only consider the $v = 0 \to v = 1$ fundamental transition, but we will also consider the $v = 0 \to v = n, \quad n \geq 2$, overtone transitions. The overtone transitions are possible because the $\Delta v = +1$ harmonic oscillator absorption selection rule can be violated since the diatomic's potential $U(r)$ is anharmonic (Figure 7.8). Both the fundamental and first few overtone transitions are schematically depicted in Figure 9.6. Notice that, as in Figure 7.7, in Figure 9.6 due to the anharmonicity of $U(r)$, the vibrational energy levels become increasingly closely spaced as the vibrational quantum number v increases. As a consequence, the true wave numbers of the overtone transitions are less than their harmonic oscillator estimates $v\bar{\omega}$, (a result which follows from the harmonic oscillator energy level formula of Equation [7.119] and the Bohr frequency condition), where $v = 2$ for the first overtone line, $v = 3$ for the second overtone line, and so on. Consequently, to accurately predict the experimental overtone wave numbers, we need a form for the vibrational energy levels E_v or equivalently the *vibrational terms* $G(v) \equiv E_v/hc$ which include anharmonicity.

By approximate solution of the Schrödinger equation (Equation 9.5) with $J = 0$ for anharmonic potentials $U(r)$, one may obtain $G(v)$ as the following power series in $v + \frac{1}{2}$:

$$G(v) = \omega_e\left(v + \frac{1}{2}\right) - \omega_e x_e\left(v + \frac{1}{2}\right)^2 + \omega_e y_e\left(v + \frac{1}{2}\right)^3 + \omega_e z_e\left(v + \frac{1}{2}\right)^4 + \cdots. \tag{9.30}$$

The first term in Equation (9.30) $\omega_e\left(v + \frac{1}{2}\right)$ pertains to a harmonic oscillator of wave number ω_e, while the remaining terms are anharmonic corrections. The wave number ω_e is however *not equal* to $\bar{\omega}$. In fact, as we will see, $\bar{\omega} < \omega_e$. The constants $\omega_e, \omega_e x_e, \omega_e y_e$, and $\omega_e z_e$ are called *spectroscopic parameters*. Note that $\omega_e x_e$, a quantity called the *anharmoniicity parameter*, is always positive (this is necessary so the spacing of the vibrational levels decreases with v), while $\omega_e y_e$ and $\omega_e z_e$ can be either positive or negative. Also typically, $\omega_e \gg \omega_e x_e \gg |\omega_e y_e| \gg |\omega_e z_e|$. For example, for $^1H^{35}Cl$ Equation (9.30) in cm^{-1} units is

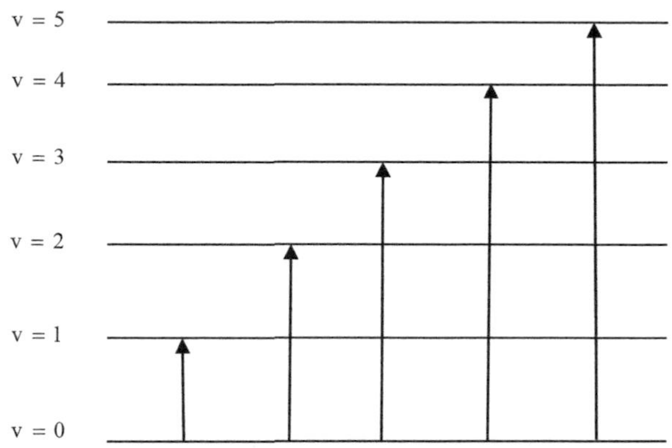

FIGURE 9.6 Schematic depiction of the fundamental $v = 0 \rightarrow v = 1$ and overtone $v = 0 \rightarrow v = n, n = 2-5$ transitions of a diatomic molecule. Because the diatomic's energy levels $E_v = hcG(v)$ become increasingly closely spaced as the vibrational quantum number v increases, the wave numbers of the overtone lines divided by v are less than the wave number $\overline{\omega}$ of the fundamental line and are a decreasing function of v.

$$G(v) = 2{,}990.9463\left(v+\frac{1}{2}\right) - 52.8186\left(v+\frac{1}{2}\right)^2 + 0.2244\left(v+\frac{1}{2}\right)^3 - 0.0122\left(v+\frac{1}{2}\right)^4 + \cdots. \quad (9.31)$$

If v is not too large, a good approximation to $G(v)$ is found by retaining only the linear and quadratic terms in Equation (9.30) to yield

$$G(v) \doteq \omega_e\left(v+\frac{1}{2}\right) - \omega_e x_e\left(v+\frac{1}{2}\right)^2. \quad (9.32)$$

Values obtained from experiment for $\omega_e, \omega_e x_e$, and other spectroscopic parameters which we will encounter later are tabulated for many molecules (Huber and Herzberg 1979). An excerpt of this tabulation is given in Table 9.1.

Next, in order to test Equation (9.32), in Table 9.2 we compare its predictions for the wave numbers of the fundamental and the first four overtone transitions of $^1H^{35}Cl$ with the corresponding observed wave numbers

$$\overline{\upsilon}^{OBS} = G(v) - G(0) \quad (9.33)$$

We will refer to the approximation to $G(v) - G(0)$ evaluated from Equation (9.32) as $\overline{\upsilon}^{ANHAR}$. Thus,

$$\overline{\upsilon}^{ANHAR} = \omega_e\left(v+\frac{1}{2}\right) - \omega_e x_e\left(v+\frac{1}{2}\right)^2 - \left(\frac{1}{2}\omega_e - \frac{1}{4}\omega_e x_e\right) \text{ or}$$

$$\overline{\upsilon}^{ANHAR} = \omega_e v - \omega_e x_e v(v+1). \quad (9.34)$$

For $^1H^{35}Cl$, from Table 9.1 Equation (9.34) becomes

$$\overline{\upsilon}^{ANHAR} = 2{,}990.95v\,\text{cm}^{-1} - 52.82v(v+1)\text{cm}^{-1}. \quad (9.35)$$

Referring to Table 9.2, we find for $^1H^{35}Cl$ that the agreement between $\overline{\upsilon}^{OBS}$ and $\overline{\upsilon}^{ANHAR}$ is excellent. Within the precision of the tabulation, $\overline{\upsilon}^{OBS}$ and $\overline{\upsilon}^{ANHAR}$ agree perfectly for the $v = 1$ fundamental line. For the $v \geq 2$ overtone lines, due to the neglect of higher order terms in Equation (9.32), small deviations build up as v increases (Problem 9.7.) Thus, for the first overtone the relative percent deviation of $\overline{\upsilon}^{OBS}$ from $\overline{\upsilon}^{ANHAR}$ is 0.053%, while for the fourth overtone it is 0.193%. In Table 9.2, we also give the harmonic oscillator approximation $\overline{\upsilon}^{HAR} = v\overline{\omega}$ to $\overline{\upsilon}^{OBS}$. Except for the $v = 0 \rightarrow v = 1$ fundamental

TABLE 9.1
Spectroscopic Parameters of Selected Ground State Diatomic Molecules (Huber and Herzberg 1979)

Molecule	r_e (pm)	ω_e (cm^{-1})	$\omega_e x_e$ (cm^{-1})	B_e (cm^{-1})	α_e (cm^{-1})	$\hat{D}_e$ (cm^{-1})
$^1H_2$	74.144	4,401.213	121.336	60.850	3.0622	4.71×10^{-2}
$^1H^2H$	74.142	3,813.15	91.65	45.655	1.986	2.605×10^{-2}
$^2H_2$	74.152	3,115.50	61.82	30.4436	1.0786	1.141×10^{-2}
$^1H^{35}Cl$	127.4552	2,990.9463	52.8186	10.593416	0.307181	5.3194×10^{-4}
$^2H^{35}Cl$	127.4581	2,145.163	27.1825	5.448794	0.113291	1.39×10^{-4}
$^1H^{19}F$	91.680	4,138.32	89.88	20.9557	0.798	21.51×10^{-4}
$^{16}O^1H$	96.966	3,737.761	84.8813	18.9108	0.7242	19.38×10^{-4}
$^7Li_2$	267.29	351.43	2.610	0.67264	0.00704	9.87×10^{-6}
$^7Li^1H$	159.57	1,405.65	23.20	7.5131	0.2132	0.8617×10^{-3}
$^{12}C_2$	124.76	1,854.71	13.340	1.81984	0.01765	6.92×10^{-6}
$^{14}N_2$	109.7685	2,358.57	14.324	1.998241	0.017318	5.76×10^{-6}
^{23}Na	307.89	159.125	0.7255	0.1547	0.00008736	5.81×10^{-7}
$^{16}O_2$	120.752	1,580.193	11.981	1.4376766	0.01593	4.839×10^{-6}
$^{19}F_2$	141.193	916.64	11.236	0.89019	0.013847	3.3×10^{-6}
$^{12}C^{16}O$	112.8323	2,169.81358	13.28831	1.93128087	0.01750441	6.12147×10^{-6}
$^{32}S_2$	188.92	725.65	2.844	0.29547	0.001570	1.90×10^{-7}
$^{35}Cl_2$	198.79	559.72	2.675	0.24399	0.00149	1.86×10^{-7}

TABLE 9.2
Comparison of Observed Wave Numbers $\bar{\upsilon}^{OBS} = G(V) - G(0)$ for the $0 \to V$ Transitions of $^1H^{35}Cl$ with the Anharmonic $\bar{\upsilon}^{ANHAR} = v\omega_e - \omega_e x_e c(v+1)$ and Harmonic $\bar{\upsilon}^{HAR} = v\bar{\omega}$ Approximations to These Wave Numbers

v	$\bar{\upsilon}^{OBS}$ (cm^{-1})	$\bar{\upsilon}^{ANHAR}$ (cm^{-1})	$\bar{\upsilon}^{HAR}$ (cm^{-1})	$\frac{\lvert\bar{\upsilon}^{OBS} - \bar{\upsilon}^{ANHAR}\rvert}{\bar{\upsilon}^{OBS}} \times 100\%$	$\frac{\lvert\bar{\upsilon}^{OBS} - \bar{\upsilon}^{HAR}\rvert}{\bar{\upsilon}^{OBS}} \times 100\%$
1	2,885.3	2,885.3	2,885.3	0%	0%
2	5,668.0	5,665.0	5,770.6	0.053%	1.810%
3	8,347.0	8,339.0	8,655.9	0.096%	3.701%
4	10,923.1	10,907.4	11,541.2	0.144%	5.659%
5	13,396.1	13,370.2	14,426.5	0.193%	7.692%

transition for which $\bar{\upsilon}^{HAR} = \bar{\omega}$ is nearly exact (since the harmonic oscillator approximations to the true v = 0 and v = 1 vibrational energy levels are nearly exact), $\bar{\upsilon}^{HAR}$ is a much poorer approximation to $\bar{\upsilon}^{OBS}$ than is $\bar{\upsilon}^{ANHAR}$. Thus, from Table 9.2, except for the fundamental transition the relative percent deviation of $\bar{\upsilon}^{HAR}$ from $\bar{\upsilon}^{OBS}$ is $\sim 35-40$ times higher than the corresponding deviation of $\bar{\upsilon}^{ANHAR}$. Consequently, as noted to accurately predict $\bar{\upsilon}^{OBS}$ for the overtone transitions, it is necessary to include anharmonicity.

We finally identify the distinction between the wave numbers $\bar{\omega}$ and ω_e. From the harmonic equation (Equation 7.125), $G(1) - G(0) = \bar{\omega}$. From the anharmonic equation (Equation 9.34),

$$G(1) - G(0) = \omega_e - 2\omega_e x_e. \tag{9.36}$$

TABLE 9.3
Vibrational Wave Numbers $\bar{\omega}$ and ω_e for Selected Diatomic Molecules

Molecule	$\bar{\omega}\left(cm^{-1}\right)$	$\omega_e\left(cm^{-1}\right)$
$^1H_2$	4,158.5	4,401.2
$^1H^{35}Cl$	2,885.3	2,990.9
$^{12}C^{16}O$	2,143.2	2,169.8
$^{16}O_2$	1,556.2	1,580.2
$^{35}Cl_2$	555.4	559.7
$^{23}Na_2$	157.7	159.1

But from Table 9.2, both the harmonic and anharmonic approximations to the fundamental line wave numbers $G(1)-G(0)$ are nearly exact. Therefore, equating these approximations gives $\bar{\omega} = \omega_e - 2\omega_e x_e$. Since $\omega_e x_e > 0$, this relation shows that $\bar{\omega} < \omega_e$. Values of $\bar{\omega}$ and ω_e obtained from the relation $\bar{\omega} = \omega_e - 2\omega_e x_e$ and from Table 9.1 are compared for several diatomic molecules in Table 9.3.

Next, we turn to a new topic, the centrifugal distortion effect that manifests itself in small deviations of the pure rotational spectrum of a diatomic molecule from the rigid rotor spectrum.

9.6 CENTRIFUGAL DISTORTION AND THE KRATZNER EQUATION

So far in our discussions of diatomic rotational motion, we have dealt exclusively with the rigid rotor approximation. Since within this approximation the vibrational coordinate r is fixed at its equilibrium value or bond length r_e, no coupling between rotations and vibrations occurs. Actually, as the molecule rotates, it is stretched by the centrifugal force acting on its atoms. This *centrifugal distortion effect* slightly increases the diatomic's bond length and moment of inertia, hence modifying its pure rotational spectrum. The rigid rotor energy levels $E_J = \dfrac{J(J+1)\hbar^2}{2I}$ or equivalently the rigid rotor *rotational terms*

$$F(J) \equiv \frac{E_J}{hc} = \frac{h}{8\pi^2 Ic} J(J+1) = \bar{B}J(J+1) \tag{9.37}$$

are consequently also modified. Since the centrifugal distortion effect becomes increasingly important as the molecule rotates faster and faster or as J increases, $F(J)$ is conveniently represented as a power series in $J(J+1)$; namely,

$$F(J) = \bar{B}J(J+1) - \bar{D}J^2(J+1)^2 + \cdots. \tag{9.38a}$$

This series expresses $F(J)$ as a rigid rotor term $\bar{B}J(J+1)$ corrected by terms $-\bar{D}J^2(J+1)^2\ldots$ which account for centrifugal distortion.

$\bar{B}$ is equal to the spectroscopic parameter B_e listed in Table 9.1 (Problem 9.8). Correspondingly, $\bar{D}$ is equal to the spectroscopic parameter $\hat{D}_e$ also listed in Table 9.1. Equation (9.38a) may thus be written as

$$F(J) = B_e J(J+1) - \hat{D}_e J^2(J+1)^2 + \cdots. \tag{9.38b}$$

From Table 9.1, we see that typically $B_e \gg \hat{D}_e$. For example, for $^1H^{35}Cl$ $B_e = 10.59\,cm^{-1}$ and $\hat{D}_e = 5.32\times10^{-4}\,cm^{-1}$, while for $^{12}C^{16}O$ $B_e = 1.93\,cm^{-1}$ and $\hat{D}_e = 6.12\times10^{-6}\,cm^{-1}$. So centrifugal distortion is a small effect.

We next derive an approximate expression for $\hat{D}_e$ known as the *Kratzner equation*. Our derivation is based on classical mechanics with quantum mechanics introduced in the last step. Our starting point is Equation (B.40) for the classical effective potential

$$U_{eff}(r) = U(r) + \frac{J^2}{2\mu r^2}. \tag{9.39}$$

where J^2 is the square of the magnitude of the diatomic's classical angular momentum vector $J = \mathbf{r} \times \mathbf{p}$. We first make the harmonic oscillator approximation of Equation (7.118) for $U(r)$. Then, $U_{eff}(r)$ has the approximate form

$$U_{eff}(r) \doteq \frac{1}{2}\mu\omega^2(r - r_e)^2 + \frac{J^2}{2\mu r^2}. \tag{9.40}$$

The two potential energy terms in Equation (9.40) act in an opposing manner. The harmonic oscillator potential $\frac{1}{2}\mu\omega^2(r - r_e)^2$ produces a force that restores r to r_e, while the centrifugal potential $J^2/2\mu r^2$ produces a repulsive force that stretches r. When the two forces balance, the diatomic is at a new equilibrium internuclear distance r_{eJ} which depends on J and which is equal to r_e when $J = 0$ and is greater than r_e when $J > 0$. The condition of balance is

$$-\left[\frac{dU_{eff}(r)}{dr}\right]_{r=r_{eJ}} = 0 \tag{9.41}$$

which using Equation (9.40) gives

$$\mu\omega^2(r_{eJ} - r_e) - \frac{J^2}{\mu r_{eJ}^3} = 0 \tag{9.42}$$

or

$$r_{eJ} = r_e + \frac{J^2}{\mu^2\omega^2 r_{eJ}^3}. \tag{9.43}$$

To estimate the energy E_J of the rotating oscillator, we will, in accord with our small amplitude vibrational excursion assumption fix r at r_{eJ} and thus neglect the vibrational kinetic energy $\frac{1}{2}\mu\dot{r}^2$ to obtain $E_J \doteq U_{eff}(r_{eJ})$. Equation (9.40) then gives

$$E_J = \frac{1}{2}\mu\omega^2(r_{eJ} - r_e)^2 + \frac{J^2}{2\mu r_{eJ}^2}. \tag{9.44}$$

Next, using Equation (9.43) it follows that $\frac{1}{2}\mu\omega^2(r_{eJ} - r_e)^2 = \frac{J^4}{2\mu^3\omega^2 r_{eJ}^6}$. Hence, E_J of Equation (9.44) may be recast as

$$E_J = \frac{J^2}{2\mu r_{eJ}^2} + \frac{J^4}{2\mu^3\omega^2 r_{eJ}^6}. \tag{9.45}$$

We may now make two approximations to Equation (9.45) which will eventually yield an expression for E_J valid to order J^4. The first approximation arises from the following approximation to Equation (9.43) $r_{eJ} \doteq r_e + \frac{J^2}{\mu^2\omega^2 r_e^3}$ yielding for the first term in Equation (9.45)

$$\frac{J^2}{2\mu r_{eJ}^2} \doteq \frac{J^2}{2\mu\left(r_e + \frac{J^2}{\mu^2\omega^2 r_e^3}\right)^2} = \frac{J^2}{2\mu r_e^2\left(1 + \frac{J^2}{\mu^2\omega^2 r_e^4}\right)^2}. \tag{9.46}$$

In the second approximation, we set $r_{eJ} = r_e$ in the second term in Equation (9.45) to yield

$$\frac{J^4}{2\mu^3\omega^2 r_{eJ}^6} \doteq \frac{J^4}{2\mu^3\omega^2 r_e^6}. \tag{9.47}$$

Given Equations (9.46) and (9.47), Equation (9.45) yields the following approximation to the energy of the diatomic:

$$E_J \doteq \frac{J^2}{2\mu r_e^2}\frac{1}{\left(1 + \frac{J^2}{\mu^2\omega^2 r_e^4}\right)^2} + \frac{J^4}{2\mu^3\omega^2 r_e^6} \tag{9.48}$$

But to order J^2 $\frac{1}{\left(1 + \frac{J^2}{\mu^2\omega^2 r_e^4}\right)} = 1 - \frac{J^2}{\mu^2\omega^2 r_e^4}$ and hence to the same level of approximation,

$$\frac{1}{\left(1 + \frac{J^2}{\mu^2\omega^2 r_e^4}\right)^2} = 1 - \frac{2J^2}{\mu^2\omega^2 r_e^4} + \frac{J^4}{\mu^4\omega_e^4 r_e^8}. \tag{9.49}$$

Then, comparing Equations (9.48) and (9.49) gives the following classical form for E_J valid to order J^4:

$$E_J = \frac{J^2}{2\mu r_e^2} - \frac{J^4}{2\mu^3\omega^2 r_e^6}. \tag{9.50}$$

E_J may be converted into a quantum form by making the classical to quantum correspondence $J^2 \leftrightarrow J(J+1)\hbar^2$. This gives the quantum formula

$$E_J = \frac{J(J+1)\hbar^2}{2\mu r_e^2} - \frac{J^2(J+1)^2\hbar^4}{2\mu^3\omega^2 r_e^6}. \tag{9.51}$$

The rotational term $F(J) = E_J/hc$ is then given by

$$F(J) = \frac{J(J+1)h}{2(2\pi)^2 cI} - \frac{J^2(J+1)^2 h^3}{2(2\pi)^4 cI^3\omega^2} \tag{9.52}$$

where we have used $I = \mu r_e^2$. Comparing Equations (9.38b) and (9.52) gives

$$B_e = \frac{h}{8\pi^2 Ic} \text{ and } \hat{D}_e = \frac{h^3}{2(2\pi)^4 cI^3\omega^2}. \tag{9.53}$$

The expression for $B_e = \bar{B}$ is our familiar result for the rotational constant. The form for $\hat{D}_e$ is a new result. The expression for $\hat{D}_e$ may be rewritten as (Problem 9.13)

$$\hat{D}_e = \frac{4B_e^3}{\omega_e^2}. \tag{9.54}$$

Equation (9.54) is the Kratzner equation. Since $B_e \ll \omega_e$, the Kratzner equation (Equation 9.54) shows in agreement with Table 9.1 that $\hat{D}_e/B_e \ll 1$ and hence as noted centrifugal distortion is a small effect. Also note that from the Kratzner equation, one may obtain an (excellent) estimate of $\hat{D}_e$ from the values of B_e and ω_e given in Table 9.1 (Problem 9.14).

Next, we treat the effects of centrifugal distortion on the pure rotational spectrum of $^{12}C^{16}O$ in its ground v = 0 vibrational state.

9.7 EFFECTS OF CENTRIFUGAL DISTORTION ON THE PURE ROTATIONAL SPECTRUM OF A DIATOMIC MOLECULE

The rotational term expression of Equation (9.38b) applies to the hypothetical situation that the molecule has zero vibrational energy. For the realistic situation where the molecule is in some vibrational state with quantum number v, due to the finite amplitude of the vibrations and the anharmonicities the average internuclear separation should be larger than r_e. The rotational constant thus should be changed from B_e to a smaller v −dependent value B_v. The centrifugal distortion constant should also be changed from $\hat{D}_e$ to a new and smaller value $\hat{D}_v$. Consequently, for the v = 0 ground state (neglecting higher order terms) Equation (9.38b) must be modified to

$$F_0(J) = B_0 J(J+1) - \hat{D}_0 J^2 (J+1)^2. \tag{9.55}$$

From Equation (9.55), the wave numbers $\bar{\upsilon}_{J\to J+1}$ of the rotational lines are given by (Problem 9.12)

$$\bar{\upsilon}_{J\to J+1} = F_0(J+1) - F_0(J) = 2B_0(J+1) - 4\hat{D}_0(J+1)^3. \tag{9.56}$$

If the centrifugal distortion term in Equation (9.56) were absent, the spectral lines would be equally spaced as for rigid rotor diatomics, with spacing $2B_0$. In fact, the lines predicted by Equation (9.56) become increasingly closely spaced as J increases (Problem 9.12). This predicted decrease in line spacing is in accord with observations as shown in Table 9.4. There we list highly accurate observed wave numbers $\bar{\upsilon}^{OBS}_{J\to J+1}$ for the first ten rotational transitions of v = 0 $^{12}C^{16}O$. We also give the line spacings. These spacings indeed decrease more and more rapidly as J increases. However, in accord with the smallness of the centrifugal distortion effect, the decreases in spacing are very small. Thus, the spacing between the $J = 0 \to J = 1$ and $J = 1 \to J = 2$ lines is $3.845\,cm^{-1}$, while the spacing between the $J = 8 \to J = 9$ and $J = 9 \to J = 10$ lines is $3.838\,cm^{-1}$. (The centrifugal distortion effect on pure rotational spectra, as shown in Problem 9.13, is however much larger for $^1H^{35}CI$.)

We next show how to determine the parameters B_0 and $\hat{D}_0$ from the experimental line positions $\bar{\upsilon}^{OBS}_{J\to J+1}$. This may be done by rewriting Equation (9.56) as

$$\frac{\bar{\upsilon}^{OBS}_{J\to J+1}}{J+1} = \frac{F_0(J+1) - F_0(J)}{J+1} = 2B_0 - 4\hat{D}_0(J+1)^2. \tag{9.57}$$

TABLE 9.4
Observed Line Positions $\bar{\upsilon}^{OBS}_{J\to J+1} = F(J+1) - F(J)$ and Line Spacings $\bar{\upsilon}^{OBS}_{J\to J+1} - \bar{\upsilon}^{OBS}_{J-1\to J}$ in the Pure Rotational Spectrum of $^{12}O^{16}O$

J	$\bar{\upsilon}^{OBS}_{J\to J+1}\left(cm^{-1}\right)$	Spacing $\left(cm^{-1}\right)$
0	3.845033	
1	7.689919	3.844886
2	11.534512	3.844592
3	15.378664	3.844152
4	19.222227	3.843563
5	23.065058	3.842831
6	26.907005	3.841948
7	30.747927	3.840921
8	34.587673	3.839746
9	38.426097	3.838424

Equation (9.57) predicts that a plot of $\frac{F_0(J+1) - F_0(J)}{J+1}$ versus $(J+1)^2$ should give a straight line with slope $-4\hat{D}_0$ and intercept $2B_0$. Such a plot based on the data of Table 9.4 is given in Figure 9.7. It is indeed linear with a negative slope, thus validating the J-dependence of $F_0(J)$ of Equation (9.55). The constants B_0 and $\hat{D}_0$ may be estimated from the plot. However, a more accurate determination may be made by least squares fitting. Making a least squares fit of Equation (9.57) to the data of Table 9.4 gives $B_0 = 1.923\,cm^{-1}$ and $\hat{D}_0 = 6.120\times10^{-6}\,cm^{-1}$. These results are in excellent agreement with the literature values of $B_0 = 1.922\,cm^{-1}$ and $\hat{D}_0 = 6.121\times10^{-6}\,cm^{-1}$.

We next return to the high-resolution IR spectrum of a diatomic molecule treated in Section 9.3 within the harmonic oscillator rigid rotor approximation. Using the concepts of Sections 9.5–9.7, we now give more advanced discussions of this spectrum. Specifically, in Problem 9.19 we compare the wave numbers of the fundamental band R- and P-branch absorption lines of $^{12}C^{16}O$ predicted

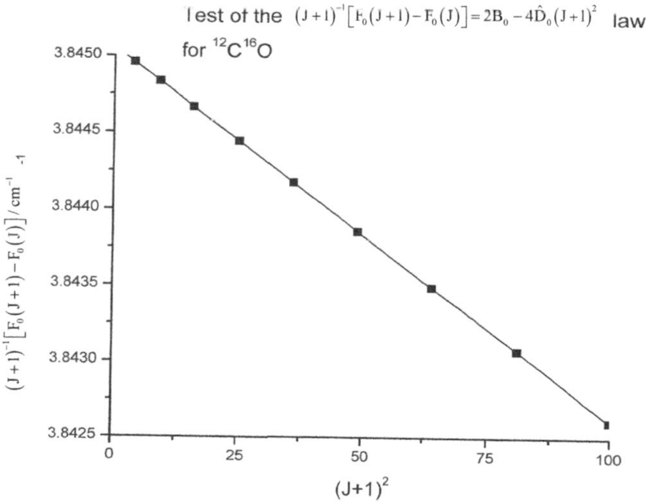

FIGURE 9.7 Verification of the linear law of Equation (9.57) for the v = 0 pure rotational spectrum of $^{12}C^{16}O$ based on the observed values in Table 9.4.

by a model that includes rotational–vibrational coupling (the dependence of rotational constant on the vibrational state) and vibrational anharmonicity with the corresponding harmonic oscillator rigid rotor wave numbers. Here, we introduce the method of *combinations and differences.* We further show how this method permits us to determine from the experimental R- and P-branch line positions of a general $v \rightarrow v'$ IR absorption spectrum the rotational and centrifugal distortion spectroscopic parameters of a diatomic molecule for both its lower (v) and upper (v') vibrational states.

9.8 THE HIGH-RESOLUTION IR ABSORPTION SPECTRUM OF A DIATOMIC MOLECULE AND THE METHOD OF COMBINATIONS AND DIFFERENCES

Recalling that both the rotational and centrifugal distortion parameters depend on the vibrational quantum state, as before we will write these parameters as B_v and $\hat{D}_v$. Then, neglecting higher order corrections, the rotational term of a diatomic molecule with vibrational quantum number v is written as the following generalization of Equation (9.55):

$$F_v(J) = B_v J(J+1) - \hat{D}_v J^2 (J+1)^2. \tag{9.58}$$

Correspondingly, neglecting higher order contributions, the vibrational term is given by Equation (9.32)

$$G(v) = \omega_e\left(v + \frac{1}{2}\right) - \omega_e x_e\left(v + \frac{1}{2}\right)^2. \tag{9.59}$$

Next, we define the *band origin* $\bar{\upsilon}_0$ of a general absorption transition $v \rightarrow v'$ by

$$\bar{\upsilon}_0 = G(v') - G(v). \tag{9.60}$$

From the previous two equations,

$$\bar{\upsilon}_0 = \omega_e(v' - v) - \left[\omega_e x_e\left(v' + \frac{1}{2}\right)^2 - \omega_e x_e\left(v + \frac{1}{2}\right)^2\right]. \tag{9.61}$$

The wave number for the absorption transition $vJ \rightarrow v'J'$ is therefore

$$\bar{\upsilon}_{v\rightarrow v', J\rightarrow J'} = F_{v'}(J') + G(v') - [F_v(J) + G(v)] = \bar{\upsilon}_0 + F_{v'}(J') - F_v(J). \tag{9.62}$$

Next, let us introduce the rotational selection rules. These are

$$J' = J + 1 \quad \text{for the R-branch} \tag{9.63a}$$

and

$$J' = J - 1 \quad \text{for the P-branch.} \tag{9.63b}$$

Further, let us denote the wave numbers of the R- and P-branch lines of the $v \to v'$ absorption band by, respectively, $R(J)$ and $P(J)$. Then, given the rotational selection rules, it follows from Equation (9.62) that

$$R(J) = \bar{\upsilon}_{v \to v', J \to J+1} = \bar{\upsilon}_0 + F_{v'}(J+1) - F_v(J) \quad (9.64a)$$

and

$$P(J) = \bar{\upsilon}_{v \to v', J \to J-1} = \bar{\upsilon}_0 + F_{v'}(J-1) - F_v(J). \quad (9.64b)$$

Notice that since $R(J)$ and $P(J)$ each depend on both $F_{v'}$ and F_v, they also depend on *both* the upper state (v') and lower state (v) rotational and centrifugal distortion parameters. This follows since from Equation (9.58) $F_{v'}(J) = B_{v'}J(J+1) - \hat{D}_{v'}J^2(J+1)^2$ and $F_v(J) = B_vJ(J+1) - \hat{D}_vJ^2(J+1)^2$. Since $R(J)$ and $P(J)$ each depend on both upper and lower state parameters, one cannot conveniently determine these parameters by fitting either the $R(J)$ or $P(J)$ wave numbers to observed spectral line positions. However, the method of combinations and differences, illustrated in Figure 9.8, separately yields the upper and lower state parameters, thus permitting the convenient determination of all parameters from experiment.

To explain the method of combinations and differences, we first consider R- and P-branch transitions for which the *initial* rotational quantum number J is the same for both types of transitions.

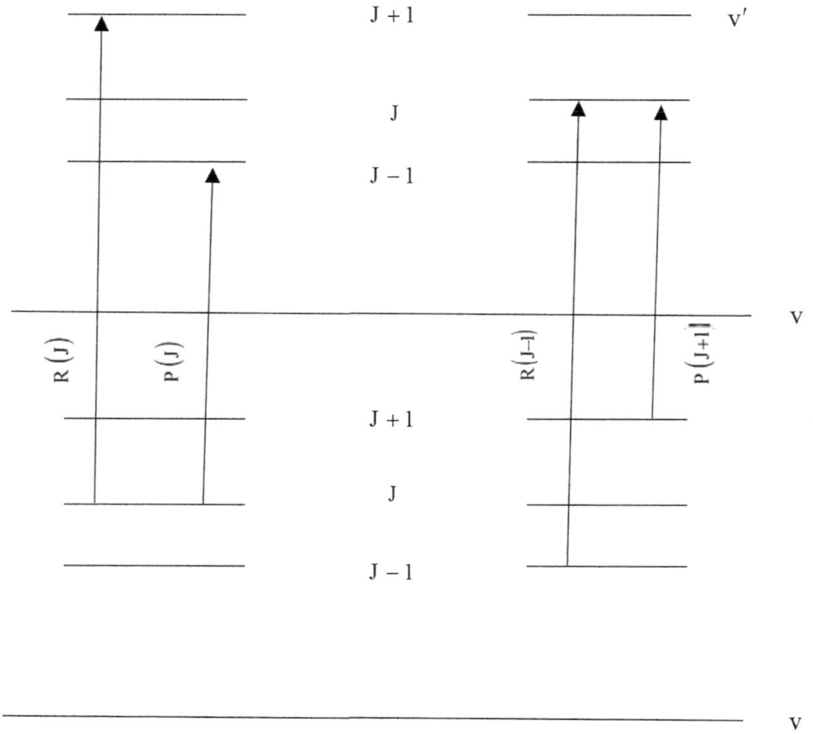

FIGURE 9.8 The method of combinations and differences. The spectroscopic parameters in the upper vibrational state $B_{v'}$ and $D_{v'}$ depend on the differences $R(J) - P(J)$ for the R- and P-branch lines with the same initial rotational quantum number J. The spectroscopic parameters in the lower vibrational state B_v and D_v depend on the differences $R(J-1) - P(J+1)$ for the R- and P-branch lines with the same final rotational quantum number J.

Then, by the rotational selection rules, for the R-branch transitions $J \rightarrow J+1$ while for the P-branch transitions $J \rightarrow J-1$. Thus, from Equations (9.64) the differences $R(J)-P(J)$ between the wave numbers of the two types of transitions are given by

$$R(J)-P(J)=\left[F_{v'}(J+1)-F_v(J)\right]-\left[F_{v'}(J-1)-F_v(J)\right]. \tag{9.65}$$

Canceling $F_v(J)$ from the above relation gives

$$R(J)-P(J)=F_{v'}(J+1)-F_{v'}(J-1). \tag{9.66a}$$

Thus, we have shown that $R(J)-P(J)$ depends only on $F_{v'}$ and hence depends only on the upper state spectroscopic parameters $B_{v'}$ and $\hat{D}_{v'}$.

To obtain an expression that depends only on the lower state parameters B_v and $\hat{D}_v$, we next consider R- and P-branch transitions for which the *final* rotational quantum number J is the same for both types of transitions. Then, by the rotational selection rules, for the R-branch transitions $J-1 \rightarrow J$, while for the P-branch transitions $J+1 \rightarrow J$. Therefore, from Equations (9.64) the differences $R(J-1)-P(J+1)$ between the wave numbers of the two types of transitions are given by

$$R(J-1)-P(J+1)=\left[F_{v'}(J)-F_v(J-1)\right]-\left[F_{v'}(J)-F_v(J+1)\right]. \tag{9.67}$$

Canceling $F_{v'}(J)$ from the above relation gives

$$R(J-1)-P(J+1)=F_v(J+1)-F_v(J-1). \tag{9.66b}$$

Thus, we have shown that $R(J-1)-P(J+1)$ depends only on F_v and hence it depends only on the lower state parameters B_v and $\hat{D}_v$.

Next, let us get explicit relationships between B_v and $\hat{D}_v$ and $B_{v'}$ and $\hat{D}_{v'}$ and IR spectral data. We first combine Equations (9.58) and (9.66a) to give

$$\begin{aligned} R(J)-P(J) &= B_{v'}(J+1)(J+2)-D_{v'}(J+1)^2(J+2)^2 \\ &\quad - B_{v'}(J-1)J+D_{v'}(J-1)^2J^2. \end{aligned} \tag{9.68}$$

Let us temporarily drop the centrifugal distortion terms. Then, the previous equation simplifies to

$$R(J)-P(J)=B_{v'}(J+1)(J+2)-B_{v'}(J-1)J$$

or (9.69a)

$$R(J)-P(J)=4B_{v'}\left(J+\frac{1}{2}\right).$$

Analogously from Equations (9.58) and (9.66b),

$$R(J-1)-P(J+1)=4B_v\left(J+\frac{1}{2}\right). \tag{9.69b}$$

Reinstating the centrifugal distortion term yields

$$R(J)-P(J)=(4B_{v'}-6D_{v'})\left(J+\frac{1}{2}\right)-8D_{v'}\left(J+\frac{1}{2}\right)^3 \tag{9.70a}$$

and

$$R(J-1)-P(J+1)=(4B_v-6D_v)\left(J+\frac{1}{2}\right)-8D_v\left(J+\frac{1}{2}\right)^3. \tag{9.70b}$$

If a sufficient number of R-branch and P-branch lines have been observed, plots can be made of $\frac{R(J)-P(J)}{J+\frac{1}{2}}$ and $\frac{R(J-1)-P(J+1)}{J+\frac{1}{2}}$ versus $\left(J+\frac{1}{2}\right)^2$. From Equations (9.70), the plots should be straight lines and the spectroscopic parameters $B_{v'}, D_{v'}$, and B_v, D_v may be found from the slopes and intercepts of these lines.

Next, let us specialize to the $v=0 \rightarrow v=1$ fundamental band transitions. For simplicity, we will assume the centrifugal distortion constants D_0 and D_1 are negligible. Then, from Equations (9.69)

$$R(J)-P(J)=4B_1\left(J+\frac{1}{2}\right) \tag{9.71a}$$

and

$$R(J-1)-P(J+1)=4B_0\left(J+\frac{1}{2}\right). \tag{9.71b}$$

According to Equations (9.71), plots of $R(J)-P(J)$ and $R(J-1)-P(J+1)$ versus $\left(J+\frac{1}{2}\right)$ should give straight lines with, respectively, slopes of $4B_1$ and $4B_0$. Using the first fourteen R-branch lines and first thirteen P-branch lines observed for $^{12}C^{16}O$ and listed in Table 9.5, such plots have been made and are given in Figure 9.9. From the respective slopes of the lines in these figures, B_1 and B_0 may be estimated. More accurate determinations may be made by least squares fitting of Equations (9.71) to the data of Table 9.5. This fitting gives $B_0 = 1.920\,cm^{-1}$ and $B_1 = 1.903\,cm^{-1}$ in very good agreement with the literature values of $B_0 = 1.923\,cm^{-1}$ and $B_1 = 1.905\,cm^{-1}$. Notice that as expected, B_1 is slightly less than B_0, reflecting the fact that the effective bond length in the $v = 1$ state is slightly longer than it is in the $v = 0$ state.

The method of combinations and differences is applied to the observed IR lines for the $v=0 \rightarrow v=2$ overtone band of $^{12}C^{16}O$ in Problem 9.14. From the least squares fits for this band, one fonds, $B_0 = 1.920\,cm^{-1}$ and $B_2 = 1.884\,cm^{-1}$. Also recall that in Section 9.7, we derived from the pure rotational spectrum of $^{12}C^{16}O$ the value $B_0 = 1.923\,cm^{-1}$. Thus, for $^{12}C^{16}O$ we have three different values of B_0 derived from three distinct experiments. These are $B_0(0-1) = 1.920\,cm^{-1}$ derived from the IR lines of the $v=0 \rightarrow v=1$ fundamental band, $B_0(0-2) = 1.920\,cm^{-1}$ derived from the IR lines of the $v=0 \rightarrow v=2$ overtone band, and B_0 (pure rotational) $= 1.923\,cm^{-1}$ derived from the lines of the $v = 0$ pure rotational spectrum. The small disagreements are probably due to ignoring centrifugal distortion in the IR calculations. The results of our least squares calculations along with comparisons to literature values are summarized in Table 9.6.

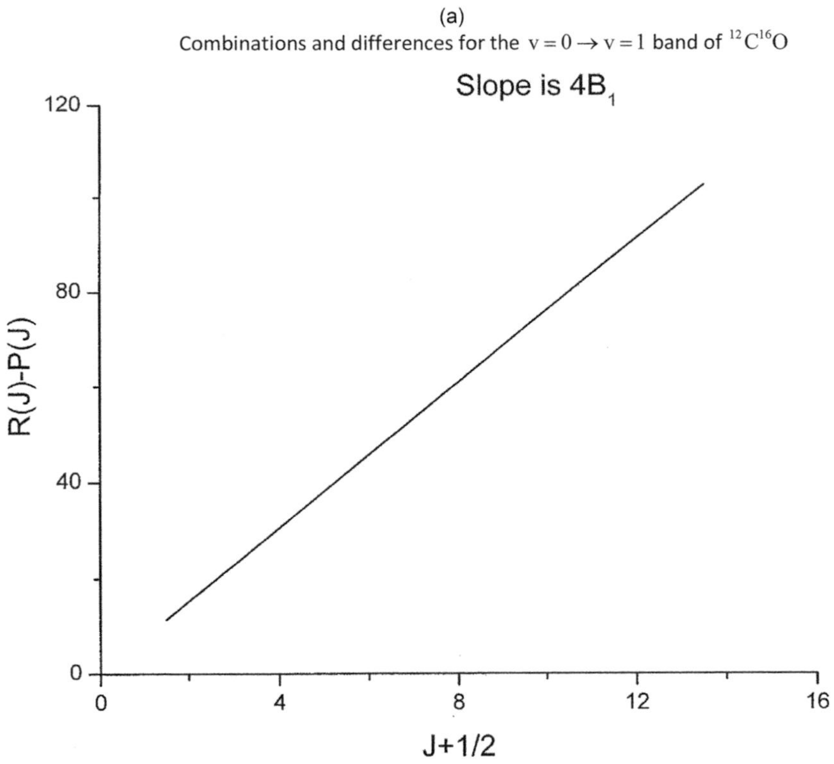

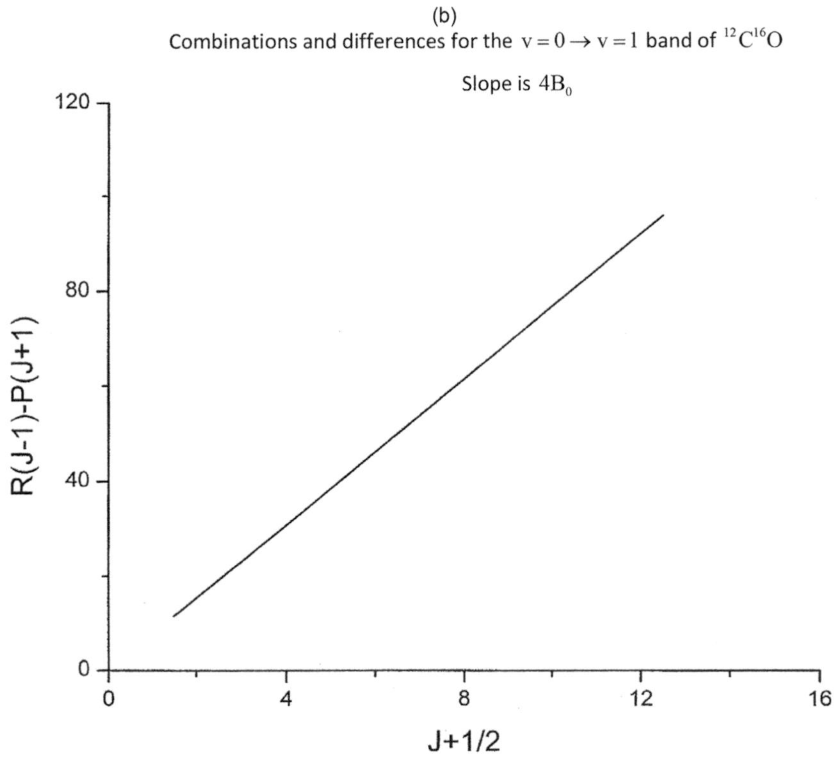

FIGURE 9.9 (a) Plot of Equation (9.71a) based on the $^{12}C^{16}O$ fundamental band IR absorption lines of Table 9.5. (b) The same as (a) except a plot of Equation (9.71b).

TABLE 9.5
Combinations and Differences $R(J)-P(J)$ and $R(J-1)-P(J+1)$ Computed from Observed R-Branch and P-Branch Lines in the Fundamental Band IR Spectrum of $^{12}C^{16}O$

J	$R(J)/cm^{-1}$	$P(J)/cm^{-1}$	$R(J)-P(J)/cm^{-1}$	$R(J-1)-P(J+1)/cm^{-1}$
0	2,147.085			
1	2,150.860	2,139.432	11.428	11.531
2	2,154.596	2,135.554	19.042	19.221
3	2,158.309	2,131.639	26.670	26.911
4	2,161.971	2,127.685	34.286	34.609
5	2,165.602	2,123.700	41.902	42.294
6	2,169.200	2,119.667	49.523	49.970
7	2,172.759	2,115.632	57.127	57.645
8	2,176.287	2,111.555	64.732	65.346
9	2,179.761	2,107.413	72.348	73.022
10	2,183.226	2,103.265	79.961	80.665
11	2,186.636	2,099.096	87.540	88.356
12	2,190.010	2,094.870	95.140	96.033
13	2,193.357	2,090.603	102.754	

TABLE 9.6
$^{12}C^{16}O$ Rotational Constants B_0, B_1, and B_2 in cm^{-1} from Least Squares Fitting of the $0-1$ and $0-2$ Lines in the Rotational–Vibrational Spectrum and of the Lines in the $v=0$ Pure Rotational Spectrum Compared with Literature Values

	$B_0(0-1)$	$B_0(0-2)$	B_0(pure rotational)	$B_1(0-1)$	$B_2(0-2)$
Fit	1.920	1.920	1.923	1.903	1.884
Literature values	1.923[a]	1.923[b]	1.922[c]	1.905[a]	1.888[b]

[a] E.K. Plyler, L.R. Blaine, and W.S. Conner, *J. Opt. Soc. America* **45**, 102 (1955).
[b] D.R. Rank, G. Skorinko, and D.P. Eastman, *J. Molec. Spectroscopy* **4**, 518 (1960).
[c] I.G. Nolt et al., *J. Molec. Spectroscopy* **12**, 274 (1987).

9.9 EVALUATION OF B_v FROM TABLE 9.1

Next, we note that the v-dependence of B_v may not only be determined from spectroscopic data but may also be found from the spectroscopic parameter α_e of Table 9.1. Namely, B_v is parameterized by the relation

$$B_v = B_e - \alpha_e\left(v+\frac{1}{2}\right). \tag{9.72}$$

Since $\alpha_e > 0$, as expected $B_e > B_0 > B_1 \cdots$. Also the deviations of B_v from B_e are small since the effective bond length r_v in vibrational state v defined by $B_v = \frac{h}{8\pi^2 \mu r_v^2 c}$ differs only slightly from r_e (Problem 9.17). This is because α_e is much smaller than B_e (Table 9.1).

Equation (9.72) is the leading term in the expansion $B_v = B_e - \alpha_e\left(v+\frac{1}{2}\right) + \gamma_e\left(v+\frac{1}{2}\right)^2 + \cdots$. An analogous parameterization of $\hat{D}_v$ is

$$\hat{D}_v = \hat{D}_e + \beta_e\left(v+\frac{1}{2}\right) + \cdots. \tag{9.73}$$

Since Equation (9.72) is expected to be valid for small v, let us compute from it B_0, B_1, and B_2 for $^{12}C^{16}O$. We need the values from Table 9.1 $B_e = 1.9313\,cm^{-1}$ and $\alpha_e = 0.0175\,cm^{-1}$. We have with these values

$$B_0 = 1.9313\,cm^{-1} - \frac{1}{2}\left(0.0175\,cm^{-1}\right) = 1.923\,cm^{-1}$$

$$B_1 = 1.9313\,cm^{-1} - \frac{3}{2}\left(0.0175\,cm^{-1}\right) = 1.905\,cm^{-1}$$

and

$$B_2 = 1.9313\,cm^{-1} - \frac{5}{2}\left(0.0175\,cm^{-1}\right) = 1.888\,cm^{-1}$$

Notice these results coincide with the literature values obtained from spectroscopic measurements listed in Table 9.6. Thus, for the v = 0,1, and 2 states of $^{12}C^{16}O$, Equation (9.72) is very accurate.

Given the parameterizations of $G(v), B_v$, and $\hat{D}_v$, the rotational-vibrational energy levels of a diatomic, which are the solutions of the rotational-vibrational Schrödinger equation (Equation 9.4) which we have been seeking, may be written as

$$\begin{aligned}\frac{E_{vJ}}{hc} &= \omega_e\left(v+\frac{1}{2}\right) - \omega_e x_e\left(v+\frac{1}{2}\right)^2 + \omega_e y_e\left(v+\frac{1}{2}\right)^3 + \omega_e z_e\left(v+\frac{1}{2}\right)^4 \\ &+ \cdots + B_e J(J+1) - \alpha_e J(J+1)\left(v+\frac{1}{2}\right) + \gamma_e J(J+1)\left(v+\frac{1}{2}\right)^2 \\ &+ \cdots - \hat{D}_e J^2(J+1)^2 - \beta_e J^2(J+1)^2\left(v+\frac{1}{2}\right) + \cdots.\end{aligned} \tag{9.74}$$

As illustrated in Table 9.1, the parameters $\omega_e, \omega_e x_e, B_e, \alpha_e$, and $\hat{D}_e$ are tabulated for many diatomic molecules (Huber and Herzfeld 1979). A widely used practical approximate energy level formula involving only these well-tabulated parameters follows from Equation (9.74) as

$$\begin{aligned}\frac{E_{vJ}}{hc} &= \omega_e\left(v+\frac{1}{2}\right) - \omega_e x_e\left(v+\frac{1}{2}\right)^2 + B_e J(J+1) \\ &- \hat{D}_e J^2(J+1)^2 - \alpha_e J(J+1)\left(v+\frac{1}{2}\right).\end{aligned} \tag{9.75}$$

We close this chapter by introducing a very useful model potential energy function, the *Morse potential.*

9.10 THE MORSE POTENTIAL

As for the harmonic oscillator potential, the Schrödinger equation (Equation 9.5) may be solved analytically for the Morse potential. However, the Morse potential is far superior to the harmonic oscillator potential. Most obviously, as for a real potential, the Morse potential is anharmonic, and in contrast to the harmonic oscillator potential, the Morse potential has a dissociation limit.

The form of the Morse potential is

$$U(r) = D_e\{1 - \exp[-\beta(r - r_e)]\}^2. \tag{9.76}$$

Note that the Morse potential involves three parameters: the classical dissociation energy D_e, the equilibrium internuclear separation r_e, and the parameter β which controls the range of the Morse potential's exponential function. For CO, the Morse parameters are

$$D_e = 1083.1\,\text{kJ mol}^{-1} \quad r_e = 112.83\,\text{pm} \quad \beta = 0.023205\,\text{pm}^{-1} \tag{9.77}$$

The Morse potential for CO is plotted in Figure 9.10. Notice that qualitatively, it appears very similar to the real HCl potential plotted in Figure B.1. Despite this apparent similarity, Morse and real potentials are qualitatively different in the $r \to \infty$ and $r \to 0$ limits. To see this, let us write Equation (9.76) as

$$U(r) = D_e\{1 - 2\exp[-\beta(r - r_e)] + \exp[-2\beta(r - r_e)]\}. \tag{9.78}$$

From Equation (9.78),

$$\lim_{r\to\infty} U(r) = D_e[1 - 2\exp(-\beta r)] \tag{9.79}$$

which shows that the Morse potential approaches D_e as $r \to \infty$ exponentially. For real diatomics, as $r \to \infty$, the approach is in contrast via an inverse power dependence. Also for real diatomics, as $r \to 0$, $U(r) \to \infty$. Instead, for the Morse potential, as $r \to 0$, $U(r)$ approaches a large but finite limit $U(r = 0)$. From Equation (9.78), this limit is

$$U(r = 0) = D_e[1 - 2\exp(\beta r_e) + \exp(2\beta r_e)]. \tag{9.80}$$

Evaluating Equation (9.80) using the CO parameters gives

$$U(r = 0) = 162.3 D_e. \tag{9.81}$$

Since for CO $U(r = 0)$ is so much larger than D_e, the fact that it is finite rather than infinite has no practical significance. This behavior is typical.

Next, let us derive a relation between the vibrational frequency υ of a Morse diatomic and its dissociation energy D_e. We will use Equation (7.117) and the relation $\upsilon = \frac{\omega}{2\pi}$ to give the general result

$$\upsilon = \frac{1}{2\pi}\left[\frac{U''(r_e)}{\mu}\right]^{1/2} \tag{9.82}$$

and then evaluate $U''(r_e)$ from Equation (9.78). We first note from Equation (9.78) that

$$U(r_e) = 0. \tag{9.83}$$

Thus, as is evident from Figure 9.10, we have chosen the zero of energy so that the potential is zero at r_e. Next, consider $U'(r_e)$. From Equation (9.78),

$$U'(r) = D_e\left\{2\beta\exp\left[-\beta(r-r_e)\right] - 2\beta\exp\left[-2\beta(r-r_e)\right]\right\}. \tag{9.84}$$

Equation (9.84) gives the expected result for a minimum

$$U'(r_e) = 0. \tag{9.85}$$

Next, differentiating Equation (9.84) with respect to r gives

$$U''(r) = D_e\left\{-2\beta^2\exp\left[-\beta(r-r_e)\right] + 4\beta^2\exp\left[-2\beta(r-r_e)\right]\right\}. \tag{9.86}$$

Then, setting $r = r_e$ yields

$$U''(r_e) = 2\beta^2 D_e. \tag{9.87}$$

Comparing Equations (9.82) and (9.87) gives the required relationship between υ and D_e

$$\upsilon = \frac{\beta}{\pi}\left(\frac{D_e}{2\mu}\right)^{1/2}. \tag{9.88}$$

Let us evaluate the above Morse form for υ for $^{12}C^{16}O$ using the values of D_e and β given in Equation (9.77) and the value of the $^{12}C^{16}O$ reduced mass $\mu = 1.139\times10^{-26}$ kg. Using the conversion factor $1\,kJ\,mol^{-1} = 1.661\times10^{-21}$ J, one finds that $D_e = 1.799\times10^{-18}$ J. Thus,

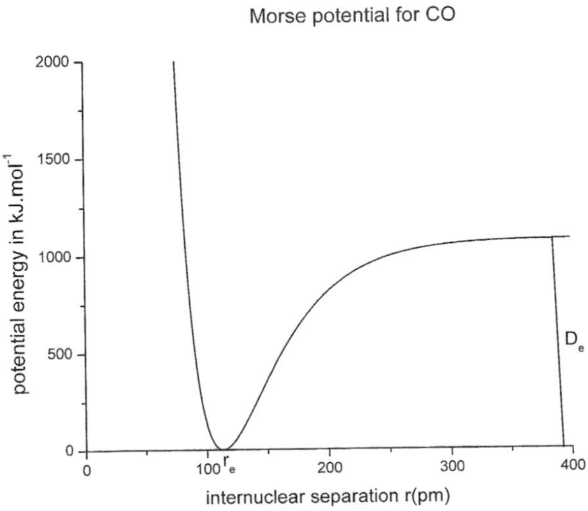

FIGURE 9.10 Morse potential calculated from Equation (9.76). The parameter values $D_e = 1{,}083\,kJ\,mol^{-1}$, $r_e = 113\,pm$, and $\beta = 0.0232\,pm^{-1}$ are those of CO.

$\left(\frac{D_e}{2\mu}\right)^{1/2} = \left[\frac{1.799\times10^{-18}}{2\times1.139\times10^{-26}}\,J\,kg^{-1}\right]^{1/2} = \left[0.7897\times10^{8}\,m^2\,s^{-2}\right]^{1/2} = 0.887\times10^{4}\,m\,s^{-1}$. Then, evaluating Equation (9.88) using $\beta = 2.321\times10^{10}\,m^{-1}$ gives $\upsilon = \frac{2.321\times10^{10}}{\pi}\times0.887\times10^{4}\,s^{-1}$ or

$$\upsilon = 6.566\times10^{13}\,s^{-1}. \tag{9.89}$$

The Morse vibrational wave number for $^{12}C^{16}O$ $\omega_0 = \frac{\upsilon}{c}$ is therefore

$$\omega_0 = 2{,}190\,cm^{-1}. \tag{9.90}$$

The corresponding experimental value is

$$\omega_{expt} = \omega_e = 2{,}170\,cm^{-1}. \tag{9.91}$$

Given the simplicity of the Morse potential, the agreement between ω_0 and ω_{expt} is satisfactory.

As noted, the Schrödinger equation (Equation 9.5) may be solved analytically for the Morse potential (Townes and Schawlow 1955, pp. 7–9). The solution is exact if $J = 0$ and approximate if $J > 0$. The energy levels derived from the solution may be written as

$$\frac{E_{vJ}}{hc} = \omega_e\left(v+\frac{1}{2}\right) - \omega_e x_e\left(v+\frac{1}{2}\right)^2 + B_e J(J+1) - \hat{D}_e J^2(J+1)^2 - \alpha_e J(J+1)\left(v+\frac{1}{2}\right). \tag{9.92}$$

Notice that the Morse levels are identical in form to the approximation to the exact energy levels of Equation (9.75). They therefore do not include the higher order corrections of Equation (9.74). The values of the Morse spectroscopic parameters found from the exact solution are

$$\omega_e = \frac{\beta}{\pi c}\left(\frac{D_e}{2\mu}\right)^{1/2} \tag{9.93}$$

$$\omega_e x_e = \frac{h\beta^2}{8\pi^2\mu c} \tag{9.94}$$

$$B_e = \frac{h}{8\pi^2\mu r_e^2 c} \tag{9.95}$$

$$\hat{D}_e = \frac{4B_e^3}{\omega_e^2} \tag{9.96}$$

$$\alpha_e = \frac{6\left(\omega_e x_e B_e^3\right)^{1/2}}{\omega_e} - \frac{6B_e^2}{\omega_e}. \tag{9.97}$$

Note that ω_e of Equation (9.93) is from Equation (9.88) identical to the Morse wave number ω_0. Thus, it will not agree perfectly with the exact value for ω_e of Table 9.1. The other Morse spectroscopic parameters except for B_e will correspondingly differ from the exact values of Table 9.1 (Problem 9.18). Also note that the Morse form for $\hat{D}_e$ of Equation (9.96) conforms to the Kratzner equation (Equation 9.54).

FURTHER READINGS

Allen, Jr. Harry C., and Paul C. Cross. 1963. *Molecular vib-rotors the theory and interpretation of high resolution infrared spectra.* New York: John Wiley and Sons, Inc.

Atkins, Peter, Julio de Paula, and Ronald Friedman. 2014. *Physical chemistry: quanta, matter, and change.* 2nd ed. Great Britain: Oxford University Press.

Atkins, Peter, and Ronald Friedman. 2011. *Molecular quantum mechanics.* 5th ed. New York: Oxford University Press.

Bernath, Peter F. 2005. *Spectra of atoms and molecules.* 2nd ed. Oxford: Oxford University Press.

Brown, John M. 1998. *Molecular spectroscopy.* Oxford: Oxford University Press.

Demtroder, Wolfgang. 2010. *Rotation and vibration of diatomic molecules. Sec. 9.5 in An introduction to atomic-, molecular-, and quantum physics.* 2nd ed. Heidelberg: Springer.

Dunham, J. L. "The energy levels of a rotating vibrator." *Phys. Rev.* 41: (1932) 721–731.

Hertel, Ingolf V. and Claus-Peter Schultz. 2015. Diatomic molecules. Chapter 3 in *Atoms, molecules, and optical physics 2: molecules and photons-spectroscopy and collisions.* Heidelberg: Springer.

Herzberg, Gerhard. 1939. *Molecular spectra and molecular structure I. diatomic molecules.* New York: Prentice-Hall, Inc.

Hollas, J. Michael. 2002. *Basic atomic and molecular spectroscopy.* Exeter, Great Britain: Wiley Interscience.

Hollas, J. Michael. 2004. *Modern spectroscopy.* 4th ed. West Sussex, England: John Wiley & Sons Ltd.

Huber, K. P and G. Herzberg. 1979. *Molecular spectra and molecular structure IV. Constants of diatomic molecules.* New York: Van Nostrand Reinhold Company.

McQuarrie, Donald A. 2008. *Quantum chemistry.* 2nd ed. Sausalito, CA: University Science Books.

Townes, C. H., and A. L. Schawlow. 1955. *Microwave spectroscopy.* New York: McGraw-Hill Book Company Inc.

PROBLEMS

9.1 Derive the form of the pseudo-one-dimensional Schrödinger equation (Equation 9.5) from Equation (9.4).

9.2 Show that the boundary conditions of Equation (9.8) follow from Equation (8.8a).

9.3 In this problem, you will show using as an example $^{12}C^{16}O$ that the rotational frequency υ_{rot} of a diatomic is much less than its vibrational frequency υ_{vib}. Then, you will establish from this fact the strong inequality $\bar{B} \ll \bar{\omega}$ of Equation (9.20). You will use that the bond length of CO $r_e = 112.83\,pm$ and the vibrational wave number of $^{12}C^{16}O$ $\bar{\omega} = 2{,}143.2\,cm^{-1}$. Using the classical to a quantum correspondence $J \leftrightarrow [J(J+1)]^{1/2}\hbar$ between the magnitude J of the classical angular momentum vector and the value J of the angular momentum quantum number, Equation (B.44) for υ_{rot} may be written in semiclassical form as $\upsilon_{rot} = \frac{[J(J+1)]^{1/2}\hbar}{4\pi\mu r_e^2}$.
(a) Taking $J = 7$, a typical value, show for $^{12}C^{16}O$ $\upsilon_{rot} = 0.811\,ps^{-1}$ and also note for $^{12}C^{16}O$ that $\upsilon_{vib} = c\bar{\omega} = 64.25\,ps^{-1}$. This shows for $^{12}C^{16}O$ that $\upsilon_{rot} \ll \upsilon_{vib}$. (b) Show that in sequence $\bar{\upsilon}_{rot} = \frac{\upsilon_{rot}}{c} = 2\bar{B}J$ and hence for $^{12}C^{16}O$ $2\bar{B}J \ll \bar{\omega}$. (c) Then, show from your last result that for $^{12}C^{16}O$ $\bar{B} \ll \bar{\omega}$, the result we set out to prove.

9.4 For $^{14}N^{16}O$, $\bar{B} = 1.67\ cm^{-1}$ and $\bar{\omega} = 1{,}876.1\,cm^{-1}$. (a) Calculate in cm^{-1} the positions of the first four harmonic oscillator rigid rotor IR R-branch lines of $^{14}N^{16}O$. (b) Repeat for the first four P-branch lines. (c) What is the position of the missing line?

9.5 Referring to Problem 9.4, (a) for $^{14}N^{16}O$ compute at $T = 300\,K$ $\frac{N_J}{N_{J=0}}$ for $J = 1-5$. (b) Using these values for $\frac{N_J}{N_{J=0}}$, sketch the ideal relative intensities of the first few R-branch and P-branch lines of the infrared spectrum of $^{14}N^{16}O$.

9.6 (a) Show from Equation (9.28) that $\frac{N_J}{N_{J=0}}$ has its maximum value for $J = J_{max}$ where $J_{max} = \left(\frac{T}{2\Theta_{rot}}\right)^{1/2} - \frac{1}{2}$. (b) Evaluate J_{max} numerically at $T = 300\,K$ for $^{1}H^{35}CI$ with $\Theta_{rot} = 15.24\,K$ and show your result is in good agreement with the maximum of $\frac{N_J}{N_0}$ in Figure 9.5.

9.7 The observed wave numbers for or the low-resolution fundamental and overtone IR lines of a diatomic molecule are given by $\bar{\upsilon}^{OBS} = G(v) - G(0)$. In Section 9.5, we approximated $\bar{\upsilon}^{OBS}$ for $^1H^{35}Cl$ using the low v limit of $G(v), \omega_e\left(v+\frac{1}{2}\right) - \omega_e x_e\left(v+\frac{1}{2}\right)^2$. In Table 9.2, we compared for $^1H^{35}Cl$ the estimate $\bar{\upsilon}^{ANHAR}$ obtained from this low v limit to $\bar{\upsilon}^{OBS}$ and found that $\bar{\upsilon}^{ANHAR} < \bar{\upsilon}^{OBS}$. In this problem, we will use a more accurate form for $G(v), \left(v+\frac{1}{2}\right)\omega_e - \omega_e x_e\left(v+\frac{1}{2}\right)^2 + \omega_e y_e\left(v+\frac{1}{2}\right)^3$ and compare the estimate $\upsilon^{(2)ANHAR}$ obtained from this more accurate form to $\bar{\upsilon}^{OBS}$ and $\bar{\upsilon}^{ANHAR}$. (a) Show $\bar{\upsilon}^{(2)ANHAR} = \bar{\upsilon}^{ANHAR} + \omega_e y_e \times v\left(v^2 + \frac{3v}{2} + \frac{3}{4}\right)$. (b) For $^1H^{35}Cl$, $\omega_e y_e = 0.22\,\text{cm}^{-1}$. From this value and from the results for $\bar{\upsilon}^{ANHAR}$ of Table 9.2, compute in cm^{-1} units $\bar{\upsilon}^{(2)ANHAR}$ for the fundamental and first four overtone lines of $^1H^{35}Cl$ and compare with $\bar{\upsilon}^{OBS}$ and $\bar{\upsilon}^{ANHAR}$. Why do you think that including the $\omega_e y_e\left(v+\frac{1}{2}\right)^3$ term overestimates the correction to $\bar{\upsilon}^{ANHAR}$?

9.8 In this problem, you will show numerically for $^{12}C^{16}O$ and $^1H^{35}Cl$ that $\bar{B} = \frac{h}{8\pi^2 Ic} = \frac{h}{8\pi^2 \mu r_e^2 c} = B_e$. Use the values of r_e listed in Table 9.1, the atomic masses $m_{^{16}O} = 15.99491\text{u}$, $m_{^1H} = 1.00783\text{u}$, and $m_{^{35}Cl} = 34.96885\text{u}$, and the values of constants $m_u = 1.6605402 \times 10^{-27}\,\text{kg}\,\text{u}^{-1}$, $h = 6.6260755 \times 10^{-34}\,\text{J s}$, and $c = 2.99792458 \times 10^{10}\,\text{cm s}^{-1}$. (a) Show for $^1H^{35}Cl$ that $\bar{B} = 10.593364\ \text{cm}^{-1}$. Compare with Table 9.1 that gives $B_e = 10.593416\ \text{cm}^{-1}$. (b) Show for $^{12}C^{16}O$ $\bar{B} = 1.931282\,\text{cm}^{-1}$. Compare with Table 9.1 that gives $B_e = 1.931281\,\text{cm}^{-1}$.

9.9 Derive the Kratzner equation (Equation 9.54) from the results given in Equation (9.53).

9.10 From the values of B_e and ω_e of Table 9.1, numerically evaluate $\hat{D}_e$ from the approximate Kratzner equation (Equation 9.54) for (a) $^1H^{35}Cl$, (b) $^{12}C^{16}O$, and (c) ^7LiH. Compare your results with the corresponding exact values for $\hat{D}_e$ given in Table 9.1. What do you conclude for these systems about the numerical accuracy of the Kratzner equation?

9.11 Prove Equation (9.56) for $\bar{\upsilon}_{J\to J+1}$ from Equation (9.55).

9.12 Show from Equation (9.56) that for the pure rotational absorption spectrum of a diatomic molecule, when centrifugal distortion is included the lines become increasingly closely spaced as J increases.

J	$\bar{\upsilon}^{OBS}_{J\to J+1}\left(\text{cm}^{-1}\right)$
0	20.878284
1	41.743895
2	62.584183
3	83.386501
4	104.138260
5	124.826909
6	145.439949
7	165.964971
8	186.389616
9	206.701655

9.13 This problem concerns the pure rotational spectrum of $v = 0\,{}^1H^{35}Cl$. You will use the following observed positions $\bar{\upsilon}^{OBS}_{J\to J+1}$ of the $J = 0 \to J = 1$ to $J = 8 \to J = 9$ spectral lines: (a) Compute the spacing of the lines in cm^{-1}. Notice that the line spacing decreases as J increases. What do you conclude about the validity of the rigid rotor approximation for ${}^1H^{35}Cl$? (b) According to Equation (9.57), a plot for $\frac{\bar{\upsilon}^{OBS}_{J\to J+1}}{J+1}$ versus $(J+1)^2$ should be linear with slope $-4\hat{D}_o$ and intercept $2B_o$. Make such a plot for ${}^1H^{35}Cl$, and from the plot, estimate for this molecule $\hat{D}_0$ and B_0. Compare your results with the literature values $B_o = 10.4402\,cm^{-1}$ and $\hat{D}_o = 5.280\times10^{-4}\,cm^{-1}$.

9.14 Consider the $v = 0 \to v = 2$ overtone transitions of ${}^{12}C^{16}O$. The observed first sixteen R-branch and first fifteen P-branch infrared spectral lines for these transitions are as follows:

J	$R(J)/cm^{-1}$	$P(J)/cm^{-1}$
0	4,263.838	
1	4,267.542	4,256.220
2	4,271.179	4,252.307
3	4,274.743	4,248.320
4	4,278.234	4,244.267
5	4,281.661	4,240.142
6	4,285.011	4,235.949
7	4,288.290	4,231.685
8	4,291.500	4,227.356
9	4,294.640	4,222.954
10	4,297.707	4,218.488
11	4,300.702	4,213.951
12	4,303.625	4,209.345
13	4,306.476	4,204.672
14	4,309.257	4,199.930
15	4,311.962	4,195.120

(a) From the data, calculate in cm^{-1} the combinations and differences $R(J)-P(J)$ and $R(J-1)-P(J+1)$. Note by specializing Equation (9.69) $R(J)-P(J)=4B_2\left(J+\frac{1}{2}\right)$ and $R(J-1)-P(J+1)=4B_0\left(J+\frac{1}{2}\right)$. (b) Plot $R(J)-P(J)$ and $R(J-1)-P(J+1)$ versus $\left(J+\frac{1}{2}\right)$. From your plots, estimate B_2 and B_0. (c) To obtain a quantitative determination of B_2, make a least squares fit of the relation $R(J)-P(J)=4B_2\left(J+\frac{1}{2}\right)$. Compare with the literature value $B_2 = 1.888\,cm^{-1}$. (d) Repeat for B_0 using the relation $R(J-1)-P(J+1)=4B_o\left(J+\frac{1}{2}\right)$. Compare with the literature value $B_0 = 1.923\,cm^{-1}$.

9.15 In the experimental ${}^1H^{35}Cl$ high-resolution IR spectrum of Figure 9.4, the R-branch lines and the P-branch lines become, respectively, increasingly narrowly spaced and increasingly widely spaced as J increases. Here, you will give a general explanation of this phenomenon.

Neglecting centrifugal distortion and higher order terms in $G(v)$, the energy levels of a diatomic are $\frac{E_{vJ}}{hc} = G(v) + B_v J(J+1)$ where $G(v) = \omega_e\left(v+\frac{1}{2}\right) - \omega_e x_e\left(v+\frac{1}{2}\right)^2$. Consider the $v == 0, v = 1$ fundamental band transitions. (a) Show that the R-branch IR spectral lines occur at wave numbers $R(J) = \bar{\omega} + (B_1 - B_0)J^2 + (3B_1 - B_0)J + 2B_1$ where from Equation (9.36) $\bar{\omega} = \omega_e - 2\omega_e x_e$. (b) Show that the P-branch lines occur at wave numbers $P(J) = \bar{\omega} + (B_1 - B_0)J^2 - (B_1 - B_0)J$. (c) Show that the spacing between R-branch lines $R(J+1) - R(J) = 2J(B_1 - B_0) + 4B_1 - 2B_0$. (d) Show that the spacing of the P-branch lines $P(J) - P(J+1) = -2J(B_1 - B_0) + 2B_0$. (e) Show from your result for $R(J+1) - R(J)$ that the spacing decreases for the R-branch lines as J increases. (f) Show from your result for $P(J) - P(J+1)$ that the spacing increases as J increases.

9.16 The high-resolution $v = 0$ pure rotational spectrum of $^1H^{35}Cl$ of Problem 9.13 gives that $B_0 = 10.440\,\text{cm}^{-1}$. Using the expression for B_v given in Problem 9.17, compute the $v = 0$ bond length r_0 in pm for $^1H^{35}Cl$.

9.17 Denote the effective internuclear separation for the vth vibrational state of a diatomic by r_v where r_v is defined by $B_v = \frac{h}{8\pi^2\mu r_v^2 c}$. (a) Using Equation (9.72) and Table 9.1, compute B_0, B_1, and B_2 in cm^{-1} for $^{16}O^1H$. (b) Compute r_0, r_1, and r_2 in pm for $^{16}O^1H$ and compare these values with $r_e = 96.966\,\text{pm}$.

9.18 Calculate $\omega_e, \omega_e x_e, B_e, \hat{D}_e$, and α_e for the Morse potential using Equations (9.92)–(9.97). Use the parameters for CO given in Equation (9.77) and the value of the reduced mass for $^{12}C^{16}O$ $\mu = 1.139 \times 10^{-26}$ kg. Compare your results with the exact values of the spectroscopic parameters for $^{12}C^{16}O$ listed in Table 9.1

9.19 In this problem, you will compare the positions of the spectral lines of the fundamental infrared band of a diatomic molecule derived from a model that includes vibrational–rotational coupling and vibrational anharmonicity with the corresponding positions derived from the harmonic oscillator rigid rotor approximation. You will take as an example the molecule $^{12}C^{16}O$ for which $\bar{\omega} = 2{,}143\,\text{cm}^{-1}$ and $\bar{B} = B_e = 1.931\,\text{cm}^{-1}$. (a) Using Equation (9.23), compute in cm^{-1} the wave numbers of the first three harmonic oscillator rigid rotor R- and P-branch fundamental band lines of $^{12}C^{16}O$. Next, you will determine the effects on the corresponding line positions of vibrational–rotational coupling and vibrational anharmonicity. You will neglect the effects of centrifugal distortion and thus will approximate the rotational term by $F(J) = B_v J(J+1)$. You will also neglect anharmonicity corrections of higher order than quadratic in $\left(v+\frac{1}{2}\right)$ and thus will approximate the vibrational term by $G(v) = \omega_e\left(v+\frac{1}{2}\right) - \omega_e x_e\left(v+\frac{1}{2}\right)^2$. Within these approximations, the vibrational–rotational energy levels are $\frac{E_{vJ}}{hc} = \omega_e\left(v+\frac{1}{2}\right) - \omega_e x_e\left(v+\frac{1}{2}\right)^2 + B_v J(J+1)$. Next, specialize to the $v = 0 \rightarrow v = 1$ fundamental band. From the form for $\frac{E_{vJ}}{hc}$, you derived in Problem 9.15 the following forms for the fundamental band the R-branch and P-branch line positions: $R(J) = \bar{\omega} + (B_1 - B_0)J^2 + (3B_1 - B_0)J + 2B_1$ and $P(J) = \bar{\omega} + (B_1 - B_0)J^2 - (B_1 + B_0)J$. (b) Show that if $B_0 = B_1 = \bar{B}$, $R(J)$ and $P(J)$ reduce to the harmonic oscillator rigid rotor line positions of Equation (9.23). For $^{12}C^{16}O$, $B_0 = 1.923\ \text{cm}^{-1}$ and $B_1 = 1.905\ \text{cm}^{-1}$. (c) Compute from the above form for $R(J)$ the first three R-branch lines in the infrared spectrum of $^{12}C^{16}O$. (d) Similarly, compute from the form for $P(J)$ the first three P-branch lines in the infrared spectrum of $^{12}C^{16}O$. (e) What do you conclude about the accuracy of the harmonic oscillator rigid rotor approximation for $^{12}C^{16}O$ at small J?

SOLUTIONS TO SELECTED PROBLEMS

9.4 (a) 1,879.4, 1882.8, 1,886.1, and 1,889.5 cm^{-1}; (b) 1,872.8, 1,869.4, 1,866.1, and 1,862.7 cm^{-1}; and (c) 1,876.1 cm^{-1}.

9.5 (a) For J = 1, 2.95; J = 2, 4.76; J = 3, 6.36; J = 4, 7.66; and J = 5, 8.64. (c) For J = 1,2.98; J = 2, 4.89; J = 3, 6.90; J = 4, 8.33; and J = 5, 10.18.

9.6 (b) $J_{max} = 3$.

9.7 $\bar{\upsilon}^{OBS}\left(cm^{-1}\right)$, $\bar{\upsilon}^{ANHAR}\left(cm^{-1}\right)$, and $\bar{\upsilon}^{(2)ANHAR}\left(cm^{-1}\right)$ are for the fundamental, respectively 2,885.3, 2,885.3, and 2,886.0; for the first overtone, 5,668.0, 5,665.0, and 5,668.4; for the second overtone, 8,347.0, 8,339.0, and 8,348.4; for the third overtone, 10,923.1, 10,907.4 and 10,927.4; and for the fourth overtone, 13,396.1, 13,370.2, and 13,406.8.

9.10 From the Kratzner equation: for $^{1}H^{35}Cl$ $\hat{D}_e = 5.3156 \times 10^{-4}\ cm^{-1}$, for $^{12}C^{16}O$ $\hat{D}_e = 6.1200 \times 10^{-6}\ cm^{-1}$, and for $^{7}Li^{1}H$ $\hat{D}_e = 0.8585 \times 10^{-3}\ cm^{-1}$. From Table 9.1: for $^{1}H^{35}Cl$ $\hat{D}_e = 5.3194 \times 10^{-4}\ cm^{-1}$, for $^{12}C^{16}O$ $\hat{D}_e = 6.1215 \times 10^{-6}\ cm^{-1}$, and for $^{7}Li^{1}H$ $\hat{D}_e = 0.8617 \times 10^{-3}\ cm^{-1}$.

9.14 (c) $B_2 = 1.884\ cm^{-1}$ and (d) $B_0 = 1.920\ cm^{-1}$.

9.16 $r_e = 127.4$ pm.

9.17 (a) $B_0 = 18.5484\ cm^{-1}$, $B_1 = 17.8242\ cm^{-1}$, and $B_2 = 17.100\ cm^{-1}$.
(b) $r_0 = 97.908$ pm, $r_1 = 99.877$ pm, and $r_2 = 101.970$ pm.

9.18 $\omega_e = 2{,}190\ cm^{-1}$, $\omega_e x_e = 12.24\ cm^{-1}$, $B_e = 1.93\ cm^{-1}$, $\hat{D}_e = 6.00 \times 10^{-6}\ cm^{-1}$, and $\alpha_e = 0.016\ cm^{-1}$.

9.19 (a) All line positions in cm^{-1}. (b) 2,147, 2,151, 2,155, 2,139, 2,135, and 2,131; (c) 2,147, 2,151, and 2,154; and (d) 2,139, 2,135, and 2,132.

10 The Hydrogen-Like Atoms

In Section 1.3, we developed the Bohr theory of the hydrogen-like atoms: H, He^{+}, L_{i}^{2+}, and so on. Here, we develop the modern quantum theory of these atoms. Again, we deal with a two-particle system comprised of an electron of mass

$$m_e = 9.10938972 \times 10^{-31}\ \mathrm{kg} \tag{10.1}$$

and of charge $-e$ where

$$e = 1.6022 \times 10^{-19}\ \mathrm{C} \tag{10.2}$$

and a nucleus of mass m_N and charge Ze where Z is the atomic number. The motion of this two-particle system is determined by an attractive Coulomb potential which is in the SI units of joules is

$$U(r) = -\frac{Ze^2}{4\pi\varepsilon_0 r} \tag{10.3}$$

where r is the distance between the electron and the nucleus and

$$4\pi\varepsilon_0 = 1.26301 \times 10^{-10}\ \mathrm{J^{-1}\,m^{-1}\,C^2} \tag{10.4}$$

is 4π times the vacuum permittivity ε_0.

With the system so defined in either classical or quantum mechanics, the problem of the hydrogen-like atoms amounts to solving for the motion of the relative coordinate vector $\mathbf{r} = \mathbf{r_e} - \mathbf{r_N} = (r, \theta, \phi)$, where $\mathbf{r}_e$ and $\mathbf{r}_N$ are the position vectors of the atom's electron and nucleus. The effective mass for the motion is the reduced mass

$$\mu = \frac{m_e m_N}{m_e + m_N}. \tag{10.5}$$

Since $m_e \ll m_N$, to an excellent approximation $\mu \doteq m_e$ and the relative coordinate vector reduces to the vector displacement of the electron from the nucleus assumed motionless. We made the approximation $\mu \doteq m_e$ in our development of the Bohr theory. Here, we will lift it. So to see the change, let us compute μ for the hydrogen atom and compare it to m_e. We need m_N which for hydrogen is nearly the mass of the proton

$$m_p = 1.6727623 \times 10^{-27}\ \mathrm{kg}. \tag{10.6}$$

Comparing Equations (10.1) and (10.6) yields

$$m_p = 1836.1538 m_e. \tag{10.7}$$

Equations (10.5) and (10.7) give for the reduced mass of hydrogen

$$\mu = \frac{1{,}836.1258\, m_e^2}{1{,}837.1258\, m_e} = 0.999456 m_e. \tag{10.8}$$

Note that the approximation $\mu \doteq m_e$ is sufficient unless very high precision is needed.

We next obtain the *radial Schrödinger equation* of a hydrogen-like atom. Its solutions are the *radial wave functions* that determines the dependence of the electron's full wave function on r.

10.1 THE RADIAL SCHRÖDINGER EQUATION

The Coulomb potential of Equation (10.3) is of the central field type. In Section 9.1, we discussed the rotational–vibrational motion of a diatomic molecule for which the potential $U(r)$ is also of the central field type. Consequently, we may write down and simplify the Schrödinger equation for a hydrogen-like atom by transcribing the results and analysis of Section 9.1.

To start, we make the following transcriptions that are in accord with the notational conventions for single electron motion: rotational angular momentum operator $\hat{J} \to$ electronic angular momentum operator $\hat{L}$, rotational angular momentum quantum numbers JM $\to$ electronic angular momentum quantum numbers ℓm, and rotational–vibrational wave functions $\Psi_{vJM}(r,\theta,\phi) \to$ hydrogen-like atom electronic wave functions $\Psi_{n\ell m}(r,\theta,\phi)$.

Note that the electronic quantum numbers n, ℓ, and m are called, respectively, the *principal, angular momentum*, and *magnetic* quantum numbers.

The Schrödinger equation for $\Psi_{n\ell m}(r,\theta,\phi)$ may now be obtained from the rotational–vibrational Schrödinger equation (Equation 9.2) for $\Psi_{vJM}(r,\theta,\phi)$ by the substitutions $\hat{J}^2 \to \hat{L}^2, vJM \to n\,\ell\,m$, and $U(r) \to -\dfrac{Ze^2}{4\pi\varepsilon_0 r}$. These substitutions yield the Schrödinger equation

$$\left[-\frac{\hbar^2}{2\mu}\frac{1}{r^2}\frac{\partial}{\partial r}\left(r^2\frac{\partial}{\partial r}\right)+\frac{\hat{L}^2}{2\mu r^2}-\frac{Ze^2}{4\pi\varepsilon_0 r}\right]\Psi_{n\ell m}(r,\theta,\phi)=E_n\Psi_{n\ell m}(r,\theta,\phi). \tag{10.9}$$

Note that we have anticipated for hydrogen-like atoms that the energies E_n are independent of the angular momentum quantum number ℓ and depend only on the principal quantum number n.

Since the Coulomb potential is a central field potential, in analogy to Equation (9.3) solutions of Equation (10.9) of the following form exist:

$$\Psi_{n\ell m}(r,\theta,\phi)=R_{n\ell}(r)Y_{\ell m}(\theta,\phi) \tag{10.10}$$

where $R_{n\ell}(r)$ is the radial wave function and $Y_{\ell m}(\theta,\phi)$ is a spherical harmonic. In analogy to the eigenvalue equations (Equation 8.108) for $\hat{J}^2$ and $\hat{J}_z$, one has that $\hat{L}^2Y_{\ell m}(\theta,\phi)=\ell(\ell+1)\hbar^2Y_{\ell m}(\theta,\phi)$ and $\hat{L}_zY_{\ell m}(\theta,\phi)=m\hbar Y_{\ell m}(\theta,\phi)$.

We may now derive the radial Schrödinger equation for $R_{n\ell}(r)$. Comparing Equations (10.9) and (10.10) and using the relation $\hat{L}^2Y_{\ell m}(\theta,\phi)=\ell(\ell+1)\hbar^2Y_{\ell m}(\theta,\phi)$ yield the radial Schrödinger equation as (compare Equation [9.4])

$$\left[-\frac{\hbar^2}{2\mu}\frac{1}{r^2}\frac{d}{dr}\left(r^2\frac{d}{dr}\right)+\frac{\ell(\ell+1)\hbar^2}{2\mu r^2}-\frac{Ze^2}{4\pi\varepsilon_0 r}\right]R_{n\ell}(r)=E_nR_{n\ell}(r). \tag{10.11}$$

The spherical harmonics $Y_{\ell m}(\theta,\phi)$ are known from Section 8.10 and are listed in Table 10.1. Thus, by Equation (10.10) the wave functions $\Psi_{n\ell m}(r,\theta,\phi)$ may be found by solving Equation (10.11) for the radial wave functions $R_{n\ell}(r)$ and then normalizing these wave functions according to the condition (derived later)

$$\int_0^\infty r^2R_{n\ell}^2(r)\,dr=1. \tag{10.12}$$

TABLE 10.1
The s, p, and d Spherical Harmonics $Y_{\ell m}(\theta,\phi)$

Type	Degeneracy $2\ell+1$	ℓ	m	$Y_{\ell m}(\theta,\phi)$
s	1	0	0	$\left(\frac{1}{4\pi}\right)^{1/2}$
p	3	1	−1	$\left(\frac{3}{8\pi}\right)^{1/2}\sin\theta\exp(-i\phi)$
			0	$\left(\frac{3}{4\pi}\right)^{1/2}\cos\theta$
			1	$-\left(\frac{3}{8\pi}\right)^{1/2}\sin\theta\exp(i\phi)$
d	5	2	−2	$\left(\frac{15}{32\pi}\right)^{1/2}\sin^2\theta\exp(-2i\phi)$
			−1	$\left(\frac{15}{8\pi}\right)^{1/2}\sin\theta\cos\theta\exp(-i\phi)$
			0	$\left(\frac{5}{16\pi}\right)^{1/2}(3\cos^2\theta-1)$
			1	$-\left(\frac{15}{8\pi}\right)^{1/2}\sin\theta\cos\theta\exp(i\phi)$
			2	$\left(\frac{15}{32\pi}\right)^{1/2}\sin^2\theta\exp(2i\phi)$

Solving Equation (10.11) also gives the energy eigenvalues E_n.

The solutions of the radial Schrödinger equation are derived in Appendix D.

We next give an overview of the results of the modern quantum theory of hydrogen-like atoms based in part on the analysis of Appendix D.

10.2 AN OVERVIEW OF THE QUANTUM THEORY OF HYDROGEN-LIKE ATOMS

In order to obtain acceptable wave functions $\Psi_{n\ell m}(r,\theta,\phi)$, the quantum numbers n,ℓ, and m must be restricted to the integer values

$$n = 1,2,\ldots \tag{10.13}$$

$$\ell = 0,1,\ldots,n-1 \tag{10.14}$$

and

$$m = -\ell,-\ell+1,\ldots,\ell-1,\ell. \tag{10.15}$$

Notice that $2\ell+1$ m values are possible for a given ℓ and n ℓ values are possible for a given n.

Also the following terminology is used. States with $\ell = 0,1,2,3,4,5,\ldots$ are called s, p, d, f, g, h,… states. The first four letters come from an old classification of atomic spectral lines as *sharp, principal, diffuse, and fundamental.* Letters after f are in alphabetical order with the exception that j is omitted.

As is familiar from general chemistry texts, Equations (10.13)–(10.15) strongly limit the possible atomic states. Thus, if n = 1, ℓ must be equal to zero, and since m is also zero, only a single s state is possible. Since n = 1, it is called a 1s state. If n = 2, ℓ can be zero or one and both s and p states are possible, called the 2s and 2p states. There are three 2p states since for $\ell = 1$ m can equal −1,0, and 1. To continue, for n = 3 there is one 3s state, three 3p states, and five 3d states, a total of nine states.

The wave functions $\Psi_{n\ell m}(r,\theta,\phi)$ are called *orbitals*. The 1s wave function $\Psi_{100}(r,\theta,\phi)$ is called the 1s orbital. The 2p wave functions $\Psi_{21-1}(r,\theta,\phi)$, $\Psi_{211}(r,\theta,\phi)$, and $\Psi_{210}(r,\theta,\phi)$ are called the 2p orbitals, and so on.

Next, let us determine the number of states with the same principal quantum number n. We just noted for n = 3 that there are $n^2 = 9$ states. This is a special case of the general result that for quantum number n, there are n^2 states. To prove this, note from Equations (10.14) and (10.15) that

$$\begin{array}{c}\text{The number of}\\ \text{states with quantum}\\ \text{number n}\end{array} = \sum_{\ell=0}^{n-1}(2\ell+1) = 1+3+5\cdots. \tag{10.16}$$

Equation (10.16) is an arithmetic series. The rule for summing an arithmetic series, however, is to add the first and last terms in the series and to multiply that sum by half the number of terms in the series. Applying this rule to Equation (10.16) gives

$$\begin{array}{c}\text{The number of}\\ \text{states with quantum}\\ \text{number n}\end{array} = \frac{n}{2}\left[1+2(n-1)+1\right] = n^2 \tag{10.17}$$

which proves the result.

Additionally, we have noted that since the energy eigenvalues E_n depend only on the quantum number n, the n^2 states with the same n value but different ℓ and m values have the same energy E_n (Figure 10.1). For this reason, the energy level E_n is n^2*-fold degenerate.*

We next turn to the form of the energies E_n. This form is

$$E_n = -\frac{Z^2e^4\mu}{2(4\pi\varepsilon_0)^2 n^2\hbar^2}, \quad n = 1,2,\ldots. \tag{10.18}$$

Notice that the form of E_n is the same as the Bohr form of Equation (1.54) if one replaces in that equation m_e by μ. Equation (10.18) may be rewritten as

$$E_n = -\frac{Z^2e^2}{2(4\pi\varepsilon_0)n^2a}, \quad n = 1,2,\ldots \tag{10.19}$$

where

$$a = \frac{4\pi\varepsilon_0}{\mu e^2}\hbar^2. \tag{10.20}$$

Replacing μ by m_e in Equation (10.20), a becomes the Bohr radius a_0 of Equation (1.56). Since a is numerically nearly identical to a_0, we will also call a the Bohr radius.

Next, let us return to the radial Schrödinger equation (Equation 10.11). Let us introduce a new function $u_{n\ell}(r) = rR_{n\ell}(r)$. Then, following steps like those that led from Equation (9.4) to Equations (9.6) and (9.7), Equation (10.11) transforms to the following pseudo-one-dimensional Schrödinger equation:

$$-\frac{\hbar^2}{2\mu}\frac{d^2}{dr^2}u_{n\ell}(r) + U_{eff}(r)u_{n\ell}(r) = E_n u_{n\ell}(r) \tag{10.21}$$

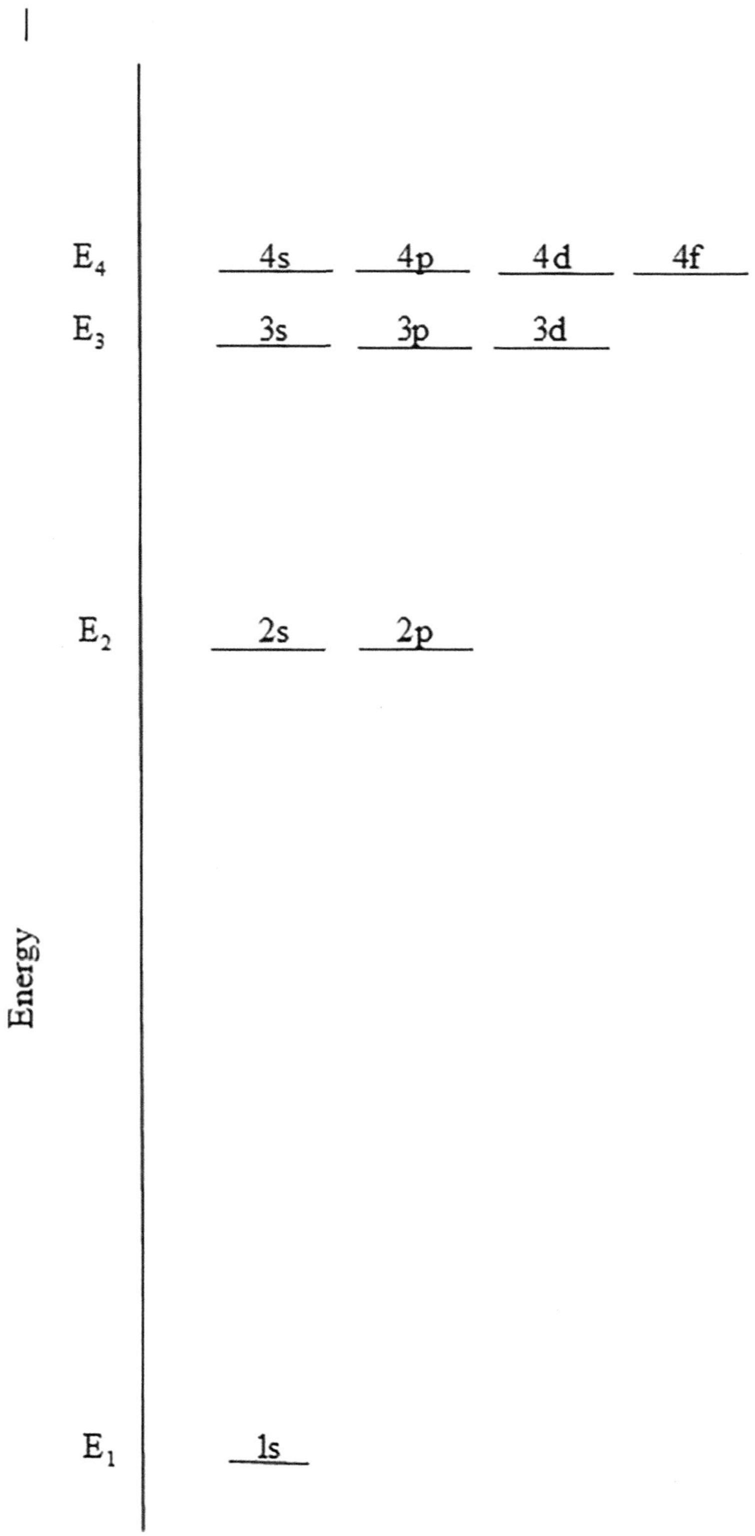

FIGURE 10.1 All hydrogen-like atom states with the same principal quantum number n have the same energy E_n.

where the effective potential $U_{eff}(r)$ is given by

$$U_{eff}(r) = \frac{\ell(\ell+1)\hbar^2}{2\mu r^2} - \frac{Ze^2}{4\pi\varepsilon_0 r}. \tag{10.22}$$

$U_{eff}(r)$ is plotted for hydrogen (Z = 1) for $\ell = 0-3$ in Figure 10.2.

We may gain insight without doing calculations into the small and large r behaviors of the radial wave functions $R_{n\ell}(r)$ from the form of $U_{eff}(r)$.

First, note that $U_{eff}(r)$ is the sum of the repulsive centrifugal potential $\frac{\ell(\ell+1)\hbar^2}{2\mu r^2}$ and the attractive Coulomb potential $-\frac{Ze^2}{4\pi\varepsilon_0 r}$. Also the centrifugal potential vanishes for s states $(\ell = 0)$ but is non-vanishing for p, d, f,... states $(\ell > 0)$. Thus, except at large r where the Coulomb potential dominates the centrifugal potential $U_{eff}(r)$ is qualitatively different for s states and for p, d, f,... states. Especially, at small r $U_{eff}(r)$ is a repulsive potential for p, d, f,... states (since the centrifugal potential dominates the Coulomb potential as $r \to 0$) and is the attractive Coulomb potential for s states. Moreover, at small r for p, d, f,... states $U_{eff}(r)$ becomes increasingly repulsive as ℓ increases. These differences in $U_{eff}(r)$ at small r lead to qualitative differences in the small r forms of the radial wave functions. As a result of these differences, the probability that an electron in a p, d, f,... state is near the nucleus decreases as ℓ increases and is qualitatively smaller than the corresponding s state probability.

In other words, a p, d, f,... electron is kept away from the nucleus because its effective potential is strongly repulsive at small r and the extent to which it is kept away increases with ℓ.

These qualitative arguments may be made quantitative by an analysis of the small r behavior of $R_{n\ell}(r)$ given in Appendix D. This analysis yields

$$\lim_{r\to 0} R_{n\ell}(r) \sim \rho^{\ell} \tag{10.23}$$

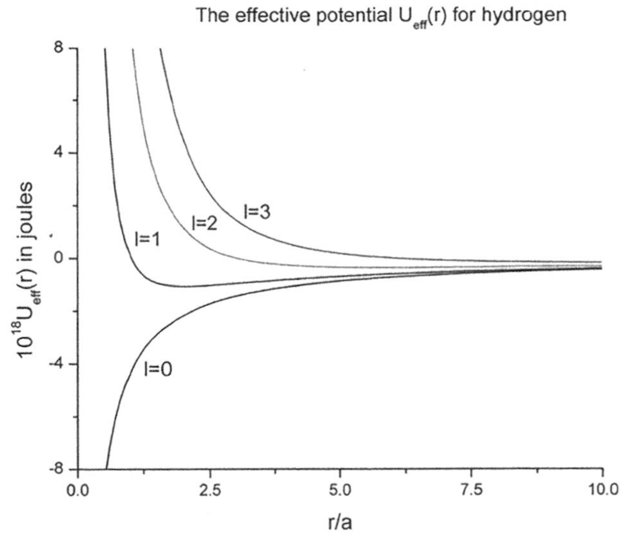

FIGURE 10.2 The hydrogen atom effective potential $U_{eff}(r) = \frac{\ell(\ell+1)\hbar^2}{2\mu r^2} - \frac{e^2}{4\pi\varepsilon_0 r}$ for $\ell = 0-3$ plotted against r/a where $a = \frac{4\pi\varepsilon_0\hbar^2}{2\mu e^2}$ is the Bohr radius. For $\ell > 0$ and for small r, $U_{eff}(r)$ is a repulsive potential that keeps an electron away from the nucleus. This results in zero probability for a p, d, f... electron to be at the nucleus.

where ρ is a dimensionless distance defined by

$$\rho = \frac{2Z}{na} r. \tag{10.24}$$

From Equation (10.23), it follows for an s state that $R_{n\ell}(0)$ is nonzero, while for a p,d,f,... state $R_{n\ell}(r)$ approaches zero as $r \to 0$ with the approach becoming increasingly rapid as ℓ increases. As a consequence, the probability that an electron is near the nucleus is large for an s state but is increasingly small for p,d,f,... states. Moreover, we will see in Section 12.9 that these behaviors are critical for the makeup of the periodic table. These rigorous conclusions are in accord with our qualitative inferences from the $U_{eff}(r)$ plots of Figure 10.2.

Next, let us consider the large r behavior of $U_{eff}(r)$. As noted, this is dominated by the Coulomb potential. Thus, we expect that at large r, the radial wave functions $R_{n\ell}(r)$ are similar for all values of ℓ. In accord with this expectation, as shown in Appendix D an analysis of Equation (10.11) gives the large r form of $R_{n\ell}(r)$ as

$$\lim_{r\to\infty} R_{n\ell}(r) \sim \exp(-\rho/2) \tag{10.25}$$

which is independent of ℓ.

We next give the general form of the normalized radial wave functions $R_{n\ell}(r)$. This form which interpolates between the $\rho \to 0$ and $\rho \to \infty$ limits of $R_{n\ell}(r)$ is

$$R_{n\ell}(r) = -\left\{\left(\frac{2Z}{na}\right)^2 \frac{(n-\ell-1)!}{2n[(n+\ell)!]^3}\right\}^{1/2} \rho^\ell L_{n+\ell}^{2\ell+1}(\rho)\exp\left(-\frac{\rho}{2}\right). \tag{10.26}$$

The functions $L_{n+\ell}^{2\ell+1}(\rho)$ are polynomials of order $n-\ell-1$ known as the associated Laguerre polynomials. $L_{n+\ell}^{2\ell+1}(\rho)$ has $n-\ell-1$ zeroes in the range $0<\rho<\infty$, and thus, $R_{n\ell}(r)$ has $n-\ell-1$ nodes.

The radial wave functions are listed in Table 10.2 for $n = 1-3$. For convenience, they are given as functions of the variable

$$\sigma = \frac{Zr}{a} \tag{10.27}$$

TABLE 10.2
Normalized Hydrogen-Like Atom Radial Wave Functions $R_{n\ell}(r)$ for $n = 1-3$[a]

N	ℓ	Type	$R_{n\ell}(r)$
1	0	1s	$2\left(\frac{Z}{a}\right)^{3/2}\exp(-\sigma)$
2	0	2s	$\frac{1}{2\sqrt{2}}\left(\frac{Z}{a}\right)^{3/2}(2-\sigma)\exp\left(-\frac{\sigma}{2}\right)$
2	1	2p	$\frac{1}{2\sqrt{6}}\left(\frac{Z}{a}\right)^{3/2}\sigma\exp\left(-\frac{\sigma}{2}\right)$
3	0	3s	$\frac{1}{\sqrt{3}}\left(\frac{2}{81}\right)\left(\frac{Z}{a}\right)^{3/2}(27-18\sigma+2\sigma^2)\exp\left(-\frac{\sigma}{3}\right)$
3	1	3p	$\frac{1}{\sqrt{6}}\left(\frac{4}{81}\right)\left(\frac{Z}{a}\right)^{3/2}(6-\sigma)\sigma\exp\left(-\frac{\sigma}{3}\right)$
3	2	3d	$\frac{1}{\sqrt{30}}\left(\frac{4}{81}\right)\left(\frac{Z}{a}\right)^{3/2}\sigma^2\exp\left(-\frac{\sigma}{3}\right)$

[a] $\sigma = \frac{Zr}{a}$ where a is the Bohr radius given by Equation (10.20).

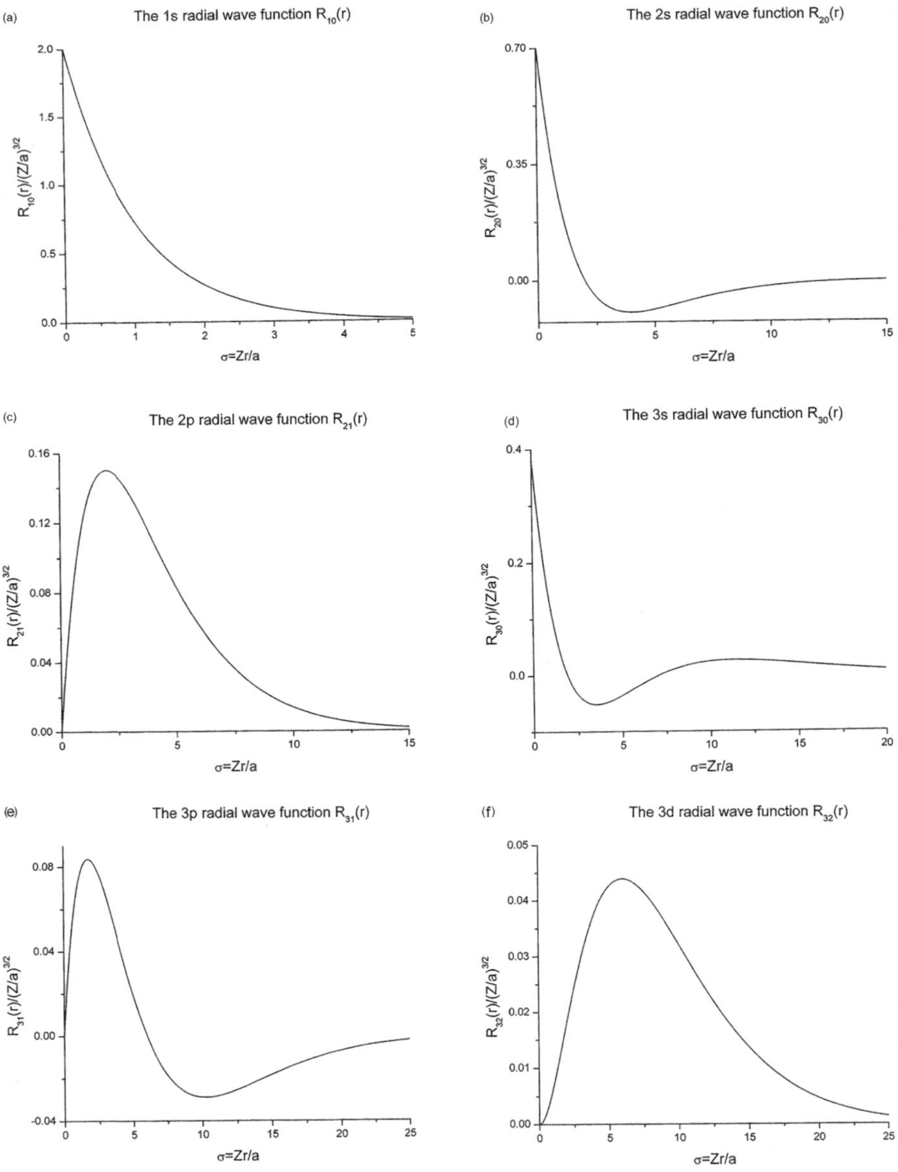

FIGURE 10.3 The hydrogen-like atom radial wave functions $R_{n\ell}(r)$ for $n = 1-3$ plotted against $\sigma = \frac{Zr}{a}$ where Z is the atomic number and a is the Bohr radius. (a) $R_{10}(r)$. (b) $R_{20}(r)$. (c) $R_{21}(r)$. (d). $R_{30}(r)$. (e) $R_{31}(r)$. (f) $R_{32}(r)$.

rather than the variable ρ of Equation (10.24). Also the radial wave functions are plotted for $n = 1-3$ in Figure 10.3. From the plots, the $n-\ell-1$ rule for the number of radial nodes is evident. For example, the 3s radial wave function $R_{30}(r)$ has two nodes, the 3p wave function $R_{31}(r)$ has one node, and the 3d wave function $R_{32}(r)$ has zero nodes.

In Table 10.3, we list the complete hydrogen-like atom wave functions $\Psi_{n\ell m}(r,\theta,\phi)$. These satisfy the orthonormality relation

$$\int_0^\infty r^2\,dr \int_0^\pi \sin\theta\,d\theta \int_0^{2\pi} d\phi\,\Psi^*_{n'\ell'm'}(r,\theta,\phi)\,\Psi_{n\ell m}(r,\theta,\phi) = \delta_{n'n}\delta_{\ell'\ell}\delta_{m'm}. \tag{10.28}$$

TABLE 10.3
Normalized Hydrogen-Like Atom Wave Functions $\Psi_{n\ell m}(r,\theta,\phi) =$ for $n = 1 - 3$[a]

n	ℓ	m	$\Psi_{n\ell m}(r,\theta,\phi)$
1	0	0	$\left(\frac{1}{\pi}\right)^{1/2}\left(\frac{Z}{a}\right)^{3/2}\exp(-\sigma)$
2	0	0	$\left(\frac{1}{4}\right)\left(\frac{1}{2\pi}\right)^{1/2}\left(\frac{Z}{a}\right)^{3/2}(2-\sigma)\exp\left(-\frac{\sigma}{2}\right)$
2	1	0	$\left(\frac{1}{4}\right)\left(\frac{1}{2\pi}\right)^{1/2}\left(\frac{Z}{a}\right)^{3/2}\sigma\exp\left(-\frac{\sigma}{2}\right)\cos\theta$
2	1	±1	$\mp\frac{1}{8}\left(\frac{1}{\pi}\right)^{1/2}\left(\frac{Z}{a}\right)^{3/2}\sigma\exp\left(-\frac{\sigma}{2}\right)\sin\theta\exp(\pm i\phi)$
3	0	0	$\frac{1}{81}\left(\frac{1}{3\pi}\right)^{1/2}\left(\frac{Z}{a}\right)^{3/2}\left(27-18\sigma+2\sigma^2\right)\exp\left(-\frac{\sigma}{3}\right)$
3	1	0	$\frac{1}{81}\left(\frac{2}{\pi}\right)^{1/2}\left(\frac{Z}{a}\right)^{3/2}(6-\sigma)\sigma\exp\left(-\frac{\sigma}{3}\right)\cos\theta$
3	1	±1	$\mp\frac{1}{81}\left(\frac{1}{\pi}\right)^{1/2}\left(\frac{Z}{a}\right)^{3/2}(6-\sigma)\sigma\exp\left(-\frac{\sigma}{3}\right)\sin\theta\exp(\pm i\phi)$
3	2	0	$\frac{1}{81}\left(\frac{1}{6\pi}\right)^{1/2}\left(\frac{Z}{a}\right)^{3/2}\sigma^2\exp\left(-\frac{\sigma}{3}\right)\left(3\cos^2\theta-1\right)$
3	2	±1	$\mp\frac{1}{81}\left(\frac{1}{\pi}\right)^{1/2}\left(\frac{Z}{a}\right)^{3/2}\sigma^2\exp\left(-\frac{\sigma}{3}\right)\sin\theta\cos\theta\exp(\pm i\phi)$
3	2	±2	$\frac{1}{162}\left(\frac{1}{\pi}\right)^{1/2}\left(\frac{Z}{a}\right)^{3/2}\sigma^2\exp\left(-\frac{\sigma}{3}\right)\sin^2\theta\exp(\pm 2i\phi)$

[a] $\sigma = \frac{Zr}{a}$ where a is the Bohr radius of Equation (10.20).

Using the orthonormality relation of the spherical harmonics

$$\int_0^{\pi}\sin\theta d\theta\int_0^{2\pi}Y_{\ell' m'}(\theta,\phi)Y_{\ell m}(\theta,\phi)d\phi = \delta_{\ell'\ell}\delta_{m'm} \tag{10.29}$$

and the factorization relation $\Psi_{n\ell m}(r,\theta,\phi) = R_{n\ell}(r)Y_{\ell m}(\theta,\phi)$ Equation (10.28) yields the following orthonormality relation for the radial wave functions:

$$\int_0^{\infty} r^2 R_{n'\ell}(r)R_{n\ell}(r)dr = \delta_{n'n}. \tag{10.30}$$

For $n' = n$, Equation (10.30) becomes the normalization condition of Equation (10.12). Also note that according to Equation (10.30), only radial wave functions with the same ℓ are orthonormal. Thus, while the 1s and 2s radial wave functions are orthogonal, the 1s and 2p functions are not.

We next turn to the spectra of hydrogen-like atoms. The Bohr theory of these spectra was developed in Section 1.3. Here, we outline some additional points concerning the spectra not dealt with in Section 1.3.

10.3 THE SPECTRA OF HYDROGEN-LIKE ATOMS

We first consider changes in the spectra which arise from dropping the assumption made in Section 1.2 that the nucleus is infinitely massive. As in Section 1.3, we consider emission spectra. These arise from transitions $n_2 \to n_1$ which change the quantum number n from an initial value n_2 to a smaller final value n_1. By combining the Bohr frequency condition as with the energy level equation (Equation 1.54) for an infinitely massive nucleus in Section 1.3, we derived Bohr's result for the wave numbers $\bar{\upsilon}_{n2\to n1}$ of the emission lines:

$$\bar{\upsilon}_{n2\to n1} = Z^2 R_\infty \left(\frac{1}{n_1^2} - \frac{1}{n_2^2} \right) \tag{10.31}$$

where

$$n_1 = 1, 2, \ldots \text{ and } n_2 = n_1 + 1, n_1 + 2, \ldots \tag{10.32}$$

and where

$$R_\infty = \frac{2\pi^2 e^4 m_e}{(4\pi\varepsilon_0)^2 h^3 c} \tag{10.33}$$

is the Rydberg constant which has the numerical value

$$R_\infty = 109{,}737.315\,\text{cm}^{-1}. \tag{10.34}$$

If one repeats the steps that led to Equations (10.31)–(10.33) with the exception that the energy level equation (Equation 10.18) valid for a nucleus of finite mass is used in place of Equation (1.54), then one obtains the following:

$$\bar{\upsilon}_{n2\to n1} = Z^2 R_N \left(\frac{1}{n_1^2} - \frac{1}{n_2^2} \right) \tag{10.35}$$

where

$$n_1 = 1, 2, \ldots \text{and}\, n_2 = n_1 + 1, n_1 + 2, \ldots \tag{10.36}$$

and where

$$R_N = \frac{2\pi^2 e^4 \mu}{(4\pi\varepsilon_0)^2 h^3 c}. \tag{10.37}$$

Equations (10.31)–(10.33) and Equations (10.35)–(10.37) differ only because m_e in R_∞ is replaced by μ in R_N. Next, let us relate R_N and R_∞. Using Equation (10.5) for μ and Equations (10.33) and (10.37) for R_∞ and R_N, we find that

$$R_N = R_\infty \frac{m_N}{m_e + m_N}. \tag{10.38}$$

Let us specialize Equation (10.38) to the hydrogen atom. Renaming R_N R_H and taking $m_N = m_p$, Equation (10.38) so specialized becomes

$$R_H = R_\infty \frac{m_p}{m_e + m_p}. \tag{10.39}$$

Evaluating Equation (10.39) using Equations (10.1) and (10.6) for m_e and m_p gives

$$R_H = R_\infty \frac{167,276,231 \times 10^{-27}\,\text{kg}}{9.10938992 \times 10^{-31}\,\text{kg} + 167,276,231 \times 10^{-27}\,\text{kg}} = \left(109,737.315\,\text{cm}^{-1}\right)(0.9994503)$$

or (10.40)

$$R_H = 109,677.582\,\text{cm}^{-1}.$$

R_H is in excellent agreement with the experimental value of the Rydberg constant

$$R = 109,677.576\,\text{cm}^{-1}. \tag{10.41}$$

showing that a near-exact spectrum is derived if nuclear motion is correctly accounted for.

The difference between R_H and R_∞ leads to small but detectable differences in the spectrum (Problems 10.5) and the ionization potential (Problem 10.6) of the hydrogen atom.

We next note that modern quantum mechanics provides a more detailed description of the transitions that give rise to spectral lines than does the Bohr theory. For example, consider the first Balmer line of hydrogen. In the Bohr theory, this line is attributed to the transition $n_3 \rightarrow n_2$ between the n = 3 and n = 2 Bohr orbits. In modern quantum mechanics, it is made up of the following transitions between n = 3 and n = 2 *orbitals*: 3s → 2p, 3p → 2s, and 3d → 2p. Notice that only some of the transitions between n = 3 and n = 2 orbitals contribute to the first Balmer line. Namely, the 3s → 2s, 3p → 2p, and 3d → 2s transitions do not contribute. The reason is that modern quantum mechanics supplies selection rules that restrict changes in ℓ and m. (There is no selection rule on n, so changes in n are unrestricted.) These selection rules are

$$\Delta\ell = \ell_2 - \ell_1 = \pm 1 \tag{10.42}$$

and

$$\Delta m = m_2 - m_1 = 0, \pm 1. \tag{10.43}$$

For the 3s → 2s and 3p → 2p transitions, $\Delta\ell = 0$, while for the 3d → 2s transition, $\Delta\ell = 2$. These transitions are thus forbidden by the selection rule on ℓ.

Table 10.4 lists the transitions allowed by the selection rules for the Lyman, Balmer, and Paschen series of hydrogen.

TABLE 10.4
Allowed Transitions $n_2 \rightarrow n_1$ for the Lyman, Balmer, and Paschen Series of Hydrogen

Series	n_1	Transitions
Lyman	1	2p → 1s, 3p → 1s,...
Balmer	2	3p → 2s, 4p → 2s,...
		3s → 2p, 4s → 2p,...
		3d → 2p, 4d → 2p,...
Paschen	3	4p → 3s, 5p → 3s,...
		4s → 3p, 5s → 3p,...
		4d → 3p, 5d → 3p,...
		4p → 3d, 5p → 3d,...
		4f → 3d, 5f → 3d,...

We next discuss the radial wave functions in more detail and introduce the important concept of the *radial distribution function*.

10.4 THE RADIAL DISTRIBUTION FUNCTIONS $P_{n\ell}(r)$

To start, let us find the location of the nodes of the 3s wave function $R_{30}(r)$. By the n-l -1 rule, $R_{30}(r)$ has two nodes. From Table 10.2, these occur at the σ values σ_1 and σ_2 which satisfy the quadratic equation

$$2\sigma^2 - 18\sigma + 27 = 0. \tag{10.44}$$

Applying the quadratic formula to Equation (10.44) gives

$$\sigma_1 = 1.902 \text{ and } \sigma_2 = 7.098. \tag{10.45a}$$

Therefore, since $\sigma = \frac{Zr}{\alpha}$, the nodes of $R_{30}(r)$ occur at

$$r_1 = 1.902\frac{a}{Z} \text{ and } r_2 = 7.098\left(\frac{a}{Z}\right). \tag{10.45b}$$

Next, according to Equation (10.30), the 1s and 2s radial wave functions obey the orthogonality relation

$$\int_0^\infty r^2 R_{10}(r) R_{20}(r) = 0 \tag{10.46}$$

Let us verify Equation (10.46). We need the forms for $R_{10}(r)$ and $R_{20}(r)$ which from Table 10.2 are

$$R_{10}(r) = 2\left(\frac{Z}{a}\right)^{3/2} \exp\left(-\frac{Zr}{a}\right) \tag{10.47}$$

and

$$R_{20}(r) = \frac{1}{2\sqrt{2}}\left(\frac{Z}{a}\right)^{3/2}\left(2 - \frac{Zr}{a}\right)\exp\left(-\frac{Zr}{2a}\right). \tag{10.48}$$

To confirm Equation (10.46), we must therefore show that the integral

$$I = \int_0^\infty r^2 \exp\left(-\frac{Zr}{a}\right)\left(2 - \frac{Zr}{a}\right)\exp\left(-\frac{Zr}{2a}\right)dr \tag{10.49}$$

vanishes. To do this, we first rewrite I as

$$I = 2\int_0^\infty r^2 \exp\left(-\frac{3}{2}\frac{Zr}{a}\right)dr - \frac{Z}{a}\int_0^\infty r^3 \exp\left(-\frac{3}{2}\frac{Zr}{a}\right)dr. \tag{10.50}$$

To proceed further, we need the standard integral (Problem 10.12)

$$\int_0^\infty x^n \exp(-kx)dx = \frac{n!}{k^{n+1}}. \tag{10.51}$$

Letting $x \to r$ and $k \to \frac{3}{2}\frac{Zr}{a}$, I may be evaluated by using Equation (10.51) to yield

$$I = \frac{2(2!)}{\left(\frac{3}{2}\frac{Z}{a}\right)^3} - \frac{Z}{a}\frac{3!}{\left(\frac{3}{2}\frac{Z}{a}\right)^4} = \frac{1}{\left(\frac{3}{2}\frac{Z}{a}\right)^3}\left(4 - \frac{6}{\frac{3}{2}}\right) = \frac{1}{\left(\frac{3}{2}\frac{Z}{a}\right)^3}(4-4) = 0. \tag{10.52}$$

We have therefore shown that I vanishes and thus verified the orthogonality of $R_{10}(r)$ and $R_{20}(r)$.

Next, we note that from the full wave function $\Psi_{n\ell m}(r,\theta,\phi)$ and the Born rule, we may obtain the probability that an electron will be in the vicinity of a point r,θ,ϕ. Often however, we want the probability that an electron will be at a distance r from the nucleus independent of the angles θ and ϕ. This leads us to the concept of the radial distribution function $P_{n\ell}(r)$. $P_{n\ell}(r)$ is defined by

$$P_{n\ell}(r)dr = \begin{array}{c}\text{probability that an electron is inside a spherical}\\ \text{shell of radius r and thickness dr}\end{array} \tag{10.53}$$

We next show that

$$P_{n\ell}(r) = r^2 R_{n\ell}^2(r). \tag{10.54}$$

To prove Equation (10.54), we first determine the probability $P(r)$ that the electron is inside a sphere of radius r which from Equation (10.10) is

$$\begin{aligned} P(r) &= \int_0^r r'^2\,dr' \int_0^\pi \sin\theta'\,d\theta' \int_0^{2\pi} d\phi' \Psi_{n\ell m}^*(r',\theta',\phi')\Psi_{n\ell m}(r',\theta',\phi') \\ &= \int_0^r r'^2 R_{n\ell}^2(r')dr' \int_0^\pi \sin\theta'\,d\theta' \int_0^{2\pi} d\phi' Y_{\ell m}^*(\theta',\phi')Y_{\ell m}(\theta',\phi'). \end{aligned} \tag{10.55}$$

Since the spherical harmonic $Y_{\ell m}(\theta,\phi)$ is normalized, Equation (10.55) reduces to

$$P(r) = \int_0^r r'^2 R_{n\ell}^2(r')dr'. \tag{10.56}$$

Similarly, the probability $P(r+dr)$ that the electron is inside a sphere of radius $r+dr$ is

$$P(r+dr) = \int_0^{r+dr} r'^2 R_{n\ell}^2(r')dr'. \tag{10.57}$$

From Equation (10.53), $P_{n\ell}(r)dr = P(r+dr) - P(r) = \int_0^{r+dr} r'^2 R_{n\ell}^2(r')dr' - \int_0^r r'^2 R_{n\ell}^2(r')dr' = \int_r^{r+dr} r'^2 R_{n\ell}^2(r')dr' = r^2 R_{n\ell}(r)\int_r^{r+dr} dr' = r^2 R_{n\ell}(r)dr$ or $P_{n\ell}(r) = r^2 R_{n\ell}(r)$. Thus, we have proven Equation (10.54).

Radial distribution functions $P_{n\ell}(r)$ are plotted for $n = 1-3$ in Figure 10.4. Notice the following from the plots. (a) The functions $P_{n\ell}(r)$ are nonnegative vanishing only at $r = 0$ and at the nodes of $R_{n\ell}(r)$, and also they approach zero as $r \to \infty$. (b) They have $n - \ell$ maxima, one more than the number of nodes of $R_{n\ell}(r)$. (c) For radial distribution functions $P_{n\ell}(r)$ with more than one maximum, the dominant maximum occurs at the largest r. (d) The radial extents of the functions $P_{n\ell}(r)$ increase with n, but for fixed n, the radial extents and the positions of the dominant maxima decrease with ℓ.

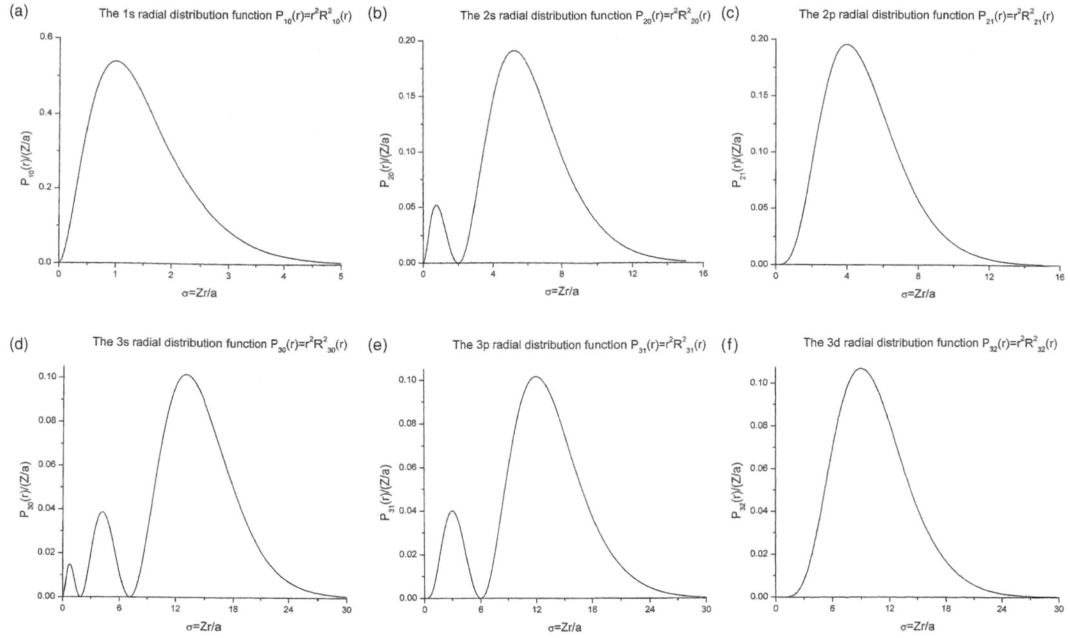

FIGURE 10.4 The hydrogen-like atom radial distribution functions $P_{n\ell}(r)=r^2R_{n\ell}(r)$ for $n=1-3$ plotted against $\sigma=\frac{Zr}{a}$. (a) $P_{10}(r)$. (b) $P_{20}(r)$. (c) $P_{21}(r)$. (d) $P_{30}(r)$. (e) $P_{31}(r)$. (f) $P_{32}(r)$.

Figure 10.4 shows that in modern quantum mechanics, an electron in a hydrogen-like atom has a nonzero probability of being at a range of distances from the nucleus. In contrast, in the Bohr theory an electron is permitted to only be at the discrete distances r_n given by Equation (1.57) as $r_n=\frac{n^2}{Z}a_o$. We next show however that the radial distribution function permits one to make contact between the modern quantum and Bohr theories of the hydrogen-like atoms.

To start, consider the 1s state. From Table 10.2 and Equation (10.54), its radial distribution function is

$$P_{10}(r)=4\left(\frac{Z}{a}\right)^3 r^2\exp\left(-\frac{2Zr}{a}\right). \tag{10.58}$$

From Figure 10.4a, $P_{10}(r)$ has a single peak. Let us find the location r_1 of that peak. This location is the point where $P_{10}(r)$ is a maximum and thus may be found by applying the condition $\left[\frac{dP_{10}(r)}{dr}\right]_{r=r_1}=0$ to Equation (10.58) to yield

$$4\left(\frac{Z}{a}\right)^3\left[2r_1\exp\left(-\frac{2Zr_1}{a}\right)-\frac{2Z}{a}r_1^{\,2}\exp\left(-\frac{2Zr_1}{a}\right)\right]$$

$$=4\left(\frac{Z}{a}\right)^3 2r_1\exp\left(-\frac{2Zr_1}{a}\right)\left(1-\frac{Zr_1}{a}\right)=0 \tag{10.59}$$

or

$$r_1=\frac{a}{Z}. \tag{10.60}$$

If we now let $\mu \to m_e$, then $a \to a_0$ and Equation (10.60) becomes $r_1 = \frac{a_0}{Z}$. which is identical to radius of the first Bohr orbit. So for the 1s state, the most probable distance r_1 of the electron from the nucleus (the position of the peak of $P_{10}[r]$) is the radius of the first Bohr orbit.

Next, let us consider the 2p state. From Table 10.2, its radial distribution function is

$$P_{21}(r) = \frac{1}{24}\left(\frac{Z}{a}\right)^5 r^4 \exp\left(-\frac{Zr}{a}\right). \tag{10.61}$$

From Figure 10.4c, $P_{21}(r)$ like $P_{10}(r)$ has only a single peak. The location r_2 of that peak may be found by applying the maximization condition $\left[\frac{dP_{21}(r)}{dr}\right]_{r=r_2} = 0$ to Equation (10.61) to yield r_2

$$\frac{2^2 a}{Z} \doteq \frac{2^2 a_0}{Z} \tag{10.62}$$

which shows that the most probable distance r_2 of a 2p electron from the nucleus is the radius of the second Bohr orbit.

Similarly, one may show that the most probable distance r_3 of a 3d electron from the nucleus is the radius of the third Bohr orbit (Problem 10.14). On the other hand, the most probable distances of 2s, 3p, and 3s electrons from the nucleus do not have simple relations to the Bohr radii.

How do the 1s, 2p, and 3d states differ from the 2s, 3p, and 3s states? Only for the former states are the radial wave functions $R_{n\ell}(r)$ nodeless (and hence the corresponding radial distribution functions have only a single peak). This follows because the number of nodes of $R_{n\ell}(r)$ is $n - \ell - 1$ and since for the 1s, 2p, and 3d states $n - \ell - 1 = 0$ or $\ell = n - 1$. The radial wave functions $R_{n,n-1}(r)$ for states with $\ell = n - 1$ are however especially simple since for such states the Laguerre polynomials are constants. Thus, one may write down a simple normalized form for $R_{n,n-1}(r)$ valid for all n (Problem D.6).

This form is

$$R_{n,n-1}(r) = \left(\frac{2}{n}\right)^{n+\frac{1}{2}} \left(\frac{Z}{a}\right)^{3/2} \frac{\sigma^{n-1} \exp\left(-\frac{\sigma}{n}\right)}{[(2n)!]^{1/2}}. \tag{10.63}$$

For n = 1, 2, and 3, Equation (10.63) reduces to the 1s, 2p, and 3d radial wave functions (Problem 10.15). Moreover, the location r_n of the *single* peak of the radial distribution function $P_{n,n-1}(r) = r^2 R^2_{n,n-1}(r)$ is identical to the radius of the nth Bohr orbit (Problem 10.16).

In summary, for a hydrogen-like atom in a state with principal quantum number n for which its radial wave function is nodeless and hence $\ell = n - 1$, the most probable distance of the electron from the nucleus is the radius $\frac{n^2 a_0}{Z}$ of the atom's nth Bohr orbit.

Next, let us turn to another measure of the typical distance of an electron from the nucleus. This is the expected or average value $\langle r \rangle_{n\ell}$ of r which is given by

$$\langle r \rangle_{n\ell} = \int_0^\infty r^2\, dr \int_0^\pi \sin\theta\, d\theta \int_0^{2\pi} d\phi\, r \Psi^*_{n\ell m}(r,\theta,\phi) \Psi_{n\ell m}(r,\theta,\phi). \tag{10.64}$$

Equation (10.64) may be reduced to an integral over the radial distribution function $P_{n\ell}(r)$. To see this, insert Equation (10.10) $\Psi_{n\ell m}(r,\theta,\phi) = R_{n\ell}(r) Y_{\ell m}(\theta,\phi)$ into Equation (10.64) and then perform the integrals over θ and ϕ using the fact that $Y_{\ell m}(\theta,\phi)$ is normalized. This gives

$$\langle r \rangle_{n\ell} = \int_0^\infty r P_{n\ell}(r)\, dr. \tag{10.65}$$

TABLE 10.5

Values of $\langle r^k \rangle_{n\ell} = \int_0^\infty r^2 R_{n\ell}^2(r) r^k dr$ for Hydrogen-Like Atoms for $k = 1, 2, -1, -2$, and -3

$$\langle r \rangle_{n\ell} = \frac{1}{2}\left[3n^2 - \ell(\ell+1)\right]\frac{a}{Z}$$

$$\langle r^2 \rangle_{n\ell} = \frac{1}{2}\left[5n^2 + 1 - 3\ell(\ell+1)\right]n^2\left(\frac{a}{Z}\right)^2$$

$$\langle r^{-1} \rangle_{n\ell} = \frac{1}{n^2}\left(\frac{Z}{a}\right)$$

$$\langle r^{-2} \rangle_{n\ell} = \frac{2}{(2\ell+1)n^3}\left(\frac{Z}{a}\right)^2$$

$$\langle r^{-3} \rangle_{n\ell} = \frac{1}{\ell\left(\ell+\frac{1}{2}\right)(\ell+1)n^3}\left(\frac{Z}{a}\right)^3$$

We will also require the more general averages $\langle r^k \rangle_{n\ell}$ where k is an integer. Following steps like those that led to Equation (10.65), we find that

$$\langle r^k \rangle_{n\ell} = \int_0^\infty r^k P_{n\ell}(r)\,dr. \tag{10.66}$$

Expressions for $\langle r^k \rangle_{n\ell}$ valid for all $n\ell$ and for $k = 1, 2, -1, -2$, and -3 are listed in Table 10.5. These expressions are derived from properties of the associated Laguerre polynomials (Pauling and Wilson 1985).

Next, let us compare the average and most probable values of r for the 1s state. From Equations (10.58) and (10.65), the average value of r is

$$\langle r \rangle_{10} = 4\left(\frac{Z}{a}\right)^3 \int_0^\infty r^3 \exp\left(-\frac{2Zr}{a}\right)dr. \tag{10.67}$$

Evaluating the integral using Equation (10.51) gives

$$\langle r \rangle_{10} = 4\left(\frac{Z}{a}\right)^3 \frac{3!}{\left(2\frac{Z}{a}\right)^4} = \frac{3}{2}\frac{a}{Z}. \tag{10.68}$$

So for a 1s state, the average value of $r = \frac{3}{2}\frac{a}{Z}$ is a little greater than its most probable value of $\frac{a}{Z}$. The average value of r for a 2p state may be found similarly as $\langle r \rangle_{21} \equiv 5\frac{a}{Z}$, which is a little larger than its most probable value of $4\frac{a}{Z}$.

We next ask another question that permits us to estimate the size of a 1s hydrogen-like atom. What is the probability $P(x)$ that a 1s electron is in a sphere centered on the nucleus of radius equal to x times the radius $\frac{a}{Z}$ of the first Bohr orbit? $P(x)$ is given by

$$P(x) = \int_0^{\frac{xa}{Z}} P_{10}(r)\,dr \tag{10.69}$$

or using Equation (10.58)

$$P(x) = 4\left(\frac{Z}{a}\right)^3 \int_0^{\frac{xa}{Z}} r^2 \exp\left(-\frac{2Zr}{a}\right) dr. \tag{10.70}$$

Making the variable change $y = \frac{Zr}{a}$ $P(x)$ becomes

$$P(x) = 4\int_0^x y^2 \exp(-2y)\,dy. \tag{10.71}$$

Integrating by parts twice, $P(x)$ may be evaluated as

$$P(x) = 1 - \exp(-2x)\left(2x^2 + 2x + 1\right). \tag{10.72}$$

Notice that $P(x)$ properly reduces in the limits $x = 0$ and $x = \infty$ to $P(0) = 0$ and $\lim_{x\to\infty} P(x) = 1$. Let us calculate $P(1), P(2)$, and $P(3)$. These are the probabilities that a 1s electron will be found in spheres with radii equal to the radius of the first Bohr orbit and twice and three times the radius of the first Bohr orbit. From Equation (10.72), these probabilities are

$$P(1) = 0.323\ P(2) = 0.762 \text{ and } P(3) = 0.938. \tag{10.73}$$

Notice that the size of a 1s hydrogen atom is $\sim 3a_0$. Next, we discuss the virial theorem. In Section 7.10, we noted that virial theorems are relationships between average kinetic energies $\langle KE\rangle$ and average potential energies $\langle PE\rangle$ which exist for many types of systems. For the harmonic oscillator, we also noted in Section 7.10 that the virial theorem is $\langle PE\rangle = \langle KE\rangle$. More generally, for a system with a potential energy function of the form $U(r) = ar^n$, the virial theorem is the relationship $\langle PE\rangle = 2n^{-1}\langle KE\rangle$. For the hydrogen-like atoms, $n = -1$, so for these atoms the virial theorem takes the form

$$\langle PE\rangle = -2\langle KE\rangle. \tag{10.74}$$

Let us verify Equation (10.74) for the 1s state. To start, we note that since the potential energy is the Coulomb potential $-\frac{Ze^2}{4\pi\varepsilon_0 r}$ for an arbitrary state,

$$\langle PE\rangle = -\frac{Ze^2}{4\pi\varepsilon_0}\left\langle \frac{1}{r}\right\rangle_{n\ell}. \tag{10.75}$$

Using Equation (10.66) for $k = -1$, Equation (10.75) becomes

$$\langle PE\rangle = -\frac{Ze^2}{4\pi\varepsilon_0}\int_0^\infty r^{-1} P_{n\ell}(r)\,dr. \tag{10.76}$$

Evaluating Equation (10.76) for the 1s state with $P_{10}(r)$ given by Equation (10.58) yields

$$\langle PE\rangle = -\frac{Ze^2}{4\pi\varepsilon_0}\int_0^\infty r^{-1} P_{10}(r) = -4\frac{Ze^2}{4\pi\varepsilon_0}\left(\frac{Z}{a}\right)^3 \int_0^\infty r \exp\left(-\frac{2Zr}{a}\right) dr. \tag{10.77}$$

Applying Equation (10.51) then gives

$$\langle PE\rangle = -\frac{Z^2 e^2}{(4\pi\varepsilon_0)a}. \tag{10.78}$$

Comparing Equation (10.78) with Equation (10.19) for E_n shows that $\langle PE \rangle = 2E_1$ or that

$$E_1 = \frac{1}{2}\langle PE \rangle. \tag{10.79}$$

However, since

$$\langle KE \rangle + \langle PE \rangle = E_1 \tag{10.80}$$

Equations (10.79) and (10.80) give that $\langle KE \rangle + \langle PE \rangle = \frac{1}{2}\langle PE \rangle$ or $\langle PE \rangle = -2\langle KE \rangle$, which is the virial theorem for hydrogen-like atoms, thus verifying that theorem for their 1s states.

We next turn to the angular dependencies of the hydrogen-like atom wave functions. This requires us to determine the real forms of these wave functions, which often are used in chemical applications (for example, when generalized to multi-electron atoms in the construction of molecular orbitals as in Section 13.4) since unlike the complex forms of Table 10.3 they are easy to plot and visualize.

10.5 THE REAL FORMS OF THE HYDROGEN-LIKE ATOM WAVE FUNCTIONS

Recall the hydrogen-like atom wave functions factorize as $\Psi_{n\ell m}(r,\theta,\phi) = R_{n\ell}(r)Y_{\ell m}(\theta,\phi)$. Their angular dependencies thus derive from the dependence of the spherical harmonics $Y_{\ell m}(\theta,\phi)$ on θ and ϕ. To display these angular dependencies, we will therefore give spherical polar plots of real versions of the spherical harmonics.

We start with the s wave functions. From Table 10.1, there is only one s spherical harmonic. It is

$$Y_{00}(\theta,\phi) = \left(\frac{1}{4\pi}\right)^{1/2}. \tag{10.81}$$

Since $Y_{00}(\theta,\phi)$ is independent of θ and ϕ, its spherical polar plot shown in Figure 10.5a is spherically symmetric.

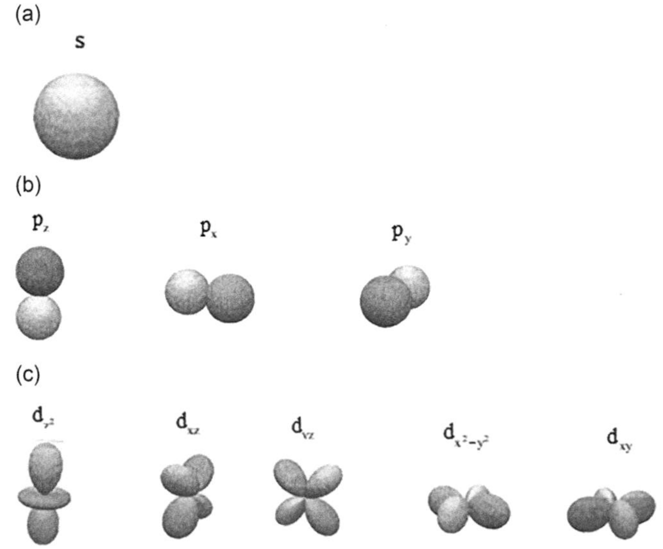

FIGURE 10.5 Spherical polar plots of the s, p, and d real spherical harmonics (see text). (a) The s spherical harmonic. (b) The p_z, p_x, and p_y spherical harmonics. (c) The d_{z^2}, d_{xz}, d_{yz}, $d_{x^2-y^2}$, and d_{xy} spherical harmonics.

Next, let us consider the p wave functions. From Table 10.1, there are three p spherical harmonics. One of them, $Y_{10}(\theta,\phi)$, which we will call p_z, is real and has the form

$$p_z = Y_{10}(\theta,\phi) = \left(\frac{3}{4\pi}\right)^{1/2}\cos\theta. \tag{10.82a}$$

Because it is real, p_z is easy to plot and to visualize. The other two p spherical harmonics

$$Y_{1-1}(\theta,\phi) = \left(\frac{3}{8\pi}\right)^{1/2}\sin\theta\exp(-i\phi) \tag{10.83a}$$

and

$$Y_{1\,1}(\theta,\phi) = -\left(\frac{3}{8\pi}\right)^{1/2}\sin\theta\exp(i\phi) \tag{10.83b}$$

are complex and therefore are not easy to plot and visualize.

To circumvent this problem, we replace $Y_{1-1}(\theta,\phi)$ and $Y_{1\,1}(\theta,\phi)$ by two real functions that we call p_x and p_y. These real functions are defined as the following linear combinations of $Y_{1-1}(\theta,\phi)$ and $Y_{1\,1}(\theta,\phi)$:

$$p_x = \sqrt{\frac{1}{2}}\left[Y_{1-1}(\theta,\phi) - Y_{1\,1}(\theta,\phi)\right] = \left(\frac{3}{4\pi}\right)^{1/2}\sin\theta\cos\phi \tag{10.82b}$$

$$p_y = i\sqrt{\frac{1}{2}}\left[Y_{1-1}(\theta,\phi) + Y_{1\,1}(\theta,\phi)\right] = \left(\frac{3}{4\pi}\right)^{1/2}\sin\theta\cos\phi \tag{10.82c}$$

where we have used $\exp[\pm i\phi] = \cos\phi \pm i\sin\phi$.

The functions p_z, p_x, and p_y are called *real spherical harmonics*. Because of their visualizability, the real spherical harmonics often replace the ordinary spherical harmonics listed in Table 10.1 in chemical applications. Like the ordinary spherical harmonics, the real spherical harmonics form an orthonormal set (Problem 10.23). Further, the real p spherical harmonics are eigenfunctions of $\hat{L}^2$ with eigenvalue $2\hbar^2$. Therefore, $R_{nl}(r)p_z$, $R_{nl}(r)p_x$, and $R_{nl}(r)p_y$ are solutions of the Schrödinger equation (Equation 10.9) and are thus eigenfunctions of the Hamiltonian operator of the atom.

The real spherical harmonics p_z, p_x, and p_y are plotted in Figure 10.5b. All three plots have the same "dumbbell" shape but are oriented, respectively, in the z, x, and y directions.

Let us now turn to the d wave functions. From Table 10.1, there are five d spherical harmonics. Only one of these, $Y_{20}(\theta,\phi)$, is real. Therefore, as we did for the p spherical harmonics, we will form real d spherical harmonics as linear combinations of the complex d spherical harmonics of Table 10.1. The usual definitions and names of the real d spherical harmonics are

$$d_{z^2} = Y_{20}(\theta,\phi) = \left(\frac{5}{16}\right)^{1/2}\left(3\cos^2\theta - 1\right) \tag{10.84a}$$

$$d_{xz} = i\sqrt{\frac{1}{2}}\left[Y_{2-1}(\theta,\phi) + Y_{21}(\theta,\phi)\right] = \left(\frac{15}{4\pi}\right)^{1/2}\sin\theta\cos\theta\sin\phi \tag{10.84b}$$

$$d_{yz} = i\sqrt{\frac{1}{2}}\left[Y_{2-1}(\theta,\phi) + Y_{21}(\theta,\phi)\right] = \left(\frac{15}{4\pi}\right)^{1/2}\sin\theta\cos\theta\sin\phi \tag{10.84c}$$

$$d_{x^2-y^2} = \sqrt{\frac{1}{2}}\left[Y_{2-2}(\theta,\phi)+Y_{22}(\theta,\phi)\right] = \left(\frac{15}{16\pi}\right)^{1/2} \sin^2\theta\cos 2\phi \tag{10.84d}$$

$$d_{xy} = i\sqrt{\frac{1}{2}}\left[Y_{2-2}(\theta,\phi)-Y_{22}(\theta,\phi)\right] = \left(\frac{15}{16\pi}\right)^{1/2} \sin^2\theta\sin 2\phi \tag{10.84e}$$

Like the real p spherical harmonics, the real d spherical harmonics form an orthonormal set, are eigenfunctions of $\hat{L}^2$, and also when multiplied by appropriate radial wave functions are eigenfunctions of the hydrogen-like atom Hamiltonian operator.

The real d spherical harmonics are plotted in Figure 10.5c. Notice that all but d_{z^2} have identical shapes and differ only in orientation.

The origin of the names of the real p and d spherical harmonics may be seen by expressing these functions in Cartesian rather than in spherical polar coordinates. Using the transformation relations between Cartesian and spherical polar coordinates $z = r\cos\theta$, $x = r\sin\theta\cos\phi$, and $y = r\sin\theta\sin\phi$, one finds the Cartesian forms

$$p_z = \left(\frac{3}{4\pi}\right)^{1/2}\frac{z}{r} \quad p_x = \left(\frac{3}{4\pi}\right)^{1/2}\frac{x}{r} \quad p_y = \left(\frac{3}{4\pi}\right)^{1/2}\frac{y}{r} \tag{10.85}$$

and

$$d_{z^2} = \left(\frac{5}{16\pi}\right)^{1/2}\frac{3z^2-r^2}{r^2} \tag{10.86a}$$

$$d_{xz} = \left(\frac{15}{4\pi}\right)^{1/2}\frac{xz}{r^2} \tag{10.86b}$$

$$d_{yz} = \left(\frac{15}{4\pi}\right)^{1/2}\frac{yz}{r^2} \tag{10.86c}$$

$$d_{x^2-y^2} = \left(\frac{15}{16\pi}\right)^{1/2}\frac{x^2-y^2}{r^2} \tag{10.86d}$$

$$d_{xy} = \left(\frac{15}{4\pi}\right)^{1/2}\frac{xy}{r^2} \tag{10.86e}$$

Notice that except for the d_{z^2} function, the names of the real p and d spherical harmonics are evident from their Cartesian forms.

In analogy to the formation of the complex complete wave functions $\Psi_{n\ell m}(r,\theta,\phi)$ from radial wave functions and complex spherical harmonics, real complete wave functions may be formed from radial wave functions and real spherical harmonics. An example is $\Psi_{3d_{x^2-y^2}}(r,\theta,\phi) \equiv R_{32}(r)\,d_{x2-y2}(\theta,\phi)$. Like the complex complete wave functions, the real complete wave functions form an orthonormal set and are eigenfunctions of the hydrogen-like atom Hamiltonian operator. The real complete wave functions are listed for $n = 1-3$ in Table 10.6.

Next, turning to a new topic, we note that due to their orbital motion, electrons in states with nonzero angular momentum $(p, d, f, \ldots$ states) behave like tiny magnets with magnetic moments **m**. In an external magnetic field **B**, the magnetic moments couple to the field, thus producing small splittings of the hydrogen-like atom energy levels E_n. These splittings in turn cause the lines in the hydrogen-like atom spectra to break into several closely spaced lines. This magnetic field effect on the spectra is known as the *Zeeman effect*.

TABLE 10.6
The Normalized Real Complete Hydrogen-Like Atom Wave Functions for n = 1 – 3[a]

$$\Psi_{1s} = \left(\frac{1}{\pi}\right)^{1/2}\left(\frac{Z}{a}\right)^{3/2}\exp(-\sigma)$$

$$\Psi_{2s} = \left(\frac{1}{4}\right)\left(\frac{1}{2\pi}\right)^{1/2}\left(\frac{Z}{a}\right)^{3/2}(2-\sigma)\exp(-\sigma/2)$$

$$\Psi_{2p_z} = \left(\frac{1}{4}\right)\left(\frac{1}{2\pi}\right)^{1/2}\left(\frac{Z}{a}\right)^{3/2}\sigma\exp(-\sigma/2)\cos\theta$$

$$\Psi_{2p_x} = \left(\frac{1}{4}\right)\left(\frac{1}{2\pi}\right)^{1/2}\left(\frac{Z}{a}\right)^{3/2}\sigma\exp\left(-\frac{\sigma}{2}\right)\sin\theta\cos\phi$$

$$\Psi_{2p_y} = \left(\frac{1}{4}\right)\left(\frac{1}{2\pi}\right)^{1/2}\left(\frac{Z}{a}\right)^{3/2}\sigma\exp\left(-\frac{\sigma}{2}\right)\sin\theta\sin\phi$$

$$\Psi_{3s} = \left(\frac{1}{81}\right)\left(\frac{1}{3\pi}\right)^{1/2}\left(\frac{Z}{a}\right)^{3/2}\left(27-18\sigma+2\sigma^2\right)\exp\left(-\frac{\sigma}{3}\right)$$

$$\Psi_{3p_z} = \left(\frac{1}{81}\right)\left(\frac{2}{\pi}\right)^{1/2}\left(\frac{Z}{a}\right)^{3/2}(6-\sigma)\sigma\exp\left(-\frac{\sigma}{3}\right)\cos\theta$$

$$\Psi_{3p_x} = \left(\frac{1}{81}\right)\left(\frac{2}{\pi}\right)^{1/2}\left(\frac{Z}{a}\right)^{3/2}(6-\sigma)\sigma\exp\left(-\frac{\sigma}{3}\right)\sin\theta\cos\phi$$

$$\Psi_{3p_y} = \left(\frac{1}{81}\right)\left(\frac{2}{\pi}\right)^{1/2}\left(\frac{Z}{a}\right)^{3/2}(6-\sigma)\sigma\exp\left(-\frac{\sigma}{3}\right)\sin\theta\sin\phi$$

$$\Psi_{3d_{z^2}} = \left(\frac{1}{81}\right)\left(\frac{1}{6\pi}\right)^{1/2}\left(\frac{Z}{a}\right)^{3/2}\sigma^2\exp\left(-\frac{\sigma}{3}\right)\left(3\cos^2\theta-1\right)$$

$$\Psi_{3d_{xz}} = \left(\frac{1}{81}\right)\left(\frac{2}{\pi}\right)^{1/2}\left(\frac{Z}{a}\right)^{3/2}\sigma^2\exp\left(-\frac{\sigma}{3}\right)\sin\theta\cos\theta\cos\phi$$

$$\Psi_{3d_{yz}} = \left(\frac{1}{81}\right)\left(\frac{2}{\pi}\right)^{1/2}\left(\frac{Z}{a}\right)^{3/2}\sigma^2\exp\left(-\frac{\sigma}{3}\right)\sin\theta\cos\theta\sin\phi$$

$$\Psi_{3d_{x^2-y^2}} = \left(\frac{1}{81}\right)\left(\frac{1}{2\pi}\right)^{1/2}\left(\frac{Z}{a}\right)^{3/2}\sigma^2\exp\left(-\frac{\sigma}{3}\right)\sin^2\theta\cos 2\phi$$

$$\Psi_{3d_{xy}} = \left(\frac{1}{81}\right)\left(\frac{1}{2\pi}\right)^{1/2}\left(\frac{Z}{a}\right)^{3/2}\sigma^2\exp\left(-\frac{\sigma}{3}\right)\sin^2\theta\sin 2\phi$$

[a] $\sigma = \frac{Zr}{a}$ where a is the Bohr radius given by Equation (10.20).

We next develop these ideas in more detail.

10.6 ORBITAL MAGNETISM AND THE ZEEMAN EFFECT

We will not give a rigorous treatment of orbital magnetism as is done, for example, by Sakurai (Sakurai 1985, p. 308). Instead, we will give a simplified argument based on classical mechanics and simple electromagnetic theory.

Consider an electron moving with uniform speed v in a circular orbit of radius r (Figure 10.6). The orbit encloses an area $A = \pi r^2$. The electron produces a current I, and thus, from classical electromagnetic theory, it is an electromagnet with magnetic moment **m** whose magnitude is given by

$$m = IA = I\pi r^2. \tag{10.87}$$

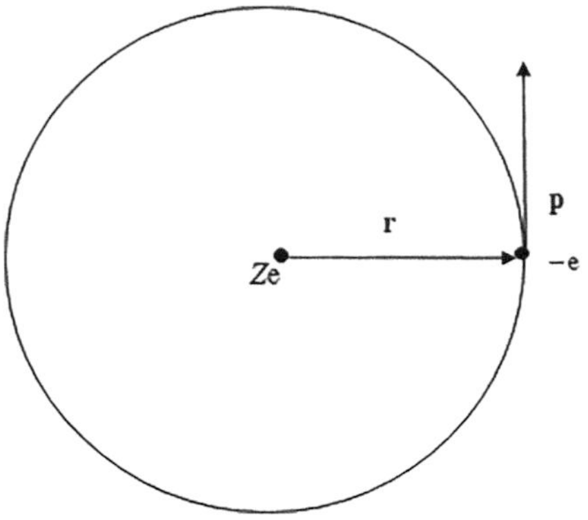

FIGURE 10.6 An electron moving in a circular orbit of radius r with orbital angular momentum $\mathbf{L} = \mathbf{r} \times \mathbf{p}$ is a tiny magnet with magnetic moment $\mathbf{m} = -\dfrac{e}{2m_e}\mathbf{L}$.

The current I is the amount of charge passing through some point on the orbit per second. Since the charge of the electron is −e,

$$I = -e\upsilon \tag{10.88}$$

where υ is the frequency of revolution of the electron. But $\upsilon = T^{-1}$, where T is the period, which is the time needed for the electron to make one revolution. Moreover, the distance to electron traverses in a period is vT = circumference of orbit= $2\pi r$, so $T = \dfrac{2\pi r}{v}$. Hence, Equation (10.88) becomes

$$I = -\frac{e}{T} = \frac{-ev}{2\pi r}. \tag{10.89}$$

Equations (10.87) and (10.89) give m as

$$m = -\frac{evr}{2}. \tag{10.90}$$

But $v = \dfrac{p}{m_e}$, where p is the magnitude of the momentum vector **p**. So $m = -\dfrac{erp}{2m_e}$. However, from Figure 10.6 **r** and **p**, the electron's position and momentum vectors, are perpendicular, so by a property of the vector cross product, rp is the magnitude L of the electron's orbital angular momentum vector $\mathbf{L} = \mathbf{r} \times \mathbf{p}$. Therefore,

$$m = -\frac{eL}{2m_e}. \tag{10.91}$$

By another property of the vector cross product, the direction of **L** is perpendicular to the plane of the orbit and is directed outward from that plane. Moreover, from classical electromagnetic theory, the direction of **m** is in the opposite of that of **L**. Thus, the scalar equation (Equation 10.91) implies the vector equation

$$\mathbf{m} = \frac{-e}{2m_e}\mathbf{L} \tag{10.92}$$

where **m** is directed inward from the plane of the orbit.

We may now introduce an external magnetic field **B**. We first note that the SI derived unit for magnetic field strength is *tesla* (T). In terms of basic units,

$$1\,\mathrm{T} = \frac{1\,\mathrm{kg}}{\mathrm{C\,s}}. \tag{10.93}$$

We next couple the magnetic field **B** to the orbital magnetic moment **m**. The classical potential energy V arising from this coupling is from classical electromagnetic theory

$$V = -\mathbf{m}.\mathbf{B}. \tag{10.94}$$

Using Equation (10.92), V becomes

$$V = \frac{e}{2m_e}\mathbf{B}.\mathbf{L}. \tag{10.95}$$

The corresponding quantum operator relation is

$$\hat{V} = \frac{e}{2m_e}\mathbf{B}.\hat{\mathbf{L}}. \tag{10.96}$$

Next, assuming **B** is aligned along the Z direction, $\hat{V}$ becomes

$$\hat{V} = \frac{e}{2m_e}B_z\hat{L}_z. \tag{10.97}$$

We may now write down the Hamiltonian operator $\hat{H}$ of a hydrogen-like atom in the presence of the magnetic field as

$$\hat{H} = \hat{H}_0 + \hat{V} \tag{10.98}$$

where $\hat{H}_0$ is the Hamiltonian operator of the atom in the absence of the field.

We next show that the hydrogen-like atom wave functions $\Psi_{n\ell m}(r,\theta,\phi)$ are eigenfunctions of $\hat{H}$. First, we note

$$\hat{H}_0\Psi_{n\ell m}(r,\theta,\phi) = E_n\Psi_{n\ell m}(r,\theta,\phi) \tag{10.99}$$

where by Equation (10.18) the energies

$$E_n = -\frac{Z^2e^4\mu}{2(4\pi\varepsilon_0)^2 n^2\hbar^2} \tag{10.100}$$

are those of the atom in the absence of the field. Also since $\hat{L}_z Y_{\ell m}(\theta,\phi) = mhY_{\ell m}(\theta,\phi)$ (a result given in Section 10.1), it follows from Equation (10.10) that

$$\hat{L}_z\Psi_{n\ell m}(r,\theta,\phi) = m\hbar\Psi_{n\ell m}(r,\theta,\phi) \tag{10.101}$$

where m can have any of the $2\ell+1$ values $-\ell, -\ell+1, \ldots, \ell-1, \ell$.

Thus, from Equation (10.97) for $\hat{V}$

$$\hat{V}\,\Psi_{n\ell m}(r,\theta,\phi) = \frac{e\hbar}{2m_e}B_z m\Psi_{n\ell m}(r,\theta,\phi). \tag{10.102}$$

So the wave functions $\Psi_{n\ell m}(r,\theta,\phi)$ are eigenfunctions of both $\hat{H}_o$ and $\hat{V}$ and hence of $\hat{H}$. In other words, they solve the Schrödinger equation

$$\hat{H}\Psi_{n\ell m}(r,\theta,\phi) = E_{nm}\,\Psi(r,\theta,\phi) \tag{10.103}$$

where the eigenvalues

$$E_{nm} = E_n + \frac{e\hbar}{2m_e} B_z m \tag{10.104}$$

are the energies of the atom in the presence of the magnetic field.

Equation (10.104) shows that the degeneracy of $\ell > 0$ states with the same n and ℓ is lifted in a magnetic field (but not for $\ell = 0$ or s states since for these states m = 0). Namely, in a magnetic field, states with the same n and ℓ but different values for mhave $2\ell + 1$ different energies, one energy for each of the $2\ell + 1$m values. This is why m is called the magnetic quantum number.

As an example, consider the 2p states. For a p state, $\ell = 1$ and so m can take on the three values $-1, 0$, or 1. Therefore, when a magnetic field is applied, according to Equation (10.104) the 2p energy level E_2 splits into three levels with energies (Figure 10.7)

$$E_{21} = E_2 + \mu_B B_z \tag{10.105a}$$

$$E_{20} = E_2 \tag{10.105b}$$

and

$$E_{2-1} = E_2 - \mu_B B_z \tag{10.105c}$$

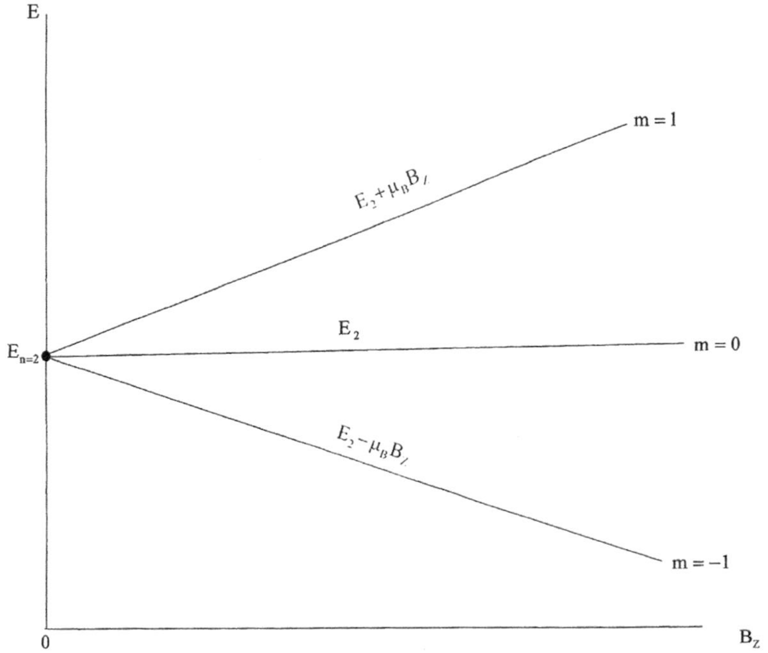

FIGURE 10.7 Splitting of the 2p hydrogen atom energy level $E_{n=2}$ in a magnetic field B_z. $\mu_B = \frac{e\hbar}{2m_e}$ is the Bohr magneton.

where

$$\mu_B = \frac{e\hbar}{2m_e} \tag{10.106}$$

is the *Bohr magneton*. The Bohr magneton has the value

$$\mu_B = 9.274 \times 10^{-24}\ \text{J}\,\text{T}^{-1}. \tag{10.107}$$

So if $B_z = 1\text{T}$, a typical value

$$\mu_B B_z = 9.274 \times 10^{-24}\ \text{J}. \tag{10.108}$$

Additionally, E_2 for hydrogen has the value

$$E_2 = -5.477 \times 10^{-19}\ \text{J}. \tag{10.109}$$

Note $|E_2| \gg \mu_B B_z$, and Equation (10.105) therefore shows that for typical values of B_z, the magnetic field effect on the 2p energy levels of hydrogen is very small.

This effect however may be observed spectroscopically. Consider, for example, the $2p \rightarrow 1s$ Lyman transition of hydrogen occurring in a magnetic field B_z. Since for Equation (10.106) the magnetic field splits the 2p energy level into three levels, three transitions are potentially possible (Figure 10.8):

$$2p_{m=1} \rightarrow 1s_{m=0} \tag{10.110a}$$

$$2p_{m=0} \rightarrow 1s_{m=0} \tag{10.110b}$$

and

$$2p_{m=-1} \rightarrow 1s_{m=0}. \tag{10.110c}$$

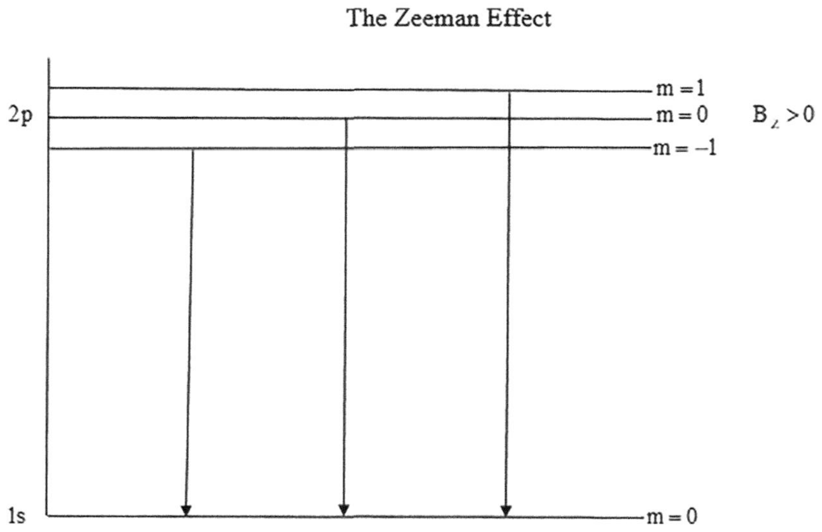

FIGURE 10.8 Zeeman effect on the $2p \rightarrow 1s$ Lyman transition of hydrogen. Due to the magnetic field splitting of the 2p energy level, the single line observed when $B_z = 0$ is split into three closely spaced lines when $B_z > 0$.

All three transitions are in accord with the selection rules of Equations (10.42) and (10.43). Thus, in a magnetic field the three transitions actually occur, and hence, 2p → 1s Lyman line will split into three lines. As noted, this phenomenon is known as the Zeeman effect.

Let us calculate the wave numbers of the three 2p → 1s Lyman lines of hydrogen. We need the value of the hydrogen atom Rydberg constant given in Equation (10.40) as $R_H = 109{,}677.582\,cm^{-1}$. We first assume $B_z = 0$. Then, the wave number of the single Lyman line is from Equation (10.35) $\bar{\upsilon}_{2p\to 1s} = \frac{1}{hc}(E_2 - E_1) = R_H\left(1 - \frac{1}{4}\right)$ or

$$\bar{\upsilon}_{2p\to 1s} = 82{,}258.187\,cm^{-1}. \qquad (10.111)$$

Next, let $B_z = 1T$. Referring to Equations (10.105a) and (10.108), the wave number of the $2p_{m=1\to 1s}$ line is $\bar{\upsilon}_{2p_{m=1}\to 1s} = \frac{1}{hc}(E_{21} - E_1) = \frac{1}{hc}(E_2 - E_1) + \frac{\mu_B(B_z = 1T)}{hc} = \bar{\upsilon}_{2p\to 1s} + \frac{9.274\times 10^{-24}\,J}{hc} = 82{,}258.187\,cm^{-1} + 0.467\,cm^{-1}$ or

$$\bar{\upsilon}_{2p_{m=1}\to 1s} = 82{,}258.654\,cm^{-1}. \qquad (10.112a)$$

Similarly,

$$\bar{\upsilon}_{2p_{m=0}\to 1s} = 82{,}258.187\,cm^{-1} \qquad (10.112b)$$

and

$$\bar{\upsilon}_{2p_{m=-1}\to 1s} = 82{,}257.720\,cm^{-1}. \qquad (10.112c)$$

Notice the Zeeman splittings of the three lines are only about $\sim 0.5\,cm^{-1}$.

However, something is wrong. Our treatment of the Zeeman effect does not agree with experiment. Ten not three 2p → 1s Zeeman lines are actually observed for hydrogen. Also even in the absence of a magnetic field, the hydrogen 2p → 1s Lyman transition gives rise to a doublet (two closely spaced lines) instead of a single line.

To understand these and other phenomena, we must modify our model for the hydrogen-like atoms to account for the fact that electrons have an intrinsic angular momentum called *spin*.

10.7 ELECTRON SPIN

The most direct evidence for the existence of electron spin is provided by the results of the Stern–Gerlach experiment (Figure 10.9). It was performed by the German physicists Otto Stern and Walther Gerlach in 1922. The date is significant since in 1922 the existence of spin was not yet suspected and moreover the modern quantum mechanics of Heisenberg, Schrödinger, and Dirac (that the contributions of the British physicist Paul Dirac are as foundational to quantum mechanics as those earlier described due to Heisenberg and Schrödinger is evident in advanced textbooks on quantum mechanics, for example the book by Cohen-Tannoudji, Diu, and Laloe) was not created until 1925–1930. The quantum theory which existed in 1922 was the old quantum theory of Sommerfeld. So the motivation of Stern and Gerlach in performing their experiment was not related to spin but rather was aimed to testing a prediction of the Sommerfeld theory concerning *orbital* angular momentum. Sommerfeld's theory predicts, like modern quantum mechanics, space quantization; that is, it predicts that the directions of angular momentum vectors are quantized. Classical theory, in contrast, predicts that angular momentum vectors can have any orientations. Stern and Gerlach performed their experiment to distinguish between these two predictions.

They chose to perform their experiment using silver atoms. We now know that for silver, the magnitude L of the electron's orbital angular momentum vector **L** is zero. But Stern and Gerlach

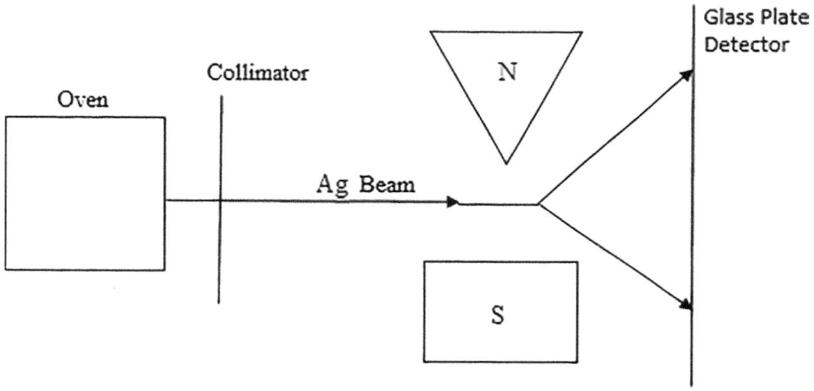

FIGURE 10.9 A schematic Stern–Gerlach apparatus

incorrectly assumed L = 1 for silver. Also while modern quantum mechanics predicts three directions of space quantization for L = 1 the Sommerfeld theory predicts two directions of space quantization for L = 1. Thus, Stern and Gerlach believed that if the Sommerfeld theory was correct, the orbital magnetic moments of silver could have only two alignments, one for each direction of space quantization while if the classical theory was correct, they would have a continuum of alignments.

To distinguish between these two predictions, Stern and Gerlach sent a beam of silver atoms through an inhomogeneous magnetic field (Figure 10.9). Because the field is inhomogeneous, the force on one end of a magnetic dipole moment is slightly greater than the force on the other end and the atom deflects. If the orientations of the magnetic moments could take on a continuum of values, the intensity pattern on the detector would have a maximum at the point of zero deflection and decrease monotonically away from that point. If the orientations of the magnetic moments were quantized, as in the way predicted by the Sommerfeld theory the initial single beam of atoms would split into two beams and two spots would appear on the detector. Stern and Gerlach found the latter result and interpreted it as confirming Sommerfeld's theory.

The correct interpretation of Stern and Gerlach's result came with the advent of modern quantum mechanics which showed silver had a single s valence electron outside a closed shell with zero orbital angular momentum (see Section 12.10) and so its orbital angular momentum was zero, not one. The observed deflection pattern could then have arisen only if the silver valence electron possessed an intrinsic angular momentum with two directions of space quantization. This intrinsic angular momentum is now known as spin.

In this way, the Stern–Gerlach experiment provided unambiguous evidence for the existence of electron spin.

Spin emerges naturally from the relativistic wave equation of Dirac. Here, we will incorporate it in an *ad hoc* manner into the nonrelativistic quantum mechanics we have been developing.

Specifically, we will develop the properties of spin angular momentum **S** in analogy to those of orbital angular momentum **L**. Thus, just as there are two orbital angular momentum quantum numbers ℓ and $m = -\ell, -\ell+1, \ldots, \ell-1, \ell$, there are two spin angular momentum quantum numbers s and $m_s = -s, -s+1, \ldots, s-1, s$. For a given value of ℓ, there are $2\ell+1$ directions of space quantization of **L**, one for each value of m. Similarly, for a given value of s, there are $2s+1$ directions of space quantization of **S**, one for each value of m_s. However, from the Stern–Gerlach experiment, we know that the number of directions of space quantization of **S** is two. So $2s+1 = 2$ or $s = \frac{1}{2}$. Thus, in contrast to ℓ which has the range $\ell = 0, 1, 2, \ldots$, s is restricted to *the single non-integral* value

$$s = \frac{1}{2}. \tag{10.113}$$

From this restriction, it follows that m_s can take on only two values

$$m_s = \frac{1}{2} \text{ and } m_s = -\frac{1}{2}. \tag{10.114}$$

If $m_s = \frac{1}{2}$, the electron is said to be in a *spin-up* state, while for $m_s = -\frac{1}{2}$, it is said to be in a *spin-down* state.

Continuing in analogy to the orbital angular momentum operator $\hat{\mathbf{L}}$, there is a spin angular momentum operator $\hat{\mathbf{S}}$. $\hat{\mathbf{S}}$ like $\hat{\mathbf{L}}$ obeys standard angular momentum commutation rules. For example, in analogy to the orbital angular momentum commutation relation (compare Equation [8.32])

$$\left[\hat{L}^2, \hat{L}_z\right] = \hat{O} \tag{10.115}$$

one has the spin angular momentum commutation rules

$$\left[\hat{S}^2, \hat{S}_z\right] = \hat{O}. \tag{10.116}$$

The commutation of $\hat{L}^2$ and $\hat{L}_z$ implies the simultaneous eigenvalue problem (compare Equations [8.108])

$$\hat{L}^2 Y_{\ell m}(\theta,\phi) = \ell(\ell+1)\hbar^2 Y_{\ell m}(\theta,\phi) \tag{10.117a}$$

and

$$\hat{L}_z Y_{\ell m}(\theta,\phi) = m\hbar\ Y_{\ell m}(\theta,\phi). \tag{10.117b}$$

Similarly, the commutation of $\hat{S}^2$ and $\hat{S}_z$ implies the simultaneous eigenvalue problem

$$\hat{S}^2\alpha = s(s+1)\hbar\alpha \text{ and } \hat{S}^2\beta = s(s+1)\hbar\beta \tag{10.118a}$$

and

$$\hat{S}_z\alpha = \frac{1}{2}\hbar\alpha \text{ and } \hat{S}_z\beta = -\frac{1}{2}\hbar\beta \tag{10.118b}$$

where α and β are the spin eigenfunctions for, respectively, the spin-up and spin-down states. Comparing Equations (10.117) and (10.118), it follows that α and β are, respectively, formally analogous to $Y_{\ell m}(\theta,\phi)$ and $Y_{\ell -m}(\theta,\phi)$. Also note that just as the magnitude of **L** is given by

$$L = \sqrt{\ell(\ell+1)}\hbar \tag{10.119}$$

the magnitude of **S** since s=1/2 is given by

$$S = \sqrt{s(s+1)}\hbar = \frac{\sqrt{3}}{2}\hbar. \tag{10.120}$$

Further, just as the orbital eigenfunctions $Y_{\ell m}(\theta,\phi)$ depend on θ and ϕ, we assume that the spin eigenfunctions $\alpha(\sigma)$ and $\beta(\sigma)$ depend on a spin variable σ, which can take on only two values $\frac{1}{2}$ (spin-up) and $-\frac{1}{2}$ (spin-down). The spin variable allows us to express the orthonormality of the spin eigenfunctions as

$$\int \alpha^*(\sigma)\alpha(\sigma)d\sigma = 1 \text{ and } \int \beta^*(\sigma)\beta(\sigma)d\sigma = 1 \tag{10.121a}$$

and

$$\int \alpha^*(\sigma)\beta(\sigma)d\sigma = 0 \tag{10.121b}$$

The integrals in Equation (10.121) are interpreted to be sums over the values of $\sigma \frac{1}{2}$ and $-\frac{1}{2}$.

Finally, just as the orbital angular momentum **L** gives rise to the magnetic moment of Equation (10.92) $\mathbf{m} = -\frac{e}{2m_e}\mathbf{L}$, the spin angular momentum **S** gives rise to the magnetic moment

$$\mathbf{m}_s = -g_e \frac{e}{2m_e}\mathbf{S} \tag{10.122}$$

where from experiment

$$g_e = 2.002319304. \tag{10.123}$$

We will take $g_e \doteq 2$. Then,

$$\mathbf{m}_s \doteq -\frac{e}{m_e}\mathbf{S}. \tag{10.124}$$

Next, note that the existence of spin requires us to modify our description of hydrogen-like atoms. This is evident, for example, by considering two 1s atoms, one with a spin-up electron and the other with a spin-down electron. The atoms are in different quantum states and thus have different wave functions. How is the difference in wave functions handled? This is done by replacing the orbitals $\Psi_{n\ell m}(r,\theta,\phi)$ by *spin orbitals* $\Psi_{n\ell mm_s}(r,\theta,\phi,\sigma)$. (Strictly speaking the subscript of the previous symbol should be $n\ell msm_s$ but the subscript s is frequently omitted since always $s = \frac{1}{2}$.) Since the space and spin motions are nearly independent, spin orbitals have nearly the forms

$$\Psi_{n\ell m\frac{1}{2}}(r,\theta,\phi,\sigma) = \Psi_{n\ell m}(r,\theta,\phi)\alpha(\sigma) \tag{10.125a}$$

and

$$\Psi_{n\ell m-\frac{1}{2}}(r,\theta,\phi,\sigma) = \Psi_{n\ell m}(r,\theta,\phi)\beta(\sigma). \tag{10.125b}$$

The spin eigenfunctions $\alpha(\sigma)$ and $\beta(\sigma)$ and the orbitals $\Psi_{n\ell m}(r,\theta,\phi)$ both form orthonormal sets and so the spin orbitals $\Psi_{n\ell mm_s}(r,\theta,\phi,\sigma)$ also form an orthonormal set. Therefore, the spin orbitals conform to the orthonormality relation

$$\int_0^\infty r^2\,dr \int_0^\pi \sin\theta\,d\theta \int_0^{2\pi} d\phi \int d\sigma \Psi^*_{n'\ell'm'm'_s}(r,\theta,\phi,\sigma)\Psi_{n\ell mm_s}(r,\theta,\phi,\sigma) = \delta_{n'n}\delta_{\ell'\ell}\delta_{m'm}\delta_{m'_s m_s}. \tag{10.126}$$

Finally, we note that there are twice as many spin orbitals as orbitals, so when spin is accounted for, the degeneracy of a hydrogen-like atom energy level with principal quantum number n is $2n^2$, not n^2.

The space and spin motions of an electron actually are not perfectly independent. They interact slightly. This interaction, which we next discuss, is called *spin–orbit coupling.*

10.8 SPIN–ORBIT COUPLING

It is found experimentally that the nonrelativistic energy level formula of Equation (10.18) does not perfectly describe the energies of hydrogen-like atoms. Rather, relativistic effects (and a small quantum electrodynamic effect called the *Lamb shift*) give rise to small shifts and splittings of the nonrelativistic energies. The effects of these shifts and splittings on spectra are called *fine structure*. The relativistic contributions increase with v/c, where v is the speed of the electron. Since v increases linearly with the nuclear charge Ze (this follows from the virial theorem), the magnitudes of the relativistic corrections increase as one proceeds down the series $H, He^+, Li^{2+}, \ldots$. As an example of the sizes of the relativistic effects, consider the n = 2 level of hydrogen. According to the nonrelativistic equation (Equation 10.18), the n = 2 energy lies $82{,}258.187\,cm^{-1}$ above the ground state energy. However, experimentally the 2s energy is shifted to $82{,}258.949\,cm^{-1}$ above the ground state energy. Moreover, the nonrelativistic 2p level splits into two levels that occur 82,258.913 and $82{,}259.279\,cm^{-1}$ above ground state.

Three relativistic terms emerge from the Dirac equation (Cohen-Tannoudji, Diu, and Laloe 1977, pp. 1213–1217). The first is a term that corrects the nonrelativistic kinetic energy for relativity. The second, which only affects the energies of s states, is called the Darwin term. The third is the spin–orbit coupling term. The kinetic energy and Darwin terms give rise to shifts in the energy levels, while the spin–orbit term splits the energy levels for p, d, f,... states with the same n. We will not attempt to compute the kinetic energy and Darwin contributions, but we will treat the spin–orbit contribution. Therefore, we will not be able to determine the experimental energy levels of the hydrogen-like atoms since they depend on the kinetic energy and Darwin shifts. But these shifts cancel when one computes the spin–orbit splittings. So we will be able to compute these splittings.

Since we are neglecting the kinetic energy and Darwin terms, we will take the Hamiltonian operator of a hydrogen-like atom to be

$$\hat{H} = \hat{H}_0 + \hat{V} \tag{10.127}$$

where $\hat{H}_0$ is the nonrelativistic Hamiltonian operator of the atom which follows from Equation (10.9) as

$$\hat{H}_0 = -\frac{\hbar^2}{2\mu}\frac{1}{r^2}\frac{\partial}{\partial r}\left(r^2\frac{\partial}{\partial r}\right) + \frac{\hat{L}^2}{2\mu r^2} - \frac{Ze^2}{4\pi\varepsilon_0 r} \tag{10.128}$$

and where $\hat{V}$ is the spin–orbit coupling operator. We next show that

$$\hat{V} = \xi(r)\hat{\mathbf{L}}\cdot\hat{\mathbf{S}} = \frac{Ze^2}{2(4\pi\varepsilon_0)m_e^2c^2r^3}\hat{\mathbf{L}}\cdot\hat{\mathbf{S}}. \tag{10.129}$$

To start, we note that the nucleus produces an electric field $\mathbf{E}$ acting on the electron which is derived from the electron–nucleus Coulomb potential as

$$\mathbf{E} = -\frac{Ze\mathbf{r}}{4\pi\varepsilon_0 r^3} \tag{10.130}$$

where $\mathbf{r}$ is electron's position vector. However, according to classical electromagnetic theory, a particle moving in an electric field $\mathbf{E}$ with velocity $\mathbf{v}$ experiences a magnetic field $\mathbf{B}$ given by

$$\mathbf{B} = \frac{\mathbf{v} \times \mathbf{E}}{c^2}. \tag{10.131}$$

Equations (10.130) and (10.131) yield since $\mathbf{v} \times \mathbf{r} = -\mathbf{r} \times \mathbf{v}$

$$\mathbf{B} = -\frac{Ze}{4\pi\varepsilon_0 c^2 r^3} \mathbf{v} \times \mathbf{r} = \frac{Ze\mathbf{L}}{4\pi\varepsilon_0 m_e c^2 r^3} \tag{10.132}$$

where the electron's angular momentum $\mathbf{L} = \mathbf{r} \times m_e \mathbf{v} = \mathbf{r} \times \mathbf{p}$. $\mathbf{B}$ couples to the spin magnetic moment of Equation (10.124) $\mathbf{m}_s = -\frac{e}{m_e}\mathbf{S}$ to give an interaction energy, analogous to that of Equation (10.94)

$$V = -\mathbf{m}_s \cdot \mathbf{B} = \frac{e}{m_e} \mathbf{S} \cdot \mathbf{B}. \tag{10.133}$$

Comparing Equations (10.132) and (10.133) gives

$$V = \frac{Ze^2}{4\pi\varepsilon_0 m_e^2 c^2 r^3} \mathbf{L} \cdot \mathbf{S} \tag{10.134}$$

The above expression is too large by a factor of two due to a relativistic effect known as *Thomas precession*. The correct expression for V is

$$V = \frac{Ze^2}{2(4\pi\varepsilon_0) m_e^2 c^2 r^3} \mathbf{L} \cdot \mathbf{S}. \tag{10.135}$$

The quantum operator that corresponds to V is just $\hat{V}$ of Equation (10.129).

Summarizing, we have shown that the Hamiltonian operator $\hat{H}$ is given by

$$\hat{H} = \hat{H}_0 + \xi(r)\hat{\mathbf{L}} \cdot \hat{\mathbf{S}} = \hat{H}_0 + \frac{Ze^2}{2(4\pi\varepsilon_0) m_e^2 c^2 r^3} \hat{\mathbf{L}} \cdot \hat{\mathbf{S}}. \tag{10.136}$$

We wish to find the eigenvalues of $\hat{H}$. We know the eigenvalues of $\hat{H}_0$. These are just the nonrelativistic energies E_n of Equation (10.18). But we are really interested in the small corrections to the energies E_n which arise from the spin–orbit coupling term $\xi(r)\hat{\mathbf{L}} \cdot \hat{\mathbf{S}}$. These corrections cannot be found exactly. However, there is a standard quantum mechanical technique for approximately determining small corrections to known energy levels. This technique is known as perturbation theory. It is the natural tool for obtaining the relativistic shifts and splittings. We will develop perturbation theory in the next chapter. Since we do not yet have perturbation theory at our disposal, we will use a simplified method to treat spin–orbit coupling. It gives results identical to those of perturbation theory. In this simplified method, we treat spin–orbit coupling in the $n\ell$th subshell by replacing $\hat{H}$ of Equation (10.136) by an *effective Hamiltonian operator*

$$\hat{H}^{\text{eff}} = \hat{H}_0 + \frac{hc}{\hbar^2} \xi_{n\ell} \hat{\mathbf{L}} \cdot \hat{\mathbf{S}}. \tag{10.137}$$

The quantity $\xi_{n\ell}$ is called the *spin–orbit coupling constant*. It has units of cm^{-1} and is defined by

$$\xi_{n\ell} = \frac{hc}{\hbar^2} \int_0^\infty r^2 \xi(r) R_{n\ell}^2(r)\,dr. \tag{10.138}$$

From Equation (10.136), $\xi(r)=\dfrac{Ze^2}{2(4\pi\varepsilon_0)m_e^{\,2}c^2r^3}$. Hence, from Equations (10.54), (10.66) and (10.138), $\xi_{n\ell}=\dfrac{hc}{\hbar^2}\dfrac{Ze^2}{2(4\pi\varepsilon_0)m_e^2c^2}\left\langle r^{-3}\right\rangle_{n\ell}$. Then, using Table 10.5 to determine $\left\langle r_{n\ell}^{-3}\right\rangle\xi_{n\ell}$ may be evaluated (setting $a=a_0$) as

$$\xi_{n\ell}=\frac{hc}{\hbar^2}\frac{Ze^2}{2(4\pi\varepsilon_0)m_e^2c^2}\left(\frac{Z}{a_0}\right)^3\frac{1}{n^3\ell(\ell+1)\left(\ell+\dfrac{1}{2}\right)}. \tag{10.139}$$

Equation (10.139) may be recast as (Problem 10.26)

$$\xi_{n\ell}=\frac{Z^4\alpha^2R_\infty}{n^3\ell(\ell+1)\left(\ell+\dfrac{1}{2}\right)} \tag{10.140}$$

where R_∞ is the Rydberg constant of Equation (10.33) and α is the *fine structure constant* defined by

$$\alpha=\frac{1}{4\pi\varepsilon_0}\frac{e^2}{\hbar c}. \tag{10.141}$$

The fine structure constant is dimensionless and has the numerical value (Problem 10.128)

$$\alpha=\frac{1}{137.036}\doteq\frac{1}{137}. \tag{10.142}$$

The fine structure constant is a direct measure of the importance of relativistic effects since it is closely related to v/c (Problem 10.29).

From Equation (10.140), the constants $\xi_{n\ell}$ are proportional to $\alpha^2R_\infty=5.33\times10^{-5}\,R_\infty$. Since from Equation (1.71) R_∞ is of the same order of magnitude as the nonrelativistic energies E_n when expressed in cm^{-1} units, the constants $\xi_{n\ell}$ are much smaller than typical values of E_n. For example, from Equations (10.33), (10.140), and (10.142) the hydrogen 2p spin–orbit coupling constant $\xi_{21}=0.2435\,cm^{-1}$ which is tiny compared to the value of the Rydberg constant 109,737 cm^{-1}. Additionally, we will see later that the spin–orbit splittings are proportional to the constants $\xi_{n\ell}$ so that the splittings are also very small relative to typical values of E_n.

Also notice from Equation (10.140) that the spin–orbit coupling constants are proportional to Z^4. So they grow rapidly with Z. Thus, for Be^{3+} with $Z=4$, $\xi_{21}=256\times0.2435\,cm^{-1}=62.34\,cm^{-1}$.

We next determine the eigenvalues of $\hat{H}^{eff}$. These are the energies of the atom with spin–orbit coupling included. To start, we introduce the total (orbital plus spin) angular momentum operator

$$\hat{\mathbf{J}}=\hat{\mathbf{L}}+\hat{\mathbf{S}}. \tag{10.143}$$

From the commutation rules for $\hat{\mathbf{L}}$ and $\hat{\mathbf{S}}$ and the fact that $\left[\hat{\mathbf{L}},\hat{\mathbf{S}}\right]=\hat{O}$, it follows that $\hat{\mathbf{J}}$ obeys standard angular momentum commutation rules. Especially,

$$\left[\hat{J}^2,\hat{J}_z\right]=\hat{O}. \tag{10.144}$$

Given Equation (10.144), the second theorem of Section 5.2, and the fact indicated at the close of Section 8.2 that the commutation rules and their consequences are identical for all types of angular momenta, $\hat{J}^2$ and $\hat{J}_z$ are expected to have common eigenfunctions with respective eigenvalues $j(j+1)\hbar^2$

and $m_j\hbar$, where m_j takes on the $2j+1$ values $-j, -j+1, \ldots, j-1, j$. Moreover, angular momentum addition theory (Cohen-Tannoudji, Diu, and Laloe 1977, chap. 10) gives that j is restricted to the values

$$j = \ell + s, \ell + s - 1, \ldots, |\ell - s|. \tag{10.145}$$

For $s = \frac{1}{2}$, this restriction becomes

$$j = \ell + \frac{1}{2}, \left|\ell - \frac{1}{2}\right|. \tag{10.146}$$

Classically, the values $\ell + \frac{1}{2}$ and $\left|\ell - \frac{1}{2}\right|$ correspond, respectively, to adding **L** and **S** parallel and antiparallel.

Next, we identify a set of mutually commuting operators that include $\hat{H}^{eff}$. To start, we note that since $\left[\hat{\mathbf{L}}, \hat{\mathbf{S}}\right] = \hat{O}$, all orbital and spin angular momentum operators commute. So $\hat{L}^2$ and $\hat{S}^2$ commute. Also $\hat{L}^2$ and $\hat{S}^2$ commute with $\hat{J}^2, \hat{J}_z$, and $\hat{H}^{eff}$ (Problem 10.30). Further, from Equation (10.144) $\hat{J}^2$ and $\hat{J}_z$ commute. Thus, $\hat{H}^{eff}, \hat{L}^2, \hat{S}^2, \hat{J}^2$, and $\hat{J}_z$ form a complete set of mutually commuting operators. ($\hat{L}_z$ and $\hat{S}_z$ are not included in the set because they do not commute with either $\hat{J}^2$ or the spin–orbit part of $\hat{H}^{eff}$.)

As a generalization of the second theorem of Section 5.2, the members of this set satisfy a simultaneous eigenvalue problem with common eigenfunctions $\Psi_{n\ell sjm_j}(r,\theta,\phi,\sigma)$. Angular momentum addition theory gives the form of the common eigenfunctions as linear superpositions of spin orbitals $\Psi_{n\ell msm_s}(r,\theta,\phi,\sigma)$ all with the same values for $n\ell$ and s but with different values for m and m_s.

The simultaneous eigenvalue problem is

$$\hat{H}^{eff}\Psi_{n\ell sjm_j} = E_{n\ell sj}\Psi_{n\ell sjm_j} \tag{10.147a}$$

$$\hat{L}^2\Psi_{n\ell sjm_j} = \ell(\ell+1)\hbar^2\Psi_{n\ell sjm_j} \tag{10.147b}$$

$$\hat{S}^2\Psi_{n\ell sjm_j} = s(s+1)\hbar^2\Psi_{n\ell sjm_j} \tag{10.147c}$$

$$\hat{J}^2\Psi_{n\ell sjm_j} = j(j+1)\hbar^2\Psi_{nsjm_j} \tag{10.147d}$$

and

$$\hat{J}_z\Psi_{n\ell sjm_j} = m_j\hbar\Psi_{n\ell sjm_j}. \tag{10.147e}$$

(*Note*: Using forethought, we have omitted the subscript m_j on E in Equation [10.147a].) We next note that $\hat{J}^2 = \hat{\mathbf{J}}\cdot\hat{\mathbf{J}} = \left(\hat{\mathbf{L}}+\hat{\mathbf{S}}\right)\cdot\left(\hat{\mathbf{L}}+\hat{\mathbf{S}}\right) = \hat{L}^2 + \hat{S}^2 + 2\hat{\mathbf{L}}\cdot\hat{\mathbf{S}}$ (where we have used the fact that $\left[\hat{\mathbf{L}}, \hat{\mathbf{S}}\right] = \hat{O}$ or $\hat{\mathbf{L}}\cdot\hat{\mathbf{S}} = \hat{\mathbf{S}}\cdot\hat{\mathbf{L}}$) so $\hat{\mathbf{L}}\cdot\hat{\mathbf{S}} = \frac{1}{2}\left(\hat{J}^2 - \hat{L}^2 - \hat{S}^2\right)$. Comparing with Equation (10.137) gives

$$\hat{H}^{eff} = \hat{H}_0 + \frac{hc}{2\hbar^2}\xi_{n\ell}\left(\hat{J}^2 - \hat{L}^2 - \hat{S}^2\right). \tag{10.148}$$

But from Equation (10.147b–d), $\left(\hat{J}^2 - \hat{L}^2 - \hat{S}^2\right)\Psi_{n\ell sjm_j} = \left[j(j+1) - \ell(\ell+1) - s(s+1)\right]\hbar^2\Psi_{n\ell sjm_j}$. Also from the superposition of spin orbitals form of $\Psi_{n\ell sjm_j}$, it follows that $H_0\Psi_{n\ell sjm_j} = E_n\Psi_{n\ell sjm_j}$.

These results together give $\hat{H}^{eff}\Psi_{n\ell sjm_j} = \left\{E_n + \frac{hc}{2}\xi_{n\ell}\left[j(j+1) - \ell(\ell+1) - s(s+1)\right]\right\}\Psi_{n\ell sjm_j}$. Then, comparing with Equation (10.147a) shows the eigenvalues of $\hat{H}^{eff}$ have the m_j-independent values

$$E_{n\ell sj} = E_n + \frac{hc}{2}\xi_{n\ell}\left[j(j+1) - \ell(\ell+1) - s(s+1)\right]. \tag{10.149}$$

Since the energy level $E_{n\ell sj}$ is independent of m_j, it is $(2j+1)$-fold degenerate.

Before going on, let us introduce *term symbols.* States with the same n, ℓ, and s values form a *term*. The $2j+1$ degenerate states with the same n, ℓ, s, and j values but different m_j values constitute a *level*. Term symbols are a convenient way of designating terms and levels.

Term symbols are of the form

$$^{2s+1}L_j. \tag{10.150}$$

L is an uppercase letter that represents the quantum number ℓ; $2s+1$ is called the *multiplicity,* and since $s = \frac{1}{2}$, it always has the value 2; and j is the total angular momentum quantum number that from Equation (10.146) takes on the values $j = \ell + \frac{1}{2}$ and $j = \left|\ell - \frac{1}{2}\right|$. The correspondence between ℓ and L is by convention

$$\begin{array}{cccccc} \ell & 0 & 1 & 2 & 3 & 4\ldots \\ L & S & P & D & F & G\ldots. \end{array} \tag{10.151}$$

Let us consider some examples. For the 2p states, $n = 2, \ell = 1, s = \frac{1}{2}$, and $j = \frac{3}{2}$ or $\frac{1}{2}$. The term symbols for these states are

$$2p\,^2P_{3/2} \text{ and } 2p\,^2P_{1/2}. \tag{10.152}$$

Notice that we have added a 2p to the left of the term symbol to indicate we are specifying 2p rather than, say, 3p or 4p states. As further examples, the term symbols for the 3d and 1s states are

$$3d\,^2D_{5/2} \text{ and } 3d\,^2D_{3/2} \tag{10.153}$$

and

$$1s\,^2S_{1/2}. \tag{10.154}$$

Let us now return to Equation (10.149). It permits the determination of the spin–orbit coupling modification of the nonrelativistic energy level E_n which splits energy levels $E_{n\ell sj}$ each with the same values for n, ℓ, and s but with different values for j into two levels. For example, for the 2p states, $n = 2, \ell = 1$, and $s = \frac{1}{2}$, but j can be either $\frac{3}{2}$ or $\frac{1}{2}$. The energy level for $j = \frac{3}{2}$ is $E_{n\ell sj} = E_{21\frac{1}{2}\frac{3}{2}}$, while for $j = \frac{1}{2}$, it is $E_{n\ell sj} = E_{21\frac{1}{2}\frac{1}{2}}$. For clarity, we will use the term symbol notation to denote $E_{21\frac{1}{2}\frac{3}{2}}$ by $E(2p\,^2P_{3/2})$ and $E_{21\frac{1}{2}\frac{1}{2}}$ by $E(2p\,^2P_{1/2})$. Then, from Equation (10.149)

$E(2p\,^2P_{3/2}) = E_2 + \frac{hc}{2}\xi_{21}\left[\frac{3}{2}\left(\frac{3}{2}+1\right) - 1(1+1) - \frac{1}{2}\left(\frac{1}{2}+1\right)\right]$ or

TABLE 10.7
The Energy Levels of Hydrogen (Moore 1972)

Configuration	Terms	j	Levels $\left(\text{cm}^{-1}\right)$
1s	^2S	1/2	0
2p	^2P	1/2	82,258.913
		3/2	82,259.279
2s	^2S	1/2	82,258.949
3p	^2P	1/2	97,492.205
		3/2	97,492.313
3s	^2S	1/2	97,492.215
3d	^2D	3/2	97,492.313
		5/2	97,492.349
4p	^2P	1/2	102,823.842
		3/2	102,823.887
4s	^2S	1/2	102,823.846
4d	^2D	3/2	102,823.887
		5/2	102,823.902
4f	^2F	5/2	102,823.902
		7/2	102,823.910

$$E\left(2p\,^2P_{3/2}\right) = E_2 + \frac{hc}{2}\xi_{21}. \qquad (10.155a)$$

Similarly, $E\left(2p\,^2P_{1/2}\right) = E_2 + \frac{hc}{2}\xi_{21}\left[\frac{1}{2}\left(\frac{1}{2}+1\right) - 1(1+1) - \frac{1}{2}\left(\frac{1}{2}+1\right)\right]$ or

$$E\left(2p\,^2P_{1/2}\right) = E_2 - hc\,\xi_{21}. \qquad (10.155b)$$

We may now calculate the spin–orbit splitting of the 2p levels. From Equations (10.155), the splitting in cm^{-1} units is

$$\text{splitting} = \frac{1}{hc}\left[E\left(2p\,^2P_{3/2}\right) - E\left(2p\,^2P_{1/2}\right)\right] = \frac{3}{2}\xi_{21}. \qquad (10.156)$$

Since evaluation of Equation (10.140) yields for Z = 1 $\xi_{21} = 0.2425\,\text{cm}^{-1}$, the above relation gives that for hydrogen the

$$\text{splitting} = 0.365\,\text{cm}^{-1}. \qquad (10.157)$$

Next, let us compare with experiment. In Table 10.7, the observed energies for the $n = 1-4$ states of hydrogen are listed with the zero of energy chosen so that the 1s energy is zero. Notice that the energy levels of the p, d, and f states with the same n are all split by spin–orbit coupling. From Table 10.7, the observed 2p state splitting is

$$82{,}259.279 - 82{,}258.913\,\text{cm}^{-1} = 0.366\,\text{cm}^{1} \qquad (10.158)$$

which is in excellent agreement with our calculated splitting $0.365\,\text{cm}^{-1}$.

While the 2p splitting is small because of it, as mentioned at the close of Section 10.6, the $2p \rightarrow 1s$ Lyman line is a detectable doublet arising from the two transitions $2p^2P_{3/2} \rightarrow 1s^2S_{1/2}$ and $2p^2P_{1/2} \rightarrow 1s^2S_{1/2}$ (Problem 10.37).

FURTHER READINGS

Atkins, Peter, Julio de Paula, and Ronald Friedman. 2014. *Physical chemistry: quanta, matter, and change.* 2nd ed. Great Britain: Oxford University Press.

Atkins, Peter, and Ronald Friedman. 2011. *Molecular quantum mechanics.* 5th ed. New York: Oxford University Press.

Beiser, Arthur. 2003. *Concepts of modern physics.* 6th ed. New York: McGraw-Hill, Inc.

Cohen-Tannoudji, Claude, Bernard Diu, and Franck Laloe. 1977. *Quantum mechanics.* Translated by Susan Reid Hemley, Nicole Ostrowsky, and Dan Ostrowsky. New York: John Wiley and Sons Inc.

Cox, P. A. 1996. *Introduction to quantum theory and atomic structure.* Oxford: Oxford University Press.

McQuarrie, Donald A. 2008. *Quantum chemistry.* 2nd ed. Sausalito, CA: University Science Books.

Moore, Charlotte E. 1972. Selected tables of atomic spectra HI, D, T. *NSRDS-NBS 3*, Section 6: A1I-1-B1T-1.

Pauling, Linus, and E. Bright Wilson Jr. 1985. *Introduction to quantum mechanics with applications to chemistry.* New York: Dover Publications, Inc.

Sakurai, J. J. 1985. *Modern quantum mechanics.* Rev. ed. Edited by San Fu Tuan. Reading, MA: Addison-Wesley Publishing Co.

Townsend, John F. 2012. *A modern approach to quantum mechanics.* 2nd ed. Mill Valley, CA: University Science Books.

Woodgate, G. K. 1980. *Elementary atomic structure.* 2nd ed. Oxford: Oxford University Press.

PROBLEMS

10.1 Show by direct substitution that the 1s radial wave function $R_{10}(r)$ solves the radial Schrödinger equation (Equation 10.11). Is your result for E_1 in accord with Equation (10.18)?

10.2 Repeat Problem 10.1 for the 2s and 2p radial wave functions. Is your result for E_2 in accord with Equation (10.18)?

10.3 (a) What is the minimum value of n for which k states exist? (b) How many k states exist for that value of n?

10.4 Show that for $\ell > 0$, the minima $r_{min}(\ell)$ of $U_{eff}(r)$ of Equation (10.22) occur at $r_{min}(\ell) = \frac{\ell(\ell+1)a}{Z}$.

10.5 Calculate the wavelength in nm for the $n_2 = 2 \rightarrow n_1 = 1$ Lyman line of hydrogen assuming that (a) the nucleus is infinitely massive and (b) the nuclear mass is m_p.

10.6 Let I be the ionization potential of hydrogen. Show that $\frac{I}{hc}$ is equal to (a) R_∞ assuming that the nucleus is infinitely massive and (b) R_H assuming that the nuclear mass is m_p. Using the conversion factor $1\,cm^{-1} = 1.239842 \times 10^{-4}$ eV, calculate I in eV assuming that (c) the nucleus is infinitely massive and (d) the nuclear mass is m_p. The experimental value of I is 13.598 eV. This value includes relativistic as well as reduced mass effects (I is the energy needed for the transition $n_1 = 1 \rightarrow n_2 = \infty$).

10.7 The masses of ^2H and ^3H nuclei the deuteron and the triton are $m_D = 3.3436 \times 10^{-27}$ kg and $m_T = 5.0074 \times 10^{-27}$ kg. Calculate in cm^{-1} (a) the Rydberg constant R_D of ^2H and (b) the Rydberg constant R_T of ^3H.

10.8 Prove that $R_\infty = \frac{m_e c \alpha^2}{2h} = \frac{e^2}{2(4\pi\varepsilon_0)a_0 hc} = \frac{\hbar}{4\pi m_e a_0^2 c}$ where $a_0 = \frac{4\pi\varepsilon_0}{m_e e^2}\hbar^2$ is the Bohr radius and $\alpha = \frac{1}{4\pi\varepsilon_0}\frac{e^2}{\hbar c}$ is the fine structure constant.

10.9 Extend Table 10.4 by giving the allowed transitions for the Brackett series ($n_1 = 4$).

10.10 The 4p radial wave function is given by $R_{41}(r) = \frac{1}{265\sqrt{15}}\left(80 - 20\sigma + \sigma^2\right)\sigma \exp\left(-\frac{\sigma}{4}\right)$. Find the locations r_1 and r_2 of the nodes of $R_{41}(r)$.

10.11 Use the result of Problem 8.15a to verify the integral of Equation (10.51).

10.12 Show that the 1s, 2s, and 2p radial wave functions are normalized.

10.13 Show that (a) the 2s and 3s and (b) 2p and 3p radial wave functions are orthogonal, while (c) the 1s and 2p radial wave functions are not orthogonal.

10.14 Show that the position of the maximum of the 3d radial distribution function coincides with the radius of the third Bohr orbit $\frac{3^2}{Z}a \doteq \frac{3^2}{Z}a_0$.

10.15 Show that for $n = 1, 2$, and 3, the radial wave function $R_{n,n-1}(r)$ of Equation (10.63) reduces to the 1s, 2p, and 3d radial wave functions.

10.16 Referring to Equation (10.63), show that the position of the maximum of the radial distribution function $P_{n,n-1}(r)$ coincides with the radius of the nth Bohr orbit $\frac{n^2 a}{Z} \doteq \frac{n^2 a_0}{Z}$.

10.17 Find the expected value of r for the 2s and 2p states. Can you rationalize your results in terms of the plots of the 2s and 2p radial distribution functions in Figure 10.4?

10.18 Find the expected value of r for the 3s, 3p, and 3d states. Can you rationalize your results in terms of the plots of the 3s, 3p, and 3d radial distribution functions in Figure 10.4?

10.19 (a) Derive from Equation (10.63) an expression for $\langle r \rangle_{n,n-1} = \int_0^\infty r P_{n,n-1}(r)\,dr = \int_0^\infty r^2 R^2_{n,n-1}(r)\, r\,dr$. (b) Show that the general expression for $\langle r \rangle_{n\ell}$ in Table 10.5 reduces to your result for $\langle r \rangle_{n,n-1}$ when $\ell = n - 1$. (c) Show that for $\ell = n - 1$, as $n \to \infty$, $\langle r \rangle_{n,n-1}$ approaches the most probable value of r, $\frac{n^2 a}{Z}$.

10.20 Prove the virial theorem for the 2s and 2p states.

10.21 Prove the virial theorem in general using the form for $\langle r^{-1} \rangle_{n\ell}$ in Table 10.5.

10.22 (a) Show that the probability $P(x)$ that a 2p electron is in a sphere centered on the nucleus of radius $x\left(\frac{a}{Z}\right)$, where $\frac{a}{Z}$ is the radius of the first Bohr orbit, is $P(x) = 1 - \exp(-x)\left(\frac{1}{24}x^4 + \frac{1}{6}x^3 + \frac{1}{2}x^2 + 1\right)$. Evaluate $P(x)$ for (b) $x = 4$, (c) $x = 6$, and (d) $x = 8$.

10.23 Show that the real spherical harmonics p_z, p_x, and p_y of Equation (10.82) form an orthonormal set.

10.24 Show that the real d spherical harmonics when written in Cartesian coordinates take form given in Equation (10.86).

10.25 This problem concerns the Zeeman splitting of the first Balmer line of hydrogen. As noted in Section 10.3, this line is due to the orbital transitions $3p \to 2s$, $3s \to 2p$, and $3d \to 2p$. These orbital transitions are made up of subtransitions in which the quantum number m changes in accord with the selection rule of Equation (10.43). For example, the subtransitions for the $3p \to 2s$ transition are $3p_{m=1} \to 2s_{m=0}$, $3p_{m=0} \to 2s_{m=0}$, and $3p_{m=-1} \to 2s_{m=0}$. Find (a) the three allowed subtransitions for the $3s \to 2p$ transition and (b) the nine allowed subtransitions for the $3d \to 2p$ transition. In a magnetic field, each subtransition gives rise to its own Zeeman line. (c) Find the wave numbers for the three lines due to the $3p \to 2s$ subtransitions. (d) Repeat for the $3s \to 2p$ subtransitions. Use Equation (10.104) written as $E_{nm} = E_n + \mu_B m B_z$ and assume $B_z = 1\text{T}$.

10.26 Derive the form for the spin–orbit coupling constants $\xi_{n\ell}$ given in Equation (10.140) from Equation (10.139).

10.27 Calculate from Equation (10.140) the spin–orbit coupling constants $\xi_{n\ell}$ in cm^{1} units for (a) the 2p, (b) 3p, (c) 3d, (d) 4p, (e) 4d, and (f) 4f states of hydrogen.

10.28 Show that the fine structure constant $\alpha = \frac{1}{4\pi\varepsilon_0}\frac{e^2}{\hbar c}$ is dimensionless and has the numerical value $\frac{1}{137.036} \doteq \frac{1}{137}$. (Hint: Calculate α^{-1}.)

10.29 This problem shows that the fine structure constant α given in Problem 10.28 is a direct measure of relativistic effects since it is closely related to v/c. We will assume a hydrogen atom in the ground state. Then, neglecting nuclear motion its energy is $E_1 = -\frac{e^4 m_e}{2(4\pi\varepsilon_0)^2\hbar^2}$ (a) Show that the virial theorem gives for the average kinetic energy of the electron $\langle KE\rangle = \frac{\langle p^2\rangle}{2m_e} = \frac{e^4 m_e}{2(4\pi\varepsilon_0)^2\hbar^2}$. (b) Defining the root mean square velocity v_{rms} by $v_{rms} = \frac{\langle p^2\rangle^{1/2}}{m_e}$, show $v_{rms} = \frac{e^2}{(4\pi\varepsilon_0)\hbar}$. (c) Finally, show $v_{rms}/c = \alpha$, thus establishing the close relationship between α and v/c.

10.30 Show $\hat{L}^2$ and $\hat{S}^2$ commute with $\hat{J}^2, \hat{J}_z$, and $\hat{H}^{eff}$.

10.31 Show $\hat{L}_z$ and $\hat{S}_z$ do not commute with either $\hat{J}^2$ or the spin–orbit part of $\hat{H}^{eff}$.

10.32 Write down the term symbols for the 8k configuration of a hydrogen-like atom.

10.33 Using the results of Problem 10.27, calculate the spin–orbit splittings for (a) the 3p and (b) 4p states of hydrogen. Compare with the observed splittings found from Table 10.7.

10.34 Show that the spin–orbit splittings for the nd and nf states are given by

(a) $\frac{1}{hc}\left[E\left(nd\,^2D_{5/2}\right) - E\left(nd\,^2D_{3/2}\right)\right] = \frac{5}{2}\xi_{n2}$

and (b) $\frac{1}{hc}\left[E\left(nf\,^2F_{7/2}\right) - E\left(nf\,^2F_{5/2}\right)\right] = \frac{7}{2}\xi_{n3}$. Using the method of Problem 10.27, calculate the spin–orbit splittings for (c) the 3d, (d) 4d, and (e) 4f states of hydrogen. Compare with the observed splittings found from Table 10.7.

10.35 Experimental energy levels for many hydrogen-like atoms are given by J.D. Garcia and J.E. *Mack J. Opt. Soc. Am.* **55**, 654 (1965). Below, we list the low-lying levels for $^4He^+$ given by

Configuration	Terms	J	Levels (cm^{-1})
1s	^2S	1/2	0
2p	^2P	1/2	329,179.275
		3/2	329,185.132
2s	^2S	1/2	329,179.744
3p	^2P	1/2	390,140.803
		3/2	390,142.538
3s	^2S	1/2	390,140.942
3d	^2D	3/2	390,142.535
		5/2	390,143.114
4p	^2P	1/2	411,477.099
		3/2	411,477.831
4s	^2S	1/2	411,477.158
4d	^2D	3/2	411,477.830
		5/2	411,478.074
4f	^2F	5/2	411,478.074
		7/2	411,478.196

Using the spin–orbit coupling constants of Problem 10.27 scaled in accord with Equation (10.140) by a factor of $2^4 = 16$, calculate the spin–orbit splittings for (a) the 2p, (b) 3p, and (c) 4p states of He^+. Compare with the observed splittings found from the data of Garcia and Mack.

10.36 Repeat Problem 10.35 for (a) the 3d, (b) 4d, and (c) 4f states of He^+.

10.37 Due to spin orbit coupling, the $2p \rightarrow 1s$ Lyman transition of hydrogen gives rise to a doublet consisting of the $2p^2P_{1/2} \rightarrow 1s^2S_{1/2}$ and $2p^3P_{3/2} \rightarrow 1s^2S_{1/2}$ lines. Using the data of Table 10.7, calculate the wavelengths λ in nm of these lines. Compare with the value of λ for the $2p \rightarrow 1s$ line found ignoring spin–orbit coupling in Problem 10.5b.

10.38 (a) Show that the uncertainty $(\Delta r)_{10} = \langle r^2 \rangle_{10} - \langle r \rangle_{10}^2$ in the distance r of the electron from the nucleus in a 1s hydrogen atom is $\frac{\sqrt{3}}{2}a$. (b) Since $a = 52.9\,pm$, show that for a 1s hydrogen atom, $\langle r \rangle_{10} = 79.4\,pm$ and $(\Delta r)_{10} = 45.8\,pm$. (c) What do you conclude from your calculations about the differences between the modern quantum and Bohr theories of the hydrogen atom?

SOLUTIONS TO SELECTED PROBLEMS

10.3 (a) 8 and (b) 15.

10.5 (a) 121.50227 nm and (b) 121.56844 nm.

10.6 (c) 13.606 eV and (d) 13.594 eV.

10.7 (a) $R_D = 109{,}707.426\ \text{cm}^{-1}$ and (b) $R_T = 109{,}717.355\ \text{cm}^{-1}$.

10.9 $5p \to 4s$, $6p \to 4s, \ldots$; $5s \to 4p$, $6s \to 4p, \ldots$; $5d \to 4p$, $6d \to 4p, \ldots$; $5p \to 4d$, $6p \to 4d, \ldots$; $5d \to 4f$, $6d \to 4f, \ldots$; $5f \to 4d$, $6f \to 4d, \ldots$; $5g \to 4f$, $6g \to 4f, \ldots$.

10.10 $r_1 = 5.528\frac{a}{Z}$ and $r_2 = 14.472\frac{a}{Z}$.

10.17 $\langle r\rangle_{2s} = 6\frac{a}{Z}$ and $\langle r\rangle_{2p} = 5\frac{a}{Z}$.

10.18 $\langle r\rangle_{3s} = \frac{27}{2}\frac{a}{Z}$, $\langle r\rangle_{3p} = \frac{25}{2}\frac{a}{Z}$, and $\langle r\rangle_{3d} = \frac{21}{2}\frac{a}{Z}$.

10.19 (a) $\langle r\rangle_{n,n-1} = \left(n^2 + \frac{n}{2}\right)\frac{a}{Z}$.

10.22 (b) 0.371, (c) 0.715, and (d) 0.900.

10.25 (a) $3s_{m=0} \to 2p_{m=1}$, $3s_{m=0} \to 2p_{m=0}$, $3s_{m=0} \to 2p_{m=-1}$.
(b) $3d_{m=2} \to 2p_{m=1}$, $3d_{m=1} \to 2p_{m=1}$, $3d_{m=1} \to 2p_{m=0}$, $3d_{m=0} \to 2p_{m=0}$, $3d_{m=0} \to 2p_{m=1}$, $3d_{m=0} \to 2p_{m=-1}$, $3d_{m=-1} \to 2p_{m=-1}$, $3d_{m=-1} \to 2p_{m=0}$, $3d_{m=-2} \to 2p_{m=-1}$.
(c) $\bar{\upsilon}_{3p_{m=1} \to 2s_{m=0}} = 15{,}233.465\ \text{cm}^{-1}$, $\bar{\upsilon}_{3p_{m=0} \to 2s_{m=0}} = 15{,}232.998\ \text{cm}^{-1}$, and $\bar{\upsilon}_{3p_{m=-1} \to 2s_{m=0}} = 15{,}232.531\ \text{cm}^{-1}$.
(d) $\bar{\upsilon}_{3s_{m=0} \to 2p_{m=1}} = 15{,}232.531\ \text{cm}^{-1}$, $\bar{\upsilon}_{3s_{m=0} \to 2p_{m=0}} = 15{,}232.998\ \text{cm}^{-1}$, and $\bar{\upsilon}_{3s_{m=0} \to 2p_{m=-1}} = 15{,}233.465\text{cm}^{-1}$.

10.27 (a) $0.2435\ \text{cm}^{-1}$, (b) $0.07214\ \text{cm}^{-1}$, (c) $0.01443\ \text{cm}^{-1}$, (d) $0.03044\ \text{cm}^{-1}$, (e) $0.006087\ \text{cm}^{-1}$, and (f) $0.002174\ \text{cm}^{-1}$.

10.32 $8k\ {}^2K_{\frac{15}{2}}$ and $8k\ {}^2K_{\frac{13}{2}}$.

10.33 (a) $0.108\ \text{cm}^{-1}$, $0.108\ \text{cm}^{-1}$ (obs.); and (b) $0.0457\ \text{cm}^{-1}$, $0.045\ \text{cm}^{-1}$ (obs.)

10.34 (c) $0.0361\ \text{cm}^{-1}$ and $0.036\ \text{cm}^{-1}$ (obs.), (d) $0.0152\ \text{cm}^{-1}$ and $0.015\ \text{cm}^{-1}$ (obs.), and (e) $0.00761\ \text{cm}^{-1}$, $0.008\ \text{cm}^{-1}$ (obs.).

10.35 (a) $5.844\ \text{cm}^{-1}$, $5.857\ \text{cm}^{-1}$ (obs.); (b) $1.731\ \text{cm}^{-1}$, $1.735\ \text{cm}^{-1}$ (obs.); and (c) $0.7305\ \text{cm}^{-1}$, $0.732\ \text{cm}^{-1}$ (obs.).

10.36 (a) $0.5772\ \text{cm}^{-1}$ and $0.579\ \text{cm}^{-1}$ (obs.); (b) $0.2435\ \text{cm}^{-1}$ and $0.244\ \text{cm}^{-1}$ (obs.), and (c) $0.1217\ \text{cm}^{-1}$ and $0.122\ \text{cm}^{-1}$ (obs.).

10.37 For the $2p\ {}^2P_{1/2} \to 1s\ {}^2S_{1/2}$ line, $\lambda = 121.56738$ nm. For the $2p\ {}^2P_{3/2} \to 1s\ {}^2S_{1/2}$ line, $\lambda = 121.56683$ nm.

11 Approximation Methods

As we have seen, the time-independent Schrödinger equation can be solved exactly for the particle-in-a-box, harmonic oscillator, rigid rotor, and hydrogen-like atom systems. For most other systems, however the Schrödinger equation cannot be solved exactly. Fortunately, approximation methods have been developed which permit one to apply quantum mechanics to a vast range of problems. Especially, we will see in Chapter 13 that approximate quantum mechanics is the indispensable tool for determining molecular electronic structures and hence treating chemical bonding and thus is of immense importance in chemistry.

In this chapter, we will develop the two most widely used methods for obtaining approximate solutions to the Schrödinger equation. These are *perturbation theory* and the *variational method.* We begin with perturbation theory.

11.1 PERTURBATION THEORY

Before developing perturbation theory, we describe a problem to which it may be applied. Specifically, we discuss the problem of obtaining anharmonic corrections to the vibrational energy levels of a diatomic molecule (see Figure 7.7). In Section 7.13, we expanded the potential energy function of the diatomic $U(r)=U(r_e+y)$ in a power series in the vibrational displacement y and terminated the expansion at quadratic order. $U(r)$ was thus approximated by the harmonic oscillator potential $\frac{1}{2}\mu\omega^2y^2$ where $\omega=\left[\frac{U''(r_e)}{\mu}\right]^{1/2}$ yielding the vibrational Hamiltonian operator of the diatomic as

$$\hat{H}_0=-\frac{\hbar^2}{2\mu}\frac{d^2}{dy^2}+\frac{1}{2}\mu\omega^2y^2. \tag{11.1}$$

$\hat{H}_0$ is a harmonic oscillator Hamiltonian operator, and thus, the Schrödinger equation $\hat{H}_0\Psi_v^{(0)}=E_v^{(0)}\Psi_v^{(0)}$ is exactly solvable. One, however, can improve on the harmonic oscillator approximation by terminating the expansion of $U(r)$ at, say, quartic rather than quadratic order in y. This yields the vibrational Hamiltonian operator as

$$\hat{H}=\hat{H}_0+\hat{V} \tag{11.2}$$

where

$$\hat{V}=\frac{1}{6}\gamma_3y^3+\frac{\gamma_4}{24}y^4 \tag{11.3}$$

with $\gamma_3=U^{(3)}(r_e)$ and $\gamma_4=U^{(4)}(r_e)$. Because of the anharmonic term $\hat{V}$ in $\hat{H}$, the Schrödinger equation $\hat{H}\Psi_v=E_v\Psi_v$ is not exactly solvable. Suppose, however, that the anharmonic corrections to the harmonic oscillator energy levels are not too large. Then, perturbation theory systematically provides accurate approximations to these corrections in terms of the known eigenfunctions and eigenvalues of $\hat{H}_0$.

We may now develop perturbation theory.

Denote the Hamiltonian operator of our system by $\hat{H}$. We wish to find a particular eigenfunction Ψ_n and the associated (assumed non-degenerate) eigenvalue E_n of $\hat{H}$. These are determined as the solutions of the Schrödinger equation

$$\hat{H}\Psi_n=E_n\Psi_n. \tag{11.4}$$

Suppose that this Schrödinger equation is not exactly solvable but that there is a closely related system whose Schrödinger equation

$$\hat{H}_0\Psi_n^{(0)} = E_n^{(0)}\Psi_n^{(0)} \tag{11.5}$$

is exactly solvable. We assume Ψ_n and E_n are very similar but not identical to $\Psi_n^{(0)}$ and $E_n^{(0)}$. Thus, we seek the small corrections to $\Psi_n^{(0)}$ and $E_n^{(0)}$ which will yield Ψ_n and E_n. Perturbation theory provides approximations to these corrections in terms of the known normalized eigenfunctions $\Psi_k^{(0)}$ and eigenvalues $E_k^{(0)}$ of $\hat{H}_0$.

To start, we break up $\hat{H}$ as follows:

$$\hat{H} = \hat{H}_0 + \hat{V}. \tag{11.6}$$

$\hat{H}_0$ is called the *unperturbed Hamiltonian operator.* $\hat{V}$, which is assumed to be much "smaller" than $\hat{H}_0$, is referred to as the *perturbation.* As examples for the problem of anharmonic corrections, the unperturbed Hamiltonian operator, $\hat{H}_0$ is the harmonic oscillator Hamiltonian operator of Equation (11.1) and the perturbation $\hat{V}$ is the anharmonic term of Equation (11.3). For the problem of spin–orbit coupling in the hydrogen-like atoms the unperturbed Hamiltonian operator is the non-relativistic hydrogen-like atom Hamiltonian operator of Equation (10.128) and the perturbation is the spin–orbit coupling operator of Equation (10.129).

To develop perturbation theory, it is convenient to define a more general Hamiltonian operator

$$\hat{H}(\lambda) = \hat{H}_0 + \lambda\hat{V} \tag{11.7}$$

where λ is a parameter with the range $0 \le \lambda \le 1$. We will see that λ is introduced to keep track of orders of perturbation theory. When $\lambda = 0$, $\hat{H}(\lambda)$ reduces to the unperturbed Hamiltonian operator $\hat{H}_0$. When $\lambda = 1$, $H(\lambda)$ becomes the actual Hamiltonian operator $\hat{H}$. We will solve the Schrödinger equation

$$\hat{H}(\lambda)\Psi_n(\lambda) = E_n(\lambda)\Psi_n(\lambda) \tag{11.8}$$

as a power series in λ and then set $\lambda = 1$ at the end of the calculation.

To carry out this solution, we expand $\Psi_n(\lambda)$ and $E_n(\lambda)$ as the following power series in λ:

$$\Psi_n(\lambda) = \Psi_n^{(0)} + \lambda\Psi_n^{(1)} + \lambda^2\Psi_n^{(2)} + \cdots \tag{11.9}$$

and

$$E_n(\lambda) = E_n^{(0)} + \lambda E_n^{(1)} + \lambda^2 E_n^{(2)} + \cdots \tag{11.10}$$

We then determine $\Psi_n^{(m)}$ and $E_n^{(m)}$ for $m = 0,1,2,\ldots$ by substituting the expansions into the Schrödinger equation (Equation 11.8). Evaluation of $\Psi_n^{(m)}$ and $E_n^{(m)}$ becomes increasingly complex as m increases. So for perturbation theory to be useful, usually the two power series must converge sufficiently rapidly that their first few terms give useful approximations at $\lambda = 1$.

Making the substitutions and using Equation (11.7) for $\hat{H}(\lambda)$ give

$$\left(\hat{H}_0 + \lambda\hat{V}\right)\left(\Psi_n^{(0)} + \lambda\Psi_n^{(1)} + \lambda^2\Psi_n^{(2)}\cdots\right) = \left(E_n^{(0)} + \lambda E_n^{(1)} + \lambda^2 E_n^{(2)} + \cdots\right)\left(\Psi_n^{(0)} + \lambda\Psi_n^{(1)} + \lambda^2\Psi_n^{(2)} + \cdots\right). \tag{11.11}$$

Multiplying out the terms in Equation (11.11) and grouping terms of the same order in λ yields (Problem 11.1)

$$\left(\hat{H}_0\Psi_n^{(0)} - E_n^{(0)}\Psi_n^{(0)}\right)\lambda^0 +$$

$$\left(\hat{H}_0\Psi_n^{(1)} + \hat{V}\Psi_n^{(0)} - E_n^{(0)}\Psi_n^{(1)} - E_n^{(1)}\Psi_n^{(0)}\right)\lambda^1 + \tag{11.12}$$

$$\left(\hat{H}_0\Psi_n^{(2)} + \hat{V}\Psi_n^{(1)} - E_n^{(0)}\Psi_n^{(2)} - E_n^{(1)}\Psi_n^{(1)} - E_n^{(2)}\Psi_n^{(0)}\right)\lambda^2 = 0.$$

where we have only kept terms to order λ^2. Since λ can be varied arbitrarily, Equation (11.12) can hold only if each term in parentheses vanishes. Thus, we have from the λ^0 term

$$\hat{H}_0\Psi_n^{(0)} = E_n^{(0)}\Psi_n^{(0)} \tag{11.13}$$

from the λ^1 term

$$\hat{H}_0\Psi_n^{(1)} + \hat{V}\Psi_n^{(0)} = E_n^{(0)}\Psi_n^{(1)} + E_n^{(1)}\Psi_n^{(0)} \tag{11.14}$$

and from the λ^2 term

$$\hat{H}_0\Psi_n^{(2)} + \hat{V}\Psi_n^{(1)} = E_n^{(0)}\Psi_n^{(2)} + E_n^{(1)}\Psi_n^{(1)} + E_n^{(2)}\Psi_n^{(0)}. \tag{11.15}$$

Additional equations emerge from the λ^3, λ^4, and so on terms. But for simplicity, we will not consider these equations here.

Equation (11.13) is just the Schrödinger equation for the unperturbed system and thus gives nothing new. However, from Equation (11.14) one may obtain the first-order correction $E_n^{(1)}$ to the unperturbed energy $E_n^{(0)}$ and the first-order correction $\Psi_n^{(1)}$ to the unperturbed wave function $\Psi_n^{(0)}$. Additionally, from Equation (11.15) one may find the second-order correction $E_n^{(2)}$ to $E_n^{(0)}$.

We will soon show that $E_n^{(1)}$ is given by

$$E_n^{(1)} = \int \left[\Psi_n^{(0)}\right]^* \hat{V}\Psi_n^{(0)} d\tau \tag{11.16}$$

where $d\tau$ is the volume element of the system. Thus, $E_n^{(1)}$ would be the expected value of the observable V if the actual wave function was the unperturbed wave function $\Psi_n^{(0)}$. Because of its simplicity, Equation (11.16) is the most useful result which emerges from perturbation theory.

The expression for $\Psi_n^{(1)}$ is (Problem 11.2)

$$\Psi_n^{(1)} = \sum_{\substack{\text{all} \\ k \neq n}} \Psi_k^{(0)} \frac{\int \left[\Psi_k^{(0)}\right]^* \hat{V}\Psi_n^{(0)} d\tau}{E_n^{(0)} - E_k^{(0)}}. \tag{11.17}$$

Note that the above sum runs over all states k of the unperturbed system except state n. This exception is necessary to avoid the divergence that would occur if the denominator $E_n^{(0)} - E_k^{(0)}$ was allowed for k = n.

Finally, the result for $E_n^{(2)}$ is (Problem 11.2)

$$E_n^{(2)} = \int \left[\Psi_n^{(0)}\right]^* \hat{V}\Psi_n^{(1)} d\tau. \tag{11.18}$$

We next recast Equation (11.18) for $E_n^{(2)}$ into a more familiar form. We do this by eliminating $\Psi_n^{(1)}$ from Equation (11.18) using Equation (11.17) to give

$$E_n^{(2)} = \sum_{\substack{\text{all}\\ k\neq n}} \frac{\left\{\int\left[\Psi_n^{(0)}\right]^* \hat{V}\Psi_k^{(0)}\, d\tau\right\}\left\{\int\left[\Psi_k^{(0)}\right]^* \hat{V}\Psi_n^{(0)}\, d\tau\right\}}{E_n^{(0)} - E_k^{(0)}}. \tag{11.19}$$

However, since $\hat{V}$ is Hermitian,

$$\int\left[\Psi_n^{(0)}\right]^* \hat{V}\Psi_k^{(0)} d\tau = \left\{\int\left[\Psi_k^{(0)}\right]^* \hat{V}\Psi_n^{(0)}\, d\tau\right\}^* \tag{11.20}$$

yielding the final form for $E_n^{(2)}$ as

$$E_n^{(2)} = \sum_{\substack{\text{all}\\ k\neq n}} \frac{\left|\int\left[\Psi_k^{(0)}\right]^* \hat{V}\Psi_n^{(0)}\, d\tau\right|^2}{E_n^{(0)} - E_k^{(0)}}. \tag{11.21}$$

We may now verify Equation (11.16) for $E_n^{(1)}$. To do this, we multiply Equation (11.14) on the left by $\left[\Psi_n^{(0)}\right]^*$ and integrate over $d\tau$. Using the normalization integral $\int\left[\Psi_n^{(0)}\right]^*\Psi_n^{(0)} d\tau = 1$, this gives

$$\int\left[\Psi_n^{(0)}\right]^* \hat{H}_0\Psi_n^{(1)} d\tau + \int\left[\Psi_n^{(0)}\right]^* \hat{V}\Psi_n^{(0)} d\tau = E_n^{(0)}\int\left[\Psi_n^{(0)}\right]^* \Psi_n^{(1)} d\tau + E_n^{(1)}. \tag{11.22}$$

Since $\hat{H}_0$ is Hermitian, we additionally have that

$$\int\left[\Psi_n^{(0)}\right]^* \hat{H}_0\Psi_n^{(1)} d\tau = \left\{\int\left[\Psi_n^{(1)}\right]^* \hat{H}_0\Psi_n^{(0)} d\tau\right\}^*. \tag{11.23}$$

But $\hat{H}_0\Psi_n^{(0)} = E_n^{(0)}\Psi_n^{(0)}$, so the previous equation becomes

$$\int\left[\Psi_n^{(0)}\right]^* \hat{H}_0\Psi_n^{(1)} d\tau = E_n^{(0)}\int\left[\Psi_n^{(0)}\right]^* \Psi_n^{(1)} d\tau. \tag{11.24}$$

With Equation (11.24), Equation (11.22) may be written as

$$E_n^{(0)}\int\left[\Psi_n^{(0)}\right]^* \Psi_n^{(1)} d\tau + \int\left[\Psi_n^{(0)}\right]^* \hat{V}\Psi_n^{(0)} d\tau = E_n^{(0)}\int\left[\Psi_n^{(0)}\right]^* \Psi_n^{(1)} d\tau + E_n^{(1)} \tag{11.25}$$

which immediately yields Equation (11.16) for $E_n^{(1)}$.

Next, let us summarize our results. From Equation (11.9) with $\lambda = 1$ and Equation (11.17), we find that the wave function to first order is

$$\Psi_n = \Psi_n^{(0)} + \Psi_n^{(1)} = \Psi_n^{(0)} + \sum_{\substack{\text{all}\\ k\neq n}} \frac{\Psi_k^{(0)}\int\left[\Psi_k^{(0)}\right]^* \hat{V}\Psi_n^{(0)}\, d\tau}{E_n^{(0)} - E_k^{(0)}}. \tag{11.26}$$

Similarly, the energy to second order is

$$E_n = E_n^{(0)} + E_n^{(1)} + E_n^{(2)} = E_n^{(0)} + \int \left[\Psi_n^{(0)}\right]^* \hat{V}\Psi_n^{(0)} d\tau + \sum_{\substack{\text{all} \\ k \neq n}} \frac{\left| \int \left[\Psi_k^{(0)}\right]^* \hat{V}\Psi_n^{(0)} d\tau \right|^2}{E_n^{(0)} - E_k^{(0)}}. \tag{11.27}$$

We next turn to some simple applications of perturbation theory. More advanced applications are given in the Problems set at the end of the chapter.

11.2 SIMPLE APPLICATIONS OF PERTURBATION THEORY

We will consider one-dimensional systems adapting to these systems our general expressions for the energy and wave function perturbation corrections. All wave functions for our systems are real and so no complex conjugates will appear in our expressions.

The first problem we treat is the determination of the first-order anharmonic correction to the harmonic oscillator approximation to the ground state vibrational energy of a diatomic molecule. The unperturbed Hamiltonian operator is the harmonic Hamiltonian operator $\hat{H}_0$ of Equation (11.1), and the perturbation is the anharmonic term $\hat{V}$ of Equation (11.3). Thus we approximate the energy as $E_0 = E_0^{(0)} + E_0^{(1)}$, where $E_0^{(0)} = \frac{1}{2}h\omega$ is the harmonic oscillator ground state energy and where

$$E_0^{(1)} = \int_{-\infty}^{\infty} \Psi_0^{(0)}(y)\hat{V}\Psi_0^{(0)}(y)dy \tag{11.28}$$

is the first-order correction to $E_0^{(0)}$. From Equation (11.3), $\hat{V} = \frac{1}{6}\gamma_3 y^3 + \frac{1}{24}\gamma_4 y^4$. Also $\Psi_0^{(0)}(y) = \left(\frac{\mu\omega}{\pi\hbar}\right)^{1/4} \exp\left(-\frac{1}{2}\frac{\mu\omega}{\hbar}y^2\right)$ is the ground state harmonic oscillator wave function of Equation (7.15) (with $m = \mu$). Thus, $E_0^{(1)} = \left(\frac{\mu\omega}{\pi\hbar}\right)^{1/2} \frac{\gamma_3}{6} \int_{-\infty}^{\infty} y^3 \exp\left(-\frac{\mu\omega}{\hbar}y^2\right)dy + \left(\frac{\mu\omega}{\pi\hbar}\right)^{1/2} \frac{\gamma_4}{24} \int_{-\infty}^{\infty} y^4 \exp\left(-\frac{\mu\omega}{\hbar}y^2\right)dy.$ The first of the two integrals vanishes since it is an integral of an odd integrand between symmetric limits (see Figure 7.6). Also from the Gaussian integral of Equation (7.45) with $\alpha = \frac{\mu\omega}{\hbar}$, the second integral has the value $\frac{3}{4}\left(\frac{\hbar}{\mu\omega}\right)^{5/2} \pi^{1/2}$. Hence, $E_0^{(1)} = \frac{\gamma_4}{32}\left(\frac{\hbar}{\mu\omega}\right)^2$ and the energy to first order is

$$E_0 = \frac{1}{2}\hbar\omega + \frac{\gamma_4}{32}\left(\frac{\hbar}{\mu\omega}\right)^2. \tag{11.29}$$

Next, consider a particle of mass m subject to a gravitational potential (Figure 11.1)

$$V(x) = \begin{cases} \frac{W}{a}x & \text{for } 0 \le x \le a \\ \infty & \text{for } x < 0 \text{ and } x > a \end{cases} \tag{11.30}$$

where $W > 0$. Notice that if $W = 0$, $\hat{V}(x)$ reduces to the potential energy operator for a particle in a box of length a.

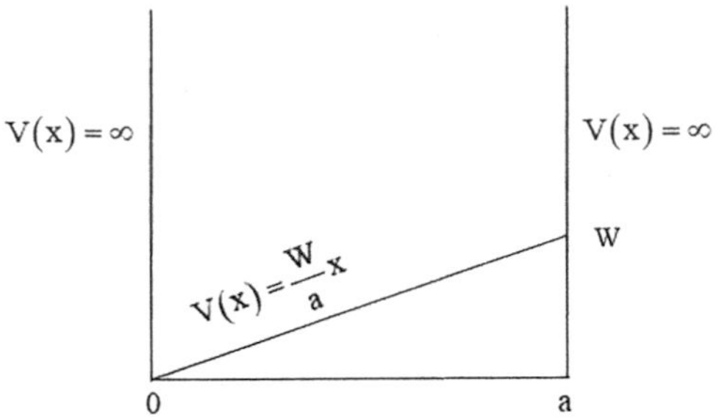

FIGURE 11.1 The potential energy function $V(x)$ of Equation (11.30).

We will use perturbation theory up to second order to find the ground state energy E_1 of the particle. We take the unperturbed Hamiltonian operator $\hat{H}_0$ to be the Hamiltonian operator of a particle in a box of length a and the perturbation to be $\hat{V} = \frac{W}{a}x$. We will need the normalized eigenfunctions $\Psi_k^{(0)}(x)$ and eigenvalues $E_k^{(0)}$ of $\hat{H}_0$. These are given by (see Section 3.4)

$$\Psi_k^{(0)}(x) = \left(\frac{2}{a}\right)^{1/2} \sin\left(\frac{k\pi x}{a}\right) \text{ and } E_k^{(0)} = \frac{k^2h^2}{8ma^2} \quad k = 1,2,\ldots. \tag{11.31}$$

To second order, the energy is $E_1 = E_1^{(0)} + E_1^{(1)} + E_1^{(2)}$, where $E_1^{(0)} = \frac{h^2}{8ma^2}$ is the particle-in-a-box ground state energy and where $E_1^{(1)}$ and $E_1^{(2)}$ are the first- and second-order perturbation corrections to $E_1^{(0)}$. Given that $\hat{V} = \frac{W}{a}x$, these corrections have the forms

$$E_1^{(1)} = \frac{W}{a}\int_0^a \Psi_1^{(0)}(x)x\Psi_1^{(0)}(x)dx \tag{11.32}$$

and

$$E_1^{(2)} = \left(\frac{W}{a}\right)^2 \sum_{k=2}^{\infty} \frac{\left[\int_0^a \Psi_k^{(0)}(x)x\Psi_1^{(0)}(x)dx\right]^2}{E_1^{(0)} - E_k^{(0)}}. \tag{11.33}$$

Evaluation of $E_1^{(1)}$ is easy since the integral $\int_0^a \Psi_1^{(0)}(x)x\Psi_1^{(0)}(x)dx$ is the particle-in-a-box ground state expected value of x which from Equation (3.56) is $\frac{a}{2}$. So

$$E_1^{(1)} = \frac{W}{2}. \tag{11.34}$$

The evaluation of $E_1^{(2)}$ proceeds as follows (also see McQuarrie 2008, Problem 8.46). First, from Equation (11.31) $E_1^{(0)} - E_k^{(0)} = \frac{h^2}{8ma^2}(1-k^2)$. $E_1^{(2)}$ may therefore be written as

$$E_1^{(2)} = -W^2 \frac{8m}{h^2} \sum_{k=2}^{\infty} \frac{H_{k1}^2}{k^2 - 1} \tag{11.35}$$

where we define H_{k1} as

$$H_{k1} = \int_0^a \Psi_k^{(0)}(x)\,x\,\Psi_1^{(0)}(x)\,dx. \tag{11.36}$$

Given Equation (11.31) for the particle-in-a-box wave functions, we may write H_{k1} as

$$H_{k1} = \frac{2}{a} I_k \tag{11.37}$$

where

$$I_k = \int_0^a x \sin\left(\frac{k\pi x}{a}\right)\sin\left(\frac{\pi x}{a}\right)dx. \tag{11.38}$$

Next, using the identity $\cos(A+B) = \cos A\cos B - \sin A\sin B$, one may verify that $\sin\left(\frac{k\pi x}{a}\right)\sin\left(\frac{\pi x}{a}\right) = \frac{1}{2}\left\{\cos\left[(k-1)\frac{\pi x}{a}\right] - \cos\left[(k+1)\frac{\pi x}{a}\right]\right\}$. Thus, I_k may be written as

$$I_k = \frac{1}{2}\int_0^a x\left\{\cos\left[(k-1)\frac{\pi x}{a}\right] - \cos\left[(k+1)\frac{\pi x}{a}\right]\right\}dx. \tag{11.39}$$

The integrals in Equation (11.39) may be evaluated by integrating by parts to give I_k as $I_k = \frac{a^2}{2\pi^2}\left\{\frac{\cos[(k-1)\pi]-1}{(k-1)^2} - \frac{\cos[(k+1)\pi]-1}{(k+1)^2}\right\}$. However, if k is even, $\cos[(k\pm1)\pi] = -1$, and if k is odd, $\cos[(k\pm1)\pi] = 1$. Thus, if k is even, $\cos[(k\pm1)\pi]-1 = -2$, and if k is odd, $\cos[(k\pm1)\pi]-1 = 0$. I_k therefore reduces to

$$I_k = \begin{cases} -\dfrac{a^2}{\pi^2}\left[\dfrac{1}{(k-1)^2} - \dfrac{1}{(k+1)^2}\right] & \text{for k even} \\ 0 & \text{for k odd} \end{cases} \tag{11.40}$$

Simplifying I_k and comparing with Equation (11.37) give H_{k1} as

$$H_{k1} = \begin{cases} -\dfrac{8a}{\pi^2}\dfrac{k}{(k-1)^2} & \text{for k even} \\ 0 & \text{for k odd} \end{cases} \tag{11.41}$$

$E_1^{(2)}$ may now be found from Equations (11.35) and (11.41) as

$$E_1^{(2)} = -\frac{64W^2}{\pi^4}\frac{8ma^2}{h^2}\sum_{\substack{k=2 \\ k\text{ even}}}^{\infty}\frac{k^2}{(k^2-1)^5} \tag{11.42}$$

Performing the above sum numerically, one finds that it converges very rapidly to the value 0.01648. Thus, our final result for $E_1^{(2)}$ is

$$E_1^{(2)} = -0.01083\frac{8ma^2}{h^2}W^2. \tag{11.43}$$

We may now write down the ground state energy of our system to second order as

$$E_1 = \frac{h^2}{8ma^2} + \frac{W}{2} - 0.01083\frac{8ma^2}{h^2}W^2. \tag{11.44}$$

Finally, introducing the dimensionless variables $w = \frac{8ma^2}{h^2}W$ and $\varepsilon_1 = \frac{8ma^2}{h^2}E_1$, our expression for E_1 simplifies to

$$\varepsilon_1 = 1 + \frac{w}{2} - 0.01083w^2. \tag{11.45}$$

Next, let us turn to another problem. Consider a particle of mass m and charge q moving in a harmonic oscillator potential $\frac{1}{2}m\omega^2y^2$ and perturbed by an electric field **E** which points in the positive y direction. We will determine by perturbation theory the corrections to the harmonic oscillator energy levels due to the electric field. The Hamiltonian operator of the particle is

$$\hat{H} = -\frac{\hbar^2}{2m}\frac{d^2}{dy^2} + \frac{1}{2}m\omega^2y^2 - \gamma y \tag{11.46}$$

where $\gamma = \alpha E_z$. We will break up $\hat{H}$ as $\hat{H} = \hat{H}_0 + \hat{V}$, where we take the unperturbed Hamiltonian operator $\hat{H}_0$ to be the harmonic oscillator Hamiltonian operator $-\frac{\hbar^2}{2m}\frac{d^2}{dy^2} + \frac{1}{2}m\omega^2y^2$ and the perturbation $\hat{V}$ to be $-\gamma y$.

We will use perturbation theory up to second order to find the energy of the vth state as $E_v = E_v^{(0)} + E_v^{(1)} + E_v^{(2)}$, where $E_v^{(0)} = \left(v + \frac{1}{2}\right)\hbar\omega$ is the vth state harmonic oscillator energy and $E_v^{(1)}$ and $E_v^{(2)}$ are the first- and second-order perturbation corrections to $E_v^{(0)}$. Since $\hat{V} = -\gamma y$, these corrections have the forms

$$E_v^{(1)} = -\gamma\int_{-\infty}^{\infty}\Psi_v^{(0)}(y)\,y\,\Psi_v^{(0)}(y)\,dy \tag{11.47}$$

and

$$E_v^{(2)} = \gamma^2\sum_{\substack{\text{all}\\ v'\neq v}}\frac{\left[\int_{-\infty}^{\infty}\Psi_{v'}^{(0)}(y)\,y\,\Psi_v^{(0)}(y)\,dy\right]^2}{E_v^{(0)} - E_{v'}^{(0)}} \tag{11.48}$$

where $\Psi_v^{(0)}(y)$ and $\Psi_{v'}^{(0)}(y)$ and $E_v^{(0)}$ and $E_{v'}^{(0)}$ are the harmonic oscillator eigenfunctions and eigenvalues.

We first show that $E_v^{(1)} = 0$. To see this, recall from Section 7.8 that $\Psi_v^{(0)}(y)$ is either an even or odd function of y. Thus, the integrand in Equation (11.47) is an odd function of y. However, an integral of an odd integrand over symmetric limits vanishes (Figure 7.6). Hence, $E_v^{(1)} = 0$.

Next, let us evaluate $E_v^{(2)}$. Defining $I_{v'v}$ by

$$I_{v'v} = \int_{-\infty}^{\infty}\Psi_{v'}^{(0)}(y)\,y\,\Psi_v^{(0)}(y)\,dy \tag{11.49}$$

and noting that $E_v^{(0)} - E_{v'}^{(0)} = \hbar\omega(v - v')$, $E_v^{(2)}$ becomes

$$E_v^{(2)} = \gamma^2 \sum_{\substack{\text{all} \\ v' \neq v}} \frac{I_{v'v}^2}{(v - v')\hbar\omega}. \tag{11.50}$$

But we evaluated $I_{v'v}$ in Section 7.14 and found that (see Equation [7.155])

$$I_{v'v} = \left(\frac{\hbar}{2m\omega}\right)^{1/2} \left[(v+1)^{1/2} \delta_{v'v+1} + v^{1/2}\delta_{v'v-1}\right]. \tag{11.51}$$

Thus, $I_{v'v}^2 = \frac{\hbar}{2m\omega}\left[(v+1)\delta_{v'v+1} + v\delta_{v'v-1}\right]$, and hence,

$$E_v^{(2)} = \frac{\gamma^2\hbar}{2m\omega} \sum_{\substack{\text{all} \\ v' \neq v}} \frac{\left[(v+1)\delta_{v'v+1} + v\delta_{v'v-1}\right]}{(v - v')\hbar\omega}. \tag{11.52}$$

Performing the sum gives

$$E_v^{(2)} = \frac{\gamma^2\hbar}{2m\omega}\left(\frac{v+1}{-\hbar\omega} + \frac{v}{\hbar\omega}\right) \tag{11.53}$$

or

$$E_v^{(2)} = -\frac{\gamma^2}{2m\omega^2}. \tag{11.54}$$

Notice that $E_v^{(2)}$ is the same for all states v.

We may now write down the energy to second order as

$$E_v = \left(v + \frac{1}{2}\right)\hbar\omega - \frac{\gamma^2}{2m\omega^2}. \tag{11.55}$$

Finally, we note that this problem may be solved exactly. The exact and second-order energies turn out to be identical (Problem 11.9).

We next turn to the variational method that is nearly universally used to determine approximate molecular wave functions.

11.3 THE VARIATIONAL METHOD

The variational method, unlike perturbation theory, is not limited to systems that are close to exactly solvable systems. However, in contrast to perturbation theory the variational method applies (with exceptions) only to the ground state.

The idea behind the variational method is very simple. Suppose we seek the ground state energy E_0 of a system whose Schrödinger equation

$$\hat{H}\Psi_0 = E_0\Psi_0 \tag{11.56}$$

is not exactly solvable. Also suppose we guess an approximation Φ to Ψ_0. Φ must be well-behaved and also must obey the boundary conditions satisfied by Ψ_0 but is not necessarily normalized. We will call Φ a *trial function*. We then form the *trial energy*

$$E_T = \frac{\int \Phi^* \hat{H} \Phi \, d\tau}{\int \Phi^* \Phi \, d\tau}. \tag{11.57}$$

According to the variational theorem whose proof hinges on a complete orthonormal expansion (Problem 11.12), E_r cannot be less than E_0. That is,

$$E_T \geq E_0. \tag{11.58}$$

If our guess was perfect so that $\Phi = \Psi_0$, then, it follows immediately from Equations (11.56) and (11.57) that E_r would equal E_0. Typically however, $\Phi \neq \Psi_0$ and E_T will be greater than E_0. Therefore, the smaller the value of E_T, the closer E_T is to E_0 and the better is the approximation Φ. In practice, we examine more than one trial function Φ and choose as the best the one which gives the lowest trial energy E_T.

Next, let us consider an example. We will apply the variational method to obtain upper bounds to the ground state energy of a one-dimensional particle-in-a-box of length a. We will consider two trial functions

$$\Phi_1(x) = x(a-x) \text{ and } \Phi_2(x) = x^2(a-x)^2. \tag{11.59}$$

Both trial functions are well-behaved and share the boundary conditions $\Psi(0) = \Psi(a) = 0$ of the exact (unnormalized) particle-in-a-box ground state wave function $\Psi(x) = \sin\left(\frac{\pi x}{a}\right)$. Thus, both are acceptable trial functions. Also both are plausible trial functions since like $\Psi(x)$ each has a single maximum at $x = \frac{a}{2}$ and each is symmetric about that maximum.

Next, let us find the trial energies for $\Phi_1(x)$ and $\Phi_2(x)$. The trial energy for $\Phi_1(x)$ is

$$E_{T_1} = \frac{\int_0^a \Phi_1(x) \hat{H} \Phi_1(x) \, dx}{\int_0^a \Phi_1(x) \Phi_1(x) \, dx}. \tag{11.60}$$

Since the one-dimensional particle-in-a-box Hamiltonian operator is $\hat{H} = -\frac{\hbar^2}{2m}\frac{d^2}{dx^2}$, $\int_0^a \Phi_1(x)\hat{H}\Phi_1(x)\,dx = \int_0^a x(a-x)\left[-\frac{\hbar^2}{2m}\frac{d^2}{dx^2}x(a-x)\right]dx = \frac{\hbar^2}{m}\int_0^a x(a-x)dx$ and thus,

$$\int_0^a \Phi_1(x)\hat{H}\Phi_1(x)\,dx = \frac{\hbar^2}{6m}a^3. \tag{11.61}$$

Also $\int_0^a \Phi_1(x)\Phi_1(x)\,dx = \int_0^a x^2(a-x)^2\,dx$, which gives

$$\int_0^a \Phi_1(x)\Phi_1(x)\,dx = \frac{1}{30}a^5. \tag{11.62}$$

Thus, $E_{T_1} = \frac{\hbar^2}{6m} a^3 \Big/ \frac{1}{30} a^5$ or

$$E_{T_1} = \frac{5\hbar^2}{ma^2} = \frac{10}{\pi^2} \frac{h^2}{8ma^2} = 1.013 \frac{h^2}{8ma^2}. \tag{11.63}$$

E_{T_1} is only 1.3% above the exact ground state energy $\frac{h^2}{8ma^2}$, so $\Phi_1(x)$ is a good trial function. Next, let us consider $\Phi_2(x)$. Evaluation of its trial energy $E_{T_2} = \int_0^a \Phi_2(x)\hat{H}\Phi_2(x)dx \Big/ \int_0^a \Phi_2(x)\Phi_2(x)dx$ gives (Problem 11.13)

$$E_{T_2} = \frac{12}{\pi^2} \frac{h^2}{8ma^2} = 1.216 \frac{h^2}{8ma^2}. \tag{11.64}$$

E_{T_2} is 21.6% above the exact energy, showing that $\Phi_1(x)$ is a substantially better trial function than $\Phi_2(x)$.

$\Phi_1(x)$, however, is not necessarily the best trial function since other functions not yet guessed may have lower trial energies. So how does one find optimal trial functions? There is a systematic procedure that partially solves this problem by permitting us to choose the best of an infinity of trial functions. Namely, one selects a function $\Phi(\lambda_1,\ldots,\lambda_n)$ which depends on a set *of variational parameters* λ and hence represents a class of trial functions, one trial function for each choice of the parameters λ. (For example, $\Phi_1[x]$ and $\Phi_2[x]$ are members of the class of trial functions $\Phi[x;\lambda] = x^\lambda [a-x]^\lambda$.) One next forms the trial energy

$$E_T(\lambda_1,\ldots,\lambda_n) = \frac{\int \Phi^*(\lambda_1,\ldots,\lambda_n)\hat{H}\Phi(\lambda_1,\ldots,\lambda_n)d\tau}{\int \Phi^*(\lambda_1,\ldots,\lambda_n)\Phi(\lambda_1,\ldots,\lambda_n)d\tau} \tag{11.65}$$

and then determines the λ values $\lambda_{1\,\min},\ldots,\lambda_{n\,\min}$ which minimize the trial energy. The best trial function of the type $\Phi(\lambda_1,\ldots,\lambda_n)$ is $\Phi(\lambda_{1\,\min},\ldots,\lambda_{n\,\min})$ since this function gives the minimum and hence the lowest value of $E_T(\lambda_1,\ldots,\lambda_n)$. This minimum value is $E_T(\lambda_{1\,\min},\ldots,\lambda_{n\,\min})$ which is the best estimate of the energy of the system obtainable from trial functions of the form $\Phi(\lambda_1,\ldots\lambda_n)$.

We next turn to some simple applications of these ideas. Much more advanced and important applications will be described in Chapters 12 and 13.

We first make a variational calculation of the ground state energy of a harmonic oscillator with Hamiltonian operator

$$\hat{H} = -\frac{\hbar^2}{2m}\frac{d^2}{dy^2} + \frac{1}{2}m\omega^2 y^2. \tag{11.66}$$

We take as our trial function

$$\Phi(y;\lambda) = \begin{cases} \lambda^2 - y^2 & \text{for } -\lambda \le y \le \lambda \\ 0 & \text{otherwise} \end{cases} \tag{11.67}$$

where λ is a variational parameter analogous to the parameters λ in Equation (11.65). Note that $\Phi(y;\lambda)$ is a parabola which like the true ground state harmonic oscillator wave function has its maximum value at $y = 0$. The trial energy $E_T(\lambda)$ is given by

$$E_T(\lambda)=\frac{\int_{-\infty}^{\infty}\Phi(y;\lambda)\hat{H}\Phi(y;\lambda)dy}{\int_{-\infty}^{\infty}\Phi(y;\lambda)\Phi(y;\lambda)dy}. \tag{11.68}$$

We will find the value λ_{min} of λ which minimizes $E_T(\lambda)$. This will give us $\Phi(y;\lambda_{min})$, the best trial function of the type $\Phi(y;\lambda)$, and also $E_T(\lambda_{min})$, the variational estimate of the ground state energy of the oscillator.

First, we must determine $E_T(\lambda)$. From Equations (11.66)–(11.68), we find (Problem 11.14)

$$\int_{-\infty}^{\infty}\Phi(y;\lambda)\hat{H}\Phi(y;\lambda)dy=\frac{4\hbar^2\lambda^3}{3m}+\frac{8m\omega^2\lambda^7}{105} \tag{11.69}$$

and

$$\int_{-\infty}^{\infty}\Phi(y;\lambda)\Phi(y;\lambda)dy=\frac{16}{15}\lambda^5. \tag{11.70}$$

Thus, $E_T(\lambda)=\left(\frac{4\hbar^2\lambda^3}{3m}+\frac{8m\omega^2\lambda^7}{105}\right)\Big/\frac{16}{15}\lambda^5$ or

$$E_T(\lambda)=\frac{5}{4}\frac{\hbar^2}{m\lambda^2}+\frac{1}{14}m\omega^2\lambda^2. \tag{11.71}$$

The condition for λ_{min} is

$$\left[\frac{dE_T(\lambda)}{d\lambda}\right]_{\lambda=\lambda_{min}}=0. \tag{11.72}$$

Applying this condition to Equation (11.71) gives $-\frac{5}{2}\frac{\hbar^2}{m\lambda_{min}^3}+\frac{1}{7}m\omega^2\lambda_{min}=0$, which yields λ_{min} as

$$\lambda_{min}=\left(\frac{35}{2}\right)^{1/4}\left(\frac{\hbar}{m\omega}\right)^{1/2}. \tag{11.73}$$

The variational estimate of the energy $E_T(\lambda_{min})$ is found by setting $\lambda=\lambda_{min}$ in Equation (11.71) for $E_T(\lambda)$. This gives $E_T(\lambda_{min})=\left[\left(\frac{5}{2}\right)\left(\frac{2}{35}\right)^{1/2}+\frac{1}{7}\left(\frac{35}{2}\right)^{1/2}\right]\frac{1}{2}\hbar\omega$ or

$$E_T(\lambda_{min})=(1.195)\frac{1}{2}\hbar\omega \tag{11.74}$$

which is 19.5% higher than the exact ground state energy $\frac{1}{2}\hbar\omega$. This is not too bad given that our trial function is a parabola and the exact ground state harmonic oscillator wave function is a Gaussian.

We next apply the variational method to the hydrogen atom. Since the ground state wave function of hydrogen is the spherically symmetric 1s orbital any reasonable trial function must also be spherically symmetric. However, when an orbital angular momentum operator operates on a spherically symmetric function, it gives zero. Thus, for the purpose of our variational calculation, the full hydrogen atom Hamiltonian operator of Equation (10.128) may be replaced by the simpler Hamiltonian operator

$$\hat{H} = -\frac{\hbar^2}{2\mu r^2}\frac{d}{dr}\left(r^2\frac{d}{dr}\right) - \frac{e^2}{4\pi\varepsilon_0 r}. \tag{11.75}$$

We will choose as our trial function the spherically symmetric Gaussian function

$$\Phi(r;\beta) = \exp(-\beta r^2) \tag{11.76}$$

where β is a variational parameter which will be chosen to minimize the trial energy

$$E_T(\beta) = \frac{\int \Phi(r;\beta)\hat{H}\Phi(r;\beta)d\mathbf{r}}{\int \Phi(r;\beta)\Phi(r;\beta)d\mathbf{r}}. \tag{11.77}$$

We will use spherical polar coordinates (r,θ,ϕ) to evaluate the integrals needed to determine $E_T(\beta)$. Since $\Phi(r;\beta)$ is independent of θ and ϕ, we may integrate over these angles to obtain $E_T(\beta)$ as

$$E_T(\beta) = \frac{\int_0^\infty r^2\Phi(r;\beta)\hat{H}\Phi(r;\beta)dr}{\int_0^\infty r^2\Phi(r;\beta)\Phi(r;\beta)dr}. \tag{11.78}$$

Evaluation of the integrals in Equation (11.78) using Equation (11.76) gives (Problem 11.15)

$$\int_0^\infty r^2\Phi(r;\beta)\hat{H}\Phi(r;\beta)dr = \frac{3\hbar^2}{16\mu\beta^{1/2}}\frac{\pi^{1/2}}{2^{1/2}} - \frac{e^2}{4\beta(4\pi\varepsilon_0)} \tag{11.79}$$

and

$$\int_0^\infty r^2\Phi(r;\beta)\Phi(r;\beta)dr = \frac{1}{8\beta^{3/2}}\frac{\pi^{1/2}}{2^{1/2}} \tag{11.80}$$

yielding $E_T(\beta)$ as

$$E_T(\beta) = \frac{3\hbar^2\beta}{2\mu} - \frac{2^{1/2}}{\pi^{1/2}}\frac{2e^2}{4\pi\varepsilon_0}\beta^{1/2}. \tag{11.81}$$

We wish to determine β_{min}, the value of β which minimizes $E_T(\beta)$. The condition for β_{min} is

$$\left[\frac{dE_T(\beta)}{d\beta}\right]_{\beta=\beta_{min}} = 0. \tag{11.82}$$

Applying this condition to Equation (11.81) gives $\frac{3\hbar^2}{2\mu} - \frac{2^{1/2}}{\pi^{1/2}}\frac{e^2}{4\pi\varepsilon_0}\beta_{min}^{-1/2} = 0$ or $\beta_{min} = \frac{8}{9\pi}\frac{\mu^2 e^4}{(4\pi\varepsilon_0)^2\hbar^4}$. However, the Bohr radius $a = \frac{4\pi\varepsilon_0}{\mu e^2}\hbar^2$. Thus, β_{min} may be written as

$$\beta_{min} = \frac{8}{9\pi}\frac{1}{a^2}. \tag{11.83}$$

The variational estimate of the energy $E_T(\beta_{min})$ may now be found by substituting Equation (11.83) for β_{min} into Equation (11.81) for $E_T(\beta)$. This yields $E_T(\beta_{min})$ as

$$E_T(\beta_{min}) = -\frac{8}{3\pi}\frac{e^2}{2(4\pi\varepsilon_0)a} = -0.849\frac{e^2}{2(4\pi\varepsilon_0)a}. \tag{11.84}$$

From Equation (10.19), the exact ground state energy of hydrogen is $-\frac{e^2}{2(4\pi\varepsilon_0)a}$. Thus, the variational energy is 15.1% too high.

Next, let us compare our optimal trial function $\Phi(r;\beta_{min}) = \exp\left(-\frac{8}{9\pi}\frac{r^2}{a^2}\right)$ with the exact 1s wave function

$$\Psi_{1s}(r) = \left(\frac{1}{\pi}\right)^{1/2}\left(\frac{1}{a}\right)^{3/2}\exp(-r/a). \tag{11.85}$$

To do this, we must normalize $\Phi(r;\beta_{min})$. Denoting the normalized form of $\Phi(r;\beta_{min})$ by $\Psi_{1s,VAR}(r)$, we find from Equations (11.76) and (11.80) both evaluated for $\beta = \beta_{min} = \frac{8}{a\pi}\frac{1}{a^2}$ that

$$\begin{aligned}\Psi_{1s,VAR}(r) &= \frac{8}{3^{3/2}\pi}\left(\frac{1}{\pi}\right)^{1/2}\left(\frac{1}{a}\right)^{3/2}\exp\left(-\frac{8}{9\pi}\frac{r^2}{a^2}\right)\\ &= 0.490\left(\frac{1}{\pi}\right)^{1/2}\left(\frac{1}{a}\right)^{3/2}\exp\left(-\frac{8}{9\pi}\frac{r^2}{a^2}\right).\end{aligned} \tag{11.86}$$

Notice that at $r = 0$, $\Psi_{1s}(r)$ is more than twice as large as $\Psi_{1s,VAR}(r)$. $\Psi_{1s}(r)$ and $\Psi_{1s,VAR}(r)$ are plotted in Figure 11.2. The two functions have different shapes and deviate significantly at small and large r.

Let us also compare the exact and variational radial distribution functions

$$4\pi r^2\Psi_{1s}^2(r) = \frac{4}{a^3}r^2\exp\left(-\frac{2r}{a}\right) \tag{11.87}$$

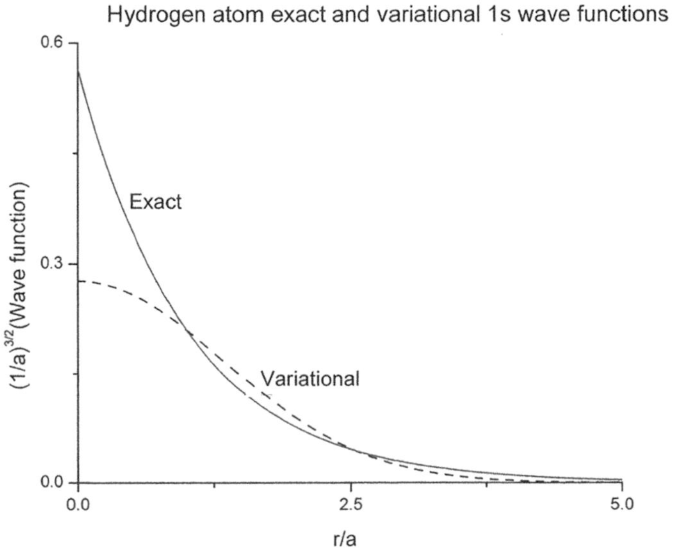

FIGURE 11.2 Exact 1s hydrogen atom wave function of Equation (11.85) compared with the Gaussian variational 1s wave function (dashed line) of Equation (11.86).

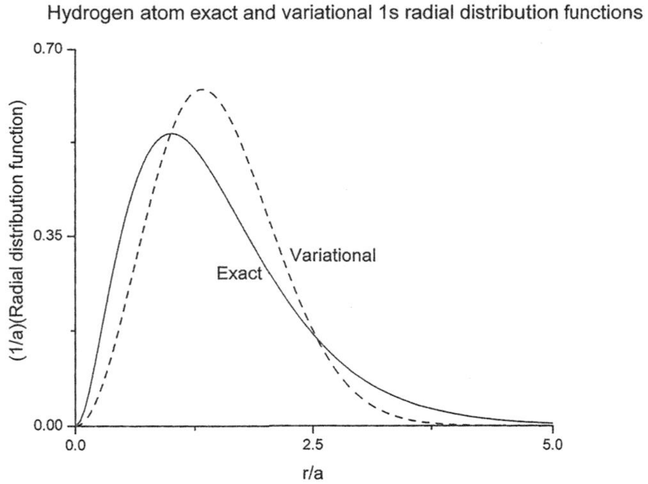

FIGURE 11.3 Exact 1s hydrogen atom radial distribution function of Equation (11.87) compared with the variational 1s radial distribution function (dashed line) of Equation (11.88).

and

$$4\pi r^2 \Psi^2_{1s,VAR}(r) = \frac{0.960}{a^3} r^2 \exp\left(-\frac{16}{9\pi}\frac{r^2}{a^2}\right). \tag{11.88}$$

Both are plotted in Figure 11.3. The plots show that the variational radial distribution function gives a poor description of the radial probability at both intermediate and large values of r. For example while from the true 1s radial distribution function, the most probable value of r is the Bohr radius a, the variational radial distribution function predicts that it is 1.33 a.

In summary, a trial function that gives a reasonable estimate of the energy may be a poor approximation to the true wave function and hence give poor approximations to other properties of the system (Problem 11.16).

We next turn to the linear variational method.

11.4 THE LINEAR VARIATIONAL METHOD

A very convenient type of trial function Φ is one that is a sum

$$\Phi(c_1,\ldots,c_n) = c_1\phi_1 + c_2\phi_2 + \ldots + c_n\phi_n \tag{11.89}$$

of n known linearly independent, but not necessarily orthonormal, *basis functions* ϕ_i. The coefficients c_i in the sum are variational parameters that are found by minimizing the trial energy

$$E_T(c_1,\ldots,c_n) = \frac{\int \Phi^*(c_1,\ldots,c_n)\hat{H}\Phi(c_1,\ldots,c_n)d\tau}{\int \Phi^*(c_1,\ldots,c_n)\Phi(c_1,\ldots,c_n)d\tau}. \tag{11.90}$$

For simplicity, we will assume the basis functions ϕ_i and the coefficients c_i are real. Then, $\Phi^* = \Phi$ and Equation (11.90) reduces to

$$E_T(c_1,\ldots,c_n) = \frac{\int \Phi(c_1,\ldots,c_n)\hat{H}\Phi(c_1,\ldots,c_n)d\tau}{\int \Phi(c_1,\ldots,c_n)\Phi(c_1,\ldots,c_n)d\tau}. \tag{11.91}$$

We will first take n = 2. Then, the trial function and trial energy become

$$\Phi(c_1,c_2)=c_1\phi_1+c_2\phi_2 \tag{11.92}$$

and

$$E_T(c_1,c_2)=\frac{\int\Phi(c_1,c_2)\hat{H}\Phi(c_1,c_2)d\tau}{\int\Phi(c_1,c_2)\Phi(c_1,c_2)d\tau}. \tag{11.93}$$

Next, let us evaluate the trial energy $E_T(c_1,c_2)$. First, note that $\int\Phi(c_1,c_2)\hat{H}\Phi(c_1,c_2)d\tau=$ $\int(c_1\phi_1+c_2\phi_2)\hat{H}(c_1\phi_1+c_2\phi_2)\,d\tau=c_1^2\int\phi_1\hat{H}\phi_1d\tau+c_1c_2\int\phi_1\hat{H}\phi_2d\tau+c_2c_1\int\phi_2\hat{H}\phi_1d\tau+c_2^2\int\phi_2\hat{H}\phi_2d\tau.$ Defining H_{ij} by

$$H_{ij}=\int\phi_i\hat{H}\phi_j\,d\tau \tag{11.94}$$

we have that $\int\Phi(c_1,c_2)\hat{H}\Phi(c_1,c_2)d\tau=c_1^2H_{11}+c_1c_2H_{12}+c_2c_1H_{21}+c_2^2H_{22}$. However, since $\hat{H}$ is Hermitian and the basis functions are real, $H_{ij}=H_{ji}$. So $\int\Phi\hat{H}\Phi d\tau$ simplifies to

$$\int\Phi(c_1,c_2)\hat{H}\Phi(c_1,c_2)d\tau=c_1^2H_{11}+2c_1c_2H_{12}+c_2^2H_{22}. \tag{11.95}$$

Similarly,

$$\int\Phi(c_1,c_2)\Phi(c_1,c_2)d\tau=c_1^2S_{11}+2c_1c_2S_{12}+c_2^2S_{22} \tag{11.96}$$

where we have defined S_{ij} by

$$S_{ij}=\int\phi_i\phi_jd\tau. \tag{11.97}$$

Thus, the trial energy of Equation (11.93) is given by

$$E_T(c_1c_2)=\frac{c_1^2H_{11}+2c_1c_2H_{12}+c_2^2H_{22}}{c_1^2S_{11}+2c_1c_2S_{12}+c_2^2S_{22}}. \tag{11.98}$$

We determine c_1 and c_2 from the minimization conditions

$$\frac{\partial E_T(c_1,c_2)}{\partial c_1}=0 \text{ and } \frac{\partial E_T(c_1,c_2)}{\partial c_2}=0. \tag{11.99}$$

To implement these conditions, it is convenient to write Equation (11.98) as

$$E_T(c_1,c_2)\left(c_1^2S_{11}+2c_1c_2S_{12}+c_2^2S_{22}\right)=c_1^2H_{11}+2c_1c_2H_{12}+c_2^2H_{22}. \tag{11.100}$$

Then, differentiating Equation (11.100) with respect to c_1 gives

$$\frac{\partial E_T(c_1,c_2)}{\partial c_1}\left(c_1^2S_{11}+2c_1c_2S_{12}+c_2^2S_{22}\right)+E_T(c_1,c_2)(2c_1S_{11}+2c_2S_{12})=2c_1H_{11}+2c_2H_{12}. \quad (11.101)$$

Since $\frac{\partial E_T}{\partial c_1}=0$, the previous equation simplifies to $E(c_1S_{11}+c_2S_{12})=c_1H_{11}+c_2H_{12}$, where we have set $E_T(c_1,c_2)$ equal to E. Rearranging then gives

$$(H_{11}-ES_{11})c_1+(H_{12}-ES_{12})c_2=0. \quad (11.102a)$$

Similarly, differentiating Equation (11.100) with respect to c_2 yields

$$(H_{12}-ES_{12})c_1+(H_{22}-ES_{22})c_2=0. \quad (11.102b)$$

Equations (11.102), which are called *secular equations*, form a pair of simultaneous homogeneous linear equations for the unknowns c_1 and c_2. From a theorem in linear algebra, these equations have a non-trivial solution (i.e., a solution other than $c_1=c_2=0$) only if the following determinantal equation is satisfied:

$$\begin{vmatrix} H_{11}-ES_{11} & H_{12}-ES_{12} \\ H_{12}-ES_{12} & H_{22}-ES_{22} \end{vmatrix}=0 \quad (11.103)$$

Equation (11.103) is referred to as a *secular determinantal equation*. When the determinant is expanded, the secular determinantal equation becomes a quadratic equation for E. The smaller root of that equation is the variational estimate of the ground state energy of the system.

Our analysis may be extended to trial functions which are sums of an arbitrary number n of basis functions. The main result is the secular determinantal equation

$$\begin{vmatrix} H_{11}-ES_{11} & H_{12}-ES_{12} & \cdots & H_{1n}-ES_{1n} \\ H_{12}-ES_{12} & H_{22}-ES_{22} & \cdots & H_{2n}-ES_{2n} \\ \vdots & \vdots & & \vdots \\ H_{1n}-ES_{1n} & H_{2n}-ES_{2n} & \cdots & H_{nn}-ES_{nn} \end{vmatrix}=0 \quad (11.104)$$

This secular determinantal equation must be solved numerically. Its smallest root is the variational estimate of the ground state energy.

We next discuss applications of the linear variational method. (Our first application closely follows the analysis of McQuarrie [2008].)

In our first application, we return to the problem of finding the ground state energy of a particle in a box of length a. Recall we earlier obtained reasonable estimates of this energy from the trial functions $x(a-x)$ and $x^2(a-x)^2$. This suggests that these functions might make good basis functions in a linear variational calculation of the ground state energy. Thus, we take as our trial function $\Phi(x;c_1,c_2)=c_1\phi_1(x)+c_2\phi_2(x)$, where

$$\phi_1(x)=x(a-x) \text{ and } \phi_2(x)=x^2(a-x)^2. \quad (11.105)$$

To set up the secular determinantal equation, we need the matrix elements $H_{ij}=\int_0^a \phi_i(x)\hat{H}\phi_j(x)dx=\int_0^a \phi_i(x)\left[-\frac{\hbar^2}{2m}\frac{d^2}{dx^2}\phi_j(x)\right]dx$ and $S_{ij}=\int_0^a \phi_i(x)\phi_j(x)dx$. A calculation of these matrix elements gives (Problem 11.17)

$$H_{11} = \frac{\hbar^2}{6m} a^3 \text{ and } S_{11} = \frac{1}{30} a^5 \tag{11.106a}$$

$$H_{12} = H_{21} = \frac{\hbar^2}{30m} a^5 \text{ and } S_{12} = S_{21} = \frac{a^7}{140} \tag{11.106b}$$

and

$$H_{22} = \frac{\hbar^2}{m}\frac{a^7}{105} \text{ and } S_{22} = \frac{1}{630} a^9. \tag{11.106c}$$

Using these results for the matrix elements, the explicit form of the secular determinantal equation (Equation 11.103) may be written down. However, if the secular determinantal equation is reexpressed in terms of the dimensionless energy ε defined by

$$E = \frac{\hbar^2}{ma^2}\varepsilon \tag{11.107}$$

its form simplifies significantly to

$$\begin{vmatrix} \left(\frac{1}{6} - \frac{1}{30}\varepsilon\right)a^3 & \left(\frac{1}{30} - \frac{1}{140}\varepsilon\right)a^5 \\ \left(\frac{1}{30} - \frac{1}{140}\varepsilon\right)a^5 & \left(\frac{1}{105} - \frac{1}{630}\varepsilon\right)a^7 \end{vmatrix} = 0 \tag{11.108}$$

Expanding the determinant yields the following quadratic equation for ε (Problem 11.18):

$$\varepsilon^2 - 56\varepsilon + 252 = 0. \tag{11.109}$$

Solving this equation, one finds two roots

$$\varepsilon = 51.065125 \text{ and } \varepsilon = 4.934875. \tag{11.110}$$

Selecting the smaller root and using Equation (11.107) give the energy as

$$E = \frac{\hbar^2}{ma^2}(4.934875) = \frac{h^2}{8ma^2}\frac{2(4.934875)}{\pi^2} \tag{11.111}$$

or

$$E = 1.0000148\frac{h^2}{8ma^2}. \tag{11.112}$$

The exact result for the ground state energy is $\frac{h^2}{8ma^2}$. So our variational energy is remarkably accurate. It is only 0.00148% too high!

We next consider a partible in a box of length a subject to a perturbing potential

$$V(x) = \frac{V_0}{a}\left(x - \frac{a}{2}\right) \tag{11.113}$$

where $V_0 > 0$. We will use the linear variational method to determine the effect of the perturbing potential on the ground state energy E of the particle. The effect of $V(x)$ on the ground state wave function is worked out in Problem 11.32.

We choose a trial function of the form $\Phi(x;c_1,c_2) = c_1\phi_1(x) + c_2\phi_2(x)$ where the basis functions

$$\phi_1(x) = \left(\frac{2}{a}\right)^{1/2} \sin\left(\frac{\pi x}{a}\right) \text{ and } \phi_2(x) = \left(\frac{2}{a}\right)^{1/2} \sin\left(\frac{2\pi x}{a}\right) \tag{11.114}$$

are the ground and first excited state particle-in-a-box wave functions. Since the basis functions are orthonormal, $S_{ij} = \int_0^a \phi_i(x)\phi_j(x)dx = \delta_{ij}$ and the secular determinantal equation (Equation 11.103) simplifies to

$$\begin{vmatrix} H_{11} - E & H_{12} \\ H_{12} & H_{22} - E \end{vmatrix} = 0 \tag{11.115}$$

Thus, we only need to calculate the matrix elements $H_{ij} = \int_0^a \phi_i(x)\hat{H}\phi_j(x)dx$ of the system's Hamiltonian operator

$$\hat{H} = -\frac{\hbar^2}{2m}\frac{d^2}{dx^2} + V(x). \tag{11.116}$$

We break up the matrix elements H_{ij} into kinetic and potential energy components. Thus, we write H_{ij} as $H_{ij} = T_{ij} + V_{ij}$, where $T_{ij} = \int_0^a \phi_i(x)\left[-\frac{\hbar^2}{2m}\frac{d^2\phi_j(x)}{dx^2}\right]dx$ and where $V_{ij} = \int_0^a \phi_i(x)V(x)\phi_j(x)dx = \int_0^a \phi_i(x)\left[\frac{V_0}{a}\left(x - \frac{a}{2}\right)\right]\phi_j(x)dx$. Since $-\frac{\hbar^2}{2m}\frac{d^2}{dx^2}$ is the particle-in-a-box Hamiltonian operator, it is easy to see that the matrix elements T_{ij} have the values

$$T_{11} = E_1 \quad T_{12} = 0 \quad \text{and} \quad T_{22} = 4E_1 \tag{11.117}$$

where $E_1 = \frac{h^2}{8ma^2}$ is the ground state energy of the unperturbed particle in a box. Also since from Equation (3.56) $\int_0^a \phi_i(x)x\phi_i(x)dx = \frac{a}{2}$, it follows that $V_{11} = V_{22} = 0$. Lastly, consider $V_{12} = \int_0^a \phi_1(x)\left[\frac{V_0}{a}\left(x - \frac{a}{2}\right)\right]\phi_2(x)dx = \frac{V_0}{a}\int_0^a \phi_1(x)x\phi_2(x)dx = \frac{2V_0}{a^2}\int_0^a x\sin\left(\frac{\pi x}{a}\right)\sin\left(\frac{2\pi x}{a}\right)dx$.

The integral may be evaluated using Equation (11.40) for $I_k = \int_0^a x\sin\left(\frac{\pi x}{a}\right)\sin\left(\frac{k\pi x}{a}\right)dx$ with k set equal to two to give $V_{12} = -\frac{16}{9}\frac{V_0}{\pi^2}$. Summarizing,

$$V_{11} = 0 \quad V_{12} = -\frac{16}{9}\frac{V_0}{\pi^2} \quad \text{and} \quad V_{22} = 0. \tag{11.118}$$

From Equations (11.117) and (11.118), the matrix elements $H_{ij} = T_{ij} + V_{ij}$ are

$$H_{11} = E_1 \quad H_{12} = -\frac{16}{9}\frac{V_0}{\pi^2} \quad \text{and} \quad H_{22} = 4E_1. \tag{11.119}$$

The secular determinantal equation (Equation 11.103) may now be written down explicitly as

$$\begin{vmatrix} E_1 - E & -\frac{16}{9}\frac{V_0}{\pi^2} \\ -\frac{16}{9}\frac{V_0}{\pi^2} & 4E_1 - E \end{vmatrix} = 0 \tag{11.120}$$

Expanding the determinant yields the following quadratic equation for ground state energy E:

$$E^2 - 5E_1E + 4E_1^2 - \left(\frac{16}{9}\frac{V_0}{\pi^2}\right)^2 = 0. \tag{11.121}$$

We next introduce the dimensionless variable λ defined by

$$\lambda = \frac{E}{E_1}. \tag{11.122}$$

The variable λ is the ratio of the perturbed and unperturbed particle-in-a-box ground state energies. Thus, it is a direct measure of the effect of the potential $V(x)$ on the ground state energy. In terms of λ, the quadratic equation simplifies to

$$\lambda^2 - 5\lambda + 4 - \mu^2 = 0 \tag{11.123}$$

where

$$\mu = \frac{16}{9\pi^2}\frac{V_0}{E_1} \tag{11.124}$$

is a dimensionless parameter that measures the strength of $V(x)$. Solving Equation (11.123) gives its smaller root as

$$\lambda = \frac{5}{2} - \left(\mu^2 + \frac{9}{4}\right)^{1/2}. \tag{11.125}$$

Notice that when $\mu^2 = 0, \lambda = 1,$ and E properly reduces to E_1. However, when $\mu^2 > 0, \lambda < 1,$ and hence $E < E_1$. Thus, the effect of $V(x)$ is to decrease the ground state energy. This is evident from Figure 11.4 which shows a rapid reduction of the ground state energy as $\frac{V_0}{E_1}$ increases. For example, for $\frac{V_0}{E_1} = 5 \quad E = 0.750E_1$, while for $\frac{V_0}{E_1} = 10 \quad E = 0.156E_1$.

FURTHER READINGS

Cohen-Tannoudji, Claude, Bernard Diu, and Franck Laloe. 1977. *Quantum mechanics*. Vol. 2. Translated by Susan Reid Hemley, Nicole Ostrowsky, and Dan Ostrowsky. New York: John Wiley & Sons Inc.

Levine, Ira N. 2013. *Quantum chemistry*. 7th ed. Upper Saddle River, NJ: Prentice-Hall.

McQuarrie, Donald A. 2008. *Quantum chemistry*. 2nd ed. Sausalito, CA: University Science Books.

Townsend, John F. 2012. *A modern approach to quantum mechanics*. 2nd ed. Mill Valley, CA: University Science Books.

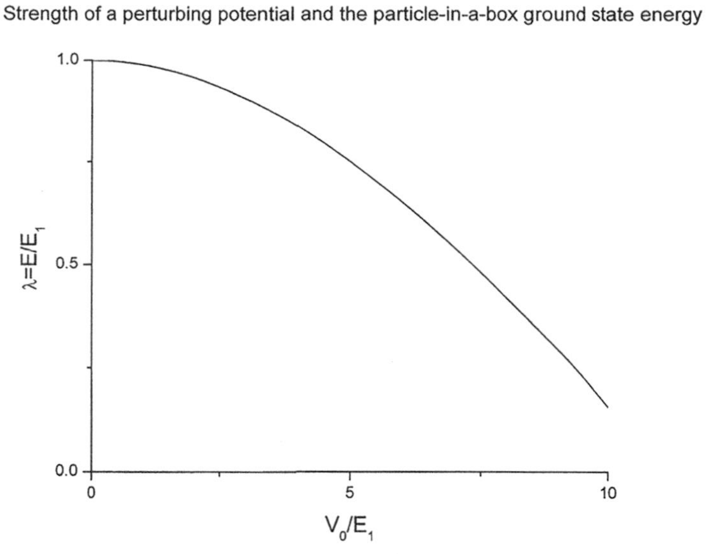

FIGURE 11.4 Effect of the strength of a perturbing potential $V(x)=\frac{V_0}{a}\left(x-\frac{a}{2}\right)$ on the ground state energy of particle in a box of length a. The ratio $\lambda=\frac{E}{E_1}$ of the perturbed and unperturbed particle-in-a-box energies is plotted against the strength parameter $\frac{V_0}{E_1}$.

PROBLEMS

11.1 Derive Equation (11.12) from Equation (11.11).

11.2 In this problem, you will verify Equations (11.17) and (11.18) for $\Psi_n^{(1)}$ and $E_n^{(2)}$. First, verify Equation (11.17). (a) Show by multiplying Equation (11.14) on the left by $\left[\Psi_k^{(0)}\right]^*$ for $k \neq n$ and integrating over $d\tau$ that $\int\left[\Psi_k^{(0)}\right]^*\hat{H}_0\Psi_n^{(1)}d\tau+\int\left[\Psi_k^{(0)}\right]^*\hat{V}\Psi_n^{(0)}d\tau=E_n^{(0)}\int\left[\Psi_k^{(0)}\right]^*\Psi_n^{(1)}d\tau$. (b) Then, show that because $\hat{H}_0$ is Hermitian and because $H_0\Psi_k^{(0)}=E_k^{(0)}\Psi_k^{(0)}$, the proceeding relation implies that $\int\left[\Psi_k^{(0)}\right]^*\Psi_n^{(1)}d\tau=\frac{\int\left[\Psi_k^{(0)}\right]^*\hat{V}\Psi_n^{(0)}d\tau}{E_n^{(0)}-E_k^{(0)}}$ for $k \neq n$. (c) Explain why $\Psi_n^{(1)}=\sum_{\text{all } k}\Psi_k^{(0)}\int\left[\Psi_k^{(0)}\right]^*\Psi_n^{(1)}d\tau$ where the sum is over all states k of the unperturbed system. (d) Show from your results in parts (b) and (c) that $\Psi_n^{(1)}=\Psi_n^{(0)}\int\left[\Psi_n^{(0)}\right]^*\Psi_n^{(1)}d\tau+\sum_{\substack{\text{all}\\ k\neq n}}\Psi_k^{(0)}\frac{\int\left[\Psi_k^{(0)}\right]^*\hat{V}\Psi_n^{(0)}d\tau}{E_n^{(0)}-E_k^{(0)}}$. However, it may be proven from Equation (11.9) by requiring to order λ $\Psi_n(x)$ be normalized that $\int\left[\Psi_n^{(0)}\right]^*\Psi_n^{(1)}d\tau=0$. Thus, your expression for $\Psi_n^{(1)}$ reduces to Equation (11.17), hence establishing that result. Next, prove Equation (11.18). (e) Noting that $\int\left[\Psi_n^{(0)}\right]^*\Psi_n^{(1)}d\tau=0$, show from Equation (11.15) that $\int\left[\Psi_n^{(0)}\right]^*\hat{H}_0\Psi_n^{(2)}d\tau+\int\left[\Psi_n^{(0)}\right]^*\hat{V}\Psi_n^{(1)}d\tau=E_n^{(0)}\int\left[\Psi_n^{(0)}\right]^*\Psi_n^{(2)}d\tau+E_n^{(2)}$. (f) Then, show that Equation (11.18) follows since $\hat{H}_0$ is Hermitian and since $\hat{H}_0\Psi_n^{(0)}=E_n^{(0)}\Psi_n^{(0)}$.

11.3 In Section 11.2, we gave a first-order perturbation theory treatment of the ground state anharmonic correction to the harmonic oscillator energy. (See Equation [11.29].) Here, you will give a parallel treatment for the first excited state. Show for this state that the energy to first order is $E_1 = \frac{3}{2}\hbar\omega + \frac{5\gamma_4}{32}\left(\frac{\hbar}{\mu\omega}\right)^2$. You will need the first excited state harmonic oscillator wave function $\Psi_1^{(0)}(y) = \left(\frac{\mu\omega}{\pi\hbar}\right)^{1/4}\left(\frac{2\mu\omega}{\hbar}\right)^{1/2} y\exp\left(-\frac{1}{2}\frac{\mu\omega}{\hbar}y^2\right)$ and the Gaussian integral of Equation (7.40).

11.4 In this problem, you will extend the first-order treatment of anharmonic corrections given for the ground and first excited states in Section 11.2 and Problem 11.3 to arbitrary states. Denote the harmonic oscillator wave function for the vth state by $\Psi_v^{(0)}(y)$. Then, from the form of the anharmonic perturbation $\hat{V} = \frac{1}{6}\gamma_3 y^3 + \frac{1}{24}\gamma_4 y^4$, the first-order correction to the unperturbed energy $E_v^{(0)} = \left(v + \frac{1}{2}\right)\hbar\omega$ is given by $E_v^{(1)} = \frac{\gamma_3}{6}I_3 + \frac{\gamma_4}{24}I_4$, where $I_3 = \int_{-\infty}^{\infty} y^3\Psi_v^{(0)}(y)\Psi_v^{(0)}(y)dy$ and $I_4 = \int_{-\infty}^{\infty} y^4\Psi_v^{(0)}(y)\Psi_v^{(0)}(y)dy$. (a) Show that $I_3 = 0$ and thus that for all states, the first-order cubic anharmonic correction $\frac{\gamma_3}{6}I_3$ vanishes. Next, consider the first-order quartic anharmonic correction $\frac{\gamma_4}{24}I_4$ and hence the evaluation of I_4. To make this evaluation, introduce the variable $\xi = \alpha^{1/2}y$ where $\alpha = \frac{\mu\omega}{\hbar}$. In terms of ξ from Equations (7.12) and (7.14) with $m \to \mu$, $I_4 = N_v^2\alpha^{-5/2}\int_{-\infty}^{\infty}\xi^4\exp(-\xi^2)H_v^2(\xi)d\xi$.

Also in terms of ξ the orthonormality relation $\int_{-\infty}^{\infty}\Psi_{v'}^{(0)}(y)\Psi_v^{(0)}(y)dy = \delta_{vv'}$ becomes $N_{v'}N_v\alpha^{-1/2}\int_{-\infty}^{\infty}\exp(-\xi^2)H_{v'}(\xi)H_v(\xi)d\xi = \delta_{vv'}$. The orthonormality relation implies the normalization condition $N_v^2\alpha^{-1/2}\int_{-\infty}^{\infty}\exp(-\xi^2)H_v^2(\xi)d\xi = 1$. (b) To proceed further, square Equation (7.105) and use the resulting expression for $\xi^4H_v^2(\xi)$ and the orthonormality relation to show that I_4 may be written as $I_4 = N_v^2\alpha^{-2}\left[v^2(v-1)^2\alpha^{-1/2}\int_{-\infty}^{\infty}\exp(-\xi^2)H_{v-2}^2(\xi)d\xi + \left(v+\frac{1}{2}\right)^2\alpha^{-1/2}\int_{-\infty}^{\infty}\exp(-\xi^2)H_v^2(\xi)d\xi + \frac{1}{16}^2\alpha^{-1/2}\int_{-\infty}^{\infty}\exp(-\xi^2)H_{v+2}^2(\xi)d\xi\right]$. (c) Simplify the expression for I_4 in part (b) using the normalization condition to show that $I_4 = \alpha^{-2}\left[v^2(v-1)^2\frac{N_v^2}{N_{v-2}^2} + \left(v+\frac{1}{2}\right)^2 + \frac{1}{16}\frac{N_v^2}{N_{v+2}^2}\right]$. (d) Use Equation (7.7) for N_v to verify that $I_4 = \left(\frac{\hbar}{\mu\omega}\right)^2\frac{3}{2}\left(v^2+v+\frac{1}{2}\right)$. Thus, show that $E_v^{(1)} = \frac{\gamma_4}{16}\left(v^2+v+\frac{1}{2}\right)\left(\frac{\hbar}{\mu\omega}\right)^2$ and that the energy to first order is therefore $E_v = \left(v+\frac{1}{2}\right)\hbar\omega + \frac{\gamma_4}{16}\left(v^2+v+\frac{1}{2}\right)\left(\frac{\hbar}{\mu\omega}\right)^2$. (e) Verify that for $v = 0$ and $v = 1$, your expression for E_v reduces to our earlier results.

11.5 Consider a harmonic oscillator with Hamiltonian $\hat{H}_0 = -\frac{\hbar^2}{2m}\frac{d^2}{dy^2} + \frac{1}{2}m\omega^2 y^2$. Denote the eigenfunctions and eigenvalues of the oscillator by $E_v^{(0)} = \left(v + \frac{1}{2}\right)\hbar\omega$ and $\Psi_v^{(0)}(y)$. Assume the oscillator is subject to a perturbation $\hat{V} = \frac{1}{2}m\omega_1^2 y^2$ where $\omega_1 \ll \omega$. This problem can of course be solved exactly, but it is instructive to treat it by perturbation theory and then compare the perturbation theory and exact results. To second order, the energy of the vth state is $E_v = E_v^{(0)} + E_v^{(1)} + E_v^{(2)}$ where $E_v^{(0)} = \left(v + \frac{1}{2}\right)\hbar\omega$, $E_v^{(1)} = \frac{1}{2}m\omega_1^2 \int_{-\infty}^{\infty} \Psi_v^{(0)}(y)y^2\Psi_v^{(0)}(y)dy$, and $E_v^{(2)} = \frac{1}{4}m^2\omega_1^4 \sum_{\substack{\text{all} \\ v' \neq v}} \frac{I_{v'v}^2}{E_v^{(0)} - E_{v'}^{(0)}}$ with $I_{v'v} = \int_{-\infty}^{\infty} \Psi_{v'}^{(0)} y^2 \Psi_v^{(0)}(y)dy$. To evaluate $E_v^{(1)}$, note that the integral $\int_{-\infty}^{\infty} \Psi_v^{(0)} y^2 \Psi_v^{(0)}(y)dy$ is the expected value of y^2 for the vth harmonic oscillator state. This expected value is given in Equation (7.68) as $\left(v + \frac{1}{2}\right)\frac{\hbar}{m\omega}$. Thus, $E_v^{(1)} = \left(v + \frac{1}{2}\right)\hbar\omega\left(\frac{1}{2}\frac{\omega_1^2}{\omega^2}\right)$. Next, we turn to $E_v^{(2)}$. Evaluation of $E_v^{(2)}$ requires the determination of $I_{v'v}$. To make this determination, introduce the variable $\xi = \alpha^{1/2}y$ where $\alpha = \frac{m\omega}{\hbar}$. The wave functions and orthonormality relation when expressed in terms of ξ become $\Psi_v^{(0)}(y) = N_v \exp\left(-\frac{1}{2}\xi^2\right)H_v(\xi)$ and $N_{v'}N_v\alpha^{-1/2}\int_{-\infty}^{\infty} \exp(-\xi^2)H_{v'}(\xi)H_v(d\xi) = \delta_{vv}$ Also when expressed in terms of ξ $I_{v'v} = \frac{\hbar}{m\omega}N_{v'}N_v\alpha^{-1/2}\int_{-\infty}^{\infty} \exp(-\xi^2)H_{v'}(\xi)\xi^2 H_v(\xi)d\xi$. (a) Show using Equation (7.106) that $I_{v'v}$ may be reexpressed as $I_{v'v} = \frac{\hbar}{m\omega}N_{v'}N_v\left[v(v-1)\alpha^{-1/2}\int_{-\infty}^{\infty} \exp(-\xi^2)H_{v'}(\xi)H_{v-2}(\xi)d\xi \right.$ $\left. + \left(v + \frac{1}{2}\right)\alpha^{-1/2}\int_{-\infty}^{\infty} \exp(-\xi^2)H_{v'}(\xi)H_v(\xi)d\xi + \frac{1}{4}\alpha^{-1/2}\int_{-\infty}^{\infty} \exp(-\xi^2)H_{v'}(\xi)H_{v+2}(\xi)d\xi.\right]$ (b) Use the orthonormality relation to simplify the proceeding expression to $I_{v'v} = \frac{\hbar}{m\omega}\left[v(v-1)\frac{N_v}{N_{v-2}}\delta_{v'v-2} + \left(v + \frac{1}{2}\right)\delta_{v'v} + \frac{1}{4}\frac{N_v}{N_{v+2}}\delta_{v'v+2}\right]$. (c) Then, show that $I_{v'v}^2 = \left(\frac{\hbar}{m\omega}\right)^2\left[v^2(v-1)^2\frac{N_v^2}{N_{v-2}^2}\delta_{v'v-2} + \left(v + \frac{1}{2}\right)^2\delta_{v'v} + \frac{1}{16}\frac{N_v^2}{N_{v+2}^2}\delta_{v'v+2}\right]$. (d) Use Equation (7.8) for N_v to rewrite $I_{v'v}^2$ as $I_{v'v}^2 = \left(\frac{\hbar}{m\omega}\right)^2\left[\frac{1}{4}v(v-1)\delta_{v'v-2} + \left(v + \frac{1}{2}\right)^2\delta_{v'v} + \frac{1}{4}(v+2)(v+1)\delta_{v'v+2}\right]$. (e) Using this form for $I_{v'v}^2$, show that $E_v^{(2)} = -\frac{1}{8}\frac{\omega_1^4}{\omega^4}\left(v + \frac{1}{2}\right)\hbar\omega$. (f) Hence, show that the energy to second order is $E_v = \left(1 + \frac{1}{2}\frac{\omega_1^2}{\omega^2} - \frac{1}{8}\frac{\omega_1^4}{\omega^4}\right)\left(v + \frac{1}{2}\right)\hbar\omega$. (g) Show that the exact energy is $E_v = \left(v + \frac{1}{2}\right)\hbar\Omega$ where $\Omega = \left(\omega^2 + \omega_1^2\right)^{1/2}$. (h) Expand

Ω as a power series in $\frac{\omega_1^2}{\omega^2}$ and verify that to order $\frac{\omega_1^4}{\omega^4}$, the exact and perturbation theory results for E_v are identical.

11.6 In this problem, you will determine to first order the ground state wave function $\Psi_1(x)$ of a particle moving in the potential of Equation (11.30). By specializing Equation (11.26), show that the first four terms in the expansion of $\Psi_1(x)$ are $\Psi_1(x) = \Psi_1^{(0)}(x) + w\left[0.0600\Psi_2^{(0)}(x) + 0.000961\Psi_4^{(0)}(x) + 0.000113\Psi_6^{(0)}(x)\right]$ where $w = \frac{8ma^2}{h^2}W$ and where $\Psi_1^{(0)}(x)$, and so on are the particle-in-a-box eigenfunctions of Equation (11.31). (Hint: Use Equation [11.41].)

11.7 Here you will determine to the second order the energy E_2 of the first excited state of a particle moving in the potential of Equation (11.30). The method you will use is a natural extension of the one employed for the ground state in Section 11.2. (a) Show that the first-order energy $E_2^{(1)} = \frac{W}{2}$. (b) Show that the second-order energy $E_2^{(1)} = \frac{256W^2}{243\pi^4}\frac{8ma^a}{h^2} - \frac{256W^2}{\pi^4}\frac{8ma^2}{h^2}\sum_{\substack{k=3\\ k\ \text{odd}}}^{\infty}\frac{k^2}{\left(k^2-4\right)^5}$. (c) Perform the sum in part (b) numerically to obtain $E_2^{(2)} = 0.003229\frac{8ma^2}{h^2}W^2$. (d) Thus, show that the energy to second order is $\varepsilon_2 = 4 + \frac{w}{2} + 0.003229w^2$ where $\varepsilon_2 = \frac{8ma^2}{h^2}E_2$ and where $w = \frac{8ma^2}{h^2}W$.

11.8 In this problem, you will determine by perturbation theory the ground state energy E of a particle in a box of length a subject to the perturbation $V(x) = \frac{V_0}{a}\left(x - \frac{a}{2}\right)$ of Equation (11.113). Then, you will compare your perturbation theory result with the variational result of Equation (11.125). (a) Taking the particle-in-a-box as the unperturbed system, show to second order that $E = \frac{h^2}{8ma^2} - 0.01083\frac{8ma^2}{h^2}V_0^2$ or $\lambda = \frac{E}{E_1} = 1 - 0.01083\left(\frac{V_0}{E_1}\right)^2$ where $E_1 = \frac{h^2}{8ma^2}$. You can do this with minimal effort if you note Equations (11.33) and (11.43). To compare with the variational result, (b) show from Equation (11.125) that for $\mu^2 << 1$, $\lambda = 1 - \frac{1}{3}\mu^2$ and thus show from Equation (11.124) that to order $\left(\frac{V_0}{E_1}\right)^2$, $\lambda = 1 - \frac{256}{243\pi^4}\left(\frac{V_0}{E_1}\right)^2 = 1 - 0.01082\left(\frac{V_0}{E_1}\right)^2$, which is in excellent agreement with the perturbation theory result.

11.9 Show that the exact eigenvalues E_υ of the Hamiltonian operator $\hat{H}$ of Equation (11.46) are the same as the perturbation theory eigenvalues of Equation (11.55). (Hint: Rewrite the term $\frac{1}{2}m\omega^2y^2 - \gamma y$ in $\hat{H}$ by completing the square.)

11.10 So far we have determined the second-order energy $E_n^{(2)}$ by performing the sum in Equation (11.21). But there is a second method for evaluating $E_n^{(2)}$. Namely, we can solve the differential equation (Equation 11.14) for the first-order wave function $\Psi_n^{(1)}$ and then perform the integral in Equation (11.18) to obtain $E_n^{(2)}$. In this problem and in Problem 11.11, you will use this second method. Here, you will recalculate the ground state second-order energy $E_0^{(2)}$ of the linearly perturbed harmonic oscillator with Hamiltonian $\hat{H}$ given in Equation (11.46). (a) To start, show by specializing Equation (11.14) that the ground state first-order wave function $\Psi_0^{(1)}(y)$ of the oscillator satisfies the differential equation (recall the first-order energy $E_0^{(1)}=0$) $\left[-\frac{\hbar^2}{2m}\frac{d^2}{dy^2}+\frac{1}{2}m\omega^2y^2\right]\Psi_0^{(1)}(y)-\gamma y\Psi_0^{(0)}(y)=\frac{1}{2}\hbar\omega\Psi_0^{(1)}(y)$ where $\Psi_0^{(0)}(y)=\left(\frac{m\omega}{\pi\hbar}\right)^{1/4}\exp\left(-\frac{1}{2}\frac{m\omega}{\hbar}y^2\right)$ is the ground state harmonic oscillator wave function. (b) Assume a solution of the differential equation of the form $\Psi_0^{(1)}(y)=\Psi_0^{(0)}(y)\phi(y)$ and then show that $\phi(y)$ satisfies the equation $-\frac{\hbar^2}{2m}\phi''(y)+h\omega y\phi'(y)-\gamma y=0$. (c) Show that the solution of this equation is $\phi(y)=\frac{\gamma}{\hbar\omega}y$ and therefore that $\Psi_0^{(1)}(y)=\frac{\gamma}{\hbar\omega}y\Psi_0^{(0)}(y)$. (d) By specializing Equation (11.18), verify that $E_0^{(2)}=-\gamma\int_{-\infty}^{\infty}\Psi_0^{(0)}(y)y\Psi_0^{(1)}(y)dy$. (e). Thus, show that $E_0^{(2)}=-\frac{\gamma^2}{\hbar\omega}\left(\frac{m\omega}{\pi\hbar}\right)^{1/2}\int_{-\infty}^{\infty}y^2\exp\left(-\frac{m\omega}{\hbar}y^2\right)dy$. (f) Finally, use the Gaussian integral of Equation (7.44) to show in agreement with Equation (11.54) that $E_0^{(2)}=-\frac{\gamma^2}{2m\omega^2}$.

11.11 This is a very challenging problem. When an atom is placed in an electric field **E**, the field induces a dipole moment $\mu=\alpha E$, where the proportionality constant α is called the *polarizability*. The polarizability is a measure of the disortability of the charge density of the atom and varies from atom to atom.

Here, you make an exact calculation of the ground state polarizability of the hydrogen atom. This polarizability is proportional to the ground state second-order energy $E_1^{(2)}$ of the atom, so you will obtain it by determining $E_1^{(2)}$ using the differential equation method of Problem 11.10.

You will need the form of the perturbation $\hat{V}$ arising from the interaction of the atom with the electric field **E.** To derive this form, note that the hydrogen atom has an instantaneous dipole moment $\mu_i=-e\mathbf{r}$ and thus $\hat{V}=-\mu_i\cdot\mathbf{E}=e\mathbf{r}\cdot\mathbf{E}$. Assume that the electric field points in the Z direction. Then, $\hat{V}=eE_zz=eE_zr\cos\theta$.

Before attempting to evaluate $E_1^{(2)}$, you will first show that the ground state first-order energy $E_1^{(1)}$ vanishes. Given the form of $\hat{V}, E_1^{(1)}=eE_z\int\Psi_1^{(0)}(r)r\cos\theta\Psi_1^{(0)}(r)d\mathbf{r}$ where $\Psi_1^{(0)}(r)=\left(\frac{1}{\pi}\right)^{1/2}\left(\frac{1}{a_0}\right)^{3/2}\exp(-r/a_0)$ is the hydrogen atom 1s orbital. (a) By evaluating the integral $\int\Psi_1^{(0)}(r)r\cos\theta\Psi_1^{(0)}(r)d\mathbf{r}$ in spherical polar coordinates, show $E_1^{(1)}=0$.

Next, turn to the determination of the ground state first-order wave function $\Psi_1^{(1)}(\mathbf{r})$.

(b) By specializing Equation (11.14) and using the forms for the hydrogen atom energies and Hamiltonian operator given in Equations (10.19) and (10.128), show that $\Psi_1^{(1)}(\mathbf{r})$ satisfies the following partial differential equation: $\left[-\frac{\hbar^2}{2m_e}\frac{1}{r^2}\frac{\partial}{\partial r}\left(r^2\frac{\partial}{\partial r}\right)+\frac{\hat{L}^2}{2m_er^2}-\frac{e^2}{4\pi\varepsilon_0 r}\right]\Psi_1^{(1)}(\mathbf{r})+eE_zr\cos\theta\Psi_1^{(0)}(r)=-\frac{e^2}{2(4\pi\varepsilon_0)a_0}\Psi_1^{(1)}(\mathbf{r})$.

(c) Assuming that $\Psi_1^{(1)}(\mathbf{r})$ has the form $\Psi_1^{(1)}(\mathbf{r})=\Phi(r)\cos\theta$ and noting that the spherical harmonic $Y_{10}(\theta,\phi)=\left(\frac{3}{4\pi}\right)^{1/2}\cos\theta$, show that $\Phi(r)$ satisfies the following ordinary differential equation: $\left[-\frac{\hbar^2}{2m_e}\frac{1}{r^2}\frac{d}{dr}\left(r^2\frac{d}{dr}\right)+\frac{\hbar^2}{m_er^2}-\frac{e^2}{4\pi\varepsilon_0 r}\right]\Phi(r)+eE_zr\Psi_1^{(0)}(r)=-\frac{e^2}{2(4\pi\varepsilon_0)a_0}\Phi(r)$.

(d) Next, define $\phi(r)$ by $\Phi(r) = \Psi_1^{(0)}(r)\phi(r)$ and then show from the differential equation for $\Phi(r)$ and the form of $\Psi_1^{(0)}(r)$ that $\phi''(r) - 2\left(\frac{1}{a_0} - \frac{1}{r}\right)\phi'(r) - \frac{2\phi(r)}{r^2} - \frac{2m_e}{\hbar^2}eE_z r = 0$. (Hint: Use the formula for the Bohr radius $a_0 = \frac{(4\pi\varepsilon_0)\hbar^2}{m_e e^2}$ to simplify your equation for $\phi[r]$.) (e) Show that the solution of the equation for $\phi(r)$ has the form $\phi(r) = Ar^2 + Br$ and further show that $A = -\frac{m_e a_0 eE_z}{2\hbar^2}$ and $B = -\frac{m_e a_0^2 eE_z}{\hbar^2}$. (f) Thus, show that $\Psi_1^{(1)}(\mathbf{r}) = -\left(\frac{m_e eE_z}{\hbar^2}\right)\left(\frac{a_0 r^2}{2} + a_0^2 r\right)\Psi_1^{(0)}(r)\cos\theta$.

Next, determine the second-order energy $E_1^{(2)}$. (g) First, specialize Equation (11.18) to show that $E_1^{(2)} = eE_z \int \Psi_1^{(0)}(r) r\cos\theta \Psi_1^{(1)}(\mathbf{r})d\mathbf{r}$. (h) Then, show from the forms for $\Psi_1^{(0)}(r)$ and $\Psi_1^{(1)}(\mathbf{r})$ that $E_1^{(2)}$ may be reduced to the following spherical polar coordinate expression: $E_1^{(2)} = -\left(\frac{m_e e^2}{\hbar^2}\right)\frac{E_z^2}{\pi a_0^3}\left[\int_0^\infty r^3\left(\frac{a_0 r^2}{2} + a_0^2 r\right)\exp\left(-\frac{2r}{a_0}\right)dr \int_0^\pi \sin\theta\cos^2\theta d\theta \int_0^{2\pi} d\phi\right]$. (i) In this expression, eliminate the factor $\frac{m_e e^2}{\hbar^2}$ using the relation $\frac{m_e e^2}{\hbar^2} = \frac{4\pi\varepsilon_0}{a_0}$ and then perform the integrals to obtain $E_1^{(2)}$ as $E_1^{(2)} = -\frac{9}{4}(4\pi\varepsilon_0)a_0^3 E_z^2$. The polarizability α is given in terms of $E_1^{(2)}$ by $E_1^{(2)} = -\frac{1}{2}\alpha E_z^2$. Thus, for the ground state of hydrogen, $\alpha = 4.5(4\pi\varepsilon_0)a_0^3$.

11.12 In this problem, you will prove the variational theorem $E_T = \frac{\int \Phi^* \hat{H}\Phi d\tau}{\int \Phi^*\Phi d\tau} \geq E_0$ where Φ is a trial function and where $\hat{H}$ and E_0 are the Hamiltonian operator and ground state energy of the system. You will need to recall that $\hat{H}$ has a complete set of orthonormal eigenfunctions Ψ_n and associated eigenvalues E_n and that Φ may be expanded in terms of the eigenfunctions as $\Phi = \sum_n c_n\Psi_n$ where $c_n = \int \Psi_n^*\Phi d\tau$. (a) By substituting the expansion of Φ into the expression for E_T, show that $E_T - E_0$ may be written as $E_T - E_0 = \sum_n |c_n|^2 (E_n - E_0) \Big/ \sum_n |c_n|^2$. (b) Then, noting that $E_n \geq E_0$, prove that $E_T \geq E_0$, which is the variational theorem.

11.13 Verify Equation (11.64) for E_{T_2}.

11.14 Verify Equations (11.69) and (11.70).

11.15 Verify Equations (11.79) and (11.80). In addition to the Gaussian integrals of Section 7.9, you will need the integral $\int_0^\infty y\exp(-\alpha y^2)dy = \frac{1}{2\alpha}$.

11.16 Determine the values of $\langle r\rangle$ and $\langle r^2\rangle$ predicted by the exact and variational 1s wave functions of Equations (11.85) and (11.86). In addition to familiar integrals of Section 7.6, you will need the integral $\int_0^\infty y^3\exp(-\alpha y^2)dy = \frac{1}{2\alpha^2}$. Your results illustrate that variational wave functions which give reasonable energies can give very poor results for the other properties.

11.17 Verify the forms for the matrix elements given in Equation (11.106). (Take note of Equations [11.61] and [11.62] and Problem 11.13.)

11.18 Derive the quadratic equation (Equation 11.109) from the determinantal equation (11.108). (Hint: Multiply the determinant through by 1,260.)

11.19 For a harmonic oscillator with Hamiltonian operator $\hat{H} = -\frac{\hbar^2}{2m}\frac{d^2}{dy^2} + \frac{1}{2}m\omega^2 y^2$, show using the results of Section 7.9 that a trial function of the form $\Phi(y;\beta) = \exp(-\beta y^2)$ where β is a variational parameter gives the exact ground state energy $E_0 = \frac{1}{2}\hbar\omega$. Explain why this is so.

11.20 For the hydrogen atom with the Hamiltonian operator of Equation (11.75), show that a trial function of the form $\Phi(r;\alpha) = \exp(-\alpha r)$ where α is a variational parameter gives the exact (fixed nucleus) ground state energy $E_1 = -\frac{e^2}{2(4\pi\varepsilon_0)a_0}$. Explain why this is so.

11.21 For a harmonic oscillator with Hamiltonian operator $\hat{H} = -\frac{\hbar^2}{2m}\frac{d^2}{dy^2} + \frac{1}{2}m\omega^2 y^2$, find the variational estimate of the ground state energy given by the trial function

$$\Phi(y;\alpha) = \begin{cases} \cos\alpha y & \text{for } -\frac{\pi}{2\alpha} \le y \le \frac{\pi}{2\alpha} \\ 0 & \text{otherwise} \end{cases}$$

where α is a variational parameter, and compare with the exact ground state energy $E_0 = \frac{1}{2}\hbar\omega$. (Hint: When calculating the trial energy, take $\cos^2\alpha y = \frac{1+\cos 2\alpha y}{2}$ and perform the integral $\int_{-\frac{\pi}{2\alpha}}^{\frac{\pi}{2\alpha}} y^2 \cos 2\alpha y\, dy$ by two integrations by parts.)

11.22 Determine using Equation (11.75) for $\hat{H}$ the variational estimate of the ground state energy of hydrogen given by the trial function $\Phi(r;\alpha) = r\exp(-\alpha r)$ where α is a variational parameter, and compare with the exact result $E_1 = -\frac{e^2}{2(4\pi\varepsilon_0)a}$.

11.23 For a quartic oscillator with Hamiltonian operator $\hat{H} = -\frac{\hbar^2}{2m}\frac{d^2}{dy^2} + \frac{1}{2}\lambda y^4$, using the results of Section 7.9 calculate the variational estimate of the ground state energy predicted by the trial function $\Phi(y;\alpha) = \exp(-\alpha y^2)$ where α is a variational parameter. Compare with the exact ground state energy $E_0 = 1.060\lambda^{1/3}\left(\frac{\hbar^2}{2m}\right)^{2/3}$.

11.24 For a three-dimensional isotropic harmonic oscillator with Hamiltonian operator $\hat{H} = -\frac{\hbar^2}{2\mu r^2}\frac{d}{dr}\left(r^2\frac{d}{dr}\right) + \frac{\hat{L}^2}{2\mu r^2} + \frac{1}{2}\mu\omega^2 r^2$, determine the variational estimate of the ground state energy predicted by the trial function $\Phi(r;\alpha) = \exp(-\alpha r)$ where α is variational parameter, and compare with the exact result $E_0 = \frac{3}{2}\hbar\omega$.

11.25 Consider a cubically perturbed harmonic oscillator with Hamiltonian operator $\hat{H} = -\frac{\hbar^2}{2\mu}\frac{d^2}{dy^2} + \frac{1}{2}\mu\omega^2 y^2 + \frac{1}{6}\gamma_3 y^3$. In this problem, you will estimate the ground state energy E_0 of the oscillator by the linear variational method taking as the basis functions $\phi_0(y) = \left(\frac{\mu\omega}{\pi\hbar}\right)^{1/4}\exp\left(-\frac{1}{2}\frac{\mu\omega}{\hbar}y^2\right)$ and $\phi_1(y) = \left(\frac{\mu\omega}{\pi\hbar}\right)^{1/2}\left(\frac{2\mu\omega}{\hbar}\right)^{1/2} y\exp\left(-\frac{1}{2}\frac{\mu\omega}{\hbar}y^2\right)$, the ground and first excited state normalized harmonic oscillator eigenfunctions.

(a)≈By calculating the matrix elements H_{11}, H_{12}, and H_{22} using the results of Section 7.9 and noting that $S_{ij} = \delta_{ij}$, derive the secular determinantal equation

$$\begin{vmatrix} \frac{1}{2}\hbar\omega - E & 2^{1/2}\frac{\gamma_3}{8}\left(\frac{\hbar}{\mu\omega}\right)^{3/2} \\ 2^{1/2}\frac{\gamma_3}{8}\left(\frac{\hbar}{\mu\omega}\right)^{3/2} & \frac{3}{2}\hbar\omega - E \end{vmatrix} = 0$$

(b) From this secular determinantal equation, show that the variational estimate E of E_0 is given by $E = \hbar\omega - \frac{\hbar\omega}{2}\left[1+\frac{1}{8}\gamma_3^2\left(\frac{\hbar}{\mu\omega}\right)^3\left(\frac{1}{\hbar\omega}\right)^2\right]^{1/2}$ (c) For $\gamma_3^2\left(\frac{\hbar}{\mu\omega}\right)^3\left(\frac{1}{\hbar\omega}\right)^2 << 1$ (small anharmonicity), show that E simplifies to $E = \frac{1}{2}\hbar\omega - \frac{1}{32}\gamma_3^2\left(\frac{\hbar}{\mu\omega}\right)^3\frac{1}{\hbar\omega} = \frac{1}{2}\hbar\omega - 0.03125\gamma_3^2\left(\frac{\hbar}{\mu\omega}\right)^3\frac{1}{\hbar\omega}$. This result is identical with that of the second-order perturbation theory treatment of the anharmonicity which gives $E_0 = \frac{1}{2}\hbar\omega - 0.03819\gamma_3^2\left(\frac{\hbar}{\mu\omega}\right)^3\frac{1}{\hbar\omega}$.

11.26 Consider a quartically perturbed harmonic oscillator with Hamiltonian operator $\hat{H} = -\frac{\hbar^2}{2\mu}\frac{d^2}{dy^2} + \frac{1}{2}\mu\omega^2 y^2 + \frac{1}{24}\gamma_4 y^4$. In this problem, you will estimate by the linear variational method the ground state energy E_0 of the oscillator taking as the basis functions $\phi_0(y) = \left(\frac{\mu\omega}{\pi\hbar}\right)^{1/4}\exp\left(-\frac{1}{2}\frac{\mu\omega}{\hbar}y^2\right)$ and $\phi_2(y) = \left(\frac{\mu\omega}{\pi\hbar}\right)^{1/4}\left(\frac{1}{2}\right)^{1/2}\left[2\left(\frac{\mu\omega}{\hbar}\right)y^2 - 1\right]\exp\left(-\frac{1}{2}\frac{\mu\omega}{\hbar}y^2\right)$, the ground and second excited state normalized harmonic oscillator eigenfunctions. (a) By calculating the matrix elements H_{11}, H_{12}, and H_{22} using the results of Section 7.9 and noting that $S_{ij} = \delta_{ij}$, derive the secular equation

$$\begin{vmatrix} \frac{1}{2}\hbar\omega + \frac{\varepsilon}{32} - E & \left(\frac{1}{2}\right)^{1/2}\frac{\varepsilon}{8} \\ \left(\frac{1}{2}\right)^{1/2}\frac{\varepsilon}{8} & \frac{5}{2}\hbar\omega + \frac{13\varepsilon}{32} - E \end{vmatrix} = 0$$

where $\varepsilon = \gamma_4\left(\frac{\hbar}{\mu\omega}\right)^2$. (b) From this secular equation, show that the variational estimate E of E_0 is given by $E = \frac{3}{2}\hbar\omega + \frac{7}{32}\varepsilon - h\omega\left[1+\frac{3}{8}\frac{\varepsilon}{\hbar\omega}+\frac{11}{286}\left(\frac{\varepsilon}{\hbar\omega}\right)^2\right]^{1/2}$. (c) Then, assume $\varepsilon \ll \hbar\omega$ (small anharmonicity) and show to lowest order in $\frac{\varepsilon}{\hbar\omega}$ that $E = \frac{1}{2}\hbar\omega + \frac{1}{32}\varepsilon$ or $E = \frac{1}{2}\hbar\omega + \frac{\gamma_4}{32}\left(\frac{\hbar}{\mu\omega}\right)^2$. Notice this result agrees with the first-order perturbation theory result of Equation (11.29).

11.27 Modify the proof of the variational theorem of Problem 11.12 to show that if a trial function Φ is orthogonal to the exact ground state wave function, then the trial energy $E_T = \dfrac{\int \Phi^* \hat{H} \Phi d\tau}{\int \Phi^* \Phi d\tau}$ is an upper bound to the energy E_1 of the first excited state of the system.

11.28 Consider a particle of mass m in a box of length a. You will make a variational calculation of the first excited state energy of the particle using the trial function $\Phi(x) = x(a-x)\left(\frac{a}{2} - x\right)$.

(a) Show using a symmetry argument that $\Phi(x)$ is orthogonal to the exact ground state particle-in-a-box wave function. Thus, by the variational theorem of Problem 11.27, the trial energy $E_T = \dfrac{\int_0^a \Phi(x)\hat{H}\Phi(x)dx}{\int_0^a \Phi(x)\Phi(x)dx} = \dfrac{\int_0^a \Phi(x)\left[-\frac{\hbar^2}{2m}\frac{d^2}{dx^2}\Phi(x)\right]dx}{\int_0^a \Phi(x)\Phi(x)dx}$ is an upper bound to the first excited state particle-in-a-box energy $E_2 = \frac{4h^2}{8ma^2}$. (b) In fact, show that $E_T = \frac{21}{\pi^2}\left(\frac{4h^2}{8ma^2}\right) = 1.0639\left(\frac{4h^2}{8ma^2}\right)$, which is only 6.39% too high.

11.29 Consider a harmonic oscillator with Hamiltonian operator $\hat{H} = -\frac{\hbar^2}{2m}\frac{d^2}{dy^2} + \frac{1}{2}m\omega y^2$. You will make a variational calculation of the first excited state energy of the oscillator using the trial function $\Phi(y;\alpha) = \begin{cases} \sin\alpha y & \text{for } -\frac{\pi}{\alpha} \le y \le \frac{\pi}{\alpha} \\ 0 & \text{otherwise} \end{cases}$ where α is a variational parameter. $\Phi(y;\alpha)$ is orthogonal to the ground state harmonic oscillator wave function $\Psi_0(y)$ since $\Phi(y;\alpha)$ is odd in y while $\Psi_0(y)$ is even in y. Thus, by the variational theorem of Problem 11.27, the optimal trial energy $E_T(\alpha_{min})$ is an upper bound to the energy $E_1 = \frac{3}{2}\hbar\omega$ of the first excited state of the oscillator. Determine $E_T(\alpha_{min})$ and compare it with $E_1 = \frac{3}{2}\hbar\omega$. (Hint: When calculating $E_T[\alpha]$, take $\sin^2\alpha y = \frac{1-\cos 2\alpha y}{2}$ and perform the integral $\int_{-\frac{\pi}{\alpha}}^{\frac{\pi}{\alpha}} y^2 \cos 2\alpha y\, dy$ by two integrations by parts.)

11.30 In Problem 11.11, neglecting nuclear motion we made an exact calculation of the ground state hydrogen atom polarizability α. As part of that calculation, we found that to first order, the wave function $\Psi(\mathbf{r})$ of the atom in the presence of an electric field E_z takes the following form: $\Psi(\mathbf{r}) = \Psi_{1s}(r) - m_e e E_z\left(\frac{a_0 r^2}{2} + a_0^2 r\right)\Psi_{1s}(r)\cos\theta$ where a_0 is the Bohr radius and where $\Psi_{1s}(r) = \left(\frac{1}{\pi}\right)^{1/2}\left(\frac{1}{a_0}\right)^{3/2}\exp(-r/a_0)$ is the hydrogen atom 1s orbital.

Here, you will make a linear variational calculation of α choosing as basis functions $\phi_1(\mathbf{r}) = \Psi_{1s}(r)$ and $\phi_2(\mathbf{r}) = \Psi_{1s}(r)r\cos\theta$. Given the form of $\Psi(\mathbf{r})$, this choice is very reasonable. To determine α, you will need to set up and solve the secular determinantal equation

$$\begin{vmatrix} H_{11} - ES_{11} & H_{12} - ES_{12} \\ H_{12} - ES_{12} & H_{22} - ES_{22} \end{vmatrix} = 0$$

where $H_{ij} = \int \phi_i(\mathbf{r})\hat{H}\phi_j(\mathbf{r})d\mathbf{r}$ and $S_{ij} = \int \phi_i(\mathbf{r})\phi_j(\mathbf{r})d\mathbf{r}$ and where $\hat{H}$ is the Hamiltonian operator of the atom in the presence of the electric field, being given by (see Problem 11.11) $\hat{H} = \hat{H}_0 + eE_z r\cos\theta$ where $\hat{H}_0 = -\frac{\hbar^2}{2m_e}\frac{1}{r^2}\frac{\partial}{\partial r}\left(r^2\frac{\partial}{\partial r}\right) + \frac{\hat{L}^2}{2m_e r^2} - \frac{e^2}{4\pi\varepsilon_0 r}$ is the hydrogen atom Hamiltonian operator neglecting nucleus motion. (a) Noting that $\phi_1(\mathbf{r}) = \Psi_{1s}(r)$ is an eigenfunction of $\hat{H}_0$ and evaluating all needed integrals in spherical polar coordinates, show that $H_{11} = E_1$ $H_{12} = eE_z a_0^2$ $H_{22} = 0$ and $S_{11} = 1$ $S_{12} = 0$ $S_{22} = a_0^2$ where $E_1 = -\frac{e^2}{2(4\pi\varepsilon_0)a_0}$ is the ground state energy of the hydrogen atom. (Hint: To evaluate H_{22}, use the relation $\frac{e^2}{4\pi\varepsilon_0} = \frac{\hbar^2}{m_e a_0}$ to first show that $\hat{H}_0\phi_2(\mathbf{r}) = -\left[\frac{1}{\pi}\right]^{1/2}\left[\frac{1}{a_0}\right]^{3/2}\frac{\hbar^2}{m_e}\left[\frac{r}{2a_0^2} - \frac{1}{a_0}\right]\exp[-r/a_0]\cos\theta$.) (b) Using your results for H_{ij} and S_{ij}, set up and expand the secular determinant equation (Equation 11.103) to obtain a quadratic equation for E, the ground state energy of the atom in the presence of the field. Then, solve the quadratic equation for its smaller root (remembering $E_1 < 0$) to obtain E as $E = \frac{1}{2}E_1 + \frac{1}{2}E_1\left(1 + \frac{4e^2E_z^2a_0^2}{E_1^2}\right)^{1/2}$. (c) Show by expanding the square root that if $eE_z a_0 \ll E_1$, then E becomes $E = E_1 + \frac{e^2E_z^2a_0^2}{E_1} = E_1 - 2(4\pi\varepsilon_0)a_0^3E_z^2$. But the polarizability α is given in terms of E by $E = E_1 - \frac{1}{2}\alpha E_z^2$. Thus, the variational estimate of α is $\alpha = 4(4\pi\varepsilon_0)a_0^3$. This estimate is in reasonable agreement with the exact result of Problem 11.11 $\alpha = 4.5(4\pi\varepsilon_0)a_0^3$.

11.31 In Problem 11.30, you made a linear variational calculation of the polarizability α of the 1s state of the hydrogen atom. Here, using the notation and definitions of Problem 11.30, you will make a second linear variational estimate of α. In this problem, however you will choose basis functions which are different from those used in Problem 11.30. Specifically, you will take as basis functions $\phi_1(\mathbf{r}) = \Psi_{1s}(r) = \left(\frac{1}{\pi}\right)^{1/2}\left(\frac{1}{a_0}\right)^{3/2}\exp(-r/a_0)$ and $\phi_2(\mathbf{r}) = \Psi_{2p_z}(\mathbf{r}) = \frac{1}{4}\left(\frac{1}{2\pi}\right)^{3/2}\left(\frac{1}{a_0}\right)^{3/2}\left(\frac{r}{a_0}\right)\exp\left(-\frac{r}{2a_0}\right)\cos\theta$, the 1s and $2p_z$ orbitals of hydrogen. The basis functions are orthonormal, so $S_{ij} = \delta_{ij}$ and the secular determinantal equation simplifies to

$$\begin{vmatrix} H_{11} - E & H_{12} \\ H_{12} & H_{22} - E \end{vmatrix} = 0$$

where $H_{ij} = \int \phi_i(\mathbf{r})\hat{H}\phi_j(\mathbf{r})d\mathbf{r}$. (a) Noting that $\Psi_{1s}(r)$ and $\Psi_{2p_z}(\mathbf{r})$ are eigenfunctions of $\hat{H}_0$, show that the matrix elements H_{ij} are given by

$$H_{11} = E_1 \quad H_{12} = \frac{eE_z a_0}{2^{1/2}} \frac{2^8}{3^5} \text{ and } H_{22} = \frac{1}{4} E_1.$$

(b) Then, using the above results, set up the secular determinantal equation, expand it to obtain a quadratic equation for E, and solve that equation for its smaller root to determine E as $E = \frac{5}{8}E_1 + \frac{3}{8}E_1\left(1 + \frac{2^{21}}{3^{12}}\frac{e^2E_z^2a_0^2}{E_1^2}\right)^{1/2}$. (c) Show that $\frac{2^{21}}{3^{12}} = 3.946$ and therefore $E = \frac{5}{8}E_1 + \frac{3}{8}E_1\left(1 + 3.946\frac{e^2E_z^2a_0^2}{E_1^2}\right)^{1/2}$. (d) Assume that the magnitude of the applied electric field E is sufficiently small that its correction to the atom's field free energy is small enough to guarantee the inequality $eE_za_0 \ll E_1$. Then, expand the square root to obtain E as $E = E_1 + 0.740\frac{e^2E_z^2a_0^2}{E_1} = E_1 - 1.480(4\pi\varepsilon_0)a_0^3E_z^2$. The polarizability α is given in terms of E by $E = E_1 - \frac{1}{2}\alpha E_z^2$. Thus, the variational estimate of α is $\alpha = 2.960(4\pi\varepsilon_0)a_0^3$. (e) Why is the estimate of α made in Problem 11.30 closer to the exact value $\alpha = 4.5(4\pi\varepsilon_0)a_0^3$ than the estimate made in this problem?

11.32 In Section 11.4, we used the linear variational method to determine the effect of the perturbing potential $V(x) = \frac{V_0}{a}\left(x - \frac{a}{2}\right)$ on the ground state energy E of a particle in a box of length a. (See Figure 11.4.) Here, we will determine the effect of V(x) on the ground state wave function of the particle. Assuming two basis functions the linear variational wave function is of the form $\Phi(x;c_1,c_2) = c_1\phi_1(x) + c_2\phi_2(x)$ where we choose $\phi_1(x)$ and $\phi_2(x)$ as the gc$_1$ round and first excited state particle-in-a-box wave functions given in Equation (11.114). Thus, to determine $\Phi(x;c_1,c_2)$, we must find the optimal values of the coefficients c_1 and c_2. We proceed as follows. We first write Φ as $\Phi(x;c_1,c_2) = c_1\left[\phi_1(x) + \frac{c_2}{c_1}\phi_2(x)\right]$, then we determine $\frac{c_2}{c_1}$, and finally, we pin down c_1 by requiring that Φ be normalized. (a) Show from Equation (11.102a) specialized to the case that $S_{ij} = \delta_{ij}$ that $\frac{c_2}{c_1} = -\frac{H_{11} - E}{H_{12}}$. (b) From Equation (11.119), $H_{11} = E_1$ and $H_{12} = -\frac{16}{9}\frac{V_0}{\pi^2}$. Using these results and further using Equations (11.122), (11.124), and (11.125), show that $\frac{c_2}{c_1}$ may be expressed solely in terms of the dimensionless parameter $\mu = \frac{16}{9\pi^2}\frac{V_0}{E_1}$ as $\frac{c_2}{c_1} = \frac{-\frac{3}{2} + \frac{3}{2}\left(1 + \frac{4}{9}\mu^2\right)^{1/2}}{\mu}$. So far our result for the wave function is thus $\Phi(x;c_1,c_2) = c_1\left\{\phi_1(x) + \left[\frac{-\frac{3}{2} + \frac{3}{2}\left(1 + \frac{4}{9}\mu^2\right)^{1/2}}{\mu}\right]\phi_2(x)\right\}$. (c) Remembering that

$\int_0^a \phi_i(x)\phi_j(x)dx = \delta_{ij}$, show from the normalization condition $\int_0^a \Phi^2(x;c_1,c_2)dx = 1$

that c_1 is given by $c_1 = \frac{\mu}{\left[2\mu^2 - \frac{9}{2}\left(1+\frac{4}{9}\mu^2\right)^{1/2} + \frac{9}{2}\right]^{1/2}}$ and therefore that the normalized ground state wave function in the presence of the potential $V(x)$ is

$$\Phi(x;c_1,c_2) = \frac{\mu}{\left[2\mu^2 - \frac{9}{2}\left(1+\frac{4}{9}\mu^2\right)^{1/2} + \frac{9}{2}\right]^{1/2}}\left\{\phi_1(x) + \left[\frac{-\frac{3}{2}+\frac{3}{2}\left(1+\frac{4}{9}\mu^2\right)^{1/2}}{\mu}\right]\phi_2(x)\right\}.$$

We may now numerically compare $\Phi(x;c_1,c_2)$ with the unperturbed particle-in-a-box ground state wave function $\phi_1(x)$ for various values of the perturbation potential strength parameter $\frac{V_0}{E_1}$. (d) Show for $\frac{V_0}{E_1} = 1$ or $\mu = 0.180$ that $\Phi(x;c_1,c_2) = 0.998\phi_1(x) + 0.0560\phi_2(x)$. (e) Show for $\frac{V_0}{E_1} = 5$ or $\mu = 0.901$ that $\Phi(x;c_1,c_2) = 0.964\phi_1(x) + 0.267\phi_2(x)$. (f) Show for $\frac{V_0}{E_1} = 10$ or $\mu = 1.80$ that $\Phi(x;c_1,c_2) = 0.906\phi_1(x) + 0.424\phi_2(x)$. (g) Set $a = 1$ and plot on the same graph $\phi_1(x)$ and $\Phi(x;c_1,c_2)$ for $\frac{V_0}{E_1} = 5$ and $\frac{V_0}{E_1} = 10$.

SOLUTIONS TO SELECTED PROBLEMS

11.16 $\langle r\rangle_{exact} = \frac{3}{2}a$, $\langle r\rangle_{variational} = 0.0185a$, $\langle r^2\rangle_{exact} = 3a^2$, and $\langle r^2\rangle_{variational} = 0.0109a^2$.

11.21 $E_T(\alpha_{min}) = \frac{1}{2}\hbar\omega\left(\frac{\pi^2-6}{3}\right)^{1/2} = 1.136\left(\frac{1}{2}\hbar\omega\right)$.

11.22 $E_T(\alpha_{min}) = \frac{3}{4}\left[-\frac{e^2}{2(4\pi\varepsilon_0)a}\right]$.

11.23 $E_T(\alpha_{min}) = 1.082\lambda^{1/3}\left(\frac{\hbar^2}{2m}\right)^{2/3}$.

11.24 $E_T(\alpha_{min}) = 1.155\left(\frac{3}{2}\hbar\omega\right)$.

11.29 $E_T(\alpha_{min}) = \frac{2}{3}\left(\frac{\pi^2}{3}-\frac{1}{2}\right)^{1/2}\frac{3}{2}\hbar\omega = 1.114\left(\frac{3}{2}\hbar\omega\right)$.

12 Electrons in Atoms

In Chapter 10, we saw that the Schrödinger equation can be solved exactly for one-electron atoms. However, for multi-electron atoms (atoms with two or more electrons), the Schrödinger equation cannot be solved exactly. Thus, as we will see in this chapter, to determine wave functions and energies for multi-electron atoms one must resort to perturbation theory or the variational method.

Before getting started, we introduce a simplified system of units known as *atomic units*. (See Table 12.1.) Atomic units are commonly used in calculations of the electronic structure of atoms and molecules.

12.1 ATOMIC UNITS

In atomic units, the basic constants $m_e, e, \hbar$, and $4\pi\varepsilon_0$ all have the value unity. This leads to significant simplifications. For example, while the hydrogen-like Hamiltonian operator if nuclear motion is neglected is in SI units

$$\hat{H} = -\frac{\hbar^2}{2m_e}\nabla^2 - \frac{Ze^2}{4\pi\varepsilon_0 r} \tag{12.1}$$

in atomic units it takes the simplified form

$$\hat{H} = -\frac{1}{2}\nabla^2 - \frac{Z}{r}. \tag{12.2}$$

As a second example, in atomic units all energies are measured as multiples of the *hartree* E_h defined by

TABLE 12.1
Atomic Units and Their SI Equivalents

Dimension	Unit	SI Value
Mass	$m_e = 1$	$9.109382 \times 10^{-31}\,kg$
Charge	$e = 1$	$1.6021765 \times 10^{-19}\,C$
Permittivity	$4\pi\varepsilon_0 = 1$	$1.11265 \times 10^{-10}\,J^{-1}\,m^{-1}\,C^2$
Angular momentum	$\hbar = 1$	$1.0545716 \times 10^{-34}\,J\,s$
Distance	$\frac{4\pi\varepsilon_0\hbar^2}{m_e e^2} = a_0 = 1$	$5.2917721 \times 10^{-11}\,m$
Energy	$\frac{e^4 m_e}{(4\pi\varepsilon_0)^2\hbar^2} = 1$	$4.359744 \times 10^{-18}\,J$
Time	$\frac{(4\pi\varepsilon_0)^2\hbar^3}{m_e e^4} = 1$	$2.4188843 \times 10^{-17}\,s$
Speed	$\frac{e^2}{4\pi\varepsilon_0\hbar} = 1$	$2.1876913 \times 10^6\,m\,s^{-1}$
Electric potential	$\frac{m_e e^3}{(4\pi\varepsilon_0)^2\hbar^2} = 1$	$27.21138\,eV$

$$E_h = 1 \text{ hartree} = \frac{e^4 m_e}{(4\pi\varepsilon_0)^2 \hbar^2} = 1. \tag{12.3}$$

Thus, while in SI units the energy levels E_n of a hydrogen-like atom neglecting nuclear motion are given by $E_n = -\frac{Z^2 e^4 m_e}{2(4\pi\varepsilon_0)^2 n^2 \hbar^2}$, in atomic units they are expressed much more simply as

$$E_n = -\frac{Z^2}{2n^2} \text{ hartree}. \tag{12.4}$$

For example, in atomic units the ground state energy of the hydrogen atom is $E_1 = -\frac{1}{2}$ hartree.

Often the abbreviation 1 hartree = 1 au (atomic unit) is used. So E_n may also be written as $E_n = -\frac{Z^2}{2n^2}$ au.

Conversion factors between the hartree and other energy units are given in Table 12.2. We will especially use the equivalence 1 hartree = 1 au = 27.21139 electron volts (eV).

We may now begin. We start with helium and the helium-like ions H^-, Li^+, Be^{2+}, and so on since these systems have only two electrons and are thus the simplest multi-electron atoms. (See Figure 12.1.)

TABLE 12.2
Conversion of Hartrees (au) to Other Energy Units

eV	J	kJmol^{-1}	cm^{-1}
27.21139	4.35974×10^{-18}	2,625.500	2.194746×10^5

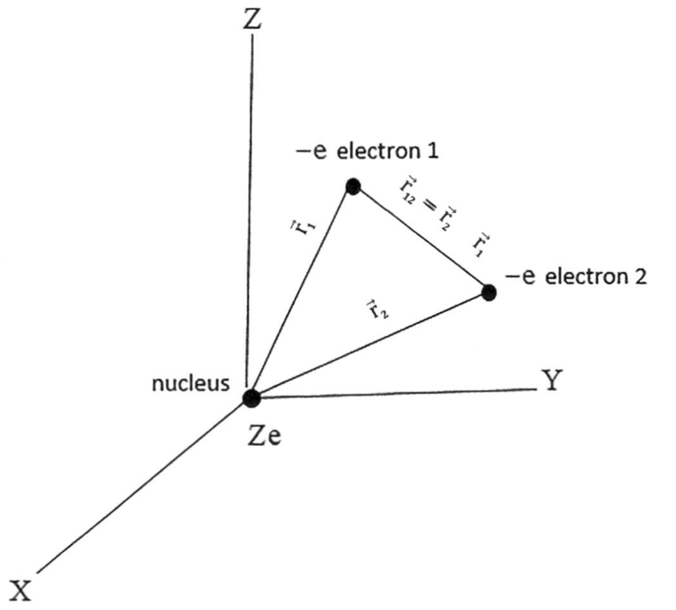

FIGURE 12.1 A helium-like atom with atomic number Z

12.2 THE HELIUM-LIKE ATOMS

Neglecting nuclear motion, the Hamiltonian operator of a helium-like atom has the form

$$\hat{H} = -\frac{\hbar^2}{2m_e}\nabla_1^2 - \frac{\hbar^2}{2m_e}\nabla_2^2 - \frac{Ze^2}{4\pi\varepsilon_0 r_1} - \frac{Ze^2}{4\pi\varepsilon_0 r_2} + \frac{e^2}{4\pi\varepsilon_0 r_{12}} \tag{12.5}$$

where Z is the atomic number, r_1 and r_2 are the distances of electrons 1 and 2 from the nucleus, and r_{12} is the distance between the two electrons. The first two terms in $\hat{H}$ are the kinetic energy operators for electrons 1 and 2, the third and fourth terms are the operators that represent the potential energies due to the attractive Coulomb interactions of the electrons and the nucleus, and the final term is the operator that represents the potential energy due to the Coulomb repulsion of the electrons. In atomic units, $\hat{H}$ simplifies to

$$\hat{H} = -\frac{1}{2}\nabla_1^2 - \frac{1}{2}\nabla_2^2 - \frac{Z}{r_1} - \frac{Z}{r_2} + \frac{1}{r_{12}}. \tag{12.6}$$

We wish to solve the Schrödinger equation

$$\left(-\frac{1}{2}\nabla_1^2 - \frac{1}{2}\nabla_2^2 - \frac{Z}{r_1} - \frac{Z}{r_2} + \frac{1}{r_{12}}\right)\Psi_0(\mathbf{r}_1,\mathbf{r}_2) = E_0\Psi_0(\mathbf{r}_1,\mathbf{r}_2) \tag{12.7}$$

for the atom's ground state wave function $\Psi_0(\mathbf{r}_1,\mathbf{r}_2)$ and ground state energy E_0. The wave function $\Psi_0(\mathbf{r}_1,\mathbf{r}_2)$ depends on the coordinates $\mathbf{r}_1$ and $\mathbf{r}_2$ of two interacting electrons and thus is a very complex function of six non-separable variables which cannot be found exactly. We will therefore obtain approximate wave functions Ψ_0 and energies E_0. We will then compare our approximate energies with the experimental ground state energy E_{expt}. E_{expt} is the negative of the energy needed to remove the electrons from the atom and thus determined from the observed first and second ionization potentials of the atom I_1 and I_2 as $E_{expt} = -(I_1 + I_2)$. For helium,

$$I_1 = 24.58739 \text{ eV and } I_2 = 54.41776 \text{ eV} \tag{12.8}$$

and so for helium

$$E_{expt} = -79.00515 \text{ eV} = -2.90339 \text{ au}. \tag{12.9}$$

Next, let us turn to the problem of finding the solution of the Schrödinger equation (Equation 12.7). This equation cannot be solved exactly because of the electron repulsion term $\frac{1}{r_{12}}$. So we will seek approximate solutions.

In the crudest approximation, we drop the electron repulsion term to obtain the zeroth-order Schrödinger equation

$$\hat{H}_0\Psi_0^{(0)}(\mathbf{r}_1,\mathbf{r}_2) = E_0^{(0)}\Psi_0^{(0)}(\mathbf{r}_1,\mathbf{r}_2) \tag{12.10}$$

where

$$\hat{H}_0 = -\frac{1}{2}\nabla_1^2 - \frac{Z}{r_1} - \frac{1}{2}\nabla_2^2 - \frac{Z}{r_2}. \tag{12.11}$$

Notice that $\hat{H}_0$ is the sum of two independent hydrogen-like atom Hamiltonian operators. Thus, the zeroth-order wave function $\Psi_0^{(0)}(\mathbf{r}_1,\mathbf{r}_2)$ is a product of two normalized hydrogen-like atom 1s orbitals $\Psi_{1s}(r)$ and the zeroth-order energy $E_0^{(0)}$ is twice the ground state hydrogen-like atom energy which by Equation (12.4) is $E_1 = -\frac{Z^2}{2}$. That is (Problem 12.6),

$$\Psi_0^{(0)}(\mathbf{r}_1,\mathbf{r}_2) = \Psi_{1s}(r_1)\Psi_{1s}(r_2) \tag{12.12}$$

where in atomic units (see first entry in Table 10.3)

$$\Psi_{1s}(r) = \left(\frac{1}{\pi}\right)^{1/2} Z^{3/2}\exp(-Zr) \tag{12.13}$$

and where

$$E_0^{(0)} = -Z^2 \text{ au.} \tag{12.14}$$

For helium, $Z = 2$ and so $E_0^{(0)} = -4$ au $= -108.87$ eV, which is 38% lower than $E_{expt} = -79.01$ eV. Clearly, the positive electron repulsion contribution to the ground state energy of helium is too great to be ignored.

The simplest way to include this contribution is via first-order perturbation theory. Thus, we take as the unperturbed Hamiltonian operator $\hat{H}_0$ and as the perturbation $\hat{V} = \frac{1}{r_{12}}$. From Equation (11.16), the first-order correction $E_0^{(1)}$ to $E_0^{(0)}$ is then

$$E_0^{(1)} = \int \Psi_0^{(0)}(\mathbf{r}_1,\mathbf{r}_2)\frac{1}{r_{12}}\Psi_0^{(0)}(\mathbf{r}_1,\mathbf{r}_2)d\mathbf{r}_1 d\mathbf{r}_2. \tag{12.15}$$

Evaluation of $E_0^{(1)}$ yields (Problem 12.8)

$$E_0^{(1)} = \frac{5}{8}Z \text{ au.} \tag{12.16}$$

Thus, the ground state energy E_0 to first order is

$$E_0^{(0)} + E_0^{(1)} = \left(-Z^2 + \frac{5}{8}Z\right) \text{au.} \tag{12.17}$$

For helium with $Z = 2$, $E_0^{(0)} + E_0^{(1)} = -\frac{11}{4}$ au $= -2.75$ au $= -74.83$ eV. This is only 5.3% higher than $E_{expt} = -79.01$ eV.

Next, we make a variational estimate of E_0. We choose as the trial function

$$\Phi(\mathbf{r}_1,\mathbf{r}_2) = \phi_{1s}(r_1)\phi_{1s}(r_2) \tag{12.18}$$

where

$$\phi_{1s}(r) = \left(\frac{1}{\pi}\right)^{1/2}(Z')^{3/2}\exp(-Z'r) \tag{12.19}$$

is a normalized 1s orbital with a variable exponent Z' which will be taken as the variational parameter. Notice that if we set $Z' = Z$, the trial function $\Phi(r_1,r_2)$ becomes identical to the zeroth-order wave function $\Psi_0^{(0)}(r_1,r_2)$. A consequence is that our variational calculation cannot give an energy higher than the first-order perturbation energy and it will give a lower energy if the optimal value of Z' differs from Z (Problem 12.10).

We next note that since $\Phi(\mathbf{r}_1,\mathbf{r}_2)$ is normalized, the trial energy $E_T(Z')$ is given by

$$E_T(Z') = \int \Phi(\mathbf{r}_1,\mathbf{r}_2)\hat{H}\Phi(\mathbf{r}_1,\mathbf{r}_2)d\mathbf{r}_1 d\mathbf{r}_2. \tag{12.20}$$

The evaluation of $E_T(Z')$ may be simplified by rewriting Equation (12.6) for $\hat{H}$ as

$$\hat{H} = -\frac{1}{2}\nabla_1^2 - \frac{Z'}{r_1} - \frac{1}{2}\nabla_2^2 - \frac{Z'}{r_2} - \frac{(Z-Z')}{r_1} - \frac{(Z-Z')}{r_2} + \frac{1}{r_{12}}. \tag{12.21}$$

Simplification arises since just as $\Psi_0^{(0)}(\mathbf{r}_1,\mathbf{r}_2)$ is an eigenfunction of $\hat{H}_0$ of Equation (12.11) with eigenvalue $-Z^2$, $\Phi(\mathbf{r}_1,\mathbf{r}_2)$ is an eigenfunction of the operator $-\frac{1}{2}\nabla_1^2 - \frac{Z'}{r_1} - \frac{1}{2}\nabla_2^2 - \frac{Z'}{r_2}$ with eigenvalue $-(Z')^2$. Consequently, $E_T(Z')$ may be written as

$$E_T(Z') = -(Z')^2 + I + J \tag{12.22}$$

where

$$I = -\int \Phi(\mathbf{r}_1,\mathbf{r}_2)\left[\frac{(Z-Z')}{r_1} + \frac{(Z-Z')}{r_2}\right]\Phi(\mathbf{r}_1,\mathbf{r}_2)d\mathbf{r}_1 d\mathbf{r}_2 \tag{12.23}$$

and where

$$J = \int \Phi(\mathbf{r}_1,\mathbf{r}_2)\frac{1}{r_{12}}\Phi(\mathbf{r}_1,\mathbf{r}_2)d\mathbf{r}_1 d\mathbf{r}_2. \tag{12.24}$$

I has the value (Problem 12.11)

$$I = 2(Z')^2 - 2ZZ'. \tag{12.25}$$

J is identical to the integral needed to evaluate $E_0^{(1)}$ if one makes the transposition $Z \rightarrow Z'$. Thus, from Equation (12.16)

$$J = \frac{5}{8}Z'. \tag{12.26}$$

Equations (12.22), (12.25), and (12.26) then yield $E_T(Z')$ as

$$E_T(Z') = (Z')^2 - 2ZZ' + \frac{5}{8}Z'. \tag{12.27}$$

The optimal value Z'_{min} of Z' is found from the minimization condition $\left[\frac{dE_T(Z')}{dZ'}\right]_{Z'=Z'_{min}} = 0$ or from Equation (12.27) $2Z'_{min} - 2Z + \frac{5}{8} = 0$ which gives Z'_{min} as

$$Z'_{min} = Z - \frac{5}{16}. \tag{12.28}$$

Inserting this value for Z'_{min} into Equation (12.27) for $E_T(Z')$ gives the variational estimate of E_0 as

$$E_T(Z'_{min}) = -Z^2 + \frac{5}{8}Z - \frac{25}{256} = -\left(Z - \frac{5}{16}\right)^2 \text{ au.} \tag{12.29}$$

For helium with $Z = 2$, $E_T(Z'_{min}) = -\left(\frac{27}{16}\right)^2$ au $= -2.8477$ au $= -77.49$ eV, which is only 1.9% higher than $E_{expt} = -79.01$ eV. Thus, for helium the variational estimate of the energy is significantly better than the energy predicted by first-order perturbation theory which recall is 5.3% too high. The reason for the large improvement is that the variational wave function approximately accounts for the effect of screening of the nuclear charge. When an electron is close to the nucleus, it experiences the full nuclear charge Z. But when it is far from the nucleus, because of the screening effect of the second electron it experiences a nuclear charge of $Z-1$. The effective nuclear charge produced by the variational treatment $Z'_{min} = Z - \frac{5}{16}$ is a compromise between these two limits.

In Table 12.3, we compare our three theoretical estimates $E_0^{(0)}$, $E_0^{(0)} + E_0^{(1)}$, and $E_T(Z'_{min})$ for the ground state energy E_0 with the experimental energies E_{expt} for the helium-like atoms H^- through C^{4+}. Reflecting the fact that the electron repulsion energy becomes less dominant as the nuclear charge increases, the percent errors for the theoretical estimates decrease as Z increases. For example, for C^{4+} $E_T(Z'_{min}) = -880.2$ eV is only 0.2% higher than $E_{expt} = -882.06$ eV.

The determinations of E_0 just described are very primitive. More sophisticated and accurate calculations of E_0 have been made. In fact, the accuracy of the best of these determinations is so high that their results cannot be meaningfully assessed by comparing with experiment. This is because experimental energies contain small contributions from the effects of relativity and nuclear motion, effects which are not included in the Schrödinger equation (Equation 12.7). Rather, for helium the E_0 values obtained from the best calculations should be compared with the value $E_0 = -2.903724375$ au (with an uncertainty of order 1 in the last digit) found by Pekeris (1959) from a very complex variational solution of Equation (12.7). Notice that Pekeris' value for E_0 differs significantly from the experimental helium ground state energy which is −2.90339 au.

Next, we describe some additional calculations of the ground state energy of helium and the helium-like ions.

The most accurate perturbation theory treatment of the ground state energies of the helium-like atoms is that of Scherr and Knight (1963) who determined E_0 to thirteenth order. Scherr and Knight gave E_0 in au as the following fourteen-term expansion in powers of Z:

TABLE 12.3
Theoretical Estimates (See Text) of the Ground State Energies of Helium-Like Atoms Compared with the Experimental Energies E_{expt}

Atom	$E_0^{(0)}$ (eV)	$E_0^{(0)} + E_0^{(1)}$ (eV)	$E_T(Z'_{min})$ (eV)	E_{expt} (eV)
H^-	−27.21	−10.21	−12.86	−14.35
He	−108.87	−74.83	−77.49	−79.01
Li^+	−244.90	−193.88	−196.55	−198.09
Be^{2+}	−435.38	−367.35	−370.02	−371.61
B^{3+}	−680.28	−595.25	−597.92	−599.59
C^{4+}	−979.61	−877.57	−880.23	−882.06

$$E_0 = -Z^2 + \frac{5}{8}Z - 0.157666405 + 0.008698991Z^{-1} - 0.000888587Z^{-2}$$
$$- 0.001036372Z^{-3} - 0.000612917Z^{-4} + \text{seven more terms.} \quad (12.30)$$

Notice that the first two terms in the above series reproduce our earlier result of Equation (12.17) for the energy to first order. In general, the sum of the first n terms of the series yields the energy to $(n-1)$th order (Problem 12.13). For example, the energy to second order is the sum of the first three terms of the series. Thus, for helium the energy to second order is $E_0 \cong -4\text{ au} + \frac{5}{4}\text{ au} - 0.157666405\text{ au} = -2.9077\text{ au}$, which is a big improvement on the energy to first order which recall is −2.75 au. Adding up all fourteen terms in the series gives the energy to thirteenth order as $E_0 = -2.90372433$ au, a result which is nearly as good as that of Pekeris.

We next turn to variational determinations of E_0 for helium. We will first assume orbital model trial functions $\Psi(\mathbf{r}_1,\mathbf{r}_2)$. These are products of one-electron wave functions or orbitals $\phi(r)$ and thus have the form

$$\Psi(\mathbf{r}_1,\mathbf{r}_2) = \phi(r_1)\phi(r_2). \quad (12.31)$$

The trial function $\Phi(\mathbf{r}_1,\mathbf{r}_2) = \phi_{1s}(r_1)\phi_{1s}(r_2)$ of Equation (12.18) is of this orbital product form.

We may however consider orbitals $\phi(r)$ of more complex form than $\phi_{1s}(r)$ which include more than a single variational parameter. The trial functions $\Psi(\mathbf{r}_1,\mathbf{r}_2)$ built from such orbitals will give lower energies than that given by $\Phi(\mathbf{r}_1,\mathbf{r}_2)$.

For example, consider the normalized *Slater-type orbitals* (STOs) $\chi_{n\ell m}(\mathbf{r})$ which have the form

$$\chi_{n\ell m}(\mathbf{r}) = \left(2\frac{\xi}{n}\right)^n \left[\frac{2\frac{\xi}{n}}{\Gamma(2n+1)}\right]^{1/2} r^{n-1}\exp\left(-\frac{\xi}{n}r\right)Y_{\ell m}(\theta,\phi) \quad (12.32)$$

where $\Gamma(x)$ is the gamma function. ($\Gamma[x]$ is a generalization of the factorial function to non-integral arguments x. For $x = 0,1,2,\ldots$, $\Gamma[x+1] = x!$.) We will choose $\phi(r)$ to be the STO

$$\chi_{n00}(r) = \frac{\left(2\frac{\xi}{n}\right)^n}{(4\pi)^{1/2}} \left[\frac{2\frac{\xi}{n}}{\Gamma(2n+1)}\right]^{1/2} r^{n-1}\exp\left(-\frac{\xi}{n}r\right). \quad (12.33)$$

Then, our trial function becomes

$$\Psi(\mathbf{r}_1,\mathbf{r}_2) = \chi_{n00}(r_1)\chi_{n00}(r_2). \quad (12.34)$$

Taking n and ξ to be variational parameters, the values of these parameters that minimize the trial energy $E_T(n,\xi)$ are $n_{min} = 0.9995$ and $\xi_{min} = 1.6116$ yielding the variational estimate of E_0 as $E_T(n_{min},\xi_{min}) = -2.8542$ au. This energy is lower than the estimate from $\Phi(\mathbf{r}_1,\mathbf{r}_2)$ which recall is $E_T(Z'_{min}) = -2.8477$ au.

One can try to further lower the energy by using increasingly complex forms for the orbital $\phi(r)$ which include more and more variational parameters. Continuing this process indefinitely will yield a sequence of energies which approach the limiting value of −2.8617 au. This value, which for reasons discussed in Section 12.3 we call the Hartree–Fock limit, is 1.44% above the exact energy of −2.9037 au. The error arises because adopting an orbital model wave function amounts to assuming that the two electrons move independently. In reality, the electrons "see" one another. So because of the Coulomb repulsion of the electrons, if electron 1 is at point $\mathbf{r}_1$ electron 2 tends to be at a point $\mathbf{r}_2$ which is distant from $\mathbf{r}_1$, thus lowering the energy. This effect is called *electron correlation*. Therefore, to obtain an energy which is lower than the Hartree–Fock limiting value, one must choose a trial function which incorporates electron correlation.

Perhaps the simplest trial function which includes electron correlation is the Eckart function

$$\Psi(\mathbf{r}_1,\mathbf{r}_2) = N\left[\exp(-Z'r_1)\exp(-Z''r_2) + \exp(-Z''r_1)\exp(-Z'r_2)\right] \tag{12.35}$$

where N is a normalization factor and Z' and Z'' are variational parameters. Since Z' and Z'' are different, the Eckart function is not of the orbital model form. The values of Z' and Z'' which minimize the trial energy $E_T(Z',Z'')$ are $Z'_{min} = 1.19$ and $Z''_{min} = 2.18$.

The variational estimate of E_0 is $E_T(Z'_{min},Z''_{min}) = -2.8757$ au, which is an improvement on the best orbital model energy of -2.8617 au. Since $\exp(-Z''_{min}r)$ decays more rapidly than $\exp(-Z'_{min}r)$, the Eckart function predicts a relatively high probability for one electron to be closer to the nucleus than the other, thus introducing correlation.

A breakthrough in the treatment of electron correlation was made by Hylleraas in 1930 who introduced trial functions which included the interelectronic distance r_{12}. Hylleraas expressed his trial functions in terms of the variables $s = r_1 + r_2$, $t = r_1 - r_2$, and $u = r_{12}$. The trial functions chosen were of the form

$$\Psi(\mathbf{r}_1,\mathbf{r}_2) = N\exp(-Z's)\left[1 + g(s,t,u)\right] \tag{12.36}$$

where N is a normalization factor, Z' is a variational parameter, and

$$g(s,t,u) = \sum_{n\ell m} c_{n\ell m}\, s^n t^\ell u^m \tag{12.37}$$

is a power series in the variables s, t, and u. Like Z', the coefficients $c_{n\ell m}$ are variational parameters.

To see how trial functions which include r_{12} introduce correlation, it is helpful to consider the simplest Hylleraas function

$$\Psi(\mathbf{r}_1,\mathbf{r}_2) = N\exp\left[-Z'(r_1 + r_2)\right](1 + cr_{12}) \tag{12.38}$$

where Z' and c are variational parameters. Since the optimal value of c is positive, for fixed r_1 and r_2 the above function increases as r_{12} increases. Thus, it predicts enhanced probabilities for configurations in which the electrons are far apart. The optimal values of Z' and c are $Z'_{min} = 1.849$ and $c_{min} = 0.364$. The corresponding variational energy is $E_T = (Z'_{min},c_{min}) = -2.8913$ au, which is a big improvement over our earlier variational estimates of E_0. An even better value for the energy, -2.9036 au,was found by Hylleraas from a trial function with ten parameters (nine terms in the power series of Equation [12.37]).

Kinoshita (1957) performed calculations similar to those of Hylleraas. Using a trial function with thirty-nine parameters, he obtained an energy of -2.9037225 au.

Finally, Pekeris using a trial function with 1078 terms obtained the energy mentioned earlier -2.903724375 au. This energy was corrected for relativistic and finite nuclear mass effects and for the Lamb shift, a small quantum electrodynamic effect. Together with an equally sophisticated calculation of the helium ion energy $E(He^+)$, Pekeris obtained the following result for the first ionization potential $I = E(He^+) - E_0$ of helium

$$I = 24.5873914 \text{ eV}. \tag{12.39}$$

TABLE 12.4
Theoretical Values for the Ground State Energy E_0 of Helium

Calculation	E_0(au)
Neglect of electron repulsion	–4.0
First-order perturbation theory	–2.75
Second-order perturbation theory	–2.9077
Thirteenth-order perturbation theory	–2.90372433
Hydrogenic orbital product wave function	–2.8477
Slater-type orbital product wave function	–2.8542
Hartree–Fock wave function	–2.861680
Eckart wave function	–2.8757
Two-parameter Hylleraas wave function	–2.8913
Ten-parameter Hylleraas wave function	–2.9036
Thirty-nine-parameter Kinoshita wave function	–2.9037225
1,078-term Pekeris wave function	–2.903724375

This theoretical result should be compared with the best available experimental value for I

$$I_{expt} = 24.5873879 \text{ eV}. \tag{12.40}$$

The percent error in the theoretical ionization potential is only 0.000014%!

The results of the calculations of E_0 described in this section are summarized in Table 12.4.

In conclusion, the problem of determining the ground state energy of helium has been solved by Hylleraas, Kinoshita, and Pekeris. Their methods however are difficult to extend to more complex systems and have not been applied to atoms with more than four electrons. Moreover, not much physical insight may be gleaned from the very complex wave functions which emerge from calculations of the Hylleraas, Kinoshita, and Pekeris types.

Thus, we return to orbital model wave functions $\Psi(\mathbf{r}_1,\mathbf{r}_2) = \phi(r_1)\phi(r_2)$. These while of limited accuracy provide great physical understanding and moreover can be found for both atoms with many electrons and for molecules. We first show for the helium-like atoms how to find the best orbitals $\phi(r)$. These are the orbitals that give the lowest energy orbital model wave functions. The best orbitals are the solutions of the *Hartree–Fock equations* that we next develop for the special case of the helium-like atoms.

12.3 THE HARTREE–FOCK EQUATIONS FOR THE HELIUM-LIKE ATOMS

We first develop the Hartree–Fock equations by an intuitive physical argument and then show how they may be rigorously derived from the variational principle. For clarity, in the intuitive argument we initially use SI units rather than atomic units.

Within the orbital model, the Born rule probability density for the position of, say, electron 2 is $P(r_2) = \phi^*(r_2)\phi(r_2)$. This form for $P(r_2)$ suggests the model shown in Figure 12.2 in which electron 2 is "smeared out" so that electron 1 at point $\mathbf{r}_1$ "sees" electron 2 as a continuous average charge density $\rho(r_2) = -e\phi^*(r_2)\phi(r_2)$ rather than as in reality a discrete charge located at a specific point. The average charge density produces an infinitesimal negative charge $dq = \rho(r_2)d\mathbf{r}_2$ in the volume element $d\mathbf{r}_2$ located at point $\mathbf{r}_2$. The repulsive Coulomb interaction of electron 1 with the charge dq gives rise to the potential energy $dV = -\frac{edq}{4\pi\varepsilon_0 r_{12}} = \frac{-e\rho(r_2)d\mathbf{r}_2}{4\pi\varepsilon_0 r_{12}} = \frac{e^2\phi^*(r_2)\phi(r_2)d\mathbf{r}_2}{4\pi\varepsilon_0 r_{12}}$ where r_{12} is the distance between points $\mathbf{r}_1$ and $\mathbf{r}_2$. The potential energy $V(r_1)$ due to the interaction of electron 1 with

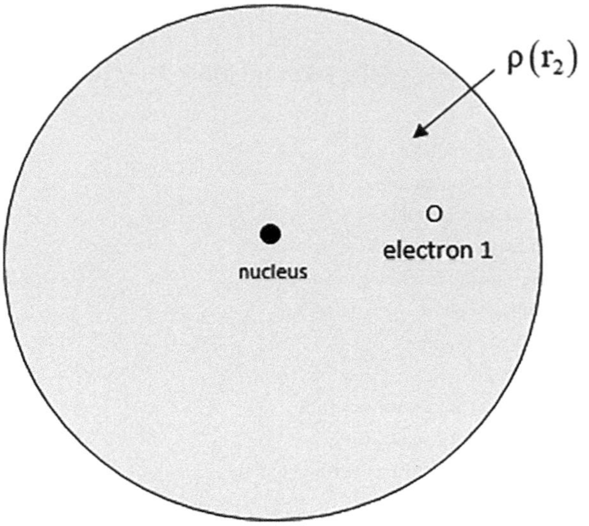

FIGURE 12.2 Electron 1 interacting with the smeared-out charge density $\rho(r_2)$ due to electron 2.

the whole charge density is found by integrating dV over all volume elements $d\mathbf{r}_2$. This integration gives $V(r_1)$ as

$$V(r_1) = \frac{e^2}{4\pi\varepsilon_0}\int \phi^*(r_2)\frac{1}{r_{12}}\phi(r_2)d\mathbf{r}_2. \tag{12.41}$$

Given this form for $V(r_1)$, the total potential energy $U(r_1)$ of electron 1 is

$$U(r_1) = -\frac{Ze^2}{4\pi\varepsilon_0 r_1} + \frac{e^2}{4\pi\varepsilon_0}\int \phi^*(r_2)\frac{1}{r_{12}}\phi(r_2)d\mathbf{r}_2 \tag{12.42}$$

where $-\dfrac{Ze^2}{4\pi\varepsilon_0 r_1}$ is the potential energy due to the interaction of electron 1 with the nucleus.

We may now return to atomic units. In these units, $U(r_1)$ is

$$U(r_1) = -\frac{Z}{r_1} + \int \phi^*(r_2)\frac{1}{r_{12}}\phi(r_2)\,d\mathbf{r}_2. \tag{12.43}$$

Since the potential energy of electron 1 is $U(r_1)$, its wave function $\phi(r_1)$ satisfies the Schrödinger equation

$$\hat{F}_1\phi(r_1) = \varepsilon_1\phi(r_1) \tag{12.44a}$$

where the Hamiltonian operator $\hat{F}_1$ is given by

$$\hat{F}_1 = -\frac{1}{2}\nabla_1^2 + U(r_1). \tag{12.44b}$$

$\hat{F}_1$ is called the *Fock operator*, and the eigenvalue ε_1 is called the *orbital energy*.

The wave function $\phi(r_2)$ of electron 2 satisfies an identical Schrödinger equation

$$\hat{F}_2\phi(r_2) = \varepsilon_2\phi(r_2) \tag{12.45a}$$

where

$$\hat{F}_2 = -\frac{1}{2}\nabla_2^2 + U(r_2) \tag{12.45b}$$

with

$$U(r_2) = -\frac{Z}{r_2} + \int \phi^*(r_1)\frac{1}{r_{12}}\phi(r_1)\,d\mathbf{r}_1. \tag{12.46}$$

Equations (12.44) and (12.45) are the Hartree–Fock equations for helium-like atom. Either one of them determines the Hartree–Fock orbital $\phi(r)$ and hence the Hartree–Fock wave function $\Psi(\mathbf{r}_1,\mathbf{r}_2) = \phi(r_1)\phi(r_2)$. Notice that the Hartree–Fock equations are one-electron equations. They are thus vastly easier to solve than the true two-electron helium-like atom Schrödinger equation (Equation 12.7). The price paid for this simplification is that the Hartree–Fock wave function does not include electron correlation.

Before describing how the Hartree–Fock equations are solved, we outline their rigorous derivation from the variational principle. This derivation proves that the Hartree–Fock orbitals give the lowest energy orbital model wave function.

We start by assuming an orbital model trial function

$$\Psi(\mathbf{r}_1,\mathbf{r}_2) = \phi(r_1)\phi(r_2). \tag{12.47}$$

The orbital $\phi(r)$ must be a well-behaved normalized function but is otherwise arbitrary. Since $\phi(r)$ is normalized, $\Psi(\mathbf{r}_1,\mathbf{r}_2)$ is also normalized and so the trial energy E is

$$\begin{aligned} E &= \int \Psi^*(\mathbf{r}_1,\mathbf{r}_2)\hat{H}\Psi(\mathbf{r}_1,\mathbf{r}_2)\,d\mathbf{r}_1\,d\mathbf{r}_2 \\ &= \int \phi^*(r_1)\phi^*(r_2)\hat{H}\phi(r_1)\phi(r_2)\,d\mathbf{r}_1\,d\mathbf{r}_2 \end{aligned} \tag{12.48}$$

where $\hat{H}$ is the helium-like atom Hamiltonian operator of Equation (12.6). We wish to find the orbital $\phi_{min}(r)$ which minimizes the trial energy E. This is a more complex variational problem than any we have dealt with so far since we are now seeking a *function* $\phi_{min}(r)$ rather than optimal variational parameters $\lambda_{1\,min}, \lambda_{2\,min}, \ldots$ which are a set of numbers.

To proceed further, we write $\hat{H}$ as

$$\hat{H} = -\frac{1}{2}\nabla_1^2 - \frac{Z}{r_1} - \frac{1}{2}\nabla_2^2 - \frac{Z}{r_2} + \frac{1}{r_{12}}. \tag{12.49}$$

Comparing Equations (12.48) and (12.49) gives the following explicit result for $E = E[\phi(r)]$:

$$\begin{aligned} E[\phi(r)] &= \int \phi^*(r_1)\left(-\frac{1}{2}\nabla_1^2 - \frac{Z}{r_1}\right)\phi(r_1)\,d\mathbf{r}_1 \\ &+ \int \phi^*(r_2)\left(-\frac{1}{2}\nabla_2^2 - \frac{Z}{r_2}\right)\phi(r_2)\,d\mathbf{r}_2 \\ &+ \int \phi^*(r_1)\,\phi^*(r_2)\frac{1}{r_{12}}\phi(r_1)\phi(r_2)\,d\mathbf{r}_1 d\mathbf{r}_2. \end{aligned} \tag{12.50}$$

We have written E as $E[\phi(r)]$ to indicate that E is a *functional* of $\phi(r)$. (A functional is a rule that converts a function into a number. For example, $E[\phi(r)]$ is a rule that converts a function $\phi[r]$ into a number E.) For each choice of $\phi(r)$, the functional $E[\phi(r)]$ yields a different value for the trial energy E. We want to find the lowest value for the trial energy. Thus, we seek a method for finding the function $\phi(r) = \phi_{min}(r)$ which minimizes the functional $E[\phi(r)]$.

To motivate our method, let us consider the simple problem of finding the number which minimizes a function $f(x)$. The value of that number x_{min} is of course given by the condition $f'(x_{min}) = 0$. Let us consider a small variation $\delta x = x - x_{min}$ of x about x_{min}. The change in $f(x)$ due to this variation is $\delta f = f(x_{min} + \delta x) - f(x_{min}) = f'(x_{min})\delta x + \frac{1}{2}f''(x_{min})(\delta x)^2 + \cdots$. We will only need to know that since $f'(x_{min}) = 0$, (a) δf is of order $(\delta x)^2$ and (b) the term in δf of order δx vanishes.

We may now return to our actual problem. Instead of considering small variations $\delta x = x - x_{min}$ of x about x_{min} and the resulting changes δf of $f(x)$, we consider small variations $\delta\phi(r) = \phi(r) - \phi_{min}(r)$ of an orbital $\phi(r)$ about $\phi_{min}(r)$ and the resulting changes δE of the energy functional $E[\phi(r)]$. We assume variations of the form $\delta\phi(r) = \lambda\chi(r)$ where $\chi(r)$ is an arbitrary well-behaved function and where λ is a small real parameter. Given this form for $\delta\phi(r)$,

$$\delta E = E[\phi_{min}(r) + \lambda\chi(r)] - E[\phi_{min}(r)]. \tag{12.51}$$

Since $E[\phi(r)]$ has its minimum value when $\phi(r) = \phi_{min}(r)$, in analogy to our discussion of the minimum of $f(x)$ (a) δE is of order λ^2 and (b) the term in δE of order λ vanishes.

Next, we will evaluate the vanishing term in δE of order λ. From Equations (12.50) and (12.51), this term is (Problem 12.15)

$$\delta E = \lambda\left\{\int d\mathbf{r}_1\chi^*(r_1)\left[-\frac{1}{2}\nabla_1^2 + U(r_1)\right]\phi(r_1) + \int d\mathbf{r}_2\chi^*(r_2)\left[-\frac{1}{2}\nabla_2^2 + U(r_2)\right]\phi(r_2)\right.$$
$$\left. + \text{complex conjugate terms}\right\} \tag{12.52}$$

where to simplify the notation we have set $\phi_{min}(r)$ equal to $\phi(r)$. In Equation (12.52), the complex conjugate terms are the complex conjugates of the terms written down explicitly and $U(r_1)$ and $U(r_2)$ are the potentials of Equations (12.43) and (12.46).

Since $\chi(r)$ can be varied arbitrarily, δE vanishes only if $\phi(r_1)$ and $\phi(r_2)$ satisfy the differential equations $\left[-\frac{1}{2}\nabla_1^2 + U(r_1)\right]\phi(r_1) = 0$ and $\left[-\frac{1}{2}\nabla_2^2 + U(r_2)\right]\phi(r_2) = 0$ (with equivalent complex conjugate equations). These differential equations are however wrong. Instead, $\phi(r_1)$ and $\phi(r_2)$ satisfy the Hartree–Fock equations which from Equations (12.44) and (12.45) are

$$\left[-\frac{1}{2}\nabla_1^2 + U(r_1)\right]\phi(r_1) = \varepsilon_1\phi(r_1) \tag{12.53a}$$

and

$$\left[-\frac{1}{2}\nabla_2^2 + U(r_2)\right]\phi(r_2) = \varepsilon_2\phi(r_2). \tag{12.53b}$$

The error arises since Equation (12.52) is the correct minimization condition for $E[\phi(r)]$ only if the orbitals $\phi(r_1)$ and $\phi(r_2)$ can be varied without any constraints. In fact, the varied orbitals are subject to the normalization constraints

$$\int \phi^*(r_1)\phi(r_1)\,d\mathbf{r}_1 = 1 \text{ and } \int \phi^*(r_2)\phi(r_2)\,d\mathbf{r}_2 = 1. \tag{12.54}$$

Thus, we must minimize $E[\phi(r)]$ subject to the constraints of Equation (12.54). There is however a standard technique for minimizing a functional subject to constraints. It is called the *method of Lagrange multipliers* (Arfken, Weber, and Harris 2012). Applying this method to our problem, we define the new functional $F[\phi(r)]$ by

$$F[\phi(r)] = E[\phi(r)] - \varepsilon_1\left[\int \phi^*(r_1)\phi(r_1)d\mathbf{r}_1 - 1\right] - \varepsilon_2\left[\int \phi^*(r_2)\phi(r_2)d\mathbf{r}_2 - 1\right] \tag{12.55}$$

where $-\varepsilon_1$ and $-\varepsilon_2$ are Lagrange multipliers which we will shortly identify as the helium-like atom's orbital energies. The change in $F[\phi(r)]$ due to small variations $\delta\phi(r) = \lambda\chi(r)$ of the orbital $\phi(r)$ is

$$\delta F = F[\phi(r) + \lambda\chi(r)] - F[\phi(r)]. \tag{12.56}$$

The Lagrange multiplier condition for minimizing $E[\phi(r)]$ subject to the normalization constraints is that the term in δF of linear order in λ must vanish. Proceeding as in the derivation of Equation (12.52), this condition gives

$$\delta F = \lambda\left\{\int d\mathbf{r}_1\chi^*(r_1)\left[-\frac{1}{2}\nabla_1^2 + U(r_1) - \varepsilon_1\right]\phi(r_1) + \int d\mathbf{r}_2\chi^*(r_2)\left[-\frac{1}{2}\nabla_2^2 + U(r_2) - \varepsilon_2\right]\phi(r_2)\infty\right.$$
$$\left. + \text{complex conjugate terms}\right\} = 0. \tag{12.57}$$

Since $\chi(r)$ may be varied arbitrarily, $\delta F = 0$ only if $\left[-\frac{1}{2}\nabla_1^2 + U(r_1) - \varepsilon_1\right]\phi(r_1) = 0$ and $\left[-\frac{1}{2}\nabla_2^2 + U(r_2) - \varepsilon_2\right]\phi(r_2) = 0$ (with equivalent complex conjugate equations). However, if we identify the Lagrange multipliers ε_1 and ε_2 with the Hartree–Fock orbital energies, these relations are just the Hartree–Fock equations (Equations 12.53). Thus, we have derived the Hartree–Fock equations from the variational principle.

We next turn to the solution of the Hartree–Fock equations. From Equations (12.43), (12.46), and (12.53), the explicit forms of the Hartree–Fock equations are

$$\left[-\frac{1}{2}\nabla_1^2 - \frac{Z}{r_1} + \int \phi^*(r_2)\frac{1}{r_{12}}\phi(r_2)d\mathbf{r}_2\right]\phi(r_1) = \varepsilon_1\phi(r_1) \tag{12.58a}$$

and

$$\left[-\frac{1}{2}\nabla_2^2 - \frac{Z}{r_2} + \int \phi^*(r_1)\frac{1}{r_{12}}\phi(r_1)d\mathbf{r}_1\right]\phi(r_2) = \varepsilon_2\phi(r_2). \tag{12.58b}$$

Equations (12.58a) and (12.58b) are equivalent both giving the same Hartree–Fock orbital and orbital energy. So we only need to consider, say, Equation (12.58a). This equation is, however, more complex than an ordinary one-electron Schrödinger equation since the Hamiltonian operator $\hat{F}_1 = -\frac{1}{2}\nabla_1^2 - \frac{Z}{r_1} + \int \phi^*(r_2)\frac{1}{r_{12}}\phi(r_2)\,d\mathbf{r}_2$ depends on the unknown orbital $\phi(r)$. Consequently, Equation (12.58a) must be solved iteratively. Namely, one makes an initial guess $\phi^{(1)}(r)$ for $\phi(r)$ and then constructs an initial Hamiltonian operator $\hat{F}_1^{(1)} = -\frac{1}{2}\nabla_1^2 - \frac{Z}{r_1} + \int\left[\phi^{(1)}(r_2)\right]^*\frac{1}{r_{12}}\phi^{(1)}(r_2)d\mathbf{r}_2$. One then solves the closed one-electron Schrödinger equation $\hat{F}_1^{(1)}\phi^{(2)}(r_1) = \varepsilon_1^{(2)}\phi^{(2)}(r_1)$ to find a second approximation $\phi^{(2)}(r)$ to $\phi(r)$ which improves on $\phi^{(1)}(r)$. One then similarly constructs a second

Hamiltonian operator $\hat{F}_1^{(2)}$ from $\phi^{(2)}(r)$, which is identical to $\hat{F}^{(1)}$ except that ϕ^1 in $\hat{F}^{(1)}$ is replaced by ϕ^2, and solves the closed Schrödinger equation $\hat{F}_1^{(2)}\phi^{(3)}(r_1) = \varepsilon_1^{(3)}\phi^{(3)}(r_1)$ to obtain a further improved third approximation $\phi^{(3)}(r)$ to $\phi(r)$. One continues this process until no further improvement occurs or in other words until the $(n+1)$th approximation $\phi^{(n+1)}(r)$ and the nth approximation $\phi^{(n)}(r)$ are (to within a prescribed tolerance) identical. Then, the iterative process has converged and the Hartree–Fock orbital $\phi(r)$ and orbital energy ε_1 are determined as $\phi^{(n)}(r)$ and $\varepsilon_1^{(n)}$. Since the converged Hartree–Fock potential $U(r)$ is unchanged by further iterations, the Hartree–Fock procedure is also called the *self-consistent field (SCF) method.*

We further note that because $U(r)$ is a spherically symmetric (central field) potential, the three-dimensional equation (Equation 12.58a) may be reduced to a one-dimensional equation for the radial part of $\phi(r)$. (See Problem 12.16.) This reduction greatly simplifies the problem of solving the Hartree–Fock equations.

While the Hartree–Fock equations for helium and other atoms were initially solved numerically, now they are more commonly solved by a *basis set* method due to Roothaan (1951). (Roothaan's procedure is universally used to solve the Hartree–Fock equations for molecules [see Section 13.5]). In Roothaan's method, the Hartree–Fock orbitals of an atom $\phi_i(\mathbf{r})$ are expanded as linear combinations

$$\phi_i(\mathbf{r}) = \sum_n c_{ni}\chi_n(\mathbf{r}) \tag{12.59}$$

of known linearly independent *basis functions* $\chi_n(\mathbf{r})$. As discussed in detail Section 13.5, the expansion coefficients c_{ni} are determined by inserting Equation (12.59) into the Hartree–Fock equations. This yields a set of equations for the coefficients which are solved by an iterative self-consistent procedure.

The basis functions typically employed for atomic Hartree–Fock calculations are the STOs $\chi_{n\ell m}(\mathbf{r})$ of Equation (12.32) for integral n. The number of STOs included in the expansion and the values of the *orbital exponents* ξ are varied until convergence is achieved. As an example, Bunge, Barrientos, and Bunge (1993) obtained the following STO representation of the helium 1s Hartree–Fock orbital $\phi(r)$:

$$\phi(r) = N_{1\xi_1}c_1\exp(-\xi_1 r) + N_{2\xi_2}c_2 r\exp\left(-\frac{\xi_2 r}{2}\right) + N_{2\xi_3}c_3 r\exp\left(-\frac{\xi_3 r}{2}\right)$$
$$+ N_{3\xi_3}c_4 r^2\exp\left(-\frac{\xi_4 r}{3}\right) \tag{12.60}$$

where the normalization factors for the STOs $N_{n\xi}$ are given by $N_{n\xi} = \left(\frac{2\xi}{n}\right)^n\left[\frac{2\xi/n}{(2n)!}\right]^{1/2}$ and where the orbital exponents and expansion coefficients ξ_m and c_m, $m = 1-4$, are listed in Table 12.5.

TABLE 12.5
Values of the Orbital Exponents ξ_m and Expansion Coefficients c_m in the STO Expansion of the Helium 1s Orbital $\phi(r)$ of Equation (12.60)[a]

m	ξ_m	c_m
1	1.4595	1.347900
2	1.7504	−0.270779
3	2.6298	−0.100506
4	5.3244	−0.001613

[a] C.F. Bunge, J.A. Barrientos, and A.V. Bunge, *At. Data Nucl. Tables* 53 113 (1993).

Next, let us consider the ground state Hartree–Fock energy E_{HF} of helium. Since the Hartree–Fock wave function is given by $\Psi(\mathbf{r}_1,\mathbf{r}_2) = \phi(r_1)\phi(r_2)$ where $\phi(r)$ is the Hartree–Fock orbital,

$$E_{HF} = \int \phi^*(r_1)\phi^*(r_2)\hat{H}\phi(r_1)\phi(r_2)d\mathbf{r}_1 d\mathbf{r}_2 \tag{12.61}$$

where $\hat{H}$ is the helium atom Hamiltonian operator. Using Equation (12.49) (with Z = 2) for $\hat{H}$, E_{HF} may be written as

$$E_{HF} = I_1 + I_2 + J_{12} \tag{12.62}$$

where

$$I_1 = \int \phi^*(r_1)\left(-\frac{1}{2}\nabla_1^2 - \frac{2}{r_1}\right)\phi(r_1)d\mathbf{r}_1 \tag{12.63a}$$

$$I_2 = \int \phi^*(r_2)\left(-\frac{1}{2}\nabla_2^2 - \frac{2}{r_2}\right)\phi(r_2)d\mathbf{r}_2 \tag{12.63b}$$

and

$$J_{12} = \int \phi^*(r_1)\phi^*(r_2)\frac{1}{r_{12}}\phi(r_1)\phi(r_2)d\mathbf{r}_1 d\mathbf{r}_2. \tag{12.64}$$

We will shortly need the following approximate interpretation of I_1 and I_2: $I_1 = I_2 \approx E(He^+)$, where $E(He^+)$ is the ground state energy of the He^+ ion. This interpretation follows since $-\frac{1}{2}\nabla^2 - \frac{2}{r}$ is the He^+ ion Hamiltonian operator and so if in Equations (12.63) the He Hartree–Fock orbital is replaced by the He^+ 1s orbital $\Psi_{1s}(r)$, then I_1 and I_2 would exactly equal $E(He^+)$. The assumption underlying the replacement, that $\Psi_{1s}(r)$ is identical to $\phi(r)$, is called the neglect of *orbital relaxation* upon ionization.

We next note that one may derive from the Hartree–Fock equations (Equation 12.58) the following results for the orbital energies (Problem 12.17):

$$\varepsilon_1 = I_1 + J_{12} \text{ and } \varepsilon_2 = I_2 + J_{12}. \tag{12.65}$$

Since within the orbital model the two electrons move independently, one might expect that $E_{HF} = \varepsilon_1 + \varepsilon_2$. This is wrong since $\varepsilon_1 + \varepsilon_2 = I_1 + I_2 + 2J_{12}$ while from Equation (12.62) $E_{HF} = I_1 + I_2 + J_{12}$. Adding the orbital energies double-counts the *Coulomb integral* J_{12}.

We may now derive for the special case of helium *Koopmans' theorem*, which is a general result which relates ionization potentials of closed shell atoms and molecules (i.e., atoms and molecules for which all orbitals are doubly occupied) to orbital energies. To start, we note that for helium, the first ionization potential I is the energy change for the process He → He^+ + a stationary electron at infinity. Thus, the exact value of I for helium is given by $I = E(He^+) - E_0$, where E_0 is the exact helium atom ground state energy. The Hartree–Fock approximation I_{HF} to I is correspondingly $I_{HF} = E(He^+) - E_{HF}$. But from Equations (12.62) and (12.65), $E_{HF} = \varepsilon_1 + I_2$. Thus, $I_{HF} = E(He^+) - \varepsilon_1 - I_2$. But we have argued that approximately $I_2 \approx E(He^+)$. Therefore within this approximation, $I_{HF} \approx -\varepsilon_1$. Assuming $I \approx I_{HF}$, we have Koopmans' theorem

$$I \approx -\varepsilon_1 \tag{12.66}$$

which states that the first ionization potential of helium is approximately equal to the negative of the helium 1s orbital energy.

Koopmans' theorem for helium is based on two assumptions: (a) Electron correlation is small so that $E_{HF} \approx E_0$ and hence $I_{HF} \approx I$ and (b) orbital relaxation is slight so that $I_2 \approx E(He^+)$.

As shown in Section 12.7, analogous assumptions may be used to derive approximate Koopmans' theorem relations between experimental ionization potentials and orbital energies for more complex closed shell atoms. Namely, one assumes that (a) electron correlation effects are small so that the Hartree–Fock approximation may be used to determine the energies of both the ion and the neutral species and (b) orbital relaxation is slight so that the Hartree–Fock orbitals of the ion may be taken as those of the neutral atom. It is found empirically that the errors due to these two assumptions tend to cancel and so Koopmans' theorem ionization potentials can be in surprisingly good agreement with experiment for both atoms and molecules.

Let us illustrate this cancellation for helium. To do this, we need the following results for helium (Bunge, Barrientos, and Bunge 1993): $\varepsilon_1 = -0.9179$ au and $E_{HF} = -2.8617$ au. Thus, for helium the Koopmans' theorem ionization potential is $I_{KOOP} = -\varepsilon_1 = 0.9179\,\text{au} = 24.98\,\text{eV}$, which differs from the experimental ionization potential $I_{expt} = 24.59$ eV by only 1.59%. We may also determine the Hartree–Fock ionization potential $I_{HF} = E(He^+) - E_{HF}$. He^+ is a hydrogen-like atom, so its ground state energy is given exactly by Equation (12.4) as $E(He^+) = -2.0$ au. Thus, $I_{HF} = -2.0\text{ au} - (-2.8617)\text{ au} = 0.8617\text{ au} = 23.45\text{ eV}$, which is in poorer agreement with experiment than I_{KOOP}. This may seem surprising since Koopmans' theorem requires both the Hartree–Fock approximation and the additional assumption that orbital relaxation may be ignored. The explanation is the cancellation of errors just noted.

So far in this chapter, we have ignored an essential feature of the behavior of electrons in atoms (and molecules too); namely, that all electrons are identical and are thus indistinguishable. We next discuss this indistinguishability and its implications for the forms of atomic and molecular wave functions.

12.4 THE INDISTINGUISHABILITY OF ELECTRONS AND THE PAULI PRINCIPLE

As discussed in Section 10.7, electrons have an intrinsic angular momentum called spin. Consequently, the wave functions Ψ and probability densities $P = \Psi^*\Psi$ of an atom with, say, N electrons depend not only on the electrons' spatial coordinates $\mathbf{r}_1,\ldots,\mathbf{r}_N$ but also on their spin coordinates $\sigma_1,\ldots,\sigma_N$. Let us denote the space and spin coordinates for electron $k = 1,\cdots,N$ collectively by $\rho_k = \mathbf{r}_k, \sigma_k$. We may then write the wave functions Ψ as $\Psi(\rho_1,\ldots,\rho_N)$ and the probability densities P as $P(\rho_1,\ldots,\rho_N)$.

As illustrated in Figure 12.3, since the electrons are indistinguishable, the probability of finding electron i at space and spin point ρ_i and electron j at space and spin point ρ_j must be identical to the probability of finding electron i at space and spin point ρ_j and electron j at space and spin point ρ_i. That is, for all i and j

$$P(\rho_1,\ldots,\rho_i,\ldots,\rho_j,\ldots,\rho_N) = P(\rho_1,\ldots,\rho_j,\ldots,\rho_i,\ldots,\rho_N). \tag{12.67}$$

The symmetry relation of Equation (12.67) is actually not restricted to electrons but holds for any set of N indistinguishable particles.

Additionally, since $P = \Psi^*\Psi$, Equation (12.67) imposes the following restriction on the wave functions:

$$\Psi(\rho_1,\ldots,\rho_i,\ldots,\rho_j,\ldots,\rho_N) = \pm\Psi(\rho_1,\ldots,\rho_j,\ldots,\rho_i,\ldots,\rho_N). \tag{12.68}$$

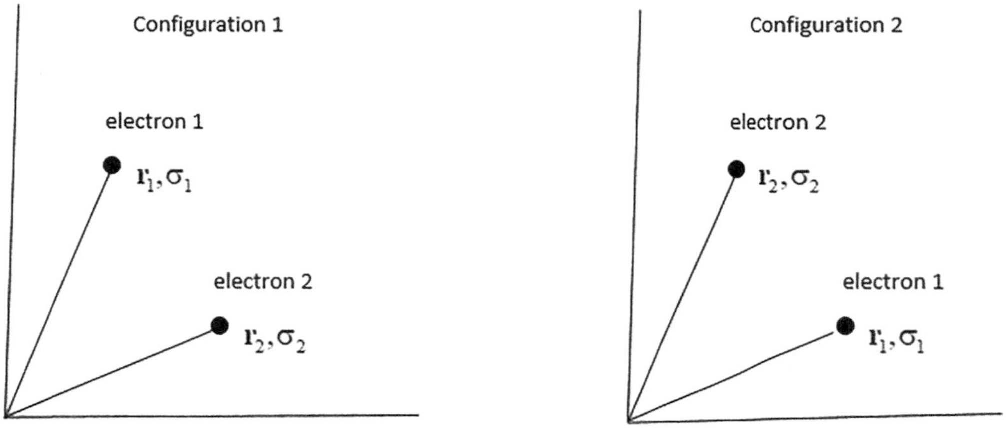

FIGURE 12.3 Since electrons 1 and 2 are indistinguishable, the probability $P(\mathbf{r}_1, \sigma_1, \mathbf{r}_2, \sigma_2)$ of configuration 1 is equal to the probability $P(\mathbf{r}_2, \sigma_2, \mathbf{r}_1, \sigma_1)$ of configuration 2.

In other words, the wave functions of any set of N indistinguishable particles must be either symmetric (positive sign) or antisymmetric (negative sign) with respect to the interchange of the space and spin coordinates of any pair of particles. It is found experimentally and is predicted theoretically that the wave functions of *fermions* or particles with half integral spin $\frac{1}{2}, \frac{3}{2}, \ldots$ are antisymmetric with respect to interchange, while the wave functions of *bosons* or particles with integral spin $0, 1, \ldots$ are symmetric with respect to interchange. Since electrons have spin $\frac{1}{2}$, they are fermions. Thus, we have arrived at the very important result:

the electronic wave functions $\Psi(\rho_1, \ldots \rho_N)$ of atoms and molecules must be antisymmetric with respect to interchange of the space and spin coordinates of any pair of electrons. (12.69)

More generally, for arbitrary N − electron systems the wave functions must conform to the condition that for all i and j

$$\Psi(\rho_1, \ldots, \rho_i, \ldots, \rho_j, \ldots, \rho_N) = -\Psi(\rho_1, \ldots, \rho_j, \ldots, \rho_i, \ldots, \rho_N). \tag{12.70}$$

The above antisymmetry requirement is referred to as the *Pauli principle* due to the Austrian physicist Wolfgang Pauli.

An important consequence of the Pauli principle is that two electrons with the same spin have zero probability of being at the same spatial point $\mathbf{r}$ and thus tend to repel one another. This effect is called *Fermi correlation*. It differs from the electron correlation due to repulsive Coulomb forces discussed earlier. For example, Fermi correlation, unlike Coulomb correlation, is included in orbital model wave functions if these properly incorporate the Pauli principle.

To see the origin of Fermi correlation, assume that electrons i and j have the same space and spin coordinates $\rho = \mathbf{r}, \sigma$. Then, evaluating Equation (12.70) for $\rho_i = \rho_j = \rho$ gives $\Psi(\rho_1, \ldots, \rho, \ldots, \rho, \ldots, \rho_N) = -\Psi(\rho_1, \ldots, \rho, \ldots, \rho, \ldots, \rho_N)$, which implies that when two electrons are at the same space and spin point $\rho = \mathbf{r}, \sigma$, their wave function vanishes. Consequently, two electrons with the same spin (or the same value of σ) have zero probability of being at the same spatial point $\mathbf{r}$.

We next return to the helium atom, but now we will properly account for electron spin and the Pauli principle.

12.5 THE PAULI PRINCIPLE AND THE 1s² GROUND STATE AND THE 1s2s EXCITED STATES OF HELIUM

Let us first consider the ground state of helium. Ignoring spin, a ground state orbital model wave function for helium is of the familiar form

$$\Psi(\mathbf{r}_1,\mathbf{r}_2) = \phi_{1s}(r_1)\phi_{1s}(r_2) \tag{12.71}$$

where $\phi_{1s}(r)$ is a normalized 1s orbital.

Next, let us introduce spin. Recalling the concepts of Section 10.7, one may construct two spin orbitals from the orbital $\phi_{1s}(r_1)$ and the spin eigenfunctions $\alpha(\sigma)$ and $\beta(\sigma)$. These spin orbitals are $\phi_{1s}(r)\alpha(\sigma)$ and $\phi_{1s}(r)\beta(\sigma)$. From these spin orbitals, one may build the following four possible helium ground state wave functions:

$$\Psi_1(\rho_1,\rho_2) = \phi_{1s}(r_1)\phi_{1s}(r_2)\alpha(\sigma_1)\alpha(\sigma_2) \tag{12.72a}$$

$$\Psi_2(\rho_1,\rho_2) = \phi_{1s}(r_1)\phi_{1s}(r_2)\beta(\sigma_1)\beta(\sigma_2) \tag{12.72b}$$

$$\Psi_3(\rho_1,\rho_2) = \phi_{1s}(r_1)\phi_{1s}(r_2)\alpha(\sigma_1)\beta(\sigma_2) \tag{12.72c}$$

and

$$\Psi_4(\rho_1,\rho_2) = \phi_{1s}(r_1)\phi_{1s}(r_2)\beta(\sigma_1)\alpha(\sigma_2) \tag{12.72d}$$

where $\rho_1 = \mathbf{r}_1,\sigma_1$ and $\rho_2 = \mathbf{r}_2,\sigma_2$ are the space and spin coordinates of electrons 1 and 2.

Let us now test the wave functions of Equation (12.72) to see if they are in accord with the Pauli principle. We note that since interchanging ρ_1 and ρ_2 in any one of these functions merely changes all subscripts 1 to subscripts 2 and vice versa, $\Psi_1(\rho_2,\rho_1) = \phi_{1s}(r_2)\phi_{1s}(r_1)\alpha(\sigma_2)\alpha(\sigma_1) = \phi_{1s}(r_1)\phi_{1s}(r_2)\alpha(\sigma_1)\alpha(\sigma_2) = \Psi_1(\rho_1,\rho_2)$. Similarly, $\Psi_2(\rho_2,\rho_1) = \Psi_2(\rho_1,\rho_2)$. Thus, both Ψ_1 and Ψ_2 are symmetric with respect to interchange of ρ_1 and ρ_2. They are therefore both ruled out as acceptable wave functions by the Pauli principle which requires antisymmetric wave functions. We similarly find that $\Psi_3(\rho_2,\rho_1) = \Psi_4(\rho_1,\rho_2)$ and $\Psi_4(\rho_2,\rho_1) = \Psi_3(\rho_1,\rho_2)$. Thus, Ψ_3 and Ψ_4 are neither symmetric nor antisymmetric and therefore also are not acceptable wave functions. Consider however the function

$$\Psi_-(\rho_1,\rho_2) = N\left[\Psi_3(\rho_1,\rho_2) - \Psi_4(\rho_1,\rho_2)\right] \tag{12.73}$$

where N is a normalization factor. From the symmetry properties of Ψ_3 and Ψ_4, it follows that

$$\Psi_-(\rho_2,\rho_1) = N\left[\Psi_3(\rho_2,\rho_1) - \Psi_4(\rho_2,\rho_1)\right] = N\left[\Psi_4(\rho_1,\rho_2) - \Psi_3(\rho_1,\rho_2)\right] = -\Psi_-(\rho_1,\rho_2).$$

Therefore, Ψ_- is antisymmetric and thus is an acceptable ground state wave function.

The antisymmetry of Ψ_- may also be seen by writing out its explicit form. From Equations (12.72c), (12.72d), and (12.73), this form is

$$\Psi_-(\rho_1,\rho_2) = N\phi_{1s}(r_1)\phi_{1s}(r_2)\left[\alpha(\sigma_1)\beta(\sigma_2) - \beta(\sigma_1)\alpha(\sigma_2)\right]. \tag{12.74}$$

The spatial part of Ψ_- is symmetric with respect to interchange of r_1 and r_2, but the spin part is antisymmetric with respect to interchange of σ_1 and σ_2 rendering the full wave function antisymmetric with respect to interchange of ρ_1 and ρ_2.

The constant N is determined by the normalization requirement

$$\int \Psi_-^*(\rho_1,\rho_2)\Psi_-(\rho_1,\rho_2)d\mathbf{r}_1 d\mathbf{r}_2 d\sigma_1 d\sigma_2 = 1. \tag{12.75}$$

Inserting Equation (12.74) into Equation (12.75) and performing the integrals over $\mathbf{r}_1$ and $\mathbf{r}_2$ noting that $\phi_{1s}(r_1)$ and $\phi_{1s}(r_2)$ are normalized yields the following condition for N:

$$N^2\int [\alpha(\sigma_1)\beta(\sigma_2)-\beta(\sigma_1)\alpha(\sigma_2)]^* [\alpha(\sigma_1)\beta(\sigma_2)-\beta(\sigma_1)\alpha(\sigma_2)]d\sigma_1 d\sigma_2 = 1. \tag{12.76}$$

From this condition and the spin orthonormality relations of Equation (10.121), it follows that $N=\frac{1}{\sqrt{2}}$. Thus, the normalized form of Ψ_- is

$$\Psi_-(\rho_1,\rho_2)=\frac{1}{\sqrt{2}}\phi_{1s}(r_1)\phi_{1s}(r_2)[\alpha(\sigma_1)\beta(\sigma_2)-\beta(\sigma_1)\alpha(\sigma_2)]. \tag{12.77}$$

In our earlier treatments of the ground state of helium, we ignored spin and thus used orbital model wave functions like $\Psi(\mathbf{r}_1,\mathbf{r}_2)$ of Equation (12.71) which are not in accord with the Pauli principle. Instead, we should have used wave functions like $\Psi_-(\rho_1,\rho_2)$ of Equation (12.77) which conform to the principle. However, disregarding the Pauli principle introduces no error for the simple problem of the helium ground state and so our earlier results are correct. This may be seen by comparing the energies determined from $\Psi(\mathbf{r}_1,\mathbf{r}_2)$ and $\Psi_-(\rho_1,\rho_2)$. These energies are

$$E=\int \Psi^*(\mathbf{r}_1,\mathbf{r}_2)\hat{H}\Psi(\mathbf{r}_1,\mathbf{r}_2)d\mathbf{r}_1 d\mathbf{r}_2 \tag{12.78}$$

and

$$E_-=\int \Psi_-^*(\rho_1,\rho_2)\hat{H}\Psi_-(\rho_1,\rho_2)d\mathbf{r}_1 d\mathbf{r}_2 d\sigma_1 d\sigma_2 \tag{12.79}$$

where $\hat{H}$ is the helium atom Hamiltonian operator. Since from Equation (12.49) $\hat{H}$ is independent of spin variables, it is readily shown that $E_- = E$ (Problem 12.18). That is, the ground state energy of helium depends only on the spatial part of the wave function of (12.77) $\phi_{1s}(r_1)\phi_{1s}(r_2)$. This result is sufficient to establish that all of our earlier results for the helium ground state (including the Hartree–Fock equations) derived ignoring spin are correct.

Next, let us consider the helium atom excited states that derive from the 1s2s electron configuration. We will form zeroth-order wave functions for these states from 1s and 2s orbitals $\phi_{1s}(r)$ and $\phi_{2s}(r)$ and from the spin eigenfunctions $\alpha(\sigma)$ and $\beta(\sigma)$. We assume that both orbitals are normalized and that $\phi_{1s}(r)$ and $\phi_{2s}(r)$ are orthogonal.

Let us first consider the spatial parts of the wave functions. We might guess that these spatial parts are $\phi_{1s}(r_1)\phi_{2s}(r_2)$ and $\phi_{1s}(r_2)\phi_{2s}(r_1)$. These guesses are wrong. Because of the indistinguishability of the electrons, it is not possible to say, for example, that electron 1 definitely occupies the 1s orbital and electron 2 definitely occupies the 2s orbital as is implied by the function $\phi_{1s}(r_1)\phi_{2s}(r_2)$. Instead, the correct zeroth-order spatial wave functions are the normalized symmetric and antisymmetric linear combinations

$$\Psi_+(\mathbf{r}_1,\mathbf{r}_2)=\frac{1}{\sqrt{2}}[\phi_{1s}(r_1)\phi_{2s}(r_2)+\phi_{1s}(r_2)\phi_{2s}(r_1)] \tag{12.80}$$

and

$$\Psi_-(\mathbf{r}_1,\mathbf{r}_2)=\frac{1}{\sqrt{2}}[\phi_{1s}(r_1)\phi_{2s}(r_2)-\phi_{1s}(r_2)\phi_{2s}(r_1)]. \tag{12.81}$$

Next, let us include spin.

Since $\Psi_+(\mathbf{r}_1,\mathbf{r}_2)$ is symmetric in the spatial variables $\mathbf{r}_1$ and $\mathbf{r}_2$, the Pauli principle requires that the spin part of the full wave function formed from $\Psi_+(\mathbf{r}_1,\mathbf{r}_2)$ be antisymmetric in the spin variables σ_1 and σ_2. There is only one normalized antisymmetric spin wave function: $\frac{1}{\sqrt{2}}[\alpha(\sigma_1)\beta(\sigma_2)-\alpha(\sigma_2)\beta(\sigma_1)]$. So there is only one full wave function with spatial part $\Psi_+(\mathbf{r}_1,\mathbf{r}_2)$:

$$\Psi_S(\rho_1,\rho_2)=\frac{1}{\sqrt{2}}[\phi_{1s}(r_1)\phi_{2s}(r_2)+\phi_{1s}(r_2)\phi_{2s}(r_1)]$$
$$\times\frac{1}{\sqrt{2}}[\alpha(\sigma_1)\beta(\sigma_2)-\alpha(\sigma_2)\beta(\sigma_1)]. \tag{12.82}$$

Correspondingly, since $\Psi_-(\mathbf{r}_1,\mathbf{r}_2)$ is antisymmetric in the spatial variables $\mathbf{r}_1$ and $\mathbf{r}_2$, the Pauli principle requires that the spin part of a full wave function formed from $\Psi_-(\mathbf{r}_1,\mathbf{r}_2)$ be symmetric in the spin variables σ_1 and σ_2. There are three normalized symmetric spin wave functions: $\alpha(\sigma_1)\alpha(\sigma_2)$, $\frac{1}{\sqrt{2}}[\alpha(\sigma_1)\beta(\sigma_2)+\beta(\sigma_1)\alpha(\sigma_2)]$, and $\beta(\sigma_1)\beta(\sigma_2)$. Thus, there are three full wave functions with spatial part $\Psi_-(\mathbf{r}_1,\mathbf{r}_2)$. These are

$$\Psi_{T_1}(\rho_1,\rho_2)=\frac{1}{\sqrt{2}}[\phi_{1s}(r_1)\phi_{2s}(r_2)-\phi_{1s}(r_2)\phi_{2s}(r_1)]\alpha(\sigma_1)\alpha(\sigma_2) \tag{12.83a}$$

$$\Psi_{T_2}(\rho_1,\rho_2)=\frac{1}{\sqrt{2}}[\phi_{1s}(r_1)\phi_{2s}(r_2)-\phi_{1s}(r_2)\phi_{2s}(r_1)]\frac{1}{\sqrt{2}}[\alpha(\sigma_1)\beta(\sigma_2)+\alpha(\sigma_2)\beta(\sigma_1)] \tag{12.83b}$$

and

$$\Psi_{T_3}(\rho_1,\rho_2)=\frac{1}{\sqrt{2}}[\phi_{1s}(r_1)\phi_{2s}(r_2)-\phi_{1s}(r_2)\phi_{2s}(r_1)]\beta(\sigma_1)\beta(\sigma_2). \tag{12.83c}$$

Since there is only one quantum state with spatial wave function $\Psi_+(\mathbf{r}_1,\mathbf{r}_2)$, we will refer to it as a *singlet state* (subscript S on the wave function). Since there are three quantum states with spatial wave function $\Psi_-(\mathbf{r}_1,\mathbf{r}_2)$, we will refer to these as *triplet states* (subscripts T_1, T_2, and T_3 on the wave functions). (The interpretation of singlet and triplet states is further developed in Problem 12.31 and Section 12.10.)

Next, let us evaluate the energies of the singlet and triplet states. Since the helium atom Hamiltonian operator $\hat{H}$ is independent of spin, as for the ground state energy the singlet and triplet energies depend only on the spatial parts of the wave functions. Thus, the energy of the singlet state is

$$E_S=\int\Psi_+^*(\mathbf{r}_1,\mathbf{r}_2)\hat{H}\Psi_+(\mathbf{r}_1,\mathbf{r}_2)d\mathbf{r}_1d\mathbf{r}_2 \tag{12.84}$$

while the energy of the three triplet states is

$$E_T=\int\Psi_-^*(\mathbf{r}_1,\mathbf{r}_2)\hat{H}\Psi_-(\mathbf{r}_1,\mathbf{r}_2)d\mathbf{r}_1d\mathbf{r}_2. \tag{12.85}$$

Using the form of $\hat{H}$ given in Equation (12.49), the forms for $\Psi_+(\mathbf{r}_1,\mathbf{r}_2)$ and $\Psi_-(\mathbf{r}_1,\mathbf{r}_2)$ given in Equations (12.80) and (12.81), and the orthonormality of the orbitals $\phi_{1s}(r)$ and $\phi_{2s}(r)$, one may show that (Problem 12.19)

$$E_S=I_{1s}+I_{2s}+J_{1s,2s}+K_{1s,2s} \tag{12.86}$$

and

$$E_T = I_{1s} + I_{2s} + J_{1s,2s} - K_{1s,2s} \tag{12.87}$$

where

$$I_{1s} = \int \phi_{1s}^*(r)\left(-\frac{1}{2}\nabla^2 - \frac{2}{r}\right)\phi_{1s}(r)\,d\mathbf{r} \tag{12.88}$$

$$I_{2s} = \int \phi_{2s}^*(r)\left(-\frac{1}{2}\nabla^2 - \frac{2}{r}\right)\phi_{2s}(r)\,d\mathbf{r} \tag{12.89}$$

$$J_{1s,2s} = \int \phi_{1s}^*(r_1)\,\phi_{2s}^*(r_2)\frac{1}{r_{12}}\phi_{1s}(r_1)\,\phi_{2s}(r_2)\,d\mathbf{r}_1 d\mathbf{r}_2 \tag{12.90}$$

and

$$K_{1s,2s} = \int \phi_{1s}^*(r_1)\,\phi_{2s}^*(r_2)\frac{1}{r_{12}}\phi_{2s}(r_1)\,\phi_{1s}(r_2)\,d\mathbf{r}_1 d\mathbf{r}_2. \tag{12.91}$$

Next, consider the interpretations of $I_{1s}, I_{2s}, J_{1s,2s}$, and $K_{1s,2s}$. First, since $-\frac{1}{2}\nabla^2 - \frac{2}{r}$ is the He^+ ion Hamiltonian operator if $\phi_{1s}(r)$ and $\phi_{2s}(r)$ were chosen as the He^+ 1s and 2s orbitals, then I_{1s} and I_{2s} would be the ground and first excited state He^+ ion energies. Next, the *Coulomb integral* $J_{1s,2s}$ is the potential energy of electrostatic repulsion of two smeared-out electrons, one with the 1s orbital charge density $-\phi_{1s}^*(r_1)\phi_{1s}(r_1)$ and the other with the 2s orbital charge density $-\phi_{2s}^*(r_2)\phi_{2s}(r_2)$. Finally, the *exchange integral* $K_{1s,2s}$ has no classical interpretation since it arises from the quantum indistinguishability of the electrons.

Let us next estimate the values of E_S and E_T. To do this, we make an approximation in the spirit of first-order perturbation theory; namely, we take $\phi_{1s}(r)$ and $\phi_{2s}(r)$ to be the He^+ 1s and 2s orbitals. Then, as noted I_{1s} and I_{2s} are the ground and first excited state He^+ ion energies. Thus, from Equation (12.4) $I_{1s} = -2.0$ au and $I_{2s} = -0.5$ au. Also for this choice of orbitals, one may show using the methods of Problems 12.8 and 12.9 that $J_{1s,2s} = \frac{34}{81}$ au and $K_{1s,2s} = \frac{32}{729}$ au. Therefore, from Equations (12.86) and (12.87) we have the estimates

$$E_S = -2.0\text{ au} - 0.5\text{ au} + \frac{34}{81}\text{ au} + \frac{32}{729}\text{ au} = -2.0364\text{ au} \tag{12.92}$$

and

$$E_T = -2.0\text{ au} - 0.5\text{ au} + \frac{34}{81}\text{ au} - \frac{32}{729}\text{ au} = -2.1241\text{ au}. \tag{12.93}$$

Notice that the energy E_T of the triplet states is lower than the energy E_s of the singlet state. This is expected since the triplet spatial wave function $\Psi_-(\mathbf{r}_1, \mathbf{r}_2)$ vanishes when $\mathbf{r}_1 = \mathbf{r}_2$ and thus predicts a lower probability than the singlet spatial wave function $\Psi_+(\mathbf{r}_1, \mathbf{r}_2)$ for the electrons to be in close proximity.

From Equations (12.92) and (12.93), we may estimate the singlet–triplet splitting $E_S - E_T$ as 0.08777 au. This estimate is in poor agreement with the experimental splitting which is 0.0293 au. Thus, while our simple treatment correctly predicts that the triplet states are lower in energy than the singlet state, it fails to predict an accurate value for the splitting.

Next, let us turn to a new topic. In our first discussions of the helium atom, we defined orbital model wave functions as products of one-electron spatial wave functions $\phi(r)$. However, this definition must be extended since, as we have seen, simple orbital product functions are not in accord with the Pauli principle and must be replaced by antisymmetric wave functions which include spin. We next show how orbital model wave functions in accord with the Pauli principle may be generally defined in terms of *Slater determinants.*

12.6 SLATER DETERMINANTS

Consider the determinant

$$D = \begin{vmatrix} a_{11} & a_{12} & \cdots & a_{1n} \\ a_{21} & a_{22} & \cdots & a_{2n} \\ \vdots & & & \\ a_{n1} & a_{n2} & \cdots & a_{nn} \end{vmatrix}. \tag{12.94}$$

D has two properties which we will need.

Property 1. Interchanging any two columns or rows of D changes its sign.
Property 2. If any two columns or rows of D are identical, then D = 0.

We can easily verify these properties for 2×2 determinants. Consider the determinant

$$D_2 = \begin{vmatrix} a_{11} & a_{12} \\ a_{21} & a_{22} \end{vmatrix} = a_{11}a_{22} - a_{12}a_{21}. \tag{12.95}$$

Interchanging the two columns of D_2 gives the determinant

$$\begin{vmatrix} a_{12} & a_{11} \\ a_{22} & a_{21} \end{vmatrix} = a_{12}a_{21} - a_{11}a_{22} \tag{12.96}$$

which is the negative of D_2. Interchanging the two rows of D_2 gives the determinant

$$\begin{vmatrix} a_{21} & a_{22} \\ a_{11} & a_{12} \end{vmatrix} = a_{21}a_{12} - a_{22}a_{11} \tag{12.97}$$

which is also the negative of D_2. Therefore, we have verified Property 1. Next, note that the following determinant with identical columns

$$\begin{vmatrix} a_{11} & a_{11} \\ a_{21} & a_{21} \end{vmatrix} = a_{11}a_{21} - a_{11}a_{21} \tag{12.98}$$

is equal to zero. Similarly, note that the following determinant with identical rows

$$\begin{vmatrix} a_{11} & a_{12} \\ a_{11} & a_{12} \end{vmatrix} = a_{11}a_{12} - a_{12}a_{11} \tag{12.99}$$

is also equal to zero. Thus, we have verified Property 2.

Next, let us return to the problem of building orbital model wave functions in accord with the Pauli principle. We will see, as first noted by Slater, that such wave functions are naturally represented as determinants since Property 1 guarantees that determinantal wave functions are antisymmetric.

To start, consider the helium atom ground state wave function $\Psi(\rho_1,\rho_2) \equiv \Psi_-(\rho_1,\rho_2)$ of Equation (12.77). It can be expressed as the following Slater determinantal wave function:

$$\tilde{\Psi}(\rho_1,\rho_2) = \frac{1}{\sqrt{2}}\begin{vmatrix} \phi_{1s}(r_1)\alpha(\sigma_1) & \phi_{1s}(r_1)\beta(\sigma_1) \\ \phi_{1s}(r_2)\alpha(\sigma_2) & \phi_{1s}(r_2)\beta(\sigma_2) \end{vmatrix}. \tag{12.100}$$

Notice that in the above determinant, the spin orbital is unchanged down a column and the electron coordinates are unchanged across a row. This is the general form of a Slater determinant.

We next prove from Property 1 that $\Psi(\rho_1,\rho_2)$ is antisymmetric. To do this, recall that interchanging ρ_1 and ρ_2 in $\Psi(\rho_1,\rho_2)$ merely changes all subscripts 1 to subscripts 2 and vice versa. Thus, referring to Equation (12.100),

$$\tilde{\Psi}(\rho_2,\rho_1) = \frac{1}{\sqrt{2}}\begin{vmatrix} \phi_{1s}(r_2)\alpha(\sigma_2) & \phi_{1s}(r_2)\beta(\sigma_2) \\ \phi_{1s}(r_1)\alpha(\sigma_1) & \phi_{1s}(r_1)\beta(\sigma_1) \end{vmatrix}. \tag{12.101}$$

But the above determinant is the negative of the determinant of Equation (12.100) since it may be derived from that determinant by interchanging its rows. Thus, we have proven the antisymmetry property $\Psi(\rho_2,\rho_1) = -\Psi(\rho_1,\rho_2)$.

We next show using Property 2 how one may derive the Pauli exclusion principle from the more fundamental Pauli principle taking helium as an example. Recall that the Pauli exclusion principle is the statement that two electrons with the same spin cannot occupy the same spatial orbital (since this would make the four quantum numbers of both electrons the same) or equivalently that two electrons cannot occupy the same spin orbital. Thus, for helium the electronic state deriving from the spin orbital configuration $1s\alpha 1s\alpha$ which puts both electrons in the spin orbital $1s\alpha = \phi_{1s}(r)\alpha(\sigma)$ is ruled out by the Pauli exclusion principle. The $1s\alpha 1s\alpha$ state is also ruled out by the Pauli principle antisymmetry requirement since the Slater determinantal wave function for that state

$$\tilde{\Psi}(\rho_1,\rho_2) = \frac{1}{\sqrt{2}}\begin{vmatrix} \phi_{1s}(r_1)\alpha(\sigma_1) & \phi_{1s}(r_1)\alpha(\sigma_1) \\ \phi_{1s}(r_2)\alpha(\sigma_2) & \phi_{1s}(r_2)\alpha(\sigma_2) \end{vmatrix} \tag{12.102}$$

vanishes because the two columns of the determinant are identical.

The proceeding argument is easily generalized. One may show for any atom that all spin orbital configurations which violate the Pauli exclusion principle by putting more than one electron in a spin orbital give rise to vanishing determinantal wave functions with identical columns (Problem 12.21).

We next move on to the lithium atom. For the lowest energy states of lithium, its three electrons all occupy the 1s orbital. These states are however forbidden by the Pauli exclusion principle since for all of them two or three electrons occupy the spin orbitals $1s\alpha = \phi_{1s}(r)\alpha(\sigma)$ or $1s\beta = \phi_{1s}(r)\beta(\sigma)$.

The true ground state of lithium arises from the degenerate spin orbital configurations $1s\alpha 1s\beta 2s\alpha$ and $1s\alpha 1s\beta 2s\beta$ which put one electron in the spin orbital $1s\alpha$, one in the spin orbital $1s\beta$, and one in either of the spin orbitals $2s\alpha$ or $2s\beta$, but for reasons discussed in Section 12.9, not in the spin orbitals $2p\alpha$ or $2p\beta$. The Slater determinantal wave function for, say, the $1s\alpha 1s\beta 2s\alpha$ state is in analogy to the helium wave function of Equation (12.100)

$$\Psi(\rho_1,\rho_2,\rho_3) = \frac{1}{\sqrt{3!}}\begin{vmatrix} \phi_{1s}(r_1)\alpha(\sigma_1) & \phi_{1s}(r_1)\beta(\sigma_1) & \phi_{2s}(r_1)\alpha(\sigma_1) \\ \phi_{1s}(r_2)\alpha(\sigma_2) & \phi_{1s}(r_2)\beta(\sigma_2) & \phi_{2s}(r_2)\alpha(\sigma_2) \\ \phi_{1s}(r_3)\alpha(\sigma_3) & \phi_{1s}(r_3)\beta(\sigma_3) & \phi_{2s}(r_3)\alpha(\sigma_3) \end{vmatrix}. \tag{12.103}$$

The wave function Ψ is normalized if $\phi_{1s}(r)$ and $\phi_{2s}(r)$ are orthonormal (Problem 12.22b). Also by extending the argument used for helium, one may show that Ψ is fully antisymmetric and thus is in accord with the Pauli principle.

Finally, let us consider atoms with an arbitrary number of electrons. Such atoms can have any spin orbital configuration which is in accord with the Pauli exclusion principle. (See Problem 12.21.) Here however, we will only consider atoms with closed shell configurations; that is, we will consider atoms like ground state helium, beryllium, neon, and argon for which all orbitals are doubly occupied by electrons with opposite spin. We restrict ourselves to closed shell atoms since the Hartree–Fock equations are simplest for such atoms. The Slater determinantal wave function for a closed shell atom with N doubly occupied orbitals $\phi_1(\mathbf{r}),\ldots,\phi_N(\mathbf{r})$ and hence 2N electrons is of the fully antisymmetric form

$$\Psi(\rho_1,\ldots,\rho_{2N}) = \frac{1}{\sqrt{(2N)!}}\begin{vmatrix} \phi_1(\mathbf{r}_1)\alpha(\sigma_1) & \phi_1(\mathbf{r}_1)\beta(\sigma_1) & \cdots & \phi_N(\mathbf{r}_1)\alpha(\sigma_1) & \phi_N(\mathbf{r}_1)\beta(\sigma_1) \\ \phi_1(\mathbf{r}_2)\alpha(\sigma_2) & \phi_1(\mathbf{r}_2)\beta(\sigma_2) & \cdots & \phi_N(\mathbf{r}_2)\alpha(\sigma_2) & \phi_N(\mathbf{r}_2)\beta(\sigma_2) \\ \vdots & \vdots & & \vdots & \vdots \\ \phi_1(\mathbf{r}_{2N})\alpha(\sigma_{2N}) & \phi_1(\mathbf{r}_{2N})\beta(\sigma_{2N}) & \cdots & \phi_N(\mathbf{r}_{2N})\alpha(\sigma_{2N}) & \phi_N(\mathbf{r}_{2N})\beta(\sigma_{2N}) \end{vmatrix} \tag{12.104}$$

Since an $n \times n$ determinant when expanded gives rise to $n!$ terms, the above Slater determinant expands into $(2N)!$ terms. A consequence is that if we choose the orbitals to be orthonormal, then Ψ is normalized; that is,

$$\int \Psi^*(\rho_1,\ldots,\rho_{2N})\Psi(\rho_1,\ldots,\rho_{2N})\,d\mathbf{r}_1 d\sigma_1 \ldots d\mathbf{r}_{2N} d\sigma_{2N} = 1. \tag{12.105}$$

For closed shell atoms, Ψ of Equation (12.104) is the general orbital model wave function in accord with the Pauli principle which we have been seeking. Thus, it is natural to ask: What are the best orbitals? Namely, what are the orbitals $\phi_1(\mathbf{r}),\ldots,\phi_N(\mathbf{r})$ which give the lowest energy Slater determinantal wave function Ψ? Since Ψ is normalized, the atom's energy is given by

$$E = \int \Psi^*(\rho_1,\ldots,\rho_{2N})\hat{H}\Psi(\rho_1,\ldots,\rho_{2N})\,d\mathbf{r}_1 d\sigma_1 \ldots d\mathbf{r}_{2N} d\sigma_{2N} \tag{12.106}$$

where $\hat{H}$ is the electronic Hamiltonian operator of the atom. The best orbitals are those which minimize E. They turn out to be the solutions of the Hartree–Fock equations for closed shell atoms which we next develop.

12.7 THE HARTREE–FOCK EQUATIONS FOR CLOSED SHELL ATOMS AND KOOPMANS' THEOREM

We start by evaluating the energy E. To do this, we need the form of the 2N-electron electronic Hamiltonian operator $\hat{H}$ which for an atom of atomic number Z is (compare Problem 12.23a)

$$\hat{H} = \sum_{i=1}^{2N}\left(-\frac{1}{2}\nabla_i^2 - \frac{Z}{r_i}\right) + \sum_{i=1}^{2N-1}\sum_{j=i+1}^{2N}\frac{1}{r_{ij}} \tag{12.107}$$

where $-\frac{1}{2}\nabla_i^2$ is the kinetic energy operator of electron i, $-\frac{Z}{r_i}$ is the potential energy due to the attractive Coulomb interaction of electron i with the nucleus, and $\frac{1}{r_{ij}}$ is the potential energy due to the Coulomb repulsion of electrons i and j. The limits of the double sum are chosen so that the Coulomb repulsion energies $\frac{1}{r_{ij}}$ between all of the electrons are included but none are double-counted and so that the infinite terms $\frac{1}{r_{ii}}$ are excluded.

Evaluation of E from Equations (12.104), (12.106), and (12.107) is complex. So we will skip the derivation and pass directly to the result. Assuming that the orbitals are orthonormal so that for $i, j = 1, \ldots, N$

$$\int \phi_i^*(\mathbf{r})\phi_j(\mathbf{r})d\mathbf{r} = \delta_{ij} \tag{12.108}$$

one finds that

$$E = 2\sum_{i=1}^{N} I_i + \sum_{i=1}^{N}\sum_{j=1}^{N}(2J_{ij} - K_{ij}) \tag{12.109}$$

where

$$I_i = \int \phi_i^*(\mathbf{r}_1)\left(-\frac{1}{2}\nabla_1^2 - \frac{Z}{r_1}\right)\phi_i(\mathbf{r}_1)d\mathbf{r}_1 \tag{12.110}$$

$$J_{ij} = \int \phi_i^*(\mathbf{r}_1)\phi_j^*(\mathbf{r}_2)\frac{1}{r_{12}}\phi_i(\mathbf{r}_1)\phi_j(\mathbf{r}_2)d\mathbf{r}_1 d\mathbf{r}_2 \tag{12.111}$$

and

$$K_{ij} = \int \phi_i^*(\mathbf{r}_1)\phi_j^*(\mathbf{r}_2)\frac{1}{r_{12}}\phi_i(\mathbf{r}_2)\phi_j(\mathbf{r}_1)d\mathbf{r}_1 d\mathbf{r}_2 \tag{12.112}$$

I_i is the "bare nucleus" energy, that is, the average energy an electron in orbital $\phi_i(\mathbf{r})$ would have in the absence of all other electrons. J_{ij} is a Coulomb integral and is equal to the repulsive potential energy of two smeared-out electrons with charge densities $-\phi_i^*(\mathbf{r}_1)\phi_i(\mathbf{r}_1)$ and $-\phi_j^*(\mathbf{r}_2)\phi_j(\mathbf{r}_2)$. K_{ij} is an exchange integral which arises from the antisymmetry of the wave function Ψ. When $i = j$, $K_{ij} = J_{ij}$ and K_{ij} may be interpreted as a potential energy. But for $i \neq j$, K_{ij} has no classical interpretation.

One may now derive the Hartree–Fock equations by minimizing E of Equation (12.109) with respect to the forms of the orbitals $\phi_i(\mathbf{r})$. This minimization must be carried out subject to the orthonormality constraints of Equation (12.108). As in our derivation of the Hartree–Fock equations for the helium-like atoms, the constrained minimization problem is solved using the method of Lagrange multipliers.

The results of the derivation are the Hartree–Fock equations for the best orbitals $\phi_1(\mathbf{r}_1), \ldots, \phi_N(\mathbf{r}_N)$ which for $i = 1, \ldots, N$ are

$$\hat{F}\phi_i(\mathbf{r}_1) = \varepsilon_i\phi_i(\mathbf{r}_1) \tag{12.113}$$

where ε_i is the ith orbital energy and where $\hat{F}$ is the *Fock operator* which is given by

$$\hat{F} = -\frac{1}{2}\nabla_1^2 - \frac{Z}{r_1} + \sum_{j=1}^{N}(2\hat{J}_j - \hat{K}_j) \tag{12.114}$$

with the *Coulomb operator* $\hat{J}_j$ and the *exchange operator* $\hat{K}_j$ being defined by

$$\hat{J}_j\phi_i(\mathbf{r}_1) = \int \frac{\phi_j^*(\mathbf{r}_2)\phi_j(\mathbf{r}_2)}{r_{12}} d\mathbf{r}_2\phi_i(\mathbf{r}_1) \tag{12.115}$$

and

$$\hat{K}_j\phi_i(\mathbf{r}_1) = \int \frac{\phi_j^*(\mathbf{r}_2)\phi_i(\mathbf{r}_2)}{r_{12}} d\mathbf{r}_2\phi_j(\mathbf{r}_1). \tag{12.116}$$

The Coulomb operator $\hat{J}_j$ is the repulsive potential energy an electron at point $\mathbf{r}_1$ experiences from a smeared-out electron with charge density $-\phi_j^*(\mathbf{r}_2)\phi_j(\mathbf{r}_2)$. The exchange terms $\hat{K}_j\phi_i(\mathbf{r})$ for $i \neq j$ have no classical interpretation since these arise from the antisymmetry of the wave function. Notice that in the exchange terms, the orbital ϕ_i appears *inside* the integral sign.

The best orbitals $\phi_1,\ldots,\phi_N$ are the solutions of the Hartree–Fock equations (Equation 12.113) with the N lowest eigenvalues ε_i. Since $\hat{F}\phi_i(\mathbf{r}_1)$ depends on the unknown orbitals $\phi_j(\mathbf{r})$, these solutions are obtained by an iterative self-consistent procedure similar to that used in Section 12.3 to solve the Hartree–Fock equations for the helium-like atoms. (Namely, one makes initial guesses for the unknown orbitals and so on.) This iterative solution may be conveniently carried out using Roothaan's basis set method (Bunge, Barrientos, and Bunge 1993).

Next, let us turn to the orbital energies ε_i. From the Hartree–Fock equations, one may show for $i = 1,\ldots,N$ that these are given by (Problem 12.24)

$$\varepsilon_i = I_i + \sum_{j=1}^{N}\left(2J_{ij} - K_{ij}\right). \tag{12.117}$$

Adding up the orbital energies for all 2N electrons gives

$$2\sum_{i=1}^{N}\varepsilon_i = 2\sum_{i=1}^{N} I_i + 2\sum_{i=1}^{N}\sum_{j=1}^{N}\left(2J_{ij} - K_{ij}\right) \tag{12.118}$$

which is not equal to the energy E of Equation (12.109). This is because summing the orbital energies double-counts the Coulomb and exchange integrals. The correct relation between the sum of the orbital energies and E is

$$2\sum_{i=1}^{N}\varepsilon_1 = E + \sum_{i=1}^{N}\sum_{j=1}^{N}\left(2J_{ij} - K_{ij}\right). \tag{12.119}$$

Before going on, we note that all of our results for the general Hartree–Fock theory properly reduce to our helium-like atom Hartree–Fock results of Section 12.3 (Problem 12.25).

We next turn to Koopmans' theorem which recall is a general result for closed shell atoms and molecules which relates orbital energies and ionization potentials. We discussed Koopmans' theorem for the helium atom in Section 12.3. Here, we will give a more rigorous and general development of the theorem. The proof of Koopmans' theorem for an arbitrary 2N–electron closed shell system is complex. So here we will prove it for the four-electron ground state beryllium atom with doubly occupied 1s and 2s orbitals. Aside from helium, beryllium is the simplest closed shell atom.

We start by specializing the energy expression of Equation (12.109) to beryllium with N = 2. Using the relations $J_{ji} = J_{ij}$, $K_{ji} = K_{ij}$, and $K_{ii} - J_{ii}$, we find the beryllium energy as

$$E = 2I_1 + 2I_2 + J_{11} + 4J_{12} - 2K_{12} + J_{22}. \tag{12.120}$$

We assume the integrals I_1, I_2, and so on are evaluated using the beryllium Hartree–Fock orbitals $\phi_1(\mathbf{r})$ and $\phi_2(\mathbf{r})$. Then, E is the Hartree–Fock energy of beryllium. Noting that since for ground state beryllium only the 1s and 2s orbitals are occupied, it follows that the subscripts 1 and 2 on the integrals in Equation (12.120) are associated with, respectively, these 1s and 2s orbitals. Given this association, it is natural to make the following definitions: $I_{1s} = I_1$, $I_{2s} = I_2$, $J_{1s,1s} = J_{11}$, $J_{1s,2s} = J_{12}$, $K_{1s,2s} = K_{12}$, and $J_{2s,2s} = J_{22}$. With these definitions, E becomes

$$E = 2I_{1s} + 2I_{2s} + J_{1s,1s} + 4J_{1s,2s} - 2K_{1s,2s} + J_{2s,2s}. \tag{12.121}$$

Additionally, because of the association, $\phi_1(\mathbf{r}) = \phi_{1s}(\mathbf{r})$ and $\phi_2(\mathbf{r}) = \phi_{2s}(\mathbf{r})$ where $\phi_{1s}(\mathbf{r})$ and $\phi_{2s}(\mathbf{r})$ are the 1s and 2s beryllium Hartree–Fock orbitals.

Given these correspondences and noting that for beryllium $Z = 4$, we have from Equation (12.110) that

$$I_{1s} = \int \phi_{1s}^*(\mathbf{r}_1)\left(-\frac{1}{2}\nabla_1^2 - \frac{4}{r_1}\right)\phi_{1s}(\mathbf{r}_1)\,d\mathbf{r}_1 \tag{12.122}$$

and

$$I_{2s} = \int \phi_{2s}^*(\mathbf{r}_1)\left(-\frac{1}{2}\nabla_1^2 - \frac{4}{r_1}\right)\phi_{2s}(\mathbf{r}_1)\,d\mathbf{r}_1. \tag{12.123}$$

Similarly, from Equations (12.111) and (12.112)

$$J_{1s,1s} = \int \phi_{1s}^*(\mathbf{r}_1)\phi_{1s}^*(\mathbf{r}_2)\frac{1}{r_{12}}\phi_{1s}(\mathbf{r}_1)\phi_{1s}(\mathbf{r}_2)\,d\mathbf{r}_1 d\mathbf{r}_2 \tag{12.124}$$

$$J_{1s,2s} = \int \phi_{1s}^*(\mathbf{r}_1)\phi_{2s}^*(\mathbf{r}_2)\frac{1}{r_{12}}\phi_{1s}(\mathbf{r}_1)\phi_{2s}(\mathbf{r}_2)\,d\mathbf{r}_1 d\mathbf{r}_2 \tag{12.125}$$

$$K_{1s,2s} = \int \phi_{1s}^*(\mathbf{r}_1)\phi_{2s}^*(\mathbf{r}_2)\frac{1}{r_{12}}\phi_{1s}(\mathbf{r}_2)\phi_{2s}(\mathbf{r}_1)\,d\mathbf{r}_1 d\mathbf{r}_2 \tag{12.126}$$

and

$$J_{2s,2s} = \int \phi_{2s}^*(\mathbf{r}_1)\phi_{2s}^*(\mathbf{r}_2)\frac{1}{r_{12}}\phi_{2s}(\mathbf{r}_1)\phi_{2s}(\mathbf{r}_2)\,d\mathbf{r}_1 d\mathbf{r}_2. \tag{12.127}$$

Next, let us consider the three-electron Be^+ ion with $Z = 4$ formed by removing, say, a spin β electron from the 2s orbital of a beryllium atom. Denote the Be^+ ion electron coordinates by $\mathbf{r}_1, \mathbf{r}_2$, and $\mathbf{r}_3$. Then, the Hamiltonian operator $\hat{H}^+$ of the ion is

$$\hat{H}^+ = -\frac{1}{2}\nabla_1^2 - \frac{4}{r_1} - \frac{1}{2}\nabla_2^2 - \frac{4}{r_2} - \frac{1}{2}\nabla_3^2 - \frac{4}{r_3} + \frac{1}{r_{12}} + \frac{1}{r_{13}} + \frac{1}{r_{23}} \tag{12.128}$$

where $r_{12} = |\mathbf{r}_1 - \mathbf{r}_2|$ and so on. Let us neglect orbital relaxation and thus assume that the Hartree–Fock orbitals of the Be^+ ion are identical to the beryllium Hartree–Fock orbitals $\phi_{1s}(\mathbf{r})$ and $\phi_{2s}(\mathbf{r})$. Then, the Hartree–Fock (single determinant) wave function of the ion becomes the following normalized Slater determinantal wave function:

$$\Psi^+(\rho_1, \rho_2, \rho_2) = \frac{1}{\sqrt{3!}}\begin{vmatrix} \phi_{1s}(\mathbf{r}_1)\alpha(\sigma_1) & \phi_{1s}(\mathbf{r}_1)\beta(\sigma_1) & \phi_{2s}(\mathbf{r}_1)\alpha(\sigma_1) \\ \phi_{1s}(\mathbf{r}_2)\alpha(\sigma_2) & \phi_{1s}(\mathbf{r}_2)\beta(\sigma_2) & \phi_{2s}(\mathbf{r}_2)\alpha(\sigma_2) \\ \phi_{1s}(\mathbf{r}_3)\alpha(\sigma_3) & \phi_{1s}(\mathbf{r}_3)\beta(\sigma_3) & \phi_{2s}(\mathbf{r}_3)\alpha(\sigma_3) \end{vmatrix}. \tag{12.129}$$

The energy of the ion E^+ is given by

$$E^+ = \int \left[\Psi^+(\rho_1,\rho_2,\rho_3)\right]^* \hat{H}^+\Psi^+(\rho_1,\rho_2,\rho_3)\,dr_1 d\sigma_1 dr_2 d\sigma_2 dr_3 d\sigma_3. \quad (12.130)$$

But one may show that (Problem 12.28)

$$E^+ = 2I_{1s} + I_{2s} + J_{1s,1s} + 2J_{1s,2s} - K_{1s,2s}. \quad (23.131)$$

Defining the Koopmans' theorem first ionization potential I_{KOOP} by

$$I_{KOOP} = E^+ - E \quad (12.132)$$

it follows from Equations (12.121) and (12.131) that

$$I_{KOOP} = -I_{2s} - 2J_{1s,2s} + K_{1s,2s} - J_{2s,2s}. \quad (12.133)$$

However, from Equation (12.117), with N = 2 the 2s orbital energy of beryllium $\varepsilon_{2s} \equiv \varepsilon_2$ is given by $\varepsilon_{2s} = I_2 + 2J_{21} - K_{21} + 2J_{22} - K_{22} = I_2 + 2J_{12} - K_{12} + J_{22}$, where we have used $K_{22} = J_{22}$, or in our new notation

$$\varepsilon_{2s} = I_{2s} + 2J_{1s,2s} - K_{1s,2s} + J_{2s,2s}. \quad (12.134)$$

Comparing Equations (12.133) and (12.134) gives Koopmans' theorem for the beryllium atom

$$I_{KOOP} = -\varepsilon_{2s} \quad (12.135)$$

which states that the approximation I_{KOOP} to the ionization potential of a beryllium 2s electron is *exactly* equal to the negative of the beryllium 2s orbital energy ε_{2s}.

The general Koopmans' theorem for a 2N –electron closed shell atom is a natural extension of Koopmans' theorem for beryllium. Suppose an electron is removed from the ith orbital of the atom to form a singly positively charged ion. Then, let us define the Koopmans' theorem ionization potential I_{KOOP} for the process by

$$I_{KOOP} = E_i^+ - E \quad (12.136)$$

where E_i^+ is the energy of the ion determined from the Slater determinantal wave function for the ion formed from the Hartree–Fock orbitals of the atom and where E is the Hartree–Fock energy of the atom. The general Koopmans' theorem is the *rigorous result*

$$I_{KOOP} = -\varepsilon_i \quad (12.137)$$

where ε_i is the orbital energy of the ith orbital, the orbital from which the electron has been removed.

The experimental ionization potential of an electron in the ith orbital will differ from $-\varepsilon_i$ since I_{KOOP} is evaluated from approximate wave functions for both the atom and the ion and thus is not equal to the true ionization potential. So the value of Koopmans' theorem ionization potentials can only be assessed by comparison with experiment. In Table 12.6, we compare several valence and core ionization potentials of the argon atom from computed orbital energies via Koopmans' theorem and from X-ray photoemission experiments. The Koopmans' theorem and experimental ionization potentials with one exception agree to within a few percent.

TABLE 12.6
Experimental I_{expt} and Koopmans' Theorem I_{KOOP} Valence and Core Electron Ionization Potentials of Argon

Electron Removed	Ion	I_{expt} (eV)[a]	I_{KOOP} (eV)	$\frac{\lvert I_{expt} - I_{KOOP} \rvert}{I_{expt}} \times 100\%$
1s	$Ar^+(1s2s^22p^63s^23p^6)$	3,206	3,227	0.66%
2s	$Ar^+(1s^22s2p^63s^23p^6)$	326	335	2.76%
2p	$Ar^+(1s^22s^22p^53s^23p^6)$	249	260	4.42%
3s	$Ar^+(1s^22s^22p^63s3p^6)$	29.2	34.8	19.2%
3p	$Ar^+(1s^22s^22p^63s^23p^5)$	15.8	16.1	1.90%

[a] D.A. Shirley et al, *Phys. Rev. B* 15 544 (1976).

We next examine the accuracy of the Hartree–Fock approximation (but not just for closed shell atoms). The Koopmans' theorem comparisons of Table 12.6 provide only a partial test of this accuracy. A far better test is provided by comparing for many atoms the Hartree–Fock and experimental ground state energies (the latter obtained as the negative of the sum of the observed first, second, third, … ionization potentials of the atom). Such a comparison is given in Table 12.7 for the atoms helium through neon. For all atoms other than helium, the Hartree–Fock energies E_{HF} and the experimental energies E_{expt} differ by only ~ 0.5%. While this may seem satisfactory, the small errors in E_{HF} are chemically significant. For example, for helium $\lvert E_{expt} - E_{HF} \rvert = 0.0416\,\text{au} = 109.2\,\text{kJ mol}^{-1}$, which is of the order of bond dissociation energies.

A more sensitive test of the accuracy of the Hartree–Fock approximation is provided by comparisons of Hartree–Fock I_{HF} and experimental I_{expt} first ionization potentials. I_{HF} is determined as $I_{HF} = E_{HF}^{ion} - E_{HF}$ where E_{HF}^{ion} and E_{HF} are, respectively, the Hartree–Fock energies of the positive ion and the neutral atom. I_{HF} and I_{expt} are compared in Table 12.8 for the atoms helium through neon.

TABLE 12.7
Experimental E_{expt} and Hartree–Fock E_{HF} Ground State Energies for the Atoms Helium through Neon

Atom	E_{expt} (au)	E_{HF} (au)[a]	$\frac{\lvert E_{expt} - E_{HF} \rvert}{E_{expt}} \times 100\%$
He	−2.9034	−2.86168	1.44%
Li	−7.4780	−7.4327	0.61%
Be	−14.6684	−14.5730	0.65%
B	−24.6581	−24.5291	0.52%
C	−37.8558	−37.6886	0.44%
N	−54.6119	−54.4009	0.39%
O	−75.1098	−74.8094	0.40%
F	−99.8071	−99.4093	0.40%
Ne	−129.0525	−128.5471	0.39%

[a] C.F. Bunge, J.A. Barrientos, and A.V. Bunge, *At. Data Nucl. Tables* 53 113 (1993).

TABLE 12.8
Experimental I_{expt} and Hartree–Fock I_{HF} First Ionization Potentials for the Atoms Helium through Neon

Atom	I_{expt} (eV)	I_{HF} (eV)[a]	$\frac{\lvert I_{expt} - I_{HF} \rvert}{I_{expt}} \times 100\%$
He	24.587	23.448	4.63%
Li	5.392	5.342	0.93%
Be	9.323	8.044	13.72
B	8.298	7.932	4.41%
C	11.260	10.786	4.21%
N	14.534	13.958	3.96%
O	13.618	11.886	12.72%
F	17.423	15.718	9.79%
Ne	21.565	19.845	7.98%

[a] C.F. Bunge, J.A. Barrientos, and A.V. Bunge, *At. Data Nucl. Tables* 53 113 (1993).

The percent errors in I_{HF} are much larger than those for E_{HF} ranging from 0.93% for lithium to 13.72% for beryllium. The average percent error in I_{HF} for the nine atoms is 6.93%. The relatively large errors arise since I_{HF} is the difference between two large quantities E_{HF}^{ion} and E_{HF} with similar values and so the errors in these quantities magnify when one forms the much smaller quantity I_{HF}.

We next introduce the concept of the *correlation energy.*

12.8 THE CORRELATION ENERGY

We noted in Section 12.2 that simple orbital product wave functions predict energies which are too high since they ignore Coulomb correlation and thus overestimate the probability that two electrons will be in close proximity. Hartree–Fock wave functions also do not include Coulomb correlation since they incorporate the effects of the interelectronic Coulomb forces in an averaged rather than in an instantaneous manner. However, since Hartree–Fock wave functions are antisymmetric, they do include some correlation; namely, they include the Fermi correlation between electrons of parallel (the same) spin. (See Problem 12.21.)

Because Hartree–Fock wave functions are not completely correlated, Hartree–Fock energies E_{HF} are higher than experimental energies E_{expt} (Table 12.7). However, experimental energies include relativistic contributions and therefore cannot be compared with non-relativistic Hartree–Fock energies to obtain a valid measure of the magnitude of electron correlation effects on atomic energies. Thus, the measure of electron correlation is not usually taken as $E_{expt} - E_{HF}$. Rather, it is conventionally taken as the correlation energy E_{corr} defined by

$$E_{corr} = E_0 - E_{HF} \tag{12.138}$$

where E_0 is the exact nonrelativistic ground state energy of the atom (which is not observable but which in principle may be obtained by exactly solving the atom's nonrelativistic Schrödinger equation). By the variational principle, $E_{HF} > E_0$. Hence, correlation energies are negative quantities.

For helium, E_{corr} may be found exactly since E_0 is known exactly from the work of Pekeris (Problem 12.29). Also very good approximations to E_{corr} may be found for three- and four-electron atoms since accurate theoretical values for E_0 are available for these systems. For more complex atoms, E_{corr} must be determined semiempirically. This is done by removing the relativistic contribution E_{rel} to E_{expt} to give the nonrelativistic energy as $E_0 = E_{expt} - E_{rel}$.

With this form for E_0, the correlation energy may be expressed as

$$E_{corr} = E_{expt} - E_{rel} - E_{HF}. \tag{12.139}$$

E_{rel} may then be estimated to yield an approximate value for E_{corr}. In this manner, Clementi (1965) obtained correlation energies for atoms and ions with up to twenty-two electrons. In Table 12.9, we reproduce his correlation energies for the atoms helium through neon and compare these energies with the corresponding values of $E_{expt} - E_{HF}$ determined from Table 12.7. If E_{rel} vanished, then from Equation (12.139) E_{corr} and $E_{expt} - E_{HF}$ would be identical. Thus, the differences between E_{corr} and $E_{expt} - E_{HF}$ are direct measures of relativistic influence on the energies of the atoms. From Table 12.9, this influence as expected from the observations at the start of Section 10.8 increases with the atomic number Z. Thus, for the atoms helium through carbon, the relativistic contributions to the energies are small, but for neon for which $E_{expt} - E_{HF} = -0.505$ au and $E_{corr} = -0.393$ au, they are appreciable.

What is more interesting is the variation of E_{corr} with Z. Since the number of electron pairs increases with Z, E_{corr} decreases monotonically with Z. But as shown in Table 12.9, the rate of decrease varies as one moves from atom to atom. As we next show, these variations may be qualitatively rationalized.

In Table 12.10, we list for the atoms lithium through neon the differences $E_{corr}(Z) - E_{corr}(Z-1)$ determined from Table 12.9 between the correlation energies of atoms with atomic numbers Z and $Z-1$. These differences will permit us to rationalize the variations of E_{corr} for the atoms in the second period of the periodic table if we assume that the contributions to E_{corr} from specific types of electron pairs (e.g., a $2s-2p$ pair) are roughly independent of Z.

First, notice from Table 12.10 that while E_{corr} for lithium is only −0.0032 au smaller than E_{corr} for helium, E_{corr} for beryllium is −0.049 au smaller than E_{corr} for lithium. This is because when one moves from helium to lithium, a 2s electron is added which is spatially separated from lithium's 1s electrons and thus is only weakly correlated with these electrons. Hence, the addition produces only a small change in E_{corr}. However, when one moves from lithium to beryllium, a 2s electron is added which is strongly correlated with beryllium's other 2s electron, and thus, the addition produces a relatively large change in E_{corr}. Continuing, when one moves from beryllium to boron, the change in E_{corr} is −0.031 au. This change is expected to be mainly due to the correlation of boron's 2p electron with its two 2s electrons. A change in E_{corr} of similar magnitude −0.033 au occurs when one moves from boron to carbon. The similarity in the changes in E_{corr} for boron and

TABLE 12.9

$E_{expt} - E_{HF}$ (See Table 12.7) Compared with the Correlation Energies E_{corr} for the Atoms Helium through Neon

Atom	$E_{expt} - E_{HF}$ (au)	E_{corr} (au)[a]
He	−0.0417	−0.0421
Li	−0.0453	−0.0453
Be	−0.0954	−0.0944
B	−0.129	−0.125
C	−0.167	−0.158
N	−0.211	−0.188
O	−0.300	−0.258
F	−0.398	−0.324
Ne	−0.505	−0.393

[a] From E. Clementi, *IBM J.* 2 (1965), Table 7.

TABLE 12.10
The Differences $E_{corr}(Z) - E_{corr}(Z-1)$ between the Correlation Energies of Atoms with Atomic Numbers Z and Z−1 for Z=3− 10

Atom	Z	$E_{corr}(Z) - E_{corr}(Z-1)/au$
Li	3	−0.0032
Be	4	−0.049
B	5	−0.031
C	6	−0.033
N	7	−0.030
O	8	−0.070
F	9	−0.066
Ne	10	−0.069

carbon may be explained as follows. The Coulomb repulsion of carbon's two 2p electrons is not expected to significantly add to E_{corr} since from Hund's first rule (see Section 12.10) these electrons occupy different p orbitals and have parallel spins and thus are kept apart by the Fermi interaction. Therefore, the change in E_{corr} for carbon like that for boron is dominated by the correlation of a newly added 2p electron with two 2s electrons. When one moves from carbon to nitrogen, the change in E_{corr} − 0.030 au is similar to the changes for boron and carbon. This similarity may be rationalized by an argument like that just given for carbon. Namely, the three 2p electrons in nitrogen occupy different p orbitals and have parallel spins, and therefore these electrons are kept apart by Fermi correlation, and hence their Coulomb interactions are not expected to contribute significantly to E_{corr}. Thus, as for boron and carbon, the change in E_{corr} for nitrogen is mainly due to the correlation of a newly added 2p electron with two 2s electrons. However, something new happens when one moves from nitrogen to oxygen. The change in E_{corr} jumps from ~ −0.30 to −0.070 au. This jump mainly arises since in oxygen one of the 2p orbitals becomes doubly occupied and the electrons in this orbital are strongly correlated since they have opposite spins and thus are not kept apart by the Fermi interaction. Jumps in E_{corr} of the similar magnitudes −0.066 and −0.069 au occur for, respectively, fluorine and neon. These jumps like the jump for oxygen are mainly due to the correlation of electrons in newly doubly occupied 2p orbitals.

In summary, we have qualitatively rationalized the variations of E_{corr} for the second-period atoms.

We next summarize how one determines the *electron configurations* of atoms. These configurations provide the bridge between the orbital model of atomic electronic structure and the empirical periodic table.

12.9 THE DETERMINATION OF ELECTRON CONFIGURATIONS OF ATOMS AND THE PERIODIC TABLE

The (Hartree–Fock) orbitals of multi-electron atoms like those of hydrogen-like atoms are specified by three quantum numbers: the principle quantum number $n = 1,2,\ldots$, the angular momentum quantum number $\ell = 0,1,\ldots,n-1$, and the magnetic quantum number $m = -\ell, -\ell+1,\ldots,\ell-1,\ell$. As for the hydrogen-like atoms, the orbitals of multi-electron atoms are denoted by $1s(n=1,\ell=0)$, $2s(n=2,\ell=0)$, $2p(n=2,\ell=1)$, and so on. For multi-electron atoms, as for the hydrogen-like atoms there are n^2 orbitals with quantum number n. These n^2 orbitals are said to comprise the nth *shell* of the atom. The $2\ell+1$ orbitals with quantum numbers n and ℓ are said to comprise a *subshell* of the nth shell. For example, an atom's three 2p orbitals comprise the 2p subshell of its second shell. Since by the Pauli exclusion principle an orbital can at most be doubly occupied, a subshell with quantum number ℓ can hold no more than $2(2\ell+1)$ electrons. Thus, the maximum numbers of electrons which can occupy an $s(\ell = 0)$, $p(\ell = 1)$, $d(\ell = 2)$, or $f(\ell = 3)$ subshell are, respectively, two, six, ten, and fourteen.

While for hydrogen-like atoms the energies of all orbitals with the same n are identical, for multi-electron atoms energies of orbitals with the same n but different ℓ differ. (Orbitals with the same n and ℓ but different m are degenerate or have the same energy.) For example, for a hydrogen-like atom the energies of the 2s and 2p orbitals are the same, but for a multi-electron atom the energy of the 2p orbitals is higher than that of the 2s orbital.

This lifting of degeneracy occurs because in a multi-electron atom the nuclear charge an electron experiences is reduced by the screening effect of the other electrons and the magnitude of this screening, and hence, its effect on the orbital energies depends on the electron's quantum number ℓ. The ℓ−dependence follows since for fixed n the radial probabilities for an s,p,d,f electron to be close to the nucleus order as (compare Figure 10.2 and accompanying discussion) $s > p > d > f$, implying that the degree of screening and hence the orbital energies order as $s < p < d < f$.

Moreover, the screening effect is so large for d and f electrons that d and f orbitals with quantum numbers $n' < n$ actually can have higher energies than s orbitals with quantum number n. For example, for multi-electron atoms 3d orbitals usually have slightly higher energies than 4s orbitals.

The energy ordering of the orbitals for most multi-electron atoms is given by *Madelung's rule* which states that:

1. *orbital energies increase with $n+\ell$*
2. *for two orbitals with the same value for $n+\ell$, the orbital with the larger n has the higher energy.*

Madelung's rule gives the following ordering of the orbital energies:

$$\begin{aligned} &1s < 2s < 2p < 3s < 3p < 4s < 3d < 4p < 5s < 4d \\ &< 5p < 6s < 4f < 5d < 6p < 7s < 5f < 6d < 7p. \end{aligned} \tag{12.140}$$

We may now determine the ground state electron configuration of any atom. This is done (as is familiar from general chemistry textbooks) by successively filling the atom's orbitals with its electrons in order of increasing orbital energies starting with the 1s orbital and keeping in mind that the maximum electron occupancies of s, p, d, and f subshells are two, six, ten, and fourteen. Introducing the designations ns^x, np^x, nd^x, and nf^x to indicate that x electrons occupy the ns, np, nd, and nf subshells (when x = 1, the superscript x is often omitted), one finds that the electron configurations of the atoms in the first and second periods of the periodic table are

$$\text{H}: 1s^1 = 1s \text{ and He}: 1s^2$$

and

$$\begin{aligned} &\text{Li}: 1s^2 2s \quad \text{Be}: 1s^2 2s^2 \quad \text{B}: 1s^2 2s^2 2p \quad \text{C}: 1s^2 2s^2 2p^2 \quad \text{N}: 1s^2 2s^2 2p^3 \\ &\text{O}: 1s^2 2s^2 2p^4 \quad \text{F}: 1s^2 2s^2 2p^5 \quad \text{and} \quad \text{Ne}: 1s^2 2s^2 2p^6. \end{aligned}$$

The first third-period atom is sodium with electron configuration $1s^2 2s^2 2p^6 3s$. This configuration may be abbreviated as [Ne]3s, where the symbol [Ne] is shorthand notation for the neon electron configuration $1s^2 2s^2 2p^6$. In this abbreviated notation, the electron configurations of the third-period atoms are

$$\text{Na}: [\text{Ne}]3s \quad \text{Mg}: [\text{Ne}]3s^2 \quad \text{Al}: [\text{Ne}]3s^2 3p \ldots \text{Ar}: [\text{Ne}]3s^2 3p^6.$$

We will use analogous abbreviated notation to designate the electron configurations of all remaining atoms. Namely, we will abbreviate the electron configuration of the noble gas atom immediately preceding the atom in question in the periodic table by its chemical symbol in brackets and then follow that abbreviation by the atom's outer electron configuration. For example, the electron configuration of bromine $1s^2 2s^2 2p^6 3s^2 3p^6 4s^2 3d^{10} 4p^5$ is abbreviated as $[Ar]4s^2 3d^{10} 4p^5$.

In Table 12.11, we list the ground state electron configurations for the atoms hydrogen through krypton. These show that for the elements listed, atoms that occur in the same group (column) of the periodic table have analogous outer electron configurations. For example, the outermost subshell

TABLE 12.11
Ground State Electron Configurations of the Elements Hydrogen through Krypton

Atomic Number Z	Element	Electron Configuration
1	Hydrogen	$1s$
2	Helium	$1s^2$
3	Lithium	$1s^2 2s$
4	Beryllium	$1s^2 2s^2$
5	Boron	$1s^2 2s^2 2p$
6	Carbon	$1s^2 2s^2 2p^2$
7	Nitrogen	$1s^2 2s^2 2p^3$
8	Oxygen	$1s^2 2s^2 2p^4$
9	Fluorine	$1s^2 2s^2 2p^5$
10	Neon	$1s^2 2s^2 2p^6$
11	Sodium	$[Ne]3s$
12	Magnesium	$[Ne]3s^2$
13	Aluminum	$[Ne]3s^2 3p$
14	Silicon	$[Ne]3s^2 3p^2$
15	Phosphorus	$[Ne]3s^2 3p^3$
16	Sulfur	$[Ne]3s^2 3p^4$
17	Chlorine	$[Ne]3s^2 3p^5$
18	Argon	$[Ne]3s^2 3p^6$
19	Potassium	$[Ar]4s$
20	Calcium	$[Ar]4s^2$
21	Scandium	$[Ar]4s^2 3d$
22	Titanium	$[Ar]4s^2 3d^2$
23	Vanadium	$[Ar]4s^2 3d^3$
24	Chromium	$[Ar]4s3d^5$
25	Manganese	$[Ar]4s^2 3d^5$
26	Iron	$[Ar]4s^2 3d^6$
27	Cobalt	$[Ar]4s^2 3d^7$
28	Nickel	$[Ar]4s^2 3d^8$
29	Copper	$[Ar]4s3d^{10}$
30	Zinc	$[Ar]4s^2 3d^{10}$
31	Gallium	$[Ar]4s^2 3d^{10} 4p$
32	Germanium	$[Ar]4s^2 3d^{10} 4p^2$
33	Arsenic	$[Ar]4s^2 3d^{10} 4p^3$
34	Selenium	$[Ar]4s^2 3d^{10} 4p^4$
35	Bromine	$[Ar]4s^2 3d^{10} 4p^5$
36	Krypton	$[Ar]4s^2 3d^{10} 4p^6$

configurations of the alkali atoms lithium, sodium, and potassium are 2s, 3s, and 4s, while those of the halogen atoms fluorine, chlorine, and bromine are $2p^5$, $3p^5$, and $4p^5$.

However, analogous outer electron configurations imply analogous chemical behavior. Thus, electron configurations that derive from only the orbital model and the Pauli exclusion principle provide the theoretical basis for the observed periodicities in the behaviors of the elements.

The electron configurations of Table 12.11 were determined from orbital energies found by spectroscopic measurements rather than from Madelung's rule. However, the spectroscopic and Madelung's rule electron configurations are the same for all atoms listed in Table 12.11 except chromium and copper. For chromium, Madelung's rule predicts the electron configuration $[Ar]4s^2 3d^4$, while the experimental electron configuration is $[Ar]4s3d^5$. The exception to Madelung's rule for chromium is usually rationalized by the notion that the half-filled 3d subshell configuration $3d^5$ is especially stable. The exception for copper is similarly rationalized by the notion that the filled 3d subshell configuration $3d^{10}$ is especially stable.

More exceptions to Madelung's rule occur for atoms heavier than krypton. But many of these cannot be simply rationalized. For example, for platinum with $Z = 78$, Madelung's rule predicts the electron configuration $[Xe]6s^2 4f^{14} 5d^8$, while the experimental electron configuration is $[Xe]6s4f^{14} 5d^9$.

We close this section by noting that for all but closed shell atoms, more than one electronic state derives from the atom's electron configuration. For example, consider carbon with the ground state electron configuration $1s^2 2s^2 2p^2$. Fifteen distinct spin orbital configurations in accord with the Pauli exclusion principle are consistent with the $1s^2 2s^2 2p^2$ configuration. Two of these are $1s\alpha 1s\beta 2s\alpha 2s\beta 2p_{+1}\alpha 2p_{+1}\beta$ and $1s\alpha 1s\beta 2s\alpha 2s\beta 2p_{+1}\alpha 2p_0\alpha$ where the subscripts 0 and +1 denote the m quantum numbers of the 2p orbitals. These two spin orbital configurations specify two different Slater determinantal wave functions with different energies and therefore are associated with two distinct electronic states.

Thus, we next describe how the electronic states of an atom may be identified from its electron configuration and how these states are labeled by *term symbols*.

12.10 THE ELECTRONIC STATES OF ATOMS AND TERM SYMBOLS

Different electronic states of an atom deriving from the same electron configuration are distinguished by different term symbols. These term symbols are a generalization of the hydrogen-like atom term symbols of Equation (10.150). Recall that the hydrogen-like atom term symbols depend on the quantum numbers ℓ, s, and j which determine, respectively, the magnitudes of the electron's orbital, spin, and total angular momentum. Similarly, the term symbols of an N-electron atom depend on quantum numbers L, S, and J which determine, respectively, the magnitudes of the atom's total orbital angular momentum $\mathbf{L} = \sum_{i=1}^{N} \boldsymbol{\ell}_i$, total spin angular momentum $\mathbf{S} = \sum_{i=1}^{N} \mathbf{s}_i$, and total electronic angular momentum $\mathbf{J} = \mathbf{L} + \mathbf{S}$, where $\boldsymbol{\ell}_i$ and $\mathbf{s}_i$ are the orbital and spin angular momenta of electron i.

The term symbols for multi-electron atoms are of the form

$$^{2S+1}\boldsymbol{L}_J \tag{12.141}$$

where $\boldsymbol{L}$ is an uppercase letter which represents the quantum number L which can take on the values 0, 1, 2, ... The correspondence between L and $\boldsymbol{L}$ is by convention

$$\begin{array}{cccccc} L & 0 & 1 & 2 & 3 & 4\ldots \\ \boldsymbol{L} & S & P & D & F & G\ldots \end{array} \tag{12.142}$$

The quantity $2S+1$ is called the *multiplicity*. We will see that S is restricted to the values $0, \frac{1}{2}, 1, \frac{3}{2}, \ldots$, so multiplicities are positive integers. States with multiplicities $1,2,3,\ldots$ are called, respectively, *singlet, doublet, triplet,* ... states. Additionally, from angular momentum addition theory (Cohen-Tannoudji, Diu, and Laloe 1977, chap. 10) the possible values of J are given for a particular pair L and S by

$$J = L+S, L+S-1, \ldots, |L-S|. \tag{12.143}$$

As an example, consider states with $L = 2$ and $S = 1$. From Equation (12.143), for such states $J = 3,2,$ or 1. Hence, the $L = 2$, $S = 1$ states are represented by three triplet D term symbols

$$^3D_3,\ ^3D_2, \text{ and } ^3D_1. \tag{12.144}$$

Each term symbol specifies $2J+1$ degenerate electronic states since for a given J the vector **J** is space quantized with $2J+1$ orientations each orientation corresponding to a different state. Thus, the total number of states specified by the three term symbols of Equation (12.144) is $2(3)+1+2(2)+1+2(1)+1=15$ states.

Additionally, states represented by term symbols with the same values for L and S but different values for J only differ slightly in energy (when Z is sufficiently small that relativistic effects are small) due to spin–orbit coupling. Often this difference is ignored and the subscripts J on the term symbols are dropped. For example, all states with $L = 2$ and $S = 1$ are often represented by the single term symbol

$$^3D. \tag{12.145}$$

Next, let us count the numbers of electronic states represented by the term symbol ^3D. For a given choice of L and S, there are $(2L+1)(2S+1)$ states since the vectors **L** and **S** have, respectively, $2L+1$ and $2S+1$ orientations each pair of orientations corresponding to a different state. Thus, the number of states represented by the ^3D term symbol is $(2L+1)(2S+1)=(4+1)(2+1)=15$ states, which, as must be true, is the same number of states represented by the term symbols $^3D_3, ^3D_2$, and $^3D_1$.

Term symbols provide a valid representation of the electronic states of atoms only if the values of L and S distinguish between different states deriving from the same electron configuration, or as it is often put if L and S are "good quantum numbers." This becomes increasingly true as Z and hence the importance of relativistic effects decrease and is found in practice to be an acceptable assumption for atoms with atomic numbers $Z \lesssim 30$.

To show this, we introduce the operators $\hat{\mathbf{L}}, \hat{L}_z$ and $\hat{\mathbf{S}}, \hat{S}_z$ which represent the total orbital angular momentum **L** and its Z-component L_z and the total spin angular momentum **S** and its Z-component S_z. These operators are defined by

$$\hat{\mathbf{L}} = \sum_{i=1}^{N} \hat{\ell}_i \tag{12.146}$$

$$\hat{L}_z = \sum_{i=1}^{N} \hat{\ell}_{zi} \tag{12.147}$$

$$\text{and } \hat{\mathbf{S}} = \sum_{i=1}^{N} \hat{\mathbf{s}}_i \tag{12.148}$$

$$\hat{S}_z = \sum_{i=1}^{N} \hat{s}_{zi} \tag{12.149}$$

where $\hat{\ell}_i$ and $\hat{\mathbf{s}}_i$ are the orbital and spin angular momentum operators of electron i and where $\hat{\ell}_{zi}$ and $\hat{s}_{zi}$ are the Z-components of these operators, with definitions analogous to those of Equations (12.147) and (12.149) for the X- and Y-components of the operators of Equations (12.146) and (12.148).

We may now outline the theoretical basis of term symbols.

We first note that all orbital angular momentum operators commute with all spin angular momentum operators. Moreover, the single electron orbital and spin angular momentum operators for the *same* electron obey standard angular momentum commutation rules. For example, for all $j = 1,\ldots,N \left[\hat{\ell}_{xj}, \hat{\ell}_{yj}\right] = i\hat{\ell}_{zj}$ and $\left[\hat{s}_{xj}, \hat{s}_{yj}\right] = i\hat{s}_{zj}$. Additionally, for both orbital and spin angular momenta, all single electron angular momentum operators for *different* electrons commute. For example, for all $i \neq j\left[\hat{\ell}_{x_i}, \hat{\ell}_{y_j}\right] = \left[\hat{\mathbf{s}}_{x_i}, \hat{\mathbf{s}}_{y_j}\right] = \hat{O}$. It then follows from Equations (12.147) and (12.149) and their X- and Y-component analogues that the total orbital and spin angular momentum operators obey standard angular momentum commutation rules. For example, $\left[\hat{L}_x, \hat{L}_y\right] = i\hat{L}_z$ and $\left[\hat{S}_x, \hat{S}_y\right] = i\hat{S}_z$. From these considerations it follows that (recall the proof of Equations [8.32] from Equation [8.31]) $\hat{L}^2, \hat{L}_z, \hat{S}^2$, and $\hat{S}_z$ form a set of mutually commuting operators.

Moreover, if relativistic effects, especially spin–orbit coupling, are sufficiently small, as is found true experimentally for atoms with atomic numbers $Z \leq 30$, the atom's true Hamiltonian operator may be approximated by its nonrelativistic Hamiltonian operator $\hat{H}$. However, an involved calculation gives that $\hat{H}$ commutes with $\hat{L}^2$ and $\hat{L}_z$. Also since $\hat{H}$ is free of spin operators, it also commutes with $\hat{S}^2$ and $\hat{S}_z$. Thus, $\hat{H}, \hat{L}^2, \hat{L}_z, \hat{S}^2$, and $\hat{S}_z$ mutually commute and therefore as an extension of the results of Section 5.2 have a complete orthonormal set of common eigenfunctions Ψ which are approximations to the atom's true relativistic stationary state wave functions. In particular, since Ψ is an eigenfunction of $\hat{L}^2$ and $\hat{S}^2$, we have in analogy to our earlier results for angular momentum eigenvalue problems that

$$\hat{L}^2\Psi = L(L+1)\Psi \tag{12.150}$$

and

$$\hat{S}^2\Psi = S(S+1)\Psi. \tag{12.151}$$

Thus, for $Z \lesssim 30$ L^2 and S^2 when the atom is in one of its true relativistic stationary states approximately have the definite values $L(L+1)$ and $S(S+1)$. Hence then different true stationary state wave functions, and the corresponding distinct atomic electron states arising from the same electron configuration can be approximately distinguished by their values for the quantum numbers L and S. When this characterization is valid, we say that the atom falls into the case of *Russell–Saunders* coupling and then the use of term symbols to represent the electronic states of atoms is justified. For very heavy atoms with $Z \gtrsim 40$, the opposite occurs and relativistic effects dominate and the system falls into the case of *$j - j$ coupling* for which L and Sare not good quantum numbers and hence for this case the use of term symbols to label electronic states (though often done) is questionable.

Additionally, we earlier argued qualitatively that each pair of quantum numbers L and S or each term symbol represents $(2L+1)(2S+1)$ electronic states. Now, we may derive this result more rigorously. Namely, since Ψ is an eigenfunction of both $\hat{L}_z$ and $\hat{S}_z$, we have that

$$\hat{L}_z\Psi = M\Psi \tag{12.152}$$

where for each L M takes on the 2L + 1 values

$$M = -L, -L+1, \ldots, L-1, L \tag{12.153}$$

and

$$\hat{S}_z\Psi = M_s\Psi \tag{12.154}$$

where for each S M_s takes on the 2S + 1 values

$$M_s = -S, -S+1, \ldots, S-1, S \tag{12.155}$$

yielding for each L, S pair $(2L+1)(2S+1)$ distinct states.

Next, note that the operators $\hat{L}_z, \hat{S}_z, \hat{\ell}_{zi}$, and $\hat{s}_{zi}$ for i = 1,…,N mutually commute and thus have common eigenfunctions Φ. We therefore have for all i that

$$\hat{L}_z\Phi = M\Phi \tag{12.156}$$

$$\hat{\ell}_{zi}\Phi = m_i\Phi \tag{12.157}$$

$$\hat{S}_z\Phi = M_s\Phi \tag{12.158}$$

and

$$\hat{s}_{zi}\Phi = m_{si}\Phi. \tag{12.159}$$

Equations (12.147), (12.156), and (12.157) then yield

$$M = \sum_{i=1}^{N} m_i. \tag{12.160}$$

Similarly, Equations (12.149), (12.158), and (12.159) give

$$M_s = \sum_{i=1}^{N} m_{si}. \tag{12.161}$$

Equations (12.160) and (12.161) will be key in the determination of term symbols from electron configurations since they permit one to obtain the total angular momentum quantum numbers M and M_s from the single electron angular momentum quantum numbers m and m_{si}, the latter being derivable from the electron configuration.

We further note from general angular momentum considerations that for i = 1,…,N, the eigenvalues of $\hat{\ell}_i^2$ and $\hat{s}_i^2$ are $\ell_i(\ell_i+1)$ and $s_i(s_i+1)$ where the quantum numbers ℓ_i and s_i determine the magnitudes of the orbital and spin angular momenta of electron i. For an electron occupying an s,p,d,f,… orbital, $\ell_i = 0,1,2,3,\ldots$ while always $s_i = \frac{1}{2}$.

Next, suppose N = 2 (two electrons) so that $\ell_i = \ell_1$ or ℓ_2 and $s_i = s_1$ or s_2. For N = 2, $\hat{\mathbf{L}} = \hat{\ell}_1 + \hat{\ell}_2$ and $\hat{\mathbf{S}} = \hat{\mathbf{s}}_1 + \hat{\mathbf{s}}_2$ and so angular momentum addition theory gives the possible values of L and S as

$$L = \ell_1+\ell_2, \ell_1+\ell_2-1, \ldots, |\ell_1-\ell_2| \tag{12.162}$$

and

$$S = s_1+s_2, s_1+s_2-1, \ldots, |s_1-s_2|. \tag{12.163}$$

Since $s_1 = s_2 = \frac{1}{2}$, for N = 2 the possible values of S are always S = 0 and S = 1.

Equations (12.162) and (12.163), like Equations (12.160) and (12.161), will be key in determining term symbols since they permit one to determine the total angular momentum quantum numbers L and S from the single electron angular momentum quantum numbers ℓ_1 and ℓ_2 and s_1 and s_2, the latter being derivable from the electron configuration.

We may now derive term symbols from electron configurations.

To start, consider the 1s2s excited state helium electron configuration. For the 1s and 2s electrons are $\ell_1 = 0(1s)$ and $\ell_2 = 0(2s)$ and $s_1 = s_2 = \frac{1}{2}$. So from Equations (12.162) and (12.163), L = 0 and as always S = 0 or 1. So using the (2L + 1)(2S + 1) rule for the number of states represented by a term symbol, the 1s2s configuration gives rise to two term symbols, ^1S which represents one singlet state and ^3S which represents three triplet states. (These states have the wave functions given in Equations [12.82] and [12.83].)

Next, let us consider the 1s2p helium electron configuration. For this configuration, $\ell_1 = 0$ and $\ell_2 = 1$. Hence, L = 1. Also again, S = 0 or 1. So the 1s2p configuration gives rise to two term symbols ^1P and ^3P which represent, respectively, three and nine distinct states.

This simple method for determining term symbols does not account for the Pauli exclusion principle, and for that reason, it fails when the two electrons are *equivalent*; that is, then they occupy the same subshell. To see why, consider the ground state helium electron configuration $1s^2$ with two equivalent electrons. The simple method predicts for the $1s^2$ electron configuration that L = 0 and S = 0 or 1 giving the same term symbols ^1S and ^3S found for the non-equivalent electron configuration 1s2s. However, for the $1s^2$ electron configuration, only the ^1S term symbol actually occurs since the ^3S term symbol includes S = 1 spin orbital configurations like $1s\alpha 1s\alpha$ which violate the Pauli exclusion principle.

The proper method for deriving term symbols for electron configurations with equivalent electrons is based on Equation (12.160) $M = \sum_{i=1}^{N} m_i$ and Equation (12.161) $M_s = \sum_{i=1}^{N} m_{si}$. Let us apply these equations to the helium ground state. For this state, there is only one spin orbital configuration in accord with the Pauli exclusion principle $1s\alpha 1s\beta$. The m and m_s quantum numbers for the spin orbital $1s\alpha$ are $m_1 = 0$ and $m_{s1} = \frac{1}{2}$, while those for the spin orbital $1s\beta$ are $m_2 = 0$ and $m_{s2} = -\frac{1}{2}$. Thus, $M = m_1 + m_2 = 0$ and $M_s = m_{s1} + m_{s2} = 0$. Therefore, for the $1s^2$ electron configuration, L = 0 and S = 0 yielding for the helium ground state the single term symbol ^1S which represents only one electronic state.

We may readily extend this argument to show that *all* closed shell electron configurations yield only the single term symbol ^1S and thus give rise to only one electronic state. Consider, for example, neon with the closed shell ground state electron configuration $1s^2 2s^2 2p^6$. There is only one spin orbital configuration in accord with the Pauli exclusion principle which is consistent with the neon electron configuration, namely, $1s\alpha 1s\beta 2s\alpha 2s\beta 2p_{+1}\alpha 2p_{+1}\beta 2p_0\alpha 2p_0\beta 2p_{-1}\alpha 2p_{-1}\beta$.

The m values for this spin orbital configuration are $m_1 = m_2 = m_3 = m_4 = 0, m_s = m_6 = +1$, $m_7 = m_8 = 0$, and $m_9 = m_{10} = -1$. Thus, $M = \sum_{i=1}^{10} m_i = 4(0) + 2(+1) + 2(0) + 2(-1) = 0$ Also for every α spin with $m_s = \frac{1}{2}$, there is a β spin with $m_s = -\frac{1}{2}$. Therefore, $M_s = \sum_{i=1}^{10} m_{si} = 0$. Since M = 0 and $M_s = 0$, L = 0 and S = 0, and therefore, the only term symbol deriving from the $1s^2 2s^2 2p^6$ electron configuration is ^1S.

Moreover, generalizing our argument for neon shows that the contributions to M and M_s and hence L and S from closed shells always vanish (in other words, both the orbital and spin angular momenta of closed shells are zero) (Problem 12.35). Consequently, when deriving the term symbols of an electron configuration, one may ignore the closed shells. For example, the term symbols of the excited state carbon electron configuration $1s^2 2s^2 2p3p$ are identical to those of the electron configuration 2p3p found by ignoring the $1s^2 2s^2$ closed shells.

Moreover, since the 2p and 3p electrons are non-equivalent, the term symbols of the 2p3p configuration may be found from Equations (12.162) and (12.163) by taking $\ell_1 = 1$ (its 2p value) and $\ell_2 = 1$ (its 3p value) and by setting $s_1 = s_2 = \frac{1}{2}$. This gives L = 2,1, or 0 and S = 1 or 0 yielding the $1s^2 2s^2 2p3p$ term symbols as ${}^3D, {}^3P, {}^3S, {}^1D, {}^1P$, and ${}^1S$.

Next, let us derive the term symbols for the carbon ground state electron configuration $1s^2 2s^2 2p^2$. Since we may ignore the $1s^2 2s^2$ closed shells, these term symbols are the same as those of the $2p^2$ electron configuration. Recalling our discussion of the helium ground state, since the two 2p electrons are equivalent, some of the 2p3p term symbols will not occur for the $2p^2$ configuration since some include spin orbital configurations that violate the Pauli exclusion principle. So as for the helium ground state, one cannot use Equations (12.162) and (12.163) to determine the term symbols. Instead, to derive the $2p^2$ term symbols, one must use the following specializations of Equations (12.160) and (12.161): $M = m_1 + m_2$ and $M_s = m_{s1} + m_{s2}$.

We thus set up the following table in which we list the fifteen distinct spin orbital configurations in accord with the Pauli exclusion principal which are consistent with the $2p^2$ electron configuration and also the M and M_s values for each spin orbital configuration:

	Spin Orbital Configuration	M	M_S
1	$2p_{+1}\alpha 2p_{+1}\beta$	2	0
2	$2p_{+1}\alpha 2p_0\alpha$	1	1
3	$2p_{+1}\alpha 2p_0\beta$	1	0
4	$2p_{+1}\alpha 2p_{-1}\alpha$	0	1
5	$2p_{+1}\alpha 2p_{-1}\beta$	0	0
6	$2p_{+1}\beta 2p_0\alpha$	1	0
7	$2p_{+1}\beta 2p_0\beta$	1	−1
8	$2p_{+1}\beta 2p_{-1}\alpha$	0	0
9	$2p_{+1}\beta 2p_{-1}\beta$	0	−1
10	$2p_0\alpha 2p_0\beta$	0	0
11	$2p_0\alpha 2p_{-1}\alpha$	−1	1
12	$2p_0\alpha 2p_{-1}\beta$	−1	0
13	$2p_0\beta 2p_{-1}\alpha$	−1	0
14	$2p_0\beta 2p_{-1}\beta$	−1	−1
15	$2p_{-1}\alpha 2p_{-1}\beta$	−2	0

Notice from the table that the largest value of M is M = 2 and that it occurs with the M_s value $M_s = 0$. Therefore, there must be states with L = 2 and S = 0 giving a ${}^1D$ term symbol. The M and M_s values of these states are $M = \pm 2, \pm 1$, and 0 and $M_s = 0$. Also notice from the table that the largest value of M_s is $M_s = 1$ and that this value occurs with the M values $M = \pm 1$ and 0. Thus, there must be states with L = 1 and S = 1 giving a ${}^3P$ term symbol. The M and M_s values of these states are $M = \pm 1$ and 0 and $M_s = \pm 1$ and 0. Eliminating fourteen states with $M = \pm 2, \pm 1, 0$ and $M_s = 0$ and $M = \pm 1, 0$ and $M_s = \pm 1, 0$ from the table leaves a single state with M = 0 and S = 0. For this state, L = 0 and S = 0 giving a ${}^1S$ term symbol.

Summarizing, the carbon $1s^2 2s^2 2p^2$ electron configuration gives rise to fifteen electronic states represented by the three term symbols ${}^1D$ (five states), ${}^3P$ (nine states), and ${}^1S$ (one state).

Next, let us determine from Equation (12.143) the J values of these term symbols. For the ${}^1D$ states, J = 2. For the ${}^3P$ states, J = 2,1, or 0. For the ${}^1S$ state, J = 0. Thus, the full term symbols for the carbon ground state are

$$^1D_2\ ^3P_2\ ^3P_1\ ^3P_0 \text{ and } ^1S_0. \tag{12.164}$$

We next turn to Hund's three rules which permit one to find for a given electron configuration the lowest energy term symbol. Hund's rules are as follows:

The term symbol with the maximum multiplicity has the lowest energy.

If two or more term symbols share the maximum multiplicity, the term symbol with the highest L has the lowest energy.

If two or more term symbols with the maximum multiplicity share the highest L, then for an atom whose outermost subshell is half-filled or less, the term symbol with the smallest J has the lowest energy, while for an atom whose outermost subshell is more than half-filled, the term symbol with the largest J has the lowest energy.

Let us apply Hund's rules to determine the ground state term symbol of carbon. From Equation (12.164), the carbon term symbols of maximum multiplicity are $^3P_2, ^3P_1$, and $^3P_0$. By Hund's first rule, the ground state term symbol must be one of these three. However, all three share the highest $L = 1$. So one must resort to Hund's third rule to determine the term symbol of lowest energy. The outermost subshell configuration of carbon is $2p^2$. So for carbon, the outermost subshell is less than half-filled. Thus, Hund's third rule gives in agreement with experiment that the ground state term symbol of carbon is $^3P_0$.

Next, let us consider the 1s2s excited state electron configuration of helium. Recall this configuration yields the term symbols ^1S and ^3S. From Hund's first rule, the lowest energy term symbol is ^3S (in agreement with Equations [12.92] and [12.93]). Similarly, for the 1s2p helium electron configuration, the lowest energy term symbol is ^3P.

Finally, for the $1s^2 2s^2 2p3p$ carbon electron configuration which yields the term symbols $^3D, ^3P, ^3S, ^1D, ^1P$, and ^1S, Hund's first and second rules identify ^3D as the lowest energy term symbol if one ignores the energy differences of ^3D states with different values for J. If these energy differences are not ignored, then one must consider all three ^3D term symbols $^3D_3, ^3D_2$, and $^3D_1$. By Hund's third rule, the lowest energy term symbol is then $^3D_1$.

FURTHER READINGS

Arfken, George B., Hans J. Weber, and Frank E. Harris. 2012. *Mathematical methods for physicists*. 7th ed. Waltham, MA: Academic Press.

Atkins, Peter and Ronald Friedman. 2011. *Molecular quantum mechanics*. 5th ed. New York: Oxford University Press.

Bunge, C.F., J.A. Barrientos, and A.V. Bunge. "Roothaan-Hartree-Fock Ground-State Atomic Wave Functions: Slater-Type Orbital Expansions and Expectation Values for $Z = 2 - 54$." *Atomic Data Nucl. Data Tables* 53 (1993): 113–162.

Clementi, E. "Ab Initio Computations in Atoms and Molecules." *IBM J.* (1965) (January): 2–19.

Cohen-Tannoudji, Claude, Bernard Diu, and Franck Laloe. 1977. *Quantum mechanics*. Vol. 1. Translated by Susan Reid, Nicole Ostrowsky, and Dan Ostrowsky. New York: John Wiley & Sons.

Kinoshita, T. "Ground State of the Helium Atom." *Phys. Rev.* 105 (1957): 1490–1502.

Levine, Ira N. 2013. *Quantum chemistry*. 7th ed. New York: Prentice-Hall.

McQuarrie, Donald A. 2008. *Quantum chemistry*. 2nd ed. Sausalito, CA: University Science Books.

Pauling, Linus, and E. Bright Wilson Jr. 1985. *Introduction to quantum mechanics with applications to chemistry*. New York: Dover Publications, Inc.

Pekeris, C. L. "1^1S and 2^3S States of Helium." *Phys. Rev.* 115 (1959): 1216–1221.

Roothaan, C.C.J. "New Developments in Molecular Orbital Theory." *Rev. Mod. Phys.* 23 (1951): 69–89.

Scherr, C.W., and R.E. Knight. "Two-Electron Atoms III. A Sixth-Order Perturbation Study of the 1^1S Ground State." *Rev. Mod. Phys.* 35 (1963): 436–442.

Slater, John C. 1960. *Quantum theory of atomic structure*. Vol. 1. New York: McGraw-Hill Book Company, Inc.

PROBLEMS

12.1 Verify that $1\,\text{hartree} = 2{,}625.5\,\text{kJ mol}^{-1}$.

12.2 Recalling from Equation (1.43) that the kinetic energy of an electron moving in a circular orbit of radius r with angular momentum of magnitude L is $\frac{L^2}{2m_e r^2}$, show that the atomic unit of speed given in Table 12.1 as $\frac{e^2}{4\pi\varepsilon_0\hbar}$ is the speed of an electron moving in the lowest energy hydrogen atom Bohr orbit.

12.3 Show that the atomic unit of time given in Table 12.1 as $\frac{(4\pi\varepsilon_0)^2\hbar^3}{m_e e^4}$ is the time needed for an electron moving with the atomic unit of speed $\frac{e^2}{4\pi\varepsilon_0\hbar}$ to traverse a distance of one Bohr radius.

12.4 Show that the speed of light in atomic units is nearly equal to 137.

12.5 Show that the Bohr magneton in atomic units is equal to $\frac{1}{2}$.

12.6 Show by applying the separation-of-variables method to solve Equation (12.10) that $\Psi_0^0(\mathbf{r}_1,\mathbf{r}_2)$ is given by Equations (12.12) and (12.13) and $E_0^{(0)}$ is given by Equation (12.14).

12.7 From Equation (12.14), for helium $E_0^{(0)} = -4\,\text{au}$, which is lower than the experimental helium ground state energy $E_{\text{expt}} = -2.9\,\text{au}$. Explain why this result does not violate the variational theorem.

12.8 Prove from Equation (12.15) that the first-order perturbation correction to the energy of a helium-like atom $E_0^{(1)} = \frac{5}{8}Z$. Proceed as follows. (a) Using the spherical polar coordinate forms $d\mathbf{r}_i = r_i^2 \sin\theta_i\, dr_i\, d\theta_i\, d\phi_i$ for i = 1 and 2 and the form for $\Psi_0^{(0)}(\mathbf{r}_1,\mathbf{r}_2)$ of Equations (12.12) and (12.13), show that

$$E_0^{(1)} = \frac{Z^6}{\pi^2}\int_0^\infty r_1^2\exp(-2Zr_1)\int_0^\infty r_2^2\exp(-2Zr_2)\int_0^{2\pi}\int_0^\pi \sin\theta_1$$

$$\times\int_0^{2\pi}\int_0^\pi \sin\theta_2\frac{1}{r_{12}}\,dr_1 dr_2 d\theta_1 d\theta_2 d\phi_1 d\phi_2.$$

(b) Next, use the following expansion of $\frac{1}{r_{12}}$ in terms of the spherical harmonics $Y_{\ell m}(\theta_1,\phi_1)$ and $Y_{\ell m}(\theta_2,\phi_2)$ $\frac{1}{r_{12}} = \sum_{\ell=0}^{\infty}\sum_{m=-\ell}^{\ell}\frac{4\pi}{2\ell+1}\frac{r_<^\ell}{r_>^{\ell+1}}Y_{\ell m}^*(\theta_1,\phi_1)Y_{\ell m}(\theta_2,\phi_2)$ where $r_<$ ($r_>$) is the smaller (larger) of r_1 and r_2 and the fact that $Y_{00}(\theta,\phi) = \left(\frac{1}{4\pi}\right)^{1/2}$ to show that

$$E_0^{(1)} = 16Z^6\sum_{\ell=0}^{\infty}\sum_{m=-\ell}^{\ell}\frac{1}{2\ell+1}\int_0^\infty r_1^2\exp(-2Zr_1)\int_0^\infty r_2^2\exp(-2Zr_2)\frac{r_<^\ell}{r_>^{\ell+1}}dr_1 dr_2$$

$$\text{x}\int_0^\pi \sin\theta_1\, d\theta_1\int_0^{2\pi}Y_{\ell m}^*(\theta_1,\phi_1)Y_{00}(\theta_1,\phi_1)\,d\phi_1\int_0^\pi \sin\theta_2\, d\theta_2\int_0^{2\pi}Y_{00}^*(\theta_2,\phi_2)Y_{\ell m}(\theta_2,\phi_2)\,d\phi_2.$$

(c) Use the orthonormality of the spherical harmonics to simplify your expression in part (b) to

$$E_0^{(1)} = 16Z^6\int_0^\infty r_1^2\exp(-2Zr_1)\,dr_1\int_0^\infty r_2^2\exp(-2Zr_2)\,dr_2\frac{1}{r_>}.$$

(d) Show that your expression in part (c) may be written as

$$E_0^{(1)} = 16Z^6\left[\int_0^\infty r_1 \exp(-2Zr_1)\,dr_1 \int_0^{r_1} r_2^2 \exp(-2Zr_2)\,dr_2 + \int_0^\infty r_1^2 \exp(-2Zr_1)\,dr_1 \int_{r_1}^\infty r_2 \exp(-2Zr_2)\,dr_2\right].$$

(e) Evaluate the above integrals over r_2 by integration by parts and simplify to show that

$$E_0^{(1)} = 16Z^6\left[-\int_0^\infty \exp(-4Zr_1)\left(\frac{1}{4Z^2}r_1^2 + \frac{r_1}{4Z^3}\right)dr_1 + \frac{1}{4Z^3}\int_0^\infty r_1 \exp(-2Zr_1)\,dr_1\right].$$

(f) Evaluate the above integrals over r_1 using Equation (10.51) to prove the required result $E_0^{(1)} = \frac{5}{8}Z.$

12.9 Using the spherical harmonic expansion method of Problem 12.8, evaluate the integral $J_{1s,2p} = \int \Psi(\mathbf{r}_1,\mathbf{r}_2)\frac{1}{r_{12}}\Psi(\mathbf{r}_1,\mathbf{r}_2)\,d\mathbf{r}_1 d\mathbf{r}_2$ where $\Psi(\mathbf{r}_1,\mathbf{r}_2) = \Psi_{1s}(\mathbf{r}_1)\Psi_{2pz}(\mathbf{r}_2)$ with $\Psi_{1s}(\mathbf{r}) = \left(\frac{1}{\pi}\right)^{1/2} Z^{3/2}\exp(-Zr)$ and $\Psi_{2pz}(\mathbf{r}) = \frac{1}{4}\left(\frac{1}{2\pi}\right)^{1/2} Z^{5/2} r \exp\left(-\frac{1}{2}Zr\right)\cos\theta.$

Note that Ψ_{1s} and Ψ_{2pz} are normalized 1s and $2p_Z$ hydrogen-like atom orbitals. (a) First, show after making the spherical harmonic expansion of $\frac{1}{r_{12}}$ that $J_{1s,2p}$ may be reduced to

$$J_{1s,2p} = \frac{Z^8}{6}\left[\int_0^\infty r_1 \exp(-2Zr_1)\,dr_1 \int_0^{r_1} r_2^4 \exp(-Zr_2)\,dr_2 + \int_0^\infty r_1^2 \exp(-2Zr_1)\,dr_1 \int_{r_1}^\infty r_2^3 \exp(-Zr_2)\,dr_2\right].$$

(b) Then, using the integrals $\int_0^{r_1} r_2^4 \exp(-Zr_2)\,dr_2 = -\left(\frac{r_1^4}{Z} + \frac{4r_1^3}{Z^2} + 12\frac{r_1^2}{Z^3} + \frac{24r_1}{Z^4} + \frac{24}{Z^5}\right)\exp(-Zr_1) + \frac{24}{Z^5}$ and $\int_{r_1}^\infty r_2^3 \exp(-Zr_2)\,dr_2 = \left(\frac{r_1^3}{Z} + \frac{3r_1^2}{Z^2} + \frac{6r_1}{Z^3} + \frac{6}{Z^4}\right)\exp(-Zr_1)$, show that $J_{1s,2p} = \frac{59}{243}Z.$

12.10 Show without doing any calculations that the optimal variational energy $E_T(Z'_{min})$ determined from Equations (12.18)–(12.20) must be lower than the first-order perturbation energy of Equation (12.17). (Hint: Your argument should be based on the variational principle.)

12.11 Prove that the integral I of Equation (12.23) has the value given in Equation (12.25).

12.12 The electron affinity of an atom X is the minimum amount of energy needed to detach an electron from its negative ion X^-. That is, the electron affinity is the energy change for the process $X^- \rightarrow X +$ a stationary electron at infinity. If the electron affinity is positive, X^- is stable, while if it is negative, X^- is unstable. The experimental electron affinity of hydrogen is 0.754 eV, showing that the ion H^- is stable. Referring to Table 12.3, calculate the variational estimate of the electron affinity of hydrogen and show that estimate incorrectly predicts that H^- is unstable.

12.13 From Equation (12.30), calculate the ground state energy E_0 of helium in au to (a) first, (b) second, (c) third, (d) fourth, (e) fifth, and (f) sixth order in perturbation theory. Compare your results for E_0 to Pekeris' result of $E_0 = -2.903724375\,\text{au}$. (g) Similarly, calculate E_0 in au and eV to sixth order in perturbation theory for the Li^+ ion. Compare with the experimental result for the ground state energy of Li^+ $E_{expt} = -198.0945\,\text{eV}$.

12.14 (a) Consider the Slater-type orbital $\chi_{n\ell m}(\mathbf{r}) = N_{n\xi} r^{n-1} \exp\left(-\frac{\xi r}{n}\right) Y_{\ell m}(\theta,\phi)$. Show for integer n that the normalization factor $N_{n\xi} = \frac{(2\xi)^n}{n}\left[\frac{2\xi/n}{(2n)!}\right]^{1/2}$. (b) Show for two normalized Slater-type orbitals $\chi_{n\ell m}(\mathbf{r})$ and $\chi_{n'\ell m}(\mathbf{r})$ with the same ξ values but with $n' \neq n$ that $\int \chi^*_{n'\ell m}(\mathbf{r})\chi_{n\ell m}(\mathbf{r})d\mathbf{r} = (n+n')!\left[\frac{1}{(2n)!(2n')!}\right]^{1/2}$ and therefore that the two orbitals are not orthogonal.

12.15 Derive Equation (12.52) for δE from Equations (12.50) and (12.51).

12.16 Show for the helium-like atoms that because the Hartree–Fock potential $U(r)$ is spherically symmetric and because the Hartree–Fock orbital $\phi(r)$ is an s-orbital, the three-dimensional Hartree–Fock equation (Equation 12.53a) may be reduced to the one-dimensional equation $\left[-\frac{1}{2}\frac{1}{r_1^2}\frac{d}{dr_1}\left(r_1^2\frac{d}{dr_1}\right) + U(r_1)\right]\phi(r_1) = \varepsilon_1\phi(r_1)$.

12.17 Derive Equation (12.65) for the orbital energies ε_1 and ε_2 from the Hartree–Fock equations (Equation 12.58). (Hint: Multiply Equation [12.58] on the left by the Hartree–Fock orbital and then integrate over all space.)

12.18 Show that E of Equation (12.78) and E_- of Equation (12.79) are equal. (Hint: Use the spin orthonormality relations of Equation [10.121].)

12.19 Verify Equation (12.86) and (12.87) for the singlet and triplet energies.

12.20 Consider the 3×3 determinant

$$D_3 = \begin{vmatrix} a_{11} & a_{12} & a_{13} \\ a_{21} & a_{22} & a_{23} \\ a_{31} & a_{32} & a_{33} \end{vmatrix}.$$

It may be expanded by the method of minors (Arfken, Weber, and Harris 2012) to give

$$D_3 = a_{11}\begin{vmatrix} a_{22} & a_{23} \\ a_{32} & a_{33} \end{vmatrix} - a_{12}\begin{vmatrix} a_{21} & a_{23} \\ a_{31} & a_{33} \end{vmatrix} + a_{13}\begin{vmatrix} a_{21} & a_{22} \\ a_{31} & a_{32} \end{vmatrix}$$

or

$$D_3 = a_{11}a_{22}a_{33} - a_{11}a_{23}a_{32} - a_{12}a_{21}a_{33} + a_{12}a_{23}a_{31} + a_{13}a_{21}a_{32} - a_{13}a_{22}a_{31}.$$

(a) Next, consider the determinant D_3' formed by interchanging the first two columns of D_3. Show that $D_3' = -D_3$.

(b) Assume the first two columns of D_3 are identical. Then, show $D_3 = 0$.

More generally, one can similarly show that interchanging any two columns or rows of D_3 changes its sign and if any two columns or rows of D_3 are identical that $D_3 = 0$.

12.21 Consider an N-electron atom with N occupied orthonormal spin orbitals $\chi_1, \chi_2, \cdots, \chi_N$. The Slater determinantal wave function for the atom is

$$\Psi(\rho_1,\rho_2,\ldots,\rho_N) = \frac{1}{\sqrt{N!}}\begin{vmatrix} \chi_1(\rho_1) & \chi_2(\rho_1) & \cdots & \chi_N(\rho_1) \\ \chi_1(\rho_2) & \chi_2(\rho_2) & \cdots & \chi_N(\rho_2) \\ \vdots & \vdots & & \vdots \\ \chi_1(\rho_N) & \chi_2(\rho_N) & \cdots & \chi_N(\rho_N) \end{vmatrix}$$

where the factor $\frac{1}{\sqrt{N!}}$ guarantees that the wave function is normalized and where for $i = 1,\ldots,N$ $\rho_i = \mathbf{r}_i,\sigma_i$ are the space and spin coordinates of electron i.

(a) Show that for spin orbital configurations which violate the Pauli exclusion principle by putting more than one electron in one or more of the spin orbitals, the above determinant has identical columns and thus the wave function Ψ vanishes. (b) Show that if two electrons, say electrons 1 and 2, have the same space and spin coordinates $\rho = \mathbf{r},\sigma$, the wave function Ψ vanishes.

(c) Explain why your result in part (b) implies that two electrons with the same spin (or the same σ) have zero probability of being at the same spatial point $\mathbf{r}$.

Your result in part (c) shows that Fermi correlation is included in antisymmetric orbital model wave functions.

12.22 (a) Referring to Problem 12.20, show that the determinantal wave function for lithium of Equation (12.103) may be expanded as

$$\begin{aligned}\Psi(\rho_1,\rho_2,\rho_3) = \frac{1}{\sqrt{3!}}\big[&\phi_{1s}(r_1)\alpha(\sigma_1)\phi_{1s}(r_2)\beta(\sigma_2)\phi_{2s}(r_3)\alpha(\sigma_3)\\ &-\phi_{1s}(r_1)\alpha(\sigma_1)\phi_{2s}(r_2)\alpha(\sigma_2)\phi_{1s}(r_3)\beta(\sigma_3)-\phi_{1s}(r_1)\beta(\sigma_1)\phi_{1s}(r_2)\alpha(\sigma_2)\phi_{2s}(r_3)\alpha(\sigma_3)\\ &+\phi_{1s}(r_1)\beta(\sigma_1)\phi_{2s}(r_2)\alpha(\sigma_2)\phi_{1s}(r_3)\alpha(\sigma_3)+\phi_{2s}(r_1)\alpha(\sigma_1)\phi_{1s}(r_2)\alpha(\sigma_2)\phi_{1s}(r_3)\beta(\sigma_3)\\ &-\phi_{2s}(r_1)\alpha(\sigma_1)\phi_{1s}(r_2)\beta(\sigma_2)\phi_{1s}(r_3)\alpha(\sigma_3)\big].\end{aligned}$$

(b) Assuming that $\phi_{1s}(r)$ and $\phi_{2s}(r)$ are orthonormal and using the spin orthonormality relations of Equation (10.121), show from your expansion of part (a) that $\Psi(\rho_1,\rho_2,\rho_3)$ is normalized. That is, show

$$\int \Psi^*(\rho_1,\rho_2,\rho_3)\Psi(\rho_1,\rho_2,\rho_3)d\mathbf{r}_1 d\sigma_1 d\mathbf{r}_2 d\sigma_2 d\mathbf{r}_3 d\sigma_3 = 1.$$

(Hint: When you multiply Ψ^* and Ψ and integrate, all cross terms vanish and the six remaining terms are all equal to one.)

12.23 Consider the four-electron beryllium atom. Denote the coordinates of its electrons by $\mathbf{r}_1,\mathbf{r}_2,\mathbf{r}_3$, and $\mathbf{r}_4$. (a) Write out the Hamiltonian operator $\hat{H}$ of beryllium.

In its ground state, beryllium has doubly occupied 1s and 2s orbitals $\phi_{1s}(r)$ and $\phi_{2s}(r)$.

(b) Assuming these orbitals are orthonormal, write down the normalized ground state Slater determinantal wave function Ψ for beryllium.

12.24 Derive Equation (12.117) for the orbital energies ε_i from the general Hartree–Fock equations (Equations 12.113–12.116). (See hint for Problem 12.17.)

12.25 Show that the results of the general Hartree–Fock theory properly reduce to our Hartree–Fock results for the helium-like atoms with N = 1. In particular, show that (a) Equation (12.109) for E reduces to Equation (12.62), (b) the general Hartree–Fock equations (Equations 12.114–12.116) reduce to Equation (12.53a), and (c) Equation (12.117) for the orbital energies ε_i reduces to Equation (12.65).

12.26 Show from Equations (12.109) and (12.117) that the Hartree–Fock energy $E = \sum_{i=1}^{N}(\varepsilon_i + I_i)$.

12.27 There are four spin orbital configurations which are consistent with the helium atom excited state 1s2s electron configuration. These are $1s\alpha 2s\alpha, 1s\alpha 2s\beta, 1s\beta 2s\alpha$, and $1s\beta 2s\beta$. (a) Write down the normalized Slater determinantal wave functions Ψ_1,Ψ_2,Ψ_3, and Ψ_4 which correspond, respectively, to these spin orbital configurations. Next, consider the helium atom singlet and triplet wave functions of Equations (12.82) and (12.83). (b) Show that $\Psi_S = \frac{1}{\sqrt{2}}(\Psi_2 - \Psi_3), \Psi_{T_1} = \Psi_1, \Psi_{T_2} = \frac{1}{\sqrt{2}}(\Psi_2 + \Psi_3)$, and $\Psi_{T_3} = \Psi_4$.

12.28 In this problem, you will prove Equation (12.132) for the energy E^+ of the Be^+ ion. First, note that the expansion of the Be^+ ion Slater determinantal wave function $\Psi^+(\rho_1, \rho_2, \rho_3)$ of Equation (12.129) is identical to the expansion of the lithium atom wave function given in Problem 12.22. (a) Thus, show that the expansion of Ψ^+ may be written as

$$\Psi^+ = \frac{1}{\sqrt{3!}}\left[\phi_{1s}(r_1)\phi_{1s}(r_2)\phi_{2s}(r_3) - \phi_{2s}(r_1)\phi_{1s}(r_2)\phi_{1s}(r_3)\right]\alpha(\sigma_1)\beta(\sigma_2)\alpha(\sigma_3)$$

$$+\frac{1}{\sqrt{3!}}\left[\phi_{2s}(r_1)\phi_{1s}(r_2)\phi_{1s}(r_3) - \phi_{1s}(r_1)\phi_{2s}(r_2)\phi_{1s}(r_3)\right]\alpha(\sigma_1)\alpha(\sigma_2)\beta(\sigma_3)$$

$$+\frac{1}{\sqrt{3!}}\left[\phi_{1s}(r_1)\phi_{2s}(r_2)\phi_{1s}(r_3) - \phi_{1s}(r_1)\phi_{1s}(r_2)\phi_{2s}(r_3)\right]\beta(\sigma_1)\alpha(\sigma_2)\alpha(\sigma_3).$$

(b) Show that because $\alpha(\sigma)$ and $\beta(\sigma)$ are orthonormal when determining E^+ from Equation (12.130) and from the form for Ψ^+ of part (a), the six cross terms vanish and the three non-vanishing terms reduce to purely spatial integrals. Additionally, it may be shown that the three non-vanishing contributions to E^+ are equal. Consequently, E^+ may be written as

$$E^+ = \frac{1}{2}\int\left[\phi_{1s}^*(r_1)\phi_{1s}^*(r_2)\phi_{2s}^*(r_3) - \phi_{2s}^*(r_1)\phi_{1s}^*(r_2)\phi_{1s}^*(r_3)\right]\hat{H}$$

$$\times\left[\phi_{1s}(r_1)\phi_{1s}(r_2)\phi_{2s}(r_3) - \phi_{2s}(r_1)\phi_{1s}(r_2)\phi_{1s}(r_3)\right]d\mathbf{r}_1 d\mathbf{r}_2 d\mathbf{r}_3.$$

(c) Use the above expression for E^+, Equation (12.131) for $\hat{H}^+$, and the orthonormality of $\phi_{1s}(r)$ and $\phi_{2s}(r)$ to derive Equation (12.131) for E^+. (Note that to derive Equation [12.131], you will have to relabel dummy variables of integration.)

12.29 Using Pekeris' exact result for the ground state energy of helium, determine in au the exact correlation energy E_{corr} for helium and compare with the approximate result of Table 12.9.

12.30 (a) Write down the electron configuration of mercury with $Z = 80$ as predicted by Madelung's rule (which is correct for mercury). (b) What are the term symbols for mercury?

12.31 Recalling Equation (10.118b), apply the operator $\hat{S}_z = \hat{S}_{z1} + \hat{S}_{z2}$ for the Z-component of the total spin angular momentum of a two-electron atom to the helium singlet and triplet wave functions of Equations (12.82) and (12.83) to show that in atomic units, (a) $\hat{S}_z\Psi_S = 0, \hat{S}_z\Psi_{T_1} = \Psi_{T_1}, \hat{S}_z\Psi_{T_2} = 0$, and (b) Explain why your results in part (a) show that for the singlet state the total spin angular momentum quantum number $S = 0$, while for the triplet states $S = 1$.

12.32 (a) Show that for scandium with ground state electron configuration $[Ne]4s^2 3d$, the term symbols deriving from this configuration are $^2D_{5/2}$ and $^2D_{3/2}$. (b) What is the ground state term symbol of scandium? (c) How many electronic states derive from the scandium ground state electron configuration?

12.33 Consider an atom whose outermost subshell is n electrons short of being completed. Then, one may show that the term symbols of that atom are identical to those of the atom with n electrons occupying the same outermost subshell. For example, the term symbols of oxygen with the electron configuration $1s^2 2s^3 2p^4$ (which is two electrons short of a completed 2p subshell) are the same as those of carbon with the electron configuration $1s^2 2s^2 2p^2$ (which has two electrons occupying the 2p subshell). (a) Given this information, show that the term symbols of fluorine with electron configuration $1s^2 2s^2 2p^5$ are $^2P_{3/2}$ and $^2P_{1/2}$. (b) What is the ground state term symbol of fluorine? (c) How many electronic states derive from the electron configuration of fluorine?

12.34 (a) Show that without including the J subscripts, the term symbols which derive from the carbon excited state electron configuration $1s^2 2s^2 2p3d$ are $^3F, ^3D, ^3P, ^1F, ^1D$, and ^1P.

(b) Show that including the J subscripts gives the term symbols $^3F_4, ^3F_3, ^3F_2, ^3D_3, ^3D_2, ^3D_1, ^3P_2, ^3P_1, ^3P_0, ^1F_3, ^1D_2$, and $^1P_1$. (c) How many electronic states derive from the $1s^2 2s^2 2p3d$ electron configuration? (d) What is the lowest energy term symbol for this configuration?

12.35 Taking gallium with electron configuration $[Ar]4s^2 3d^{10} 4p$ as an example, show that the contributions to M and M_s from closed shells vanish.

12.36 Using the same method we employed in Section 12.10 to obtain the term symbols of the $1s^2 2s^2 2p^2$ electron configuration of carbon, show that the term symbols for the $[Ar]4s^2 3d^2$ configuration of titanium are $^3F, ^3P, ^1G, ^1D$, and ^1S.

12.37 Hund's first and second rules provide a very simple way to find the lowest energy term symbol of an atom with a single open subshell. Namely, fill the open subshell as follows. First, add spin α electrons to the subshell orbitals in order of decreasing m quantum number starting with the orbital of maximum m. If the subshell becomes filled with spin α electrons to continue, fill the orbitals in the same order (maximum m first and then decreasing m) with spin β electrons. After the filling process is completed, determine M and M_s for the resulting subshell spin orbital configuration and note that $L = M$ and $S = M_s$ to determine the term symbol. For example, for oxygen with the open 2p subshell configuration $2p^4$, the filling process produces the 2p subshell spin orbital configuration $2p_{+1}\alpha 2p_0\alpha 2p_{-1}\alpha 2p_{+1}\beta$ yielding L as $L = M = 1 + 0 - 1 + 1 = 1$ and S as $S = M_s = \frac{1}{2} + \frac{1}{2} + \frac{1}{2} - \frac{1}{2} = 1$ giving the ground state term symbol as ^3P. Using this method, show that the ground state term symbol of (a) nitrogen is ^4S, (b) fluorine is ^2P, (c) iron is ^5D, and (d) nickel is ^3F. (Refer to Table 12.11.)

SOLUTIONS TO SELECTED PROBLEMS

12.7 The Hamiltonian operator $\hat{H}_0$ is not equal to the true Hamiltonian operator $\hat{H}$.

12.12 Variational estimate of the electron affinity of $H = -0.75$ eV.

12.13 (a) −2.75 au, (b) −2.90767 au, (c) −2.90332 au, (d) −2.90354 au, (e) −2.90367 au, (f) −2.90371 au, and (g) −7.27991 au = −198.0965 eV.

12.23 (a) $\hat{H} = -\frac{1}{2}\nabla_1^2 - \frac{4}{r_1} - \frac{1}{2}\nabla_2^2 - \frac{4}{r_2} - \frac{1}{2}\nabla_3^2 - \frac{4}{r_3} - \frac{1}{2}\nabla_4^2 - \frac{4}{r_4} + \frac{1}{r_{12}} + \frac{1}{r_{13}} + \frac{1}{r_{14}} + \frac{1}{r_{23}} + \frac{1}{r_{24}} + \frac{1}{r_{34}}$.

(b)
$$\Psi = \frac{1}{\sqrt{4!}}\begin{vmatrix} \phi_{1s}(r_1)\alpha(\sigma_1) & \phi_{1s}(r_1)\beta(\sigma_1) & \phi_{2s}(r_1)\alpha(\sigma_1) & \phi_{2s}(r_1)\beta(\sigma_1) \\ \phi_{1s}(r_2)\alpha(\sigma_2) & \phi_{1s}(r_2)\beta(\sigma_2) & \phi_{2s}(r_2)\alpha(\sigma_2) & \phi_{2s}(r_2)\beta(\sigma_2) \\ \phi_{1s}(r_3)\alpha(\sigma_3) & \phi_{1s}(r_3)\beta(\sigma_3) & \phi_{2s}(r_3)\alpha(\sigma_3) & \phi_{2s}(r_3)\beta(\sigma_3) \\ \phi_{1s}(r_4)\alpha(\sigma_4) & \phi_{1s}(r_4)\beta(\sigma_4) & \phi_{2s}(r_4)\alpha(\sigma_4) & \phi_{2s}(r_4)\beta(\sigma_4) \end{vmatrix}.$$

12.27 (a)

$$\Psi_1 = \frac{1}{\sqrt{2}}\begin{vmatrix} \phi_{1s}(r_1)\alpha(\sigma_1) & \phi_{2s}(r_1)\alpha(\sigma_1) \\ \phi_{1s}(r_2)\alpha(\sigma_2) & \phi_{2s}(r_2)\alpha(\sigma_2) \end{vmatrix} \quad \Psi_2 = \frac{1}{\sqrt{2}}\begin{vmatrix} \phi_{1s}(r_1)\alpha(\sigma_1) & \phi_{2s}(r_1)\beta(\sigma_1) \\ \phi_{1s}(r_2)\alpha(\sigma_2) & \phi_{2s}(r_2)\beta(\sigma_2) \end{vmatrix}$$

$$\Psi_3 = \frac{1}{\sqrt{2}}\begin{vmatrix} \phi_{1s}(r_1)\beta(\sigma_1) & \phi_{2s}(r_1)\alpha(\sigma_1) \\ \phi_{1s}(r_2)\beta(\sigma_2) & \phi_{2s}(r_2)\alpha(\sigma_2) \end{vmatrix} \quad \Psi_4 = \frac{1}{\sqrt{2}}\begin{vmatrix} \phi_{1s}(r_1)\beta(\sigma_1) & \phi_{2s}(r_1)\beta(\sigma_1) \\ \phi_{1s}(r_2)\beta(\sigma_2) & \phi_{1s}(r_2)\beta(\sigma_2) \end{vmatrix}$$

12.29 For helium, E_{corr} is exactly –0.042044 au.
12.30 (a) $[Xe]6s^2 4f^{14} 5d^{10}$ and (b) $^1S_0$.
12.32 (b) $^2D_{3/2}$ and (c) 10.
12.33 (b) $^2P_{3/2}$ and (c) 6.
12.34 (c) 60 and (d) $^3F_2$.

13 Molecular Electronic Structure and Chemical Bonding

We began this textbook by describing in Chapter 1 the early major developments of quantum mechanics. We started with the seminal contribution of Planck who initiated the quantum theory by introducing the revolutionary hypothesis that the resonators in the walls of a blackbody radiation cavity have discrete rather than continuous energies. Then, in Chapters 2 and 4–6 we laid out the physical and mathematical foundations of quantum mechanics. Next, in Chapters 7–12 we made increasingly advanced applications of these foundations. This chapter is the culmination of all we have done so far for here we develop the theory of molecular electronic structure and especially describe the greatest achievement of molecular quantum mechanics, the elucidation of the origin and nature of the chemical bond.

We begin by introducing the *Born–Oppenheimer approximation* which is foundational to most applications of quantum mechanics to chemistry.

13.1 THE BORN–OPPENHEIMER APPROXIMATION AND THE ELECTRONIC AND NUCLEAR SCHRÖDINGER EQUATIONS

The Born–Oppenheimer approximation was formulated by the German physicist Max Born and the American physicist J. Robert Oppenheimer in 1927 just one year after the discovery of the Schrödinger equation. The Born–Oppenheimer approximation yields two major results.

It shows that the rotational–vibrational Schrödinger equation of Equations (8.4) and (8.6) which depends only on nuclear coordinates may be derived from a more fundamental equation called the electron–nuclear Schrödinger equation which depends on the coordinates of *all* of the particles of a molecular system, electrons as well as nuclei. This result puts the theory of molecular rotations and vibrations described in Chapters 7–9 on a rigorous foundation.

It supplies a tractable Schrödinger equation called the electronic Schrödinger equation which unlike the intractable electron nuclear Schrödinger equation governs only the motion of the electrons. The electronic Schrödinger equation provides the foundation for all treatments of molecular electronic structure and chemical bonding and therefore forms the basis for all developments in subsequent sections of this chapter.

The principle underlying the Born–Oppenheimer approximation is that the masses m_N of all atomic nuclei are much greater than the electron mass m_e. Thus, even for the lightest nucleus the hydrogen atom nucleus with mass m_p (the proton mass) $m_p = 1{,}836 m_e$. For the heavier nuclei which occur in most molecules, the nuclear masses can be as much as tens of thousands of times greater than the electron mass.

Because of this enormous mass disparity, a molecule's electrons move much more rapidly than its nuclei (Problem 13.1). Consequently, the electronic wave function of a molecule can nearly instantaneously adjust to changes in the molecule's configuration (the positions of its nuclei) and thus can almost perfectly follow these changes. What this implies is that the electronic wave function of a molecule with *moving nuclei* which reach a specific configuration at some arbitrary time t is nearly identical to the electronic wave function the molecule would have if its nuclei were *fixed* at that configuration.

These qualitative considerations are quantified by the rigorous analysis of Born and Oppenheimer. According to Pauling and Wilson (Pauling and Wilson 1985, p. 261) who studied Born and Oppenheimer's original German-language paper, the derivation of Born and Oppenheimer is

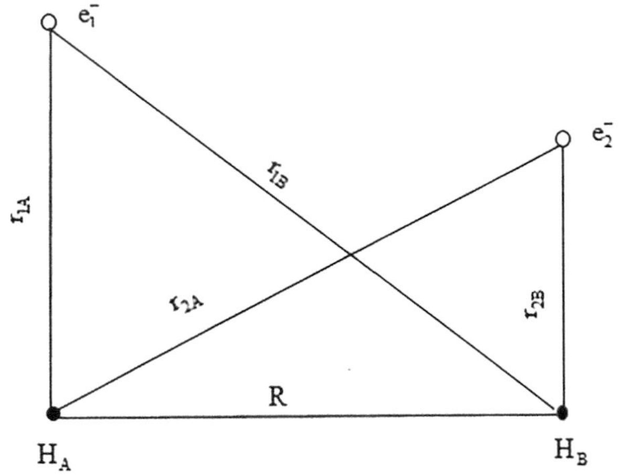

FIGURE 13.1 Sketch of molecular hydrogen H_2.

fearsomely complex. But the results of the derivation are simple and easy to understand. So here following Pauling and Wilson, we will omit the derivation and pass directly to its results.

To keep the notation simple, we will only describe the Born–Oppenheimer theory for the simplest neutral molecule, the two-electron hydrogen molecule H_2 sketched in Figure 13.1. The discussion however extends to more complex molecules in a straightforward manner (Problems 13.2 and 13.3).

To distinguish between the two hydrogen atoms of H_2, we will label the nuclei of these atoms by the subscripts A and B. Similarly, we will label the two electrons by the subscripts 1 and 2. Given these labels, we will denote the H_2 internuclear separation by $R = |\mathbf{R}_A - \mathbf{R}_B|$ where $\mathbf{R}_A$ and $\mathbf{R}_B$ are the position vectors of the nuclei of hydrogen atoms A and B. Similarly, we will denote the distance between the two electrons by $r_{12} = |\mathbf{r}_1 - \mathbf{r}_2|$ where $\mathbf{r}_1$ and $\mathbf{r}_2$ are the position vectors of electrons 1 and 2. Finally, we will denote the distance between electron 1 and nucleus A by $r_{1A} = |r_1 - \mathbf{R}_A|$ with, as shown in Figure 13.1, analogous designations for the other three electron–nuclear distances.

With these definitions, the H_2 electron–nuclear Hamiltonian operator $\hat{H}_{en}$ is in SI units

$$\hat{H}_{en} = -\frac{\hbar^2}{2m_p}\nabla_A^2 - \frac{\hbar^2}{2m_p}\nabla_B^2 - \frac{\hbar^2}{2m_e}\nabla_1^2 - \frac{\hbar^2}{2m_e}\nabla_2^2 - \frac{e^2}{4\pi\varepsilon_0}\frac{1}{r_{1A}} - \frac{e^2}{4\pi\varepsilon_0}\frac{1}{r_{1B}} - \frac{e^2}{4\pi\varepsilon_0}\frac{1}{r_{2A}} - \frac{e^2}{4\pi\varepsilon_0}\frac{1}{r_{2B}} + \frac{e^2}{4\pi\varepsilon_0}\frac{1}{R} + \frac{e^2}{4\pi\varepsilon_0}\frac{1}{r_{12}}. \tag{13.1}$$

The first two terms in $\hat{H}_{en}$ are the operators which represent the kinetic energies of nuclei A and B, the next two terms represent the kinetic energies of electrons 1 and 2, the next four terms represent the attractive Coulomb potentials between electrons 1 and 2 and nuclei A and B, and finally, the last two terms represent the repulsive Coulomb potentials between, respectively, the two nuclei and the two electrons. When expressed in atomic units, $\hat{H}_{en}$ simplifies to

$$\hat{H}_{en} = -\frac{1}{2m_p}\nabla_A^2 - \frac{1}{2m_p}\nabla_B^2 - \frac{1}{2}\nabla_1^2 - \frac{1}{2}\nabla_2^2 - \frac{1}{r_{1A}} - \frac{1}{r_{1B}} - \frac{1}{r_{2A}} - \frac{1}{r_{2B}} + \frac{1}{R} + \frac{1}{r_{12}} \tag{13.2}$$

where now $m_p = 1{,}836$ is the mass of a hydrogen atom nucleus expressed in atomic units (multiples of the electron mass m_e).

We next denote the electron–nuclear stationary state wave functions and energies of H_2 by $\Psi_{n_A,n_B,n}$ and $E_{n_A,n_B,n}$ where the interpretations of the quantum numbers n_a, n_b and n will be clear shortly. These wave functions and energies are the solutions of the H_2 electron–nuclear Schrödinger equation which from Equation (13.2) for $\hat{H}_{en}$ is

$$\left(-\frac{1}{2m_p}\nabla_A^2 - \frac{1}{2m_p}\nabla_B^2 - \frac{1}{2}\nabla_1^2 - \frac{1}{2}\nabla_2^2 - \frac{1}{r_{1A}} - \frac{1}{r_{1B}} - \frac{1}{r_{2A}} - \frac{1}{r_{2B}} + \frac{1}{R} + \frac{1}{r_{12}}\right)$$
$$\times \Psi_{n_A,n_B,n}(\mathbf{R}_A,\mathbf{R}_B,\mathbf{r}_1,\mathbf{r}_2) = E_{n_A,n_B,n}\Psi_{n_A,n_B,n}(\mathbf{R}_A,\mathbf{R}_B,\mathbf{r}_1,\mathbf{r}_2). \quad (13.3)$$

The H_2 electron–nuclear Schrödinger equation (Equation 13.3), like the electron–nuclear Schrödinger equations for all other molecules, is intractable because it governs the motion of two different kinds of particles, electrons and nuclei. Moreover, it cannot be simplified in a straightforward manner. This is because the attractive electron–nuclear potential terms like $-\frac{1}{r_{1A}}$ in $\hat{H}_{en}$ render it impossible to exactly separate the Schrödinger equation (Equation 13.3) into electronic and nuclear parts. Because of this non-separability, the wave functions $\Psi_{n_A,n_B,n}$ cannot be rigorously expressed as products of electronic and nuclear wave functions and the energies $E_{n_A,n_B,n}$ cannot be rigorously written as sums of electronic and nuclear energies.

However, Born and Oppenheimer showed that while the wave functions $\Psi_{n_A,n_B,n}$ cannot be exactly factorized because of the electron–nuclear mass disparity to a very high degree of approximation, they can be so factorized. Specifically, Born and Oppenheimer showed that the wave functions $\Psi_{n_A,n_B,n}$ can be nearly exactly factorized as

$$\Psi_{n_A,n_B,n}(\mathbf{R}_A,\mathbf{R}_B,\mathbf{r}_1,\mathbf{r}_2) \doteq \Psi_n(\mathbf{r}_1,\mathbf{r}_2;R)\Psi_{n_A,n_B}(\mathbf{R}_A,\mathbf{R}_B). \quad (13.4)$$

Ψ_n and Ψ_{n_A,n_B} are, respectively, called the electronic and nuclear wave functions, while n, n_a, and n_B are the quantum numbers which distinguish between these wave functions.

Notice that the electronic wave functions Ψ_n not only depend on the electron coordinates $\mathbf{r}_1$ and $\mathbf{r}_2$ but also depend on the nuclear coordinates $\mathbf{R}_A$ and $\mathbf{R}_B$ through their dependence on $R = |\mathbf{R}_A - \mathbf{R}_B|$. The nuclear wave functions Ψ_{n_A,n_B} however depend only on the nuclear coordinates. These features extend to all molecules.

By substituting the factorized form for $\Psi_{n_A,n_B,n}$ of Equation (13.4) into the electron–nuclear Schrödinger equation (Equation 13.3) and by making further simplifications based on the electron–nuclear mass disparity, Born and Oppenheimer showed that Ψ_n and Ψ_{n_A,n_B} are the solutions of two independent and tractable Schrödinger equations called, respectively, the electronic and nuclear Schrödinger equations. Both Schrödinger equations are of prime importance. The electronic Schrödinger equation provides the foundation for all treatments of molecular electronic structure. The nuclear Schrödinger equation is the basis for all treatments of molecular vibrations and rotations and their associated spectroscopies.

We first discuss the electronic Schrödinger equation. It governs the motion of the two electrons conditional that the nuclei are fixed with internuclear separation R. The electronic Schrödinger equation is

$$\left(-\frac{1}{2}\nabla_1^2 - \frac{1}{2}\nabla_2^2 - \frac{1}{r_{1A}} - \frac{1}{r_{1B}} - \frac{1}{r_{2A}} - \frac{1}{r_{2B}} + \frac{1}{r_{12}} + \frac{1}{R}\right)\Psi_n(\mathbf{r}_1,\mathbf{r}_2;R) = E_n(R)\Psi_n(\mathbf{r}_1,\mathbf{r}_2;R). \quad (13.5)$$

where $E_n(R)$ is the energy of the electronic state with quantum numbers n.

Next, let us write the electronic Schrödinger equation as $\hat{H}_e\Psi_n(\mathbf{r}_1,\mathbf{r}_2;R) = E_n(R)\Psi_n(\mathbf{r}_1,\mathbf{r}_2;R)$ where $\hat{H}_e = -\frac{1}{2}\nabla_1^2 - \frac{1}{2}\nabla_2^2 - \frac{1}{r_{1A}} - \frac{1}{r_{1B}} - \frac{1}{r_{2A}} - \frac{1}{r_{2B}} + \frac{1}{r_{12}} + \frac{1}{R}$ is the H_2 electronic Hamiltonian operator. Notice that $\hat{H}_e$ is identical to the electron–nuclear Hamiltonian operator $\hat{H}_{en}$ of Equation (13.2) except that the nuclear kinetic energy operators $-\frac{1}{2m_p}\nabla_A^2$ and $-\frac{1}{2m_p}\nabla_B^2$ included in $\hat{H}_{en}$ are absent in $\hat{H}_e$. This absence is expected since because of the electron–nuclear mass disparity, the H_2 nuclei may be taken as fixed or motionless. Consequently, by the classical–quantum correspondence the nuclear kinetic energy terms in the classical Hamiltonian H_e corresponding to $\hat{H}_e$ vanish, implying again by the classical–quantum correspondence that the nuclear kinetic energy operators are absent from $\hat{H}_e$.

Moreover, since the nuclei are fixed, the internuclear separation R enters the electronic Schrödinger equation as a constant parameter rather than as a variable like either of the electron position vectors $\mathbf{r}_1$ or $\mathbf{r}_2$. Therefore, as R varies over some range, $\Psi_n(\mathbf{r}_1,\mathbf{r}_2;R)$ becomes a sequence of functions of $\mathbf{r}_1$ and $\mathbf{r}_2$, one function for each value of R in the range. Because of this behavior, $\Psi(\mathbf{r}_1,\mathbf{r}_2;R)$ is referred to as a *parametric function* of R.

Next, let us turn to the nuclear Schrödinger equation which has the form

$$\left[-\frac{1}{2m_p}\nabla_A^2-\frac{1}{2m_p}\nabla_B^2+U_n(R)\right]\Psi_{n_A,n_B}(R_A,R_B)=E_{n_A,n_B}\Psi_{n_A,n_B}(\mathbf{R}_A,\mathbf{R}_B). \tag{13.6}$$

The nuclear Schrödinger equation governs the molecule's translational, rotational, and vibrational motions. The uninteresting translational motions may be removed from the nuclear Schrödinger equation by expressing it in terms of center-of-mass and relative coordinates. (Recall Section 8.1.) The nuclear Schrödinger equation so expressed provides the foundation for studying molecular rotational–vibrational motions.

In the nuclear Schrödinger equation, E_{n_A,n_B} is the nuclear energy of the molecule when it is in the state with nuclear quantum numbers n_A and n_B and $U_n(R)$ is the potential energy function of the molecule when it is in the electronic state with quantum numbers n. Since $U_n(R)$ depends on electronic state, we have the very important result that a molecule's rotational–vibrational motions depend on its electronic state. Additionally, when the molecule is in its electronic ground state, the nuclear Schrödinger equation (Equation 13.6) with translational motions removed reduces to the rotational–vibrational Schrödinger equation of Equations (8.4) and (8.6), thus putting that key equation on a rigorous foundation.

We have just noted that a molecule's potential energy function $U_n(R)$ depends on its electronic state. A related but much more powerful statement is one of the most consequential results of the Born–Oppenheimer theory. Namely, *a molecule's potential energy function* $U_n(R)$ *is identical to its electronic energy* $E_n(R)$. Stated concisely,

$$U_{\mathbf{n}}(R)=E_{\mathbf{n}}(R). \tag{13.7}$$

This result has two very important consequences.

Most importantly, determining $E_n(R)$ for a broad range of R values by solving the electronic Schrödinger equation (Equation 13.5) for each R on the range is the standard method for determining molecular potentials $U_n(R)$ *ab initio* (from first principles) rather than empirically by, say, adopting a Morse potential.

Since $E_n(R)$ is the electronic energy conditional that the nuclei are fixed, it is independent of the nuclear masses and therefore is the same for all isotopes of an element. Consequently, $U_n(R)$ is also isotope independent. This result confirms and generalizes Equation (7.128) for the isotope independence of ground electronic state potentials.

We finally note that very importantly for chemistry, both the Born–Oppenheimer equation (Equation 13.7) and its two consequences extend to polyatomic molecules. For polyatomics however since their configurations depend on the position vectors $\mathbf{R}_1,\mathbf{R}_2,\ldots$ of all of their nuclei their potentials, called

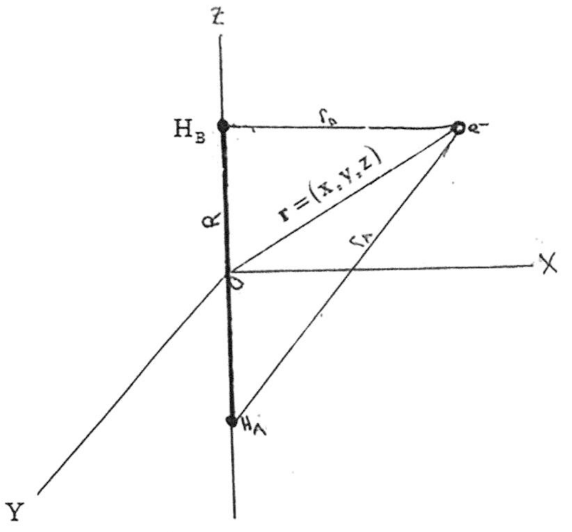

FIGURE 13.2 Sketch of the hydrogen molecular ion H_2^+.

potential energy surfaces, are complex multi-dimensional functions $U_n(\mathbf{R}_1, \mathbf{R}_2, \ldots)$ of all nuclear position vectors, not simple one-dimensional functions $U_n(R)$ of the internuclear separation R as is true for diatomic molecule.

This concludes our discussion of the Born–Oppenheimer approximation. We next turn to the main topic of this chapter, molecular electronic structure and chemical bonding. We start with the simplest molecule, the instructive one-electron hydrogen molecular ion H_2^+ sketched in Figure 13.2.

13.2 MOLECULAR ORBITALS AND THE HYDROGEN MOLECULAR ION H_2^+

H_2^+ was first studied theoretically within the framework of the old quantum theory touched on in Section 1.4. The old quantum theory failed qualitatively to explain the properties of H_2^+. The old quantum theory also failed qualitatively to explain the line spectra of the helium atom. These disastrous failures hastened both the demise of the old quantum theory and its replacement by modern quantum mechanics.

Proper understanding of H_2^+ only came in the late 1920s and early 1930s just after the advent of the Schrödinger equation and the Born–Oppenheimer approximation. Before we describe both exact and approximate treatments of H_2^+, we explain why the chemically uninteresting H_2^+ ion is worthy of intensive study.

Recall from Chapter 12 that the exact atomic orbital wave functions of the one-electron hydrogen atom provide the conceptual basis for atomic orbital electron configurations and approximate atomic orbital Slater determinantal wave functions for atoms of arbitrary complexity. Quite analogously, the exact *molecular orbital* wave functions of the one-electron H_2^+ ion provide the conceptual basis for equally useful molecular orbital electron configurations and approximate molecular orbital Slater determinantal wave functions for a vast range of molecules. This is why study of the H_2^+ ion is worthwhile particularly within the molecular orbital framework which at present is almost universally used in both qualitative and quantitative treatments of molecular electronic structure.

Before going further, we compare and contrast atomic and molecular orbitals. Both are conceptually very similar. This is because both are one-electron wave functions which give rise to probability clouds within which the electron does not have a definite position but rather is characterized by a Born rule probability density. While conceptually atomic and molecular orbitals are very similar, the two types of orbitals differ radically in one essential way. Namely, while atomic orbitals give rise to probability clouds which restrict an electron to the vicinity of a single atom, the probability clouds of molecular orbitals are spread out rendering it possible for an electron occupying a molecular orbital to be found anywhere in the molecule rather than being restricted in position to, say, the region of a chemical bond (as is predicted by the Lewis electron dot model or the quantum valence bond theory).

We may now turn to the determination of the molecular orbitals and properties of the H_2^+ ion. Because the ion has only one electron, its electronic Schrödinger equation may be solved exactly, and therefore, within the Born–Oppenheimer approximation the attributes of H_2^+ may be found exactly. We first outline the exact ground state solution of the H_2^+ electronic Schrödinger equation initially carried out by Burrau in 1927 and slightly later improved by Hylleraas. We then note that the method of the exact solution cannot be extended to molecules with more than one electron. Consequently, approximate methods that can be extended are needed. We therefore next introduce the simplest form of an extendable approximate method for determining molecular orbitals called the linear combination of atomic orbitals–molecular orbital (LCAO-MO) procedure. We then apply our simplest LCAO-MO method to determine the ground and first excited state molecular orbitals, potential energy functions, and selected properties of H_2^+ and compare its predictions for the ground state with the results of the exact solution.

Both the exact and LCAO-MO determinations start with the H_2^+ Born–Oppenheimer electronic Hamiltonian operator

$$\hat{H} = -\frac{1}{2}\nabla^2 - \frac{1}{r_A} - \frac{1}{r_B} + \frac{1}{R} \tag{13.8}$$

where $-\frac{1}{2}\nabla^2$ is the electron's kinetic energy operator, $-\frac{1}{r_A} - \frac{1}{r_B}$ are the operators which represent the attractive Coulomb interactions between the electron and, respectively, hydrogen atom nucleus A and nucleus B, and $\frac{1}{R}$ is the operator which represents the repulsive Coulomb potential between the two nuclei which are fixed with internuclear separation R.

Further, note from Figure 13.2 that we have chosen the coordinate system used to reference the electron motion so that the two nuclei lie on its Z-axis and so that its origin occurs at the midpoint of the ion's bond axis. Also note from Figure 13.2 that the position vector of the electron relative to this origin is denoted by $\mathbf{r} = (x, y, z)$.

Thus, we will write the H_2^+ electronic wave functions as $\Psi_{\mathbf{n}}(\mathbf{r};R)$ where $\mathbf{n}$ denotes the electronic quantum numbers. Also for brevity, we will write the ion's electronic energies as $E_{\mathbf{n}}$ rather than as $E_{\mathbf{n}}(R)$. With these designations, the H_2^+ electronic Schrödinger equation is from Equation (13.8) for $\hat{H}$

$$\left(-\frac{1}{2}\nabla^2 - \frac{1}{r_A} - \frac{1}{r_B} + \frac{1}{R}\right)\Psi_{\mathbf{n}}(\mathbf{r};R) = E_{\mathbf{n}}\Psi_{\mathbf{n}}(\mathbf{r};R). \tag{13.9}$$

The exact solution of the above Schrödinger equation is carried out in terms of a set of generalized coordinates denoted by λ, μ, and ϕ and known as *elliptical coordinates* rather than in terms of the electron's more familiar Cartesian coordinates x, y, and z. This is because the Schrödinger equation (Equation 13.9) is separable in elliptical coordinates but not in Cartesian coordinates. The elliptical coordinates are defined by

$$\lambda = \frac{r_A + r_B}{R} \quad \mu = \frac{r_A - r_B}{R} \quad \text{and} \quad \phi = \tan^{-1}\left(\frac{y}{x}\right) \tag{13.10}$$

where r_A and r_B are the respective distances between the electron and nuclei A and B.

From these definitions, the ranges of the elliptical coordinates follow as (Problem 13.4)

$$1 \le \lambda < \infty \quad -1 \le \mu \le 1 \quad \text{and} \quad 0 \le \phi \le 2\pi. \tag{13.11}$$

The volume element dr = dxdydz when expressed in elliptical coordinates is

$$\mathbf{dr} = \frac{R^3}{8}\left(\lambda^2 - \mu^2\right)d\lambda d\mu d\phi. \tag{13.12}$$

This volume element formula may be established by the Jacobian determinant method described in Problem 8.1. Thus, we write

$$\mathbf{dr} = dxdydz = Jd\lambda d\mu d\phi \tag{13.13}$$

where J, the Jacobian of the transformation $(x,y,z) \rightarrow (\lambda,\mu,\phi)$, is the determinant

$$J = \begin{vmatrix} \frac{\partial x}{\partial \lambda} & \frac{\partial y}{\partial \lambda} & \frac{\partial z}{\partial \lambda} \\ \frac{\partial x}{\partial \mu} & \frac{\partial y}{\partial \mu} & \frac{\partial z}{\partial \mu} \\ \frac{\partial x}{\partial \phi} & \frac{\partial y}{\partial \phi} & \frac{\partial z}{\partial \phi} \end{vmatrix}. \tag{13.14}$$

For the present problem, it is more convenient to determine the Jacobian J^{-1} for the inverse transformation $(\lambda,\mu,\phi) \rightarrow (x,y,z)$. J^{-1} is the determinant

$$J^{-1} = \begin{vmatrix} \frac{\partial \lambda}{\partial x} & \frac{\partial \lambda}{\partial y} & \frac{\partial \lambda}{\partial z} \\ \frac{\partial \mu}{\partial x} & \frac{\partial \mu}{\partial y} & \frac{\partial \mu}{\partial z} \\ \frac{\partial \phi}{\partial x} & \frac{\partial \phi}{\partial y} & \frac{\partial \phi}{\partial z} \end{vmatrix}. \tag{13.15}$$

Since from the transformation equations (Equation 13.10) λ and μ depend on r_A and r_B, to evaluate J^{-1} from Equation (13.15) we need relations which give r_A and r_B in terms of x, y, and z. The required relations are (Problem 13.5)

$$r_A^2 = x^2 + y^2 + \left(z - \frac{R}{2}\right)^2 \quad \text{and} \quad r_B^2 = \left(x^2 + y^2\right) + \left(z + \frac{R}{2}\right)^2. \tag{13.16}$$

Eventually, from Equations (13.10), (13.15), and (13.16) we find that (Problem 13.6)

$$J^{-1} = \frac{8}{R^3\left(\lambda^2 - \mu^2\right)} \tag{13.17}$$

and hence

$$J = \frac{R^3\left(\lambda^2 - \mu^2\right)}{8}. \tag{13.18}$$

The result that we sought to establish, Equation (13.12), then follows from Equations (13.13) and (13.18).

We may now outline the exact ground state solution of the H_2^+ electronic Schrödinger equation (Equation 13.9). We first use the transformation equations (Equation 13.10) to express this Schrödinger equation in elliptical coordinates as

$$-\frac{1}{2}\left\{\frac{\partial}{\partial\lambda}\left[\lambda^2-1\right]+\frac{\partial}{\partial\mu}\left[\left(1-\mu^2\right)\frac{\partial\Psi}{\partial\mu}\right]+\left(\frac{1}{\lambda^2-1}-\frac{1}{\mu^2-1}\right)\frac{\partial^2\Psi}{\partial\phi^2}+R\lambda\right\}=\frac{R^2}{4}\left(\lambda^2-\mu^2\right)\left(E-\frac{1}{R}\right) \tag{13.19}$$

where we abbreviate the ground state wave function $\Psi(\lambda,\mu,\phi;R)$ by Ψ and the ground state energy $E(R)$ by E.

We next assume that Ψ separates as

$$\Psi = \Lambda(\lambda)M(\mu)\Phi(\phi). \tag{13.20}$$

Inserting this form into Equation (13.19) and making the separation-of-variables argument yields the following independent ordinary differential equations for Λ, M, and Φ:

$$\frac{d}{d\lambda}\left[\left(\lambda^2-1\right)\frac{d\Lambda(\lambda)}{d\lambda}\right]+\left[-\alpha\lambda^2+2\beta\lambda-\frac{m^2}{\lambda^2-1}+\beta\right]\Lambda(\lambda)=0 \tag{13.21}$$

$$\frac{d}{d\mu}\left[\left(1-\mu^2\right)\frac{dM(\mu)}{d\mu}\right]+\left[\alpha\mu^2-\frac{m^2}{1-\mu^2}-\beta\right]M(\mu)=0 \tag{13.22}$$

and

$$\frac{d^2\Phi(\phi)}{d\phi^2}=-m^2\Phi(\phi). \tag{13.23}$$

The constants m, α, and β in the above differential equations arise from the application of the separation-of-variables procedure. The constant α is especially important since it determines the energy E from the relation $\alpha = -4\pi^2 m_e\left(E - R^{-1}\right)$.

All three constants are determined by the requirement that the functions Λ, M, and Φ be well-behaved.

To obtain Λ, M, and Φ, one proceeds as follows. First, one solves the Φ equation subject to the single-valuedness condition $\Phi(\phi+2\pi)=\Phi(\phi)$ which guarantees that Φ is well-behaved. This solution yields $\Phi(\phi)=\exp(im\phi)$ where m is restricted to the integer values $m = 0, \pm 1, \ldots$. For the ground state, $m = 0$. One then notes that the Λ and M equations are not solvable analytically even for $m = 0$ and therefore must be solved numerically. Numerical solution yields $\Lambda(\lambda)$ and $M(\mu)$ as inconvenient numerical tabulations. In the course of this numerical solution, one obtains α and hence E.

These steps solve the problem since they yield Λ, M, and Φ and hence the R-dependent H_2^+ ground state wave function Ψ and they also yield the R-dependent ground state energy $E(R)=E$. Further, since from the Born–Oppenheimer equation (Equation 13.7) $U(R)=E(R)$, they additionally yield the exact H_2^+ ground state potential $U(R)$.

While many H_2^+ ground state properties may be derived from $U(R)$, we will here focus on only two properties. These are the equilibrium internuclear separation or bond length R_e and the classical dissociation energy D_e. R_e follows immediately as the position of the minimum of $U(R)$. D_e may be obtained from $U(R)$ via the relation $D_e = U(\infty) - U(R_e)$. $U(\infty)$ is the potential energy of H_2^+ when it is dissociated and is therefore equal to -0.5 au, the sum of the energy -0.5 au of a 1s H atom and the energy 0.0 au of a free proton. Thus, $D_e = -0.5\,\text{au} - U(R_e)$. From the exact form of $U(R)$, one then finds for ground state H_2^+ that $D_e = 0.10264\,\text{au}\left(269.5\,\text{kJ}\,\text{mol}^{-1}\right)$ and $R_e = 105.7\,\text{pm}$, results which agree perfectly with experiments

We next note that experimental potential energy curves are plotted in Figure 13.3 for the ground and several low-lying excited states of H_2^+. Notice in the plots we have chosen the arbitrary zero of energy so that as $R \to \infty$ the ground and first excited state potentials approach the energy -0.5 au of a dissociated H_2^+ ion. This choice of zero of energy suitably adopted for other molecules is conventional for electronic structure problems but differs from our earlier choice $U(R_e) = 0$ for molecular vibrations (Figures 7.7 and 7.8).

Also note from Figure 13.3 that the ground state potential energy curve has a deep well indicating correctly that ground state H_2^+ is chemically bonded. Additionally, notice that the first excited state potential energy curve is purely repulsive, showing that first excited state H_2^+ is not chemically bonded. These features will persist in all of our later approximate treatments of H_2^+.

The exact ground state solution of the H_2^+ electronic Schrödinger equation (Equation 13.9) yields the wave function Ψ as an inconvenient numerical tabulation and provides little or no physical insight, and all importantly, the method of solution cannot be extended to molecules with more than one electron.

Thus, we next develop and apply the most primitive LCAO-MO theory to study the ground and first excited states of H_2^+. Unsurprisingly, this primitive theory yields H_2^+ ground state attributes which are in poor accord with those of the exact solution. Yet even our most primitive LCAO-MO theory has a number of virtues. Most importantly, in agreement with experiment it predicts that H_2^+ is chemically bonded in the ground state but not in the first excited state. Also in contrast to the exact solution, it is relatively simple and straightforward, yields analytic rather than numerical wave functions, and provides a great deal of physical understanding. Additionally and very importantly, as we will see later, it can be systematically improved to eventually yield sophisticated models which predict H_2^+ properties which are essentially in perfect agreement with those obtained from the exact solution.

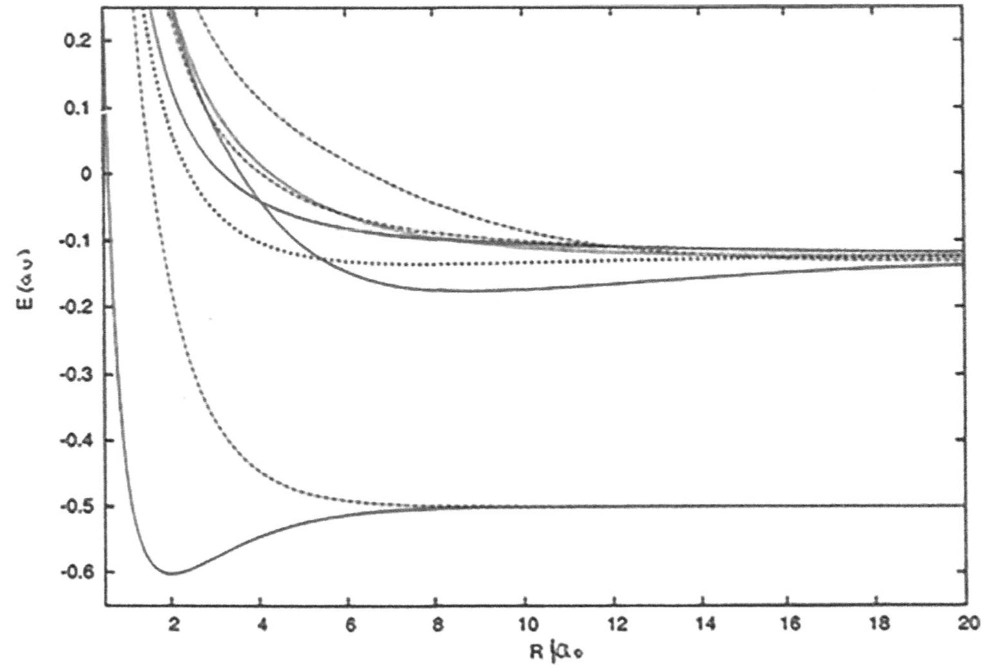

FIGURE 13.3 Ground and low-lying excited state potential energy curves of H_2^+.

Our simplest LCAO-MO theory is an application of the linear variational method developed in Section 11.4 with only two, the minimal number, of basis functions included. Thus, our ground and first excited state molecular orbitals which are denoted by $\Psi_{\pm}$ are the two term expansions.

$$\Psi_{\pm}(\mathbf{r};R) = C_{A\pm}\phi_A(\mathbf{r}) + C_{B\pm}\phi_B(\mathbf{r}) \tag{13.24}$$

where $\phi_A(\mathbf{r})$ and $\phi_B(\mathbf{r})$ are the basis functions and $C_{A\pm}$ and $C_{B\pm}$ are the expansion coefficients (with for simplicity their $R-$ dependence suppressed).

The following question immediately arises: What is the best choice of basis functions? To answer this question, we note that if the H_2^+ electron was localized on hydrogen atom A, the wave functions $\Psi_{\pm}$ would coincide with the normalized 1s hydrogen atom orbital centered on nucleus A which in atomic units is

$$\Psi_{1s}(r_A) = \left(\frac{1}{\pi}\right)^{1/2} \exp(-r_A). \tag{13.25a}$$

Similarly, if the electron was localized on hydrogen atom B, the wave functions $\Psi_{\pm}$ would be identical to the normalized 1s hydrogen atom orbital centered on nucleus B

$$\Psi_{1s}(r_B) = \left(\frac{1}{\pi}\right)^{1/2} \exp(-r_B). \tag{13.25b}$$

These considerations suggest that the best choice of basis functions is $\Psi_{1s}(r_A)$ and $\Psi_{1s}(r_B)$. With this choice, the expansion of Equation (13.24) becomes the linear combination of atomic orbitals

$$\Psi_{\pm}(\mathbf{r};R) = C_{A\pm}\Psi_{1s}(r_A) + C_{B\pm}\Psi_{1s}(r_B). \tag{13.26}$$

To proceed further, let us evaluate the LCAO-MO energy which (with its $R-$dependence and $\pm$ subscripts suppressed) is

$$E = \frac{\int \Psi_{\pm}(\mathbf{r};R)\hat{H}\Psi_{\pm}(\mathbf{r};R)d\mathbf{r}}{\int \Psi_{\pm}(\mathbf{r};R)\Psi_{\pm}(\mathbf{r};R)d\mathbf{r}} \tag{13.27}$$

where $\hat{H}$ is the H_2^+ electronic Hamiltonian operator of Equation (13.8). Next, inserting Equation (13.26) for $\Psi_{\pm}$ into Equation (13.27) for E and noting the normalization of the 1s orbitals yields

$$E = \frac{C_{A\pm}^2 H_{AA} + C_{A\pm}C_{B\pm}(H_{AB} + H_{BA}) + C_{B\pm}H_{BB}}{C_{A\pm}^2 + C_{B\pm}^2 + 2C_{A\pm}C_{B\pm}S}. \tag{13.28}$$

Before giving the forms of H_{AA} and so forth, we note that because hydrogen atoms A and B are identical, $H_{BB} = H_{AA}$, and because $\hat{H}$ is Hermitian and the 1s orbitals are real, $H_{AB} = H_{BA}$. Therefore, the proceeding expression simplifies to

$$E = \frac{C_{A\pm}^2 H_{AA} + 2C_{A\pm}C_{B\pm}H_{BA} + C_{B\pm}^2 H_{AA}}{C_{A\pm}^2 + C_{B\pm}^2 + 2C_{A\pm}C_{B\pm}S} \tag{13.29}$$

We many now give the forms for H_{AA}, H_{BA}, and S. These are

$$H_{AA} = \int \Psi_{1s}(r_A)\hat{H}\Psi_{1s}(r_A)d\mathbf{r} \tag{13.30}$$

$$H_{BA} = \int \Psi_{1s}(r_B)\hat{H}\Psi_{1s}(r_A)d\mathbf{r} \tag{13.31}$$

and

$$S = \int \Psi_{1s}(r_B)\Psi_{1s}(r_A)d\mathbf{r}. \tag{13.32}$$

S is called the *overlap integral*. This is because its integrand is non-vanishing only in the region where both $\Psi_{1s}(r_A)$ and $\Psi_{1s}(r_B)$ are non-vanishing, that is, only the region important for chemical bonding where the two 1s orbitals overlap. In the once-popular but now largely obsolete valence bond theory, the magnitude of S was taken as a measure of the strength of the chemical bond between atoms A and B.

According to the linear variational method, the next step in the determination of E and $\Psi_\pm$ is to derive secular equations for $C_{A\pm}$ and $C_{B\pm}$. They are derived by applying the linear variational minimization conditions

$$\frac{\partial E}{\partial C_{A\pm}} = 0 \quad \text{and} \quad \frac{\partial E}{\partial C_{B\pm}} = 0 \tag{13.33}$$

to Equation (13.29) for E and then following steps like those that led from Equations (11.99) to Equations (11.102). This yields the secular equations

$$(H_{AA} - E)C_{A\pm} + (H_{BA} - ES)C_{B\pm} = 0 \tag{13.34a}$$

and

$$(H_{BA} - ES)C_{A\pm} + (H_{AA} - E)C_{B\pm} = 0. \tag{13.34b}$$

The secular equations (Equation 13.34) comprise a set of two linear homogeneous equations for $C_{A\pm}$ and $C_{B\pm}$. These equations have non-trivial solutions only if the secular determinantal equation

$$\begin{vmatrix} H_{AA} - E & H_{BA} - ES \\ H_{BA} - ES & H_{AA} - E \end{vmatrix} = 0 \tag{13.35}$$

holds. Expanding the determinant yields a quadratic equation for E. Denote its smaller root by E_+ and its larger root by E_-. One finds these roots as (Problem 13.7)

$$E_+ = \frac{H_{AA} + H_{BA}}{1 + S} \tag{13.36a}$$

and

$$E_- = \frac{H_{AA} - H_{BA}}{1 - S}. \tag{13.36b}$$

The next step is to determine $C_{A\pm}$ and $C_{B\pm}$. To accomplish this, we first set $E = E_+$ and then set $E = E_-$. Next, we use Equations (13.36) to successively eliminate both values of E from either of the secular equations (Equation 13.34). After simplifying, we find that $C_{A+} = C_{B+}$ and $C_{A-} = -C_{B-}$ (Problem 13.8a). Lastly, we compare these proportionality relations with the normalization conditions on $\Psi_\pm$ which from Equation (13.26) is

$$\int \Psi_\pm(r;R)\Psi_\pm(r;R)dr = C_{A\pm}^2 + C_{B\pm}^2 + 2C_{A\pm}C_{B\pm}S = 1 \tag{13.37}$$

to obtain our final results for $C_{A\pm}$ and $C_{B\pm}$ (Problem 13.8b)

$$C_{A_+} = \left[\frac{1}{2(1+S)}\right]^{1/2} \quad \text{and} \quad C_{B_+} = \left[\frac{1}{2(1+S)}\right]^{1/2} \tag{13.38a}$$

and

$$C_{A_-} = \left[\frac{1}{2(1-S)}\right]^{1/2} \quad \text{and} \quad C_{B_-} = -\left[\frac{1}{2(1-S)}\right]^{1/2}. \tag{13.38b}$$

From Equations (13.26) and (13.38), the H_2^+ LCAO-MO molecular orbitals are

$$\Psi_+(\mathbf{r};R) = \left[\frac{1}{2(1+S)}\right]^{1/2} \Psi_{1s}(r_A) + \left[\frac{1}{2(1+S)}\right]^{1/2} \Psi_{1s}(r_B) \tag{13.39a}$$

and

$$\Psi_-(\mathbf{r};R) = \left[\frac{1}{2(1-S)}\right]^{1/2} \Psi_{1s}(r_A) - \left[\frac{1}{2(1-S)}\right]^{1/2} \Psi_{1s}(r_B). \tag{13.39b}$$

Notice that both Ψ_+ and Ψ_- are normalized and also that these functions are orthogonal. Therefore, Ψ_+ and Ψ_- comprise an orthonormal pair.

Before continuing the quantitative development of our simplest LCAO-MO theory, we make some qualitative inferences based on the theory about the origin and nature of the chemical bonding of H_2^+. Our inferences are founded on qualitative analyses of the Born rule probability densities $P_+ = \Psi_+\Psi_+$ and $P_- = \Psi_-\Psi_-$. From Equations (13.39), these probability densities are

$$P_+(\mathbf{r};R) = \frac{1}{2(1+S)}\Psi_{1s}^2(r_A) + \frac{1}{2(1+S)}\Psi_{1s}^2(r_B) + \frac{1}{1+S}\Psi_{1s}(r_A)\Psi_{1s}(r_B) \tag{13.40a}$$

and

$$P_-(\mathbf{r};R) = \frac{1}{2(1-S)}\Psi_{1s}^2(r_A) + \frac{1}{2(1-S)}\Psi_{1s}^2(r_B) - \frac{1}{1-S}\Psi_{1s}(r_A)\Psi_{1s}(r_B). \tag{13.40b}$$

We begin with P_+. We show from the form of P_+ that when the H_2^+ ion is in the Ψ_+ state, it is chemically bonded. For this reason, Ψ_+ is referred to as a *bonding molecular orbital.* To start, we note that the first two terms on the right-hand side of Equation (13.40a) for P_+ determine the probabilities that the ion's electron is localized on either hydrogen atom nucleus A or hydrogen atom nucleus B. It is the third right-hand term in Equation (13.40a) which is the important one. It gives the correction to the localized electron densities of the first two terms which is needed since the electron may be found in the internuclear region, that is, between the two nuclei. The essential point is that the third term is positive. This is because the factor $\frac{1}{1+S}\Psi_{1s}(r_A)\Psi_{1s}(r_B)$ is positive since 1s orbitals are strictly positive rendering both the overlap factor $\Psi_{1s}(r_A)\Psi_{1s}(r_B)$ and its integral $S = \int \Psi_{1s}(r_A)\Psi_{1s}(r_B)d\mathbf{r}$ positive. The positivity of the third term produces a buildup of the internuclear electron density and by total charge conservation a consequent depletion of the electron density near the two H_2^+ nuclei (Figure 13.4 bottom). There are two competing effects of this shift of the electron density due to the third term. First, the energy of the ion is lowered because of the attractive Coulomb interaction between the internuclear electron density and the electron-depleted nuclei, the lowering being enhanced by the depletion. Second, the energy of the ion is raised by the repulsive Coulomb interaction between the two electron-depleted nuclei, the raising being enhanced by the depletion. It turns out that the former factor dominates. So the net effect of the shift is to lower the energy of the ion relative to the value it would have if the third term was absent and the electron was localized. Further, the energy lowering of a bound ion implies that such an ion has a lower energy than a dissociated ion since the latter has no energy lowering internuclear electron density. This inference stated in terms of the ion's potential energy function means that for an ion in the Ψ_+ state, its potential has a well, and hence, an ion in that state is chemically bonded.

The origin of this chemical bonding is the *constructive interference* of the ion's two 1s orbitals when the ion is in the Ψ_+ state. This is evident from the plot of the Ψ_+ molecular orbital given in Figure 13.4. This plot shows that the constructive interference increases the magnitude of Ψ_+ in the internuclear region. This increase produces the buildup of internuclear electron density which is responsible for the chemical bonding of a H_2^+ ion in the Ψ_+ state.

We next turn to the Ψ_- state. We will show qualitatively that the form of $P_- = \Psi_-\Psi_-$ implies that an H_2^+ ion in the Ψ_- state is not chemically bonded. For this reason, Ψ_- is called an *anti-bonding molecular orbital.* To start, we note that similarly to what we found for P_+, the first two right-hand terms in Equation (13.40b) for P_- merely determine the probabilities that the electron is localized on either of the two nuclei. So again it is the third right-hand term which is the important one since it determines the internuclear electron density. The essential point is that this third term is negative because the factor $\frac{1}{1-S}\Psi_{1s}(r_A)\Psi_{1s}(r_A)$ is positive. This is so since the overlap factor $\Psi_{1s}(r_A)\Psi_{1s}(r_B)$ is positive and since $S < 1$ (because $S = 1$ only occurs for the impossible situation that the two nuclei are at the same position). The result of this negativity of the third term is an electron density depletion in the internuclear region and a consequent electron density buildup in the regions near the H_2^+ ion's nuclei (Figure 13.4 top). Then, the following arguments similar to those just given for a Ψ_+ ion the electron density shift due the third term renders the energy of a bound Ψ_- H_2^+ ion higher than that of a dissociated ion. In potential energy language, the potential of a Ψ_- ion is purely repulsive, and hence, a Ψ_- ion is not chemically bound.

The origin of the non-bonding of a H_2^+ ion in the Ψ_- state is the *destructive interference* of the two 1s orbitals of the ion when it is in that state. This is evident from the plot of the Ψ_- molecular orbital given in Figure 13.4. The plot shows that not only does the destructive inference reduce the magnitude of Ψ_- in the internuclear region, but it also produces a nodal place in Ψ_- halfway between the two nuclei. This reduction of the magnitude of Ψ_- in the internuclear region enhanced

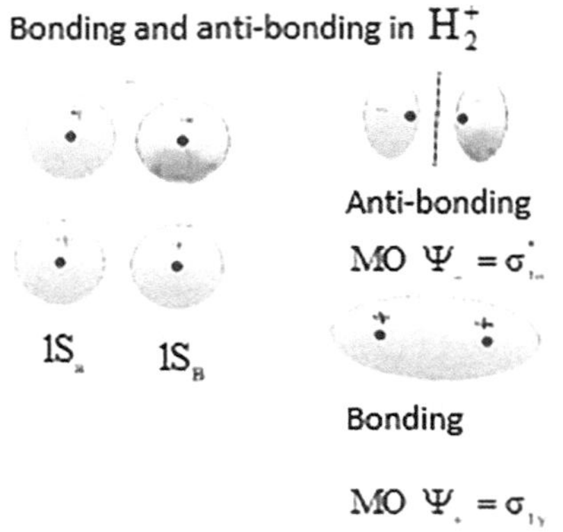

FIGURE 13.4 Lowest energy bonding and anti-bonding molecular orbitals of H_2^+.

by the nodal plane causes the depletion of the internuclear electron density which is responsible for the non-bonding of an H_2^+ ion in the Ψ_- state.

Finally, we note that our qualitative considerations show that Ψ_+ is the ground state H_2^+ molecular orbital, while Ψ_- is its first excited state molecular orbital.

This concludes our qualitative discussion of the chemical bonding of H_2^+.

Before returning to the quantitative development of our simplest LCAO-MO theory for H_2^+, we introduce some terminology commonly used in molecular orbital theory. We take the molecular orbitals Ψ_+ and Ψ_- as our main examples. Since these molecular orbitals are cylindrically symmetric about the bond axis (Figure 13.4), we call both σ molecular orbitals. To distinguish between these two σ molecular orbitals, we give them distinct labels. Thus, we denote Ψ_+ and Ψ_- by, respectively, σ_{1g} and σ_{1u}^*. The star on σ_{1u}^* and the absence of a star on σ_{1g} mean, respectively, that these orbitals are anti-bonding and bonding molecular orbitals. The subscript g on σ_{1g} and the subscript u on σ_{1u}^* mean that these molecular orbitals have respective even and odd symmetries with respect to inversion through the origin of the coordinate system shown in Figure 13.2. (The subscripts g and u are abbreviations for the German words gerade meaning even and ungerade meaning odd.) The subscript 1 on σ_{1g} and σ_{1u}^* means that these are the lowest energy σ molecular orbitals of, respectively, even and odd inversion symmetry. (Similarly, the second lowest energy σ molecular orbitals are denoted by σ_{2g} and σ_{2u}^* and so on.) Analogous labels will be used in Section 13.4 to distinguish between distinct π molecular orbitals.

We may now return to the quantitative development of our simplest LCAO-MO theory of H_2^+. We start by obtaining more explicit forms for E_+ and E_- of Equations (13.36), the respective energies of the Ψ_+ and Ψ_- H_2^+ states. From Equation (13.8) for $\hat{H}$, Equations (13.25) for the 1s orbitals, and Equations (13.30)–(13.32) for H_{AA}, H_{BA}, and S, these more explicit forms are (Problem 13.9)

$$E_+ = -0.5\,\text{au} + \frac{J+K}{1+S} \tag{13.41a}$$

and

$$E_- = -0.5\,\text{au} + \frac{J-K}{1-S} \tag{13.41b}$$

where $-0.5\,\text{au}$ is the energy of a 1s hydrogen atom, where

$$J = \frac{1}{R} - j \tag{13.42}$$

and

$$K = \frac{S}{R} - k \tag{13.43}$$

and where j, k, and S are the integrals

$$j = \int \frac{\Psi_{1s}(r_A)\Psi_{1s}(r_A)}{r_B}\,d\mathbf{r} = \frac{1}{\pi}\int \frac{\exp(-2r_A)}{r_B}\,d\mathbf{r} \tag{13.44}$$

$$k = \int \frac{\Psi_{1s}(r_A)\Psi_{1s}(r_B)}{r_B}\,d\mathbf{r} = \frac{1}{\pi}\int \frac{\exp[-(r_A+r_B)]}{r_B}\,d\mathbf{r} \tag{13.45}$$

and

$$S = \int \Psi_{1s}(r_A)\Psi_{1s}(r_B)\,d\mathbf{r} = \frac{1}{\pi}\int \exp[-(r_A+r_B)]\,d\mathbf{r}. \tag{13.46}$$

The integrals J and j are called *Coulomb integrals*, and the integrals K and k are called *resonance integrals*.

The integrals j, k, and S are most conveniently evaluated in elliptical coordinates. The first step in making these evaluations is to transform Equations (13.44)–(13.46) for j, k, and S into elliptical coordinate integrals using the transformation equations (Equations 13.10–13.12). One finds these elliptical coordinate integrals as (Problem 13.10)

$$j=\frac{R^2}{2}\int_1^\infty \exp(-R\lambda)d\lambda\int_{-1}^1 \exp(-R\mu)(\lambda+\mu)d\mu \tag{13.47}$$

$$k=\frac{R^2}{2}\int_1^\infty \exp(-R\lambda)d\lambda\int_{-1}^1 (\lambda+\mu)d\mu \tag{13.48}$$

and

$$S=\frac{R^3}{4}\int_1^\infty \exp(-R\lambda)d\lambda\int_{-1}^1 (\lambda^2-\mu^2)d\mu. \tag{13.49}$$

The final step is to evaluate these integrals (the evaluations only requiring integrations by parts) to yield (Problem 13.11)

$$j=\frac{1}{R}-\left(1+\frac{1}{R}\right)\exp(-2R) \tag{13.50}$$

$$k=(1+R)\exp(-R) \tag{13.51}$$

and

$$S=\left(1+R+\frac{1}{3}R^2\right)\exp(-R). \tag{13.52}$$

Next, we determine J and K. From Equations (13.42), (13.43), and (13.50)–(13.52),

$$J=\left(1+\frac{1}{R}\right)\exp(-2R) \tag{13.53}$$

and

$$K=\exp(-R)\left(\frac{1}{R}-\frac{2}{3}R\right). \tag{13.54}$$

The energies E_+ and E_- may now be found from Equations (13.41) and (13.52)–(13.54) as

$$E_+=-0.5\,\text{au}+\frac{\left(1+\frac{1}{R}\right)\exp(-2R)+\left(\frac{1}{R}-\frac{2}{3}R\right)\exp(-R)}{1+\left(1+R+\frac{1}{3}R^2\right)\exp(-R)} \tag{13.55a}$$

and

$$E_-=-0.5\,\text{au}+\frac{\left(1+\frac{1}{R}\right)\exp(-2R)-\left(\frac{1}{R}-\frac{2}{3}R\right)\exp(-R)}{1-\left(1+R+\frac{1}{3}R^2\right)\exp(-R)}. \tag{13.55b}$$

From the Born–Oppenheimer equation (Equation 13.7), the electronic energies E_+ and E_- are the potential energy functions for the Ψ_+ and Ψ_- states. Like the exact H_2^+ ground and first excited state potentials plotted in Figure 13.3, as $R \to \infty$, both potentials approach the H_2^+ dissociation limit -0.5 au.

J, K, and S are plotted as functions of R in Figure 13.5. Notice that both J and K diverge as $R \to 0$ and approach zero as $R \to \infty$. Also note that K has a negative well for $R > \left(\frac{3}{2}\right)^{1/2}$. This well may be related to the chemical bonding of H_2^+ when it is in the Ψ_+ state. Further notice that S decays smoothly to zero from its $R = 0$ value of unity as $R \to \infty$. (Actually, S never attains the value unity since this value requires the impossible situation that the two nuclei occupy the same position.)

The LCAO-MO potentials of Equations (13.55) are plotted as functions of R in Figure 13.6. These plots provide a complete quantitative validation of our earlier qualitative inferences. First, the plots show that the Ψ_+ and Ψ_- states with potentials E_+ and E_- are, respectively, the H_2^+ ground and first excited states. Also as predicted by our qualitative analysis, E_+ has a well implying ground state chemical bonding and E_- is purely repulsive proving first excited state non-bonding.

Moreover, from the minimum of E_+ one finds the simplest LCAO-MO H_2^+ ground state classical dissociation energy and bond length as $D_e = 0.0683$ au and $R_e = 132$ pm. These approximate results are as expected in poor agreement with those of the exact solution of the Schrödinger Equation (13.9), $D_e = 0.10264$ au and $R_e = 106$ pm. But at least our simplest LCAO-MO theory correctly predicts that ground state H_2^+ is chemically bonded even though it severely underestimates the strength of the bond and it also correctly predicts that first excited state H_2^+ is not chemically bonded.

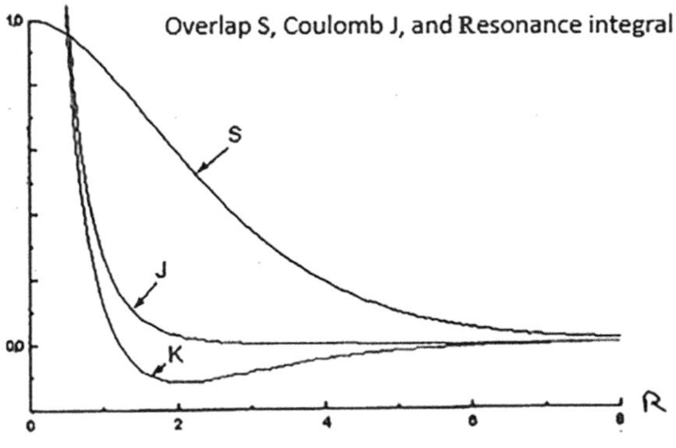

FIGURE 13.5 Plots of Equations (13.52)–(13.54) for the H_2^+ integrals S, J, and K.

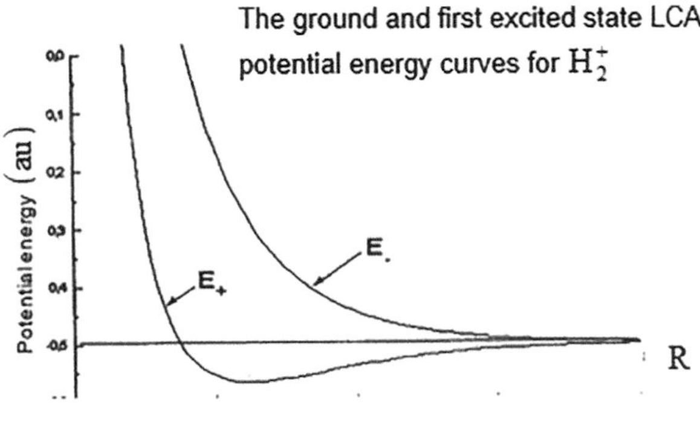

FIGURE 13.6 Plot of Equations (13.55) for the simplest LCAO-MO theory ground E_+ and first excited state E_- potential energy curves for H_2^+.

Despite its correct predictions about the ion's bonding and anti-bonding, because of its poor quantitative agreement with key exact results, our simplest LCAO-MO theory clearly needs major improvement. Fortunately, as noted earlier it may be systematically improved.

The most obvious flaw in our most primitive LCAO-MO theory is that while as $R \to \infty$ the natural choices of basis functions are the hydrogen atom 1s orbitals, chosen for all R in the primitive theory, for arbitrary R a better choice of basis functions are the normalized hydrogen-like 1s orbitals with variable Z

$$\Psi_{1s}(r_A;Z) = \left(\frac{1}{\pi}\right)^{1/2} Z^{3/2} \exp(-Zr_A) \tag{13.56a}$$

and

$$\Psi_{1s}(r_B;Z) = \left(\frac{1}{\pi}\right)^{1/2} Z^{3/2} \exp(-Zr_B). \tag{13.56b}$$

With this choice of basis functions, the simplest LCAO-MO form of Equation (13.26) for the molecular orbitals of the ion generalizes to

$$\Psi_{\pm}(\mathbf{r};R,Z) = C_{A_{\pm}}(Z)\Psi_{1s}(r_A;Z) + C_{B_{\pm}}(Z)\Psi_{1s}(r_B;Z). \tag{13.57}$$

To determine the optimal molecular orbitals $\Psi_{\pm}(\mathbf{r};R,Z)$ and their associated energies $E_{\pm}(Z)$, we will later variationally choose a unique value of Z for each value of R.

Next, paralleling the steps that lead from Equations (13.26) to (13.36), we obtain the variable Z theory molecular orbital energies as (Problem 13. 12)

$$E_{+}(Z) = \frac{H_{AA}(Z) + H_{BA}(Z)}{1 + S(Z)} \tag{13.58a}$$

and

$$E_{-}(Z) = \frac{H_{AA}(Z) - H_{BA}(Z)}{1 - S(Z)} \tag{13.58b}$$

where as generalizations of Equations (13.30) to (13.32)

$$H_{AA}(Z) = \int \Psi_{1s}(r_A;Z)\hat{H}\Psi_{1s}(r_A;Z)d\mathbf{r} \tag{13.59}$$

$$H_{BA}(Z) = \int \Psi_{1s}(r_B;Z)\hat{H}\Psi_{1s}(r_A;Z)d\mathbf{r} \tag{13.60}$$

and

$$S(Z) = \int \Psi_{1s}(r_A;Z)\Psi_{1s}(r_B;Z)d\mathbf{r} \tag{13.61}$$

and where $\hat{H}$ is again the H_2^+ electronic Hamiltonian operator of Equation (13.8).

Similarly, we obtain the final forms of the molecular orbitals $\Psi_{\pm}(\mathbf{r};R,Z)$ as the following generalization of Equations (13.39) (Problem 13.13):

$$\Psi_{+}(r;R,Z) = \left\{\frac{1}{2[1+S(Z)]}\right\}^{1/2} \Psi_{1s}(r_A;Z) + \left\{\frac{1}{2[1+S(Z)]}\right\}^{1/2} \Psi_{1s}(r_B;Z) \tag{13.62a}$$

and

$$\Psi_{-}(\mathbf{r};R,Z)=\left\{\frac{1}{2[1-S(Z)]}\right\}^{1/2}\Psi_{1s}(r_A;Z)-\left\{\frac{1}{2[1-S(Z)]}\right\}^{1/2}\Psi_{1s}(r_B;Z). \qquad (13.62b)$$

To proceed further, it will prove advantageous to rearrange $\hat{H}$ of Equation (13.8) as

$$\hat{H}=-\frac{1}{2}\nabla^2-\frac{Z}{r_A}-\frac{1}{r_B}+\frac{(Z-1)}{r_A}+\frac{1}{R}. \qquad (13.63)$$

The advantage arises since $\left(-\frac{1}{2}\nabla^2-\frac{Z}{r_A}\right)\Psi_{1s}(r_A;Z)$ –0.5Z^2 au which permits $\hat{H}\Psi_{1s}(r_A;Z)$ to be written in the convenient form

$$\hat{H}\Psi_{1s}(r_A;Z)=-0.5Z^2\text{au}\Psi_{1s}(r_A;Z)-\frac{1}{r_B}\Psi_{1s}(r_A;Z)+\frac{Z-1}{r_A}\Psi_{1s}(r_A;Z)+\frac{1}{R}\Psi_{1s}(r_A;Z). \qquad (13.64)$$

From Equations (13.56), (13.59)–(13.61), and (13.64), one may derive the following more explicit forms for $E_{+}(Z)$ and $E_{-}(Z)$ of Equations (13.58) (Problem 13.14):

$$E_{+}(Z)=-0.5Z^2\text{au}+\frac{J(Z)+K(Z)}{1+S(Z)} \qquad (13.65a)$$

and

$$E_{-}(Z)=-0.5Z^2\text{au}+\frac{J(Z)-K(Z)}{1-S(Z)} \qquad (13.65b)$$

where

$$J(Z)=\frac{1}{R}-j(Z)+(Z-1)\bar{j}(Z) \qquad (13.66a)$$

and

$$K(Z)=\frac{S(Z)}{R}-k(Z)+(Z-1)\bar{k}(Z). \qquad (13.66b)$$

The integrals $j(Z), k(Z), S(Z), \bar{j}(Z)$, and $\bar{k}(Z)$ are given by

$$j(Z)=\frac{1}{\pi}Z^3\int\frac{\exp(-2Zr_A)}{r_B}d\mathbf{r} \qquad (13.67)$$

$$k(Z)=\frac{1}{\pi}Z^3\int\frac{\exp[-Z(r_A+r_B)]}{r_B}d\mathbf{r} \qquad (13.68)$$

$$S(Z)=\frac{1}{\pi}Z^3\int\exp[-Z(r_A+r_B)]d\mathbf{r} \qquad (13.69)$$

$$\bar{j}(Z)=\frac{1}{\pi}Z^3\int\frac{\exp(-2Zr_A)}{r_A}d\mathbf{r} \qquad (13.70)$$

and

$$\bar{k}(Z) = \frac{1}{\pi} Z^3 \int \frac{\exp[-Z(r_A + r_B)]}{r_A} d\mathbf{r}. \tag{13.71}$$

The integrals $j(Z), k(Z)$, and $S(Z)$ may be found without direct elliptical coordinate integration by a shortcut. This shortcut involves relating the elliptical coordinate forms of these integrals written in terms of the variable $\rho = RZ$ to the elliptical coordinate forms for j, k, and S given in Equations (13.47)–(13.49). The results for $j(Z), k(Z)$, and $S(Z)$ then follow from those for j, k, and S of Equations (13.50)–(13.52) as (Problem 13.15)

$$j(Z) = Z\left[\frac{1}{RZ} - \left(1 + \frac{1}{RZ}\right)\exp(-2RZ)\right] \tag{13.72}$$

$$k(Z) = Z(1 + RZ)\exp(-RZ) \tag{13.73}$$

and

$$S(Z) = \left(1 + RZ + \frac{1}{3}R^2Z^2\right)\exp(-RZ). \tag{13.74}$$

The integrals $\bar{j}(Z)$ and $\bar{k}(Z)$ may be found by direct elliptical coordinate integration as (Problem 13.16)

$$\bar{j}(Z) = Z \tag{13.75}$$

and

$$\bar{k}(Z) = Z(1 + RZ)\exp(-RZ). \tag{13.76}$$

We may now determine $J(Z)$ and $K(Z)$ of Equations (13.66) from Equations (13.72)–(13.76) as

$$J(Z) = \left(Z + \frac{1}{R}\right)\exp(-2RZ) + Z(Z - 1) \tag{13.77a}$$

and

$$K(Z) = \left[\frac{1}{R} - Z + Z^2\left(1 - \frac{5}{3}R\right) + RZ^3\right]\exp(-RZ). \tag{13.77b}$$

The energies $E_+(Z)$ and $E_-(Z)$ then follow from their forms of Equations (13.65), from Equation (13.74) for $S(Z)$, and from Equations (13.77) for $J(Z)$ and $K(Z)$ as

$$E_+(Z) = -0.5Z^2\text{au} + \frac{\left(Z + \frac{1}{R}\right)\exp(-2RZ) + Z(Z-1) + \left[\frac{1}{R} - Z + Z^2\left(1 - \frac{5}{3}R\right) + RZ^3\right]\exp(-RZ)}{1 + \left(1 + RZ + \frac{1}{3}R^2Z^3\right)} \tag{13.78a}$$

and

$$E_-(Z) = -0.5Z^2\text{au} + \frac{\left(Z+\frac{1}{R}\right)\exp(-2RZ)+Z(Z-1)-\left[\frac{1}{R}-Z+Z^2\left(1-\frac{5}{3}R\right)+RZ^3\right]\exp(-RZ)}{1-\left(1+RZ+\frac{1}{3}R^2Z^3\right)} \tag{13.78b}$$

Notice that for $Z = 1$, $E_+(Z)$ and $E_-(Z)$ properly reduce to the energies E_+ and E_-, given in Equations (13.55), of our simplest LCAO-MO theory.

The R – dependent optimal value of Z is defined as the value which minimizes the ground state energy $E_+(Z)$ and thus is determined from the condition

$$\frac{\partial E_+(Z)}{\partial Z} = 0. \tag{13.79}$$

To conveniently implement the above minimization condition, let us write $E_+(Z)$ as

$$E_+(Z) = -0.5Z^2 + \frac{X(Z)}{Y(Z)} \tag{13.80}$$

where from Equation (13.78a)

$$X(Z) = Z(Z-1) + \left(Z+\frac{1}{R}\right)\exp(-2ZR) + \left(\frac{1}{R}-Z+Z^2-\frac{5}{3}Z^2R+Z^3R\right)\exp(-ZR) \tag{13.81}$$

and

$$Y(Z) = 1 + \left(1+ZR+\frac{1}{3}Z^2R^2\right)\exp(-ZR). \tag{13.82}$$

From Equation (13.80), the energy minimization condition of Equation (13.79) may be conveniently expressed as (Problem 13.18)

$$X(Z)\frac{\partial Y(Z)}{\partial Z} - Y(Z)\frac{\partial X(Z)}{\partial Z} = Z. \tag{13.83}$$

Next, from Equation (13.81) if follows that

$$\frac{\partial X(Z)}{\partial Z} = (2Z-1) - (1+2ZR)\exp(-2ZR) - \left(2-2Z+\frac{7}{3}ZR-2Z^2R-\frac{5}{3}Z^2R^2+Z^3R^3\right)\exp(-ZR) \tag{13.84}$$

while from Equation (13.82) we have that

$$\frac{dY(Z)}{dZ} = -\frac{1}{3}\left(ZR^2+Z^2R^3\right)\exp(-ZR). \tag{13.85}$$

Inserting the above results for $X, Y, \frac{\partial X}{\partial Z}$, and $\frac{\partial Y}{\partial Z}$ into the energy minimization equation (Equation 13.83) yields after a tedious but straightforward calculation a complex transcendental equation in Z and R (Problems 13.19a and b). This equation implicitly determines for each R the optimal Z.

Solving the transcendental equation on a computer yields the optimal Z as a numerical function of R. The optimal Z so obtained is plotted versus R in Figure 13.7. Notice from the figure that at $R = 0$, the optimal Z has the He^+ ion value 2, while as $R \rightarrow \infty$, the optimal Z correctly approaches its H atom value 1. For intermediate values of R, the optimal Z-versus-R curve smoothly interpolates between these two limiting values.

For each R, the value of Z found by solving the transcendental equation may be eliminated from Equation (13.78a) to obtain E_+ solely as a function of R. However, from Equation (13.7) this function is the variable Z theory approximation to the H_2^+ ground state potential. The minimum of this potential, which occurs when the optimal $Z = 1.238$, determines the variable Z theory results for D_e and R_e as $D_e = 0.08651\,\text{au}$ and $R_e = 106\,\text{pm}$. Unsurprisingly, these variable Z theory results are in better agreement with the exact values $D_e = 0.10264\,\text{au}$ and $R_e = 105.7\,\text{pm}$ than are the results $D_e = 0.06483\,\text{au}$ and $R_e = 132\,\text{pm}$ of the most primitive theory. Further, note that while the variable Z and exact values for R_e are in excellent agreement, the variable Z prediction for D_e is too small. Consequently, the variable Z theory seriously underestimates the strength of the H_2^+ chemical bond but not as severely as does the most primitive theory.

From these comments, it is clear that further improvements of our LCAO-MO treatment of H_2^+ are necessary.

The most obvious way to make this improvement is to replace the 1s orbital basis functions used so far by more elaborate basis functions which are linear superpositions of 1s,2p,3d,... orbitals.

The convergence of D_e and R_e as the basis functions become increasingly elaborate is documented by McQuarrie (McQuarrie 2008, Table 10.2). The most sophisticated of these basis functions is a linear superposition of 1s, $2p_z$, and $3d_{z^2}$. Slater-type orbitals (STOs) each with a variationally chosen exponent Z. This basis function is

$$\phi = \Psi_{1s}(Z = 1.2458) + C_1\Psi_{2p_z}(Z = 1.2152) + C_2\Psi_{3d_{z^2}}(Z = 1.333) \tag{13.86}$$

where C_1 and C_2 are determined variationally. The LCAO-MO treatment based on this basis function yields $D_e = 0.1020\,\text{au}$ and $R_e = 106\,\text{pm}$, results in excellent agreement with the exact values $D_e = 0.10264\,\text{au}$ and $R_e = 105.7\,\text{pm}$.

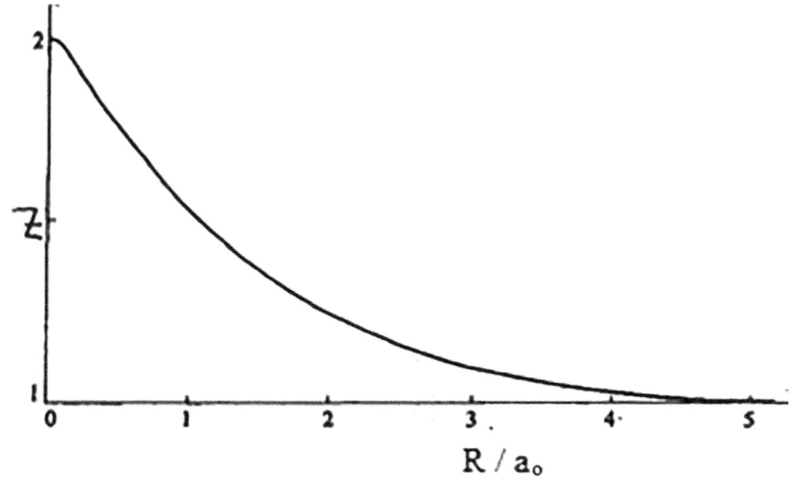

FIGURE 13.7 Exponent Z versus internuclear separation R found by numerically solving the variable Z theory transcendental equation.

In summary, our most primitive LCAO-MO theory can be systematically improved to give sophisticated LCAO-MO models which yield results in excellent agreement with those obtained from the exact solution of the H_2^+ electronic Schrödinger equation.

This concludes our discussion of H_2^+. We next move to the simplest neutral molecule, two-electron molecular hydrogen H_2.

13.3 THE ELECTRONIC STRUCTURE OF GROUND STATE MOLECULAR HYDROGEN H_2

In this section, we will describe an essentially exact calculation of the ground state potential energy function of H_2 and successively sophisticated approximate determinations based on molecular orbitals of the main features of this potential. Both the exact and approximate treatments are founded on the H_2 electronic Schrödinger equation (Equation 13.5) which we write as

$$\hat{H}_e \Psi_n(\mathbf{r}_1,\mathbf{r}_2;R) = E_n(R)\Psi_n(\mathbf{r}_1,\mathbf{r}_2;R) \tag{13.87}$$

where $\hat{H}_e$ is the H_2 electronic Hamiltonian operator given by

$$\hat{H}_e^+ = -\frac{1}{2}\nabla_1^2 - \frac{1}{2}\nabla_2^2 - \frac{1}{r_{1A}} - \frac{1}{r_{1B}} - \frac{1}{r_{2A}} - \frac{1}{r_{2B}} + \frac{1}{r_{12}} + \frac{1}{R}. \tag{13.88}$$

To start, recall that in Section 12.2, we saw that since helium has only two electrons, an essentially exact result may be found for its infinite nuclear mass nonrelativistic ground state energy E_0.

This result for two-electron helium suggests that an essentially exact result may be found for the infinite nuclear mass nonrelativistic ground state potential energy function of two-electron molecular hydrogen. Important progress along these lines, especially remarkable since it was made before the invention of computers, occurred in the early 1930s and is due to the American physicists James and Coolidge. However, truly major breakthroughs, made possible by the advent of modern digital computing, were made in the 1960s by the Polish physicists W. Kolos and L. Wolniewicz, henceforth referred to as K and W. Building on the methodology and calculational results described in earlier papers (Kolos and Wolniewicz 1964,1965), in 1968 K and W made an incredibly accurate calculation of the ground state Born–Oppenheimer potential energy function $U(R)$ of H_2.

Before describing K and W's method of calculation and results for $U(R)$, we note that in their 1968 paper, they also determined some spectroscopic properties of several isotopes of molecular hydrogen including H_2,HD, and D_2. While we will not give their results for these spectroscopic properties (which are tabulated in tables III–VII of their 1968 paper), we will summarize their calculational method since it goes beyond *all* other treatments of molecular electronic structure by including approximate non-Born–Oppenheimer corrections for nuclear motion. Because of this inclusion, K and W's results for the spectroscopic properties are nearly exact.

The most important spectroscopic properties dealt with by K and W are vibrational–rotational quantum dissociation energies $D_0(v,J)$, where v and J are the vibrational and rotational quantum numbers of the molecular hydrogen isotopes. For $v = J = 0$, $D_0(v,J)$ reduces to the familiar quantum dissociation energy D_0 defined in Equation (7.112) (Problem 13.20). For an arbitrary diatomic molecule, $D_0(v,J)$ is defined as $D_0(v,J) = E_{diss} - E_{vJ}$ where E_{diss} is the isotope-independent (Problem 13.21) energy of a dissociated diatomic (e.g., −1.0 au for the molecular hydrogen isotopes) and E_{vJ} is the isotope-dependent energy of the diatomic when it is in its vJth vibrational–rotational state. Therefore, to calculate $D_0(v,J)$ for the molecular hydrogen isotopes, one must determine each isotope's vibrational–rotational energy levels E_{vJ}. With one alteration, K and W determined these levels by solving the H_2 nuclear Schrödinger equation (Equation 13.6) (with the nuclear kinetic energy mass factors changed to be appropriate for each isotope). The novel and important alteration of this procedure introduced by K and W is that they replaced the Born–Oppenheimer potential

$U(R)$ in the nuclear Schrödinger equation by a slightly different potential which includes approximate corrections for nuclear motion. The vibrational–rotational levels E_{vJ} found by solving the resulting new nuclear Schrödinger equation thus include non-Born–Oppenheimer corrections for nuclear motion. In this way, the quantum vibrational–rotational dissociation energies $D_0(v,J)$ were corrected for breakdown of the Born–Oppenheimer approximation.

We may now return to K and W's determinations of $U(R)$. These determinations were variational. Specifically, in all of their papers (1964, 1965, and 1968) they employed variational trial functions which consisted of multi-term expansions in basis functions expressed in elliptical coordinates, each basis function including the interelectronic separation r_{12} explicitly. In their definitive 1968 paper, K and W used a 100-term trial function which after energy minimization yielded a 100-term wave function.

In Table 13.1, we tabulate for seventeen values of R K and W's 100-term wave function results for $U(R)$. These results were found by variationally solving (with the trial function just noted) the Schrödinger equation (Equation 13.87) to obtain for the seventeen values of R the H_2 ground state electronic energy function $E(R)$ which by Equation (13.7) is equal to the potential $U(R)$.

K and W found that the H_2 bond length $R_e = 1.4011\ a_0 = 74.1439$ pm. While K and W did not evaluate the H_2 classical dissociation energy D_e, its value may be easily found from the results of Table 13.1. The evaluation proceeds in analogy to the determination of the exact value of D_e for H_2^+ described in Section 13.2. Specifically, the exact value of D_e for H_2 is given by the relation $D_e = U(\infty) - U(R_e)$ where $U(\infty)$ = energy of two isolated hydrogen atoms $= -1.0$ au. Thus, from Table 13.1 $D_e = -1.0\ \text{au} + 1.17447498\ \text{au}$ or $D_e = 0.17447498$ au. K and W's theoretical values for R_e and D_e agree with the measured values $R_e = 74.14$ pm and $D_e = 0.174$ au to within experimental error.

TABLE 13.1
H_2 Ground State Potential Energy Function $U(R)$ Computed from a 100-Term Wave Function

$R(a_0)$	$U(R)$(au)
1.0	−1.12453881
1.2	−1.16493435
1.3	−1.17234623
1.39	−1.17445190
1.4	−1.17447477
1.4011	−1.17447498
1.41	−1.17446041
1.5	−1.17285408
1.6	−1.16858212
1.8	−1.15506752
2.0	−1.13813155
2.2	−1.12013055
2.4	−1.10242011
2.6	−1.08578740
2.8	−1.07067758
3.0	−1.05731738
3.2	−1.04578647

Source: Excerpted from Table 2 of W. Kolos and L. Wolniewicz *J. Chem. Phys.* 49 404 (1968).

K and W's elegant work however has one very severe limitation. The methods of K and W cannot be extended to molecules with more than two electrons. Thus, we next describe approximate treatments of H_2 based on molecular orbitals which can be extended.

To start, we note that in analogy to the atomic orbital electron configuration $1s^2$ of ground state helium, the molecular orbital electron configuration of ground state molecular hydrogen is σ_{1g}^2. In the crudest approximation, the molecular orbital σ_{1g} is taken as the ground state H_2^+ wave function $\Psi_+(\mathbf{r};R)$ of Equation (13.39a). That is, we take

$$\sigma_{1g}(\mathbf{r};R) = \sigma_{1g} = \left[\frac{1}{2(1+S)}\right]^{1/2} \Psi_{1s}(r_A) + \left[\frac{1}{2(1+S)}\right]^{1/2} \Psi_{1s}(r_B). \tag{13.89}$$

The crudest molecular orbital approximation to the normalized H_2 wave function corresponding to the ground state electron configuration σ_{1g}^2 is thus the Slater determinant

$$\Psi_0(\rho_1,\rho_2;R) = \frac{1}{\sqrt{2}} \begin{vmatrix} \sigma_{1g}(\mathbf{r}_1;R)\alpha(\sigma_1) & \sigma_{1g}(\mathbf{r}_1;R)\beta(\sigma_1) \\ \sigma_{1g}(\mathbf{r}_2;R)\alpha(\sigma_2) & \sigma_{1g}(\mathbf{r}_2;R)\beta(\sigma_2) \end{vmatrix} \tag{13.90}$$

where $\rho_1 = \mathbf{r}_1,\sigma_1$ and $\rho_2 = \mathbf{r}_2,\sigma_2$. Multiplying out the Slater determinant gives

$$\Psi_0(\rho_1,\rho_2;R) = \frac{1}{\sqrt{2}}\sigma_{1g}(\mathbf{r}_1;R)\sigma_{1g}(\mathbf{r}_2;R)\left[\alpha(\sigma_1)\beta(\sigma_2) - \beta(\sigma_2)\sigma(\mathbf{r}_1)\right]. \tag{13.91}$$

The crudest molecular orbital approximation to the H_2 ground state energy $E_0(R)$ is

$$E_0(R) = \int \Psi_0(\rho_1,\rho_2;R)\hat{H}_e\Psi_0(\rho_1,\rho_2;R)\,\mathbf{d}\rho_1\mathbf{d}\rho_2 \tag{13.92}$$

where $\hat{H}_e$ is the H_2 electronic Hamiltonian operator of Equation (13.88). But since $\hat{H}_e$ is free of spin coordinates, $E_0(R)$ simplifies to

$$E_0(R) = \int \Psi_0(\mathbf{r}_1,\mathbf{r}_2;R)\hat{H}_e\Psi_0(\mathbf{r}_1,\mathbf{r}_2;R)\,d\mathbf{r}_1 d\mathbf{r}_2 \tag{13.93}$$

where

$$\Psi_0(\mathbf{r}_1,\mathbf{r}_2;R) = \sigma_{1g}(\mathbf{r}_1;R)\sigma_{1g}(\mathbf{r}_2;R) \tag{13.94}$$

is the spatial part of $\Psi_0(\rho_1,\rho_2;R)$. Finally, from the Born–Oppenheimer equation (Equation 13.7) the crudest approximation $U_0(R)$ to the H_2 ground state potential energy function is identical to $E_0(R)$ and is thus given by

$$U_0(R) = \int \Psi_0(\mathbf{r}_1,\mathbf{r}_2;R)\hat{H}_e\Psi_0(\mathbf{r}_1,\mathbf{r}_2;R)\,d\mathbf{r}_1 d\mathbf{r}_2. \tag{13.95}$$

Before considering $E_0(R)$ and $U_0(R)$, let us first examine in more detail $\Psi_0(\mathbf{r}_1,\mathbf{r}_2;R)$. From Equations (13.89) and (13.94), the explicit form of $\Psi_0(\mathbf{r}_1,\mathbf{r}_2;R)$ is

$$\begin{aligned}\Psi_0(\mathbf{r}_1,\mathbf{r}_2;R) = {} & \frac{1}{2(1+S)}\left[\Psi_{1s}(r_{1A})\Psi_{1s}(r_{2B}) + \Psi_{1s}(r_{1B})\Psi_{1s}(r_{2A})\right] \\ & + \frac{1}{2(1+S)}\left[\Psi_{1s}(r_{1A})\Psi_{1s}(r_{2A}) + \Psi_{1s}(r_{1B})\Psi_{1s}(r_{2B})\right].\end{aligned} \tag{13.96}$$

Next, let us interpret the terms in $\Psi_0(\mathbf{r}_1,\mathbf{r}_2;R)$. The first two terms $\Psi_{1s}(r_{1A})\Psi_{1s}(r_{2B})$ and $\Psi_{1s}(r_{1B})\Psi_{1s}(r_{2A})$ describe the hypothetical situation for which electrons 1 and 2 are separately localized on nucleus A and nucleus B. These terms when added together form a zeroth-order wave function for a covalently bonded H_2 molecule. They are therefore referred to as *covalent terms.* In contrast, the third and fourth terms $\Psi_{1s}(r_{1A})\Psi_{1s}(r_{2A})$ and $\Psi_{1s}(r_{1B})\Psi_{1s}(r_{2B})$ describe the hypothetical situation for which both electrons are, respectively, localized on nucleus A or nucleus B. Each of these terms forms a zeroth-order wave function for a hypothetical ionic H_2 molecule each comprised of a hydride (H^-) ion bound to a proton. For this reason, they are called *ionic terms.* One may determine the H_2 dissociation limit predicted by $\Psi_0(\mathbf{r}_1,\mathbf{r}_2;R)$ by inserting Equation (13.96) for this wave function into Equation (13.95) for $U_0(R)$ and then evaluating the limit $U_0(R \to \infty)$. It turns out that because of the ionic terms in $\Psi_0(\mathbf{r}_1,\mathbf{r}_2;R)$, this dissociation limit is significantly higher than the true H_2 dissociation limit −1.0 au, the energy of two 1s hydrogen atoms. Thus, our simplest molecular orbital treatment of H_2 fails disastrously as $R \to \infty$.

This disaster cannot be removed merely by choosing a more sophisticated form for σ_{1g} rather than the crude form of Equation (13.89). Rather, the disaster is a consequence of the single-determinant form for $\Psi(\rho_1,\rho_2;R)$. It thus reflects an inherent defect of molecular orbital theory when applied to H_2.

To graphically illustrate these points, let us consider the purely covalent approximate H_2 ground state wave function

$$\Psi_{VB}(\mathbf{r}_1,\mathbf{r}_2;R) = N\left[\Psi_{1S}(r_{1A})\Psi_{1S}(r_{2B}) + \Psi_{1S}(r_{1B})\Psi_{1S}(r_{2A})\right] \tag{13.97}$$

where N is a normalization factor (Problem 13.23). Since Ψ_{VB} lacks ionic terms, we expect and will soon show graphically that it predicts the correct H_2 dissociation limit. The wave function Ψ_{VB} was introduced in 1927 by the German physicists Walter Heitler and Fritz London in their pioneering first application of quantum mechanics to molecules. Heitler and London found that Ψ_{VB} predicted that ground state H_2 is chemically bonded, thus providing the first evidence that quantum mechanics could account for chemical bonding.

The subscript VB on the Heitler–London wave function is included because this wave function is the first and simplest example of a *valence bond* wave function. Valence bond wave functions were further developed and widely applied in the 1930s, especially by the American chemist Linus Pauling, and are still discussed in many textbooks. However, outside of the present discussion we will not pursue valence bond wave functions since they have been largely superseded by molecular orbital wave functions and are therefore not an important part of modern research in molecular quantum mechanics.

To proceed further, in analogy to Equation (13.95) for the H_2 ground state molecular orbital potential energy function $U_0(R)$, the H_2 ground state valence bond potential is given by

$$U_{VB}(R) = \int \Psi_{VB}(\mathbf{r}_1,\mathbf{r}_2;R)\hat{H}_e\Psi_{VB}(\mathbf{r}_1,\mathbf{r}_2;R)d\mathbf{r}_1 d\mathbf{r}_2 \tag{13.98}$$

$U_0(R)$ and $U_{VB}(R)$ may be found explicitly by performing the integrals in Equations (13.95) and (13.98) (a very formidable task in the pre-computer age during which the integrals were first evaluated). $U_0(R)$, $U_{VB}(R)$, and the exact H_2 ground state potential are compared in Figure 13.8. Notice that for R near R_e, both the molecular orbital and the valence bond potentials are in qualitative agreement with the exact potential. However, because its determining wave function is purely covalent, the valence bond potential properly dissociates into two 1s hydrogen atoms and thus has the correct H_2 dissociation limit −1.0 au. In contrast, the molecular orbital potential because its determining wave function is an ionic–covalent mixture incorrectly and disastrously dissociates into a higher energy state. Evidently the higher energy state is comprised of one 1s hydrogen atom and a hydride ion+proton ion pair. The energy of the hydrogen atom is −0.5 au. Since a proton has

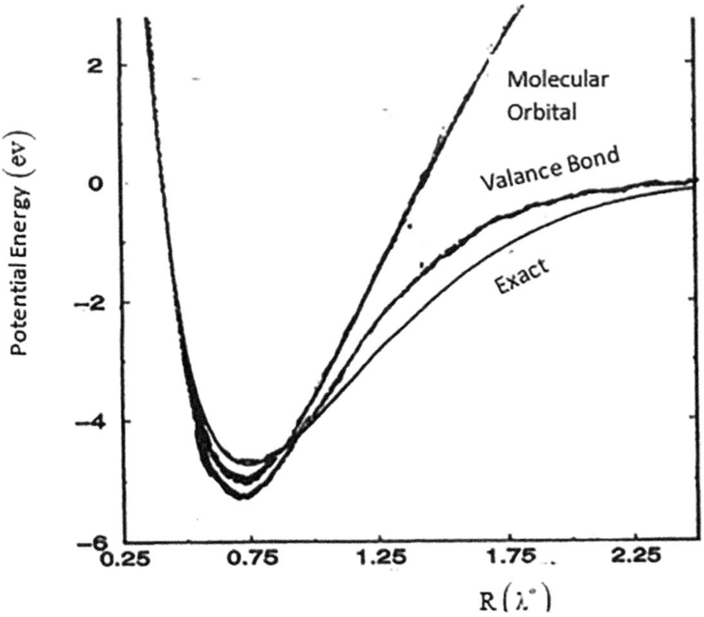

FIGURE 13.8 Exact, simplest molecular orbital, and valence bond potential energy function for H_2. Notice that the molecular orbital potential disastrously does not dissociate into two 1s H atoms.

zero electronic energy, the energy of the ion pair is just the energy of the hydride ion –0.028 au. Thus, the dissociation limit of the molecular orbital potential $U_0(R)$ is too high having the value $-0.5au - 0.028\ au = -0.5028\ au > -1.0\ au$.

We next note that the molecular orbital classical dissociation energy cannot be meaningfully evaluated. This is so since the standard formula for this dissociation energy $D_e = U(\infty) - U(R_e)$ is invalid for the molecular orbital model for H_2 since the molecular orbital value of $U(\infty)$ is unphysical due to the incorrect dissociation limit of $U_0(R)$.

However, the simplest molecular orbital theory value for R_e may be meaningfully determined since $U_0(R)$ is physical near R_e (Figure 13.8). One finds however that the simplest molecular orbital theory prediction for $R_e = 84.33\text{pm}$ is in poor agreement with the exact KW value $R_e = 74.14\text{pm}$. One might try for improvement by basing the molecular orbital treatment on the variable Z H_2^+ molecular orbital of Equation (13.62a) (with Z chosen by minimizing the H_2 ground state energy $E_0[R]$) rather than on the crudest H_2^+ molecular orbital of Equation (13.39a). This yields the significantly improved value $R_e = 72.29\text{pm}$. However, the dissociation limit of the variable Z molecular orbital potential is identical to the unphysical limit of the crudest molecular orbital potential $U_0(R)$. This exemplifies the point made earlier, that a single-determinant form for the H_2 ground state wave function $\Psi_0(\rho_1,\rho_2;R)$ is qualitatively flawed irrespective of the choice of the form of the molecular orbital σ_{1g}.

Thus, we next turn to ground state H_2 wave functions $\Psi_0(\rho_1,\rho_2;R)$ called *configuration interaction* (CI) wave functions which are built from more than one Slater determinant. CI wave functions are linear superpositions of two of more Slater determinants D_1, D_2, and thus have the form

$$\Psi_1(\rho_1,\rho_2;R) = C_1D_1 + C_2D_2 + \cdots. \tag{13.99}$$

where the CI expansion coefficients are determined variationally. The leading determinant D_1 is of the molecular orbital form of Equation (13.90) where the molecular orbital σ_{1g} can be of arbitrary sophistication. Thus, CI wave functions reduce to molecular orbital wave functions if all coefficients

$C_2, C_3, \ldots$ vanish. Hence, CI wave functions are a generalization of and an improvement on molecular orbital wave functions.

The great virtue of taking $\Psi_0(\rho_1, \rho_2; R)$ to be a CI wave function is that if the non-leading determinates $D_2, D_3, \ldots$ are chosen judiciously, then $\Psi_0(\rho_1, \rho_2; R)$ determines a form for the ground state H_2 potential $U_0(R)$ which properly dissociates into two 1s hydrogen atoms.

However, as we will see in Section 13.5, CI wave functions can do much more than eliminate the disastrous H_2 dissociation energy predicted by molecular orbital models as well as analogous catastrophes which occur for some other molecules.

Specifically, CI expansions are one of the premier *post-Hartree–Fock* methods which can be applied to a vast range of molecules, not just those with molecular orbital catastrophes. Post-Hartree–Fock wave functions transcend single-determinant Hartree–Fock wave functions and thus capture at least part of the correlation energy.

We next show how CI wave functions predict the correct H_2 dissociation limit. To start, recall that in molecular orbital theory, the H_2 ground state wave function is a single normalized Slater determinant. In contrast, in the simplest CI theory the ground state H_2 wave function is the following linear superposition of two normalized Slater determinants:

$$\Psi_0(\rho_1, \rho_2; R) = C_1 \frac{1}{\sqrt{2}} \begin{vmatrix} \sigma_{1g}(\mathbf{r}_1; R)\alpha(\sigma_1) & \sigma_{1g}(\mathbf{r}_1; R)\beta(\sigma_1) \\ \sigma_{1g}(\mathbf{r}_2; R)\alpha(\sigma_2) & \sigma_{1g}(\mathbf{r}_2; R)\beta(\sigma_2) \end{vmatrix}$$

$$+ C_2 \frac{1}{\sqrt{2}} \begin{vmatrix} \sigma_{1u}^*(\mathbf{r}_1; R)\alpha(\sigma_1) & \sigma_{1u}^*(\mathbf{r}_1; R)\beta(\sigma_1) \\ \sigma_{1u}^*(\mathbf{r}_2; R)\alpha(\sigma_2) & \sigma_{1u}^*(\mathbf{r}_2; R)\beta(\sigma_2) \end{vmatrix} \quad (13.100)$$

In analogy to our choice of σ_{1g} in the simplest molecular orbital theory, in the simplest CI theory we choose σ_{1g} and σ_{1u}^* to be the crudest ground and first excited state H_2^+ molecular orbitals. That is, we choose (see Equations [13.39])

$$\sigma_{1g}(\mathbf{r}; R) = \sigma_{1g}$$

$$= \left[\frac{1}{2(1+S)}\right]^{1/2} \Psi_{1s}(r_A) + \left[\frac{1}{2(1+S)}\right]^{1/2} \Psi_{1s}(r_B) \quad (13.101)$$

and

$$\sigma_{1u}^*(\mathbf{r}; R) = \sigma_{1u}^*$$

$$= \left[\frac{1}{2(1-S)}\right]^{1/2} \Psi_{1s}(r_A) - \left[\frac{1}{2(1-S)}\right]^{1/2} \Psi_{1s}(r_B). \quad (13.102)$$

Multiplying out the determinants in Equation (13.100) yields

$$\Psi_0(\rho_1, \rho_2; R) = \Psi_0(\mathbf{r}_1, \mathbf{r}_2; R) \frac{1}{\sqrt{2}} [\alpha(\sigma_1)\beta(\sigma_2) + \alpha(\sigma_2)\beta(\sigma_1)] \quad (13.103)$$

where

$$\Psi_0(\mathbf{r}_1, \mathbf{r}_2; R) = C_1 \sigma_{1g}(\mathbf{r}_1; R)\sigma_{1g}(\mathbf{r}_2; R) + C_2 \sigma_{1u}^*(\mathbf{r}_1; R)\sigma_{1u}^*(\mathbf{r}_2; R) \quad (13.104)$$

is the spatial part of $\Psi_0(\rho_1, \rho_2; R)$.

We may now prove that the above 2B CI form for $\Psi_0(\mathbf{r}_1,\mathbf{r}_2;R)$ implies that H_2 properly dissociates into two 1s hydrogen atoms. To make this proof, we first use Equations (13.101) and (13.102) to eliminate σ_{1g} and σ^*_{1u} from Equation (13.104) to yield

$$\Psi_0(\mathbf{r}_1,\mathbf{r}_2;R)=\left[\frac{C_1}{2(1+S)}-\frac{C_2}{2(1-S)}\right]\left[\Psi_{1s}(r_{1A})\Psi_{1s}(r_{2B})+\Psi_{1s}(r_{1B})\Psi_{1s}(r_{2A})\right]$$

$$+\left[\frac{C_1}{2(1+S)}+\frac{C_2}{2(1-S)}\right]\left[\Psi_{1s}(r_{1A})\Psi_{1s}(r_{2A})+\Psi_{1s}(r_{2B})\Psi_{1s}(r_{2B})\right]. \quad (13.105)$$

Next, let $R \to \infty$. In this limit, $S \to 0$ and the above expression simplifies to

$$\lim_{R\to\infty}\Psi_0(\mathbf{r}_1,\mathbf{r}_2;R)=\left(\frac{C_1-C_2}{2}\right)\left[\Psi_{1s}(r_{1A})\Psi_{1s}(r_{2B})+\Psi_{1s}(r_{1B})\Psi_{1s}(r_{2A})\right]$$

$$+\left(\frac{C_1+C_2}{2}\right)\left[\Psi_{1s}(r_{1A})\Psi_{1s}(r_{2A})+\Psi_{1s}(r_{2B})\Psi_{1s}(r_{1B})\right]. \quad (13.106)$$

The next step is to determine the expansion coefficients C_1 and C_2 by variational minimization of the CI ground state energy $E_0(R)$. Since $\hat{H}_e$ is independent of spin coordinates, this energy is given by

$$E_0(R)=\int \Psi_0(\mathbf{r}_1,\mathbf{r}_2;R)\hat{H}_e\Psi_0(\mathbf{r}_1,\mathbf{r}_2;R)d\mathbf{r}_1 d\mathbf{r}_2 \quad (13.107)$$

But it turns out that as $R \to \infty$, the variational values of C_1 and C_2 are

$$\lim_{R\to\infty} C_1 = \frac{1}{\sqrt{2}} \text{ and } \lim_{R\to\infty} C_2 = -\frac{1}{\sqrt{2}}. \quad (13.108)$$

Inserting these values into Equation (13.106) yields that in the $R \to \infty$ dissociation limit, the H_2 ground state CI wave function reduces to

$$\lim_{R\to\infty}\Psi_0(\mathbf{r}_1,\mathbf{r}_2;R)=\frac{1}{\sqrt{2}}\left[\Psi_{1s}(r_{1A})\Psi_{1s}(r_{2B})+\Psi_{1s}(r_{1B})\Psi_{1s}(r_{2A})\right].$$

Notice that the ionic terms have dropped out of $\lim_{R\to\infty}\Psi_0(\mathbf{r}_1,\mathbf{r}_2;R)$ rendering it a normalized purely covalent wave function. Thus, like the purely covalent valence bond wave function Ψ_{VB}, $\lim_{R\to\infty}\Psi_0(\mathbf{r}_1,\mathbf{r}_2;R)$ correctly predicts that ground state H_2 dissociates into two 1s hydrogen atoms.

In summary, the use of the simplest CI ground state H_2 wave function eliminates the H_2 dissociation catastrophe predicted by molecular orbital theory.

We next note that the simplest CI approximation to the H_2 ground state potential $U_0(R)$ may be found from Equation (13.107) for $E_0(R)$ since $U_0(R)=E_0(R)$. From the minimum of $U_0(R)$, we find the simplest CI estimates $D_e = 1.19$au and $R_e = 88.87$pm. Both estimates are in mediocre to poor agreement with the exact K and W values $D_e = 1.174$au and $R_e = 74.14$pm. Substantial improvements occur if one bases the CI treatment on optimized variable Z molecular orbitals. One finds for this CI model that $D_e = 1.148$au and $R_e = 75.66$pm.

As expected and as illustrated by McQuarrie (McQuarrie 2008, Table 10.9), further improvements result if one expands the CI model to include more than two determinants, with D_e and R_e converging to the K and W results as more and more determinants are included. For example, from a five-determinant expansion McLean, Weiss, and Yoshimine (1960) found that $D_e = 1.167$ au and $R_e = 74.14$ pm, while from a thirty-three-determinant expansion Hagstrom and Schull (1963) found the essentially exact results $D_e = 1.174$ au and $R_e = 74.14$ pm.

This concludes our discussion of H_2.

We next turn to a synopsis of qualitative molecular orbital theory.

13.4 A SYNOPSIS OF QUALITATIVE MOLECULAR ORBITAL THEORY

Our discussion of qualitative molecular orbital theory is limited to a synopsis since the topic is well covered in many textbooks more elementary than ours. For example, qualitative molecular orbital theory is described in the general chemistry text of Silberberg (2006, sec. 11.3) and the introductory physical chemistry text of Atkins, de Paula, and Friedman (2014, topic 24). While our synopsis of qualitative molecular theory is only a review, we include it because the theory's simplicity and usefulness make uniquely clear (clearer than the earlier discussions in this chapter) the power and elegance of the molecular orbital concept.

Qualitative molecular orbital theory applies to all diatomic molecules. But here we will restrict our discussion to the simplest case of homonuclear diatomics formed from atoms in the first and second periods of the periodic table, the atoms hydrogen through neon.

First and foremost, qualitative molecular orbital theory correctly predicts from the molecular orbital electron configuration of a diatomic molecule the number of chemical bonds which bind the molecule. It also permits one to qualitatively rationalize the variation from molecule to molecule of properties such as bond lengths and classical or quantum dissociation energies which correlate with the number of these chemical bonds. Additionally, qualitative molecular orbital theory is the only simple theory that correctly predicts the magnetic properties of diatomic molecules, that is, whether a molecule is diamagnetic (zero electronic magnetic moment) or paramagnetic (nonzero electronic magnetic moment).

The method of qualitative molecular orbital is very similar to the familiar atomic orbital electron filling procedure used to find the electron configurations of atoms (Section 12.9). First, for the diatomic in question one writes down all of the molecular orbitals which can potentially be filled by the molecule's electrons. (To illustrate, in the closely similar atomic case the potential orbitals of a ground state carbon atom are $1s, 2s, 2p_x, 2p_y$, and $2p_z$.) Then, starting with the lowest energy molecular orbital the potential molecular orbitals are filled in order of increasing energy by the molecule's electrons taking into account the restrictions imposed by the Pauli exclusion principle and for degenerate molecular orbitals those required by Hund's maximum multiplicity rule (Section 12.10).

After all of the molecule's electrons have been used to fill the molecular orbitals, one can then immediately write down the molecule's electron configuration. From this configuration, one may count the number of electrons occupying both bonding and anti-bonding molecular orbitals.

From this count, one may determine the molecule's bond order which is defined by

$$\text{bond order} = \frac{1}{2}\big(\text{number of electrons occupying bonding molecular orbitals}$$

$$- \text{number of electrons occupying anti-bonding molecular orbitals}\big). \quad (13.109)$$

The number of chemical bonds binding the molecule is equal to its bond order. From the electron configuration, one may also immediately obtain the magnetism of the molecule. If it has no unpaired electrons, it is diamagnetic. If it has one or more unpaired electrons, it is paramagnetic.

We will assume throughout this section that all molecular orbitals are built from atomic orbitals by the simplest LCAO-MO prescription. For example, we approximate the σ_{1g} and σ^*_{1u} molecular orbitals by those of Equations (13.101) and (13.102).

To illustrate these principles, we start with the simplest case of diatomic molecules formed from the two first-period atoms hydrogen and helium. These include H_2^+, H_2, He_2^+, and He_2. The potential molecular orbitals for these molecules are σ_{1g} and σ^*_{1u}. Since H_2^+ has only one electron, that electron fills the lower energy σ_{1g} molecular orbital giving the H_2^+ electron configuration as σ_{1g}. Similar filling of the potential molecular orbitals of H_2, He_2^+ and He_2 gives for their respective electron configurations $\sigma_{1g}^2, \sigma_{1g}^2\sigma_{1u}^*$, and $\sigma_{1g}^2\sigma_{1u}^{*2}$. H_2^+ has one bonding electron and zero anti-bonding electrons, and therefore its bond order is $\frac{1}{2}(1+0)=\frac{1}{2}$, showing that bond orders can be half-integral as well as integral. The H_2^+ bond order does not mean that H_2^+ is held together by half a chemical bond, but rather means that it is weakly chemically bonded. Similarly, from their electron configurations the bond orders of H_2, He_2^+, and He_2 are, respectively, $1, \frac{1}{2}$, and 0. This means that in agreement with experiment, H_2 is singly bonded, He_2^+ is weakly bonded, and as is expected, He_2 is not a stable molecule.

Let us next compare the bond lengths R_e and the classical dissociation energies D_e of H_2^+ and H_2. Since the chemical bond of H_2 is stronger than that of H_2^+, we expect that D_e for H_2 is larger than that for H_2^+ and that R_e for H_2 is smaller than that for H_2^+. These qualitative inferences are completely confirmed by comparing K and W's exact results for H_2 $D_e = 0.1774\,\text{au}$ and $R_e = 74.143\,\text{pm}$ with the results of the exact solution of the H_2^+ Schrödinger equation $D_e = 0.1026\,\text{au}$ and $R_e = 105.7\,\text{pm}$. These results illustrate the great power of qualitative molecular orbital theory. Namely, with essentially zero labor, the theory can correctly predict qualitative trends which can only be established quantitatively by detailed or even massive calculations.

Next, let us turn to the predictions of molecular magnetism made by qualitative molecular orbital theory. From their electron configurations, H_2^+ and He^+ each have one unpaired electron, while H_2 has no unpaired electrons. Thus, in agreement with experiment the theory predicts that H_2^+ and He_2^+ are paramagnetic while H_2 is diamagnetic.

Next, let us move on to homonuclear diatomics formed from second-period atoms, the atoms lithium through neon. For all diatomics formed from atoms of the second period or higher, the σ_{1g} and σ_{1u}^* molecular orbitals are doubly occupied and therefore contribute zero to both the bond orders and magnetism of these molecules. Thus, we will exclude these two molecular orbitals from the group of potential molecular orbitals of the molecules we will consider.

The potential molecular orbitals which we need are identified from simple general chemistry atomic orbital overlap arguments as $\sigma_{2g}, \sigma_{2u}^*, \sigma_{3g}, \sigma_{3u}^*, \pi_{1u}^{(1)}, \pi_{1u}^{(2)}, \pi_{1g}^{(1)*}$, and $\pi_{1g}^{(2)*}$. The σ_{2g} and σ_{2u}^* molecular orbitals are built as LCAO-MO linear combinations of 2s atomic orbitals. The σ_{3g} and σ_{3u}^* molecular orbitals are similarly built from $2p_z$ atomic orbitals. The $\pi_u^{(1)}$ and the $\pi_{1u}^{(2)}$ molecular orbitals are degenerate as are the $\pi_g^{(1)*}$ and $\pi_g^{(2)*}$ molecular orbitals. Both pairs are similarly built from $2p_x$ and $2p_y$ atomic orbitals. Notice that the bonding and anti-bonding σ molecular orbitals have, respectively, even (g) and odd (u) inversion symmetry, while the opposite is true for the π molecular orbitals.

To obtain the valence electron configurations of the diatomics in question, we need the energy ordering of their potential molecular orbitals. These orderings may be determined either from photoelectron spectroscopy or from the solutions of the Hartree–Fock–Roothaan equations, a computationally convenient formulation of the Hartree–Fock equations discussed extensively in the next section.

It turns out that for O_2 and F_2, *only* the energies of their σ_{3g} molecular orbitals are slightly less than the energies of their degenerate $\pi_{1u}^{(1)}$ and $\pi_{1u}^{(2)}$ molecular orbitals (Atkins, de Paula, and Friedman 2014, Fig. 24.11). For this reason, the energy ordering of the molecules Li_2 through N_2 differs from that of the molecules O_2 and F_2. Specifically, for Li_2 through N_2 the energy ordering is

$$\sigma_{2g} < \sigma_{2u}^* < \pi_{1u}^{(1)} = \pi_{1u}^{(2)} < \sigma_{3g} < \pi_{1g}^{(1)*} = \pi_{1g}^{(2)*} < \sigma_{3u}^* \tag{13.110}$$

while the energy ordering for O_2 and F_2 is

$$\sigma_{2g} < \sigma_{2u}^* < \sigma_{3g} < \pi_{1u}^{(1)} = \pi_{1u}^{(2)} < \pi_{1g}^{(1)*} = \pi_{1g}^{(2)*} < \sigma_{3u}^*. \tag{13.111}$$

With these energy orderings, we may determine the valence electron configurations, chemical bonding, and magnetism of the molecules Li_2 through F_2. We will take as examples N_2 and O_2. More examples are given as problems at the end of the chapter (Problems 13.23–13.26).

N_2 has ten valence electrons. Its potential molecular orbitals are listed in Equation (13.110) in order of increasing energy. Filling these molecular orbitals in that order taking into account the Pauli and Hund restrictions yields the N_2 valence electron configuration $\sigma_{2g}^2\sigma_{2u}^{*2}\pi_{1u}^{(1)2}\pi_{1u}^{(2)2}\sigma_{3g}^2$. The bond order of N_2 is thus $\frac{1}{2}(2+2+2+2-2)=3$. Therefore, N_2 has a triple bond. Its chemical bonding is in fact stronger than that of any other homonuclear diatomic molecule. Also N_2 has no unpaired electrons and consequently is diamagnetic. These molecular orbital theory predictions of the chemical bonding and magnetism of N_2 are in perfect agreement with experiment.

Before the advent of molecular orbital theory, O_2 had presented a paradox. This is because the earlier Lewis electron dot and valence bond theories, which in contradiction to molecular orbital theory are based on the assumption that electrons are localized to the region of chemical bonds, both failed qualitatively to predict either the chemical bonding or the magnetism of O_2. Molecular orbital theory resolved this paradox, one of its greatest successes.

To see this, we first determine the valence electron configuration of O_2. O_2 has twelve valence electrons, and its energy-ordered potential molecular orbitals are listed in Equation (13.111). Filling these molecular orbitals in order of increasing energy taking account of the Pauli and Hund restrictions yields the O_2 valence electron configuration $\sigma_{2g}^2\sigma_{2u}^{*2}\sigma_{3g}^2\pi_{1u}^{(1)2}\pi_{1u}^{(2)2}\pi_{1g}^{(1)*}\pi_{1g}^{(2)*}$. The bond order of O_2 is therefore $\frac{1}{2}(2+2+2+2-2-1-1)=2$. O_2 has two unpaired electrons, arising from the enforcement of Hund's maximum multiplicity rule. Thus qualitative molecular orbital theory predicts that O_2 is doubly bonded and paramagnetic. Both of these predictions of molecular orbital theory are in perfect accord with experiment, thus resolving the paradox.

We next note that since N_2 is triply bonded while O_2 is doubly bonded, N_2 is held together more tightly than O_2. This difference has implications for several properties of O_2 and N_2. For example, we expect that the quantum dissociation energy D_0 of N_2 is larger than that of O_2 while the bond length R_e of N_2 is shorter than that of O_2. These qualitative predictions of molecular orbital theory are borne out quantitatively by experiments. Thus, the measured values of D_0 for N_2 and O_2 are, respectively, 945 and 498 kJ mol^{-1}, while the experimental values of R_e for N_2 and O_2 are, respectively, 110 and 121 pm. Thus, we again see that with almost zero effort, qualitative molecular orbital theory correctly predicts trends which can only be established quantitatively by laborious experiments or by the kind of large-scale quantum calculations described in the next section.

This concludes our synopsis of qualitative molecular orbital theory.

We next move on to the final topic of this textbook *ab initio* (from first principles) methods. These methods underlay modern quantitative molecular orbital and post-Hartree–Fock treatments of molecules.

13.5 *AB INITIO* MOLECULAR QUANTUM MECHANICS

The goal of *ab initio* theory is to solve on a computer the Schrödinger equation for molecules of chemical interest and thus to compute important molecular properties and to simulate real chemistry.

Ab initio treatments do not replace experimental chemistry. Rather, they supplement it by permitting one to better interpret experimental results and to study chemical phenomena not readily accessible in the laboratory, for example, the structures, bonding, and properties of exotic species.

Solving the Schrödinger equation for molecules of chemical interest presents severe mathematical difficulties. So the solutions are always approximate (Hehre, Redom, Schleyer, and Pople 1986). In general, the more sophisticated the approximation, the greater is its accuracy, but also the greater is its computational cost. Thus, one must always compromise between higher accuracy and higher cost or lower accuracy and lower cost. The choice depends on the complexity of the molecule treated, the accuracy to which its properties are needed, and the power of the available computational facilities.

We begin our discussion of *ab initio* methods by developing a basis set formulation of the Hartree–Fock equations, already touched on in Section 12.3, known as the *Hartree–Fock–Roothaan* equations. These equations were derived in 1951 by the Dutch-American physicist and theoretical chemist Clemens Roothaan (1951). It is a tribute to Roothaan's insight that he realized as early as 1951 when scientific computation was still in its infancy that the Hartree–Fock equations when expressed in a basis set could be far more efficiently solved on a computer than the conventional Hartree–Fock equations. In fact, because of their relative ease of solution, the Hartree–Fock–Roothaan equations have completely replaced the standard Hartree–Fock equations as a tool in *ab initio* treatments of molecular electronic structure.

The Hartree–Fock–Roothaan equations are the workhorse of *ab initio* molecular quantum mechanics. For many molecules, their solutions alone are sufficient to satisfactorily compute a host of molecular properties. However, for some molecules and properties the solutions of the Hartree–Fock–Roothaan equations are not sufficiently accurate and must be corrected for the effects of electron correlation by post-Hartree–Fock methods. But even for these cases, the Hartree–Fock–Roothaan solutions provide the departure point for the post-Hartree–Fock calculations.

We will derive the Hartree–Fock–Roothaan equation by first generalizing the single-nucleus atomic Hartree–Fock equations (Equations 12.113–12.116) to multi-nucleus molecular Hartree–Fock equations. Then, we will express the molecular equations in a basis set.

To satisfy electron pair atomic bonding requirements, in contrast to most ground state atoms most ground state molecules are closed shell systems. Such systems are treated by solving the *restricted* Hartree–Fock–Roothaan equations. For open shell molecules like free radicals, one instead optimally solves the more complex *unrestricted* Hartree–Fock–Roothaan equations (Hehre, Radom, Schleyer, and Pople 1986, sec. 2.62). Here for simplicity, we will deal only with closed shell molecules and thus develop and apply the restricted Hartree–Fock–Roothaan equations.

We will assume closed shell molecules with N doubly occupied molecular orbitals ϕ_k and hence 2N electrons and with M nuclei. We will adopt the following notation. We employ the subscripts i, j,... to distinguish between different electrons and the subscripts s, t,... to distinguish between distinct nuclei. Thus, the position vectors of two arbitrary electrons will be denoted by $\mathbf{r}_i$ and $\mathbf{r}_j$ and the position vectors of two arbitrary nuclei will be written as $\mathbf{R}_s$ and $\mathbf{R}_t$. Also we will collectively denote the nuclear position vectors by $\mathbf{R} = \mathbf{R}_1, \ldots, \mathbf{R}_M$. With this notation, the molecule's nuclear configuration-dependent kth molecular orbital is written as $\phi_k(\mathbf{r}_i;\mathbf{R})$.

A normalized determinantal wave function of a closed shell molecule is the following natural generalization of the closed shell atomic wave function of Equation (12.104):

$$\Psi(\rho_1,\cdots,\rho_{2N};R) =$$

$$\frac{1}{\sqrt{(2N)!}}\begin{vmatrix} \phi_1(\mathbf{r}_1;\mathbf{R})\alpha(\sigma_1) & \phi_1(\mathbf{r}_1;\mathbf{R})\beta(\sigma_1) & \cdots & \phi_N(\mathbf{r}_1;\mathbf{R})\alpha(\sigma_1) & \phi_N(\mathbf{r}_1;\mathbf{R})\beta(\sigma_1) \\ \phi_1(\mathbf{r}_2;\mathbf{R})\alpha(\sigma_2) & \phi_1(\mathbf{r}_2;\mathbf{R})\beta(\sigma_2) & \cdots & \phi_N(\mathbf{r}_2;\mathbf{R})\alpha(\sigma_2) & \phi_N(\mathbf{r}_2;\mathbf{R})\beta(\sigma_2) \\ \vdots & & & & \\ \phi_1(\mathbf{r}_{2N};\mathbf{R})\alpha(\sigma_{2N}) & \phi_1(\mathbf{r}_{2N};\mathbf{R})\beta(\sigma_{2N}) & \cdots & \phi_N(\mathbf{r}_{2N};\mathbf{R})\alpha(\sigma_{2N}) & \phi_N(\mathbf{r}_{2N};\mathbf{R})\beta(\sigma_{2N}) \end{vmatrix} \tag{13.112}$$

where $\rho_i = \mathbf{r}_i, \sigma_i$.

The electronic energy $E(\mathbf{R})$ of the molecule is given in terms of its determinantal wave function Ψ and its electronic Hamiltonian operator $\hat{H}$ by

$$E(\mathbf{R}) = \int \Psi^*(\rho_1,\ldots,\rho_{2N};\mathbf{R})\hat{H}\Psi(\rho_1,\ldots,\rho_{2N};\mathbf{R})d\mathbf{r}_1 d\sigma_1 \cdots d\mathbf{r}_{2N} d\sigma_{2N}. \tag{13.113}$$

$\hat{H}$ is the following generalization of the atomic electronic Hamiltonian operator of Equation (12.107):

$$\hat{H} = \sum_{i=1}^{2N}\left(-\frac{1}{2}\nabla_i^2 - \sum_{s=1}^{M}\frac{Z_s}{r_{is}}\right) + \sum_{i=1}^{2N-1}\sum_{j=i+1}^{2N}\frac{1}{r_{ij}} + \sum_{s=1}^{M-1}\sum_{t=s+1}^{M}\frac{Z_s Z_t}{R_{st}}. \quad (13.114)$$

The meanings of the non-atomic quantities in $\hat{H}$ are as follows. Z_s is the atomic number of nucleus s, $r_{is} = |\mathbf{r}_i - \mathbf{R}_s|$ is the distance between electron i and nucleus s, and $R_{st} = |\mathbf{R}_s - \mathbf{R}_t|$ is the distance between nucleus s and nucleus t. Thus, the non-atomic energy operators in $\hat{H}$ have the following interpretations: $-\frac{Z_s}{r_{is}}$ is the attractive Coulomb potential between electron i and nucleus s_j and $\frac{Z_s Z_t}{R_{st}}$ is the repulsive Coulomb potential between nucleus s and nucleus t. Also the limits of the two double sums in Equation (13.114) are chosen so that all Coulomb repulsion terms are counted but are only counted once and so that infinite terms like $\frac{1}{r_{it}}$ and $\frac{1}{R_{ss}}$ are excluded.

Using properties of determinants, one may show by taking suitable linear combinations of the molecular orbitals $\phi_k(\mathbf{r};R)$ that they may be orthogonalized without changing the value of the molecule's full Slater determinantal wave function Ψ. The molecular orbitals can of course also be normalized. Thus, we can and will choose the molecular orbitals to form an orthonormal set satisfying for i and $j = 1,\ldots,N$ the orthonormality relation

$$\int \phi_i(\mathbf{r};R)\phi_j(\mathbf{r};R)d\mathbf{r} = \delta_{ij}. \quad (13.115)$$

The next step is to derive an expression for the electronic energy $E(\mathbf{R})$ of the molecule which generalizes the atomic electronic energy expression of Equations (12.109)–(12.112). Inserting Equation (13.112) for Ψ and Equation (13.114) for $\hat{H}$ into Equation (13.113) for $E(\mathbf{R})$ and performing a complex calculation yield $E(\mathbf{R})$ as

$$E(\mathbf{R}) = 2\sum_{i=1}^{N} I_i(\mathbf{R}) + \sum_{i=1}^{N}\sum_{j=1}^{N}\left[2J_{ij}(\mathbf{R}) - K_{ij}(\mathbf{R})\right] + \sum_{s=1}^{M-1}\sum_{t=1+s}^{M}\frac{Z_s Z_t}{R_{st}} \quad (13.116)$$

where the final term in the above expression is the molecule's total nuclear repulsion energy, a constant for a fixed nuclear configuration, and where

$$I_i(\mathbf{R}) = \int \phi_i^*(\mathbf{r}_1;\mathbf{R})\left(-\frac{1}{2}\nabla_1^2 - \sum_{s=1}^{M}\frac{Z_s}{r_{1s}}\right)\phi_i(\mathbf{r}_1;\mathbf{R})d\mathbf{r}_1 \quad (13.117)$$

$$J_{ij}(\mathbf{R}) = \int \phi_i^*(\mathbf{r}_1;\mathbf{R})\phi_j^*(\mathbf{r}_2;\mathbf{R})\frac{1}{r_{12}}\phi_i(\mathbf{r}_1;\mathbf{R})\phi_j(\mathbf{r}_2;\mathbf{R})d\mathbf{r}_1 d\mathbf{r}_2 \quad (13.118a)$$

and

$$K_{ij}(\mathbf{R}) = \int \phi_i^*(\mathbf{r}_1;\mathbf{R})\phi_j^*(\mathbf{r}_2;\mathbf{R})\frac{1}{r_{12}}\phi_i(\mathbf{r}_2;\mathbf{R})\phi_j(\mathbf{r}_1;\mathbf{R})d\mathbf{r}_1 d\mathbf{r}_2. \quad (13.118b)$$

The interpretations of the molecular energy terms $I_i(\mathbf{R})$, $J_{ij}(\mathbf{R})$, and $K_{ij}(\mathbf{R})$ are analogous to those of the atomic energy terms I_i, J_{ij}, and K_{ij} given after Equation (12.112).

The best molecular orbitals $\phi_i(\mathbf{r};\mathbf{R})$ are the solutions of the molecular Hartree–Fock equations with the N lowest orbital energies ε_i. In analogy to the atomic case, the molecular Hartree–Fock equations are derived by minimizing $E(\mathbf{R})$ with respect to the forms of the orbitals $\phi_i(\mathbf{r};\mathbf{R})$ subject to the orthonormality constraint of Equation (13.115). This minimization process gives the molecular Hartree–Fock equations as the following generalization of the atomic Hartree–Fock equations (Equations 12.113–12.116):

$$\hat{F}\phi_i(\mathbf{r}_1;\mathbf{R}) = \varepsilon_i\phi_i(\mathbf{r}_1;\mathbf{R}) \tag{13.119}$$

where $\hat{F}$ is the molecular Fock operator defined by

$$\hat{F} = -\frac{1}{2}\nabla_1^2 - \sum_{s=j}^{M}\frac{Z_s}{r_{1s}} + \sum_{j=1}^{N}\left(2\hat{J}_j - \hat{K}_j\right). \tag{13.120}$$

$\hat{J}_j$ and $\hat{K}_j$ are Coulomb and exchange operators which in analogy to the corresponding atomic operators are defined by

$$\hat{J}_j\phi_i(\mathbf{r}_1;\mathbf{R}) = \int\left[\frac{\phi_j^*(\mathbf{r}_2;\mathbf{R})\phi_j(\mathbf{r}_2;\mathbf{R})d\mathbf{r}_2}{r_{12}}\right]\phi_i(\mathbf{r}_1;\mathbf{R}) \tag{13.121}$$

and

$$\hat{K}_j\phi_i(\mathbf{r}_1;\mathbf{R}) = \int\left[\frac{\phi_j^*(\mathbf{r}_2;\mathbf{R})\phi_i(\mathbf{r}_2;\mathbf{R})d\mathbf{r}_2}{r_{12}}\right]\phi_j(\mathbf{r}_1;\mathbf{R}). \tag{13.122}$$

Notice that $\hat{F}\phi_i$ through the terms $\hat{J}_j\phi_i$ and $\hat{K}_j\phi_i$ depends on the unknown molecular orbitals ϕ_j. So like the atomic Hartree–Fock equations, the molecular Hartree–Fock equations must be solved by an iterative self-consistent procedure.

It will prove convenient to rewrite the molecular Hartree–Fock equations as

$$\hat{F}\phi_i(\mathbf{r}_1;\mathbf{R}) = \hat{F}_1\phi_i(\mathbf{r}_1;\mathbf{R}) + \sum_{j=1}^{N}\left\{2\left[\int\frac{\phi_j^*(\mathbf{r}_2;\mathbf{R})\phi_j(\mathbf{r}_2;\mathbf{R})}{r_{12}}d\mathbf{r}_2\right]\phi_i(\mathbf{r}_1;\mathbf{R}) - \left[\int\frac{\phi_j^*(\mathbf{r}_2;\mathbf{R})\phi_i(\mathbf{r}_2;\mathbf{R})}{r_{12}}d\mathbf{r}_2\right]\phi_j(\mathbf{r}_1;\mathbf{R})\right\} = \varepsilon_i\phi_i(\mathbf{r}_1;\mathbf{R}) \tag{13.123}$$

where we have defined the one-electron Fock operator $\hat{F}_1$ by

$$\hat{F}_1 = -\frac{1}{2}\nabla_1^2 - \sum_{s=1}^{M}\frac{Z_s}{r_{1s}}. \tag{13.124}$$

The next step in the derivation of the Hartree–Fock–Roothaan equations is to express the molecular Hartree–Fock equations in a basis set. To do this, we expand the N Hartree–Fock molecular orbitals $\phi_i(\mathbf{r};\mathbf{R})$ in terms of a set of $n \geq N$ *real linearly independent* but *non-orthogonal* known basis functions $\chi_\mu(\mathbf{r})$. This expansion is

$$\phi_i(\mathbf{r}_1;\mathbf{R}) = \sum_{\mu=1}^{n} C_{i\mu}(\mathbf{R})\chi_\mu(\mathbf{r}). \tag{13.125}$$

Notice that there are nN expansion coefficients $C_{i\mu}$ since there are N molecular orbitals ϕ_i and each is expanded in terms of n basis functions χ_μ.

It is important to note that the basis functions $\chi_\mu(\mathbf{r})$ are independent of $\mathbf{R}$ and therefore are the same for all molecular configurations. (For example, the basis functions are the same for all isomers of a multi-isomeric molecule.) The configuration dependence of the molecular orbital $\phi_i(\mathbf{r};\mathbf{R})$ enters solely through the $\mathbf{R}$ – dependence of the expansion coefficients $C_{i\mu}(\mathbf{R})$.

So far we have not specified the forms of the basis functions $\chi_\mu(\mathbf{r})$. This will be done later. For now, we only require that the basis functions approach a complete set as their number $n \to \infty$. Also we have not yet said how the expansion coefficients $C_{i\mu}(\mathbf{R})$ are determined. We will soon see that they are determined as the solutions of the Hartree–Fock–Roothaan equations. Moreover, when the expansion coefficients are so determined, $\phi_i(\mathbf{r};\mathbf{R})$ found from the expansion of Equation (13.125) is an approximation, called the Hartree–Fock–Roothaan molecular orbital, to the molecule's true ith Hartree–Fock molecular orbital. The Hartree–Fock–Roothaan orbital becomes the true Hartree–Fock orbital as the basis set approaches completeness.

To improve the agreement between the Hartree–Fock–Roothaan and the true Hartree–Fock molecular orbitals, one must improve the basis set. In practice, this is done in two different ways. Either one increases the number n of the basis functions, or one makes a more judicious choice of the basis functions.

We next use Equation (13.125) to expand each of the N molecular orbitals ϕ in the molecular Hartree–Fock equations (Equation 13.123) in terms of the n basis functions χ. Recalling that the basis functions are real, the required expansions are

$$\phi_i(\mathbf{r}_1;\mathbf{R}) = \sum_{\beta=1}^{n} C_{i\beta}(\mathbf{R})\chi_\beta(\mathbf{r}_1) \tag{13.126a}$$

$$\phi_j^*(\mathbf{r}_2;\mathbf{R}) = \sum_{\delta=1}^{n} C_{j\delta}^*(\mathbf{R})\chi_\delta(\mathbf{r}_2) \tag{13.126b}$$

$$\phi_j(\mathbf{r}_2;\mathbf{R}) = \sum_{\varepsilon=1}^{n} C_{j\varepsilon}(\mathbf{R})\chi_\varepsilon(\mathbf{r}_2) \tag{13.126c}$$

$$\phi_i(\mathbf{r}_2;\mathbf{R}) = \sum_{\phi=1}^{n} C_{i\phi}(\mathbf{R})\chi_\phi(\mathbf{r}_2) \tag{13.126d}$$

and

$$\phi_j(\mathbf{r}_1;\mathbf{R}) = \sum_{\gamma=1}^{n} C_{j\gamma}(\mathbf{R})\chi_\gamma(\mathbf{r}_1). \tag{13.126e}$$

Using Equations (13.126) to eliminate the molecular orbitals ϕ from the molecular Hartree–Fock equations (Equation 13.123), one finds that these equations are now expressed in terms of the basis functions χ as

$$\begin{aligned}\sum_{\beta=1}^{n} C_{i\beta}(\mathbf{R})\hat{F}\chi_\beta(\mathbf{r}_1) = {} & \sum_{\beta=1}^{n} C_{i\beta}(\mathbf{R})\hat{F}_1\chi_\beta(\mathbf{r}_1) + \sum_{j=1}^{N}\Bigg\{2\sum_{\delta=1}^{n}\sum_{\varepsilon=1}^{n}\sum_{\beta=1}^{n} C_{j\delta}^*(\mathbf{R})C_{j\varepsilon}(\mathbf{R})C_{i\beta}(\mathbf{R}) \\ & \times\left[\int \frac{\chi_\delta(\mathbf{r}_2)\chi_\varepsilon(\mathbf{r}_2)d\mathbf{r}_2}{r_{12}}\right]\chi_\beta(\mathbf{r}_1) - \sum_{\delta=1}^{n}\sum_{\phi=1}^{n}\sum_{\gamma=1}^{n} C_{j\delta}^*(\mathbf{R})C_{j\gamma}(\mathbf{R})C_{i\phi}(\mathbf{R}) \\ & \times\left[\int \frac{\chi_\delta(\mathbf{r}_2)\chi_\phi(\mathbf{r}_2)}{r_{12}}d\mathbf{r}_2\right]\chi_\delta(\mathbf{r}_1)\Bigg\} = \varepsilon_i\sum_{\beta=1}^{n} C_{i\beta}(\mathbf{R})\chi_\beta(\mathbf{r}_1).\end{aligned} \tag{13.127}$$

Next, let us change the dummy summation indices ϕ and ε in Equation (13.127) to, respectively, β and γ. Making this change and rearranging, one finds that Equation (13.127) greatly simplifies to (Problem 13.27)

$$\sum_{\beta=1}^{0} C_{i\beta}(\mathbf{R})\hat{F}\chi_\beta(\mathbf{r}_1) = \sum_{\beta=1}^{n} C_{i\beta}(\mathbf{R})\hat{F}_1\chi_\beta(\mathbf{r}_1) + P(\mathbf{R})\left\{2\left[\int \frac{\chi_\delta(\mathbf{r}_2)\chi_\gamma(\mathbf{r}_2)d\mathbf{r}_2}{r_{12}}\right]\sum_{\beta=1}^{n} C_{i\beta}(\mathbf{R})\chi_\beta(\mathbf{r}_1)\right.$$

$$\left. -\left[\int \frac{\chi_\delta(\mathbf{r}_2)\chi_\beta(\mathbf{r}_2)d\mathbf{r}_2}{r_{12}}\right]\sum_{\beta=1}^{n} C_{i\beta}(\mathbf{R})\chi_\gamma(\mathbf{r}_1)\right\} = \varepsilon_i \sum_{\beta=1}^{n} C_{i\beta}(\mathbf{R})\chi_\beta(\mathbf{r}_1) \qquad (13.128)$$

where

$$P(\mathbf{R}) = \sum_{j=1}^{N}\sum_{\delta=1}^{n}\sum_{\gamma=1}^{n} C^*_{j\delta}(\mathbf{R})C_{j\gamma}(\mathbf{R}). \qquad (13.129)$$

The molecular orbitals ϕ may be and typically are chosen to be real. With this choice, from Equation (13.125) the expansion coefficients are also real and hence $C^*_{j\delta}(\mathbf{R}) = C_{j\delta}(\mathbf{R})$ and $P(\mathbf{R})$ becomes the real quantity

$$P(\mathbf{R}) = \sum_{j=1}^{N}\sum_{\delta=1}^{n}\sum_{\gamma=1}^{n} C_{j\delta}(\mathbf{R})C_{j\gamma}(\mathbf{R}). \qquad (13.130)$$

We are now nearly done with our rather complex derivation of the Hartree–Fock–Roothaan equations. To complete it, we multiply Equation (13.128) on the left by $\chi_\alpha(\mathbf{r}_1)$, integrate over $\mathbf{r}_1$, and rearrange to yield (Problem 13.28)

$$\sum_{\beta=1}^{n} F_{\alpha\beta}C_{i\beta}(\mathbf{R}) = \sum_{\beta=1}^{n} F_{1\alpha\beta}C_{i\beta}(\mathbf{R})$$

$$+P(\mathbf{R})\sum_{\beta=1}^{n}\iint\left[\frac{2\chi_\delta(\mathbf{r}_2)\chi_\gamma(\mathbf{r}_2)\chi_\alpha(\mathbf{r}_1)\chi_\beta(\mathbf{r}_1) - \chi_\delta(\mathbf{r}_2)\chi_\beta(\mathbf{r}_2)\chi_\alpha(\mathbf{r}_1)\chi_\gamma(\mathbf{r}_1)}{r_{12}}\right]d\mathbf{r}_1\,d\mathbf{r}_2 C_{i\beta}(\mathbf{R})$$

$$= \varepsilon_i \sum_{\beta=1}^{n} S_{\alpha\beta}C_{i\beta}(\mathbf{R}) \qquad (13.131)$$

where the matrix elements

$$F_{\alpha\beta} = \int \chi_\alpha(\mathbf{r}_1)\hat{F}\chi_\beta(\mathbf{r}_1)d\mathbf{r}_1 \qquad (13.132)$$

$$F_{1\alpha\beta} = \int \chi_\alpha(\mathbf{r}_1)\hat{F}_1\chi_\beta(\mathbf{r}_1)d\mathbf{r}_1 \qquad (13.133)$$

and

$$S_{\alpha\beta} = \int \chi_\alpha(\mathbf{r}_1)\chi_\beta(\mathbf{r}_1)d\mathbf{r}_1. \qquad (13.134)$$

Equation (13.131) with the definitions of the matrix elements $F_{\alpha\beta}, F_{1\alpha\beta}$, and $S_{\alpha\beta}$ comprise the Hartree–Fock–Roothaan equations. They are a set of nN coupled equations for the nN expansion coefficients $C_{i\beta}(\mathbf{R})$. Solving for these coefficients and then using Equation (13.125) yields the result we have been seeking, the molecule's N Hartree–Fock–Roothaan approximate molecular orbitals for the basis set in question.

Before going on, we note that a completely different method for deriving the Hartree–Fock–Roothaan equations may be developed from the expression for the molecule's electronic energy $E(\mathbf{R})$ given in Equations (13.116)–(13.118b). (See Problem 13.29.)

We next turn to the solution of the Hartree–Fock–Roothaan equations. For simplicity, we adopt a method of solution which is more elementary, but probably less accurate, than the types of methods used in practice (Hehre, Radom, Schleyer, and Pople 1986, secs 3.2 and 3.32).

Because from Equation (13.130) the factor $P(\mathbf{R})$ is highly non-linear in the expansion coefficients, the Hartree–Fock–Roothaan equations are also highly non-linear in these coefficients. For this reason (as is true for even the methods of solution used in practice), the Hartree–Fock–Roothaan equations cannot be solved directly but must be solved iteratively. We next outline our iterative solution.

To start, we rewrite the Hartree–Fock–Roothaan equations (Equation 13.131) compactly as

$$\sum_{\beta=1}^{n}\left[F_{1\alpha\beta}+P(\mathbf{R})I_{\delta\gamma\alpha\beta}\right]C_{i\beta}(\mathbf{R})=\varepsilon_i\sum_{\beta=1}^{n}S_{\alpha\beta}C_{i\beta}(\mathbf{R}) \tag{13.135}$$

where the quantities

$$I_{\delta\gamma\alpha\beta}=\iint\left[\frac{2\chi_\delta(\mathbf{r}_2)\chi_\gamma(\mathbf{r}_2)\chi_\alpha(\mathbf{r}_1)\chi_\beta(\mathbf{r}_1)-\chi_\delta(\mathbf{r}_2)\chi_\beta(\mathbf{r}_2)\chi_\alpha(\mathbf{r}_1)\chi_\gamma(\mathbf{r}_1)}{r_{12}}\right]d\mathbf{r}_1\,d\mathbf{r}_2 \tag{13.136}$$

are referred to as *two-electron integrals*. These are a large number n^4 of these integrals. So their evaluation (for some sets of basis functions) is one of the most expensive parts of a Hartree–Fock–Roothaan calculation. Fortunately, the two-electron integrals only need to be evaluated once in a Hartree–Fock–Roothaan solution since they remain unchanged in every iterative step of the solution.

We next outline these iterative steps. We use the notational convention that first, second, ... iteration quantities are labeled by, respectively, the superscripts $1,2,\cdots$.

We first note that since an iterative solution is needed only because $P(\mathbf{R})$ is non-linear in the expansion coefficients in the first iteration, we replace $P(\mathbf{R})$ by a function $P^{(1)}(\mathbf{R})$ which is linear in these coefficients. The simplest choice for $P^{(1)}(\mathbf{R})$ is unity. With this choice from Equation (13.135), the first iteration form of the Hartree–Fock–Roothaan equations is

$$\sum_{\beta=1}^{n}\left[F_{1\alpha\beta}+I_{\delta\gamma\alpha\beta}\right]C_{i\beta}^{(1)}(\mathbf{R})=\varepsilon_i^{(1)}\sum_{\beta=1}^{n}S_{\alpha\beta}C_{i\beta}^{(1)}(\mathbf{R}). \tag{13.137}$$

The above equation comprises a set of nN coupled linear equations for the coefficients $C_{i\beta}^{(1)}(\mathbf{R})$. These may be readily solved on a computer to obtain the first iteration approximations $C_{i\beta}^{(1)}(\mathbf{R})$ to the nN exact expansion coefficients $C_{i\beta}(\mathbf{R})$. The now-known quantities $C_{i\beta}^{(1)}(\mathbf{R})$ may be used in Equation (13.130) to obtain a second iteration form $P^{(2)}(\mathbf{R})$ for $P(\mathbf{R})$. With this form for $P^{(2)}(\mathbf{R})$, one may immediately write down the second iteration Hartree–Fock–Roothaan equations as

$$\sum_{\beta=1}^{n}\left[F_{1\alpha\beta}+P^{(2)}(\mathbf{R})I_{\delta\gamma\alpha\beta}\right]C_{i\beta}^{(2)}(\mathbf{R})=\sum_{\beta=1}^{n}S_{\alpha\beta}C_{i\beta}^{(2)}(\mathbf{R}). \tag{13.138}$$

However, since $P^{(2)}(\mathbf{R})$ is a known constant, Equation (13.138) again comprises a set of nN coupled linear equations which can be readily solved to obtain the nN second iteration expansion coefficients $C_{i\beta}^{(2)}(\mathbf{R})$. From these, one may obtain $P^{(3)}(\mathbf{R})$, the third iteration approximation to $P(\mathbf{R})$.

Continuing in this manner, one eventually obtains $P^{(n)}(\mathbf{R})$ and $P^{(n+1)}(\mathbf{R})$, the nth and $(n+1)$ th approximations to $P(\mathbf{R})$. If these two approximations are identical (to within a prescribed tolerance), the iterative procedure has converged and the nN quantities $C_{i\beta}^{(n)}(\mathbf{R})$ are the converged Hartree–Fock–Roothaan expansion coefficients $C_{i\beta}(\mathbf{R})$. Then, we have solved the Hartree–Fock–Roothaan equations and have accomplished our ultimate goal of obtaining the N Hartree–Fock–Roothaan approximate molecular orbitals $\phi_i(\mathbf{r};\mathbf{R})$.

This concludes our discussion of the Hartree–Fock–Roothaan equations.

Except for a brief overview of density functional theory, a radically different approach to electronic structure problems not based on the Hartree–Fock–Roothaan equations, in the remainder of this section we will describe the most common basis sets used to implement both the Hartree–Fock–Roothaan procedure and two of the most widely used post-Hartree–Fock methods, CI and Moller–Plesset (MP) perturbation theory. We then apply these three quantum methods to compute several properties of a number of molecules. We especially compare the results of these computations with experiment and also examine how the quality of the results depends on both the sophistication of the basis set and the accuracy of the quantum method.

Our discussion in much of the remainder of this section draws heavily on the monograph, cited earlier, by Hehre, Radom, Schleyer, and Pople. This back provides a relatively simple introduction to *ab initio* theory and also includes a vast number of applications of the theory.

The book is mainly based on the pioneering research of the American theoretical chemist John A. Pople. Over a period of years, Pople and his research group converted bare bones formal developments like the Hartree–Fock–Roothaan equations, CI, and MP perturbation theory into a powerful practical methodology which permits the *ab initio* treatment of a vast range of molecules and chemical phenomena. Pople and his coworkers made their methodology accessible to broad segments of the chemical community by encoding it to create their very widely used GAUSSIAN series of computer programs. The GAUSSIAN programs are so user-friendly that experimental chemists with only very limited knowledge of quantum mechanics can use the programs to study theoretically chemical phenomena that they are simultaneously observing in their laboratories. The GAUSSIAN programs are even used in some introductory physical chemistry courses to give undergraduates hands-on experience with *ab initio* computation.

For his many major contributions, Pople was awarded the 1998 Nobel Prize in chemistry. Pople shared the Nobel Prize with the American solid-state physicist Walter Kohn. Kohn and his collaborators the physicists Pierre Hohenberg and Lu Ju Sham developed a radically new approach to electronic structure problems called *density functional theory.*

Density functional theory was born in a remarkable paper by Hohenberg and Kohn (1964). In this paper, Hohenberg and Kohn did hardly less than develop a new formulation of quantum mechanics. In this new formulation, the $3N-$ variable ground state wave function $\Psi(\mathbf{r}_1,\ldots,\mathbf{r}_N)$ of an $N-$ electron system is replaced by a much simpler quantity, the system's three-variable ground state electron density $\rho(\mathbf{r})$ defined by

$$\rho(\mathbf{r}) = N\int \Psi^*(\mathbf{r},\mathbf{r}_2,\ldots,\mathbf{r}_N)\Psi(\mathbf{r},\mathbf{r}_2,\ldots,\mathbf{r}_N)d\mathbf{r}_2\ldots d\mathbf{r}_N. \tag{13.139}$$

Non-rigorously stated, Hohenberg and Kohn then argued that since the density $\rho(\mathbf{r})$ may be determined from the wave function $\Psi(\mathbf{r}_1,\mathbf{r}_2\ldots,\mathbf{r}_N)$, the wave function may be determined from the density. But since the values A of all system observables A may be determined (at least probabilistically) from the system wave function, these values may also be determined from the system's density. We write this statement symbolically as

$$A = A[\rho(\mathbf{r})] \tag{13.140}$$

which is read as the value A of an observable A is a functional A[ ] of the density $\rho(\mathbf{r})$. (A functional, as noted in Section 12.3, is a mathematical operation which converts a function, here $\rho[\mathbf{r}]$, into a number, here the value A of an observable A.)

Next, Hohenberg and Kohn specialized the observable A to the system's electronic energy E. Then, Equation (13.140) specializes to

$$E = E[\rho(\mathbf{R})]. \tag{13.141}$$

In analogy to the expected decomposition of the electronic energy observable E, Hohenberg and Kohn decomposed the electronic energy density functional as $E[\ ] = T[\ ] + V[\ ] + U_{coul}[\ ] + U_{exch}[\ ] + U_{corr}[\ ]$. Dropping the brackets on E[ ] and so on for brevity, the decomposition of the energy functional E[ ] is written as

$$E = T + V + U_{coul} + U_{exch} + U_{corr}. \tag{13.142}$$

The functionals T, V, U_{coul}, U_{exch}, and U_{corr} convert $\rho(\mathbf{r})$ into, respectively, the following observables: the system's kinetic energy T, its energy V of Coulomb attraction between its electrons and nuclei, its energy U_{coul} of Coulomb repulsion between its electrons, its non-classical exchange energy U_{exch}, and its electron correlation energy U_{corr}. The forms of the density functionals V and U_{coul} are easily found (Problem 13.30). But the forms of the functionals T, U_{exch}, and U_{corr} are unknown.

We may now turn to two theorems proven by Hohenberg and Kohn which are central to density functional theory.

Theorem 1

While the functional V is specific to the system since it depends on the nuclear configuration, remarkably the functional $T + U_{coul} + U_{exch} + U_{corr}$ is a *universal functional;* that is, it is the *same* density functional for *all* electron systems.

Hohenberg and Kohn's first theorem of course does not mean that the value of the observable $T + U_{coul} + U_{exch} + U_{corr}$ is the same for all electron systems. It only means that the value of $T[\rho(\mathbf{r})] + U_{coul}[\rho(\mathbf{r})] + U_{exch}[\rho(\mathbf{r})] + U_{corr}[\rho(\mathbf{r})]$ is the same. The value of the observable $T + U_{coul} + U_{exch} + U_{corr}$ can vary widely from system to system since the wave function $\Psi(\mathbf{r}_1, \mathbf{r}_2, \ldots, \mathbf{r}_N)$ and hence the density $\rho(\mathbf{r})$ are radically system dependent.

Hohenberg and Kohn's second theorem is the following variational principle.

Theorem 2

If one guesses a trial density $\bar{\rho}(\mathbf{r})$ and if as before E is the electronic energy observable of the system and E[ ] is the corresponding energy density functional, then

$$E[\bar{\rho}(\mathbf{r})] \geq E \tag{13.143}$$

with equality holding only if the guess is perfect, that is, if $\bar{\rho}(\mathbf{r}) = \rho(\mathbf{r})$.

The Hohenberg–Kohn analysis while beautiful and strikingly original still does not provide a basis for determining the electronic energy of a molecular system. This is because the components T, U_{exch}, and U_{corr} of the energy functional E are unknown.

This roadblock was partially soon overcome by Kohn and Sham in another remarkable paper (Kohn and Sham 1965). First, these authors developed an accurate approximation to the kinetic energy functional T, thus leaving only the exchange energy functional U_{exch} and the correlation energy functional U_{corr} unknown. Then, building on the foundation of Hohenberg and Kohn, Kohn and Sham developed a formal framework which was further developed by later workers into a powerful method for computationally determining the energy, structure, and properties of any multi-electron system, especially solid-state and molecular systems.

The main result of the formal framework of Kohn and Sham is a Hartree-like equation, now known as the *Kohn–Sham equation* for the solid-state or molecular orbitals of a fictitious non-interacting N electron system whose electron density $\rho(\mathbf{r})$ is identical to that of the true interacting electron system. (Hartree equations are an obsolete form of Hartree–Fock equations for which the Hartree–Fock exchange terms $\hat{K}_j\phi_i(\mathbf{r}_1;\mathbf{R})$ of Equation [13.122] are absent.)

The Kohn–Sham equation differs from a Hartree equation only in that it includes two extra terms. The first term requires the exchange energy functional U_{exch}, while the second term requires the correlation energy functional U_{corr}. If suitable approximations can be found to these two strictly unknown density functionals, the Kohn–Sham equation becomes closed and can be solved by an iterative procedure to obtain both the orbitals of the fictitious non-interacting electron system and the electron density $\rho(\mathbf{r})$ of the true system.

The great virtue of the Kohn–Sham equation is that while the computational effort needed to find its solution is comparable to that needed to obtain a Hartree–Fock–Roothaan solution, the Kohn–Sham energy includes a correlation energy correction to the Hartree–Fock energy. The accuracy of the correction depends on the quality of the approximate density functionals used in the Kohn–Sham computation. The fact that the Kohn–Sham equation delivers an approximate correlation energy at relatively low cost is the basis of the wide use of density functional methods in solid-state physics, chemistry, and related disciplines like material science and nanotechnology.

In summary, the Kohn–Sham theory radically transforms the electronic structure problem in several disciplines from the very complex problem of solving the electronic Schrödinger equation for many interacting electron systems to the hopefully simpler problem of developing high-quality approximate exchange and correlation energy density functionals for these systems.

The first implementations of this radically new approach to electronic structure problems occurred in solid-state physics. It was first realized that crude density functionals for solids could be developed from the exactly known form for U_{exch} and the approximately known form for U_{corr} of the model homogeneous electron gas system. These crudest density functionals could be greatly improved by introducing simple corrections for the inhomogeneity of the electron gas. Widespread use of the Kohn–Sham theory based on these electron gas functionals then began in solid-state physics in the 1970s.

Because molecular electronic structure problems require more sophisticated density functionals than the solid-state electron gas functionals, widespread application of the Kohn–Sham theory to chemical problems did not occur until the 1980s and 1990s.

It was soon found that with sufficiently high-quality density functionals, accurate correlation energies could be computed for many molecules at much lower cost from the Kohn–Sham equation than from sophisticated post-Hartree–Fock methods. Thus, some might surmise that density functional methods have replaced the more traditional procedures of Pople and others. This is not the case. Both density functional and more traditional methodologies are widely used in current chemical research.

There are several reasons why this is so. First, there is no systematic and rigorous way to choose the needed density functionals U_{exch} and U_{corr}. Choosing density functionals is as much of an art as a science. As seriously in contrast to the approximations of the traditional theories which in principle can be systematically improved to exactness, there is no well-defined way to improve approximate density functionals.

Despite their wide use in chemistry, we will not further pursue density functional methods. This is both because of their inherent limitations, two of which have just been mentioned, and more importantly because density functional theory when developed in detail is not particularly easy to understand by readers of this book since it differs so radically from the standard quantum mechanics we have been developing.

Thus, in the rest of this section we will develop and apply the more traditional Pople-type *ab initio* methods.

To start, recall that the Hartree–Fock–Roothaan equations are based on expansions of molecular orbitals $\phi_i(\mathbf{r};\mathbf{R})$ in terms of a set of real non-orthogonal molecular basis functions $\chi_\mu(\mathbf{r})$. We have however not yet said much about the nature of the basis functions. We now remedy this omission.

We first note that the basis functions used in *ab initio* molecular electronic structure calculations are (nearly) always functions which are centered on one of the nuclei of the molecule under study. Thus, the argument $\mathbf{r}$ of an acceptable basis function $\chi_\mu(\mathbf{r})$ is the vector displacement of a generic electron from the nucleus on which $\chi_\mu(\mathbf{r})$ is centered.

To proceed further, we determine the mathematical forms of the acceptable basis functions $\chi_\mu(\mathbf{r})$. To do this, we first make the following guess for these forms. We assume that they are identical to the forms of the members of a selected set of atomic basis functions (the functions used in atomic orbital expansions analogous to the molecular orbital expansions of Equation [13.125]). To choose the set of selected atomic basis functions, we require that its members optimally resemble nuclear centered normalized atomic orbitals. The normalized STOs, touched on in Section 12.3, more closely resemble such atomic orbitals than do the members of any other set of normalized well-behaved functions. Thus, we choose the normalized STOs as our selected set of atomic basis functions and hence as our first guess of the set of acceptable molecular basis functions $\chi_\mu(\mathbf{r})$.

The normalized STOs $\chi_{ngm}(\mathbf{r})$ are given by (compare Equation [12.32] for integer n)

$$\chi_{n\ell m}(\mathbf{r})=\left(\frac{2\varsigma_n}{n}\right)^n\left\{\left(\frac{2\varsigma_n}{n}\right)\left[\frac{1}{(2n)!}\right]\right\}^{1/2} r^{n-1}\exp\left(\frac{-\varsigma_n r}{n}\right)Y_{\ell m}(\theta,\phi). \qquad (13.144)$$

The STO quantum numbers $n\ell m$ have the same interpretations and ranges as the hydrogen-like atom orbital quantum numbers $n\ell m$. The parameters ς_n are called *STO orbital exponents*. Their interpretations are that the ratios $\frac{\varsigma_n}{n}$ govern the radial extents of the STOs with principle quantum number n. (Note that unlike the hydrogen-like atom orbital exponents, which have the constant value $\frac{Z}{a_0}$, the STO orbital exponents ς_n depend on the quantum number n).

We may now turn to our actual problem, determining the forms of the molecular basis functions $\chi_\mu(\mathbf{r})$. Our first guess that the functions $\chi_\mu(\mathbf{r})$ may be taken as the STOs $\chi_{n\ell m}(\mathbf{r})$ is clearly wrong. This is because the STOs due to their spherical harmonic factors are complex functions while the basis functions $\chi_\mu(\mathbf{r})$ are real functions.

The next simplest guess for the molecular basis functions $\chi_\mu(\mathbf{r})$ is that they are the normalized real versions of the STOs. This choice is very reasonable since the real STOs strongly resemble the real forms of atomic orbitals (Section 10.5) rendering real STO expansions of molecular orbitals analogous to standard LCAO-MO expansions. Thus, we will select the normalized real STOs as our choice of the basis functions $\chi_\mu(\mathbf{r})$.

The first few normalized real STOs are (Problem 13.31)

$$\chi_{1s}(\varsigma_1;\mathbf{r})=\left(\frac{\varsigma_1^2}{\pi}\right)^{1/2}\exp\left(\frac{-\varsigma_1 r}{1}\right) \qquad (13.145a)$$

$$\chi_{2s}(\varsigma_2;\mathbf{r})=\left(\frac{\varsigma_2^5}{96\pi}\right)^{1/2} r\exp\left(\frac{-\varsigma_2 r}{2}\right) \qquad (13.145b)$$

$$\chi_{2p_x}(\varsigma_2;\mathbf{r})=\left(\frac{\varsigma_2^5}{36\pi}\right)^{1/2} x\exp\left(\frac{-\varsigma_2 r}{2}\right) \tag{13.145c}$$

$$\chi_{2p_y}(\varsigma_2;\mathbf{r})=\left(\frac{\varsigma_2^5}{36\pi}\right)^{1/2} y\exp\left(\frac{-\varsigma_2 r}{2}\right) \tag{13.145d}$$

$$\chi_{2p_z}(\varsigma_2;\mathbf{r})=\left(\frac{\varsigma_2^5}{36\pi}\right)^{1/2} z\exp\left(\frac{-\varsigma_2 r}{2}\right) \tag{13.145e}$$

and so on.

The parameters $\varsigma_1, \varsigma_2, \ldots$ are referred to as *real STO orbital exponents,* and the ratios $\frac{\varsigma_n}{n}$ give the radial extents of the real STOs with principle quantum numbers $n = 1(\text{1s real STO})$, $n = 2(\text{2s and 2p real STO's})$, $n = 3(\text{3s,3p, and 3d real STO's})$, and so on.

Because the real STOs strongly resemble the real forms of atomic orbitals, they are theoretically satisfactory as molecular basis functions and were so used in early *ab initio* studies of diatomic molecules. But real STOs while satisfactory theoretically suffer from a fatal practical flaw. This flaw is as follows. Namely, if real STOs are chosen as the molecular basis functions $\chi_\mu(\mathbf{r})$, the evaluation of the two-electron integrals of Equation (13.136), an evaluation which is an essential component of both Hartree–Fock–Roothaan and post-Hartree–Fock calculations, cannot be performed analytically but rather must be performed numerically. Such numerical evaluations however are slow and subject to serious numerical errors for even diatomic molecules. For polyatomic molecules, the numerical integrations are nearly intractable. This is because for polyatomics evaluation of many of the two-electron integrals requires integration over products of basis functions centered on three or four different nuclear sites. The four-center two-electron integrals especially were long viewed as nearly insurmountable roadblocks to *ab initio* treatments of the polyatomic molecules of primary interest in chemistry.

Ab initio quantum mechanics was however saved as a tool for studying chemical problems by a breakthrough contribution due to the British physicist S.F. Boys (1950). Boys suggested non-intuitively that the atomic orbital-like real STOs be replaced by new basis functions which do not at all resemble atomic orbitals. These new basis functions are Gaussian functions, now commonly called Gaussian-type orbitals (GTOs). Boys' great insight in choosing GTOs soon became apparent when he derived the monumental result that for GTO basis functions all two-electron integrals, including the notorious four-center two-electron integrals, could be performed analytically to yield closed-form expressions for the integrals which could be conveniently evaluated numerically.

The first calculations based on Boys' GTO basis functions were performed in the 1960s apparently by the Pople group. It was soon established by these workers that with GTO basis functions, *ab initio* calculations on a wide range of molecules were feasible. The Pople group then adopted GTOs for all of their subsequent research. This is why their series of computer programs are named GAUSSIAN. GTOs were soon adopted by other researchers and are now universally used as the basis functions for *ab initio* molecular (but not atomic) calculations.

The GTOs are products of Gaussian functions of the electron-nuclear separation r and either the Cartesian components x, y, z of the electron position vector $\mathbf{r}$ or multiples of these components. The first ten normalized GTOs are (Problem 13.32)

$$g_s(\alpha;r^2)=\left(\frac{2\alpha}{\pi}\right)^{1/4} \exp(-\alpha r^2) \tag{13.146a}$$

$$g_x(\alpha;r^2)=\left(\frac{128\alpha^5}{\pi^3}\right)^{1/4} x\exp(-\alpha r^2) \tag{13.146b}$$

$$g_y(\alpha;r^2)=\left(\frac{128\alpha^5}{\pi^3}\right)^{1/4} y\exp(-\alpha r^2) \tag{13.146c}$$

$$g_z(\alpha;r^2)=\left(\frac{128\alpha^5}{\pi^3}\right)^{1/4} z\exp(-\alpha r^2) \tag{13.146d}$$

$$g_{xx}(\alpha;r^2)=\left(\frac{2048\alpha^7}{9\pi^3}\right)^{1/4} x^2\exp(-\alpha r^2) \tag{13.146e}$$

$$g_{yy}(\alpha;r^2)=\left(\frac{2048\alpha^7}{9\pi^3}\right)^{1/4} y^2\exp(-\alpha r^2) \tag{13.146f}$$

$$g_{zz}(\alpha;r^2)=\left(\frac{2048\alpha^7}{9\pi^3}\right)^{1/4} z^2\exp(-\alpha r^2) \tag{13.146g}$$

$$g_{xy}(\alpha;r^2)=\left(\frac{2048\alpha^7}{\pi^3}\right)^{1/4} xy\exp(-\alpha r^2) \tag{13.146h}$$

$$g_{xz}(\alpha;r^2)=\left(\frac{2048\alpha^7}{\pi^3}\right)^{1/4} xz\exp(-\alpha r^2) \tag{13.146i}$$

and

$$g_{yz}(\alpha;r^2)=\left(\frac{2048\alpha^7}{\pi^3}\right)^{1/4} yz\exp(-\alpha r^2). \tag{13.146j}$$

Notice that the first four GTOs g_s, g_x, g_y, and g_z have the s, p_x, p_y, and p_z angular symmetries of the real forms of atomic orbitals and of real STOs. However, none of the six higher order GTOs g_{xx} and so on have proper atomic orbital angular symmetry. This is not a serious problem since linear combinations of the six higher GTOs can be constructed which have correct s, p, d angular symmetries (Problem 13.33).

There is however one remaining important question concerning the use of GTOs as basis functions. Atomic orbitals and real STOs have an exponential r-dependence while GTOs have a gaussian r-dependence. These differences qualitatively spoil the resemblances between GTOs and both atomic orbitals and real STOs at both small r and large r.

At small r, GTOs lack the cusps of s-type atomic orbitals (see Figure 10.3) and real STOs since these cusps arise from the $r = 0$ discontinuities of the first derivatives of exponentially decaying functions. At large r, GTOs decay much more slowly than atomic orbitals (Figure 11.2) or real STOs. These qualitative differences in form between GTOs and atomic orbitals are serious since, as we have emphasized, the members of a good set of basis functions must resemble atomic orbitals.

Despite this deficiency of GTOs, we cannot abandon them since they are the key to solving the two-electron integral problem.

What is often done is to replace the single or *primitive* GTOs of Equations (13.146) by *contracted* GTOs or GTO contractions which are linear combinations of the primitives chosen to approximately simulate the forms of real STOs. While the simulations are not perfect (but can be systematically improved, at increasing cost, by including more and more primitives in the contractions), the errors from the imperfections are of small importance compared to the vast advantages arising from the use of GTO rather than real STO basis functions.

We are now finally ready to describe actual *ab initio* molecular calculations. Each calculation requires a quantum method (e.g., the Hartree–Fock–Roothaan method, CI, or one of the MP perturbation theory methods) and a *basis set*.

For our purposes, a basis set is a collection of n well-behaved linearly independent functions which can be used to expand molecular orbitals $\Phi_i(\mathbf{r};\mathbf{R})$ according to Equation (13.125). For the hypothetical case of a complete basis set $(n = \infty)$, all of the molecule's molecular orbitals can be expanded exactly. For the realistic case of an incomplete basis set $(n < \infty)$, only some of the molecular orbitals can be expanded and each can only be expanded approximately. Since many linearly independent well-behaved sets of functions which can expand molecular orbitals exist, many distinct basis sets also exist (Helgaker, Jorgensen, and Olsen 2004, chap. 6). Here however, we will only consider the two basis sets useful for *ab initio* molecular calculations. These are the real STO and the GTO basis sets.

Many GTO basis sets are available in the literature. Here however, we will focus on the widely used basis sets developed starting in the 1960s by Pople (Hehre, Radom, Schleyer, and Pople 1986, chap. 4). Pople's basis sets form a comprehensive hierarchy designed to treat *ab initio* problems at increasingly high levels of sophistication and accuracy.

We begin with Pople's *minimal basis sets.* These are Pople's earliest, crudest, least accurate, but also least costly basis sets. While minimal basis sets are very important historically, they are so primitive that they currently are not used to produce publication quality research but now are mainly employed to give qualitative preliminary results.

Minimal basis sets are strictly defined as the collection of a molecule's nuclear centered atomic orbitals which can accommodate the electrons of each of the molecule's atoms.

Before showing how to construct minimal basis sets for arbitrary molecules, we first explain the perhaps murky concept of an accommodating atomic orbital. We start with the simplest case of the atoms in the first period of the periodic table, H and He. Both of these atoms have only a single accommodating orbital, the 1s orbital. This is because the 1s orbital is sufficient to accommodate both the single electron of H and the two electrons of He. Next, let us move on to the second-period atoms Li and Be and B through Ne. The accommodating orbitals of Li and Be are 1s and 2s. This is so since the 1s orbital can accommodate the two core electrons of both atoms and since the 2s orbital can accommodate both the single valence electron of Li and the two valence electrons of Be. But something new happens for the next second-period atom B with three valence electrons. These three electrons cannot be accommodated by the 2s orbital because of the Pauli exclusion principle. Thus, B must have an accommodating 2p orbital. However, since the $2p_x, 2p_y$, and $2p_z$ orbitals are degenerate, we will add all three to accommodate the third valence electron of B. Therefore, the orbitals needed to accommodate all five electrons of B are $1s, 2s, 2p_x, 2p_y$, and $2p_z$. These orbitals also accommodate all of the electrons of the remaining second-period atoms. So the accommodating orbitals of the second-period atoms are 1s and 2s for Li and Be and $1s, 2s, 2p_x, 2p_y$, and $2p_z$ for B through Ne.

But there is one additional complication. For free Li and Be atoms, the 2s orbital is lower in energy than the three degenerate 2p orbitals (Section 12.9). However, for Li and Be atoms in a molecular environment the orbital energy ordering may reverse and so the valence electrons of Li and Be in a molecular environment may be found to occupy 2p as well as 2s orbitals. To account for this, the accommodating orbitals of Li and Be must be changed from 1s and 2s to $1s, 2s, 2p_x, 2p_x, 2p_y$, and $2p_z$. With this change, the accommodating orbitals for *all* second-period atoms Li through Ne are $1s, 2s, 2p_x, 2p_y$, and $2p_z$.

These arguments may be easily extended to the atoms in the higher periods of the periodic table. For example, for the fourth period the accommodating orbitals for K and Ca are $1s, 2s, 2p_x, 2p_y, 2p_z, 3s, 3p_x, 3p_y, 3p_z, 4s, 4p_x, 4p_y$, and $4p_z$ while for Sc through Kr they are $1s, 2s, 2p_x, 2p_y, 2p_z, 3s, 3p_x, 3p_y, 3p_z, 3d_{xy}, 3d_{xz}, 3d_{yz}, 3d_{z^2}, 3d_{x^2-y^2}, 4s, 4p_x, 4p_y$, and $4p_z$.

We may now determine the minimal basis set for any molecule (Problem 13.34). It is just the collection of accommodating atomic orbitals for each atom in the molecule. For example, for HCl the accommodating orbital for H is $1s_H$ while the accommodating orbitals for Cl are $1s_{Cl}, 2s_{Cl}, 2p_{xCl}, 2p_{yCl}, 2p_{zCl}, 3s_{Cl}, 3p_{xCl}, 3p_{yCl}$, and $3p_{zCl}$. Therefore, the minimal basis set for HCl is $1s_H, 1s_{Cl}, 2s_{Cl}, 2p_{xCl}, 2p_{yCl}, 2p_{zCl}, 3s_{Cl}, 3p_{xCl}, 3p_{yCl}$, and $3p_{zCl}$, a total of ten atomic orbital basis functions. For the polyatomic molecules of primary interest, the number of atomic basis functions in their minimal basis sets can be much larger than ten.

Pople approximated the atomic orbitals which comprise a minimal basis set by real STOs. From Table 4.1 of Hehre, Radom, Schleyer, and Pople, each real STO is specified by an orbital exponent $\varsigma_{\beta n}$, where β denotes the atom on which the real STO is centered and where n is the principle quantum number of the atomic orbital corresponding to the real STO. A large tabulation of $\varsigma_{\beta n}$ values exists (Hehre, Radom, Schleyer, and Pople 1986, table 4.1). From this tabulation, real STO minimal basis sets may be constructed for a host of molecules.

The next step in our development of the minimal basis set theory is to convert computationally intractable real STO minimal basis sets into approximate computationally tractable GTO minimal basis sets. To make this conversion, in a general way we introduce new notation for both real STO and GTO basis functions. First, we write each real STO as $\chi_{\beta n\ell X}(\varsigma_{\beta n};\mathbf{r})$ where n and ℓ are the principle and angular momentum quantum numbers of the real STO and X is an index which distinguishes between real STOs with different angular symmetries. For example, for, respectively, $s-, p_x-, p_y-$, and p_z −type real STOs $X = 1, x, y$, and z.

In this new notation, the $1s_H$ member of the HCl minimal basis set is written as $\chi_{H1\theta1}(\varsigma_{H1})$, while the $3p_{zCl}$ member is written as $\chi_{Cl31Z}(\varsigma_{Cl3};\mathbf{r})$.

Next, let us turn to the GTOs which we write as $g_Y(\alpha_{\beta Y};r^2)$ where again β is an index which tells which atom the GTO is centered on and where Y is an index which distinguishes between different GTO types. From Equations (13.146), Y can take on the values s, x, y, z, xx, and so on. For example, the normalized s-type GTO centered on a Cl atom is denoted by $g_s(\alpha_{Cls};r^2)$.

We will also use the symbol Y for a summation index with the possible values $Y = 1, 2, \ldots, m, \ldots$. The value $Y = m$ indicates the mth GTO type. Thus, $Y = 1, 2, 3, \ldots$ indicate, respectively, the GTO types s, x, y....

We may now make the STO–GTO conversion which is the critical step of the minimal basis set theory. Specifically, we approximate each member $\chi_{\beta n\ell X}(\varsigma_{\beta n};\mathbf{r})$ of a molecule's real STO minimal basis set by a simulating contracted GTO built as a linear superposition of K=1, 2, ... GTO primitives $g_Y(\alpha_{\beta Y};r^2)$. That is, we approximate $\chi_{\beta n\ell X}$ as the GTO contraction

$$\chi_{\beta n\ell X}(\varsigma_{\beta n};\mathbf{r}) \doteq \varsigma_{\beta n}^{3/2}\sum_{Y=1}^{K} C_{\beta n\ell XY}\, g_Y(\alpha_{\beta Y};r^2) \qquad (13.147)$$

(The origin of the factor $\varsigma_{\beta n}^{3/2}$ is explained on pp. 68–69 of Hehre, Radom, Schleyer, and Pople.) The minimal basis set theory which emerges from Equation (13.147) is in Pople's terminology called STO-KG. The quality of the theory increases with the number K of GTO primitives g_Y included in the GTO contraction. In actual calculations, $K = 2-6$. So in practice, the least and most accurate minimal basis set theories are STO-2G and STO-6G.

To go on, we note that the real STOs $\chi_{\beta n\ell X}$ comprise a basis set which for appropriate choices of $\beta n\ell X$ is the molecule's minimal basis set (Problem 13.35). This minimal basis set may be used to expand the molecule's molecular orbitals ϕ_i by the following adaptation of Equation (13.125):

$$\phi_i(\mathbf{r};\mathbf{R}) = \sum C_{i\beta n\ell X}(\mathbf{R})\chi_{\beta n\ell X}(\varsigma_{\beta n};\mathbf{r}) \qquad (13.148)$$

where the sum is over all $\beta n\ell X$ values for which the real STOs $\chi_{\beta n\ell X}$ are members of the molecule's minimal basis set. The expansion coefficients $C_{i\beta n\ell X}$ may apparently be found in the usual manner as solutions of the Hartree–Fock–Roothaan equations. With these solutions, Equation (13.148) yields the real STO minimal basis set approximations to the molecule's Hartree–Fock molecular orbitals. This scheme however fails since the two-election integrals in the minimal basis set Hartree–Fock–Roothaan equations are over the real STOs $\chi_{\beta n\ell X}$ and are therefore intractable.

The way out of this impasse is provided by Equation (13.147) which implies that a real STO minimal basis set may be approximated by a collection of GTO contractions, one contraction for each member of the real STO minimal basis set. This collection may be used as a basis set to form new Hartree–Fock–Roothaan equations. These new equations are solvable since their two-election integrals are over the new basis set of GTO contractions and hence may be reduced to analytically evaluatable integrals over GTO primitives. The solutions of the new Hartree–Fock–Roothaan equations yield the STO-KG approximations to the molecule's already approximate real STO minimal basis set Hartree–Fock molecular orbitals. From these STO-KG molecular orbitals, one may construct an approximate Slater determinantal wave function Ψ for the molecule. From Ψ, one may obtain approximations to all of the molecule's Hartree–Fock properties. These approximate molecular properties are the central results of the STO-KG theories. We however are not yet finished with our outline of the STO-KG procedure. This is since we still do not know the values of the expansion coefficients $C_{\beta n\ell XY}$ and the GTO exponents $\alpha_{\beta Y}$ which are needed since they determine the new basis set of GTO contractions from Equation (13.147) which eventually yield the STO-KG molecular orbitals. To obtain the needed parameters in practice, one uses the least squares fitting method to determine these parameters by fitting the unknown right-hand side of Equation (13.147) to its known left-hand side (Hehre, Radom, Schleyer, and Pople 1986, sec. 4.31a).

Next, let us look at some results of the STO-KG theories. In Table 13.2, we give for formaldehyde the convergence in K for $K = 2 - 6$ of the electronic energy E(au) and three equilibrium geometrical properties. The theory converges slowly for E yielding the values in au of –109.0244 $(K = 2)$, –113.1611 $(K = 4)$, and –113.4408 $(K = 6)$. The formaldehyde geometrical properties converge much more rapidly. For example, for $K = 4$ the CO bond length (pm), the CH bond length (pm), and the HCH bond angle (degrees) have the respective values 121.6, 109.9, and 114.5, while for $K = 6$ these properties have the very similar values 121.6, 109.8, and 114.5.

The relative computation times of the STO-KG theories increases rapidly with the number of primitive GTO basis functions per atom used in the computations and hence with the number of atoms and electrons in the molecule. For large molecules, the STO-3G theory seems to be close to an optimal compromise between accuracy and cost.

Finally, while the formaldehyde equilibrium geometry converges quite rapidly with K, this does not imply that the STO-KG models are satisfactory theories of molecular electronic structure. In

TABLE 13.2
Formaldehyde Energy E(au), Equilibrium Bond Lengths r_e (pm) and Bond Angles θ_e (Degrees) Computed from STO-KG Minimal Basis Set Theories for $K = 2 - 6$

Property	K = 2	K = 3	K = 4	K = 5	K = 6
E	−109.0244	−112.3525	−113.1611	−113.3752	−113.4408
r_e (CO)	122.0	121.7	121.6	121.6	121.6
r_e (CH)	111.0	110.1	109.9	109.8	109.8
θ_e (HCH)	113.2	114.5	114.5	114.8	114.8

Source: Adapted from Table 4.2 of W.J. Hehre, L. Radom, P.v.R Schleyer, and J. A. Pople. 1986. *Ab initio molecular orbital theory*. Irvine, CA: John Wiley & Sons.

fact, the frequent gross disagreement of their predictions with experiment shows that they are actually quite poor approximations.

So we next move on to more accurate, but of course, also more costly, Pople theories. We first note that the STOs used to build the STO-KG minimal basis set theories have dropped out of the more advanced theories. In these, the basis sets are comprised solely of GTO primitives and molecular properties are determined from Hartree–Fock–Roothaan equations derived from purely GTO basis sets.

We begin with the simplest improvement of the minimal basis set theories. This improvement is based on the realization that an important source of error in these models arises because they are based on GTOs with *fixed* orbital exponents. Because of these fixed exponents, the GTOs cannot properly change their radial extents in response to changes in molecular environments. The remedy for this problem is to replace each valence GTO (core GTOs have little influence on chemical bonding) by two GTOs of the same symmetry but with very different orbital exponents and hence very different radial extents. We refer to GTO with the smaller radial extent and hence the larger exponent as *compact,* and we call the GTO with larger radial extent and hence the smaller exponent *diffuse*. Mixing these two GTOs by the Hartree–Fock–Roothaan method produces a GTO of intermediate radial extent with this extent dependent on the molecular environment (Hehre, Radom, Schleyer, and Pople 1986, pp. 73 and 75).

The doubling of the number of valence GTOs to adjust radial extents produces what are called *split-valence* basis sets. The two most important *split-valence* basis sets are named in Pople's terminology the 6-21G basis set, which is designed to treat smaller molecules, and the simpler and less costly 3-21G basis set, which is designed to treat larger molecules. In the 6-21G basis set, all inner shell atomic orbitals of the molecule are modeled by contracted GTOs each comprised of six (thc 6 in 6-21G) compact GTO primitives, while the valence atomic orbitals are approximated both by contracted GTOs each comprised of two (the 2 in 6-21G) compact GTO primitives and by a single (the 1 in 6-21G) diffuse GTO primitive (Problem 13.36). Having both compact and diffuse valence GTOs in the basis set is how the 6-21G model properly accounts for variable radial valence orbital extents.

The 3-21G and 6-21G basis sets differ only in that for the former the inner shell atomic orbitals are modeled by three rather than six compact GTO primitives.

Comparisons of several properties of fluoromethane computed from the 3-21G and 6-21G basis sets are available (Hehre, Radom, Schleyer, and Pople 1986, Table 4.3). Despite the greater size of the 6-21G basis set for all properties other than the energy, the predictions of the two models are nearly identical.

Finally, we have not yet said how the primitive-to-contracted GTO expansion coefficients or the GTO primitive orbital exponents are determined in the 3-21G and 6-21G theories. The procedure is variational energy minimization (Hehre, Radom, Schleyer, and Pople 1986, p. 77). Since the details are complex, we will omit them.

We next turn to a still more sophisticated (and costly) group of Pople basis sets. These are designed to treat the phenomenon of *orbital polarization*, a feature of molecular electronic structure that we have not yet dealt with. These new basis sets are called polarization basis sets. In Pople's notation, polarization basis sets are labeled by one or two star superscripts.

The most widely used polarization basis sets are the 6-31G* and 6-31G** basis sets. Both are built from the 6-31G split-valence basis set. The 6-31G basis set is identical to the 6-21G basis set except that for the former there are three rather than two compact GTOs modeling each valence atomic orbital. Before specifying the above polarization basis sets, we explain the concept of orbital polarization. We earlier stated that the charge of a molecular basis function, normally a GTO, is centered on one of the molecule's nuclei. This is actually an oversimplification. For some molecules, for example, highly polar molecules, the center of charge of a GTO is slightly displaced away from a molecular nucleus (Hehre, Radom, Schleyer, and Pople 1986, sec 4.3.3). This slight displacement of GTO charge density is the phenomenon of orbital polarization.

The best way to treat orbital polarization is to add primitive GTOs to a parent split-valence basis set which are of higher angular symmetry than the GTOs in the parent set. (See schematic illustrations of this concept on p.81 of Hehre, Radom, Schleyer, and Pople). These augmented split-valence basis sets are the polarization basis sets.

For example, the 6-31G* polarization basis set is constructed by adding to the 6-31G basis set six compact d-type GTO primitives for each heavy (non-first period) atom in the molecule. The 6-31G** basis set is built by augmenting the 6-31G* basis set to include a diffuse p-type GTO primitive for each hydrogen and helium atom in the molecule (Problem 13.37).

A very sophisticated but costly polarization basis set is the 6-311G** basis set. This set is formed by augmenting the parent 6-311G split-valence basis set (Problem 13.38) with five compact d-type GTO primitives for every second-period atom in the molecule and a diffuse p-type GTO primitive for each hydrogen atom in the molecule (Problem 13.39).

Next, let us look at some results. In Table 13.3, we compare with experiment equilibrium bond lengths $r_e\,(\text{pm})$ and equilibrium bond angles θ_e (degrees), determined from Hartree–Fock–Roothaan equation calculations for several of the basis sets we have discussed for eight small molecules. Globally, the agreement with experiment increases in the rough order polarization basis sets > split-valence basis sets > STO-KG basis sets. For example, for the LiH bond length in pm the detailed results are 151.8 (STO-3G), 164.0 (3-21G), 163.6 (6-31G*), 162.3 (6-31G**), 160.7 (6-311G**), and 159.2 (expt).

So far all of the results we have presented are based solely on the Hartree–Fock–Roothaan equations and therefore do not include the effects of electron correlation. Next, we turn to post-Hartree–Fock quantum methods which yield the shifts in the values of Hartree–Fock–Roothaan molecular properties due to electron correlation. We will focus on the two simplest post-Hartree–Fock methods, MP perturbation theory and CI.

MP perturbation theory is the standard time-independent perturbation theory developed in Section 11.1 with the unperturbed Hamiltonian operator chosen as the molecular Fock operator $\hat{F}$ of Equation (13.120). The first, second, third, fourth … orders of MP perturbation theory are denoted by MP1, MP2, MP3, MP4, …. The MP1 wave function is the molecule's Hartree–Fock–Roothaan wave function. Therefore, MP1 perturbation theory is of no interest since it returns only Hartree–Fock–Roothaan molecular properties. However, electron correlation shifts occur at the MP2, MP3, MP4 … orders of perturbation theory. So these are the levels of theory used in actual calculations.

We will not go into the details of MP perturbation theory since, especially for higher orders, they are complex (Hehre, Radom, Schleyer, and Pople 1986, sec 2.9.3). Suffice to say that to implement

TABLE 13.3
Hartree–Fock–Roothaan Equilibrium Bond Lengths $r_e\,(\text{pm})$ and Bond Angles θ_e (Degrees) Computed Using Increasingly Sophisticated Basis Sets Compared with Experiment

Molecule	STO-3G	3-21G	6-31G*	6-31G**	6-311G**	Expt.
$H_2\,(r_e)$	71.2	73.5	73.0	73.2	73.8	74.2
$LiH\,(r_e)$	151.0	164.0	163.6	162.3	160.7	159.1
$CH_4\,(r_e)$	108.3	108.3	108.4	108.4	108.4	109.2
$NH_3\,(r_e)$	103.3	100.3	100.2	100.2	100.2	101.2
$NH_3\,(\theta_e)$	104.2	112.4	107.2	107.2	107.4	106.7
$H_2O\,(r_e)$	99.0	96.7	94.7	94.7	94.1	95.5
$H_2O\,(\theta_e)$	100.00	107.6	105.5	105.9	105.4	104.5
$HF\,(r_e)$	95.6	93.7	91.1	90.1	89.6	91.7

Source: Excerpted from Table 6.1 of W. J. Hehre, L. Radom, P. v. R Schleyer, and J. A. Pople. 1986. *Ab initio molecular orbital theory*. Irvine, CA: John Wiley & Sons.

each order of MP perturbation theory, one requires a basis set and so both the order and the basis set must be denoted to specify an MP treatment. The notation used for such a specification is illustrated by the following example. An MP2 calculation based on a 6-31G* basis set is denoted by MP2/6-31G*. Analogous notation is used for other quantum methods. Thus, when confusion is possible, a full specification of a Hartree–Fock–Roothaan treatment is used. For example, the full specification of a Hartree–Fock–Roothaan calculation employing a 6-21G basis set is denoted by HF/6-21G.

Next, we look at an application of MP perturbation theory. In Table 13.4, we compare with experiment the results for the equilibrium geometric properties $r_e\,(\mathrm{pm})$ and θ_e (degrees) for several small molecules obtained from an MP4/6-311G** computation. This is a very advanced treatment since MP4 is the highest order MP perturbation theory commonly used and since 6-311G** is a very sophisticated basis set. So it is not surprising that the theoretical and experimental results shown in Table 13.4 are in excellent agreement (error $\leq 5\%$) for all properties of the listed molecules. These results illustrate how accurate *ab initio* treatments can be if sufficiently advanced theoretical models are employed

We next turn to a second post-Hartree–Fock method CI. The principles underlying CI are very simple (Hehre, Radom, Schleyer, and Pople 1986, secs. 2.91 and 2.92).

Namely, there are $n > N$ (where recall n is the number of basis functions) solutions of the Hartree–Fock–Roothaan equations, not just the N low-energy solutions, the molecular orbitals $\phi_1,\ldots,\phi_N$ used to construct the Hartree–Fock–Roothaan Slater determinantal wave function Ψ_0. We will call the additional solutions *virtual molecular orbitals* and denote them by $\chi_1,\ldots,\chi_{n-N}$.

One may modify the Hartree–Fock–Roothaan determinantal wave function Ψ_0 by replacing in the determinant one or more of its 2N spin orbitals $\phi_i\alpha$ or $\phi_i\beta$ by virtual spin orbitals $\chi_i\alpha$ or $\chi_i\beta$. Let us illustrate this replacement for the two-electron H_2 molecule with $N = 1$. For H_2, the Hartree–Fock–Roothaan wave function

$$\Psi_0 = \frac{1}{\sqrt{2}}\begin{vmatrix} \phi_1(\mathbf{r}_1;R)\alpha(r_1) & \phi_1(\mathbf{r}_1;R)\beta(\sigma_1) \\ \phi_1(\mathbf{r}_2;R)\alpha(r_2) & \phi_1(\mathbf{r}_2;R)\beta(\sigma_2) \end{vmatrix} \tag{13.149}$$

TABLE 13.4
Sophisticated MP4/6-311G Computations of Equilibrium Bond Lengths $r_e\,(\mathrm{pm})$ and Bond Angles θ_e (Degrees) Compared with Experiment**

Molecule	Theory	Experiment
$H_2\,(r_e)$	74.2	74.2
$LiH\,(r_e)$	159.7	159.6
$CH_4\,(r_e)$	109.4	109.2
$NH_3\,(r_e)$	101.7	101.2
$NH_3\,(\theta_e)$	105.6	106.7
$H_2O(r_e)$	95.9	95.8
$H_2O(\theta_e)$	102.4	104.5
$HF\,(r_e)$	91.3	91.7

Source: Adapted from Table 6.3 of W. J. Hehre, L. Radom, P. v. R. Schleyer, and J. A. Pople. 1986. *Ab initio molecular orbital theory*. Irvine, CA: John Wiley & Sons.

where ϕ_1 is the $H_2\sigma_{1g}$ Hartree–Fock–Roothaan molecular orbital. We first illustrate the replacement procedure for single substitutions. For example, making the replacement $\phi_1\alpha \rightarrow \chi_1\alpha$ in Ψ_0 yields the singly substituted determinant

$$\Psi_1 = \frac{1}{\sqrt{2}}\begin{vmatrix} \chi_1(\mathbf{r}_1;R)\alpha(\sigma_1) & \phi_1(\mathbf{r}_1;R)\beta(\sigma_1) \\ \chi_1(\mathbf{r}_2;R)\alpha(\sigma_2) & \phi_1(\mathbf{r}_2;R)\beta(\sigma_2) \end{vmatrix}. \tag{13.150}$$

This replacement process may be carried out for all n – N virtual orbitals yielding the singly substituted determinants $\Psi_1,\ldots,\Psi_{n-N}$ where, for example,

$$\Psi_m = \frac{1}{\sqrt{2}}\begin{vmatrix} \chi_m(\mathbf{r}_1;R)\alpha(\sigma_1) & \phi_1(\mathbf{r}_1;R)\beta(\sigma_1) \\ \chi_m(\mathbf{r}_2;R)\alpha(\sigma_2) & \phi_1(\mathbf{r}_2;R)\beta(\sigma_2) \end{vmatrix}. \tag{13.151}$$

One may now form a linear combination of the Hartree–Fock–Roothaan wave function Ψ_0 and the n – N singly substituted determinants $\Psi_1,\ldots,\Psi_{n-N}$. This linear combination is

$$\Psi_{CIS} = C_0\Psi_0 + C_1\Psi_1 + \cdots + C_{n-N}\Psi_{n-N}. \tag{13.152}$$

The expansion coefficients in Equation (13.152) may be obtained by variational minimization of the H_2 energy. With this determination, Ψ_{CIS} becomes a proper normalized wave function of H_2. Since it is a CI wave function derived from single orbital replacements $\phi_m\alpha \rightarrow \chi_m\alpha$, it is referred to as a CI singles or CIS wave function. The CIS theory is not restricted to H_2 but can be readily generalized to any closed shell molecule.

Ψ_{CIS} apparently delivers an electron correlation contribution to the properties of a molecule since it includes the determinants $\Psi_1,\ldots,\Psi_{n-N}$ which correct the Hartree–Fock–Roothaan wave function Ψ_0. This is however not the case. Rather, a quantum theorem shows that unobviously, Ψ_{CIS} predicts Hartree–Fock–Roothaan molecular properties.

So to obtain electron correlation corrections to Hartree–Fock–Roothaan properties, one must use the next simplest CI theory called CI doubles or more briefly CID. CID and CIS differ only in that the non-Hartree–Fock determinants in the CI expansion are formed by replacing the spin orbitals in the Hartree–Fock–Roothaan wave function Ψ_0 by two (rather than one) sets of virtual spin orbitals. As an example for H_2, a typical non-Hartree–Fock–Roothaan CID determinant is

$$\Psi_{12} = \frac{1}{\sqrt{2}}\begin{vmatrix} \chi_1(\mathbf{r}_1;R)\alpha(\sigma_1) & \chi_2(\mathbf{r}_1;R)\beta(\sigma_1) \\ \chi_1(\mathbf{r}_2;R)\alpha(\sigma_2) & \chi_2(\mathbf{r}_2;R)\beta(\sigma_2) \end{vmatrix}. \tag{13.153}$$

We next note that the results of a CI calculation depend on the basis set used to solve the Hartree–Fock–Roothaan equations since the forms of the Hartree–Fock–Roothaan ground state and virtual molecular orbitals are basis set dependent. So to describe a CI calculation, one must specify the basis set used. An example of the notation used for this specification is CID/6-31G*, which means that the treatment is a CI doubles computation which employs a 6-31G* basis set.

Next, let us look at the results of some MP and CID 6-31G* calculations and compare these results with those of Hartree–Fock–Roothaan calculations for the 6-31G* basis set and experiment. The results in question given in Table 13.5 are for r_e (pm) and $\theta_\uparrow$ (degrees) for several small molecules. Notice that the results predicted by the MP2, MP3, and MP4 perturbation theories are very similar and are also very similar to those of the CID treatment. These results show that at least for the molecules listed, predicted results for molecular geometries are very insensitive to the choice of post-Hartree–Fock method and do not improve significantly if one increases that order of the MP perturbation theory. Also the Hartree–Fock–Roothaan and post-Hartree–Fock results differ only slightly showing that at least for the listed molecules molecular geometries depend only slightly

TABLE 13.5
Hartree–Fock–Roothaan and Correlated Electron 6-31G* Basis Set Results for Equilibrium Bond Lengths r_e (pm) and Bond Angles θ_e (Degrees) Compared with Experiment

Molecule	Hartree–Fock–Roothaan	MP2	MP3	MP4	CID	Expt.
H_2 (r_e)	73.0	73.8	74.2	74.4	74.6	74.2
LiH (r_e)	163.6	164.0	164.3	164.8	164.9	159.6
CH_4 (r_e)	108.4	109.0	109.1	109.4	109.1	109.2
NH_3 (r_e)	100.2	101.7	101.7	102.1	101.6	101.2
NH_3 (θ_e)	107.2	106.3	106.2	105.8	106.3	106.7
H_2O (r_e)	94.7	96.9	96.7	97.0	96.6	95.8
H_2O (θ_e)	105.5	104.0	104.5	103.9	104.3	104.5
HF (r_e)	91.1	93.4	93.2	93.5	93.1	91.7

Source: Excerpted from Table 6.2 of W. J. Hehre, L. Radom, P. v. R Schleyer, and J. A. Pople. 1986. *Ab initio molecular orbital theory*. Irvine, CA: John Wiley & Sons.

on electron correlation effects. Finally, all of the theoretical results agree well with experiment. This shows that even at the relatively low level of HF/6-31G* theory, molecular geometries are reasonably predicted, at least for the listed molecules.

FURTHER READINGS

Atkins, Peter, Julio de Paula, and Ronald Friedman. 2014. *Physical chemistry, quanta, matter and change*. 2nd ed. New York: W.H. Freeman and Company.

Boys, S.F. "Electronic Wave Functions. I. A General Method of Calculation for Stationary States of Any Molecular System." *Proc. Royal Soc. (London)*. A200 (1950): 542–554.

Cohen-Tannoudji, Claude, Bernard Diu, and Franck Laloe. 1977. *Quantum mechanics*. Vol. 2. Translated by Susan Reid Hemley, Nicole Ostrowsky, and Dan Ostrowsky. New York: John Wiley & Sons.

Hagstrom, S., and H. Schull. "The Nature of the Two-Electron Chemical Bond III. Natural Orbitals for H_2." *Rev. Mod. Phys*. 35 (1963): 624.

Hehre, Warren J., Leo Radom, Paul v.R. Schleyer, and J.A. Pople. 1986. *Ab initio molecular orbital theory*. Irvine, CA: John Wiley & Sons.

Helgaker, Trygve, Paul Jorgensen, and Jeppe Olsen. 2012. *Molecular electronic-structure theory*. Hoboken, NJ: John Wiley & Sons Inc.

Hohenberg, Pierre and Walter Kohn. "Inhomogeneous Electron Gas." *Phys. Rev*. 136 (1964): B864–B870.

Kohn, Walter and Lu Ju Sham. "Self-Consistent Equations Including Exchange and Correlation Effects." *Phys. Rev*. 140 (1965): A1133–A1138.

Kolos, W. and L. Wolniewicz. "Accurate Adiabatic Treatment of the Ground State of H_2." *J. Chem. Phys*. 41 (1964): 3663.

Kolos, W. and L. Wolniewicz. "Potential Energy Curves for the $X\sum_g^+, b\sum_u^+$, and $C^1\Pi_u^+$ States of the Hydrogen Molecule." *J. Chem. Phys*. 43 (1965): 2429.

Kolos, W. and L. Wolniewicz. "Improved Theoretical Ground-State Energy of the Hydrogen Molecule." *J. Chem. Phys*. 49 (1968): 404–410.

McLean, A.D., A. Weiss, and M. Yoshimine. "Configuration Interaction in the Hydrogen Molecule." *Rev. Mod. Phys*. 32 (1960): 211.

McQuarrie, Donald A. 2008. *Quantum chemistry*. 2nd ed. Mill Valley, CA: University Science Books.

Pauling, Linus, and E. Bright Wilson Jr. 1985. *Introduction to quantum mechanics with applications to chemistry*. New York: Dover Publications, Inc.

Roothaan, C.C.J. "New Developments in Molecular Orbital Theory." *Rev. Mod. Phys*. 23 (1951): 69–89.

Silberberg, Martin S. 2006. *Chemistry: the molecular nature of matter and change*. 4th ed. Boston, MA: McGraw-Hill.

PROBLEMS

13.1 Compute at $T-300\,K$ the average speed $v=\left(\frac{3kT}{m}\right)^{1/2}$ of both an electron with mass $m_e=9.109\times10^{-31}$ kg and a nucleus with mass $m_N=9m_p$ where $m_p=1{,}836m_e$ is the mass of a proton. Note $k=1.381\times10^{-23}\,JK^{-1}$ is Boltzmann's constant.

13.2 Write down for benzene C_6H_6 the electron–nuclear Hamiltonian operator $\hat{H}_{en}$ in atomic units. Use summation notation.

13.3 Write down the nuclear Schrödinger for ethene $H-C\equiv C-H$. Denote the position vectors of the hydrogen and carbon nuclei by, respectively, $\mathbf{R}_{H1}$ and $\mathbf{R}_{H2}$ and $\mathbf{R}_{C1}$ and $\mathbf{R}_{C2}$.

13.4 Show that the ranges of the elliptical coordinates given in Equation (13.11) follow from their definitions of Equation (13.10).

13.5 Prove both relations given in Equation (13.16).

13.6 Derive Equation (13.17). (For help on this difficult problem, see the text by Cohen-Tannoudji, Diu, and Laloe, pp. 1171-2, referenced at the end of this chapter.)

13.7 Derive Equations (13.36).

13.8 (a) Show $C_{A+}=C_{B+}$ and $C_{A-}=-C_{B-}$. (b) Prove Equation (13.38).

13.9 Derive Equations (13.41)–(13.46).

13.10 Derive Equations (13.47)–(13.49).

13.11 Derive Equations (13.50)–(13.52).

13.12 Derive Equations (13.58)–(13.61).

13.13 Derive Equation (13.62).

13.14 Derive Equations (13.65)–(13.71).

13.15 Derive Equations (13.72)–(13.74) by the shortcut method described in the text.

13.16 Derive Equations (13.75) and (13.76).

13.17 Show that the average ground state kinetic energy $\langle T\rangle$ of H_2^+ is within the variable Z molecular orbital theory $\langle T\rangle=\frac{\frac{1}{2}-\frac{1}{2}S(Z)+k(Z)}{1+S(Z)}$.

13.18 Derive Equation (13.83).

13.19 (a) Derive the transcendental equation which determines the optimal Z as a function of R. (b) Show from this equation that at $R=0$, the optimal Z is the He^+ value 2, while as $R\to\infty$, the optimal Z approaches the H value 1.

13.20 Show that for $v=J=0$, $D_0(v,J)\equiv E_{diss}-E_{vJ}=D_0$ of Equation (7.120).

13.21 Explain why E_{diss} is isotope independent.

13.22 Derive the form of the Heitler–London normalization factor N in Equation (13.97).

13.23 Determine the valence electron configurations, bond orders, and magnetism of the homonuclear diatomics Li_2 through F_2.

13.24 Assuming that their molecular orbital energy ordering is that of F_2, determine the valence electron configurations, bond orders, and magnetism of F_2^+, F_2, and F_2^-.

13.25 Repeat Problem 13.24 for B_2^+, B_2, and B_2^-.

13.26 (a) Show that qualitative molecular orbital theory predicts that C_2 and O_2 are both doubly chemically bonded. (b) Experimentally, the bond dissociation energies of C_2 and O_2 are, respectively, $D_0=348\,kJ\,mol^{-1}$ and $D_0=145\,kJ\,mol^{-1}$. Reconcile this experimental result with your theoretical result in part (a).

13.27 Derive Equation (13.128) from Equation (13.127).

13.28 Derive Equation (13.131) from Equation (13.128).

13.29 Derive the Hartree–Fock–Roothaan equations by minimizing Equations (13.116)–(13.118b) for the electronic energy $E(R)$.

13.30 Derive explicit expressions for the density functionals $V=V[\ ]$ and $U_{coul}=U_{coul}[\ ]$.

13.31 Show that the first three real STOs of Equation (13.149) are normalized.
13.32 Show that the GTOs of Equation (13.146a–d) are normalized (Hint: Recall Section 7.9 on Gaussian integrals.)
13.33 (a) Show that the linear combinations of GTOs $\frac{1}{2}\left(g_{zz}-g_{xx}-g_{yy}\right)$ and $\left(\frac{3}{4}\right)^{1/2}\left(g_{xx}-g_{yy}\right)$ have d-type angular symmetry. (b) Show that the linear combination of GTOs $\left(\frac{3}{4}\right)^{1/2}\left(g_{xx}-g_{yy}\right)$ has s-type angular symmetry.
13.34 What are the atomic orbital minimal basis sets for the molecules (a) Be F_2 and (b) CO_2?
13.35 What are the values of $\beta n\ell$ and X for each member of the real STO minimal basis set for HF? (Hint: First derive the atomic orbital minimal basis set for HF.)
13.36 For formaldehyde, how many GTOs are included in (a) the 3-21G and (b) 6-21 basis sets?
13.37 For fluoromethane, how many GTOs are included in (a) the 6-31G* and (b) 6-31G** basis sets?
13.38 For water, how many GTOs are included in the 6-311G basis set?
13.39 For water, how many GTOs are included in the 6-311G** basis set?

SOLUTIONS TO SELECTED PROBLEMS

13.1 Average speed of electron $=1.168\times10^{5}\,\mathrm{m\,s^{-2}}$. Average speed of nucleus $=9.09\times10^{2}\,\mathrm{m\,s^{-1}}$.
13.2 Denoting the carbon and hydrogen atoms by, respectively, the subscripts C_i and H_i for benzene $C_6H_6\hat{H}_{en}=-\frac{1}{2m_C}\sum_{i=1}^{6}\nabla_{C_i}^2-\frac{1}{2m_p}\sum_{i=1}^{6}\nabla_{Hi}^2-\sum_{i=1}^{12}\sum_{j=1}^{6}\frac{1}{\left(\mathbf{r}_i-\mathbf{R}_{Hj}\right)}$

$$-\sum_{i=1}^{12}\sum_{j=1}^{6}\frac{1}{\left(\mathbf{r}_i-\mathbf{R}_{H_j}\right)}-\frac{1}{2}\sum_{i=1}^{12}\nabla_i^2-\sum_{i=1}^{12}\sum_{j>i}^{12}\frac{1}{\left(\mathbf{r}_{ij}\right)}+\sum_{i=1}^{6}\sum_{j>i}^{6}\frac{1}{\left(\mathbf{R}_{C_i}-\mathbf{R}_{C_j}\right)}+\sum_{i=1}^{6}\sum_{j>i}^{6}\frac{1}{\left(\mathbf{R}_{H_i}-\mathbf{R}_{H_j}\right)}.$$

13.3 Denoting the first and second hydrogen atoms by the subscripts 1 and 2 and the first and second carbon atoms also by the subscripts 1 and 2, the nuclear Schrödinger equation for ethane $H-C\equiv C-H$ is $\left[-\frac{1}{2m_p}\nabla_{H1}^2-\frac{1}{2m_p}\nabla_{H2}^2-\frac{1}{2m_C}\nabla_{C1}^2-\frac{1}{2m_C}\nabla_{C2}^2\right]$ $\times\Psi_m\left(\mathbf{R}_{H1},\mathbf{R}_{H2},\mathbf{R}_{C1},\mathbf{R}_{C2}\right)+U_n\left(\mathbf{R}_{H1},\mathbf{R}_{H2},\mathbf{R}_{C1},\mathbf{R}_{C2}\right)=E_m\Psi_m\left(\mathbf{R}_{H1},\mathbf{R}_{H2},\mathbf{R}_{C1},\mathbf{R}_{C2}\right)$.
13.21 The electronic energies of atoms are isotope independent if one neglects the tiny effects of nuclear motion and relativity.
13.22 $N=\left(\frac{1}{2}\right)^{1/2}\left(1+S_{AB}^2\right)^{1/2}$ where $S_{AB}=\int\Psi_{1s}(r_{1A})\Psi_{1s}(r_{1B})d\mathbf{r}_1$.
13.23 $Li_2;\sigma_{2g}^2,1$, diamagnetic; $Be_2;\sigma_{2g}^2\sigma_{2u}^{*2},0$, diamagnetic; $B_2:\sigma_{2g}^2\sigma_{2u}^{*2}\pi_{1u}^{(1)}\pi_{1u}^{(2)},1$, paramagnetic; $C_2:\sigma_{2g}^2\sigma_{2u}^{*2}\pi_{1u}^{(1)2}\pi_{1u}^{(2)2},2$, diamagnetic; $N_2:\sigma_{2g}^2\sigma_{2u}^{*2}\pi_{1u}^{(1)2}\pi_{1u}^{(2)2}\sigma_{3g}^2,3$, diamagnetic; $O_2:\sigma_{2g}^2\sigma_{2u}^{*2}\sigma_{3g}^2\pi_{1u}^{(1)2}\pi_{1u}^{(2)2}\pi_{1g}^{(2)*}\pi_{1g}^{(2)*},2$, paramagnetic; and $F_2:\sigma_{2g}^2\sigma_{2u}^{*}\sigma_{3g}^2\pi_{1u}^{(1)2}\pi_{2u}^{(2)2}\pi_{1g}^{(1)*2}\pi_{1g}^{(2)*2},1$, diamagnetic.
13.25 $B_2^+:\sigma_{2g}^2\sigma_{2u}^{*2}\pi_{1u}^{(1)},\frac{1}{2}$, paramagnetic; B_2: the same as in Problem 13.23; and $B_2^-:\sigma_{2g}^2\sigma_{2u}^{*2}\pi_{1u}^{(1)2}\pi_{1u}^{(2)},\frac{3}{2}$, paramagnetic.
13.26 (a) Electron configuration of $C_2:\sigma_{2g}^2\sigma_{2u}^{*2}\pi_{1u}^{(1)2}\pi_{1u}^{(2)2}BO=2$ or C_2 doubly chemically bonded. Electron configuration of $O_2:\sigma_{2g}^2\sigma_{2u}^{*2}\sigma_{3g}^2\pi_{1u}^{(1)2}\pi_{1u}^{(2)2}\pi_{1g}^{(1)*}\pi_{1g}^{(2)*}BO=2$ or O_2 doubly chemically bonded. (b) σ_{3g}^2 bond is stronger than the $\pi_{1g}^{(1)*}\pi_{1g}^{(2)*}$ anti-bond.

13.30 $V[\ ]=-\sum_{S=1}^{M}\int\frac{Z_S d\mathbf{r}}{|\mathbf{R}_S-\mathbf{r}|};U_{coul}[\ ]=\int\int\frac{\rho(\mathbf{r}_1)}{|\mathbf{r}_1-\mathbf{r}|}d\mathbf{r}_1 d\mathbf{r}.$

13.34 (a) Atomic orbital minimal basis set for $BeF_2 = BeFF'$: $1s_{Be}, 2s_{Be}, 2p_{xBe}, 2p_{yBe} 2p_{zBe}$, $1s_P, 2s_P, 2p_{xP}, 2p_{yP}, 2p_{zP}, 1s_{P'}, 2s_{P'}, 2p_{xP'}, 2p_{yP'}, 2p_{xP'}$. (b) Atomic orbital minimal basis set for $CO_2 = COO'$: $1s_C, 2s_C, 2p_{xC}, 2p_{yC}, 2p_{zC}, 1s_O, 2s_O, 2p_{xO}, 2p_{yO}, 2p_{zO}, 1s_{O'}, 2s_{O'}, 2p_{xO'}$, $2p_{yO'}, 2p_{zO'}$.

13.35 $H101(1s_H), F101(1s_F), F201(2s_F), F21x(2p_{xF}), F21y(2p_{yF}), F21z(2p_{zF})$.

13.36 (a) 30 and (b) 36.

13.37 (a) 72 and (b) 75.

13.38 (a) 38 and (b) 37.

13.39 44.

Appendix A
Elements of Classical Dynamics

Classical mechanics is the theory of motion developed by Isaac Newton in the last third of the seventeenth century. Classical mechanics successfully describes the motion of particles whose mass is vastly greater than the mass of an electron. But for small particles like atoms and molecules, it fails and we must abandon it for quantum mechanics. However, many concepts and results that may be developed relatively simply in classical mechanics prove useful in the more complex setting of quantum mechanics. Thus, in this appendix we give a synopsis of the elements of classical mechanics.

Before starting, we note that classical mechanics concerns itself with the motion of *dynamical systems*. For the present purposes, a dynamical system may be regarded as either a single point particle or a collection of such particles. Given this definition, the central problem of classical mechanics is easy to state. Determine the position of each particle in the system as a function of time. That is, determine the *trajectory* of the system.

We begin with the simplest dynamical systems: those comprised of a single particle of mass m moving in one dimension (Figure A.1). We will denote the coordinate of the particle by x. Thus, our problem is to determine the particle's trajectory $x(t)$.

A.1 A SINGLE PARTICLE MOVING IN ONE DIMENSION

The trajectory $x(t)$ is determined by Newton's second law of motion $F = ma$. Noting that the acceleration $a(t)$ of the particle at time t is $\frac{d^2x(t)}{dt^2}$ and writing the force F explicitly as $F_x[x(t)]$, we have the following Newton equation of motion for $x(t)$:

$$m\frac{d^2x(t)}{dt^2} = F_x[x(t)]. \tag{A.1}$$

For brevity, we will write the first and second derivatives of functions of time $A(t)$ as $\dot{A}(t) = \frac{dA(t)}{dt}$ and $\ddot{A}(t) = \frac{d^2A(t)}{dt^2}$. When rewritten using this notation, the equation of motion becomes

$$m\ddot{x}(t) = F_x[x(t)]. \tag{A.2}$$

We will later need the following alternative form of the equation of motion:

$$\dot{p}_x(t) = F_x[x(t)] \tag{A.3}$$

where $p_x(t) = m\dot{x}(t)$ is the momentum of the particle.

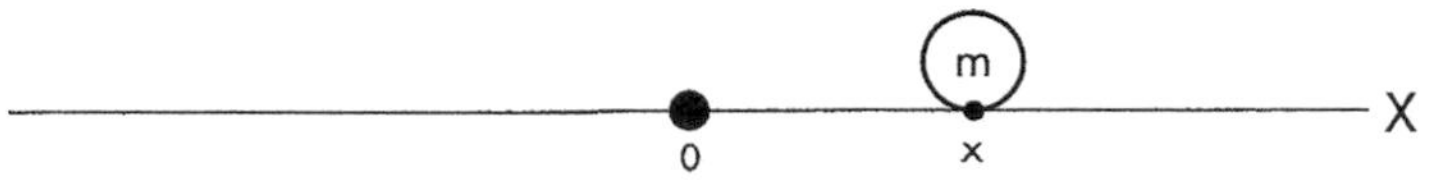

FIGURE A.1 A particle of mass m and coordinate x moving in one dimension.

We next note that since a trajectory can start at any initial position $x \equiv x(t=0)$ with any initial momentum $p_x \equiv p_x(t=0)$, a particle subject to a force $F_x(x)$ can execute an infinity of trajectories. Each distinct choice for x and p_x, called the *initial conditions*, generates a unique trajectory. Mathematically, the infinity of trajectories arises since Equation (A.2) is a second-order differential equation and so its general solution involves two integration constants. Each choice of these constants yields a different trajectory. Usually, the integration constants are determined from the initial conditions.

To illustrate these points, consider the simplest case of free particle motion. For this motion, $F_x(x)=0$ and Equation (A.2) reduces to $\ddot{x}(t)=0$. This equation of motion has the general solution $x(t)=At+B$, where A and B are integration constants which we determine as follows. Since $x(t=0)=x$, $B=x$, while since $m\dot{x}(t=0)=p_x$, $A=\frac{p_x}{m}$. Thus, the particular free particle trajectory that starts at x with momentum p_x is

$$x(t)=m^{-1}p_xt+x. \tag{A.4}$$

Next, we turn to the harmonic oscillator and harmonic oscillator motion. The harmonic oscillator is an idealized rather than a real system. Yet it is of great impotance in chemistry as a model for the molecular vibrational motions that underlie infrared spectroscopy. (See Chapters 7 and 9.)

A real system that approximates a harmonic oscillator is a particle of mass m bound by a spring as depicted in Figures A.2. In Figure A.2a, we show the particle at its *equilibrium position* x_e. At x_e, the force on the particle $F_x(x_e)=0$. So if the particle is initially at rest at x_e, it remains at x_e for all times $t>0$. Figures A.2b and c show the particle with *vibrational displacements* $y=x-x_e$ from x_e. An essential point is that for the particle to undergo oscillatory motion, $F_x(x)$ must resist vibrational displacements. Thus, for a positive vibrational displacement y (spring stretch), $F_x(x)$ must point in the negative X-direction with the converse being true for a negative vibrational displacement (spring compression). Mathematically, $F_x(x)$ must satisfy the conditions

$$F_x(x)<0 \quad \text{for } y>0 \text{ and } F_x(x)>0 \quad \text{for } y<0. \tag{A.5}$$

Equation (A.7) is summarized by the statement that $F_x(x)$ is a *restoring force*.

The simplest restoring force has the *Hooke's law* form

$$F_x[x]=-k(x-x_e)=-ky \tag{A.6}$$

where the *force constant* k is a property of the spring whose magnitude increases with the spring'd "stiffness." The Hooke's law force gives rise to harmonic oscillator motion, and thus, the harmonic oscillator equation of motion is

$$m\ddot{x}(t)=-k[x(t)-x_e]. \tag{A.7}$$

This equation of motion is more conveniently written in terms of $y=x-x_e$ as

$$m\ddot{y}(t)=-ky(t). \tag{A.8}$$

To proceed further, we introduce the *circular frequency* ω defined by

$$\omega=\left(\frac{k}{m}\right)^{1/2} \tag{A.9}$$

or $k=m\omega^2$. When rewritten in terms of ω, Equation (A.8) becomes

$$\ddot{y}(t)=-\omega^2y(t). \tag{A.10}$$

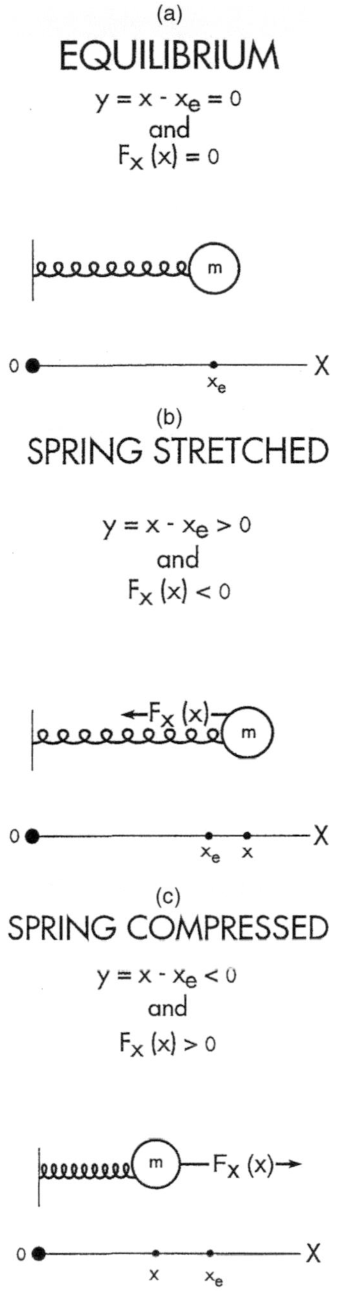

FIGURE A.2 A particle of mass m and coordinate x bound by a spring which produces a restoring force $F_x(x)$ on the particle which resists vibrational displacements $y = x - x_e$ from the equilibrium position x_e. (a) Particle is at equilibrium, $y = 0$ and $F_x(x) = 0$. (b) Spring is stretched, $y > 0$ and $F_x(x) < 0$. (c) Spring is compressed, $y < 0$ and $F_x(x) > 0$.

From Equation (A.10), it is evident that ω has frequency units of s^{-1} or hertz (Hz). As may be seen by direct substitution, the general solution of Equation (A.10) is of the form $y(t) = A\cos\omega t + B\sin\omega t$, where A and B are arbitrary constants. A and B are determined by imposing the initial conditions $y(t=0) = y$ and $p_y(t=0) = m\dot{y}(t=0) = p_y$ to give the harmonic oscillator trajectories as (Problem A.1)

$$y(t) = y\cos\omega t + \frac{1}{m\omega}p_y \sin\omega t. \tag{A.11}$$

Notice that the trajectories are *periodic* (i.e., they repeat themselves) satisfying for positive integer n $y(t+nT) = y(t)$ where

$$T = \frac{2\pi}{\omega} \tag{A.12}$$

is a time interval called the *period*. T is the number of seconds required for the harmonic oscillator to make one complete vibration. Thus, the vibrational frequency υ of the oscillator defined by

$$\upsilon = \begin{matrix}\text{number of complete vibrations} \\ \text{made by the oscillator per second}\end{matrix} \tag{A.13}$$

is the inverse of the period and therefore is given by

$$\upsilon = T^{-1} = \frac{\omega}{2\pi}. \tag{A.14}$$

Equation (A.14) shows that the circular frequency ω is 2π times the actual frequency υ.

We next note that nearly all of the forces dealt with in this text are *conservative forces*. That is, they are derived from a *potential energy function* $U(x)$ by the relation

$$F_x(x) = -\frac{dU(x)}{dx}. \tag{A.15}$$

From Equations (A.2) and (A.15), the equation of motion for a particle subject to a conservative force is

$$m\ddot{x}(t) = -\frac{dU[x(t)]}{dx(t)}. \tag{A.16}$$

Notice that $U(x)$ is undetermined to within a constant since adding a constant to $U(x)$ does not change $F_x(x)$ and hence the constant has no effect on the trajectories. Choosing the constant is referred to as *setting the zero of energy*. We will here set the zero of energy by taking the constant to be zero. Then, for a free particle since $F_x(x) = 0$ $U(x)$ is also zero. For a harmonic oscillator, $F_x(x) = -k(x - x_e) = -m\omega^2(x - x_e) = -m\omega^2 y$, so

$$U(x) = \frac{1}{2}m\omega^2(x - x_e)^2 = \frac{1}{2}m\omega^2 y^2. \tag{A.17}$$

Next, we introduce within the context of classical mechanics the concept of an *observable*.

This concept is extended to quantum mechanics in Section 4.2.

A.2 OBSERVABLES AND CONSTANTS OF THE MOTION

A classical observable is defined as any quantity whose value is determined by the system's trajectory $x(t)$. The basic observables are the position observable $x(t)$, which is the position of the particle at time t, and the momentum observable $p_x(t) = m\dot{x}(t)$, which is the momentum of the particle at time t. (Note that we use the same symbol $x[t]$ for two different quantities, the position observable which is the position at time t and the trajectory which is the position as a function of time. A similar comment holds for $p_x[t]$.) Complex observables are functions $A[x(t), p_x(t)]$ of the basic observables $x(t)$ and $p_x(t)$. Examples of complex observables are the kinetic energy observable $T[p_x(t)] = \frac{p_x^2(t)}{2m}$ and the potential energy observable $U[x(t)]$. The most important complex observable in classical mechanics is the Hamiltonian observable or simply the Hamiltonian defined by

$$H[x(t), p_x(t)] = \frac{p_x^2(t)}{2m} + U[x(t)]. \tag{A.18}$$

We will see that in classical mechanics, the Hamiltonian is equal to the (conserved) energy E of the system.

We next note that since $x(t)$ and $p_x(t)$ change as the particle moves along its trajectory, one expects the values of observables $A[x(t), p_x(t)]$ to depend on time. This is true, for example, for the kinetic and potential energy observables. However, for many systems there exist special observables whose values are independent of time and thus satisfy

$$A[x(t), p_x(t)] = A(x, p_x). \tag{A.19}$$

Such special observables are called *constants of the motion*, and statements like Equation (A.19) are called *conservation laws*. As a simple example for a free particle from Equation (A.4), $p_x(t) = m\dot{x}(t) = p_x$, and thus, $p_x(t)$ is a constant of the motion. The corresponding conservation law is called the law of conservation of linear momentum.

Conservation of linear momentum holds only for free particles, but a much more general conservation law exists which holds for all systems subject to conservative forces. This is the law of conservation of the Hamiltonian

$$H[x(t), p_x(t)] = H(x, p_x) \tag{A.20}$$

more commonly called the law of conservation of energy E. Recalling that $H(x, p_x) = \frac{p_x^2}{2m} + U(x)$, the law of conservation of energy takes the form

$$\frac{p_x^2(t)}{2m} + U[x(t)] = \frac{p_x^2}{2m} + U(x) \equiv E. \tag{A.21}$$

In Problem A.2, Equation (A.21) is proven generally for one-dimensional conservative force systems. Here, we will only verify it for the harmonic oscillator. To do this, using Equation (A.17) for the harmonic oscillator potential energy function and noting that $p_y = m\dot{y} = p_x$, we write the harmonic oscillator Hamiltonian as

$$H(y, p_y) = \frac{p_y^2}{2m} + \frac{1}{2}m\omega^2 y^2. \tag{A.22}$$

Thus, we will show that $H[y(t), p_y(t)] = H(y, p_y)$ or $\frac{p_y^2(t)}{2m} + \frac{1}{2}m\omega^2 y(t) = \frac{p_y^2}{2m} + \frac{1}{2}m\omega^2 y^2$. We start with Equation (A.11) for the harmonic oscillator trajectory $y(t)$. Namely, $y(t) = \cos\omega t\, y + \frac{1}{m\omega}\sin\omega t\, p_y$, which gives $p_y(t) = m\dot{y}(t)$ as $p_y(t) = -m\omega \sin\omega t\, y + \cos\omega t\, p_y$. These results for $y(t)$ and $p_y(t)$ give $\frac{1}{2}m\omega^2 y^2(t) = \frac{1}{2}m\omega^2 y^2\cos^2\omega t + \frac{p_y^2}{2m}\sin^2\omega t + \omega y\, p_y \sin\omega t\cos\omega t$ and $\frac{p_y^2(t)}{2m} = \frac{1}{2}m\omega^2 y^2 \sin^2\omega t + \frac{p_y^2}{2m}\cos^2\omega t - \omega y p_y \sin\omega t\cos\omega t$. Therefore, $\frac{p_y^2(t)}{2m} + \frac{1}{2}m\omega^2 y^2(t)$

$= \frac{p_y^2}{2m} + \frac{1}{2}m\omega^2 y^2$, and thus, we have verified the law of conservation of energy for the harmonic oscillator.

This law provides the basis for a powerful method for qualitatively studying trajectories which is used in Section 7.5 to contrast the classical and quantum descriptions of harmonic oscillator motion. We next describe this method.

A.3 ENERGY CONSERVATION AND THE QUALITATIVE STUDY OF TRAJECTORIES

The principle underlying the method is very simple. Writing the kinetic energy of the particle $\frac{p_x^2(t)}{2m}$ as $\frac{1}{2}mv_x^2(t)$ where $v_x(t) = \dot{x}(t)$ is the particle's velocity, the energy conservation law of Equation (A.21) may be expressed as

$$E = \frac{1}{2}mv_x^2(t) + U[x(t)]. \tag{A.23}$$

Since E is a constant, the following is evident from Equation (A.23). If along the particle's trajectory the potential energy $U(x)$ decreases, its kinetic energy and hence its velocity increase and vice versa. Thus, a plot of $U(x)$ reveals how the velocity of the particle changes as its position changes (Figure A.3). Therefore, valuable information about the motion may be obtained without solving Newton's equation of motion but by merely examining a plot of $U(x)$.

Let us apply these ideas to the harmonic oscillator. From Equation (A.17), for the harmonic oscillator potential energy function, Equation (A.23) specializes to

$$E = \frac{1}{2}mv_x^2(t) + \frac{1}{2}m\omega^2[x(t) - x_e]^2. \tag{A.24}$$

Using Equation (A.24), we may now qualitatively study the motions of the oscillator. We assume that at $t = 0$, the particle is at its equilibrium position x_e and moving to the right with velocity $v_x(t = 0) = v_x$. From Equation (A.24), these initial conditions determine the energy as

$$E = \frac{1}{2}mv_x^2. \tag{A.25}$$

To proceed further, we next discuss the *turning points* of the oscillator. In order that the particle execute vibratory motion, it must be moving to the right at some times and to the left at others. Thus, positions x^* called turning points must exist at which the particle changes direction. At a turning point x^*, the velocity v_x of the particle vanishes, and therefore, all of its energy is potential energy. Thus, from Equation (A.24) the turning points are determined by the condition

$$E = \frac{1}{2}m\omega^2(x^* - x_e)^2. \tag{A.26}$$

Equation (A.26) is satisfied for two values of x^*:

$$x_R = x_e + \left(\frac{2E}{m\omega^2}\right)^{1/2} \text{ and } x_L = x_e - \left(\frac{2E}{m\omega^2}\right)^{1/2} \tag{A.27}$$

where x_R and x_L are called, respectively, the right-hand and left-hand turning points. (See Figure A.3.)

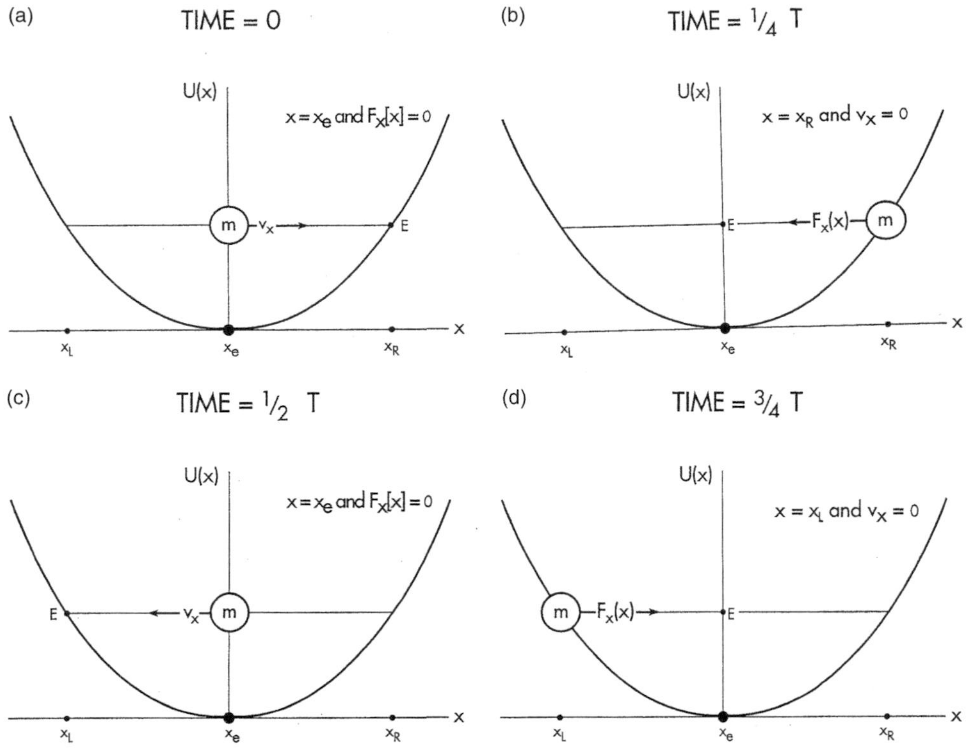

FIGURE A.3 Energy conservation study of the harmonic motion of a particle of mass m. The energy of the particle is E, its equilibrium position is x_e, and its right- and left-hand turning points are x_L and x_R. Only the first oscillation is shown since the motion is periodic with period T. The system begins (a) the first, (b) second, (c) third, and (d) final quarter cycles of the oscillation.

We next qualitatively follow the motion of the particle for one complete vibration. We divide the vibration into four quarter cycles each of duration $\frac{1}{4}T$, where recall T is the period. As noted, we assume that when the motion begins at $t = 0$, $x(t) = x_e$ and $v_x(t) > 0$. Hence, initially all of the particle's energy is kinetic energy. Then, the motion may be described as follows (Figure A.3).

Quarter cycle 1. Initially, the particle is moving to the right and the restoring force $F_x(x_e) = 0$. Consequently, the particle continues to move to the right, and hence, its potential energy increases. Its kinetic energy thus decreases, and the particle slows down eventually halting at the right-hand turning point x_R with all of its energy in the form of potential energy.

Quarter cycle 2. Initially, the particle is motionless at x_R and the restoring force $F_x(x_R) < 0$ points in the negative X-direction. Consequently, the particle changes direction and moves to the left, and hence, its potential energy decreases. Its kinetic energy thus increases, and the particle speeds up eventually returning to its initial speed at the equilibrium position x_e with all of its energy in the form of kinetic energy.

Quarter cycle 3. Initially, the particle is moving to the left and the restoring force $F(x_e) = 0$. Consequently, the particle continues to move to the left, and hence, its potential energy increases. Its kinetic energy thus decreases, and the particle slows down eventually halting at the left-hand turning point x_L with all of its energy in the form of potential energy.

Quarter cycle 4. Initially, the particle is motionless at x_L and the restoring force $F_x(x_L) > 0$ points in the positive X-direction. Consequently, the particle changes direction and moves to the right, and hence, its potential energy decreases. Its kinetic energy thus increases, and the particle speeds up eventually returning to its initial speed at the equilibrium position x_e with all of its energy in the form of kinetic energy and a complete vibration of the oscillator has occurred.

Thus, we have obtained a qualitative picture of harmonic oscillator motion merely by examining a plot of the harmonic oscillator potential energy function. The real power of our method is that it may be applied to systems for which Newton's equation of motion is not exactly solvable.

This concludes our discussion of single-degree-of-freedom systems, that is, systems described by a single coordinate x. We next turn to several-degree-of-freedom systems.

A.4 SEVERAL-DEGREE-OF-FREEDOM SYSTEMS

We begin with a single particle of mass m moving in three dimensions (Figure A.4). The position of the particle is specified by its three Cartesian coordinates x, y, z and its momentum is specified by its three Cartesian components of momentum $p_x = m\dot{x}$, $p_y = m\dot{y}$, and $p_z = m\dot{z}$. The potential energy function of the particle $U(x,y,z)$ depends on all three coordinates x, y, z, and the trajectory of the particle $x(t),y(t),z(t)$ is determined by the following three equations of motion which generalize Equation (A.16):

$$m\ddot{x}(t) = -\frac{\partial U[x(t),y(t),z(t)]}{\partial x(t)} \quad m\ddot{y}(t) = -\frac{\partial U[x(t),y(t),z(t)]}{\partial y(t)}$$
$$m\ddot{z}(t) = -\frac{\partial U[x(t),y(t),z(t)]}{\partial z(t)}. \tag{A.28}$$

Since the motion of $x(t)$ depends on that of $y(t)$ and $z(t)$ and so on, the equations of motion are *coupled* or *non-separable* and must be solved simultaneously. Additionally, to determine a particular trajectory, six initial conditions – $x(t=0)=x$, $y(t=0)=y$, $z(t=0)=z$, $p_x(t=0)=p_x$, $p_y(t=0)=p_y$, and $p_z(t=0)=p_z$ – must be imposed.

The particle has six basic observables: the position observables $x(t),y(t),z(t)$ and the momentum observables $p_x(t),p_y(t),p_z(t)$. Complex observables $A[x(t),y(t),z(t),p_x(t),p_y(t),p_z(t)]$ are functions of the basic observables. The most important complex observable is the Hamiltonian

$$H[x(t),y(t),z(t),p_x(t),p_y(t),p_z(t)]$$
$$= \frac{p_x^2(t)}{2m} + \frac{p_y^2(t)}{2m} + \frac{p_z^2(t)}{2m} + U[x(t),y(t),z(t)] \tag{A.29}$$

which is a constant of the motion equal to the energy E of the particle (Problem A.3).

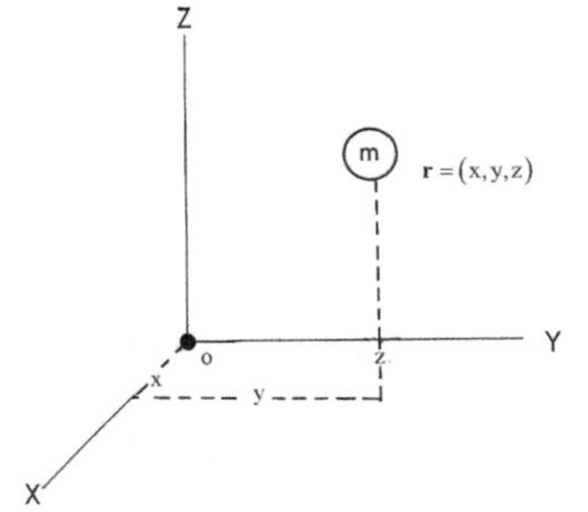

FIGURE A.4 A particle of mass m and coordinates x, y, and z moving in three dimensions.

For several-degree-of-freedom systems, the forms for the equations of both classical and quantum mechanics greatly simplify, looking (deceptively) no more complex than the corresponding single-degree-of freedom equations, when written in *vector notation* which we next introduce.

While a scalar is a single number, a vector $\mathbf{v} = (v_x, v_y, v_z)$ is a triple of three numbers v_x, v_y, and v_z called the *components* of v. As an example, the coordinates x,y,z of a single particle moving in three dimensions in vector notation are denoted by the single position vector $\mathbf{r} = (x,y,z)$. Thus, the potential energy function $U = U(x,y,z)$ in vector notation has the simple form $U = U(\mathbf{r})$. The three-dimensional equations of motion when written in vector notation greatly simplify looking no more complicated than the equation of motion of a particle moving in one dimension. Noting that two vectors are equal if all of their components are equal, the equations of motion (Equation A.28) may be written as $m[\ddot{x}(t),\ddot{y}(t),\ddot{z}(t)] = \left\{-\frac{\partial U[x(t),y(t),z(t)]}{\partial x(t)}, -\frac{\partial U[x(t),y(t),z(t)]}{\partial y(t)}, -\frac{\partial U[x(t),y(t),z(t)]}{\partial z(t)}\right\}$ or $m\ddot{\mathbf{r}}(t) = -\left[\frac{\partial}{\partial x(t)}\frac{\partial}{\partial y(t)}\frac{\partial}{\partial z(t)}\right]U[\mathbf{r}(t)]$. Defining the *vector operator* $\frac{\partial}{\partial \mathbf{r}}$ by $\frac{\partial}{\partial \mathbf{r}} = \left(\frac{\partial}{\partial x}, \frac{\partial}{\partial y}, \frac{\partial}{\partial z}\right)$, the equations of motion take the following form in vector notation:

$$m\ddot{\mathbf{r}}(t) = -\frac{\partial U[\mathbf{r}(t)]}{\partial \mathbf{r}[t]}. \tag{A.30}$$

The forms for observables also simplify in vector notation. The position and momentum observables become $\mathbf{r}(t) = [x(t),y(t),z(t)]$ and $\mathbf{p}(t) = [p_x(t),p_y(t),p_z(t)]$. Complex observables are functions $A[\mathbf{r}(t),\mathbf{p}(t)]$ of $\mathbf{r}(t)$ and $\mathbf{p}(t)$. Especially, the Hamiltonian (Equation A.29) is given in vector notation as

$$H[\mathbf{r}(t),\mathbf{p}(t)] = \frac{\mathbf{p}(t)\cdot\mathbf{p}(t)}{2m} + U[\mathbf{r}(t)] \tag{A.31}$$

where to obtain Equation (A.31) we have used the definition of the *dot product* **v.w** of two vectors $\mathbf{v} = (v_x, v_y, v_z)$ and $\mathbf{w} = (w_x, w_y, w_z)$; namely, $\mathbf{v}\cdot\mathbf{w} = v_x w_x + v_y w_y + v_z w_z$.

We finally consider a general system comprised of N particles each moving in three dimensions. When expressed in scalar form, the equations of classical mechanics are cumbersome for a general system. Thus, we will give these equations in vector form. Denote the masses, position vectors, and momentum vectors for the particles by $m_1,\ldots,m_N,\mathbf{r}_1(t),\ldots,\mathbf{r}_N(t)$, and $\mathbf{p}_1(t),\ldots,\mathbf{p}_N(t)$ where for $i = 1,\ldots,N$ $\mathbf{r}_i(t) = [x_i(t),y_i(t),z_i(t)]$ and $\mathbf{p}_i(t) = [p_{x_i}(t),p_{y_i}(t),p_{z_i}(t)]$. The potential energy function of the system depends on the coordinates of all of the particles and thus is the form $U[\mathbf{r}_1(t),\ldots,\mathbf{r}_N(t)]$. Defining for $i = 1,\ldots,N$ the vector operators $\frac{\partial}{\partial \mathbf{r}_i(t)}$ by $\frac{\partial}{\partial \mathbf{r}_i(t)} = \left[\frac{\partial}{\partial x_i(t)}, \frac{\partial}{\partial y_i(t)}, \frac{\partial}{\partial z_i(t)}\right]$, the equations of motion for the system are the following straightforward extension of Equation (A.30):

$$m_1\ddot{\mathbf{r}}_1(t) = -\frac{\partial U[\mathbf{r}_1(t),\ldots,\mathbf{r}_N(t)]}{\partial \mathbf{r}_1(t)},\ldots,m_N\ddot{\mathbf{r}}_N(t) = -\frac{\partial U[\mathbf{r}_1(t),\ldots,\mathbf{r}_N(t)]}{\partial \mathbf{r}_N(t)}. \tag{A.32}$$

Equation (A.32) forms a set of 3N coupled equations of motion which must be solved subject to the 6N initial conditions $\mathbf{r}_i(t=0) = \mathbf{r}_i$ and $\mathbf{p}_i(t=0) = \mathbf{p}_i$ for $i = 1,\ldots,N$.

Observables may be defined straightforwardly. The basic observables are the position and momentum observables $\mathbf{r}_1(t),\ldots,\mathbf{r}_N(t)$ and $\mathbf{p}_1(t),\ldots,\mathbf{p}_N(t)$. Complex observables are functions

$A[\mathbf{r}_1(t),\ldots,\mathbf{r}_N(t),\mathbf{p}_1(t),\ldots,\mathbf{p}_N(t)]$ of the basic observables. The most important complex observable is the Hamiltonian

$$H[\mathbf{r}_1(t),\ldots,\mathbf{r}_N(t),\mathbf{p}_1(t),\ldots,\mathbf{p}_N(t)] = \sum_{i=1}^{N} \frac{\mathbf{p}_i(t)\cdot\mathbf{p}_i(t)}{2m_i} + U[\mathbf{r}_1(t),\ldots,\mathbf{r}_N(t)] \tag{A.33}$$

which is a constant of the motion equal to the energy E of the system (Problem A.4).

A.5 GENERALIZED COORDINATES AND HAMILTON'S FORMULATION

The equations of both classical and quantum mechanics are much more difficult to solve for three-dimensional systems than for one-dimensional systems. In classical mechanics, for example, the equation of motion for a one-dimensional system, Equation (A.16), is a single differential equation for $x(t)$, while the equation of motion for a three-dimensional system, Equation (A.28), is a set of three coupled or non-separable differential equations for $x(t), y(t)$, and $z(t)$ which must be solved simultaneously. However, in both classical and quantum mechanics, non-separability problems may sometimes be effectively dealt with by expressing the fundamental equations in terms of a suitable set of non-Cartesian or *generalized coordinates.*

For example, suppose the potential energy function U for a particle moving in three dimensions is a *central field potential.* That is, suppose U depends only on the distance $r = (x^2 + y^2 + z^2)^{1/2}$ of the particle from the origin and thus is of the form $U(x,y,z) = U\left[(x^2 + y^2 + z)^{1/2}\right] = U(r)$. Then, the three-dimensional classical and quantum fundamental equations simplify genuinely (not apparently as for the case of expressing the equations in vector notation) when expressed in terms of a set of generalized coordinates which include r, known as spherical polar coordinates. (See Problem A.9.) We describe the simplifications arising from the use of spherical polar coordinates in classical three-dimensional central field motion in Appendix B and in quantum three-dimensional central field motion in Chapters 8–10.

Here, to illustrate the value of generalized coordinates as simply as possible, we consider two-dimensional central field motion. That is, we consider a particle of mass m moving in the XY-plane with Cartesian coordinates x, y subject to the two-dimensional central field potential $U(x,y) = U\left[(x^2 + y^2)^{1/2}\right] = U(r)$, where $r = (x^2 + y^2)^{1/2}$ is the distance of the particle from the origin. The Cartesian coordinate equations of motion of the particle are

$$m\ddot{x}(t) = -\frac{\partial U\left\{\left[x^2(t) + y^2(t)\right]^{1/2}\right\}}{\partial x(t)} \qquad m\ddot{y}(t) = -\frac{\partial U\left\{\left[x^2(t) + y^2(t)\right]^{1/2}\right\}}{\partial y(t)}. \tag{A.34}$$

Since U depends on both x and y, these equations are coupled or non-separable and therefore must be solved simultaneously. However, since U depends only on r, one expects simplifications if the equations of motion are expressed in terms of a set of generalized coordinates which include r. An appropriate set is the polar coordinates r and ϕ (Figure A.5) with ranges $0 \le r \le \infty$ and $0 \le \phi \le 2\pi$. These are defined in terms of the Cartesian coordinates x, y by the *transformation relations*

$$r = (x^2 + y^2)^{1/2} \qquad \phi = \tan^{-1}(y/x) \tag{A.35}$$

or equivalently by the inverse transformation relations

$$x = r\cos\phi \qquad y = r\sin\phi. \tag{A.36}$$

From the transformation relations, it is evident that $r(t)$ and $\phi(t)$ determine $x(t)$ and $y(t)$ and vice versa. Thus, $r(t), \phi(t)$ and $x(t), y(t)$ are equivalent representations of the trajectory. We will see however that the $r(t), \phi(t)$ representation is far more easily found, exemplifying the heart of the value of generalized coordinates in both classical and quantum mechanics.

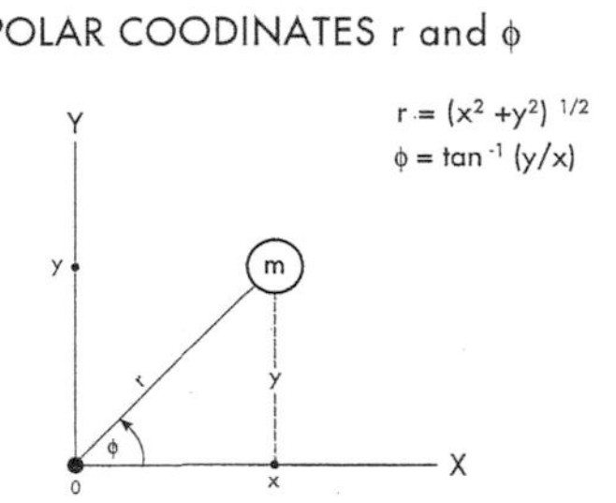

FIGURE A.5 Polar coordinate description of the two-dimensional motion of a particle of mass m.

To start, we assume that Newton's equations of motion retain their Cartesian forms when expressed in polar coordinates. If this is so, the polar coordinate equations of motion are

$$m\ddot{r}(t) = -\frac{\partial U[r(t)]}{\partial r(t)} \quad m\ddot{\phi}(t) = -\frac{\partial U[r(t)]}{\partial \phi(t)} = 0. \tag{A.37}$$

Notice that these equations of motion in sharp contrast to the Cartesian equations (Equation A.34) are not coupled. Rather, they are both of independent one-dimensional form. Moreover, $\phi(t)$ obeys a free particle equation of motion. Thus, apparently the use of polar coordinates has yielded great simplification. Unfortunately, our polar coordinate equations of motion (Equation A.37) are wrong. This is because Newton's equations of motion do not retain their Cartesian forms when expressed in generalized coordinates.

Thus, we turn to Hamilton's formulation of classical mechanics. Hamilton's formulation while equivalent to Newton's yields equations of motion which retain their Cartesian forms when expressed in generalized coordinates. Hamilton's formulation is thus much more convenient than Newton's when one uses generalized coordinates.

We will not develop Hamilton's equations rigorously (since this task is both lengthy and advanced and is accomplished in every textbook on classical mechanics). Instead, we will write down these equations for a one-dimensional system and show their equivalence to Newton's equations. Then, by analogy to the one-dimensional equations, we will "derive" Hamilton's equations for two-dimensional central field motion.

We first note that the Hamiltonian for a one-dimensional system may be written in two forms. These are the *Cartesian velocity form*

$$H[\dot{x}(t), x(t)] = \frac{1}{2} m\dot{x}^2(t) + U[x(t)] \tag{A.38}$$

and the *Cartesian momentum form*

$$H[x(t), p_x(t)] = \frac{p_x^2(t)}{2m} + U[x(t)]. \tag{A.39}$$

Since $p_x(t) = m\dot{x}(t)$, the two forms are equivalent.

To develop Hamilton's equations, we adopt the following mathematical viewpoint. We view $H[x(t), \dot{x}(t)]$ as a function of the pair of *independent* variables $x(t)$ and $\dot{x}(t)$. Similarly, we view $H[x(t), p_x(t)]$ as a function of the pair of independent variables $x(t)$ and $p_x(t)$. Given this viewpoint, we may take partial derivatives of $H[x(t), \dot{x}(t)]$ with respect to either $x(t)$ or $\dot{x}(t)$ and we may take partial derivatives of $H[x(t), p_x(t)]$ with respect to either $x(t)$ or $p_x(t)$.

For a one-dimensional system, Hamilton's formulation is comprised of three relations. The first is Hamilton's definition of the momentum

$$p_x(t)=\frac{\partial H[x(t),\dot{x}(t)]}{\partial \dot{x}(t)} \tag{A.40}$$

and the second two are Hamilton's equations of motion

$$\dot{x}(t)=\frac{\partial H[x(t),p_x(t)]}{\partial p_x(t)} \tag{A.41a}$$

and

$$\dot{p}_x(t)=-\frac{\partial H[x(t),p_x(t)]}{\partial x(t)} \tag{A.41b}$$

Notice that while in Newton's formulation the equation of motion $m\ddot{x}(t)=-\frac{dU[x(t)]}{dx(t)}$ is a single second-order differential equation for $x(t)$, in Hamilton's formulation the equation of motion is a pair of coupled first-order differential equations for $x(t)$ and $p_x(t)$.

The equivalence of the Hamilton and Newton formulations is easily established for a one-dimensional system. With Equation (A.38) for $H[\dot{x}(t),x(t)]$, Hamilton's momentum defining Equation (A.40) gives $p_x(t)=m\dot{x}(t)$, which is Newton's definition of the momentum. With Equation (A.39) for $H[x(t),p_x(t)]$, Hamilton's equations of motion give $\dot{x}(t)=\frac{p_x(t)}{m}$, which is again $p_x(t)=m\dot{x}(t)$, and $\dot{p}_x(t)=-\frac{dU[x(t)]}{dx(t)}=F_x[x(t)]$, which is Newton's equation of motion, is the form of Equation (A.3).

We may now apply Hamilton's formulation to two-dimensional central field motion. We proceed via the following six steps:

1. Write down the Cartesian velocity form of the Hamiltonian as

$$H[x(t),y(t),\dot{x}(t),\dot{y}(t)]=\frac{1}{2}m\dot{x}^2(t)+\frac{1}{2}m\dot{y}^2(t)+U\left\{\left[x^2(t)+y^2(t)\right]^{1/2}\right\}. \tag{A.42}$$

2. Using the transformation relations of Equation (A.36), express $H[x(t),y(t),\dot{x}(t),\dot{y}(t)]$ in terms of $r(t)$ and $\phi(t)$ and the *generalized velocities* $\dot{r}(t)$ and $\dot{\phi}(t)$ as (Problem A.5)

$$H\left[r(t),\phi(t),\dot{r}(t),\dot{\phi}(t)\right]=\frac{1}{2}m\dot{r}^2(t)+\frac{1}{2}mr^2(t)\dot{\phi}^2(t)+U[r(t)]. \tag{A.43}$$

3. In analogy to Equation (A.40), define the *generalized momenta* $p_r(t)$ and $p_\phi(t)$ as

$$p_r(t)=\frac{\partial H\left[r(t),\phi(t),\dot{r}(t),\dot{\phi}(t)\right]}{\partial \dot{r}(t)} \quad p_\phi(t)=\frac{\partial H\left[r(t),\phi(t),\dot{r}(t),\dot{\phi}(t)\right]}{\partial \dot{\phi}(t)} \tag{A.44}$$

and then compare Equations (A.43) and (A.44) to obtain $p_r(t)$ and $p_\phi(t)$ as

$$p_r(t)=m\dot{r}(t) \quad p_\phi(t)=mr^2(t)\dot{\phi}(t). \tag{A.45}$$

4. Use Equation (A.45) to eliminate $\dot{r}(t)$ and $\dot{\phi}(t)$ from Equation (A.43) to obtain the generalized momentum form of the Hamiltonian as

$$H[r(t),\phi(t),p_r(t),p_\phi(t)] = \frac{p_r^2(t)}{2m} + \frac{p_\phi^2(t)}{2mr^2(t)} + U[r(t)]. \tag{A.46}$$

5. In analogy to Equation (A.41), write down Hamilton's equations of motion as

$$\dot{r}(t) = \frac{\partial H[r(t),\phi(t),p_r(t),p_\phi(t)]}{\partial p_r(t)} \tag{A.47a}$$

$$\dot{\phi}(t) = \frac{\partial H[r(t),\phi(t),p_r(t),p_\phi(t)]}{\partial p_\phi(t)} \tag{A.47b}$$

and

$$\dot{p}_r(t) = -\frac{\partial H[r(t),\phi(t),p_r(t),p_\phi(t)]}{\partial r(t)} \tag{A.47c}$$

$$\dot{p}_\phi(t) = -\frac{\partial H[r(t),\phi(t),p_r(t),p_\phi(t)]}{\partial \phi(t)}. \tag{A.47d}$$

6. Obtain the explicit forms of Hamilton's equations of motion from Equations (A.46) and (A.47) as

$$\dot{r}(t) = \frac{p_r(t)}{m} \tag{A.48a}$$

$$\dot{\phi}(t) = \frac{p_\phi(t)}{mr^2(t)} \tag{A.48b}$$

$$\dot{p}_r(t) = -\frac{\partial}{\partial r(t)}\left\{\frac{p_\phi^2(t)}{2mr^2(t)} + U[r(t)]\right\} \tag{A.48c}$$

and since $H[r(t),\phi(t),p_r(t),p_\phi(t)]$ is independent of $\phi(t)$,

$$\dot{p}_\phi(t) = 0. \tag{A.48d}$$

From Equation (A.48), we can see the simplifications that arise from the use of polar coordinates. First, from Equation (A.48d) we see that $p_\phi(t)$ is independent of t so that

$$p_\phi(t) = p_\phi(0) \equiv p_\phi. \tag{A.49}$$

In other words, $p_\phi = mr^2\dot{\phi}(t)$ is a constant of the motion. We will show later that the corresponding conservation law is the law of conservation of angular momentum which holds generally for central field motion. Second, from Equations (A.48a) and (A.49), Equation (A.48c) simplifies to the following equation of motion for $r(t)$:

$$m\ddot{r}(t) = -\frac{dU_{eff}[r(t)]}{dr(t)} \tag{A.50}$$

where

$$U_{eff}[r(t)] = \frac{p_\phi^2}{2mr^2(t)} + U[r(t)]. \tag{A.51}$$

Notice that since p_ϕ is a constant, $U_{eff}(r)$ depends only on r. Because of this, Equation (A.50) for $r(t)$ is identical in form to the equation of motion $m\ddot{x}(t) = -\frac{dU[x(t)]}{dx(t)}$ for a one-dimensional system. For this reason, $U_{eff}(r)$ is called the effective potential energy function or *effective potential* for the r motion. $U_{eff}(r)$ is actually not a true potential energy function since the term $\frac{p_\phi^2}{2mr^2}$ is the kinetic energy arising for the motion of the ϕ coordinate. Despite this, it is often very useful to view $U_{eff}(r)$ as a true potential. In any case, determining $r(t)$ by solving Equation (A.50) is no more difficult than finding a one-dimensional trajectory $x(t)$.

Finally, since $p_\phi(t) = p_\phi$, Equation (A.48b) simplifies to

$$\dot{\phi}(t) = \frac{p_\phi}{mr^2(t)}. \tag{A.52}$$

Thus, once $r(t)$ is found by solving Equation (A.50), $\phi(t)$ may be easily determined by simple integration of Equation (A.52). Performing this integration gives $\phi(t)$ as

$$\phi(t) = \phi + m^{-1}p_\phi \int_0^t r^{-2}(\tau)d\tau \tag{A.53}$$

where $\phi = \phi(t = 0)$.

Our analysis shows that the equations of motion for $r(t)$ and $\phi(t)$ nearly separate (since they may be solved sequentially as opposed to simultaneously). Hence, determining the trajectory as $r(t), \phi(t)$ is much simpler than determining it as $x(t), y(t)$.

Summarizing the two-dimensional central field problem illustrates in a simple manner that in classical mechanics, non-separability problems can sometimes be largely solved by expressing the equations of motion in terms of a suitable set of generalized coordinates. Analogous simplifications occur in the more complex context of quantum mechanics.

We next turn to angular momentum.

A.6 ANGULAR MOMENTUM

We earlier defined the dot product of two arbitrary vectors $\mathbf{v} = (v_x, v_y, v_z)$ and $\mathbf{w} = (w_x, w_y, w_z)$ by $\mathbf{v} \cdot \mathbf{w} = v_x w_x + v_y w_y + v_z w_z$. We next introduce their *cross product* $\mathbf{v} \times \mathbf{w}$ which is a second way of multiplying $\mathbf{v}$ and $\mathbf{w}$. While the dot product is a scalar, the cross product is defined as the vector

$$\mathbf{u} = \mathbf{v} \times \mathbf{w} \tag{A.54}$$

whose components u_x, u_y, and u_z are given by

$$u_x = v_y w_z - v_z w_y \quad u_y = v_z w_x - v_x w_z \quad u_z = v_x w_y - v_y w_z. \tag{A.55}$$

We may now define the angular momentum of a particle of mass m moving in three dimensions. The angular momentum $\mathbf{J}(t) = [J_x(t), J_y(t), J_z(t)]$ is a vector observable defined in terms of the particle's basic observables $\mathbf{r}(t)$ and $\mathbf{p}(t)$ as the cross product

$$\mathbf{J}(t) = \mathbf{r}(t) \times \mathbf{p}(t). \tag{A.56}$$

From the definition of the cross product just given, the components of $\mathbf{J}(t)$ are

$$J_x(t) = y(t)p_z(t) - z(t)p_y(t) \quad J_y(t) = z(t)p_x(t) - x(t)p_z(t) \quad J_z(t) = x(t)p_y(t) - y(t)p_x(t). \tag{A.57}$$

We will also need the *magnitude* $J(t)$ of $\mathbf{J}(t)$ defined by

$$J(t) = \left[J_x^2(t) + J_y^2(t) + J_z^2(t)\right]^{1/2}. \tag{A.58}$$

Next, let us specialize to two-dimensional motion in the XY-plane. For two-dimensional motion, $z(t) = 0$, and hence, $p_z(t) = m\dot{z}(t) = 0$. It then follows from Equation (A.57) that $J_x(t) = J_y(t) = 0$, and thus, for two-dimensional motion

$$\mathbf{J}(t) = \left[0, 0, J_z(t)\right] \tag{A.59}$$

and

$$J(t) = J_z(t). \tag{A.60}$$

However, writing $J_z(t)$ as $J_z(t) = m\left[x(t)\dot{y}(t) - y(t)\dot{x}(t)\right]$ and using the transformation relations of Equation (A.36), one may show that $J_z(t)$ may be expressed in polar coordinates as (Problem A.6)

$$J_z(t) = mr^2(t)\dot{\phi}(t) = p_\phi(t) \tag{A.61}$$

where we have used Equation (A.48b).

Next, let us assume two-dimensional central field motion. However for this type of motion, from Equation (A.49), $p_\phi(t) = p_\phi$ is a constant of the motion. Therefore from Equation (A.61), $J_z(t)$ is also a constant of the motion and thus satisfies

$$J_z(t) = J_z(t = 0) \equiv J_z. \tag{A.62}$$

Equations (A.59), (A.60), and (A.62) imply that $\mathbf{J}(t)$ and $J(t)$ are also constants of the motion. That is,

$$\mathbf{J}(t) = \mathbf{J}(t = 0) \equiv \mathbf{J} \tag{A.63}$$

and

$$J(t) = J(t = 0) \equiv J. \tag{A.64}$$

Equation (A.63) is the law of conservation of angular momentum for two-dimensional central field motion.

Additionally, it follows from our discussion that

$$p_\phi(t) = p_\phi = J. \tag{A.65}$$

Given Equation (A.65), the Hamiltonian of Equation (A.46) simplifies to

$$H = \frac{p_r^2(t)}{2m} + \frac{J^2}{2mr^2(t)} + U\left[r(t)\right]. \tag{A.66}$$

Similarly, one has for the effective potential of Equation (A.51)

$$U_{eff}\left[r(t)\right]=\frac{J^2}{2mr^2(t)}+U\left[r(t)\right]. \tag{A.67}$$

Additionally, Equations (A.53) and (A.65) give $\phi(t)$ as

$$\phi(t)=\phi+m^{-1}J\int_0^t r^{-2}(\tau)d\tau. \tag{A.68}$$

These results for $H, U_{eff}\left[r(t)\right]$ and $\phi(t)$ are used in Appendix B.

Finally, we note that our definition of vectors $\mathbf{v}=\left(v_x v_y v_z\right)$ as triples of scalars v_x, v_y, and v_z is equivalent to the more common definition of vectors $\vec{v}$ as "arrows" with magnitude and direction. This is because the components v_x, v_y, and v_z of a vector $\mathbf{v}$ are just the projections of the corresponding arrow $\vec{v}$ on the X-, Y-, and Z-axes. This one-to-one correspondence between arrows and vectors permits a geometric interpretation of vectors. For example, the magnitude $v=\left(v_x^2+v_y^2+v_z^2\right)^{1/2}$ of a vector $\mathbf{v}$ is interpreted as the length of the corresponding arrow $\vec{v}$. Similarly, the direction of $\mathbf{v}$ is defined as the direction of $\vec{v}$. This geometric interpretation of vectors is used in the plots of Appendix B.

FURTHER READINGS

Goldstein, Herbert, Charles Poole, and John Safko. 2002. *Classical mechanics.* 3rd ed. San Francisco, CA: Addison Wesley.

Pauling, Linus, and E. Bright Wilson Jr. 1985. *Introduction to quantum mechanics with applications to chemistry.* New York: Dover Publications, Inc.

Taylor, John R. 2005. *Classical mechanics.* Sausalito, CA: University Science Books.

Thornton, Stephen T., and Jerry B. Marion. 2004. *Classical dynamics of particles and systems.* 5th ed. Belmont, CA: Brooke/Cole.

PROBLEMS

A.1 (a) Show that the general solution of the harmonic oscillator equation of motion $\ddot{y}(t)=-\omega^2 y(t)$ is $y(t)=A\cos\omega t+B\sin\omega t$ where A and B are arbitrary constants. (b) Show that the particular solution that satisfies the initial conditions $y(t=0)=y$ and $p_y(t=0)=p_y$ is given by Equation (A.11).

A.2 In this problem, you will prove the conservation of energy law for a system comprised of a particle of mass m moving in one dimension subject to an arbitrary conservative force $F_x(x)=-\frac{dU(x)}{dx}$. That is, you will prove Equation (A.21). (a) First, show that $m\dot{v}_x(t)v_x(t)=-\frac{dU\left[x(t)\right]}{dx(t)}\frac{dx(t)}{dt}=-\frac{dU\left[x(t)\right]}{dt}$. (b) Next, show that $m\dot{v}_x(t)v_x(t)=\frac{d}{dt}\left[\frac{p_x^2(t)}{2m}\right]$. (c) Thus, show that $\frac{d}{dt}\left\{\frac{p_x^2(t)}{2m}+U\left[x(t)\right]\right\}=0$. (d) Show that this last result implies Equation (A.21).

A.3 Prove the energy conservation law for a particle of mass m moving in three dimensions. You will need the result $\frac{dU\left[x(t),y(t),z(t)\right]}{dt}=\frac{\partial U\left[x(t),y(t),z(t)\right]}{\partial x(t)}\frac{dx(t)}{dt}+\frac{\partial U\left[x(t),y(t),z(t)\right]}{\partial y(t)}\frac{dy(t)}{dt}+\frac{\partial U\left[x(t),y(t),z(t)\right]}{\partial z(t)}\frac{dz(t)}{dt}$, which may be proven from partial differential calculus.

A.4 Prove the energy conservation law for N particles with masses $m_1,\dots,m_N$ moving in three dimensions. You will need the following vector generalization of the result for $\frac{\partial U[x(t),y(t),z(t)]}{dt}$ given in Problem A.3: $\frac{dU[\mathbf{r}_1(t),\dots,\mathbf{r}_N(t)]}{dt} = \sum_{i=1}^{N} \frac{\partial U[\mathbf{r}_1(t),\dots,\mathbf{r}_N(t)]}{\partial \mathbf{r}_i(t)} \cdot \frac{d\mathbf{r}_i(t)}{dt}$.

A.5 Using the transformation relations between Cartesian coordinates and polar coordinates of Equation (A.36), derive Equation (A.43) from Equation (A.42).

A.6 Using Equation (A.36), prove $J_z(t) = m[x(t)\dot{y}(t) - y(t)\dot{x}(t)] = mr^2(t)\dot{\phi}(t) = p_\phi(t)$.

A.7 Assume the one-dimensional harmonic oscillator formulas of Section A.1 provide a valid description of the vibrations of the $^{12}C^{16}O$ molecule if one takes $m = 1.14 \times 10^{-26}$ kg and $k = 1{,}860\ \text{N m}^{-1}$. Compute (a) the vibrational frequency υ in Hz of $^{12}C^{16}O$ and (b) the time T in fs it takes $^{12}C^{16}O$ to make one vibration.

A.8 Consider a particle of mass m moving in one dimension subject to the gravitational force $F_x(x) = -mg$ where g is the gravitational constant and where $x \geq 0$ is the height of the particle above the earth's surface. (a) Obtain the general solution of Newton's equation of motion for the particle. (b) Show that the particular trajectory generated by the initial conditions $x(t=0) = x$ and $p_x(t=0) = p_x$ is $x(t) = x + m^{-1}p_x t - \frac{1}{2}gt^2$. (c) Show that if the particle is initially moving up $(p_x > 0)$, it reaches a maximum height $x^* = x + \frac{1}{2}\frac{p_x^2}{gm^2}$ at time $t^* = \frac{p_x}{gm}$. (d) Noting that the potential energy function of the particle is $U(x) = mgx$, show from the result for $x(t)$ you obtained in part (b) that the energy E of the particle is a constant of the motion. (e) Using the energy conservation arguments of Section A.3, infer from a plot of $U(x)$ the qualitative form of the particle's trajectories. Especially, show in agreement with the result of part (c) that a particle with $p_x > 0$ has a turning point at the maximum height $x^x = x + \frac{1}{2}\frac{p_x^2}{gm^2}$.

A.9 Every dynamical system has many sets of generalized coordinates. For example, two valid sets of generalized coordinates for a particle moving in three dimensions are the cylindrical coordinates ρ, χ, z and the spherical polar coordinates r, θ, ϕ. These are defined in terms of the particle's Cartesian coordinates x, y, z by the respective transformation relations $\rho = (x^2 + y^2)^{1/2}$, $\chi = \tan^{-1}\left(\frac{y}{x}\right)$, and $z = z$ and $r = (x^2 + y^2 + z^2)^{1/2}$, $\theta = \cos^{-1}\left[\frac{z}{(x^2 + y^2 + z^2)^{1/2}}\right]$, and $\phi = \tan^{-1}\left(\frac{y}{x}\right)$. Prove that (a) the Cartesian coordinates when expressed in terms of the cylindrical coordinates are given by $x = \rho\cos\chi$, $y = \rho\sin\chi$, and $z = z$, (b) the Cartesian coordinates are determined in terms of the spherical polar coordinates by $x = r\sin\theta\cos\phi$, $y = r\sin\theta\sin\phi$, and $z = r\cos\theta$, and (c) the cylindrical coordinates are determined in terms of the spherical polar coordinates by $\rho = r\sin\theta$, $\chi = \phi$, and $z = r\cos\theta$.

A.10 Consider a one-dimensional system comprised of a particle of mass m and coordinate x subject to a force $F_x(x)$. (a) Prove that the time derivative of $\dot{F}_x[x(t)]$ are functions of the particle's basic observables $x(t)$ and $p_x(t)$ and thus are observables. (b) Similarly, show that an arbitrary time derivative of the force $\frac{d^n}{dt^n}F_x[x(t)]$ is an observable.

A.11 Consider a particle of mass m whose potential energy function is of the form $U = U(x,y,z) = U\left[\left(x^2+y^2\right)^{1/2}, z\right]$. The motion of the system is best analyzed using the cylindrical coordinates ρ, χ, z of Problem A.9. This is because when expressed in terms of the cylindrical coordinates, U is of the form $U = U(\rho, z)$ and thus is independent of the cylindrical coordinate χ. In this problem, you will treat the motion of the particle using Hamilton's equations expressed in cylindrical coordinates. You will need the transformation relations between the Cartesian and cylindrical coordinates given in Problem A.9. (a) Following the six-step procedure for implementing Hamilton's formulation described in Section A.5, show that ρ, χ, and z obey the following equations of motion:

$$m\ddot{\rho}(t) = -\frac{\partial}{\partial \rho(t)}\left\{\frac{p_\chi^2}{2m\rho^2(t)} + U[\rho(t), z(t)]\right\}$$

$$\dot{\chi}(t) = \frac{p_\chi}{m\rho^2(t)}$$

$$m\ddot{z}(t) = -\frac{\partial U[\rho(t), z(t)]}{\partial z(t)}$$

where $p_\chi = m\rho^2(t)\dot{\chi}(t)$ is a constant of motion. (b) Why are the equations of motion simpler in cylindrical coordinates than in Cartesian coordinates?

A.12 Particles undergoing two-dimensional central field motion can rotate about the origin in circular orbits. These orbits are characterized by a period T_{rot}, which is the time needed for the particle to make one complete revolution. T_{rot} depends on the form of the central field potential $U(r)$. In this problem, you will determine T_{rot} for a potential of the form $U(r) = -\frac{c}{r}$, where c is a positive constant. This form describes the gravitational attraction of the earth to the sun and the Coulomb attraction of an electron to a proton. (a) First, show from Equations (A.50) and (A.67) that the condition for a circular orbit of radius r is $r^3U'(r) = \frac{J^2}{m}$. (b) Next, show from Equation (A.68) specialized a circular orbit of radius r that $\frac{J}{mr^2}T_{rot} = 2\pi$ or $T_{rot} = \frac{2\pi mr^2}{J}$. (c) Specialize the general condition of part (a) to potentials of the form $U(r) = -\frac{c}{r}$ to obtain the following relation between J and $r: J = (mc)^{1/2} r^{1/2}$. (d) Thus, show from the result for T_{rot} in part(b) that for the case of a gravitational potential, this result is equivalent Kepler's third law of planetary motion when specialized to circular orbits. It states that if the planet is in a circular orbit, the period of revolution of a planet around the sun (or the length of a planet's year) is proportional to the $\frac{3}{2}$ power of the distance of the planet from the sun.

SOLUTIONS TO SELECTED PROBLEMS

A.7 (a) $\upsilon = 6.43 \times 10^{13}$ Hz and (b) 15.6 fs.

Appendix B
Classical Theory of Diatomic Rotational–Vibrational Motions

Microwave and infrared spectra reflect changes in the rotational and vibrational energies of molecules. Such spectra therefore require for their interpretation an understanding of molecular rotational and vibrational motions.

In this appendix, we present a classical analysis of these motions for diatomic molecules based on the introduction to classical mechanics given in Appendix A. We present this analysis for three reasons. (a) The classical theory of rotational–vibrational motions is much simpler than the corresponding quantum theory developed in Chapters 8 and 9 and thus the classical treatment is an instructive introduction to the quantum theory. (b) The classical theory leads to simple physical pictures that are hard to discern in the quantum theory. (c) Some of the results of the classical theory greatly help in the development of the quantum theory of diatomic rotational and vibrational motions as described in Chapter 8.

To start, we note that the force on an isolated molecule is conservative and therefore is derived from a potential energy function. Thus, we begin with a discussion of *diatomic potential energy functions.*

B.1 DIATOMIC POTENTIAL ENERGY FUNCTIONS

We first observe that diatomic potentials U depend only on the distance r between the two nuclei which is referred to as the internuclear separation. In Figure B.1, we plot as a typical example $U = U(r)$ for the HCl molecule. Note from the plot that at large $r, U(r)$ gives rise to attractive forces between the atoms that reflect chemical bonding, and at small r, it gives rise to repulsive forces due to electrostatic interactions. At $r = r_e$, these attractive and repulsive forces balance and $U(r)$ has its minimum value. Thus, r_e is the internuclear separation at which the molecule is maximally stable. It is called the *equilibrium internuclear separation* and is equal to the bond length of the molecule.

Next, note that since we have set the zero of energy (i.e., chosen the arbitrary constant in $U[r]$) so that $U(r_e) = 0,\ U(r)$ is the negative of the work required to bring the diatomic from r_e to r (Problem B.1). Given this interpretation of $U(r), D_e$ in Figure B.1 is the energy needed to bring the molecule from r_e to $r = \infty$. That is, D_e is the energy needed to pull apart the diatomic, and it is thus called the *classical dissociation energy.* (As shown in Figure 7.7, the true dissociation energy D_0 is slightly less than D_e due to the quantum zero point energy.) Similarly, $U(r)$ for $r \ll r_e$ is the energy required to bring the atoms from r_e to a point of close proximity. This energy is very large because of the strong short-range repulsive forces between the atoms and so $U(r)$ rises very steeply for $r < r_e$.

Finally, we note from Figure B.1 that the effective force on the molecule $F_r(r) = -\dfrac{dU(r)}{dr}$ has the following properties:

$$F_r = 0 \text{ for } y = 0 \text{ and } F_r(r) < 0 \quad \text{for } y > 0 \text{ and } F_r(r) > 0 \quad \text{for } y < 0 \tag{B.1}$$

where $y = r - r_e$. Recalling the discussion of Equation (A.5), $F_r(r)$ is a restoring force and thus produces oscillations of r about r_e. The oscillations are just the molecular vibrations of the diatomic studied in Chapter 7.

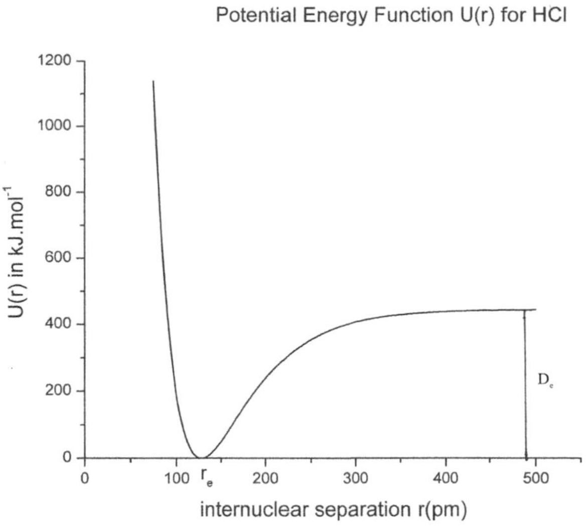

FIGURE B.1 The hydrogen chloride potential energy function $U(r)$. The classical dissociation energy $D_e = 445.1\,\text{kJ}\,\text{mol}^{-1}$ and the equilibrium internuclear separation $r_e = 127.5\,\text{pm}$.

Given this discussion of the potential energy function $U(r)$, we may now turn to the classical analysis of diatomic motion. This analysis will be performed in generalized coordinates. We next discuss the need for generalized coordinates and describe the first stage in the development of these coordinates.

B.2 BEGINNING OF THE DEVELOPMENT OF GENERALIZED COORDINATES FOR DIATOMIC MOTION

We consider the diatomic molecule shown in Figure B.2 with atomic masses m_1 and m_2 and atomic position vectors $\mathbf{r}_1 = (x_1, y_1, z_1)$ and $\mathbf{r}_2 = (x_2, y_2, z_2)$. Specializing Equation (A.32), the equations of motion of the molecule are

$$m_1\ddot{\mathbf{r}}_1(t) = -\frac{\partial U[\mathbf{r}_1(t), \mathbf{r}_2(t)]}{\partial \mathbf{r}_1(t)} \text{ and } m_2\ddot{\mathbf{r}}_2(t) = -\frac{\partial U[\mathbf{r}_1(t), \mathbf{r}_2(t)]}{\partial \mathbf{r}_2(t)} \tag{B.2}$$

where the potential energy function of the diatomic $U(\mathbf{r}_1, \mathbf{r}_2) = U(r)$ depends only on the internuclear separation $r = |\mathbf{r}_1 - \mathbf{r}_2|$ and hence may be written as $U(r)$. These vector equations (Equation B.2) are equivalent to six scalar equations of motion for the six atomic Cartesian coordinates x_1, y_1, z_1 and x_2, y_2, z_2. The scalar equations of motion are non-separable since the potential $U(r) = U\left\{\left[(x_1 - x_2)^2 + (y_1 - y_2)^2 + (z_1 - z_2)^2\right]^{1/2}\right\}$ depends on all six Cartesian coordinates. However, since $U(r)$ depends only on r, we expect simpler equations of motion if we transform to a set of generalized coordinates which include r.

So far, the only systems we have treated using generalized coordinates are the two-dimensional central field systems dealt with in Section A.5. To bridge the gap between these two-degree-of-freedom systems and six-degree-of-freedom diatomics, we build up the generalized coordinate treatment in stages. Thus, we first restrict ourselves to molecules comprised of one light and one heavy atom. An example is the $^1H^{127}I$ molecule with atomic masses

$$m(^1H) = 1.0084\text{u and } m(^{127}I) = 126.904\text{u}. \tag{B.3}$$

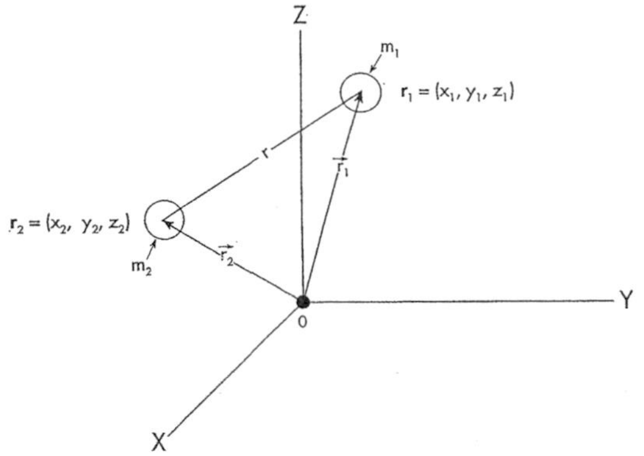

FIGURE B.2 A diatomic molecule with atomic masses m_1 and m_2, atomic position vectors $\mathbf{r}_1$ and $\mathbf{r}_2$, and internuclear separation r.

In the first approximation, the motion of the heavy atom $\left({}^{127}\text{I for } {}^{1}\text{H}^{127}\text{I}\right)$ may be ignored and it thus may be held fixed at the origin. Then, only the light atom with position vector $\mathbf{r} = (x, y, z)$ moves and the molecule becomes a three-degree-of-freedom system (Figure B.3). The motion of this system is the three-dimensional central field motion of the light atom in the potential $U(r)$, where $r = \left(x^2 + y^2 + z^2\right)^{1/2}$ is now the distance of the light atom from the origin. For convenience, we will call this three-dimensional central field system "hydrogen iodide" or "HI."

However, further simplification is possible. Because the angular momentum **J** is conserved for a particle undergoing classical three-dimensional central field motion, this motion is planar (Goldstein, Poole, and Safko 2002, p. 72). The plane of the motion may be taken as the XY-plane. Thus, the motion of "HI" may be reduced to the two-degree-of-freedom problem of two-dimensional central field motion occurring in the XY-plane (Figure B.4). This type of motion however was treated in Sections A.5 and A.6. Thus, one may adapt the results of these sections to determine the classical rotational–vibrational motions of "HI."

This determination is vastly simpler than the determination of the quantum rotational–vibrational motions of real diatomics given in Chapters 8 and 9. Thus, two-dimensional "HI" may be viewed as an instructive prototype system for the study of molecular rotations and vibrations.

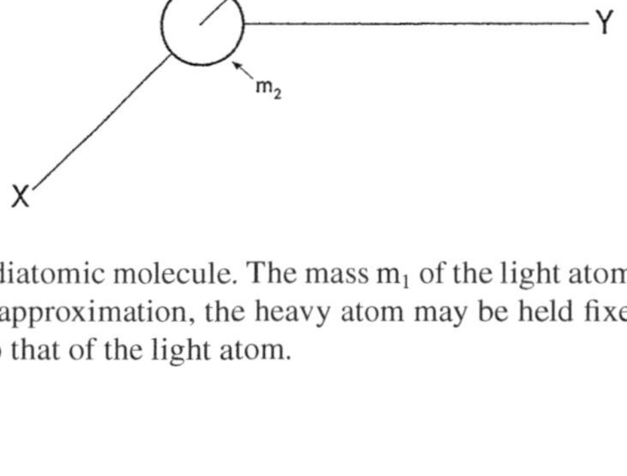

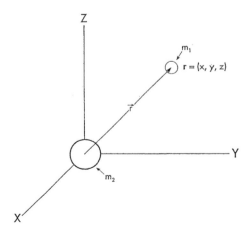

FIGURE B.3 A light–heavy diatomic molecule. The mass m_1 of the light atom is much less than the mass m_2 of the heavy atom. In the first approximation, the heavy atom may be held fixed at the origin and the motion of the diatomic then reduces to that of the light atom.

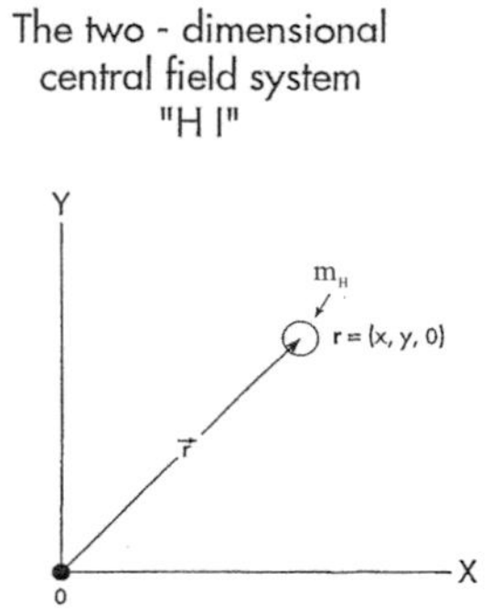

FIGURE B.4 The two-dimensional central field system "HI", the simplest model for diatomic rotational–vibrational motions.

B.3 TWO-DIMENSIONAL "HYDROGEN IODIDE": A PROTOTYPE SYSTEM FOR MOLECULAR ROTATIONS AND VIBRATIONS

The motions of two-dimensional "HI" are very easy to understand if as shown in Figure B.5 we represent the position vector $\mathbf{r}$ of the ^{1}H atom in terms of its polar coordinates r, ϕ. From Figure B.5a, the rotational motions of the ^{1}H atom are changes in the polar coordinate ϕ or equivalently changes in the direction of $\mathbf{r}$. From Figure B.5b, the vibrational motions of the ^{1}H atom are changes in its distance from the origin which is the polar coordinate r or equivalently changes in the magnitude of $\mathbf{r}$. Thus, r and ϕ are the vibrational and rotational coordinates of "HI."

Given these interpretations of r and ϕ, one may use the results of Sections A.5 and A.6 to obtain the vibrational and rotational trajectories of "HI." Denoting the mass of the ^{1}H atom by m_H from Equations (A.50) and (A.67), the vibrational trajectory $r(t)$ is the solution of the following equation of motion:

$$m_H\ddot{r}(t) = -\frac{dU_{eff}[r(t)]}{dr(t)} \tag{B.4}$$

where the effective potential is given by

$$U_{eff.}[r(t)] = \frac{J^2}{2m_H r^2(t)} + U[r(t)]. \tag{B.5}$$

J in Equation (B.5) is now interpreted as the magnitude of the "HI" rotational angular momentum vector $\mathbf{J} = \mathbf{r}(t) \times \mathbf{p}(t) = m_H \mathbf{r}(t) \times \dot{\mathbf{r}}(t)$. Next, from Equation (A.68) the rotational trajectory $\phi(t)$ is given by

$$\phi(t) = \phi + m_H^{-1} J \int_0^t r^{-2}(\tau)d\tau. \tag{B.6}$$

Equations (B.4)–(B.6) determine the classical rotational-vibrational motions of "HI."

We will also be interested in the rotational–vibrational Hamiltonian of "HI." From Equation (A.66), this Hamiltonian is

$$H = \frac{p_r^2(t)}{2m_H} + \frac{J^2}{2m_H r^2(t)} + U[r(t)] \tag{B.7}$$

(a)

The polar coordinate ϕ as an "HI" rotational coordinate

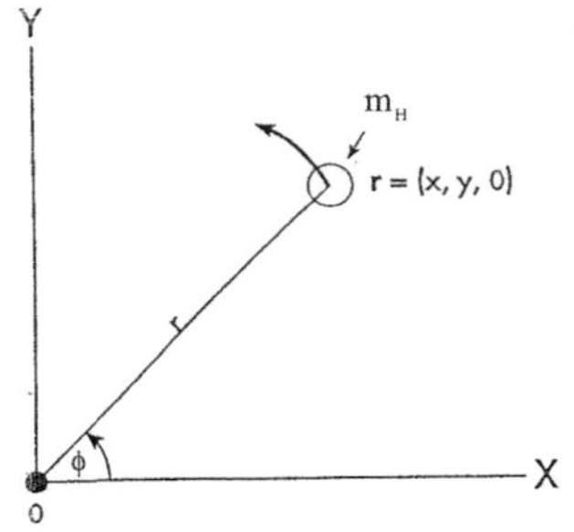

(b)

The polar coordinate r as an "HI" vibrational coordinate

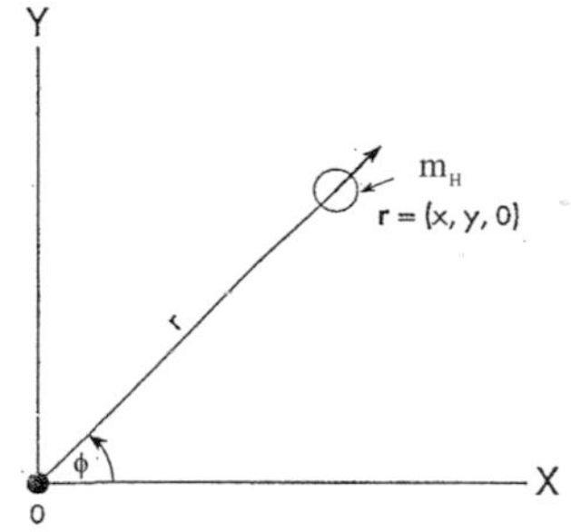

FIGURE B.5 The polar coordinates ϕ and r as "HI" rotational and vibrational coordinates. (a) The rotational motions of "HI" are changes in ϕ, and thus, ϕ is the rotational coordinate of "HI." (b) The vibrational motions of "HI" are changes in r, and thus, r is the vibrational coordinate of "HI."

where $p_r(t) = m_H\dot{r}(t)$ is the vibrational momentum. The three terms in the Hamiltonian H are the vibrational kinetic energy $\frac{p_r^2(t)}{2m_H}$, the rotational kinetic energy $\frac{J^2}{2m_H r^2(t)}$, and the potential energy $U[r(t)]$.

Finally, we note that from Equation (B.6), one may estimate the rotational period T_{rot} of "HI." This estimate is possible since often the maximum vibrational displacement $y = r - r_e$ of the internuclear separation r from r_e is small. If this is so, one may fix $r(\tau)$ at r_e. Then, the ^1H atom executes circular motion about the origin with radius r_e and constant angular velocity $\dot{\phi}(t)$ and T_{rot} is the time needed for the atom to make one revolution. Mathematically, one sets $r(\tau)$ equal r_e in Equation (B.6) to obtain the following approximation for $\phi(t)$:

$$\phi(t) \doteq \phi + \frac{J}{m_H r_e^2} t. \tag{B.8}$$

The condition for T_{rot} is $\phi(T_{rot}) = \phi + 2\pi$, which from Equation (B.8) yields T_{rot} as

$$T_{rot} = \frac{2\pi m_H r_e^2}{J}. \tag{B.9}$$

Additionally, from Equation (B.9) the number of rotations made per second or rotational frequency $\upsilon_{rot} = T_{rot}^{-1}$ is given by

$$\upsilon_{rot} = \frac{J}{2\pi m_H r_e^2}. \tag{B.10}$$

Next, we note that while in classical mechanics three-dimensional central field motion may be reduced to two-dimensional central field motion, this is not true in quantum mechanics. Thus, to compare classical and quantum results, we next study the motion of "HI" as the three-degree-freedom problem of three-dimensional central field motion.

B.4 THREE-DIMENSIONAL "HYDROGEN IODIDE"

Three-dimensional "HI" along with the generalized coordinates r, θ, and ϕ which will be used to analyze its motion is depicted in Figure B.6. The coordinates r, θ, and ϕ are known as the spherical polar coordinates. They are defined in terms of the Cartesian coordinates x, y, and z of the ^1H atom by the transformation relations

$$r = \left(x^2 + y^2 + z^2\right)^{1/2} \quad \theta = \cos^{-1}\left[\frac{z}{\left(x^2 + y^2 + z^2\right)^{1/2}}\right] \quad \phi = \tan^{-1}\left(\frac{y}{x}\right). \tag{B.11}$$

The inverse transformation relations are

$$x = r\sin\theta\cos\phi \quad y = r\sin\theta\sin\phi \quad z = r\cos\theta. \tag{B.12}$$

Note that the central field potential $U(r)$ is independent of the coordinates θ and ϕ and depends only on the coordinate r. For this reason, the "HI" equations of motion simplify when expressed in terms of the spherical polar coordinates.

The spherical polar coordinates have additional advantages. Namely, from Figure B.6 the rotational motions of the ^1H atom are changes in the spherical polar angles θ and ϕ or equivalently changes in the direction of the ^1H atom position vector **r**. Similarly, the vibrational motions of "HI" are changes in the spherical polar coordinate r or equivalently changes in the magnitude of **r**. Thus, r is the vibrational coordinate of three-dimensional "HI" and θ and ϕ are its rotational coordinates.

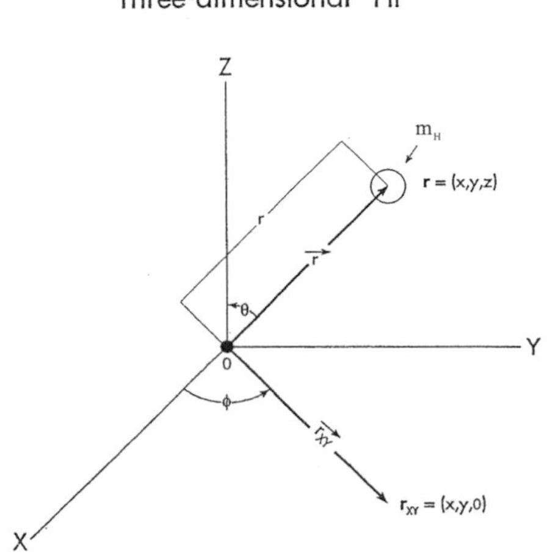

FIGURE B.6 Spherical polar coordinates r, θ, and ϕ of the ^1H atom position vector **r** of three-dimensional "HI."

We next turn to the generalized coordinate treatment of three-dimensional "HI." We start with the Cartesian velocity form for its Hamiltonian

$$H = \frac{1}{2}m_H\dot{x}^2(t) + \frac{1}{2}m_H\dot{y}^2(t) + \frac{1}{2}m_H\dot{z}^2(t) + U\left\{\left[x^2(t) + y^2(t) + z^2(t)\right]^{1/2}\right\}. \tag{B.13}$$

Using the transformation relations of Equation (B.12), one may convert the Cartesian velocity form of H into the generalized velocity form (Problem B.2)

$$H = \frac{1}{2}m_H\dot{r}^2(t) + \frac{1}{2}m_H r^2(t)\dot{\theta}^2(t) + \frac{1}{2}m_H r^2(t)\sin^2\theta(t)\dot{\phi}^2 + U[r(t)]. \tag{B.14}$$

We may now obtain the generalized momenta $p_r(t), p_\theta(t)$, and $p_\phi(t)$ conjugate (associated) to with r, θ, and ϕ from Equation (B.14) for H using Hamilton's prescription

$$p_r(t) = \frac{\partial H}{\partial \dot{r}(t)} \quad p_\theta(t) = \frac{\partial H}{\partial \dot{\theta}(t)} \quad \text{and} \quad p_\phi(t) = \frac{\partial H}{\partial \dot{\phi}(t)}. \tag{B.15}$$

This gives the generalized momenta as

$$p_r(t) = m_H\dot{r}(t), \quad p_\theta(t) = m_H r^2(t)\dot{\theta}(t), \quad p_\phi(t) = m_H r^2(t)\sin^2\theta(t)\dot{\phi}(t). \tag{B.16}$$

Using these results for the generalized momenta to eliminate $\dot{r}(t), \dot{\theta}(t)$, and $\dot{\phi}(t)$ from Equation (B.14) gives the generalized momentum form of the Hamiltonian as

$$H = \frac{p_r^2(t)}{2m_H} + \frac{1}{2m_H r^2(t)}\left[p_\theta^2(t) + \frac{p_\phi^2(t)}{\sin^2\theta(t)}\right] + U[r(t)]. \tag{B.17}$$

Given Equation (B.17), we may now write down Hamilton's equations of motion (Problem B.3). These show that while $p_\phi(t)$ is a constant of the motion, $p_\theta(t)$ is not. As a consequence, the "HI" equations of motion are more complex in three than in two dimensions. For example, in two dimensions the equation of motion for $r(t)$, Equation (B.4), is of independent one-dimensional form. The corresponding equation of motion in three dimensions

$$m_H\ddot{r}(t) = -\frac{\partial}{\partial r(t)}\left\{\frac{1}{2m_H r^2(t)}\left[p_\theta^2(t) + \frac{p_\phi^2(t)}{\sin^2\theta(t)}\right] + U[r(t)]\right\} \tag{B.18}$$

is in contrast not closed since it depends on $\theta(t)$ and $p_\theta(t)$ as well as on $r(t)$.

We next show however that $\left[p_\theta^2(t) + \frac{p_\phi^2(t)}{\sin^2\theta(t)}\right]$ is a constant of the motion, and hence, Equation (B.18) may be reduced to independent one-dimensional form. This simplification is possible since one may show that the angular momentum

$$\mathbf{J}(t) = \mathbf{r}(t)\times\mathbf{p}(t) = m_H\mathbf{r}(t)\times\dot{\mathbf{r}}(t) \tag{B.19}$$

is conserved for three-dimensional central field motion (Taylor 2005, pp. 90–91). Therefore, the magnitude $J(t) = J$ of $\mathbf{J}(t)$ is a constant of the motion, and hence J^2 is also a constant of the motion. However, when expressed in terms of spherical polar coordinates, J^2 has the form (Problem B.5)

$$J^2 = p_\theta^2(t) + \frac{p_\phi^2(t)}{\sin^2\theta(t)}. \tag{B.20}$$

Consequently, the first two terms in square brackets on right-hand side of Equation (B.20) as promised earlier is a constant of the motion and hence as also promised earlier, Equation (B.18) reduces to an independent one-dimensional equation of motion:

$$m_H \ddot{r}(t) = -\frac{dU_{eff}[r(t)]}{dr(t)} \tag{B.21}$$

where the effective potential is given by

$$U_{eff}[r(t)] = \frac{J^2}{2m_H r^2(t)} + U[r(t)]. \tag{B.22}$$

Similarly, the rotational–vibrational Hamiltonian of Equation (B.17) simplifies to

$$H = \frac{p_r^2}{2m_H} + \frac{J^2}{2m_H r^2(t)} + U[r(t)]. \tag{B.23}$$

Note that the basic equations (Equations B.21–B.23) for three-dimensional "HI" are identical in form to the corresponding equations for two-dimensional "HI" (Equations B.4, B.5 and B.7).

Next, let us turn to real diatomic molecules.

B.5 THE MOTIONS OF REAL DIATOMIC MOLECULES: RELATIVE AND CENTER-OF-MASS COORDINATES

Consider a diatomic molecule with atomic masses m_1 and m_2 and atomic position vectors $\mathbf{r}_1$ and $\mathbf{r}_2$ (Figure B.2). A typical diatomic is the $^{12}C^{16}O$ molecule with atomic masses $m(^{12}C) = 12u$ and $m(^{16}O) = 15.9994u$. These masses are comparable. So in contrast to $^1H^{127}I$, for typical diatomics it is not a permissible first approximation to hold one of the atoms fixed at the origin. Rather, for typical diatomics both atoms move appreciably during rotations and vibrations. Additionally, the molecule as a whole translates through space.

Thus, to determine the motion of a real diatomic, one must find the trajectories $\mathbf{r}_1(t)$ and $\mathbf{r}_2(t)$ of both atoms. This requires solving the six coupled equations of motion of Equation (B.2). Fortunately, this six-degree-of-freedom problem may be greatly simplified if one transforms from the six atomic Cartesian coordinates $\mathbf{r}_1 = (x_1, y_1, z_1)$ and $\mathbf{r}_2 = (x_2, y_2, z_2)$ to the set of six generalized coordinates $\mathbf{r} = (x, y, z)$ and $\mathbf{R} = (X, Y, Z)$ depicted in Figure B.7 and defined by the transformation relations

$$\mathbf{r} = \mathbf{r}_1 - \mathbf{r}_2 \tag{B.24}$$

and

$$\mathbf{R} = \frac{m_1\mathbf{r}_1 + m_2\mathbf{r}_2}{m_1 + m_2}. \tag{B.25}$$

The generalized coordinates $\mathbf{r}$ are called the *relative coordinates*, while the generalized coordinates $\mathbf{R}$ are referred to as the *center-of-mass coordinates*. The transformation relations that are the inverses of Equations (B.24) and (B.25) are

$$\mathbf{r}_1 = \frac{m_2}{m_1 + m_2}\mathbf{r} + \mathbf{R} \tag{B.26}$$

and

$$\mathbf{r}_2 = -\frac{m_1}{m_1 + m_2}\mathbf{r} + \mathbf{R}. \tag{B.27}$$

We next express the Hamiltonian H of the diatomic in terms of **r** and **R**. We start with the Cartesian velocity form of H which in vector notation is

$$H=\frac{1}{2}m_1\dot{r}_1(t)\cdot\dot{r}_1(t)+\frac{1}{2}m_2\dot{r}_2(t)\cdot\dot{r}_2(t)+U[r(t)]. \tag{B.28}$$

(where we have used the fact emphasized in Section B.1 that diatomic potentials U = U(r) depend only on the molecule's internuclear distance r.) Eliminating $\dot{r}_1(t)$ and $\dot{r}_2(t)$ from Equation (B.28) using Equations (B.26) and (B.27) yields the generalized velocity form of H as (Problem B.6)

$$H=\frac{1}{2}M\dot{\mathbf{R}}(t)\cdot\dot{\mathbf{R}}(t)+\frac{1}{2}\mu\dot{\mathbf{r}}(t)\cdot\dot{\mathbf{r}}(t)+U[r(t)] \tag{B.29}$$

where

$$M=m_1+m_2 \tag{B.30}$$

is the total mass of the molecule and where

$$\mu=\frac{m_1m_2}{m_1+m_2} \tag{B.31}$$

is a quantity with the dimensions of mass called the *reduced mass*. For a light–heavy diatomic with atomic masses $m_1 << m_2$, the reduced mass μ is close to the mass m_1 of the light atom. Thus, for $^1H^{127}I$ with atomic masses $m(^1H)=1.008u$ and $m(^{127}I)=126.904u$ $\mu=1.000u$, which is only slightly less than $m(^1H)$. But for $^{12}C^{16}O$ with the comparable atomic masses $m(^{12}C)=12u$ and $m(^{16}O)=15.9994u$ $\mu=6.0857u$, which is not close to either atomic mass. As a third example, for a homonuclear diatomic molecule, $m_1=m_2=m$, and thus, $\mu=\frac{1}{2}m$. So for example, for $^{12}C^{12}C$ $\mu=6u$, which is not close to $m(^{12}C)=12\mu$. Moreover, as illustrated by our examples, in general μ is less than either m_1 or m_2. This is why μ is called the reduced mass.

To continue our analysis, we next derive the generalized momentum form of To do this, we first apply Hamilton's prescription to Equation (B.29) to give the generalized momenta $\mathbf{P}(t)$ and $\mathbf{p}(t)$ for the center-of-mass and relative coordinate motions as

$$\mathbf{P}(t)=\frac{\partial H}{\partial\dot{\mathbf{R}}(t)}=M\dot{\mathbf{R}}(t)\text{ and }\mathbf{p}(t)=\frac{\partial H}{\partial\dot{\mathbf{r}}(t)}=\mu\dot{\mathbf{r}}(t). \tag{B.32}$$

We then use Equation (B.32) to eliminate $\dot{\mathbf{R}}(t)$ and $\dot{r}(t)$ from Equation (B.29) to obtain the generalized momentum form of H

$$H=\frac{\mathbf{P}(t)\cdot\mathbf{P}(t)}{2M}+\frac{\mathbf{p}(t)\cdot\mathbf{p}(t)}{2\mu}+U[r(t)]. \tag{B.33}$$

However, the internuclear separation r is the magnitude of the relative coordinate vector **r**. Thus, the potential $U(r)$ depends on **r** but is independent of **R**. This leads to great simplification since from Hamilton's equations of motion the motions of **R** and **r** move independently. Therefore, the six-degree-of-freedom problem has been reduced to two independent three-degree-of-freedom problems: the problem of the motion of **R** and the problem of the motion of **r**. Moreover, the former problem is very simple since the motion of **R** is that of a fictitious free particle of mass M (Problem B.7). Additionally, the latter problem has in effect already been solved (in the context of three-dimensional "HI") since the motion of **r** is that of a fictitious particle of mass μ subject to the three-dimensional central field potential $U(r)$.

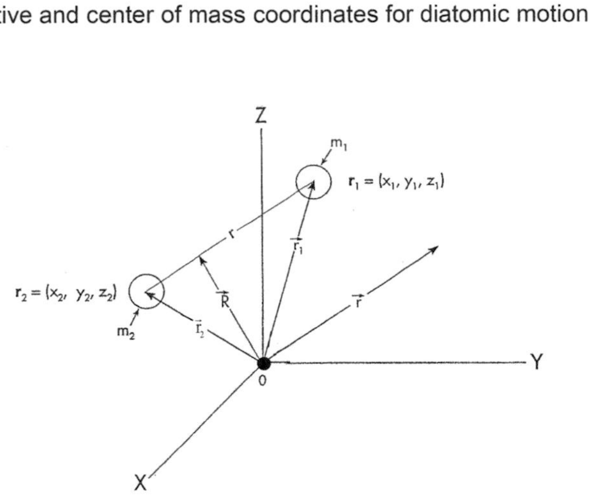

FIGURE B.7 Relative and center-of-mass coordinates $\mathbf{r} = \mathbf{r}_1 - \mathbf{r}_2$ and $\mathbf{R} = \dfrac{m_1\mathbf{r}_1 + m_2\mathbf{r}_2}{m_1 + m_2}$ of a diatomic molecule.

We further note that for a free particle, the linear momentum is a constant of the motion. (See Section A.2.) Consequently,

$$\mathbf{P}(t) = \mathbf{P}(t = 0) = \mathbf{P}. \tag{B.34}$$

Thus, the velocity $\mathbf{V} = M^{-1}\mathbf{P}$ of the fictitious free particle is constant. Therefore, one may choose a reference frame moving with the constant velocity $\mathbf{V}$. In this frame, $\mathbf{V} = \mathbf{P} = \mathbf{0}$ and the Hamiltonian of Equation (B.33) simplifies to

$$H = \frac{\mathbf{p}(t) \cdot \mathbf{p}(t)}{2\mu} + U[r(t)]. \tag{B.35}$$

H of Equation (B.35) is the rotational–vibrational Hamiltonian of the diatomic.

It describes the three-dimensional central field motion of a particle of mass μ in the potential $U(r)$. It is therefore identical to the Cartesian momentum form of the Hamiltonian of three-dimensional "HI" if one makes the following two substitutions. (a) The position vector $\mathbf{r} = (x,y,z)$ of the ^{1}H atom in the "HI" Hamiltonian is replaced by the relative coordinate vector $\mathbf{r} = (x,y,z)$. (b) The mass m_H of the ^{1}H atom in the "HI" Hamiltonian is replaced by the reduced mass μ of the diatomic.

Given these correspondences, we may write down the basic results for the rotational–vibrational motions of real diatomics by transcribing the results derived in Section B.4 for three-dimensional "HI." To do this, we represent the relative coordinate vector $\mathbf{r}$ in terms of its spherical polar coordinates r, θ, and ϕ (Figure B.8). In analogy to our interpretation of the spherical polar coordinates of "HI," r is the diatomic's vibrational coordinate and θ and ϕ are its rotational coordinates. By transcribing Equations (B.16) and (B.17), we may now write down the following spherical polar form for the diatomic's rotational–vibrational Hamiltonian:

$$H = \frac{p_r^2}{2\mu}(t) + \frac{1}{2\mu r^2(t)}\left[p_\theta^2(t) + \frac{p_\phi^2(t)}{\sin^2\theta(t)}\right] + U[r(t)] \tag{B.36}$$

where $p_r(t) = \mu\dot{r}(t), p_\theta(t) = \mu r^2(t)\dot{\theta}(t)$, and $p_\phi(t) = \mu r^2(t)\sin^2\theta(t)\dot{\phi}(t)$.

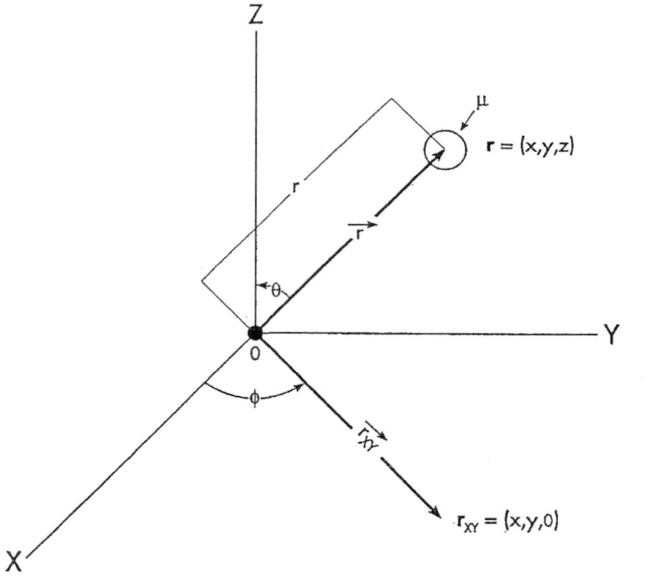

FIGURE B.8 Spherical polar coordinates r,θ, and ϕ of the relative coordinate vector **r** of a diatomic molecule.

H of Equation (B.36) may be reduced to the following form:

$$H = \frac{p_r^2(t)}{2\mu} + \frac{J^2}{2\mu r^2(t)} + U[r(t)] \tag{B.37}$$

which is analogous to the "HI" equation (Equation B.23). J in Equation (B.37) is the magnitude of the diatomic's conserved rotational angular momentum vector

$$\mathbf{J}(t) = \mathbf{r}(t) \times \mathbf{p}(t) = \mu\mathbf{r}(t) \times \dot{\mathbf{r}}(t). \tag{B.38}$$

Similarly, the equation of motion of the diatomic's vibrational coordinate $r(t)$ is in analogy to the "HI" equations (Equations B.21 and B.22)

$$\mu\ddot{r}(t) = -\frac{dU_{eff}[r(t)]}{dr(t)} \tag{B.39}$$

where the effective potential is given by

$$U_{eff}[r(t)] = \frac{J^2}{2\mu r^2(t)} + U[r(t)]. \tag{B.40}$$

We next use the classical to quantum correspondence $J^2 \leftrightarrow J(J+1)\hbar^2$ between the magnitude J of the classical rotational angular momentum vector **J** and the value J of the rotational angular momentum quantum number to write $U_{eff}[r(t)]$ in semiclassical form as

$$U_{eff}[r(t)] = \frac{J(J+1)\hbar^2}{2\mu r^2(t)} + U[r(t)]. \tag{B.41}$$

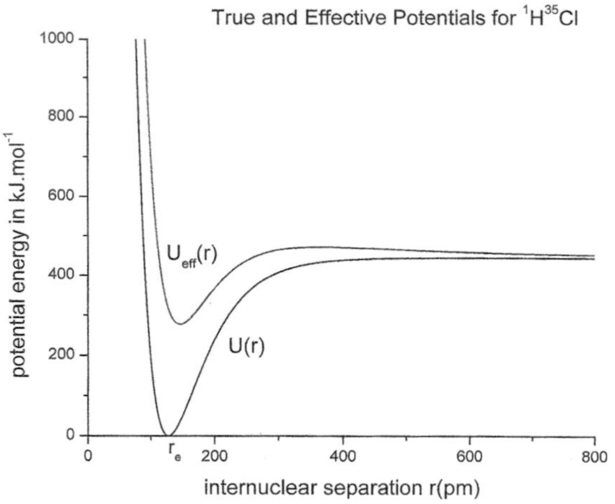

FIGURE B.9 True $U(r)$ and effective $U_{eff}(r)$ potentials for the $^{1}H^{35}Cl$ molecule. $U_{eff}(R)$ is determined from Equation (B.41) with $U(r)$ taken as the HCl potential of Figure B.1 and with the rotational angular momentum quantum number $J = 50$.

$U_{eff}(r)$ determined from Equation (A.41) is plotted in Figure B.9 for the $^{1}H^{35}Cl$ molecule for rotational angular momentum quantum number $J = 50$. For comparison purposes, the true potential $U(r)$, which is $U_{eff}(r)$ for a non-rotating molecule with $J = 0$, is also plotted. These plots show that molecular rotation renders the well-depth of $U_{eff}(r)$ smaller than that of $U(r)$ and also shifts the equilibrium internuclear separation to a value larger than r_e. Additionally, $U_{eff}(r)$ has a maximum at $r_e \doteq 291\text{pm}$. Because of this maximum, $U_{eff}(r)$ has a *centrifugal barrier* at large r.

We next note that valuable information about the vibrational trajectories $r(t)$ may be obtained without solving the equation of motion of Equation (B.39) but by merely examining a plot of $U_{eff}(r)$ like that of Figure B.9. The basis of this observation is that the energy conservation method of Section A.3 for the qualitative study of one-dimensional trajectories $x(t)$ may be extended to vibrational trajectories $r(t)$. This follows because H of Equation (B.37) is equal to the conserved rotational–vibrational energy E of the diatomic and thus may be written as

$$E = \frac{1}{2}\mu v_r^2(t) + U_{eff}\left[r(t)\right] \tag{B.42}$$

where $v_r(t) = \dot{r}(t)$ is the vibrational velocity and $U_{eff}\left[r(t)\right]$ is given by Equation (B.40). Equation (B.42) shows that as $U_{eff}\left[r(t)\right]$ increases, $v_r(t)$ decreases, and vice versa, so a plot of $U_{eff}(r)$ shows how $v_r(t)$ varies along a trajectory. Additionally, a plot of $U_{eff}(r)$ locates the turning points of the trajectories $r(t)$. Thus, such a plot reveals, for example, how the centrifugal barrier can prevent breakup of molecules with energies greater than the dissociation energy.

Finally, we note that since the classical three-dimensional central field motion of the relative coordinate **r** may be reduced to two-dimensional central field motion, the "HI" results for the rotational period T_{rot} and the rotational frequency υ_{rot} of Equations (B.9) and (B.10) may be transcribed to real diatomics as

$$T_{rot} = \frac{2\pi\mu r_e^2}{J} \tag{B.43}$$

and

$$\upsilon_{rot} = \frac{J}{2\pi\mu r_e^2}. \tag{B.44}$$

FURTHER READINGS

Goldstein, Herbert, Charles Poole, and John Safko. 2002. *Classical mechanics*. 3rd ed. San Francisco, CA: Addison Wesley.

Pauling, Linus, and E. Bright Wilson Jr. 1985. *Introduction to quantum mechanics with applications to chemistry*. New York: Dover Publications, Inc.

Taylor, John R. 2005. *Classical mechanics*. Sausalito, CA: University Science Books.

Thornton, Stephen T., and Jerry B. Marion. 2004. *Classical dynamics of particles and systems*. 5th ed. Belmont, CA: Brooke/Cole.

PROBLEMS

B.1 For a particle moving in one dimension subject to a force $F_x(x)$, the standard expression for the work W required to displace the particle from x_1 to x_2 is $W = \int_{x_1}^{x_2} F_x(x)dx$. For a particle moving in a three-dimensional central field potential $U(r)$, this expression generalizes to $W = \int_{r_1}^{r_2} F_r(r)dr$, where $F_r(r) = -\frac{dU(r)}{dr}$ is the effective force acting on the particle. Show that if the zero of energy is set so that $U(r_e) = 0$, where r_e is the point at which $U(r)$ has its minimum value, then $U(r)$ is the negative of the work required to displace the particle from r_e to r.

B.2 Using the transformation relations of Equation (B.12), derive the generalized velocity form of the Hamiltonian of three-dimensional "HI" of Equation (B.14) from its Cartesian velocity form given in Equation (B.13).

B.3 (a) Starting with Equation (B.17) and following the prescription of Section A.5, (a) write down Hamilton's equations of motion for three-dimensional "HI". From these equations of motion, show that (b) $p_\phi(t)$ is a constant of the motion, (c) $p_\theta(t)$ is not a constant of the motion, and (d) Equation (B.18) holds.

B.4 Show that when expressed in spherical polar coordinates, the components $J_x(t)$, $J_y(t)$, and $J_z(t)$ of the rotational angular momentum vector $\mathbf{J}(t)$ defined in Equation (B.19) have the forms $J_x(t) = -[\sin\phi(t)p_\theta(t) + \cos\phi(t)\cot\theta(t)p_\phi(t)]$, $J_y(t) = [\cos\phi(t)p_\theta(t) - \sin\phi(t)\cot\theta(t)p_\phi(t)]$, and $J_z(t) = p_\phi(t)$.

B.5 Show from the results of Problem B.4 that $J^2(t) = J_x^2(t) + J_y^2(t) + J_z^2(t) = p_\theta^2(t) + \frac{p_\phi^2(t)}{\sin^2\theta(t)}$.

B.6 Derive Equation (B.29) with M and μ given by Equations (B.30) and (B.31) from Equation (B.28) using the transformation relations of Equations (B.26) and (B.27).

B.7 (a) Prove from Equation (B.33) and Hamilton's equation of motion $\dot{\mathbf{P}}(t) = -\frac{\partial H}{\partial \mathbf{R}(t)}$ that $\mathbf{P}(t)$ is a constant of the motion and thus conforms to Equation (B.34). (b) Prove from this result that $\mathbf{R}(t) = M^{-1}\mathbf{P}t + \mathbf{R}$ and thus comparing with Equation (A.4) that the center of mass coordinate $\mathbf{R}$ moves like a free particle of mass M.

B.8 For a diatomic molecule with atomic masses m_1 and m_2, show that the reduced mass $\mu = \frac{m_1 m_2}{m_1 + m_2}$ is less than either m_1 or m_2.

B.9 Consider a diatomic molecule with atomic masses m_1 and m_2 and atomic position vectors $\mathbf{r}_1$ and $\mathbf{r}_2$. Denote its relative coordinate and center of mass coordinate vectors by $\mathbf{r}$ and $\mathbf{R}$. Assume a coordinate system has been chosen that is moving with the velocity $\mathbf{V}$ of the center of mass. Further assume that the center of mass is at the origin of the system so that $\mathbf{R} = \mathbf{0}$. (a) Then, show for a diatomic with $m_2 \gg m_1$ that $\mathbf{r}_1 \approx \mathbf{r}$ and $\mathbf{r}_2 \approx \mathbf{0}$. (b) Explain why this result and the fact that $\mu \approx m_2$ if $m_2 \gg m_1$ demonstrate that for light–heavy diatomic molecules, the "HI" model of Figure B.3 approximately emerges from the rigorous treatment of diatomic motion of Section B.5.

B.10 In this problem, you will determine the rotational period T_{rot} for the $^1H^{19}F$ molecule. To do this, you will use the correspondence $J \leftrightarrow [J(J+1)]\hbar^{1/2}$ between the magnitude J of the classical rotational angular momentum vector **J** and the value J of the angular momentum quantum number to write Equation (B.43) in semiclassical form as $T_{rot} = \dfrac{2\pi\mu r_e^2}{J\hbar \leftrightarrow [J(+1)]^{1/2}\hbar}$. To apply this relation to calculate T_{rot} for $^1H^{19}F$, you will need the atomic masses of the ^1H and ^{19}F atoms $m(^1H) = 1.0078\,u$ and $m(^{19}F) = 18.9984\,u$ and the HF bond length $r_e = 91.68\,pm$. You will also need the conversion factor $m_u = 1.66054 \times 10^{-27}\,kg\,u^{-1}$. Taking J = 2 a typical value, calculate T_{rot} is ps for $^1H^{19}F$.

B.11 Show from energy conservation arguments based on Equation (B.42) that a diatomic molecule with J = 0 spends more time in vibrational stretches than in vibrational compressions and that the time difference increases as the energy E increases.

B.12 In this problem, you will make a plot of the effective potential $U^{eff}(r)$ like the one in Figure B.9 for the $^1H^{127}I$ molecule. You will then use this plot and Equation (A.42) to qualitatively discuss the vibrational trajectories r(t) of $^1H^{137}I$. You will need the $^1H^{137}I$ reduced mass that is $\mu(^1H^{127}I) = 1.66 \times 10^{-27}$ kg. For the true potential U(r), you will use the empirical Morse form $U(r) = D_e\{1 - \exp[-\beta(r - r_e)]\}^2$ where D_e is the classical dissociation energy, r_e is the equilibrium internuclear separation, and β is a parameter that controls the range of the Morse potential. For HI, the Morse parameters have the values $D_e = 294.9\,kJ\,mol^{-1}$, $r_e = 160.9\,pm$, and $\beta = 0.018\,pm^{-1}$.

(a) Plot on the same graph the Morse potential U(r) and $U_{eff}(r)$ of Equation (B.41). Assume the rotational quantum number J = 50. (b) From your plots and Equation (B.42), qualitatively discuss how molecular rotation affects the vibrational trajectories of $^1H^{127}I$. (c) From your plot of $U_{eff}(r)$ and Equation (B.42), explain how the centrifugal barrier can prevent dissociation of molecules with energy greater than D_e. (d) Similarly, qualitatively discuss collisions of the ^1H and ^{127}I atoms and how the centrifugal barrier can drastically alter the outcome of these collisions.

SOLUTIONS TO SELECTED PROBLEMS

B.10 $T_{rot} = 0.845$ ps.

Appendix C
Solution of the Time-Independent Schrödinger Equation for the One-Dimensional Harmonic Oscillator

Here, we solve the time-independent Schrödinger equation (Equations 7.8) for the one-dimensional harmonic oscillator to obtain the results of Section 7.3. Especially, we derive the harmonic oscillator energy level formula of Equations (7.10) and the forms for the Hermite polynomials $H_n(x)$ given Table 7.1. We in essence follow the treatment of Pauling and Wilson (1985) which is an application of the Frobenius series method (Arfken, Weber, and Harris 2012) for solving differential equations.

C.1 THE TIME-INDEPENDENT SCHRÖDINGER EQUATION IN REDUCED VARIABLES

To start, we write the harmonic oscillator Schrödinger equation (Equation 7.8) as

$$\frac{d^2\Psi}{dy^2} + \frac{2m}{\hbar^2}\left(E - \frac{1}{2}m\omega^2 y^2\right)\Psi = 0. \tag{C.1}$$

Defining

$$\lambda = \frac{2m}{\hbar^2}E \tag{C.2}$$

and

$$\alpha = \frac{m\omega}{\hbar} \tag{C.3}$$

the Schrödinger equation may be rewritten as

$$\frac{d^2\Psi}{dy^2} + \left(\lambda - \alpha^2 y^2\right)\Psi = 0. \tag{C.4}$$

In terms of the new variable

$$\xi = \alpha^{1/2} y \tag{C.5}$$

Equation (C.4) becomes

$$\frac{d^2\Psi}{d\xi^2} + \left(\frac{\lambda}{\alpha} - \xi^2\right)\Psi = 0. \tag{C.6}$$

We must solve Equation (C.6) subject to the boundary conditions

$$\lim_{\xi \to \pm\infty} \Psi = 0. \tag{C.7}$$

C.2 FACTORIZATION OF THE WAVE FUNCTION AND HERMITE'S EQUATION

For reasons that will be clear shortly, it will be advantageous to determine the rough form of Ψ as $\xi \to \pm\infty$. In this limit, Equation (C.6) becomes

$$\frac{d^2\Psi}{d\xi^2} = \xi^2\Psi(\xi). \tag{C.8}$$

A solution of Equation (C.8) valid as $\xi \to \pm\infty$ and in accord with the boundary conditions is

$$\Psi \sim \exp\left(-\frac{\xi^2}{2}\right). \tag{C.9}$$

For functions H which are well-behaved one may legitimately write the wave function as

$$\Psi = \exp\left(\frac{-\xi^2}{2}\right)H(\xi). \tag{C.10}$$

Inserting Equation (C.10) into Equation (C.6) gives the following differential equation for $H(\xi)$ called Hermite's differential equation (Problem C.1):

$$H''(\xi) - 2\xi H'(\xi) + \left(\frac{\lambda}{\alpha} - 1\right)H(\xi) = 0. \tag{C.11}$$

Now, we can see why it was advantageous to find the $\xi \to \infty$ form of $\Psi \sim \exp\left(-\frac{1}{2}\xi^2\right)$. Factoring out this component of Ψ yielded a differential equation (Equation C.11) that is much simpler to solve than the original equation (Equation C.6). For example, we will solve Equation (C.11) as a power series in ξ. The corresponding solution of Equation (C.6) requires a much more complex power series that accounts for the dependence of Ψ on the $\exp\left(-\frac{\xi^2}{2}\right)$ factor.

C.3 POWER SERIES SOLUTION OF HERMITE'S EQUATION

Equation (C.11) is Hermite's differential equation. We will solve it by the power series method. Thus, we assume that $H(\xi)$ may be expanded as the power series

$$H(\xi) = a_0 + a_1\xi + a_2\xi^2 + \cdots = \sum_{n=0}^{\infty} a_n\xi^n \tag{C.12}$$

and then, we substitute the series into the Hermite equation (Equation C.11) to find the coefficients $a_0, a_1, \ldots$. To do this, we require the derivatives

$$H'(\xi) = \sum_{n=0}^{\omega} na_n\xi^{n-1} \tag{C.13}$$

and

$$H''(\xi) = \sum_{n=0}^{\omega} n(n-1)a_n\xi^{n-2}. \tag{C.14}$$

We may rewrite $H''(\xi)$ as (Problem C.2)

$$H''(\xi) = \sum_{n=0}^{\infty} (n+2)(n+1)a_{n+2}\xi^{n}. \tag{C.15}$$

Then, inserting Equations (C.12), (C.13), and (C.15) into Equation (C.11) gives

$$\sum_{n=0}^{\infty}\left[(n+2)(n+1)a_{n+2} - 2na_n + \left(\frac{\lambda}{\alpha} - 1\right)a_n\right]\xi^n = 0. \tag{C.16}$$

Equation (C.16) can only be obeyed for all ξ if for each n the term in square brackets vanishes, that is, if

$$(n+2)(n+1)a_{n+2} = \left(2n - \frac{\lambda}{\alpha} + 1\right)a_n. \tag{C.17}$$

Rewriting Equation (C.17) as

$$a_{n+2} = \frac{\left(2n - \frac{\lambda}{\alpha} + 1\right)}{(n+2)(n+1)}a_n \tag{C.18}$$

we have a *recursion relation* for the coefficients a_n. Setting $n = 0$, the recursion relation gives a_2 in terms of a_0. Setting $n = 2$, it gives a_4 in terms of a_2 and, hence, also gives a_4 in terms of a_0. Continuing it will give all a_n with even n in terms of a_0. Similarly, it will give all a_n with odd n in terms of a_1. Thus, if we set $a_1 = 0$, $H(\xi)$ of Equation (C.12) will be an even series proportional to a_0, and if we set $a_0 = 0$, it will be an odd series proportional to a_1.

We next show that for most values of λ,

$$\lim_{\xi\to\pm\infty} H(\xi) \sim \exp(\xi^2). \tag{C.19}$$

C.4 DIVERGENCE OF $\Psi(\xi)$ FOR MOST VALUES OF λ

Let us return to the power series of Equation (C.12). For large $|\xi|$, it is dominated by terms $a_n\xi^n$ with large n. Thus, we will consider the large n limit of the recursion relation of Equation (C.18). This limit is

$$\lim_{n\to\infty} a_{n+2} = \lim_{n\to\infty} \frac{(2n - \frac{\lambda}{\alpha} + 1)a_n}{(n+2)(n+1)} = \frac{2}{n}a_n. \tag{C.20}$$

Similarly,

$$\lim_{n\to\infty} a_{n+4} = \lim_{n\to\infty} \frac{2}{n}a_{n+2} = \left(\frac{2}{n}\right)^2 a_n \tag{C.21}$$

and so on. Thus, the large $|\xi|$ limit of the power series of Equation (C.12) for $H(\xi)$ may be written as

$$\lim_{\xi\to\pm\infty} H(\xi) = \text{negligible terms} + \left[\xi^n + \frac{2}{n}\xi^{n+2} + \left(\frac{2}{n}\right)^2 \xi^{n+4} + \cdots\right] a_n. \tag{C.22}$$

Next, consider the function $\exp(\xi^2)$. It has the power series expansion

$$\exp(\xi^2) = 1 + \frac{\xi^2}{1!} + \frac{\xi^4}{2!} + \cdots + \frac{\xi^n}{(n/2)!} + \frac{\xi^{n+2}}{[(n+2)/2]!} + \cdots. \tag{C.23}$$

Alternatively, we may write the expansion as

$$\exp(\xi^2) = 1 + b_2\xi^2 + \cdots b_n\xi^n + b_{n+2}\xi^{n+2} + \cdots. \tag{C.24}$$

Comparison of Equations (C.23) and (C.24) shows that $b_n = 1/(n/2)!$ and $b_{n+2} = 1/[(n+2)/2]!$ so $\lim_{n\to\infty} \frac{b_{n+2}}{b_n} = \lim_{n\to\infty} \frac{\left(\frac{n}{2}\right)!}{\left(\frac{n}{2}+1\right)!} = \lim_{n\to\infty} \frac{1}{\left(\frac{n}{2}+1\right)} = \frac{2}{n}$. Thus, as $n \to \infty$,

$$b_{n+2} = \frac{2}{n} b_n. \tag{C.25}$$

Similarly, as $n \to \infty$,

$$b_{n+4} = \frac{2}{n} b_{n+2} = \left(\frac{2}{n}\right)^2 b_n \tag{C.26}$$

and so on.

Then, we have from Equations (C.24)–(C.26) that

$$\lim_{\xi\to\pm\infty} \exp(\xi^2) = \text{negligible terms} + \left[\xi^n + \left(\frac{2}{n}\right)\xi^{n+2} + \left(\frac{2}{n}\right)^2 \xi^{n+4} + \cdots\right] b_n. \tag{C.27}$$

However, the non-negligible terms in the series of Equations (C.22) and (C.27) are proportional to one another. So we conclude that as $\xi \to \infty$,

$$H(\xi) \sim \exp(\xi^2). \tag{C.28}$$

Equations (C.10) and (C.28) give for the large $|\xi|$ form of the wave function

$$\lim_{\xi\to\pm\infty} \Psi \sim \exp\left(\frac{-\xi^2}{2}\right) \exp(\xi^2) = \exp\left(\frac{\xi^2}{2}\right). \tag{C.29}$$

Equation (C.29) is not an acceptable form for Ψ. For example, it is in disagreement with the boundary conditions $\lim_{\xi\to\pm\infty} \Psi = 0$. Thus, to obtain an acceptable wave function, we impose the boundary conditions. This will yield quantized energies.

C.5 IMPOSITION OF THE BOUNDARY CONDITIONS AND QUANTIZED ENERGIES

To obtain an acceptable wave function, we require that the power series in Equation (C.12) terminate at the $(v+1)$th term to give the polynomial

$$H(\xi)=a_0+a_1\xi+\cdots+a_v\xi^v. \tag{C.30}$$

Then, $\lim_{\xi\to\pm\infty}\Psi=\lim_{\xi\to\pm\infty}\exp\left(\frac{-\xi^2}{2}\right)(\text{polynomial})=0$ and Ψ obeys the boundary conditions. Next, consider the requirement for termination. Let $n\to v$ and rewrite the recursion equation (Equation C.18) as

$$a_{v+2}=\frac{2v-\frac{\lambda}{\alpha}+1}{(v+2)(v+1)}a_v. \tag{C.31}$$

The condition for termination at the $(v+1)$th term is $a_{v+2}=0$. Thus, termination occurs only for special values λ_v of λ which from Equation (C.31) satisfy the condition that $2v-\frac{\lambda_v}{\alpha}+1=0$ or equivalently

$$\lambda_v=2\left(v+\frac{1}{2}\right)\alpha. \tag{C.32}$$

Using Equations (C.2) and (C.3) for λ and α shows that Equation (C.32) holds only for values of the energy E_v such that

$$E_v=\left(v+\frac{1}{2}\right)\hbar\omega. \tag{C.33}$$

Since v can equal $0,1,2,\ldots$ Equation (C.33) is the formula for the quantized energies of the harmonic oscillator given without proof in Equation (7.10). Thus, we have derived Equation (7.10) for E_v.

C.6 DERIVATION OF THE FORMS OF THE HERMITE POLYNOMIALS

Next, let us derive the forms of the Hermite polynomials given in Table 7.1. For definiteness, we will consider the even polynomial of order v

$$H_v(\xi)=a_0+\xi^2a_2+\cdots+a_{v-2}\xi^{v-2}+a_v\xi^v. \tag{C.34}$$

We will use the recursion relation $a_{n+2}=\frac{2n-\frac{\lambda}{\alpha}+1}{(n+2)(n+1)}a_n$ to determine the coefficients $a_2,a_4,\ldots$ in Equation (C.34) in terms of a_0 and then make a separate determination of a_0. First, note that since we want the polynomial of order v, we must have that $a_{v+2}=0$. From Equation (C.31), this condition only holds if $-\frac{\lambda}{\alpha}+1=-2v$. Setting $\frac{-\lambda}{\alpha}+1=-2v$ in the recursion relation for a_{n+2} gives

$$a_{n+2}=\frac{-2(v-n)}{(n+2)(n+1)}a_n. \tag{C.35}$$

From Equation (C.35),

$$a_2 = \frac{-2v}{2!} a_0 \tag{C.36}$$

$$a_4 = \frac{-2(v-2)}{(4)(3)} a_2 = \frac{(-1)^2 2^2 v(v-2)}{4!} a_0 \tag{C.37}$$

$$a_6 = \frac{-2(v-4)}{(6)(5)} a_4 = \frac{(-1)^3 2^3 v(v-2)(v-4)}{6!} a_0 \tag{C.38}$$

$$\vdots$$

$$a_m = \frac{(-1)^{m/2} 2^{m/2} v(v-2)\cdots(v-m+2)}{m!} a_0 \tag{C.39}$$

and so on.

The form of $H_v(\xi)$ of Equation (C.34) is thus

$$H_v(\xi) = \left[1 - \frac{2v}{2!}\xi^2 + \frac{2^2 v(v-2)}{4!}\xi^4 - \frac{2^3 v(v-2)(v-4)}{6!}\xi^6 + \cdots\right] a_0. \tag{C.40}$$

Next, let us determine a_0. This is done by using the convention that the coefficient a_v in Equation (C.34) for $H_v(\xi)$ is taken as

$$a_v = 2^v \tag{C.41}$$

and then determining a_v in terms of a_0. To make this determination, we set m = v in Equation (C.39) to obtain a_v as

$$a_v = (-1)^{v/2} 2^{v/2} \frac{v(v-2)\ldots(2)}{v!} a_0. \tag{C.42}$$

There are $v/2$ factors in the product Thus, we may write this product as

$$v(v-2)\cdots(2) = 2^{v/2}\left(\frac{v}{2}\right)\left(\frac{v}{2}-1\right)\cdots(1) = 2^{v/2}\left(\frac{v}{2}\right)!. \tag{C.43}$$

Equations (C.42) and (C.43) then give a_v as

$$a_v = \frac{(-1)^{v/2} 2^v \left(\frac{v}{2}\right)!}{v!} a_0. \tag{C.44}$$

Comparing Equations (C.41) and (C.44) determines a_0 as

$$a_0 = (-1)^{v/2} \frac{v!}{\left(\frac{v}{2}\right)!}. \tag{C.45}$$

Thus, Equations (C.40) and (C.45) give the final form for the even Hermite polynomials of order $v = 0, 2, 4\ldots$ as

$$H_v(\xi) = (-1)^{v/2}\frac{v!}{\left(\frac{v}{2}\right)!}\left[1 - \frac{2v}{2!}\xi^2 + \frac{2^2 v(v-2)}{4!}\xi^4 - \frac{2^3 v(v-2)(v-4)\xi^6}{6!} + \cdots\right]. \quad (C.46)$$

The odd Hermite polynomials of order $v = 1, 3, 5\ldots$ are correspondingly given by (Problem C.3)

$$H_v(\xi) = (-1)^{(v-1)/2}\frac{2(v!)}{\left(\frac{v-1}{2}\right)!}\left[\xi - \frac{2(v-1)}{3!}\xi^3 + 2^2\frac{(v-1)(v-3)}{5!}\xi^5 + \cdots\right]. \quad (C.47)$$

Setting $v = n$ and $\xi = x$ in Equations (C.46) and (C.47) gives the Hermite polynomials $H_n(x)$ listed in Table 7.1.

FURTHER READINGS

Arfken, George B., Hans J. Weber, and Frank E. Harris. 2012. *Mathematical methods for physicists.* 7th ed. Waltham, MA: Academic Press.

Pauling, Linus, and E. Bright Wilson Jr. 1985. *Introduction to quantum mechanics with applications to chemistry.* New York: Dover Publications, Inc.

PROBLEMS

C.1 Derive Equation (C.11) from Equations (C.6) and (C.10).

C.2 Derive Equation (C.15) from Equation (C.14).

C.3 Prove Equation (C.47).

C.4 Consider the differential equation $\frac{d^2}{dx^2}f(x) = -k^2 f(x)$. Assume $f(x) = a_0 + a_1 x + \cdots + a_p x^p + \cdots$. (a) Inserting this form for $f(x)$ into the differential equation, derive the recursion relation $a_{m+2} = \frac{-k^2 a_m}{(m+1)(m+2)}$. Consider an even power series for which $a_1 = 0$ and hence by the recursion relation $a_1, a_3, \ldots$ all vanish and $f(x) = a_0 + a_2 x^2 + a_4 x^6 + \cdots$. (b). Show from the recursion relation that (i) $a_2 = \frac{-k^2 a_0}{2!}$, (ii) $a_4 = \frac{-k^2 a_2}{(4)(3)} = (-1)^2\frac{k^4 a_0}{4!}$, and (iii) $a_6 = \frac{-k^2}{(6)(5)}a_4 = \frac{(-1)^3 k^6}{6!}a_0$. Thus, show that the mth term is $a_m = \frac{(-1)^{m/2} k^m}{m!}a_0$. (c) From your results in part (b), write out the explicit form for $f(x)$ as a power series. Thus, show that for the even power series, $f(x) = a_0 \cos kx$. (d) Repeat for the odd power series. Find a_3, a_5, a_7, and a_m in terms of a_1, and thus, show that for the odd power series, $f(x) = a_1 \sin kx$.

C.5 Obtain expressions for $H_0(x), H_2(x)$, and $H_4(x)$ from Equation (C.46) and for $H_1(x), H_3(x)$, and $H_5(x)$ from Equation (C.47), and compare your results with those listed in Table 7.1.

Appendix D
Solution of the Radial Schrödinger Equation for Hydrogen-Like Atoms

Here, we describe a method of solution of the hydrogen-like atom radial Schrödinger equation (Equation 10.11). We follow in main outline the approach of Townsend (2012). As in Chapter 10, we introduce the functions $u_{n\ell}(r) = rR_{n\ell}(r)$. This yields the radial Schrödinger equation in the form given in Equations (10.21) and (10.22); namely,

$$-\frac{\hbar^2}{2\mu}\frac{d^2u_{n\ell}(r)}{dr^2} + \frac{\ell(\ell+1)\hbar^2}{2\mu r^2}u_{n\ell}(r) - \frac{Ze^2}{4\pi\varepsilon_0 r}u_{n\ell}(r) = E_n u_{n\ell}(r). \tag{D.1}$$

In Equation (D.1), we anticipate that the energy E_n depends only on the principal quantum number n, a result proven in this appendix.

D.1 THE RADIAL EQUATION IN DIMENSIONLESS FORM

To start, we rewrite Equation (D.1) in dimensionless form. Thus, we introduce the dimensionless distance ρ defined by

$$\rho = \left[\frac{8\mu(-E_n)}{\hbar^2}\right]^{1/2} r. \tag{D.2}$$

(The reason for the minus sign in Equation [D.2] is that for the bound states considered here, E_n is negative.) We will express Equation (D.1) in terms of ρ using Equation (D.2) rewritten as

$$r = \left[\frac{\hbar^2}{8\mu(-E_n)}\right]^{1/2} \rho. \tag{D.3}$$

Writing $u_{n\ell}(r)$ as $u_{n\ell}(\rho)$, we have from Equation (D.3) that

$$\frac{-\hbar^2}{2\mu}\frac{d^2}{dr^2}u_{n\ell}(r) = 4E_n\frac{d^2}{d\rho^2}u_{n\ell}(\rho). \tag{D.4}$$

Next,

$$\frac{\ell(\ell+1)\hbar^2}{2\mu r^2} = -\frac{4E_n\ell(\ell+1)}{\rho^2}. \tag{D.5}$$

Finally,

$$-\frac{Ze^2}{4\pi\varepsilon_0 r} = -\frac{Ze^2}{4\pi\varepsilon_0\hbar}\frac{\left[8\mu(-E_n)\right]^{1/2}}{\rho}. \tag{D.6}$$

Equations (D.4)–(D.6) give Equation (D.1) in terms of ρ as

$$4E_n \frac{d^2}{d\rho^2} u_{n\ell}(\rho) - \frac{4E_n \ell(\ell+1)}{\rho^2} - \frac{Ze^2}{4\pi\varepsilon_0 \hbar} \frac{[8\mu(-E_n)]^{1/2}}{\rho} u_{n\ell}(\rho) = E_n u_{n\ell}(\rho). \tag{D.7}$$

Dividing Equation (D.7) by $4E_n$ yields

$$\frac{d^2}{d\rho^2} u_{n\ell}(\rho) - \frac{\ell(\ell+1)}{\rho^2} u_{n\ell}(\rho) + \frac{\lambda}{\rho} u_{n\ell}(\rho) - \frac{1}{4} u_{n\ell}(\rho) = 0 \tag{D.8}$$

where

$$\lambda = \frac{Ze^2}{4\pi\varepsilon_0 \hbar}\left(\frac{\mu}{-2E_n}\right)^{1/2}. \tag{D.9}$$

Both ρ and λ are dimensionless quantities (Problem D.1). Thus, Equation (D.8) is the radial Schrödinger equation (Equation D.1) written in dimensionless form.

Next, solving Equation (D.9) for E_n gives

$$E_n = -\frac{Z^2 e^4 \mu}{2(4\pi\varepsilon_0)^2 \lambda^2 \hbar^2}. \tag{D.10}$$

E_n of Equation (D.10) is identical to Equation (10.18) for the energy levels of a hydrogen-like atom if λ is restricted to the integral values $n = 1, 2, \ldots$. We will prove later that λ is so restricted.

D.2 THE ASYMPTOTIC LIMITS OF $u_{n\ell}(\rho)$

Next, let us determine the $\rho \to 0$ and $\rho \to \infty$ forms from $u_{n\ell}(\rho)$. Let us assume that as $\rho \to 0$, $u_{n\ell}(\rho) = A\rho^s$, where A is an arbitrary constant and s is to be determined. Inserting this form into Equation (D.8) gives the $\rho \to 0$ relation

$$s(s-1)\rho^{s-2} - \ell(\ell+1)\rho^{s-2} + \lambda\rho^{s-1} - \frac{1}{4}\rho^s = 0. \tag{D.11}$$

The ρ^{s-2} terms in the above expression dominate as $\rho \to 0$. Thus, as $\rho \to 0$, Equation (D.11) reduces to $s(s-1)\rho^{s-2} - \ell(\ell+1)\rho^{s-2} = 0$, which yields the following condition for s:

$$s(s-1) - \ell(\ell+1) = 0. \tag{D.12}$$

Equation (D.12) has two solutions: $s = -\ell$ and $s = \ell + 1$. The $s = -\ell$ solution gives $\lim_{\rho \to 0} u_{n\ell}(\rho) = A\rho^{-\ell}$. This solution must be rejected since it yields a radial wave function $R_{n\ell}(r) = r^{-1} u_{n\ell}(\rho)$ which diverges as $\rho \to 0$. The $s = \ell + 1$ solution yields

$$\lim_{\rho \to 0} u_{n\ell}(\rho) = A\rho^{\ell+1}. \tag{D.13}$$

This is correct solution giving in agreement with Equation (10.23) that $\lim_{r \to 0} R_{n\ell}(r) \sim \rho^{\ell}$.

Next, let us determine the $\rho \to \infty$ form of $u_{n\ell}(\rho)$. This is found from the $\rho \to \infty$ limit of Equation (D.8) which is

$$\frac{d^2 u_{n\ell}(\rho)}{d\rho^2} - \frac{1}{4} u_{n\ell}(\rho) = 0. \quad \text{(D.14)}$$

Equation (D.14) has two solutions: $\lim_{\rho\to\infty} u_{n\ell}(\rho) = B\exp(-\rho/2)$ and $\lim_{\rho\to\infty} u_{n\ell}(\rho) = C\exp\left(+\frac{\rho}{2}\right)$, where B and C are arbitrary constants. The $\exp(\rho/2)$ solution must be rejected since it predicts that $R_{n\ell}(r)$ grows without bound as $r \to \infty$. Thus, the correct $\rho \to \infty$ form of $u_{n\ell}(\rho)$ is

$$\lim_{\rho\to\infty} u_{n\ell}(\rho) = B\exp(-\rho/2). \quad \text{(D.15)}$$

We next assume a solution of Equation (D.8) valid for all ρ of the form

$$u_{n\ell}(\rho) = L(\rho)\rho^{\ell+1}\exp(-\rho/2) \quad \text{(D.16)}$$

where $L(\rho)$ is a function that interpolates between the $\rho \to 0$ and $\rho \to \infty$ limits of $u_{n\ell}(\rho)$. We will determine $L(\rho)$ by inserting Equation (D.16) into Equation (D.8) to derive a differential equation for $L(\rho)$ and then by solving that equation by a power series method similar to that used to solve the Hermite equation in Appendix C.

D.3 DERIVATION OF THE DIFFERENTIAL EQUATION FOR $L(\rho)$

We next derive the differential equation for $L(\rho)$. Using Equation (E.16), we first evaluate the term $\frac{d^2}{d\rho^2} u_{n\ell}(\rho)$ in Equation (D.8) as (Problem D.2)

$$\frac{d^2}{d\rho^2} u_{n\ell}(\rho) = \exp(-\rho/2)\Big\{\rho^{\ell+1}L''(\rho) + \left[(2\ell+2)\rho^{\ell} - \rho^{\ell+1}\right]L'(\rho)$$
$$+\left[\ell(\ell+1)\rho^{\ell-1} - (\ell+1)\rho^{\ell} + \frac{1}{4}\rho^{\ell+1}\right]L(\rho)\Big\} \quad \text{(D.17)}$$

Next, let us consider the remaining terms in Equation (D.8). Using Equation (D.16) to evaluate these terms, we have that

$$-\frac{\ell(\ell+1)}{\rho^2} u_{n\ell}(\rho) + \frac{\lambda}{\rho} u_{n\ell}(\rho) - \frac{1}{4} u_{n\ell}(\rho) =$$
$$\exp(-\rho/2)\left[-\ell(\ell+1)\rho^{\ell-1} + \lambda\rho^{\ell} - \frac{1}{4}\rho^{\ell+1}\right]L(\rho). \quad \text{(D.18)}$$

Combining Equations (D.8), (D.17), and (D.18) and dividing through by $\rho^{\ell+1}\exp(-\rho/2)$ yield the differential equation for $L(\rho)$ as

$$L''(\rho) + \left[\frac{2\ell+2}{\rho} - 1\right]L'(\rho) + \left(\frac{\lambda}{\rho} - \frac{\ell+1}{\rho}\right)L(\rho) = 0. \quad \text{(D.19)}$$

We next solve Equation (D.19) by the power series method.

D.4 SOLUTION OF THE DIFFERENTIAL EQUATION FOR $L(\rho)$ BY THE POWER SERIES METHOD

Thus, we expand $L(\rho)$ as the power series

$$L(\rho)=\sum_{q=0}^{\infty} c_q \rho^q. \tag{D.20}$$

We will also need the expansions of the derivatives $L''(\rho)$ and $L'(\rho)$. These follow from Equation (D.20) as

$$L'(\rho)=\sum_{q=0}^{\infty} c_q q \rho^{q-1} \tag{D.21}$$

and

$$L''(\rho)=\sum_{q=0}^{\infty} c_q q(q-1)\rho^{q-2}. \tag{D.22}$$

Inserting these expansions into Equation (D.19) and grouping terms with like powers of ρ gives

$$\sum_{q=0}^{\infty}\left[q(q-1)+(2\ell+2)q\right]c_q\rho^{q-2}+\sum_{q=0}^{\infty}\left[-q+\lambda-(\ell+1)\right]c_q\rho^{q-1}=0. \tag{D.23}$$

However, the first sum in Equation (D.23) may be recast as (Problem D.3) $\sum_{q=0}^{\infty}\left[q(q-1)+(2\ell+2)q\right]c_q\rho^{q-2}=\sum_{q=0}^{\infty}\left[q(q+1)+(2\ell+2)(q+1)\right]c_{q+1}\rho^{q-1}$.

Thus, Equation (D.23) may be rewritten as

$$\sum_{q=0}^{\infty}\left\{\left[q(q+1)+(2\ell+2)(q+1)\right]c_{q+1} +\left[-q+\lambda-(\ell+1)\right]c_q\right\}\rho^{q-1}=0. \tag{D.24}$$

Equation (D.24) can only be true for all values of ρ if for each q

$$\left[q(q+1)+(2\ell+2)(q+1)\right]c_{q+1} +\left[-q+\lambda-(\ell+1)\right]c_q=0. \tag{D.25}$$

Solving Equation (D.25) for c_{q+1} gives the recursion relation

$$c_{q+1}=\frac{(q-\lambda+\ell+1)}{q(q+1)+(2\ell+2)(q+1)}c_q. \tag{D.26}$$

D.5 Exp(ρ) DIVERGENCE OF THE POWER SERIES FOR L(ρ) FOR ARBITRARY λ

The situation is now similar to what we found for the harmonic oscillator in Appendix C. For arbitrary λ, the recursion relation gives a form for $L(\rho)$ which diverges for $\rho \to \infty$ as $\exp(\rho)$. From Equation (D.16), $u_{n\ell}(\rho)$ then unacceptably diverges for $\rho \to \infty$ as $\exp(\rho/2)$. However, if λ is an integer, the series terminates so that $L(\rho)$ is a polynomial and $u_{n\ell}(\rho)$ of Equation (D.16) is an acceptable solution of Equation (D.8).

We next show this. For large ρ, the power series expansion of $L(\rho)$ of Equation (D.20) is dominated by terms $c_q\rho^q$ with large q. Thus, we consider the large q limit of the recursion relation of Equation (D.26) which is

$$\lim_{q\to\infty} c_{q+1} = \frac{1}{q}c_q. \tag{D.27}$$

Similarly,

$$\lim_{q\to\infty} c_{q+2} = \frac{1}{q}\lim_{q\to\infty} c_{q+1} = \frac{1}{q^2}c_q. \tag{D.28}$$

Continuing, the large ρ limit of the power series for $L(\rho)$ may be written as

$$\lim_{\rho\to\infty} L(\rho) = \text{negligible terms} + \left(\rho^q + \frac{1}{q}\rho^{q+1} + \frac{1}{q^2}\rho^{q+2} + \cdots\right)c_q. \tag{D.29}$$

Next, let us consider the power series expansion of $\exp(\rho)$ which is

$$\exp(\rho) = 1 + \frac{\rho}{1!} + \frac{\rho^2}{2!} + \cdots + \frac{\rho^q}{q!} + \frac{\rho^{q+1}}{(q+1)!} + \cdots. \tag{D.30}$$

We may rewrite this expansion as

$$\exp(\rho) = 1 + b_1\rho + \cdots + b_q\rho^q + b_{q+1}\rho^{q+1}\cdots. \tag{D.31}$$

Note that $b_q = \frac{1}{q!}$ and $b_{q+1} = \frac{1}{(q+1)!}$. Hence, $\lim_{q\to\infty}\frac{b_{q+1}}{b_q} = \lim_{q\to\infty}\frac{q!}{(q+1)!} = \lim_{q\to\infty}\frac{1}{q+1} = \frac{1}{q}$. Thus,

$$\lim_{q\to\infty} b_{q+1} = \frac{1}{q}b_q. \tag{D.32}$$

Continuing,

$$\lim_{q\to\infty} b_{q+2} = \frac{1}{q}\lim_{q\to\infty} b_{q+1} = \frac{1}{q^2}b_q \tag{D.33}$$

and so on.

Thus, the large ρ limit of the power series for $\exp(\rho)$ is

$$\lim_{\rho\to\infty} \exp(\rho) = \text{negligible terms} + \left(\rho^q + \frac{1}{q}\rho^{q+1} + \frac{1}{q^2}\rho^{q+2} + \cdots\right)b_q. \tag{D.34}$$

However, the non-negligible terms in the power series of Equations (D.29) and (D.34) are proportional. Thus, we conclude

$$\lim_{\rho\to\infty} L(\rho) \sim \lim_{\rho\to\infty} \exp(\rho). \tag{D.35}$$

Comparing Equations (D.16) and (D.35) shows that as $\rho \to \infty$, $u_{n\ell}(\rho)$ unacceptably diverges as $\exp(\rho/2)$.

We next choose λ so that the power series for $L(\rho)$ terminates giving an acceptable form for $u_{n\ell}(\rho)$.

D.6 TERMINATION OF THE POWER SERIES FOR $L(\rho)$ AND QUANTIZED ENERGIES

An acceptable form for $u_{n\ell}(\rho)$ occurs only if the power series for $L(\rho)$ of Equation (D.20) terminates for some value of $q = k = 0,1,2,\ldots$. Then, $L(\rho)$ becomes the following polynomial of order k:

$$L(\rho) = \sum_{q=0}^{k} c_q \rho^q \tag{D.36}$$

rendering

$$u_{n\ell}(\rho) = \rho^{\ell+1} \sum_{q=0}^{k} c_q \rho^q \exp(-\rho/2) \tag{D.37}$$

an acceptable function which properly approaches zero as $\rho \to \infty$. The termination of Equation (C.36) can only occur if $c_{k+1} = 0$. However, from the recursion relation of Equation (D.26) evaluated for $q = k$

$$c_{k+1} = \frac{(k - \lambda + \ell + 1)c_k}{k(k+1) + (2\ell + 2)(k+1)} \tag{D.38}$$

it follows that $c_{k+1} = 0$ only if

$$\lambda = k + \ell + 1. \tag{D.39}$$

Since k and ℓ are both restricted to the values $0,1,\ldots$, it follows from Equation (D.39) that λ must be an integer n that can only take on the values

$$\lambda = n = 1,2,\ldots. \tag{D.40}$$

This restriction on the allowed values of λ implies that the energies E_n of a hydrogen-like atom are quantized. Namely, comparing Equations (D.10) and (D.40) gives the quantized hydrogen-like atom energy level formula

$$E_n = -\frac{Z^2 e^4 \mu}{2(4\pi\varepsilon_0)^2 n^2 \hbar^2} \quad \text{where } n = 1,2,\ldots \tag{D.41}$$

of Equation (10.18).

Additionally, since $L(\rho)$ is a polynomial of order k, it has k zeros that can be shown to be real and positive (Problem D.4). Consequently, $u_{n\ell}(\rho)$ has k nodes occurring at the zeros of $L(\rho)$. However, since $\lambda = n$, from Equation (D.39)

$$k = n - \ell - 1. \tag{D.42}$$

Thus, $u_{n\ell}(r)$ and hence $R_{n\ell}(r)$ have $n-\ell-1$ nodes. Also note from Equation (A.42) that $\ell = n-1-k$. Thus, since $k = 0,1,\ldots$, the maximum value of ℓ is $n-1$, and hence, ℓ has the range $\ell = 0,1,\ldots n-1$.

We may now derive a convenient form for ρ. Using Equation (D.41) to eliminate E_n in Equation (D.2) for ρ gives that $\rho = \frac{2Z}{n}\frac{\mu e^2}{(4\pi\varepsilon_0)\hbar^2}r$. However, from Equation (10.20), the Bohr radius $a = \frac{4\pi\varepsilon_0}{\mu e^2}\hbar^2$. Thus,

$$\rho = \frac{2Z}{na}r. \tag{D.43}$$

We next derive the forms of some of the radial wave functions $R_{n\ell}(r)$.

D.7 THE RADIAL WAVE FUNCTIONS $R_{n\ell}(r)$

We first obtain the 1s radial wave function $R_{10}(r)$. To start, we note that since for the 1s state $n = 1$ and $\ell = 0$, for this state from Equation (D.42) $k = 0$ and hence from Equation (D.36) $L(\rho)$ is a constant c_0. Then, Equation (D.16) gives $u_{10}(\rho)$ as

$$u_{10}(\rho) = c_0\rho\exp(-\rho/2). \tag{D.44}$$

Thus, since for $n = 1$ $\rho = \frac{2Z}{a}r$, $u_{10}(\rho) = u_{10}(r)$ becomes $u_{10}(r) = N_{10}r\exp\left(-\frac{Zr}{a}\right)$, where the normalization factor $N_{10} = \frac{2Z}{a}c_0$. Hence, $R_{10}(r) = r^{-1}u_{10}(r)$ is given by

$$R_{10}(r) = N_{10}\exp\left(-\frac{Zr}{a}\right). \tag{D.45}$$

N_{10} is determined by the normalization condition of Equation (10.12)

$$\int_0^\infty r^2R_{n\ell}^2(r)dr = 1 \tag{D.46}$$

which gives $N_{10}^2\int_0^\infty r^2\exp\left(-\frac{2Zr}{a}\right)dr = 1$. Performing the integral using Equation (10.51) yields N_{10} as $N_{10} = 2\left(\frac{Z}{a}\right)^{3/2}$. Hence, the normalized 1s radial wave function is

$$R_{10}(r) = 2\left(\frac{Z}{a}\right)^{3/2}\exp(-\sigma) \tag{D.47}$$

where

$$\sigma \equiv \frac{Zr}{a}. \tag{D.48}$$

Next, let us derive the form of the 2p radial wave function $R_{21}(r)$. For the 2p state, $n = 2$ and $\ell = 1$, and we again find that $k = 0$ and hence $L(\rho)$ is a constant c_0. Thus, from Equation (D.16)

$$u_{21}(\rho) = c_0\rho^2 \exp(-\rho/2). \tag{D.49}$$

Since for $n = 2\ \rho = \frac{Zr}{a}$, from Equation (D.49) $u_{21}(\rho) = u_{21}(r) = N_{21}r^2 \exp\left(-\frac{Zr}{2a}\right)$, where $N_{21} = c_0\left(\frac{Z}{a}\right)^2$. Thus, $R_{21}(r) = r^{-1}u_{21}(r)$ is given by

$$R_{21}(r) = N_{21} r \exp\left(-\frac{Zr}{2a}\right). \tag{D.50}$$

Determining N_{10} using Equation (D.46) gives the normalized 2p radial wave function as

$$R_{21}(r) = \frac{1}{2\sqrt{6}}\left(\frac{Z}{a}\right)^{3/2} \sigma \exp\left(-\frac{\sigma}{2}\right). \tag{D.51}$$

A similar evaluation of the 3d radial wave function $R_{32}(r)$ gives (Problem D.5)

$$R_{32}(r) = \frac{1}{\sqrt{30}}\left(\frac{4}{81}\right)\left(\frac{Z}{a}\right)^{3/2} \sigma^2 \exp\left(-\frac{\sigma}{3}\right). \tag{D.52}$$

For the radial wave functions that we have determined – $R_{10}(r)$, $R_{21}(r)$, and $R_{32}(r)$ – ℓ has its maximum value $n - 1$. From Equation (D.42), however, for all radial wave functions with $\ell = n - 1$, $k = 0$, and thus, $L(\rho)$ is equal to a constant. Thus, such radial wave functions have an especially simple form. In fact, a general result valid for all n for the radial wave function $R_{n\ell=n-1}(r)$ of a state with $\ell = n - 1$ may be developed. It takes the normalized form (Problem D.6)

$$R_{n\ell=n-1}(r) = \frac{\left(\frac{2}{n}\right)^{n+\frac{1}{2}}\left(\frac{Z}{a}\right)^{3/2} \sigma^{n-1} \exp\left(-\frac{\sigma}{n}\right)}{\left[(2n)!\right]^{1/2}}. \tag{D.53}$$

For $n = 1,2$, or 3, $R_{n\ell=n-1}(r)$ reduces to our results for $R_{10}(r)$, $R_{21}(r)$, and $R_{32}(r)$. Moreover, the radial wave functions for higher lying $\ell = n - 1$ states (like the 4f state, 5g state, 6h state, and so on) may be found by specializing the form for $R_{n\ell=n-1}(r)$ (Problem D.7).

Next, let us consider states for which $\ell \neq n - 1$. For such states, $k \neq 0$ and $L(\rho)$ is not a constant. The simplest example of an $\ell \neq n - 1$ state is the 2s state. For this state, $n = 2, \ell = 0$, and hence, from Equation (D.42), $k = 1$. Thus, $L(\rho)$ is a polynomial of order 1 with the form

$$L(\rho) = c_0 + c_1\rho. \tag{D.54}$$

From the recursion relation of Equation (D.26) evaluated for $q = 0, \ell = 0$, and $\lambda = n = 2$, it follows that $c_1 = -\frac{1}{2}c_0$, and hence,

$$L(\rho) = \frac{1}{2}c_0(2 - \rho). \tag{D.55}$$

Comparing Equations (D.16) and (D.55) then gives $u_{20}(\rho)$ as

$$u_{20}(\rho) = \frac{1}{2}c_0(2 - \rho)\rho \exp\left(-\frac{\rho}{2}\right) \tag{D.56}$$

and thus, the 2s radial wave function is

$$R_{20}(r) = N_{20}\left(2 - \frac{Zr}{a}\right)\exp\left(-\frac{Zr}{2a}\right) \tag{D.57}$$

where $N_{20} = \frac{Z}{2a}c_0$. Determining N_{20} using Equation (D.46) gives the normalized form of $R_{20}(r)$ as

$$R_{20}(r) = \frac{1}{2\sqrt{2}}\left(\frac{Z}{a}\right)^{3/2}(2-\sigma)\exp\left(-\frac{\sigma}{2}\right). \tag{D.58}$$

Proceeding similarly for the 3p state gives its normalized radial wave function as (Problem D.8)

$$R_{31}(r) = \frac{1}{\sqrt{6}}\left(\frac{4}{81}\right)\left(\frac{Z}{a}\right)^{3/2}(6-\sigma)\sigma\exp\left(-\frac{\sigma}{3}\right). \tag{D.59}$$

FURTHER READINGS

Cohen-Tannoudji, Claude, Bernard Diu, and Franck Laloe. 1977. *Quantum mechanics*. Vol. 1. Translated by Susan Reid Hemley, Nicole Ostrowsky, and Dan Ostrowsky. New York: John Wiley & Sons.

Griffiths, David J. 2014. *Introduction to quantum mechanics*. 2nd ed. Edinburgh Gate, England: Pearson Education Limited.

Merzbacher, Eugen. 1961. *Quantum mechanics*. New York: John Wiley & Sons, Inc.

Pauling, Linus, and E. Bright Wilson Jr. 1985. *Introduction to quantum mechanics with applications to chemistry*. New York: Dover Publications, Inc.

Townsend, John F. 2012. *A modern approach to quantum mechanics*. 2nd. ed. Mill Valley, CA: University Science Books.

PROBLEMS

In Problems D.5, D.6, and D.8, you will need the integral $\int_0^\infty x^n \exp(-kx)dx = \frac{n!}{k^{n+1}}$.

D.1 Show from Equations (D.2) and (D.9) that ρ and λ are dimensionless. Recall that the units of $4\pi\varepsilon_0$ are J^{-1}. m^{-1}. C^2.

D.2 Verify Equation D.17. (Hint: Recall for arbitrary functions $f[x]$ and $g[x]$ that $\frac{df[x]g[x]}{dx} = f'[x]g[x] + f[x]g'[x]and\frac{d^2f[x]g[x]}{dx^2} = f''[x]g[x] + 2f'[x]g'[x] + f[x]g''[x]$.)

$c_{q+1}\rho^{q-1}$.

D.3 Show that the sum $\sum_{q=0}^{\infty}\left[q(q-1)+(2\ell+2)q\right]c_q\rho^{q-2} = \sum_{q=0}^{\infty}\left[q(q+1)+(2\ell+2)(q+1)\right]$

D.4 A polynomial of order n is an expression of the form $a_nx^n + a_{n-1}x^{n-1} + \cdots + a_1x + a_0$. Such a polynomial has n zeros. Illustrate this by (a) writing down a polynomial of order 1 and finding its one zero in terms of a_0 and a_1 and (b) writing down a polynomial of order 2 and finding its two zeros in terms of a_0, a_1, and a_2.

D.5 Derive the form for $R_{32}(r)$ given in Equation (D.52).

D.6 Derive Equation (D.53) for $R_{n\ell=n-1}(r)$.

D.7 Obtain the form of the 4f radial wave function $R_{43}(r)$ by specializing Equation (D.51) for $R_{n-1}(r)$ to the case $n = 4$.

D.8 Derive the form for the 3p radial wave function $R_{31}(r)$ given in Equation (D.59). Proceed according to the following steps. (a) Show that $L(\rho)=c_0\left(1-\frac{1}{4}\rho\right)$. (b) Then, show that $u_{31}(r)=\left(\frac{2Z}{3a}\right)^2\frac{1}{6}c_0r^2\left(6-\frac{Zr}{a}\right)\exp\left(-\frac{Zr}{3a}\right)$. (c) Thus, show that $R_{31}(r)=N_{31}(6-\sigma)\sigma\exp\left(-\frac{\sigma}{3}\right)$ where $\sigma=\frac{Zr}{a}$ and where N_{31} is a normalization factor proportional to c_0. (d) Show that the normalization condition for $R_{31}(r)$, $\int_0^\infty r^2R_{31}^2(r)dr=1$, may be written as $\left(\frac{a}{Z}\right)^3\int_0^\infty \sigma^2\left[R_{31}(r)\right]^2 d\sigma=1$. (e) Thus, show that $N_{31}=\left(\frac{Z}{a}\right)^{3/2}I^{-1/2}$ where $I=\int_0^\infty \sigma^4(6-\sigma)^2\exp\left(-\frac{2\sigma}{3}\right)d\sigma$. (f) Show that (g) Thus, show that $N_{31}=\frac{1}{\sqrt{6}}\frac{4}{81}\left(\frac{Z}{a}\right)^{3/2}$ and therefore that in agreement with Equation (D.59),

$$R_{31}(r)=\frac{1}{\sqrt{6}}\left(\frac{4}{81}\right)\left(\frac{Z}{a}\right)^{3/2}(6-\sigma)\sigma\exp\left(-\frac{\sigma}{3}\right).$$

SOLUTIONS TO SELECTED PROBLEMS

D.4 (a) a_1x+a_0, zero at $x=-\frac{a_0}{a_1}$; and (b) $a_2x^2+a_1x+a_0$, zeros at $x=\frac{-a_2\pm\sqrt{a_1^2-4a_2a_0}}{2a_2}$.

D.7 $R_{43}(r)=\frac{1}{\sqrt{105}}\frac{1}{768}\left(\frac{Z}{a}\right)^{3/2}\sigma^3\exp\left(-\frac{\sigma}{4}\right)$.

Index

Note: **Bold** page numbers refer to tables and *Italic* page numbers refer to figures.

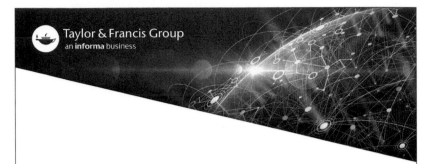

9781498733991